APPLIED MACHINE LEARNING

APPLIED MACHINE LEARNING

Dr. M. Gopal

Ex-Professor
Indian Institute of Technology Delhi

New York Chicago San Francisco
Athens London Madrid
Mexico City Milan New Delhi
Singapore Sydney Toronto

Library of Congress Control Number: 2018966932

Applied Machine Learning

1 2 3 4 5 6 QVS 23 22 21 20 19

ISBN 978-1-260-45684-4
MHID 1-260-45684-6

The sponsoring editor for this book was Lara Zoble and the production supervisor was Lynn M. Messina. The art director for the cover was Jeff Weeks.

This book was previously published by McGraw-Hill Education (India) Private Limited, New Delhi, copyright 2018.

This book is printed on acid-free paper.

Dedicated

to my granddaughter

Astika

the most precious gift life has given me

ABOUT THE AUTHOR

M.Gopal, an Ex-Professor of IIT Delhi, is a globally known academician with excellent credentials as author, teacher, and researcher. He is the author/co-author of five books on *Control Engineering.* His books are used worldwide, and some of them have been translated into Chinese and Spanish. McGraw-Hill Education, Singapore, has published his books for the Asia Pacific market; and McGraw-Hill Education, USA, has published for the US market. His latest contribution as an author is the book *Applied Machine Learning*, which is in your hand.

As a teacher, his potential is being used globally through a video course (http://www.youtube.com/iit); one of the most popular courses on YouTube by the IIT faculty. Millions of students, teachers, researchers, and practitioners used his books and the video course for knowledge acquisition over the past more than four decades.

A recognized researcher in the area of Machine Learning, he is the author/co-author of over 150 research papers; the key contributions have been published in high impact factor journals. He has supervised 16 doctoral research projects (seven of them in the machine learning area), and two projects are in progress. His current research interests are in the areas of Machine Learning and Intelligent Control.

M.Gopal holds B.Tech (Electrical), M.Tech (Control Systems), and PhD degrees from BITS, Pilani. His teaching and research stints span more than four decades at prestigious institutes like IIT Delhi (about three decades), IIT Bombay, BITS Pilani, and MNIT Jaipur.

He has been associated with Shiv Nadar University since 2012.

CONTENTS

PREFACE

Over the past two decades, the field of *Machine Learning* has become one of the mainstays of information technology. Many successful machine learning applications have been developed, such as: machine vision (image processing) in the manufacturing industry for automation in assembly line, biometric recognition, handwriting recognition, medical diagnosis, speech recognition, text retrieval, natural language processing, and so on. Machine learning is so pervasive today that you probably use it several times a day, without knowing it. Examples of such "ubiquitous" or "invisible" usage include search engines, customer-adaptive web services, email managers (spam filters), computer network security, and so on. We are rethinking on everything we have been doing, with the aim of doing it differently using tools of machine learning for better success.

Many organizations are routinely capturing huge volumes of historical data describing their operations, products, and customers. At the same time, scientists and engineers are capturing increasingly complex datasets. For example, banks are collecting huge volumes of customer data to analyze how people spend their money; hospitals are recording what treatments patients are on, for which periods (and how they respond to them); engine monitoring systems in cars are recording information about the engine in order to detect when it might fail; world's observatories are storing incredibly high-resolution images of night sky; medical science is storing the outcomes of medical tests from measurements as diverse as Magnetic Resonance Imaging (MRI) scans and simple blood tests; bioinformatics is storing massive amounts of data with the ability to measure gene expression in DNA microarrays, and so on. The field of machine learning addresses the question of how best to use this historical data to discover general patterns and improve the process of making decisions.

Terminology in the field of *learning* is exceptionally diverse, and very often similar concepts are variously named. In this book, the term *machine learning* has been mostly used to describe various concepts, though the terms: *artificial intelligence, machine intelligence, pattern recognition, statistical learning, data mining, soft computing, data analytics* (when applied in business contexts), also appear at various places.

There have been important advances in the theory and algorithms that form the foundations of machine learning field. The goal of this text book is to present basic *concepts* of the theory, and a wide range of *techniques* (algorithms) that can be applied to a variety of problems. There are many machine learning algorithms not included in this book, that can be quite effective in specific situations. However, almost all of them are some adaptation of the algorithms included in this book. Self-learning will easily help to acquire the required knowledge.

Basically, there are two approaches for understanding machine learning field. In one approach, we treat machine learning techniques as a 'black box', and focus on understanding the problems (tasks) of interest: matching these tasks to machine learning tools and assessing the quality of the output. This gives us hands-on experience with machine learning from practical case studies. Subsequently, we delve into the components of this black box by examining machine learning algorithms (a theoretical principle-driven exposition is necessary to be effective in machine learning). The second approach starts with the theory; this is then followed by hands-on experience.

The approach into the field of machine learning taken in this book has been the second one. We have focussed on machine learning theory. For hands-on experience, we propose to provide a platform through self-study machine learning projects.

In this book on "Applied Machine Learning", the reader will get not only the theoretical underpinnings of learning, but also gain the practical know-how needed to quickly and powerfully apply these techniques to challenging problems: learning how to conceptualize a problem, knowing how to represent the data, selecting and tuning algorithms, being able to interpret results properly, doing an effective analysis of results to make strategic decisions. Recognizing that most ideas behind machine learning are wonderfully simple and straightforward, the book presents machine learning concepts and techniques in a non-rigorous mathematical setting, with emphasis on effective methodology for using machine learning to solve practical problems. It is a comprehensive textbook on the subject, covering broad array of topics with more emphasis on the techniques (algorithms) that have been profitably employed, thus exploiting the available knowledge base.

Machine learning draws on concepts and techniques from many fields, including computational statistics (a discipline that aims at the design of algorithms for implementing statistical methods on computers), artificial intelligence, information theory, mathematical optimization, biology, cognitive science, and control theory. The primary goal of this book is to provide a broad-based single source introduction to the field. It introduces basic concepts from various fields as the need arises, focussing on just those concepts most relevant to machine learning. Though the required material has been given in the book, some experience with probability, statistics, and linear algebra will be useful.

The *first-generation* machine learning algorithms covered in this book, have been demonstrated to be of significant value in a variety of real-world applications with numeric features. But these algorithms also have significant limitations, for example, although some learning algorithms are available to classify images, text or speech, we still lack effective algorithms for learning from data that is represented by a combination of these various media. Also most learning algorithms perform acceptably well on datasets with tens of thousands of training examples, but many important datasets are significantly larger. The volume and diversity (structured/unstructured) of data available on the Internet and corporate Intranets is extremely large and growing rapidly. Scaling to complex, extremely large datasets—the *big data analytics*—is probably the most debated current issue. Given these and other limitations, and the strong commercial interest despite them, we might well expect the next decade to produce an order of magnitude advancement in the state of the art. *Deep learning* algorithms are emerging as very powerful *next-generation* tools. Like most other areas of technology, data mining exists on a shifting landscape; not only is the old part of the landscape being redefined, but new areas of interest always loom ahead.

All learning algorithms are explained so that the student can easily move from the equations in the book to computer programs. Proliferation of free software that makes machine learning easier to implement, will also be helpful in the project work. The diversity of machine learning libraries means that there is likely to be an option available of what language or environment a student uses.

There are many machine learning websites that give information on available machine learning software. Some of the popular *software sources* are R, SAS, Python, Weka, MATLAB, Excel, and Tableau.

This book does not promote any specific software. We have included a large number of examples, but we use illustrative datasets that are small enough to allow the reader to follow what is going on without the help of software. Real datasets are far too large to show this. Datasets in the book are chosen not to illustrate actual large-scale practical problems, but to help the reader understand what the different techniques do, how they work, and what their range of application is. This explains why a heavy focus on project work is a necessity. Each project must handle a large-scale practical problem. Use of domain knowledge to formulate the problem in machine learning setting, and interpretation of the results given by machine learning algorithms are important ingredients of training the students, in addition to the training on machine learning software. This book on 'Applied Machine Learning' provides necessary ingredients for practice—the concepts and the techniques —but the actual *practice* will follow through project work on real-life problems.

In a university setting, this book provides an introductory course for undergraduate students in computer science and all engineering degree programs. Such an introductory course will require a properly selected subset of techniques covered in the book. The course design must have a heavy focus on project work, so that when a student has completed the course, he/she should be fully prepared to attack new problems using machine learning.

Postgraduate students and Ph.D. research scholars will find in this book a useful initial exposure to the subject, before they go for highly theoretical depth in the specific areas of their research.

The book is aimed at a large professional audience as well: engineers, scientists, and business managers; with machine learning and deep learning predicted to be the next 'grand slam' in technology, professionals in almost all fields will need to know at least the basics of machine learning.

I hope that the reader will share my excitement on the subject of machine learning, and will find the book useful.

M. Gopal
mgopal.iitd@gmail.com

ACKNOWLEDGEMENTS

I am deeply indebted to many people who, directly or indirectly, are responsible for this book coming into being. I am grateful to Dr. Ashwani Kumar (Professor, Indian Institute of Management, Lucknow) for his support in my endeavor and for helping me in including an overview of Business Decision Making concepts in the book.

I thank my undergraduate and postgraduate students from Indian Institute of Technology Delhi, especially my former PhD students with research projects in machine learning: Dr. Laxmidhar Behera (Professor, IIT Kanpur), Dr. Rajen Bhatt (Director, Machine Learning Division, Qeexo, Pittsburgh), Dr. Alok Kanti Deb (Associate Professor, IIT Kharagpur), Dr. Rajneesh Sharma (Associate Professor, Netaji Subhash Institute of Technology, Delhi), Dr. M. Arunkumar (Lead Scientist, Target Corporation India, Bengaluru), Dr. Hitesh Shah (Professor, G.H. Patel College of Engineering and Technology, Vallabh Vidyanagar, Gujarat), and Dr. Ehtesham Hassan (Research Scientist, Innovation Labs, Tata Consultancy Services India), whose desire for new knowledge was both a challenge and an inspiration to me.

Ashish Kushwaha, currently working with me for his PhD, has been of great help in completing this project; my sincere thanks to him.

M. Gopal
mgopal.iitd@gmail.com

APPLIED MACHINE LEARNING

Chapter

1

INTRODUCTION

1.1 TOWARDS INTELLIGENT MACHINES

Human beings have always dreamt of creating machines with human-like traits. In today's world of technology, there are machines that have matched several human functions or even outdone them with extraordinary capacity and abilities. Robots in manufacturing, mining, agriculture, space, ocean exploration, and health sciences, are only a few examples.

These machines are, however, enslaved by commands. One of the tenets of recent research in robotics and systems science is that intelligence can be cast into a machine. This is, perhaps, an ultimate challenge to science—to create *intelligent* machines that emulate human *intelligence*.

Human intelligence possesses robust attributes with complex sensory, control, affective (emotional processes), and cognitive (thought processes) aspects of information processing and decision making. Biological neurons, over one hundred billion in number, in our central nervous system (CNS), play a key role in these functions. Essentially, CNS acquires information from the external environment through various natural sensory mechanisms such as vision, hearing, touch, taste, and smell. It integrates the information and provides appropriate interpretation through cognitive computing. The cognitive process then advances further towards some attributes such as learning, recollection, and reasoning, which results in appropriate actions through muscular control.

Recent progress in information-based technology has significantly broadened the capabilities and application of computers. Traditionally, computers have been mainly used for the storage and processing of numerical data. If we wish to emulate in a machine (compute), some of the cognitive functions (learning, remembering, reasoning, perceiving, etc.) of humans, we have to generalize the definition of information and develop new mathematical tools and hardware that must deal with the simulation and processing of cognitive information. Mathematics, as we know it today, was developed for the understanding of physical processes, whereas the process of cognition does not necessarily follow these mathematical laws. Then what is *cognitive mathematics*? This is a difficult and challenging question to answer. However, scientists have realized that if we re-examine some of the 'mathematical aspects' of our 'thinking process' and 'hardware aspects' of 'the neurons'—the principle element of the brain—we may succeed to some extent in the emulation process.

Biological neuronal processes are enormously complex, and the progress made in the understanding of the field through experimental observations is limited and crude. Nevertheless, it is true that this limited understanding of the biological processes has provided a tremendous impetus to the emulation of certain human learning behaviors, through the fields of mathematics and systems science. In neuronal information processing, there are a variety of complex mathematical operations and mapping functions involved, that, in synergism, act as a parallel-cascade computing structure. As system scientists, our objective is that, based upon this limited understanding of the brain, we create an intelligent cognitive system that can aid humans in various decision-making tasks. New computing theories under the category of *neural networks* have been evolving. Hopefully, these new computing methods with the neural network architecture as the basis will be able to provide a *thinking machine*—a low-level cognitive machine for which the scientists have been striving for so long.

The cognitive functions of brain, unlike the computational functions of the computer, are based upon *relative grades* of information acquired by the neural sensory systems. The conventional mathematical tools, whether deterministic or probabilistic, are based upon some absolute measure of information. Our natural sensors acquire information in the form of relative grades rather than in absolute numbers. The 'perceptions' and 'actions' of the cognitive process also appear in the form of relative grades. *The theory of fuzzy logic*, which is based upon the notion of graded membership, provides mathematical power for the emulation of the higher-order cognitive functions—the thought and perception process. A union of these two developing disciplines—neural networks and fuzzy logic—may strongly push the theory of independent field of cognitive information.

The subject of machine intelligence is in an exciting state of research and we believe that we are slowly progressing towards the development of truly intelligent machines. The present day versions of intelligent machines are not truly intelligent; however, the loose usage of the term 'intelligent' acts as a reminder that we have a long way to go.

Needs, Motivations, and Rationale

The combination of computing and communication has given rise to a society that feeds on information. In this information age, everyone strongly believes that information is power, and a must for success. Therefore, it is natural that computers have been collecting large amounts of information. Information repositories are collecting a myriad of information capturing operational experience in diverse fields. The collected information may be categorized on the basis of nature of experience:

- Experimental data (examples, samples, measurements, records, patterns or observations).
- Structured human knowledge (experience, expertise, heuristics) expressed in linguistic form (IF-THEN rules).

Computers capture and store tetrabytes of data across the world, daily. For instance, banks are creating records of how people spend their cash; hospitals are documenting details of the treatment given to patients, in terms of the duration of treatment, the response of the patients to the medicines, and so on. Similarly, engine monitoring systems in automobiles keep track of the working of the engine so that any malfunction can be detected; observatories around the world store high-resolution images of the night sky; medical science stores the results of medical tests from measurements as diverse as magnetic resonance imaging (MRI) scans and simple blood tests; bioinformatics stores huge amounts of data capable of measuring gene expression in DNA microarrays, along with protein transcription data and phylogenetic trees relating species to each other; and so on.

In man-machine control systems, an experienced process operator employs, consciously or subconsciously, a set of IF-THEN rules to control a process. The operator estimates the *important* process variables (*not* in numerical form, rather in *linguistic* graded form) at discrete time instants, and based on this information she/he manipulates the control signal. Intelligent machines based on the human experience in linguistic form have been very useful in process industry.

Lot of potential exists in business and finance. Decisions in these areas are often based on human induction, common sense, and experience, rather than availability of data. The scene is, however, changing very fast. The significant role of intelligent machines based on experience data in decision making is being realized. These machines are proving to be very useful in business decision-making.

Machine intelligence, in this modern era, has actually become an industry. Performing something useful with the stored data is the challenge. If the computers installed in banks can find out details related to the spending patterns of customers, is it possible for them to also detect credit card fraud promptly? If the data available with hospitals is shared, will it be possible to identify quickly treatments that do not work, and those that are expected to work? Will it be possible for an intelligent automobile to provide an early warning in case of an impending malfunction so that timely action can be taken? Is it possible to replace the process operators with intelligent machines?

The needs of the industry are motivating research in machine intelligence with focus on rationality: a machine is rational if it does the 'right thing' given what it knows; it acts so as to achieve the best outcome in terms of needs of the industry (or, when there is uncertainty, the best expected outcome). We will be concerned with the aspects of machine intelligence that serve the immediate needs of the industry.

Soft Computing/Machine Learning

In the conventional approach to solving decision problems, the variables related to the problem are identified (input or condition variables and output or action variables); the relationships among the variables are expressed in terms of mathematical (like algebraic, differential (ordinary/ partial), difference, integral or functional) equations, that fit our prior knowledge on the problem and cover the observed data. Actions (decisions) are given by analytical or numerical solutions of these equations. The statistical tools consider description of data in terms of probability measure rather than a deterministic function; the estimation methods yield the decisions. *Whenever devising mathematical/statistical model is feasible using a reasonable number of equations that can solve the given problem in a reasonable time, at a reasonable cost, and with reasonable accuracy, there is no need to look for an alternative.*

A long-standing tradition in science gives more respect to those approaches that are quantitative, formal, and precise. The validity of this tradition has been challenged by the emergence of new desires (problems, needs). Many of these problems do not lend themselves to precise solutions within the framework of conventional mathematical/statistical tools; for instance, problems of recognition of handwriting, speech, images; financial forecasting; weather forecasting; etc.

It is worth emphasizing that these and similar problems do not fit well within the framework of conventional mathematical tools—that does not lead to the conclusion that such problems cannot be solved. In fact, human intelligence has been routinely solving such problems and even more complex problems. The human intelligence can process millions of visual, acoustic, olfactory, tactile, and motor data, and it shows astonishing abilities to learn from experience, generalize from learned rules, recognize patterns, and make decisions. Human intelligence performs these tasks as well as it can, using ad hoc solutions (heuristics), approximations, low precision, or less generality, depending on the problem to be solved.

Compared to conventional mathematical tools, the solutions obtained using human intelligence are suboptimal and inexact. Though such solutions are imprecise, very often they yield results within the range of 'acceptability'. In fact, for some problems, even when precise solutions can be obtained, we settle for imprecise solutions within the range of acceptability, to save the cost involved in obtaining precise solutions.

The present-day scene is much different from yesterday; we now have ocean of data to be processed. Humans are unable to extract useful information from them. Computers of today can store this data and analyze it. However, to lead to meaningful analysis, human-like abilities need to be incorporated into software solutions. This, in fact, is the essence of *machine learning*.

Machine learning solutions are also mathematical in nature; however, these tools are different from conventional mathematical tools. With machine learning, a new mathematical theory has emerged which is built on the foundation of human faculties of learning, memorizing, adapting, and generalizing (recognizing the similarity between different situations, so that decisions that worked in one situation could be tried in another).

The basic premises of machine learning are as follows:

- *The real world is pervasively imprecise and uncertain.*
- *The precision and certainty carry a cost.*

The guiding principle of machine learning, which follows from these premises is as follows:

- *Exploit tolerance for imprecision, uncertainty, and partial truth, to achieve tractability, robustness, and low solution costs.*

Both the premises and the guiding principle differ strongly from those in classical mathematical tools (*hard computing*) which require precision and certainty. The machine learning mathematical tools (*soft computing*) exploit tolerance for imprecision (inherent in human reasoning) when necessary. Efficient soft computing techniques that are qualitative, informal, and approximate are now routinely solving problems that do not lend themselves to precise solutions within the framework of classical hard computing techniques; and also soft computing is supplementing/replacing 'the best for sure' hard computing solutions with 'good enough with high probability' low-cost solutions.

In this book, our primary focus is on learning from experimental data. Learning from *structured human knowledge* (IF-THEN rules) will be taken up in Chapter 6. Rule-based methods have found modest use in pattern recognition problems; nonetheless, they are gainfully being employed in process industry and manufacturing for function approximation problems. Also, the potential of success is very high in some business and finance applications.

1.2 WELL-POSED MACHINE LEARNING PROBLEMS

The field of machine learning is concerned with the question of how to construct computer programs that improve their performance at some task through experience. Machine learning is about making computers modify or adapt their actions (whether the task is making predictions, or controlling a robot) so that these actions get more accurate with experience, where accuracy is measured by how well the chosen actions reflect the correct ones. Put more precisely [1],

A computer program is said to learn from experience with respect to some class of tasks and performance measure, if the performance at the tasks, as measured by performance measure, improves with the experience.

In general, to have a well-defined learning problem, we must identify these three features:

- The learning task
- The measure of performance
- The task experience

The key concept that we will need to think about for our machines is *learning from experience*. Important aspects of 'learning from experience' behavior of humans and other animals embedded in machine learning are *remembering, adapting*, and *generalizing*.

- **Remembering and Adapting:** Recognizing that last time in a similar situation, a certain action (that resulted in this output) was attempted and had worked; therefore, it should be tried again or this same action failed in the last attempt in a similar situation, and so something different should be tried.
- **Generalizing:** This aspect is regarding recognizing *similarity* between different situations. This makes learning useful because we can use our knowledge in situations unseen earlier.

Given a situation not faced earlier, recognizing similarity with the situations faced earlier, we take a decision for the new situation—a generalizing capability of animal learning.

Machine learning concerns getting computers to *alter* or *adapt* their actions in a way that those actions improve in terms of accuracy, with experience. Machine learning, like animal learning, relies heavily on the notion of *similarity* in its search for valuable knowledge in data.

The computer program is the 'machine' in our context. The computer program is designed employing learning from the task experience. Equivalently, we say that the machine is *trained* using task experience, or machine *learns* from task experience. The terms: learning machine, learning algorithm, learned knowledge, all refer to a computer program design with respect to the assigned task.

In case of any software system, understanding the inputs and outputs is of greater importance than being aware of what takes place in between, and machine learning does just that. The input is defined by the learning task. Four different types of learning tasks appear in the real-world applications (details given later in Section 1.7). In *classification learning*, the expectation is that the machine will learn a technique of classifying examples of measurements/observations. In *association learning*, any relation between observations is required, not merely association capable of predicting a specific *class* value. In *clustering*, groups of observations that belong together are sought. In *regression*, the output to be predicted is not a discrete class but a continuous numeric quantity.

The classification and regression tasks are carried out through the process of *directed/supervised* learning. For the examples of measurements/observations, the outcome is known 'a priori'; for classification problems, the outcome is the class to which the example belongs; and for regression problems, the outcome is the numeric value on the approximating curve that fits the data. The other form of learning is *undirected/unsupervised*, wherein the outcome is not known 'a priori'; clustering and association learning belong to this category, as we shall see in later chapters.

The experience with which the machine will be *trained* (from which the machine will *learn*) may be available in the form of data collected in databases. Most of the information that surrounds us, manifests itself in the form of data that can be as basic as a set of measurements or observations, characterized by vectors with numerical values; or may be in forms which are more difficult to characterize in the form of numerical vectors—set of images, documents, audio clips, video clips, graphs, etc. For different forms of raw data (text, images, waveforms, and so forth), it is common to represent data in standard fixed length vector formats with numerical values. Such abstractions typically involve significant loss of information, yet they are essential for a well-defined learning problem.

Thus, though the *raw data* is an agglomerated mass that cannot be fragmented accurately into individual experience examples characterized by numerical vectors—yet it is very useful for learning many things. This book is about simple, practical methods of machine learning, and we focus on situations where input can be supplied in the form of individual experience examples in the form of numerical vectors.

Numerical form of data representation allows us to deal with patterns geometrically, and thus we shall study *learning algorithms* using linear algebra and analytic geometry (refer to Section 1.9). Characterizing the similarity of patterns in state space can be done through some form of *metric* (distance) measure: distance between two vectors is a measure of similarity between two corresponding patterns. Many measures of 'distance' have been proposed in the literature.

In another class of machine learning problems, the input (experience) is available in the form of *nominal* (or *categorical*) data, described in linguistic form (not numerical). For nominal form of data, there is no natural notion of similarity. Each learning algorithm based on nominal data employs some *nonmetric* method of similarity.

In an alternative learning option, there is no training dataset, but human knowledge (experience, expertise, heuristics) is available in linguistic form. This form of human knowledge, when properly structured as a set of IF-THEN rules, can be embedded into a learning machine. In the subsequent chapters, we will discuss learning algorithms (machines) that accept explicit prior structured knowledge as an input. These algorithms accept a long list of IF-THEN rules as training experience instead of or in addition to the training dataset.

One must typically pick up the learning algorithm that performs best according to some type of criterion. This criterion can be formulated in many different ways, but should ideally relate to the intended use of the learning machine, i.e., the learning task in hand. We will discuss the various commonly used performance criteria for different learning tasks in the next chapter.

Having described the *input* to the software system, let us now look at the *output* description. The output of an algorithm represents the learned knowledge. This knowledge is in the form of a *model* of the structural patterns in the data. The model is deployed by the user for decision-making; it gives the prediction with respect to the assigned task for measurements/observations not in the task experience; a good model will *generalize* well to observations unseen by the machine during training.

A block diagrammatic representation of a learning machine is shown in Fig. 1.1.

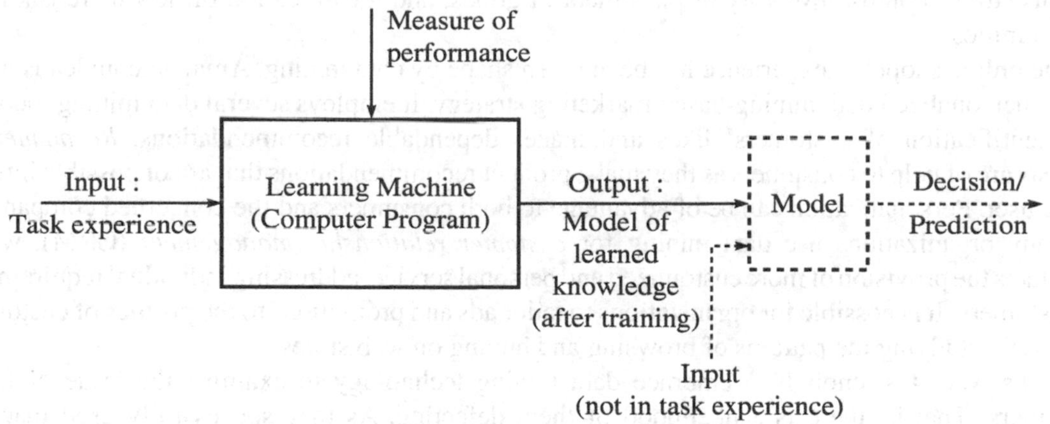

Figure 1.1 A block diagrammatic representation of a learning machine

1.3 EXAMPLES OF APPLICATIONS IN DIVERSE FIELDS

Machine learning is a growing technology used to *mine knowledge from data* (popularly known as *data mining* field (Section 1.8)). Wherever data exists, things can be learned from it. Whenever there is excess of data, the mechanics of learning must be automatic. Machine learning technology is meant for automatic learning from voluminous datasets.

The range of tasks handled by machine learning is fast expanding. Applications emerge not from machine learning experts, nor from the data itself, but from people who work with the data and the problems from which it arises.

In fact, data mining exists in multiple aspects of our daily lives, but we may or may not realize it. Instances of such 'ubiquitous' or 'invisible' data mining are web services that adapt to customers, search engines, e-mail managers, and so on.

Google is by far the most popular and extensively used of all search engines. It offers access to information from billions of web pages, which have been indexed on its server. We type some keywords pertaining to our topic of interest, and Google throws back a list of websites on that topic, mined, indexed, and organized by a set of data mining algorithms including *PageRank*. The popularity of Google has even led to a new verb in the English language, 'google', which means 'to search for (something) on the Internet using any comprehensive search engine'.

While going through the results of our Google query, many different advertisements show up relating to our query. To tailor ads to match the interests of the users is a strategy by Google and is one of the typical services that every Internet search provider tries to offer. Mining *information on the World Wide Web is an area that is fast growing, almost exploding*.

Data mining is also actively applied in the area of marketing and sales. In these spheres, organizations have huge volumes of accurately recorded data, which is potentially very valuable. One instance is Wal-Mart, with innumerable customers visiting its innumerable stores on a weekly basis. These stores obtain the data pertaining to their products and analyze the same with the help of data mining software. This permits the identification of customers' purchase patterns at various stores, control inventory of placement of goods, and identification of new merchandizing opportunities.

The online shopping experience has been given shape by data mining. Amazon.com leads in the use of personalized data mining-based marketing strategy. It employs several data mining methods for identification of customers' likes and makes dependable recommendations. *Recommender systems* are of help to consumers as they make product recommendations that are of possible interest to the user. Personalization can be of advantage to both consumers and the concerned company.

Many organizations use data mining for *customer relationship management* (CRM), which facilitates the provision of more customized and personal service, addressing individual requirements of customers. It is possible for organizations to tailor ads and promotions to the profiles of customers by closely studying the patterns of browsing and buying on web stores.

Banks were fast enough to embrace data mining technology to examine the issue of fickle customers. That is, there is a likelihood of them defecting. As they successfully used machine learning to assess credit, it was possible to reduce *customer attrition*. Cellular phone companies handle *churn* by identifying behavioral patterns that gain from new services, and then promoting such services in order to retain their customer base. These days, it is common to get a phone call from one's credit card company about unusual or questionable spending behavior. Credit card companies detect fraud and wrong usage with the help of data mining, and manage to save billions of dollars annually.

Data mining has greatly impacted the ways in which people use computers. On getting on to the Internet, for instance, let us say we feel like checking our email. Unknown to us, many irritating emails have already been noticed using *spam filters* that use machine learning to identify spam.

Computer network security is a continually rising issue. While protectors keep hardening networks, operating systems, and applications, attackers keep discovering weak spots in all these areas. Systems for detecting intrusions are able to detect unusual patterns of activity. Data mining is being applied to this issue in an attempt to find out semantic connections among attacker traces in computer network data. *Privacy-preserving data mining* assists in protecting privacy-sensitive information, such as credit card transaction records, healthcare records, personal financial records, biological features, criminal/justice investigations, and ethnicity.

Of late, huge data collection and storage technologies have altered the landscape of scientific data analysis. Major examples include applications which involve natural resources, the prediction of floods and droughts, meteorology, astronomy, geography, geology, biology, and other scientific and engineering data. Machine learning/data mining is present in all of these examples.

To arouse the reader's curiosity about machine learning, we examine some application domains, wherein machine learning is present and is yielding encouraging results.

Machine Vision: It is a field where pattern recognition has been applied with major successes. A machine vision system captures images through a camera and analyzes these to be able to describe the image. A machine vision system is applicable in the manufacturing industry, for automating the assembly line or for automated visual inspection. For instance, in inspection, manufactured objects on a moving conveyor may pass the inspection station, where the camera stands, and it has to be established whether a flaw or fault exists. Therefore, images have to be analyzed online, and a pattern recognition system has to categorize the objects into the 'defect' or 'non-defect' category. A robot arm can then put the objects in the right place.

With digital cameras and optical scanner becoming commonplace accessories, medical imaging technology producing detailed physiological measurements, laser scanners capturing 3D environments, satellites, and telescopes bringing pictures of earth and distant stars, there has been a flood of images. Researchers are actively working on the task of analyzing these images for various purposes.

Biometric Recognition: It has been made clear by decades of research in pattern recognition that the level of visual understanding and recognition that humans exhibit cannot be matched by computer algorithms. Certain problems, such as biometric recognition (fingerprints identification, face and gesture recognition, etc.) are being handled with success, but general purpose image-representation systems are still not visible on the horizon.

Handwriting Recognition: It is another area where pattern recognition can be applied, with major consequences in automation and information handling. Take first the simpler problem of printed *character recognition*. The commercially available Optical Character Recognition or OCR system has a light source, a document transport, as well as a detector. At the point where the light-sensitive detector gives output, light intensity variation is translated into 'numbers'. On this image array, image processing and pattern recognition methods are applied to identify the characters—that is, to categorize each character into the correct 'letter', 'number', and 'punctuation' class.

In addition to the printed character recognition systems, there is a lot of interest in handwriting recognition systems. A typical commercial application of such a system is in the machine reading of bank cheques. Another application is in automatic mail sorting machines for postal code identification in post offices.

Today, the tendency is to create and develop machines that possess human-like skills. *Pen computers*, for instance, is an area of huge potential interest: entry of data done not via the keyboard but through writing. Online handwriting-recognition systems have potential to offer a solution.

Medical Diagnosis: It also uses pattern recognition. Doctors make use of it while making diagnostic decisions. The ultimate diagnosis is, of course, made by the doctor. Computer-aided diagnosis has been applied to, and is of interest for, a range of medical data—X-rays, computed tomographic images, ultrasound images, electrocardiograms (ECGs), and electroencephalograms (EEGs).

Alignment of Biological Sequences: Alignment of sequences is done on the basis of the fact that all living organisms are related by evolution. This means, nucleotide (DNA, RNA) and amino acid (proteins) sequences of species that have evolved close to each other, should display more similarities. An alignment is the procedure of lining up sequences to obtain a maximum identity level, which also expresses the level of similarity between sequences. Biological sequence analysis is significant in *bioinformatics* and modern biology.

Drug Design: It is usually based on a long and expensive process involving complex chemical experiments to check whether or not a particular chemical compound could be a good candidate for a specific drug, which would be a positive result involving further clinical experiments. For several years, a new scheme based on computational simulations has been emerging. The general idea is to assess the feasibility of a chemical compound for the synthesis of the drug with a predictive model based on a database of previous experiments.

Speech Recognition: It is an area that has been well researched. Speech is the most natural means by which humans share, convey and exchange information. Intelligent machines that recognize spoken information can be used in numerous applications, for example, to help control machines by talking to them—entering data into a computer via a microphone. Speech recognition can enhance our ability to communicate with the deaf and dumb.

Text Mining: It concerns indentification of patterns in text. The procedure involves analysis of text for extraction of useful information for specific purposes.
The way information available on the Web and on corporate intranets, digital libraries, and news wires is spread or propagated, is overwhelming. Integration of this information into the decision-making process, at a fast pace, is essential in order to help businesses stay competitive in today's market. Text mining has reached the industrial world and is helping to exploit knowledge that, due to its shear size, is often beyond human consumption. Typical jobs for mining text databases are—classification of documents into predefined classes, grouping together of similar documents, and identifying documents that fulfill the criteria/specifications of a query.

Natural Language Processing: Ever since the computer age dawned, computer science research has been attempting to understand human language. In 1950, immediately following the first invention of the computer, Alan Turing, one of the greatest computer scientists of the twentieth century, suggested a test for computer intelligence. In a paper titled "*Computing Machinery and Intelligence*", he introduced this machine. Over sixty years later, computers could perform extraordinary actions that Alan Turing probably never imagined could be possible.

Language is obviously a critical component of how people communicate and how information is stored in the business world and beyond. The goal of Natural Language Processing (NLP) is to analyze, understand, and generate languages that humans use naturally so that eventually a computer will 'naturally' be able to interpret what the other person is saying. Voice automation is just starting with robot vacuum cleaners that respond to cleaning orders; telephones and household appliances that obey voice commands. Other NLP applications include spell checking and grammar checking, translation from one human language to another, searching the documents, summarizing the documents, and human-computer interfaces. However, natural language processing is still mostly a future application area.

Fault Diagnostics: Preventive upkeep of motors and generators and other electromechanical devices, can delay malfunctions that may otherwise interrupt industrial procedures. Typical defects or flaws include misalignment of shaft, mechanical slackening, defective bearings, and unbalanced pumps. Over a thousand different types of devices may be used in a chemical plant—from small pumps to huge turbo-alternators. Machine learning is extremely helpful in such applications.

Machine learning technologies are increasingly facilitating the real-time monitoring of bridges and highways. The state (health) of a bridge is continually assessed on the basis of inputs from a series of sensor networks and the processing of the received data.

Load Forecasting: It is quite essential to establish future power demand in the electricity supply industry. In fact, the earlier the demand is known, the better. Precise estimates can be made with the help of machine learning methods for the maximum and minimum load for each hour, day, month, season, and year. Utility companies create important economics in setting the operating reserve, maintenance scheduling, fuel inventory management, and so on.

Control and Automation: A quiet revolution is ongoing in the manufacturing world which is changing the look of factories. Computers are controlling and monitoring manufacturing processes with a high degree of automation, facilitated by machine learning techniques. The computer control includes control of all types of processes such as *Computerized Numerical Control* (CNC), welding, electrochemical machining, etc., and control of *industrial robots*. High degree of automation is applied in today's *Flexible Manufacturing Systems* (FMS) that can be readily rearranged to handle new market requirements. Flexible manufacturing systems, combined with automatic assembly and product inspection on one hand, and CAD/CAM system on the other, are the basic components of the modern Computer Integrated Manufacturing System.

Sophisticated processes of process/chemical industry involve tasks like acquisition of process data (i.e., collection of instantaneous values of individual process variables, and status messages of plant control facilities (valves, pumps, motors, etc.)); processing of collected data; plant hardware monitoring, system check and diagnosis, closed-loop control and logic functions. Processing of data helps provide optimal plant work conditions—this often requires tweaking control parameters. For example, separating crude oil from natural gas is an essential prerequisite to oil refinement, and controlling the separation process is a tricky job. Machine learning is being used to create rules for setting the parameters. Machine learning plays a significant rule in modern Computer Integrated Process Systems.

It is essential to remind the readers here that whenever reasonably accurate mathematical model of a manufacturing/chemical process can be derived, controllers are designed using these models, and not by machine learning techniques which are based on empirical models. Machine learning aids in achieving automation of the overall manufacturing system/processing system. Hybridization of model-based control with machine learning techniques achieves strong robustness in presence of parameter variations. This is true for many other scientific and engineering applications as well.

Business Intelligence: It is essential for businesses to be able to comprehend the commercial control of their organization well, in terms of customer base, market, supply and resources, and competition. Business Intelligence (BI) technologies offer not only historical and current information but also predictive views of business operations. Data mining is the fundamental core of business intelligence. In the absence of data mining, many businesses may be unable to effectively perform market analyses, compare customer feedback on similar products, find the strengths and weaknesses of their competitors, retain extremely valuable customers, and arrive at intelligent business decisions. Several customized data mining tools have been created to cater to domain-specific applications in many areas—finance, retail industry, telecommunications, health care, insurance, and so on. Section 9.6 will give an overview of these applications.

Machine learning/data mining is omnipresent, as can be seen from these examples. We could go on and on with such scenarios. We may now conclude that machine learning/data mining is an empirical technology that has applications in all knowledge domains: economics, social science, physics, chemistry, biology, computer engineering, electrical engineering, mechanical engineering, chemical engineering, civil engineering, business management, and others. Initial exploration of applications in a specific domain may be done by exploiting information available through search engines. Hopefully, this will provide enough fodder to the reader for moving on.

The reader may go a little deeper in these applications to ascertain that conventional mathematical approach is not feasible. We re-emphasize the fact that wherever an accurate mathematical model can be built, machine learning approach need not be used.

1.4 DATA REPRESENTATION

Experience in the form of raw data is a source of learning in many applications (human knowledge in linguistic form is an additional learning source). Raw data require some *preprocessing* (discussed in detail in later chapters) with respect to the class of tasks. This leads to an *information system*, that represents the knowledge in the raw data (input to the learning machine shown in Fig. 1.1) used for decision making.

The information-system data (*representation-space data*) may be stored in data warehouse (refer to Section 9.3). *Data warehousing* provides integrated, consistent, and cleaned data to machine learning algorithms. However, machine learning is not confined to analysis of data accessed online from data warehouses. For many applications, we can assume availability of data in a *flat file*, which is a simple *data table*. In fact, all through the book, we have worked on data tables, postponing warehousing discussion to Chapter 9.

Information system is a form of data table $\mathcal{D}$; each row of the table represents a measurement/observation, and each column gives the values of an *attribute* of the information system for all measurements/observations. Different terms have been used to name the rows depending on the context of application. Some commonly used terms are: *instances, examples, samples, measurements, observations, records, patterns, objects, cases, events*. Similarly, different terms have been used to name the columns; *attributes* and *features* being the most common.

An example information system dataset, given in Table 1.1, is simply a set of patients specified by three attributes/features: Headache, Muscle-pain, and Temperature. These attributes are described in linguistic form. *Nominal representation* (linguistic form) of knowledge is common for problems with small datasets. For problems with large datasets, each attribute has a *numerical* value (real) for the observation. In Table 1.1, patient index gives observations.

Table 1.1 An example dataset

Patient index	Headache	Muscle-pain	Temperature	Flu
1	no	yes	high	yes
2	yes	no	high	yes
3	yes	yes	very high	yes
4	no	yes	normal	no
5	yes	no	high	no
6	no	yes	very high	yes

For directed/supervised learning problems, an outcome for each observation is known *a priori*. This knowledge is expressed by one distinguished attribute, called the *decision attribute*. Information systems of this kind are called *decision systems*.

The last column in Table 1.1 represents a decision attribute with respect to the task to categorize patients into two classes: {Flu: yes}, {Flu: no}; Flu is the decision attribute with respect to the *condition attributes*: Headache, Muscle-pain, and Temperature.

In a decision system data file, we represent input as N instances $s^{(1)}, s^{(2)}, \ldots, s^{(N)}$; each is an example of the concept to be learned. Each individual instance $s^{(i)}$; $i = 1, 2, \ldots, N$, that provides the input to the machine learning tool, is characterized by its values for a fixed predefined set of n features/attributes $x_1, x_2, \ldots, x_n$ (x_j; $j = 1, 2, \ldots, n$). The instances are rows of data table, and features are the columns. A template for such a data table is shown in Table 1.2.

Table 1.2 A template for data table

Features x_j / Instances $s^{(i)}$	x_1	x_2	…	x_n	Decision y
$s^{(1)}$			…		
$s^{(2)}$			…		
⋮			…		⋮
$s^{(N)}$			…		

The value of the attribute for a specific instance is a measurement/representation of the quantity to which the attribute refers. There is a wide distinction between *numeric* and *nominal* quantities. Numeric attributes, often referred to as *continuous attributes*, adopt values that are either real numbers or integer values. The word 'continuous' is often wrongly used in this context; integer-valued attributes are not continuous mathematically speaking. Nominal attributes take on values in a prespecified finite set of possibilities, and are at times referred to as *categorical* or *discrete*.

In Table 1.2, the training experience is available in the form of N examples: $s^{(i)} \in S$; $i = 1, 2, \ldots, N$; where S is a set of possible instances.

We specify an instance by a fixed number n of attributes/features x_j ; $j = 1, 2, \ldots, n$. We can visualize each instance with n numerical features as a point in n-dimensional state space $\Re^n$:

$$\mathbf{x} = [x_1 \; x_2 \cdots x_n]^T \in \Re^n \tag{1.1}$$

The set $\mathbf{X}$ is a finite set of feature vectors $\mathbf{x}^{(i)}$ for all the possible instances. We can visualize $\mathbf{X}$ as a region in the state space $\Re^n$ to which the instances belong, i.e., $\mathbf{X} \subset \Re^n$. Note that $\mathbf{x}^{(i)}$ is a representation of $s^{(i)}$, and $\mathbf{X}$ is the *representation space*.

The pair $(S, \mathbf{X})$ constitutes the information system where S is a non-empty set of instances, and $\mathbf{X}$ is a non-empty set of features; we have represented instances by the index 'i', and features by the index 'j':

$$\begin{aligned} &\{s^{(i)}; i = 1, 2, \ldots, N\} \in S \\ &\{\mathbf{x}^{(i)}; i = 1, 2, \ldots, N\} \in \mathbf{X} \\ &\{x_j^{(i)}; j = 1, 2, \ldots, n\} = \mathbf{x}^{(i)} \end{aligned} \tag{1.2}$$

Features x_j ; $j = 1, 2, \ldots, n$, may be viewed as state variables, and feature vector $\mathbf{x}$ as state vector in n-dimensional state space.

With every feature x_j, we associate a set $V_{x_j} \in \Re$ of its *values*, called the *domain* of x_j ; $j = 1, 2, \ldots, n$.

Therefore, $V_{x_j}^{(i)} \in V_{x_j}$; $i = 1, 2, \ldots, N$.

Attribute x_j may take a finite value from a finite set of d_j discrete values $v_{1x_j}, v_{2x_j}, \ldots, v_{d_j x_j}$. In such a case,

$$V_{x_j} = \{v_{1x_j}, v_{2x_j}, \ldots, v_{d_j x_j}\} = \{v_{lx_j}; l = 1, 2, \ldots, d_j\} \tag{1.3}$$

$$\text{and } v_{lx_j}^{(i)} \in V_{x_j};\ i = 1, 2, \ldots, N.$$

The tuple $(S, \mathbf{X}, Y)$ constitutes a decision system where a finite set of condition attributes $x_j \in \mathbf{X}$, and the decision attribute (output) $y \in Y$.

We can visualize Y as a one-dimensional region in state space, i.e., $Y \subset \Re$.

In a data table for decision systems, rows correspond to instances belonging to S, and columns correspond to attributes/features belonging to $\mathbf{X} \cup Y$.

With decision attribute y, we associate a set V_y of its values. For multi-class (M classes) classification problems, the domain of y is given by the set:

$$V_y = \{1, 2, \ldots, M\} \tag{1.4a}$$

The output $y^{(i)}$ for a pattern $s^{(i)}$ will take a value from the set V_y, which represents the class to which the pattern belongs.

We may express attribute y as,

$$\begin{aligned} y \in \{y_1, y_2, \ldots, y_M\} &= \{y_q ; q = 1, \ldots, M\} \\ \{y_1, y_2, \ldots, y_M\} &= \{1, 2, \ldots, M\} \end{aligned} \tag{1.4b}$$

For regression (numeric prediction) problems,

$$V_y \in \Re \tag{1.5}$$

1.4.1 Time Series Forecasting

In this section, we describe the data structure of forecasting (predicting the future) problems. Some examples of such problems are predicting the closing price of IBM stock, the daily value of the Euro to U.S. dollar exchange rate, the future demand for power as far in advance as possible in the electricity supply industry, etc. For such problems, the data naturally falls into a *time series*, and therefore these are *time series prediction* problems.

For financial time series, if we are able to predict the next value, or even whether the series is heading up or down, it has a tremendous advantage over other competitors. Sound predictions of foreign exchange rates are increasingly important to financial managers of multinational companies. With accurate estimates of load demand, electricity supply industry can make significant economics in areas such as setting the operating reserves, maintenance scheduling, and fuel inventory management.

Time series data is *sequential*—a sequence of observations is measured over time; each observation is indexed by a time variable t. The measurements are often made at fixed time intervals so that without loss of generality, t can be taken as an integer.

Some powerful statistical methods to develop models for time-series data are well known, for example, ARIMA (Auto-Regressive, Integrated, Moving Average) model. Our concern here is with NARMA (Nonlinear, Auto-Regressive, Moving Average) model [199]. In the extensive literature on modeling dynamic systems, it has been proved, after making some moderate assumptions, that any nonlinear, discrete, time-invariant system can always be represented by a NARMA model:

$$y(t+1) = f(y(t), y(t-1), \ldots, y(t-n)) \tag{1.6a}$$

where $y(t)$ is the output signal at time t, and $y(t-1)$, $y(t-2)$, …, represent past values of this signal. Equation (1.6a) is a simple deterministic version of NARMA model (there are no noise terms in it). These models assume that the system is dependent on itself (autoregressive part). In realistic cases, the nonlinear function $f(\cdot)$ is very complex and generally unknown. Our concern here is with inferences from the past and current observations, leading to learning based inductive techniques.

In the next chapter, we will see that the basic assumption regarding data for machine learning applications is that the instances that constitute a sample (dataset) are *iid* (independently and identically drawn). This assumption is, however, not valid for applications where successive instances are dependent. Processes where there are sequence of observations cannot be modeled as simple probability distributions. We have used NARMA model for time-series forecasting applications. MDP (Markovian Decision Process) model will be used in Appendix B for reinforcement learning applications.

Typically, successive events or inputs that affect the time series, are serially correlated; this causes a time series pattern that can give some hint of the future. We will model time series by assuming that there is a relationship between the value at a given time t, and the values at the earlier times $t-1$, $t-2$, $t-3$, and so on, for as many *lags* as required.

The input-output structure for *univariate* time series is shown in Fig. 1.2(a). The model gives predictions of response variable y in terms of current and past values of y. It captures the following functional relation:

$$\hat{y}(t+k) = f(y(t), y(t-1), \ldots, y(t-n)) \tag{1.6b}$$

$y(t+k)$ is unknown. We want the model to predict its value. $\hat{y}(t+k)$ is the prediction of $y(t+k)$; this is *k-step ahead prediction.*

The number of *lags*, n, define the attributes/features for prediction. It is expected that more the past information (lags) used to predict the future, more accurate the prediction. It is, however, not true in general. Computational complexity, redundant information (noise), etc., introduced by large number of features are to be avoided, as we shall see in later chapters. For k-step ahead prediction problem, $n \geq k$.

Time series prediction is typically a regression problem. Machine learning techniques provide a data-driven approach that can capture linear/nonlinear data structures without prior assumption about the underlying relationship in a particular problem. To build a machine learning model, we require n (a design variable) features, and past and current experience in terms of N (a design variable) feature vectors. For the response variable y, the time series data is given by,

$$\{y(t), y(t-1), y(t-2), \ldots\} \tag{1.7a}$$

The n features $x_1, x_2, \ldots, x_n$:

$$x_1 = y(t-1), x_2 = y(t-2), x_3 = y(t-3), \ldots, x_n = y(t-n)$$

The past N measurements/observations:

$$\mathbf{x} = [x_1 \;\; x_2 \cdots x_n]^T : \text{Feature vector}$$

$$\begin{aligned}
\mathbf{x}^{(1)} &= \{x_1^{(1)}, x_2^{(1)}, \ldots, x_n^{(1)}\} = \{y(t-1), y(t-2), \ldots, y(t-n)\} \\
\mathbf{x}^{(2)} &= \{x_1^{(2)}, x_2^{(2)}, \ldots, x_n^{(2)}\} = \{y(t-2), y(t-3), \ldots, y(t-n-1)\} \\
&\vdots \\
\mathbf{x}^{(N)} &= \{x_1^{(N)}, x_2^{(N)}, \ldots, x_n^{(N)}\} = \{y(t-N), y(t-N-1), \ldots, y(t-n-N+1)\}
\end{aligned} \tag{1.7b}$$

Experience data:

$$\{\mathbf{x}^{(i)}, y^{(i)}\};\; i = 1, 2, \ldots, N,$$

$$y^{(1)} = y(t), y^{(2)} = y(t-1), \ldots, y^{(N)} = y(t-N+1) \tag{1.7c}$$

Using experience data, we construct a machine learning model.

Note that the time series data is not limited to data from just a single time series. It can include inputs from multiple time series. For instance, to predict the value of the Euro to U.S. dollar exchange rate, other time-series information might be included, such as volume of the previous day's transactions, the U.S. dollar to Japanese yen exchange rate, the closing value of the stock exchange, and the day of the week.

The input-output structure for *multivariate* time series model is shown in Fig. 1.2(b). The model explains the variation in response variable y in terms of the variations in inputs from time series. z_1, z_2, …, z_m; where each time series z_l; $l = 1, \ldots, m$, is given by,

$$\{z_l(t), z_l(t-1), z_l(t-2), \cdots\} \tag{1.8a}$$

The lagged data $z_1(t-1), \ldots, z_1(t-L_1), \ldots, z_m(t-1), \ldots, z_m(t-L_m)$ are used to predict output $y(t+k)$; where k is some prediction time-step into the future, and $L_1, L_2, \ldots, L_m$ represent the number of lags in time series z_l; $l = 1, \ldots, m$, used for constructing the model.

For the l time series, we define l variables z_l; $l = 1, \ldots, m$. For each variable z_l, we define the lagged data $z_l(t-1), \ldots, z_l(t-L_l)$. The feature vector $\mathbf{x}$ for the dataset is then given by,

$$\{x_1, x_2, \ldots, x_n\} = \{z_1(t-1), \ldots, z_1(t-L_1), z_2(t-1), \ldots, z_2(t-L_2), \ldots, z_m(t-1), \ldots, z_m(t-L_m)\} \tag{1.8b}$$

$x_1 = y(t-1)$
$x_2 = y(t-2)$
$\vdots$
$x_n = y(t-n)$
Predictor
$\hat{y}(t+k)$

(a) Input-Output structure for a univariate time series data

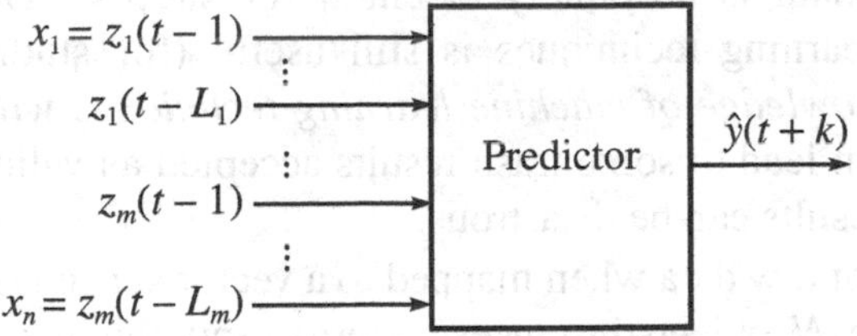

(b) Input-Output structure for multivariate time series data

Figure 1.2

1.4.2 Datasets for Toy (Unrealistically Simple) and Realistic Problems

Various datasets have the tendency to reveal new issues and challenges, and it is interesting and instructive to consider a range of problems while considering learning techniques. The set of problems stated in this book is unrealistically simple. To apply machine learning seriously would mean involving innumerable individual cases. However, when explaining what algorithms do and the manner in which they work, we require examples capable of capturing the essence of the problems, but at the same time, basic enough to be comprehensible in terms of detail.

The datasets employed in this book, for instance, are proposed to be 'academic' in nature—they will facilitate understanding of what is going on.

Some realistic fielded applications of learning techniques have been given in Section 1.3, and many more will be covered later in Chapter 9. It is obvious that the knowledge base required for handling such applications is not just how learning algorithms work, but much more than that; for example, domain knowledge for the application, open source/commercial software usage or writing code in any appropriate language from the equations given in the book, etc. In a university setting, students must be trained for realistic applications through the project work.

Description of datasets for some realistic problems is given at the end of the book. This will help the reader appreciate better the real-world environment for machine learning.

1.5 DOMAIN KNOWLEDGE FOR PRODUCTIVE USE OF MACHINE LEARNING

The productive use of machine learning is not just a matter of finding some data and then blindly applying machine learning algorithms to it. Available commercial work-benches make that easy to do, even with little apparent understanding of what these algorithms do. The usefulness of such results is questionable.

The datasets used in this book are *toy* problems. They are not applications of machine learning which have actually been put to use. They have a limited objective of providing an understanding of what machine learning algorithms do.

The productive use of machine learning is in designing a learning system that is put to work. The design of such learning systems requires, in addition to the knowledge of what various machine learning algorithms do, a deep knowledge of the application domain. The fielded applications listed in Section 1.3 give an idea of various problem domains. Many of them are speculative research domains, but some of the applications in these domains have been put to productive use employing machine learning techniques.

Knowledge of the domain is absolutely essential for success. Domain knowledge without knowledge of machine learning techniques is still useful (for studies based on non-machine learning methods), but *knowledge of machine learning techniques without domain knowledge is of no productive use*; it can lead to some trash results accepted as valid conclusions, and strategic decisions based on such results can be disastrous.

We have seen earlier that raw data when mapped to a vector space is $N \times n$ matrix (*data matrix*); the N rows represent the N objects/instances/patterns, and the n columns represent features/attributes. For many practical applications, the features/attributes are numeric in nature. Mapping of raw data to vector spaces requires appropriate processing of the data.

Today, raw data are no longer restricted to numerical measurements/observations only. Machine intelligence is capable of dealing with multimedia data: image, audio, text. Conversion of multimedia raw data to vector space is a tedious task requiring in-depth knowledge in the application area (refer to Section 9.5).

The problems of *feature generation* and *feature selection* must be addressed at the outset of any machine learning system design. The key is to choose features that

- are computationally feasible;
- lead to 'good' machine-learning success; and
- reduce the problem data into a manageable amount of information without discarding valuable (or vital) information.

Generation of features for patterns/objects in a machine learning problem is very much application dependent. Although similarities among various applications do exist, there are also big differences. Domain knowledge plays a crucial role in generating features that will subsequently be fed to a machine learning algorithm. Each feature must carry valuable information with respect to the machine learning task. Also, two features carrying 'good' information when treated separately, may be highly mutually correlated; in that case, there is little gain in including both of them in the feature vector.

In general, the compactness of feature vector in a data matrix is a desirable characteristic. There are many reasons for the necessity of compact feature vector. Computational complexity is the obvious one. The major reason is imposed by the required *generalization* properties of the resulting learning machine. As we shall see in the next chapter, for a finite and usually limited number, N, of learning patterns (statisticians give us procedures to learn with some precision how many patterns, N, we should need to achieve a given degree of reliability with a given dataset and a selected machine learning technique (discussed in next chapter)), keeping the number of features as small as possible is in line with our desire to design learning machines with good *generalization* capabilities (a large number of features is directly translated into a large number of machine parameters).

The requirement of a compact set of features is thus crucial in machine learning task. If we select features with little information with respect to the machine learning task, the subsequent design of the machine would lead to poor generalization, and hence poor performance. On the other hand, if information-rich features are not selected, the resulting data matrix will not represent the raw data (and hence, available knowledge) for the application in hand. Naturally, such a machine does not possess practical value for the application.

The usual procedure is to initially select a large number of features which are expected to carry rich information. Domain knowledge plays a significant role in this step of feature generation. The next step is to select a compact set of features by systematically reducing the features of the initial large set. Although automated methods for feature reduction are available (Chapter 7), it is important to ensure that the reduced set given by these methods does not weed out information-rich features, and also does not retain too much of irrelevant and redundant features. As we shall see in Chapter 7, domain knowledge integrated with the automated methods leads to better results.

The more data we tackle, the greater the chances of encountering erroneous values, emerging from measurement error, data-entry error, and so on. If the erroneous value lies in the same range as the remaining data, it may bear no harm. But, if it lies beyond the range of data (e.g., a misplaced decimal), it may substantially impact some of the machine learning processes we intend to employ. The values lying far away from the bulk of the data are known as *outliers*. As we shall see in Chapter 7, certain rules help identify outliers, but there are no statistical rules to help us find out whether such outliers are caused by an error. The answer may lie in domain knowledge.

Some datasets consist of variables with a very large number of *missing values*. In that case, dropping data patterns with missing values will only result in massive data loss. Imputing the missing values (Chapter 7) may also be of no use as the imputations are done on the basis of a small number of existing data patterns. Another option is to study how important the attribute is. If it is not important, it can be removed. If the attribute is crucial, the ideal solution is to invest in procuring the missing data. Domain knowledge is rather important in such scenarios.

We can go on citing examples of the importance of domain knowledge for productive use of machine learning. As we will learn when we progress with the book, probably at every stage of the design cycle for various machine learning techniques, domain knowledge plays a significant role.

1.6 DIVERSITY OF DATA: STRUCTURED/UNSTRUCTURED

Generally, digital information can be categorized into two classes—structured and unstructured. Studies have recently revealed that 70–80 per cent of all the data available to corporations today is unstructured, and this figure is only increasing at a rapid rate.

Let us look at the types of data from a layman's point of view.

Usually, traditional data sources exist in the structured realm, which means, traditional data follows a predefined format, which is unambiguous and is available in a predefined order. For instance, for the stock trade, the first field received should be the date in an *MM/DD/YYYY* format. Next could be an account number in a 12-digit data format, followed by stock symbol—three to five digit character field, and so on. Structured data fits into a fixed file format (data table of the form shown in Table 1.2).

We rarely have any control on unstructured data sources, for instance, text streams from a social media site. It is not possible to ask users to follow specific standards of grammar or sentence ordering or vocabulary. We will only get what we get when people post something. This amounts to capturing everything possible, and worrying later about what matters. This makes such data different from traditional data. The 'big data' problems (to be introduced in Chapter 9) are faced with this type of data that we get largely from unstructured data sources.

For the structured data, *first-generation data mining algorithms* (we have used the term 'first-generation', although 'next generation' is still envolving and is yet to stabilize) are in effective use in various kinds of applications wherein it becomes possible to derive numeric and categorical features. For many applications listed in Section 1.3, numeric/categorical features can be extracted. As we shall see in Section 9.5, limited analysis objectives can be realized for applications involving text, image, and audio raw data by extracting numeric features from text/image/audio data, that is, some structured patterns found in unstructured datasets lead to some useful analysis using first-generation machine learning algorithms. However, the need to go beyond this state of the art (dealing with unstructured text/image/audio/mixed multimedia data) is still mostly a research challenge.

Diversity of data, leading us to unstructured domain in a big way, poses a big research challenge: mining sequence data, mining graphs and networks, mining spatiotemporal data, mining cyber-physical data, mining multimedia data, mining web data, mining data streams, and other issues. In addition to complexity in data, the volume of data is too massive. Scaling to complex, extremely large datasets—*the big data*—is probably the most debated current research issue. In Section 9.7, we provide an overview of the evolving database technologies for higher levels of scalability, and the evolving machine learning techniques as a result of new scalability. The *next-generation machine learning* is on an evolving platform today.

Though big data is difficult to handle, it has come into prominence because of the long-term strategic value as well as immediate and tactical value associated with it. Perhaps the most exciting thing about big data is not what it will do for a business by itself, but what it will do for a business when combined with an organization's other data. It is critically important that organizations do not develop a big data strategy that is distinct from their traditional data strategy. To succeed, organizations need to develop a cohesive strategy that is not a distinct 'stand alone' concept. Rather big data must be simply another facet of an enterprise data strategy.

This book is mostly concerned with traditional (structured) data, and the first generation machine learning algorithms.

1.7 FORMS OF LEARNING

In the broadest sense, any method that incorporates information from experience in the design of a machine employs *learning*. A learning method depends on the type of experience from which the machine will learn (with which the machine will be trained). The type of available learning experience can have a significant impact on the success or failure of the learning machine.

The field of machine learning usually distinguishes four forms of learning: supervised learning, unsupervised learning, reinforcement learning, and learning based on natural processes—evolution, swarming, and immune systems.

1.7.1 Supervised/Directed Learning

The machine is designed by exploiting the *a priori* known information in the form of 'direct' training examples consisting of observed values of system states (input vectors): $\mathbf{x}^{(1)}, \ldots, \mathbf{x}^{(N)}$, and the response (output) to each state: $y^{(1)}, \ldots, y^{(N)}$.

The 'supervisor' has, thus, provided the following data:

$$\begin{aligned} \mathcal{D} &= \{s^{(i)}, y^{(i)}\}; i = 1, \ldots, N \\ s^{(i)} &= \mathbf{x}^{(i)} : \{x_1^{(i)}, x_2^{(i)}, \ldots, x_n^{(i)}\} \end{aligned} \tag{1.9}$$

The dataset $\mathcal{D}$ is used for inferring a model of the system.

If the data $\mathcal{D}$ lies in the region $\mathbf{X}$ of the state space $\Re^n$ ($\mathbf{X} \subset \Re^n$), then $\mathbf{X}$ must be fully representative of situations over which our machine will later be used. Choice of features/attributes $x_j; j = 1, \ldots, n$, significantly affects the output.

There are two types of tasks for which supervised/directed learning is used: classification (pattern recognition) and regression (numeric prediction).

Classification: Training data $\{\mathbf{x}, y\}$ are the input-output data; $\mathbf{x}$ is an input vector with n features $x_j; j = 1, \ldots, n$, as its components and output y is a discrete class y_q ; $q = 1, \ldots, M$. In classification tasks, the goal is to predict the output values for new inputs (i.e., deciding which of the M classes each new vector $\mathbf{x}$ belongs to) based on training from examples of each class.

Regression: Training data $\{\mathbf{x}, y\}$ are the input-output data; $\mathbf{x}$ are the *regressors*, and y is a continuous numeric quantity. Regression task consists of *fitting a function* to the input-output data with the goal of predicting output values (numeric) for new inputs.

Classification and regression tasks arise in many applications, such as, signal processing, optimization, modeling and identification, control, and many business applications. In fact, most of the applications listed in Section 1.3 employ supervised learning. Our major focus in the book will be on this form of learning.

1.7.2 Unsupervised/Undirected Learning

Another form of machine learning tasks is when output $y^{(i)}$ is not available in training data. In this type of problem, we are given a set of feature vectors $\mathbf{x}^{(i)}$, and the goal is to unravel the underlying *similarities*.

Two different types of learning tasks frequently appear in the real-world applications of unsupervised learning.

Cluster Analysis: Cluster analysis is employed to create groups or *clusters* of similar records on the basis of many measurements made for these records. A primary issue in clustering is that of defining 'similarity' between feature vectors $\mathbf{x}^{(i)}$; $i = 1, 2, \ldots, N$, representing the records. Another important issue is the selection of an algorithmic scheme that will cluster (group) the vectors based on the accepted similarity measure (we will take up these issues in Chapter 7).

Clustering jobs emerge in several applications. Biologists, for instance, make use of classes and subclasses to organize species. A popular use of cluster analysis in marketing is for *market segmentation*: customers are segmented on the basis of demographic and transaction history information, and a marketing strategy is tailored for each segment. In finance, cluster analysis is used to create *balanced portfolios*: choosing securities from various clusters on the basis of financial performance variables, such as return, volatility, and so on. Internet search engines make use of clustering methods to group together queries that users submit, which can then be employed to improve search algorithms.

Other application domains for cluster analysis are remote sensing, image segmentation, image and speech coding, and many more.

After cluster patterns have been detected, it is the responsibility of the investigator to interpret them and decide whether they are useful.

Association Analysis: Association analysis uses unsupervised learning to discover patterns in the data where no target is specified earlier. It is up to human interpretation to make sense of the patterns.

Association analysis emerged with the study of customer transaction databases to establish an association between purchases of different items/services on offer. This common area of application is known as *market basket analysis*, which studies customers' purchase patterns for products that are bought together. This application is widely encountered in online *recommender systems*, where customers considering buying a product(s) are shown other products that are often bought along with the desired product(s), for example, display from Amazon.com.

Other application domains for association analysis are medical diagnosis, scientific data analysis, web mining, and many more.

We will discuss association analysis procedure in Chapter 9.

1.7.3 Reinforcement Learning

Reinforcement learning is founded on the concept that if an action is followed by a satisfactory state of affairs, or by an improved state of affairs (according to some properly defined way), then the inclination to produce that action becomes stronger, i.e., *reinforced*. This idea can be extended to permit action choices to be dependent on state information, which then brings in the aspect of feedback. A reinforcement learning system, therefore, is a system that via interaction with its environment *enhances its performance* by obtaining *feedback* in the form of a scalar reward (or penalty)—a *reinforcement signal*, that is indicative of the suitability of the response. The learning system is not instructed with regard to what action has to be taken. Instead, it is expected to find out which actions produce the maximum reward by trying them. The actions may influence not only the *immediate reward* but also the next situation, and through that all subsequent rewards.

The two aspects—trial-and-error search, and *cumulative reward*—are the two significant distinguishing attributes of reinforcement learning. Even though the early performance may fail to be up to the mark, with adequate interaction with the environment, it will ultimately learn an effective strategy for maximizing cumulative reward.

The problem of reinforcement learning is the most general of the three categories. A purely unsupervised learning agent cannot learn what to do, because it has no information as to what constitutes a desirable state or a correct action. In supervised learning, the agent can predict the result of action and can tune the action that leads to the desirable state. In reinforcement learning, the state-action model is not available. The agent has to take the actual action and learn from the results of the action taken—the state to which the action has driven the system. This method of learning is thus concerned with optimizing decisions, rather than predictions.

The reinforcement learning problem covers tasks such as learning to control a mobile robot, learning to optimize operations in factories, and learning to play board games. Reinforcement learning algorithms are related to *dynamic programming* algorithms frequently used to solve optimization problems.

The subject of reinforcement learning is introduced in Appendix B. The interested reader may also find reference [33] useful, where the focus is on reinforcement learning solutions to control problems: the controller (agent) has a set of sensors to observe the state of the controlled process (environment); the learning task is to learn a control strategy (policy) for choosing control signals (actions) that achieve minimization of a performance measure (maximization of cumulative reward). Reinforcement learning systems do not depend upon models of the environment because they learn through trial-and-error experience with the environment.

1.7.4 Learning Based on Natural Processes: Evolution, Swarming, and Immune Systems

Some learning approaches take inspiration from nature for the development of novel problem-solving techniques. The thread that ties together learning based on evolution process, swarm intelligence, and immune systems is that all have been applied successfully to a variety of optimization problems.

Optimization may not appear to be like a machine learning task, but optimization techniques are commonly used as part of machine learning algorithms.

Evolutionary Computation

It derives ideas from evolutionary biology to develop search and optimization methods that help solve complicated problems. Evolutionary biology essentially states that a population of individuals possessing the ability to reproduce and exposed to genetic variation followed by selection, gives rise to new populations, which are fitted to their environment. Computational abstraction of these processes gave rise to the so called *evolutionary algorithms*. The primary streams of evolutionary computation are *genetic algorithms*, *evolution strategies*, *evolutionary programming*, and *genetic programming*. Even though differences exist among these models, they all present the fundamental traits of an evolution process.

Swarm Intelligence

Swarm intelligence is a feature of systems of unintelligent agents with inadequate individual abilities, displaying collectively intelligent behavior. It includes algorithms derived from the collective behavior of social insects and other animal societies.

The primary lines of research that can be recognized within swarm intelligence are:

(i) Based on social insects (Ant Colony Optimization)

(ii) Based on the ability of human societies to process knowledge (Particle Swarm Optimization)

Although the resulting models are rather different in sequence of steps and sources of inspiration, they share some common properties. They are both dependent on a population (Colony or Swarm) of individuals (social insects or particles) possessing the ability of direct or indirect interaction not only with each other but also with the environment.

Ant Colony Optimization (ACO): Ants, seemingly small simple creatures, cooperate to solve complex problems, such as the most effective route to a source of food, that seem well beyond the ability of individual members of the hive or colony.

Particle Swarm Optimization (PSO): The particle swarm algorithm is motivated, among other things, by the creation of a simulation of human social behavior—the quality of human societies to process knowledge. Particle swarm considers a population of individuals possessing the ability to interact with the environment and one another. Therefore, population-level behaviors will arise from individual interactions. Although the approach was initially inspired by particle systems and the collective behavior of some animal societies, the algorithm primarily emphasizes on its social adaptation of knowledge.

Artificial Immune Systems

All living beings possess the ability to resist disease-causing agents or *pathogens* in the form of bacteria, viruses, parasites, and fungi. The main role of the *immune system* is to act as a shield for the body, protecting it from infections caused by pathogens.

An Artificial Immune System (AIS) replicates certain aspects of the natural immune system, which is primarily applied to solve pattern-recognition problems and cluster data. The natural immune system has an extraordinary ability to match patterns, employed to differentiate between foreign cells making an entry into the body (referred to as *antigen*) and the cells that are part of the body. As the natural immune system faces and handles antigens, it exhibits its adaptive nature: the immune system memorizes the structure of these antigens to ensure a speedier response to the antigens in the future.

In this book, we have mostly used calculus-based optimization methods. Genetic Algorithms have been introduced in Appendix A, but learning based on swarm intelligence, and immune systems have not been included. The interested readers may find references [26–29] useful.

1.8 MACHINE LEARNING AND DATA MINING

Various organizations, as a matter of routine, capture massive amounts of historical data, which describe their operations, products, and customers. At the same time, scientists and engineers in various fields capture datasets that are growing in terms of complexity. The field of *data mining* resolves the issue of the best way of using this historical data to find out the general patterns and improve the decision-making process [34].

The swelling interest in data mining is due to the convergence of many latest trends:

(i) The reducing prices of large data storage devices
(ii) The increasing convenience of data collection over networks
(iii) The reducing cost of computational power, which enables the use of computationally-intensive techniques to analyze data, and develop robust and efficient techniques of extraction of (predictive) models from data.

With the growth of data and the machines capable of searching, the opportunities for data mining will only grow manifold. With the growing complexity of the world, and the alarming rate at which data is being generated, only data mining will be able to explain the patterns that underline it. Data that has been intelligently analyzed is a valuable resource, which can lead to new insights and, in commercial settings, to competitive benefits.

The set of techniques for extracting (predictive) models from data constitutes the field of *machine learning*. Historically, data mining was born from machine learning as a research field concentrated on issues emerging by the examination of real-world applications. Research concentrated on commercial applications, and business issues of data analysis tend to use more of data mining techniques. However, the two fields happen to be related—they are both concerned with the data analysis which aims to discover informative patterns. The two also share methods and algorithms.

Looking forward, the main challenge ahead is applications. Applications will come not from machine learning experts, but from the people who work with the data and problems from which the data arises. Machine learning research will respond to the challenges thrown by new applications and will create new opportunities in decision making.

Data mining is thus a practical field and involves learning in a practical and not in a theoretical sense. Machine learning provides a technical base for data mining.

Even though machine learning forms the core of the data mining process, there are other steps involved data mining, including construction and maintenance of the database, data formatting and cleansing, data visualization and summarization, the use of human expert knowledge to formulate the inputs to the machine learning algorithm and evaluate the empirical regularities it discovers, and determining how to deploy the results. Machine learning is thus an essential component in the data mining process, but is not data mining by itself.

The field of data mining has already produced practical applications in such areas as analysis of medical outcomes, detection of credit card frauds, prediction of customer buying behavior, prediction of the personal interest of web users, and optimization of manufacturing processes. We now have machine learning algorithms that have been demonstrated to be of significant value in a variety of real-world data-mining applications. Many companies across the globe are now offering commercial implementations of these algorithms (see www.kdnuggets.com), along with other efficient interfaces to commercial databases and well-designed user interfaces.

1.9 BASIC LINEAR ALGEBRA IN MACHINE LEARNING TECHNIQUES

The field of machine learning is a well-founded discipline, composed of several classical mathematics areas. One could say that machine learning is nothing but value-added applied mathematics and statistics, although this statement may be valid for many other fields as well. Here, 'value-added' primarily means modern computer-based applications of standard and novel mathematical and statistical techniques.

This book assumes the working knowledge of applied mathematics and statistics. Whenever the concepts and techniques presented in this book require more than the knowledge the reader possesses, he/she is advised to acquire from other sources the required level of mathematical/ statistical preparedness.

To help the reader refresh his/her memory, some help is provided in this book. The present section gives the basics of linear algebra. In Section 3.2, we present basic statistics. The basics of other required mathematical tools will be presented throughout the book as and when the requirement arises.

This section is intended to be a concise summary of facts from linear algebra. It also serves to define the notation and terminology which are, regrettably, not entirely standard [33, 35].

An n-dimensional *column vector* $\mathbf{x}$ and its transpose $\mathbf{x}^T$ (an n-dimensional *row vector*) can be written as,

$$\mathbf{x} = \begin{bmatrix} x_1 \\ x_2 \\ \vdots \\ x_n \end{bmatrix}; \; \mathbf{x}^T = [x_1 \; x_2 \cdots x_n] \tag{1.10}$$

$x_j; j = 1, \ldots, n$, are the elements of the vector.

We denote the $N \times n$ (rectangular) matrix $\mathbf{A}$ and its $n \times N$ transpose $\mathbf{A}^T$ as,

$$\mathbf{A} = \begin{bmatrix} a_{11} & a_{12} & \cdots & a_{1n} \\ a_{21} & a_{22} & \cdots & a_{2n} \\ \vdots & \vdots & & \vdots \\ a_{N1} & a_{N2} & \cdots & a_{Nn} \end{bmatrix}; \; \mathbf{A}^T = \begin{bmatrix} a_{11} & a_{21} & \cdots & a_{N1} \\ a_{12} & a_{22} & \cdots & a_{N2} \\ \vdots & \vdots & & \vdots \\ a_{1n} & a_{2n} & \cdots & a_{Nn} \end{bmatrix} \tag{1.11}$$

A has N rows and n columns; a_{ij} denotes $(i, j)^{\text{th}}$ element, i.e., the element located in i^{th} row and j^{th} column.

Vectors may also be viewed as rectangular matrices. **x** is thus an $n \times 1$ matrix, and $\mathbf{x}^T$ is $1 \times n$ matrix.

Note that we have used lower case italic letters for scalars, lower case bold non-italic letters for vectors, and upper case bold non-italic letters for matrices.

When $n = N$, i.e., when the number of columns is equal to the number of rows, the matrix is said to be a *square matrix* of order n. A square matrix is called *symmetric* when its entries obey $a_{ij} = a_{ji}$.

We will be mostly be concerned with vectors/matrices that have real elements, i.e., vector $\mathbf{x} \in \Re^n$; $x_j \in \Re$, where $\Re^n$ is an n-dimensional real vector space.

Matrix **A** can be expressed in terms of its columns/rows. For example, a square matrix

$$\mathbf{A} = [\mathbf{a}_1 \quad \mathbf{a}_2 \cdots \mathbf{a}_n];\ \mathbf{a}_j = \begin{bmatrix} a_{1j} \\ a_{2j} \\ \vdots \\ a_{nj} \end{bmatrix} = j^{\text{th}} \text{ column in } \mathbf{A} \tag{1.12}$$

$$\mathbf{A} = \begin{bmatrix} \boldsymbol{\alpha}_1 \\ \boldsymbol{\alpha}_2 \\ \vdots \\ \boldsymbol{\alpha}_n \end{bmatrix};\ \boldsymbol{\alpha}_i = [\alpha_{i1} \quad \alpha_{i2} \cdots \alpha_{in}] = i^{\text{th}} \text{ row in } \mathbf{A} \tag{1.13}$$

A *diagonal matrix* is a square matrix whose elements off the principal diagonal are all zeros.

$$\mathbf{A} = \begin{bmatrix} a_{11} & 0 & \cdots & 0 \\ 0 & a_{22} & \cdots & 0 \\ \vdots & \vdots & & \vdots \\ 0 & 0 & \cdots & a_{nn} \end{bmatrix} \tag{1.14}$$

A particularly important matrix is the *identity matrix* **I**—an $n \times n$ (square) diagonal matrix whose principal diagonal entries are all 1's, and all other entries zeros.

$$\mathbf{I} = \begin{bmatrix} 1 & 0 & \cdots & 0 \\ 0 & 1 & \cdots & 0 \\ \vdots & \vdots & & \vdots \\ 0 & 0 & \cdots & 1 \end{bmatrix} \tag{1.15}$$

A *null matrix* **0** is a matrix whose elements are all equal to zero.

$$\mathbf{0} = \begin{bmatrix} 0 & 0 & \cdots & 0 \\ 0 & 0 & \cdots & 0 \\ \vdots & \vdots & & \vdots \\ 0 & 0 & \cdots & 0 \end{bmatrix} \tag{1.16}$$

Some Properties of Transpose

$$\begin{aligned} &\text{(i)}\ (\mathbf{A}^T)^T = \mathbf{A} \\ &\text{(ii)}\ (k\mathbf{A})^T = k\mathbf{A}^T;\ k \text{ is a scalar} \\ &\text{(iii)}\ (\mathbf{A}+\mathbf{B})^T = \mathbf{A}^T + \mathbf{B}^T \\ &\text{(iv)}\ (\mathbf{AB})^T = \mathbf{B}^T\mathbf{A}^T \\ &\text{(v)}\ \text{For any matrix } \mathbf{A},\ \mathbf{A}^T\mathbf{A} \text{ and } \mathbf{A}\mathbf{A}^T \text{ are both symmetric} \\ &\text{(vi)}\ \text{When a square matrix } \mathbf{A} \text{ is symmetric, } \mathbf{A} = \mathbf{A}^T \end{aligned} \tag{1.17}$$

Addition of vectors and of matrices is component by component.

The *product* **AB** of an $N \times n$ matrix **A** by an $n \times p$ matrix **B** (number of columns of **A** must be equal to number of rows of **B**; compatibility requirement for multiplication) is an $N \times p$ matrix **C**:

$$\begin{gathered} \mathbf{C} = \mathbf{AB} \quad \text{or} \quad c_{ij} = \sum_{r=1}^{n} a_{ir}\, b_{rj};\ i = 1, \ldots, N;\ j = 1, \ldots, p. \\ \mathbf{AB} \neq \mathbf{BA};\ (\mathbf{AB})\mathbf{C} = \mathbf{A}(\mathbf{BC});\ (\mathbf{A}+\mathbf{B})\mathbf{C} = \mathbf{AC} + \mathbf{BC} \end{gathered} \tag{1.18}$$

Determinant of a Matrix

Determinants are defined for square matrices only. The determinant of an $n \times n$ matrix **A**, written as $|\mathbf{A}|$, is a scalar-valued function of **A**. The calculation of the determinant is simple in low dimensions, but a bit more involved in high dimensions. If **A** is itself a scalar (i.e., 1×1 matrix), then $|\mathbf{A}| = \mathbf{A}$. If **A** is 2×2 matrix, then $|\mathbf{A}| = a_{11}\, a_{22} - a_{21}\, a_{12}$. The determinant of a general square matrix can be computed by expansion by *minors*. For an $n \times n$ matrix **A**, we define *minor* m_{ij} to be the $(n-1) \times (n-1)$ matrix obtained by deleting i^{th} row and j^{th} column of **A**. The *cofactor* c_{ij} of the element a_{ij} is defined by the equation,

$$c_{ij} = (-1)^{i+j}\, m_{ij} \tag{1.19a}$$

Selecting an arbitrary row k, $|\mathbf{A}|$ is given by,

$$|\mathbf{A}| = \sum_{j=1}^{n} a_{kj}\, c_{kj} \tag{1.19b}$$

Similarly, expansion can be carried out with respect to any arbitrary column l, to obtain

$$|\mathbf{A}| = \sum_{i=1}^{n} a_{il}\, c_{il} \tag{1.19c}$$

Expansion by minors reduces the evaluation of $n \times n$ determinant down to evaluation of a string of $(n-1) \times (n-1)$ determinants.

Some properties of determinants are:

$$\begin{aligned}
&\text{(i)}\ |\mathbf{AB}| = |\mathbf{A}|\,|\mathbf{B}| \\
&\text{(ii)}\ |\mathbf{A}^T| = |\mathbf{A}| \\
&\text{(iii)}\ |k\mathbf{A}| = k^n|\mathbf{A}|;\ \mathbf{A} \text{ is an } n \times n \text{ matrix and } k \text{ is a scalar} \\
&\text{(iv)}\ \text{The determinant of any diagonal matrix is the product of its diagonal elements}
\end{aligned} \tag{1.20}$$

A square matrix is called *singular* if the associated determinant is zero; it is called *nonsingular* if associated determinant is nonzero.

The *rank* $\rho(\mathbf{A})$ of a matrix $\mathbf{A}$ is the dimension of the largest array in $\mathbf{A}$ with a nonzero determinant.

Inverse and Pseudoinverse Matrices

So long as its determinant does not vanish (nonsingular matrix), the inverse of an $n \times n$ matrix $\mathbf{A}$, denoted $\mathbf{A}^{-1}$, is the $n \times n$ matrix such that,

$$\mathbf{AA}^{-1} = \mathbf{A}^{-1}\mathbf{A} = \mathbf{I} \tag{1.21}$$

The *adjoint* of $\mathbf{A}$, written $\mathbf{A}^+$, is the matrix whose $(i, j)^{\text{th}}$ entry is the $(j, i)^{\text{th}}$ cofactor of $\mathbf{A}$.

Given these definitions, we can write the inverse of $\mathbf{A}$ as,

$$\mathbf{A}^{-1} = \frac{\mathbf{A}^+}{|\mathbf{A}|} \tag{1.22}$$

Some properties of matrix inverse are:

$$\begin{aligned}
&\text{(i)}\ (\mathbf{A}^{-1})^{-1} = \mathbf{A} \\
&\text{(ii)}\ (\mathbf{A}^T)^{-1} = (\mathbf{A}^{-1})^T \\
&\text{(iii)}\ (\mathbf{AB})^{-1} = \mathbf{B}^{-1}\mathbf{A}^{-1} \\
&\text{(iv)}\ |\mathbf{A}^{-1}| = \frac{1}{|\mathbf{A}|} \\
&\text{(v)}\ |\mathbf{P}^{-1}\mathbf{A}\,\mathbf{P}| = |\mathbf{A}|
\end{aligned} \tag{1.23}$$

If $\mathbf{A}$ is not square (or if $\mathbf{A}^{-1}$ does not exist because $\mathbf{A}$ is singular), we typically use instead the *pseudoinverse* of $\mathbf{A}$. If $\mathbf{A}^T\mathbf{A}$ is nonsingular square matrix, the pseudoinverse of $\mathbf{A}$ is defined as $(\mathbf{A}^T\mathbf{A})^{-1}\,\mathbf{A}^T$.

Inner and Outer Product

The *inner* (*scalar*) product of two n-dimensional vectors $\mathbf{x}$ and $\mathbf{w}$ is a scalar a:

$$a = \mathbf{x}^T\mathbf{w} = \mathbf{w}^T\mathbf{x} \tag{1.24}$$

The *outer* product of **x** and **w** is a matrix:

$$\mathbf{A} = \mathbf{x}\mathbf{w}^T = \begin{bmatrix} x_1 \\ x_2 \\ \vdots \\ x_n \end{bmatrix} [w_1 \; w_2 \cdots w_n] = \begin{bmatrix} x_1w_1 & x_1w_2 & \cdots & x_1w_n \\ x_2w_1 & x_2w_2 & \cdots & x_2w_n \\ \vdots & \vdots & & \vdots \\ x_nw_1 & x_nw_2 & \cdots & x_nw_n \end{bmatrix} \tag{1.25}$$

If the dimensions of **x** and **w** are not the same, then **A** is not a square matrix.

Any two vectors which have a zero scalar product are said to be *orthogonal* vectors. A set of vectors is said to be orthogonal, if and only if, every two vectors from the set are orthogonal.

Vector Norms

Norms are (positive) scalars and are used as measures of length, size, distance, and so on, depending on the context. For the vector

$$\mathbf{x} = \begin{bmatrix} x_1 \\ x_2 \\ \vdots \\ x_n \end{bmatrix} \tag{1.26a}$$

the Euclidean vector norm, $\|\mathbf{x}\|$, is defined by,

$$\|\mathbf{x}\| = (x_1^2 + x_2^2 + \cdots + x_n^2)^{1/2} = (\mathbf{x}^T\mathbf{x})^{1/2} \tag{1.26b}$$

The *p-norm* of vector **x**:

$$\|\mathbf{x}\|_p = \left(\sum_{j=1}^{n} |x_j|^p \right)^{1/p} \tag{1.26c}$$

Mostly $p = 1, 2$ or ∞.

$$\|\mathbf{x}\|_1 = \sum_{j=1}^{n} |x_j| \quad \text{(absolute value)}$$

$$\|\mathbf{x}\|_2 = \left(\sum_{j=1}^{n} x_j^2 \right)^{1/2} \quad \text{(Euclidean norm)} \tag{1.27}$$

$$\|\mathbf{x}\|_\infty = \max |x_j|$$

$\|\mathbf{x}\| \geq 0$ if $\mathbf{x} \neq \mathbf{0}$; $\|k\mathbf{x}\| = |k| \, \|\mathbf{x}\|$ for any scalar k.

A *unit vector* is, by definition, a vector whose norm is unity. Any nonzero vector $\mathbf{x}$ can be normalized to form a unit vector:

$$\text{unit vector} = \frac{\mathbf{x}}{\|\mathbf{x}\|} \tag{1.28}$$

A set of vectors is said to be *orthonormal* if, and only if, the set is orthogonal and each vector in the orthogonal set is a unit vector.

Orthonormal Matrix

Let $\mathbf{A}$ be an $n \times n$ matrix

$$\mathbf{A} = [\mathbf{a}_1 \quad \mathbf{a}_2 \cdots \mathbf{a}_n] \tag{1.29}$$

where $\mathbf{a}_j$ is the jth *column* vector. The set of vectors $\{\mathbf{a}_j\}$ is said to be orthonormal, if and only if, every two vectors from the set are orthonormal. When the set of vectors $\{\mathbf{a}_j\}$ is orthonormal, the matrix $\mathbf{A}$ is said to be an orthonormal matrix.

An important property of an orthonormal matrix $\mathbf{A}$ *is that its inverse is equal to its transpose,* that is,

$$\mathbf{A}^{-1} = \mathbf{A}^T \tag{1.30}$$

Let us examine $(i,j)^{\text{th}}$ element of the matrix $\mathbf{A}^T\mathbf{A}$: $\mathbf{a}_i^T\,\mathbf{a}_j$. Since the vectors $\mathbf{a}_i$ and $\mathbf{a}_j$ are orthonormal, their inner product is zero; the only exception is an inner product of one particular column with itself.

$$\mathbf{a}_i^T\,\mathbf{a}_j = \begin{cases} 1\,;\, i = j \\ 0\,;\, i \neq j \end{cases}$$

Therefore,

$$\mathbf{A}^T\mathbf{A} = \mathbf{I} \tag{1.31}$$

Since $\mathbf{A}^{-1}\,\mathbf{A} = \mathbf{I}$, it follows that

$$\mathbf{A}^{-1} = \mathbf{A}^T$$

Linearly Independent Vectors

Consider a set of n vectors $\{\mathbf{x}_1, \mathbf{x}_2, \ldots, \mathbf{x}_n\}$, each of which has n components. If there exists a set of n scalars α_i, at least one of which is not zero, which satisfies

$$\alpha_1\mathbf{x}_1 + \alpha_2\mathbf{x}_2 + \cdots + \alpha_n\mathbf{x}_n = \mathbf{0} \tag{1.32}$$

then the set of vectors $\{\mathbf{x}_i\}$ is said to be *linearly dependent*.

If,

$$\alpha_1\mathbf{x}_1 + \alpha_2\mathbf{x}_2 + \cdots + \alpha_n\mathbf{x}_n = \mathbf{0} \tag{1.33}$$

implies that each $\alpha_i = 0$, then $\{\mathbf{x}_i\}$ are *linearly independent* vectors.

Eigenvalues and Eigenvectors

Given an $n \times n$ matrix $\mathbf{A}$, a very important class of linear equations is of the form

$$\mathbf{Av} = \lambda \mathbf{v}$$

for scalar λ, which can be rewritten as,

$$(\mathbf{A} - \lambda \mathbf{I})\mathbf{v} = \mathbf{0} \tag{1.34}$$

where $\mathbf{I}$ is the identity matrix, and $\mathbf{0}$ is the zero vector.

For a given λ, this is a set of n homogeneous equations in n unknowns—the components in vector $\mathbf{v}$.

There are two questions of interest with regard to Eqn (1.34):

(i) whether a solution to Eqn (1.34) exists; and
(ii) if the answer to the first question is yes, how many linearly independent solutions occur?

We have the following answers to the two questions [33].

(i) For Eqn (1.34) to have a nontrivial solution, rank of the matrix $(\mathbf{A} - \lambda \mathbf{I})$, denoted $\rho(\mathbf{A} - \lambda \mathbf{I})$, must be less than n, or equivalently

$$|\mathbf{A} - \lambda \mathbf{I}| = 0 \tag{1.35a}$$

On expanding the determinant, we get,

$$\lambda^n + \alpha_1 \lambda^{n-1} + \cdots + \alpha_{n-1}\lambda + \alpha_n = 0 \tag{1.35b}$$

This equation is called the *charateistic equation* of matrix $\mathbf{A}$, and its roots are *characteristic roots*, or *eigenvalues* of matrix $\mathbf{A}$.

(ii) The number of linearly independent solutions to Eqn (1.35b), is equal to $(n - \rho(\mathbf{A} - \lambda \mathbf{I}))$, called the *nullity* of matrix $\mathbf{A}$.

We restrict our discussion to distinct eigenvalues $\lambda = \lambda_i$; $i = 1, \ldots, n$. That is, we assume that there are n distinct (and not repeated) roots of the characteristic equation of a given matrix $\mathbf{A}$. For distinct eigenvalue λ_i, there is one, and only one, linearly independent solution to Eqn (1.35b).

This solution is called the *eigenvector* $\mathbf{v} = \mathbf{e}_i$ of matrix $\mathbf{A}$ associated with the eigenvalue $\lambda = \lambda_i$. When the eigenvalues are not distinct, the concept of *generalized eigenvectors* becomes applicable [33].

How does one go about finding eigenvalues and eigenvectors of a given $n \times n$ matrix. Unfortunately, it is only easy for $n = 2$ or at most 3 [33]. The usual way to find eigenvectors is by some complex iterative method, which is beyond the scope of our presentation. We will rather rely on a useful package that does it all for us.

We assume that for a given $n \times n$ matrix $\mathbf{A}$, we have found the eigenvalues

$$\lambda = \lambda_1, \lambda_2, \ldots, \lambda_n \tag{1.36a}$$

and corresponding eigenvectors

$$\mathbf{v} = \mathbf{e}_1, \mathbf{e}_2, \cdots, \mathbf{e}_n \tag{1.36b}$$

The vectors $\mathbf{e}_1, \mathbf{e}_2, \ldots, \mathbf{e}_n$ are linearly independent and the nonsingular matrix

$$\mathbf{E} = [\mathbf{e}_1 \;\; \mathbf{e}_2 \cdots \mathbf{e}_n] \tag{1.36c}$$

transforms **A** into diagonal form:

$$\mathbf{A}\,\mathbf{e}_i = \lambda_i \mathbf{e}_i$$

Therefore,

$$\mathbf{A}[\mathbf{e}_1 \quad \mathbf{e}_2 \cdots \mathbf{e}_n] = [\mathbf{e}_1 \quad \mathbf{e}_2 \cdots \mathbf{e}_n]\begin{bmatrix} \lambda_1 & 0 & \cdots & 0 \\ 0 & \lambda_2 & \cdots & 0 \\ \vdots & \vdots & & \vdots \\ 0 & 0 & \cdots & \lambda_n \end{bmatrix} \tag{1.37a}$$

This can be equivalently expressed as,

$$\mathbf{AE} = \mathbf{E\Lambda} \tag{1.37b}$$

where $\mathbf{\Lambda}$ is the diagonal matrix having eigenvalues of matrix **A** as its diagonal elements.

Premultiplying both sides by $\mathbf{E}^{-1}$, we get,

$$\mathbf{E}^{-1}\mathbf{A}\mathbf{E} = \mathbf{\Lambda} \tag{1.37c}$$

So far we have considered a square matrix **A**. Let us now extend our results to *symmetric matrix* **A**.

A *symmetric matrix always has orthonormal set of eigenvectors*. This can be shown as follows. Let λ_1 and λ_2 be distinct eigenvalues associated with eigenvectors $\mathbf{e}_1$ and $\mathbf{e}_2$, respectively. The inner product of vectors $\mathbf{e}_1$ and $\mathbf{e}_2$ is given by,

$$\mathbf{e}_1^T\mathbf{e}_2 = \mathbf{e}_2^T\mathbf{e}_1$$

Since,

$$\begin{aligned}(\lambda_1\,\mathbf{e}_1^T)\mathbf{e}_2 &= (\mathbf{A}\,\mathbf{e}_1)^T\mathbf{e}_2 \\ &= \mathbf{e}_1^T\,\mathbf{A}^T\mathbf{e}_2 \\ &= \mathbf{e}_1^T\,\mathbf{A}\mathbf{e}_2;\ \text{since } \mathbf{A} \text{ is symmetric} \\ &= \mathbf{e}_1^T(\lambda_2\,\mathbf{e}_2)\end{aligned}$$

we have,

$$(\lambda_1 - \lambda_2)\,\mathbf{e}_1^T\mathbf{e}_2 = 0$$

Since $\lambda_1 \neq \lambda_2$, we have,

$$\mathbf{e}_1^T\mathbf{e}_2 = 0$$

Thus, the eigenvectors of a symmetric matrix are orthogonal/orthonormal. This means that **E** is an orthonormal matrix; so by the result $\mathbf{E}^T = \mathbf{E}^{-1}$, we get (refer to Eqn (1.37c)),

$$\mathbf{E}^T\mathbf{A}\mathbf{E} = \mathbf{\Lambda} \tag{1.38}$$

1.10 RELEVANT RESOURCES FOR MACHINE LEARNING

The machine learning algorithms we discuss in the coming chapters have their origins in different domains: statistics, cognitive science, computer vision, signal processing, control, artificial intelligence, natural evolution. The research in these different domains followed different paths in the past with different emphases. In this book, the aim is to incorporate these emphases together to give a unified treatment of the problems and the proposed solutions.

The reader may like to start with the classic books on the subject [1–5]. These titles give a deeper view of the origins of machine learning and its relationship with different domains of original research. Coming to the 'applied' aspect of the subject, there are many books currently available. Some representative titles are [6–16]; the focus in these books varying over different domains. The popular titles [17–20] primarily deal with data mining for the business world—application of machine learning algorithms to the business optimization problems. Some other titles on data mining which the reader may find useful to refer to are [21–25]. Other subgroup of books is on computational intelligence [26–29] focusing more on neural networks, fuzzy logic systems, and evolutionary algorithms.

Reinforcement learning methods are applied to applications where the output of the system is a sequence of *actions*. In such cases, a single action (decision) is not important; what is important is the *policy* that is the sequence of correct actions to reach the goal. The learning algorithm should be able to assess the goodness of policies and learn from past action sequences to be able to generate a good policy. The field of reinforcement learning is developing rapidly, and we may expect to see some impressive results in the near future.

Sutton and Barto [30] discuss all the aspects of reinforcement learning, learning algorithms and several applications. A comprehensive presentation is given in [31]. Recent work on reinforcement learning applied to robotics is given in [32]. Author's presentation of the subject is given in [33].

Many useful tutorial presentations on machine learning (including video tutorial lectures) are available over the Internet. On-line courses on different modules of machine learning are also being offered.

Research in machine learning is distributed over journals and conferences from different fields. Popular journals which contain machine learning papers are: *Machine Learning*; *Journal of Machine Learning Research*; *Neural Computation*; *Neural Networks*; *Knowledge Discovery and Data Mining; IEEE Transactions on* (*Neural Networks and Learning Systems; Pattern Analysis and Machine Intelligence; Knowledge and Data Engineering; Systems, Man, and Cybernetics*).

Journals of artificial intelligence, pattern recognition, signal processing, image processing, and those with a focus on statistics, also publish machine learning research papers.

Major conferences on machine learning are: *Neural Information Processing systems (NIPS), Uncertainty in Artificial Intelligence (UAI), International Conference on Machine Learning (ICML), European Conference on Machine Learning (ECML), and Computational Learning Theory (COLT)*. Conferences on pattern recognition, artificial intelligence, neural networks, fuzzy logic, robotics and data mining, include machine learning in their spectrum of research areas.

Most recent papers by machine learning researchers are accessible over the Internet.

UCI Repository, at http://archive.ics.uci.edu/ml, contains a large number of datasets frequently used by machine learning researchers for bench-marking purposes. Another resource is the *Kaggle*

Repository, at https://www.kaggle.com. In addition to these, there are also repositories for particular applications, for example, computational biology, speech recognition, image classification, text classification, and so forth. New and larger datasets are constantly being added to these repositories. These datasets in public domain are used repeatedly by researchers while tailoring a new algorithm. Many authors/organizations also make their data available over the web.

Datasets available in public domain are mostly speculative research projects and not production systems.

Students who are studying a course on machine learning, must attempt a set of machine-learning experiments. Public-domain datasets for the experiments and the relevant research papers using these datasets, are the important resource of gaining confidence in the available tools and techniques of machine learning.

Another resource for machine learning experiments is the codes for machine learning algorithms. Before developing their own codes for the algorithms, it may be helpful using available codes first. Many of the authors of machine learning research make their codes available over the web. There are also free software toolboxes and packages implementing various machine learning algorithms. Among these, Weka, R, Rapid Miner, Python are especially noteworthy. Popular commercial software vendors are MATLAB, SAS.

Many books on machine learning/data mining/predictive analytics are written with the dual objective of (i) explaining the concepts and techniques of machine learning, and (ii) training the readers in the usage of one of the software packages.

Chapter

2

SUPERVISED LEARNING: RATIONALE AND BASICS

2.1 LEARNING FROM OBSERVATIONS

We consider the following setting in this chapter to present the basics of learning theory.

There is some set S of possible patterns/observations/samples over which various output functions may be defined. The training experience is available in the form of N patterns $s^{(i)} \in S$; $i = 1, 2, \ldots, N$. We specify a pattern by a fixed number n of attributes/features x_j; $j = 1, 2, \ldots, n$; where each feature has real numerical value for the pattern. The domain of x_j is given by a set $V_{x_j} \in \Re$ of its values. A data pattern $s^{(i)}$ has the feature-value set $\{x_1^{(i)}, \ldots, x_n^{(i)}\}$, where $x_j^{(i)} \in V_{x_j}$. We can visualize each pattern with n numerical features as a point in n-dimensional state space $\Re^n$:

$$\mathbf{x} = [x_1 \; x_2 \; \cdots \; x_n]^T \in \Re^n$$

The set $\mathbf{X}$ is the finite set of feature vectors $\mathbf{x}^{(i)}$ for all the N patterns. We can visualize $\mathbf{X}$ as a region of the state space $\Re^n$ to which the patterns belong, i.e., $\mathbf{X} \subset \Re^n$. Note that $\mathbf{x}^{(i)}$ is the representation of $s^{(i)}$; and $\mathbf{X}$ is the *representation space*.

For supervised learning problems, the decision attribute (output) y is known *a priori* for each pattern in the set S. For multiclass classification (pattern recognition) problems, the domain of y is given by the set $V_y = \{1, 2, \ldots, M\}$. The output $y^{(i)}$ for a pattern $s^{(i)}$ will take the value from the set V_y, which represents the class to which the pattern belongs. Thus, in classification problems, each pattern $s^{(i)}$ is associated with an output $y^{(i)} \in V_y$. For regression (numeric prediction) problems, $V_y \in \Re$, and each pattern $s^{(i)}$ is associated with an output $y^{(i)} \in \Re$. We can visualize the output belonging to one-dimensional region Y of the state space.

We assume that different patterns $\mathbf{x}^{(i)}$ in $\mathbf{X}$ may be encountered with different frequencies. A convenient way to model this is to assume that there is some unknown probability distribution that defines the probability of encountering each pattern $\mathbf{x}^{(i)}$ in $\mathbf{X}$. The training examples are provided to the learning machine by a trainer who draws each pattern independently according to the distribution, and who then forwards the pattern $\mathbf{x}$ along with its true (observed) output y to the learning machine. The training experience is in the form of data $\mathcal{D}$ that describes how the system behaves over its entire range of operation.

$$\mathcal{D}: \{\mathbf{x}^{(i)}, y^{(i)}\}; i = 1, 2, \ldots, N \tag{2.1}$$

In general, learning is most reliable when the training examples follow a distribution similar to that of future patterns (unseen by the machine during training) which the machine will receive; the machine will be required to give estimated output values $\hat{y}$ for these patterns. If the training experience consists of data that lies in a region of the state space, then that region must be fully representative of situations over which the trained machine will later be used. The most current theory of machine learning rests on the crucial assumption that the distribution of training examples is identical to the distribution of unseen examples.

Assume that the data $\mathcal{D}$ are independently drawn and identically distributed (commonly referred to as *iid* data) samples from some unknown probability distribution represented by the probability density function $p(\mathbf{x}, y)$ (refer to Section 3.2). Assume a machine defined by a function

$$f: \mathbf{X} \rightarrow Y$$

that maps from $\mathbf{x} \in \mathbf{X}$ to $y \in Y$, where $\mathbf{X}$ is the *input space* and Y is the *output space* of the function. When $f(\cdot)$ is selected, the machine is called a *trained machine* that gives estimated output value

$$\hat{y} = f(\mathbf{x})$$

for a given pattern $\mathbf{x}$. To assess the success of learning, we need an evaluation criterion. The evaluation criteria are generally based on the consequences of the decision made by the machine, namely—the errors, profits or losses, penalties or rewards involved. In supervised learning, quite often used criterion is a minimization criterion involving potential loss into a classification/regression decision made.

From the set of possible learning machines (functions), we want to select the *optimal* one that minimizes the loss. We can define the set of learning machines by a function $f(\mathbf{x}, \mathbf{w})$, where $\mathbf{w}$ contains adjustable parameters. Thus, the set of parameters $\mathbf{w}$ becomes the subject of learning. A *loss function* $L(y, f(\mathbf{x}, \mathbf{w}))$ is a measure of the error between the actual output y and estimated output

$$\hat{y} = f(\mathbf{x}, \mathbf{w}) \tag{2.2}$$

Suppose we observe a particular $\mathbf{x}$ and we contemplate taking decision $f(\mathbf{x}, \mathbf{w})$ which yields the output $\hat{y}$ (the estimated output). If the true value of the output is y, by definition, we will incur the loss $L(y, f(\mathbf{x}, \mathbf{w}))$. If $p(\mathbf{x}, y)$ represents the joint probability density function of the data, then expected loss associated with decision $f(\mathbf{x}, \mathbf{w})$ is merely (refer to Section 3.2)

$$\mathbb{E}[L(y, f(\mathbf{x}, \mathbf{w}))] = \int_{\mathbf{X} \times Y} L(y, f(\mathbf{x}, \mathbf{w}))\, p(\mathbf{x}, y)\, d\mathbf{x}\, dy = R(\mathbf{w}) \tag{2.3}$$

$d\mathbf{x}\,dy$ is our notation for the $(n + 1)$-space volume element, and the integral extends over the entire joint space created by feature space (input space) $\mathbf{X}$ and output space Y.

A *risk* refers to a predictable loss in decision-theoretic terminology, and $R(\mathbf{w})$ is known as the *risk function*. On coming across a specific observation $\mathbf{x}$, the expected loss can be minimized by choosing a decision function that minimizes the risk function. If tractable, this process actually provides optimum performance.

Stated formally, *our problem is to find a decision function* $f(\mathbf{x}, \mathbf{w})$ *against* $p(\mathbf{x}, y)$ *that minimizes the risk function* $R(\mathbf{w})$.

Thus, we could design optimal classifiers/numeric predictors if we know the joint probability density function $p(\mathbf{x}, y)$. In machine learning applications, this type of information pertaining to the probabilistic structure of the problem is hardly found. Typically, we simply have some vague general knowledge regarding the situation together with several *design samples* or training data—specific representatives of the patterns we wish to categorize/regress. The issue, therefore, is to seek a way to employ this information to design or train the learning machine. Different processes that have been established to tackle such problems will be considered throughout the book.

Empirical-Risk-Minimization

The learning problem defined before is, in fact, intractable, since we do not know $p(\mathbf{x}, y)$ explicitly. The risk function representing *true risk*, given by Eqn (2.3), cannot be calculated. All that is available is a training dataset of examples drawn by independent sampling a $(\mathbf{X} \times Y)$ space according to some unknown probability distribution. Therefore, a learning machine can at best guarantee that the estimated output values $\hat{y}$ fit the true values y over the training data.

With dataset (2.1) being the only source of information, the risk function given by Eqn (2.3) must be approximated by the *empirical risk*, $R_{emp}(\mathbf{w})$:

$$R_{emp}(\mathbf{w}) \triangleq \frac{1}{N} \sum_{i=1}^{N} L(y^{(i)}, f(\mathbf{x}^{(i)}, \mathbf{w})) \tag{2.4}$$

The empirical risk given by Eqn (2.4), replaces average over $p(\mathbf{x}, y)$ by an average over the training sample.

In classification problems, probably the simplest and quite often used criterion involves counting the misclassification error (Section 2.8); if a pattern $\mathbf{x}$ is classified wrongly, we incur loss 1, otherwise there is no penalty. The *loss function for classification problems*

$$L(y, f(\mathbf{x}, \mathbf{w})) = \begin{cases} 0 & \text{if } y = f(\mathbf{x}, \mathbf{w}) = \hat{y} \\ 1 & \text{otherwise} \end{cases} \tag{2.5}$$

where $\mathbf{x}$ denotes the pattern, y denotes observation (true output) and $f(\mathbf{x}, \mathbf{w}) = \hat{y}$ is the estimation. The minimum of the loss function is 0.

When estimating the real-valued quantities (regression problems), it is usually the size of the difference $(y - f(\mathbf{x}, \mathbf{w}))$, i.e., the amount of misestimation (misprediction), which is used to determine the quality of the estimate. The popular choice (Section 2.7) is to minimize the sum of squares of the residuals $(y - f(\mathbf{x}, \mathbf{w}))$. In most cases, the *loss function for the regression problems* will be of the type

$$L(y, f(\mathbf{x}, \mathbf{w})) = \tfrac{1}{2}(y - f(\mathbf{x}, \mathbf{w}))^2 \tag{2.6}$$

It appears as if (2.4) is the answer to our problems and all that remains to be done is to find a suitable function $f(\cdot)$ out of the set of all possible functions (it is an infinite set, theoretically) that minimizes R_{emp}. Unfortunately, this problem of finding the minimum of the empirical risk is an *ill-posed problem*. Here, the 'ill-posed' characteristic of the problem is due to the infinite number of possible solutions to the problem. We will show this with the help of some simple examples.

Before we take up the examples, a relook at the *learning task* in hand will be helpful. Given a set of training examples $\{\mathbf{x}^{(i)}, y^{(i)}\}$; $i = 1, 2, \ldots, N$, the learning task requires estimating/predicting outputs $\hat{y} = f(\mathbf{x}, \mathbf{w})$ for the data beyond the training data, i.e., for the data the machine has not seen during its training phase. The only information available about $f(\cdot)$ is its true value over the training examples. Our aim is to use the machine for prediction of the output (required for decision making) for the data for which the true output is not known.

The key assumption for reliable learning is that the training examples and future examples (unseen by the machine during training) are drawn randomly and independently from some population of examples with the same (though unknown) probability distribution (the data is *iid*—independently and identically distributed). If this assumption is met, we can claim that a learning machine giving 100% accuracy on training data will give high accuracy on unseen data (data beyond the training sample). However, this assumption is not guaranteed. In fact, machine learning tasks based on real-world data are unlikely to find the *iid* data assumption tenable. The real-world data tend to be incomplete, noisy, and inconsistent (we will consider these aspects of real-world data in later chapters).

Let us now return to the examples.

Consider a regression problem. Suppose we are given empirical observations:

$$((x^{(1)}, y^{(1)}), \ldots, (x^{(N)}, y^{(N)})) \in \mathbf{X} \times Y$$

where for simplicity, we take $\mathbf{X} \in \Re$, and $Y \in \Re$. Figure 2.1 shows a plot of such a dataset (data points indicated by o). Note that all functions that *interpolate* these data points will result in zero value for R_{emp}. Figure 2.1 shows two out of infinite many different interpolating functions of training data pairs that are possible. Each of the two interpolants results in $R_{emp} = 0$, but at the same time, none is a good model of the true underlying dependency between $\mathbf{X}$ and Y, because both the functions perform very poorly outside the training inputs (indicated as ×). Thus, interpolating functions that result in zero empirical risk, can mislead. There are many other approximating functions (learning machines) that will minimize the empirical risk (*training error*) but not necessarily the true (expected) risk.

Consider now a classification problem. In two-class pattern recognition, we seek to infer a function

$$f(\cdot) : \mathbf{X} \rightarrow \{\pm 1\}, \text{ i.e., } Y = \{\pm 1\}$$

Figure 2.2 shows a simple 2-dimensional example of a pattern recognition problem. The task is to separate solid dots from circles by finding a function that takes +1 on the dots and –1 on the circles. The classification function shown by dashed line in Fig. 2.2 separates all training points correctly.

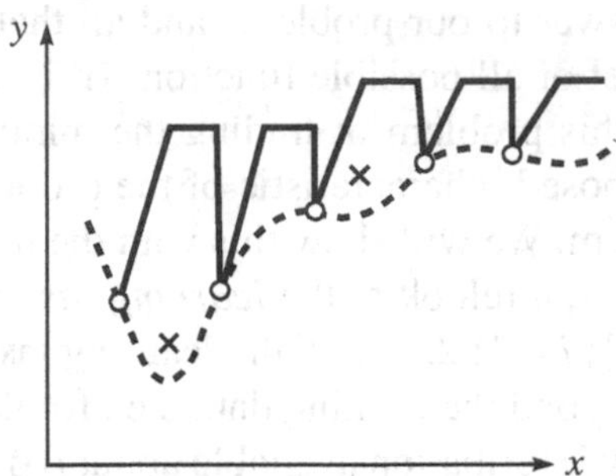

Figure 2.1 A simple regression example

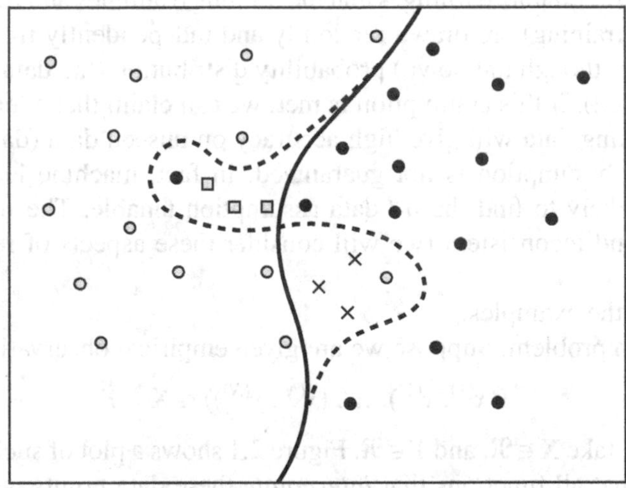

Figure 2.2 A simple classification example

From this figure, it is, however, unclear whether the same would hold true for other points. For instance, we take some *test* points for which the output is known but these points have not been used for training. The test points indicated by × are known to have +1 output and those indicated by □ have –1 output. We see from Fig. 2.2 that a classifier with even zero training error may not be able to get all test points right. We need some compromise (e.g., the decision boundary shown by solid line in Fig. 2.2) which gets most of the test points right; such a classifier may not possess zero training error.

Minimization of training error given by the empirical risk function in (2.4) is, thus, not a solution to our learning task. Minimization of true error (errors of prediction for total data, that includes both training and unseen patterns) given by the risk function in (2.3) is our learning task, but this problem is intractable with $p(\mathbf{x}, y)$ unknown explicitly.

A generic question about machine learning is:

What is it that we need for empirical risk minimization to lead to a successful solution to the learning task?

Lacking any further information except the training samples (which is the situation for most of the complex decision-making problems we face today), the solution lies in *inductive inference*.

Inductive Learning

The task of inductive learning is this:

Given a collection of examples $(\mathbf{x}^{(i)} f(\mathbf{x}^{(i)}))$; $i = 1, 2, \ldots, N$, of a function $f(\mathbf{x})$, return a function $h(\mathbf{x})$ that *approximates* $f(\mathbf{x})$.

In the statistical literature, the approximating function $h(\mathbf{x})$ is called *hypothesis function*.

The *true function* $f(\mathbf{x})$ correctly maps the input space $\mathbf{X}$ (of the entire data) to the output space Y. This function is not known in a real-world decision-making problem. Our satisfaction on an hypothesis function (*machine learning algorithm*) $h(\mathbf{x})$ with high accuracy on the given collection of examples (learning examples for the machine/examples for training the machine) of $f(\mathbf{x})$ could be premature because *central aim* of designing a machine is to suggest decisions when presented with *novel* patterns (unseen by the machine during training).

From a conceptual point of view, it is not easy to tell whether any particular $h(\cdot)$ is a good approximation of $f(\cdot)$. A good approximation will *generalize* well—that is, will predict novel patterns correctly. *Generalization performance is, thus, the fundamental problem in inductive learning.*

How do we judge the generalization performance of an algorithm? We can estimate such performance by means of a *test dataset*, sampled independently as is the training set. The *off-training set error*—the error on points *not* in the training set, will be used as a measure of generalization performance. Details will appear later in this chapter.

The assumption in inductive learning is that the ideal hypothesis related to unseen patterns is the one *induced* by the observed training data. Inductive learning techniques obtain a general hypothesis by seeking *empirical regularities* over the training examples. These regularities induce the approximation of the mapping function well over the observed examples. It generalizes from the specific training examples, hypothesizing a general function covering these examples.

When we confront various hypotheses (*hypothesis space*) we also encounter certain generic questions:

If we are only interested in the generalizing performance, is there a reason for the preference of one hypothesis over another? If a specific hypothesis outperforms another over some data sample, will the hypothesis necessarily be more accurate? Is it possible for a hypothesis to be better on the whole?

No Free Lunch Theorem [4] highlights the unpleasant fact that if the aim is to achieve perfect generalization performance, there are no reasons—independent of context or usage—to prefer one learning algorithm over another. If one algorithm appears to outperform another in a specific situation, it is as a result of its fit to the specific learning problem, *not* the general supremacy of the algorithm. Even if the algorithms are widely used and grounded in terms of theory, they will not perform well on certain problems.

It is clear from the No Free Lunch Theorem that in absence of knowledge pertaining to the relevant learning domains, we should not favor any learning algorithm over another. According to the *Ugly Duckling Theorem* [4], in absence of assumptions about the learning domains there is no privileged or 'best' feature representation even the idea of similarity between patterns is dependent on assumptions we make related to the learning domains. These assumptions may be correct or incorrect.

In certain fields, rigid conservation laws exist—such as conservation of energy, charge, and momentum; and constraint laws—such as the second law of thermodynamics. These apply irrespective of how many forces are at play or their configuration. Do analogous results exist in machine learning? Certainly not. Unfortunately, no rigid results in machine learning exist that can apply to all situations, irrespective of the learning algorithm selected, the number of patterns and their distribution, and the nature of the learning task. Awareness of the learning domains, and experience with a wide range of algorithms is the greatest insurance to solve new learning problems.

Many results that look for ways to quantify the 'match' between a learning algorithm and the problem addressed by it will appear in the forthcoming discussions. In classical statistics, this problem has been examined in terms of *bias* and *variance*. The two terms are not independent, as is clear from Section 2.2.

Fragments of *statistical learning theory* have begun to emerge, providing answers to generic questions about machine learning within particular problem settings. *VC* (Vapnik-Chervonenkis) model is mainly concerned with *hypothesis space complexity*. PAC (Probably-Accurately-Correct) model answers questions on *sample complexity* (training-data size), and *computational complexity*. We will present these results in Section 2.3.

As we shall see in this chapter, such results are of theoretical interest. Practitioners, by and large, are depending on *heuristic* (trial-and-error) search process. Trial-and-error effort can, of course, be reduced using knowledge of the learning domains, and experience with broad range of learning algorithms. Frequent empirical 'successes' of philosophical principle of *Occam's razor*, and *minimum description length principle*, have given some tools in the hands of practitioners.

Section 2.4 uses intuitive platform of discussion on Occam's razor principle and *overfitting avoidance*. The discussion on minimum description length principle will be taken up in Chapter 3.

The discussion given in this book is without much mathematical rigor or statistical justification. Our aim is to give main insights of *learning theory* in a fairly simple manner.

2.2 BIAS AND VARIANCE

Consider performing the following experiment. We first collect a random sample $\mathcal{D}$ of N independently drawn patterns from the distribution $p(\mathbf{x}, y)$, and then measure the *sample error/training error/approximation error* from Eqn (2.4), using loss function (2.5) for classification problem or (2.6) for regression problem. Let us denote the approximation error based on data sample $\mathcal{D}$ and hypothesis h as 'error$_\mathcal{D}[h]$'. If we repeat the experiment several times, each time drawing a different sample $(\mathbf{x}^{(i)}, y^{(i)})$ of size N, we would expect to see different values of error$_\mathcal{D}[h]$, on the basis of the random differences in the way the different samples of size N are made up. We say in such cases that error$_{\mathcal{D}_j}[h]$, the result of the jth experiment, is a *random variable*.

Imagine that we were to run K such experiments, measuring the random variables error$_{\mathcal{D}_j}[h]$; $j = 1, 2, \ldots, K$. The average over the K experiments:

$$\text{error}_\mathcal{D}[h] = \mathbb{E}_\mathcal{D}\{\text{error}_{\mathcal{D}_j}[h]\} \tag{2.7}$$

where $\mathbb{E}_\mathcal{D}\{\cdot\}$ denotes the expectation or ensemble average.

Bias and variance are most easily understood in the context of regression (numeric prediction). It is convenient to consider the particular case of a hypothesis function trained using the risk function (2.4), although our conclusions will be much more general.

Let us suppose that there is a true (yet unknown) function $f(\mathbf{x})$ possessing continuous-valued output y with noise, and we try to estimate this function on the basis of N samples in set $\mathcal{D}_j$ generated from $f(\mathbf{x})$.

The regression function estimated is denoted $h(\mathbf{x};\ \mathcal{D}_j)$, and we are interested in the dependence of this approximation on the training set $\mathcal{D}_j$. Due to random variations in the selection of datasets $\mathcal{D}_j;\ j = 1, \ldots, K$, for some datasets, the approximation will be excellent, while for other datasets of the same size, approximation will be poor. The natural measure of the effectiveness of the estimator can be expressed as its mean-square deviation (refer to Eqn (2.6)) from the desired optimal: $[h(\mathbf{x};\ \mathcal{D}_j) - f(\mathbf{x})]^2$.

A measure of how close the mapping function $h(\mathbf{x};\ \mathcal{D}_j)$ is to the desired one is, therefore, given by the error function,

$$\text{error}_{\mathcal{D}_j}[h] = [h(\mathbf{x};\ \mathcal{D}_j) - f(\mathbf{x})]^2 \tag{2.8}$$

The value of this quantity will depend on the particular dataset $\mathcal{D}_j$ on which it is trained. We write the average over the complete ensemble of datasets as,

$$\text{error}_{\mathcal{D}}[h] = \mathbb{E}_{\mathcal{D}}\{[h(\mathbf{x};\ \mathcal{D}_j) - f(\mathbf{x})]^2\} \tag{2.9}$$

A non-zero error can arise for essentially two reasons:

1. It may be that the hypothesis function $h(\cdot)$ is on average, different from the regression function $f(\mathbf{x})$. This is called *bias*.
2. It may be that the hypothesis function is very sensitive to the particular dataset $\mathcal{D}_j$, so that for a given $\mathbf{x}$, it is larger than the required value for some datasets, and smaller for other datasets. This is called *variance*.

We can make the decomposition into bias and variance explicit by writing (2.8) in somewhat different, but mathematically equivalent form.

$$\begin{aligned}
\text{error}_{\mathcal{D}_j}[h] &= [h(\mathbf{x};\ \mathcal{D}_j) - f(\mathbf{x})]^2 \\
&= [h(\mathbf{x};\ \mathcal{D}_j) - \mathbb{E}_{\mathcal{D}}\{h(\mathbf{x};\ \mathcal{D}_j)\} + \mathbb{E}_{\mathcal{D}}\{h(\mathbf{x};\ \mathcal{D}_j\} - f(\mathbf{x})]^2 \\
&= [h(\mathbf{x};\ \mathcal{D}_j) - \mathbb{E}_{\mathcal{D}}\{h(\mathbf{x};\ \mathcal{D}_j)\}]^2 + [\mathbb{E}_{\mathcal{D}}\{h(\mathbf{x};\ \mathcal{D}_j)\} - f(\mathbf{x})]^2 + \\
&\quad 2[h(\mathbf{x};\ \mathcal{D}_j) - \mathbb{E}_{\mathcal{D}}\{h(\mathbf{x};\ \mathcal{D}_j)\}]\,[\mathbb{E}_{\mathcal{D}}\{h(\mathbf{x};\ \mathcal{D}_j)\} - f(\mathbf{x})]
\end{aligned} \tag{2.10}$$

In order to compute the expression in (2.10), we take the expectation of both sides over the ensemble of datasets $\mathcal{D}$.

$$\begin{aligned}
\text{error}_{\mathcal{D}}[h] &= \mathbb{E}_{\mathcal{D}}\{[h(\mathbf{x};\ \mathcal{D}_j) - f(\mathbf{x})]^2\} \\
&= \mathbb{E}_{\mathcal{D}}\{[h(\mathbf{x};\ \mathcal{D}_j) - \mathbb{E}_{\mathcal{D}}\{h(\mathbf{x};\ \mathcal{D}_j)\}]^2\} + \mathbb{E}_{\mathcal{D}}\{[\mathbb{E}_{\mathcal{D}}\{h(\mathbf{x};\ \mathcal{D}_j)\} - f(\mathbf{x})]^2\} \\
&\quad + 2\mathbb{E}_{\mathcal{D}}\{[h(\mathbf{x};\ \mathcal{D}_j) - \mathbb{E}_{\mathcal{D}}\{h(\mathbf{x};\ \mathcal{D}_j)\}]\,[\mathbb{E}_{\mathcal{D}}\{h(\mathbf{x};\ \mathcal{D}_j)\} - f(\mathbf{x})]\}
\end{aligned} \tag{2.11}$$

Note that the third term on the right hand side of (2.11) vanishes (Help: $\mathbb{E}_{\mathcal{D}}\{f(\mathbf{x})\} = f(\mathbf{x})$; $\mathbb{E}_{\mathcal{D}}\{\mathbb{E}_{\mathcal{D}}\{h(\mathbf{x}; \mathcal{D}_j)\}\} = \mathbb{E}_{\mathcal{D}}\{h(\mathbf{x}; \mathcal{D}_j)\}$) and we are left with,

$$\begin{aligned} \text{error}_{\mathcal{D}}[h] &= \mathbb{E}_{\mathcal{D}}\{[h(\mathbf{x}; \mathcal{D}_j) - f(\mathbf{x})]^2\} \\ &= \underbrace{[\mathbb{E}_{\mathcal{D}}\{h(\mathbf{x}; \mathcal{D}_j)\} - f(\mathbf{x})]^2}_{(\text{bias})^2} + \underbrace{\mathbb{E}_{\mathcal{D}}\{[h(\mathbf{x}; \mathcal{D}_j) - \mathbb{E}_{\mathcal{D}}\{h(\mathbf{x}; \mathcal{D}_j)]^2\}}_{\text{variance}} \end{aligned} \tag{2.12}$$

The *bias* measures the level to which the average (over all datasets) of the hypothesis function is different from the desired function $f(\mathbf{x})$. The *variance* is a measure of the level to which the hypothesis function $h(\mathbf{x}; \mathcal{D}_j)$ is sensitive to the specific selection of the dataset.

The true (but unknown) function we wish to approximate is $f(\mathbf{x})$. We have dataset $\mathcal{D}$ of N samples having continuous-valued output y with noise: $y = f(\mathbf{x}) + \varepsilon$ where ε represents noise. We seek to find an approximate function. $\hat{y} = h(\mathbf{x}; \mathbf{w})$ such that mean-square-error is minimized ($\mathbf{w}$ represents adjustable parameters of hypothesis function $h(\cdot)$). To assess the effectiveness of the learned model $h(\cdot)$, we consider $\mathcal{D}_j$; $j = 1, \ldots, K$, training sets.

From Eqn (2.12), we observe that the bias term reflects the approximation error that the hypothesis $h(\cdot)$ is expected to have on average when trained on datasets of same finite size. In practice, it is often necessary to iterate in order to adjust the parameters of the hypothesis function to the problem specifics, so as to reduce the approximation error. In general, *higher the complexity* of the hypothesis function (more flexible function with large number of adjustable parameters), the *lower is the approximation error*.

From Eqn (2.12), we also observe that the variance term reflects the capability of the trained model on a data sample to *generalize* to other data samples. Low variance means that the estimate of $f(\mathbf{x})$ based on a data sample does not change much on the average as the data sample varies. Unfortunately, the *higher the complexity* of the hypothesis function (which results in low bias/low approximation error), the *higher is the variance*.

We see that there is a natural trade-off between bias and variance. Procedures with increased flexibility to adopt to the training data tend to have lower bias but higher variance. Simpler (inflexible; less number of free parameters) models tend to have higher bias but lower variance. To minimize the overall mean-square-error, we need a hypothesis that results in *low bias* and *low variance*. This is known as *bias-variance dilemma* or *bias-variance trade-off* [35].

The design of a good hypothesis through the study of trade-off between bias and variance in the context of a finite data-sample size is one of the main themes of modern automatic learning theory. In general, finding an optimal bias-variance trade-off is hard, but acceptable solutions can be found. New statistical learning techniques are being developed with promising results. The next section introduces this approach. This is then followed by procedures for avoiding overfitting, and heuristic search methods.

Example 2.1

Suppose that the true target function is of quadratic form in one variable with noise: $y = x^2 + \varepsilon$; ε is noise term; x takes on values on the interval $(0,5)$ [3]. Assume that a data sample is generated from this model.

Consider a linear hypothesis function; slope and intercept are the two adjustable parameters of the linear curve that we fit to the training data. If the experiment is repeated many times, it can be seen that estimates are scattered around $\hat{y} = 10.81$, given $x = 3$. According to the original model, the value of y, given $x = 3$, is scattered around $y = 9$. This difference is the *bias* (a sort of persistent error because of our choice of a simpler model), i.e., a linear model instead of a more complex polynomial model. The bias is depicted in Fig. 2.3(a).

Now, let us fit a complex model, say a 5th order polynomial, having many adjustable parameters. The distribution of the estimate $\hat{y}$ at $x = 3$ from different random samples will look like the one shown in Fig. 2.3(b). We note that the bias is now zero but variability in prediction is very high. We see from Fig. 2.3(a) that when we fit a model which is simpler (compared to the model which generated the data) bias is high but variability is less, whereas when we fitted a model of higher order, bias is negligible but variance is very high. This is bias-variance dilemma.

Figure 2.4 shows how bias and variance vary with the complexity of the model. For increasing complexity of the model, the bias component of the error decreases, but the variance component increases. For a certain model complexity, the total error is minimal. Different models result in different bias-variance trade-off curves.

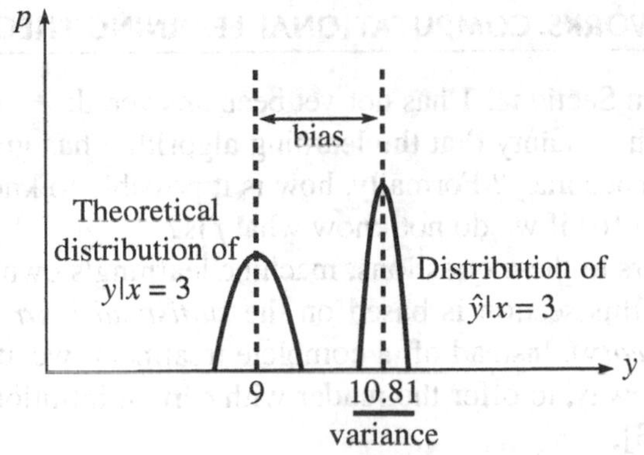

(a) Linear curve fitting

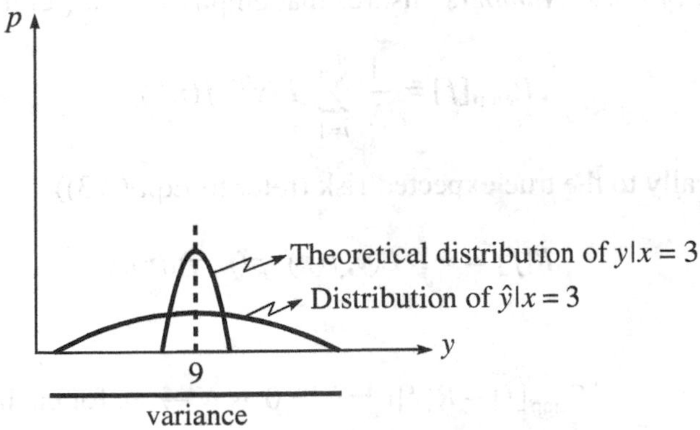

(b) Higher-order polynomial fitting

Figure 2.3 Illustration of bias and variance in the domain of regression

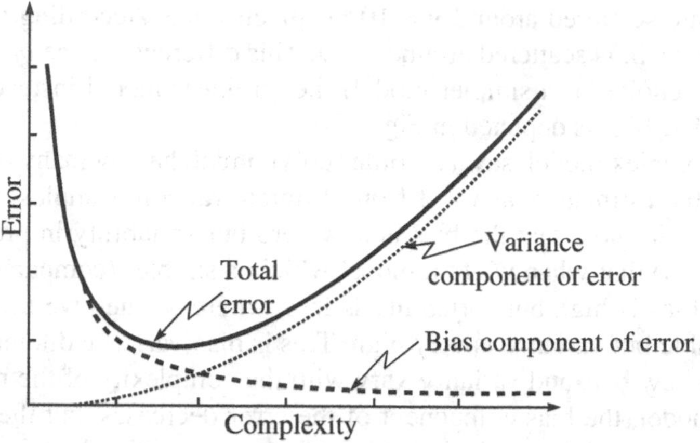

Figure 2.4 Bias-variance trade-off

2.3 WHY LEARNING WORKS: COMPUTATIONAL LEARNING THEORY

The main question posed in Section 2.1 has not yet been answered:

Is it possible to say with certainty that the learning algorithm has given rise to a theory, which can predict the future with accuracy? Formally, how is it possible to know that the hypothesis $h(\cdot)$ is close to the true function $f(\cdot)$ if we do not know what f is?

Till we know the answers to these questions, machine learning's own success will be baffling.

The approach taken in this section is based on the *statistical learning theory* (also known as *computational learning theory*). Instead of a complete treatment, we attempt to provide the main insights in a non-technical way, to offer the reader with certain intuition as to how various pieces of the puzzle fit together [3].

What is required for empirical risk minimization over a given dataset $(\mathbf{x}^{(i)}, y^{(i)})$; $i = 1, \ldots, N$, to work (bias and variance low at the same time)? Let us examine the possibility of success as $N \to \infty$ [53]. The classical *Law of Large Numbers* ensures that empirical risk (refer to Eqn (2.4))

$$R_{\text{emp}}[f] = \frac{1}{N}\sum_{i=1}^{N} L(y^{(i)}, f(\mathbf{x}^{(i)}))$$

converges probabilistically to the true expected risk (refer to Eqn (2.3))

$$R[f] = \int_{\mathbf{X}\times Y} L(y, f(\mathbf{x}))\, p(\mathbf{x}, y)\, d\mathbf{x}\, dy$$

as $N \to \infty$:

$$|R_{\text{emp}}[f] - R[f]| \xrightarrow{P} 0 \text{ as } N \to \infty, \text{ for each fixed } f. \tag{2.13}$$

This means that $\forall\, \delta > 0$, we have (P denotes the probability)

$$\lim_{N\to\infty} P\{|R_{\text{emp}}[f] - R[f]| > \delta\} = 0 \tag{2.14}$$

This statement applies as well for the parameters **w** which define function f (f: $f(\mathbf{x}, \mathbf{w})$).

$$\lim_{N\to\infty} P\{|R_{emp}[\mathbf{w}] - R[\mathbf{w}]| > \delta\} = 0 \;\forall\, \delta > 0 \tag{2.15}$$

At first sight, it seems that empirical risk minimization should work (as $N \to \infty$) in contradiction to our fears expressed in bias-variance dilemma. Unfortunately, it does not work. The convergence in probability does not guarantee that the function f^{emp} that minimizes $R_{emp}[f]$ converges to the true function f^{opt} that minimizes $R[f]$.

Simplified depiction of the convergence of empirical risk to true risk is given in Fig. 2.5. x-axis gives one-dimensional representation of function class, and y-axis gives risk (error). Downward arrow indicates that $R_{emp}[f]$ converges to $R[f]$ for fixed f as $N \to \infty$. f^{opt} is the function that gives best attainable true risk, $R[f^{opt}]$. There is no guarantee that as $N \to \infty, f \to f^{opt}$.

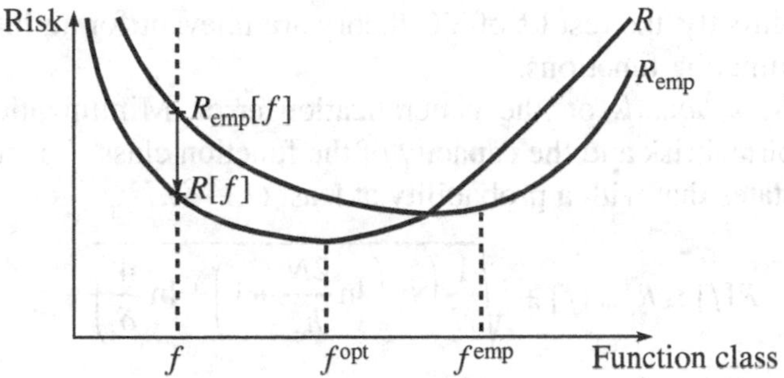

Figure 2.5 Convergence of empirical risk to true risk

We have to do more than just solving empirical risk minimization problem. What is the catch?

What is needed is *asymptotic consistency* or *uniform convergence*. This property of consistency is defined in the key learning theorem for bounded loss functions [37, 38], which states that empirical risk minimization principle is consistent if and only if *empirical risk converges uniformly to the true risk* in the following probabilistic sense:

$$\lim_{N\to\infty} P\{\sup_{f\in\mathcal{F}} |R[f] - R_{emp}[f]| > \delta\} = 0 \;\forall\, \delta > 0 \tag{2.16}$$

$\mathcal{F}$ is the set of functions that the learning machine can implement.

Equivalently, in terms of parameters of function f, we can express asymptotic consistency condition as,

$$\lim_{N\to\infty} P\{\sup_{\mathbf{w}} |R[\mathbf{w}] - R_{emp}[\mathbf{w}]| > \delta\} = 0 \;\forall\, \delta > 0 \tag{2.17}$$

sup denotes the supremum (least upper bound) of the set $S = |R(\mathbf{w}) - R_{emp}(\mathbf{w})|$; it is defined by the smallest element b such that $b \geq s$ for all $s \in S$.

Equation (2.17) asserts that the consistency is determined by the function from the set of approximating functions $\mathcal{F}$ that gives a worst-case measure (maximum value of $(R - R_{emp})$), i.e., that provides the largest error between the empirical risk and the true risk. Consistency, and thus learning, crucially depends on the set of functions $\mathcal{F}$. If we consider the set of all possible functions, learning is not possible.

It turns out that *without restricting the set of admissible functions, empirical risk minimization is not consistent.*

We now address whether there are properties of learning machines, i.e., of sets of functions, which *ensure* uniform convergence of risks.

VC (Vapnik and Chervonenkis) Theory [38] shows that it is imperative to restrict the set of functions $\mathcal{F}$ from which f is chosen to one that has a *capacity* suitable for the amount of available training data. The best-known capacity concept of VC theory is *VC Dimension*; it restricts the function class $\mathcal{F}$ to a finite one for the purpose of bounding (2.16).

Despite the fact that the VC dimension is very important, the unfortunate reality is that its analytic estimations can be used only for the simplest sets of functions. The calculation of VC dimension for nonlinear function classes is very difficult task, if possible at all. Also, the nature of learning problem (classification/regression) affects estimation complexity. But even when the VC dimension cannot be calculated directly, the results of VC theory are relevant for an introduction of *structure* on the class of approximating functions.

The VC theory offers *bounds* on the generalization error. Minimization of these bounds is dependent on the empirical risk and the capacity of the function class. A result from VC theory for binary classification states that with a probability at least $(1 - \delta)$,

$$R[f] \leq R_{\text{emp}}[f] + \sqrt{\frac{1}{N}\left(h_c\left(\ln\frac{2N}{h_c}+1\right)+\ln\frac{4}{\delta}\right)} \tag{2.18}$$

In case of binary classification, the VC dimension (capacity) h_c of a set $\mathcal{F}$ of functions is defined as the number of points that can be separated (shattered) in all possible ways.

Bounds like (2.18) can be used to justify induction principle different from the empirical risk minimization principle. Vapnik and Chervonenkis proposed minimizing the right hand side of the bound (2.18), rather than the empirical risk. The confidence term in the present case, $\sqrt{\frac{1}{N}\left(h_c\left(\ln\frac{2N}{h_c}+1\right)+\ln\frac{4}{\delta}\right)}$, then ensures that the chosen function not only leads to small risk, but also comes from a function class with small capacity. The capacity term is a property of the function class $\mathcal{F}$, and not of any individual function f. Thus, the bound cannot simply be minimized over choices of f. Instead, we introduce a so-called *structure* on $\mathcal{F}$, and minimize over elements of the structure. This leads to an induction principle called *structural risk minimization.* We leave out the technicalities involved; the main idea is as follows:

The function class $\mathcal{F}$ is decomposed into a nested sequence of subsets of increasing capacity. The structural risk minimization principle picks a function f^* which has small training error (R_{emp}), and comes from an element of the structure that has low capacity h_c; thus minimizing a risk bound of the type (2.18). This is graphically depicted in Fig. 2.6.

Structural risk minimization is a novel inductive principle for learning from finite training data sets. It is very useful for dealing with small samples. The basic idea is to choose from a large number of candidate learning machines, a machine of the right complexity to describe training data pairs. This is done by restricting the hypothesis space of approximating functions and simultaneously controlling their complexity.

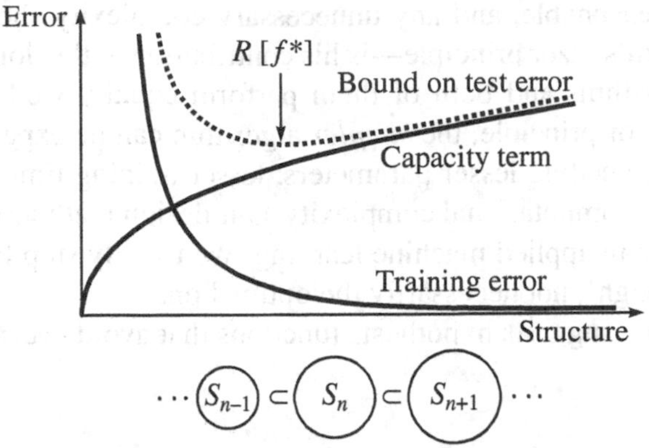

Figure 2.6 Structural risk minimization

Another popular formalism of computational learning theory is the *Probably-Approximately-Correct* (PAC) learning. Let us examine *sample complexity* in PAC framework.

The sample complexity of a learning algorithm is defined as the smallest number of samples required for learning by $\mathcal{F}$, that achieves a given approximation accuracy ε with a probability $(1 - \delta)$. Accuracy ε is defined as follows:

$$\text{error}_P[f^{\text{emp}}] \leq \varepsilon + \text{error}_{\mathcal{D}}[f^{\text{emp}}] \tag{2.19}$$

where $\mathcal{D}$ is the set of examples available to the machine, and P represents the probability distribution over the entire set of patterns.

In terms of cardinality of $\mathcal{F}$, denoted $|\mathcal{F}|$, the sample complexity of a learning algorithm N_{PAC} (smallest number of training examples required for learning), is at most

$$N_{\text{PAC}} = \frac{1}{2\varepsilon^2}\left(\ln|\mathcal{F}| + \ln\left(\frac{1}{\delta}\right)\right) \tag{2.20}$$

For most hypothesis spaces for binary classification, $N \geq N_{\text{PAC}}$ gives a good result. Bounds for other problems have also been found.

Note that the bound (2.20) can be a substantial overestimate. The weakness of this bound is mainly due to $|\mathcal{F}|$ term. In fact, much tighter bound is possible using VC dimension:

$$N \geq \frac{1}{\varepsilon}\left[4\log_2\left(\frac{2}{\delta}\right) + 8h_c\log_2\left(\frac{13}{\varepsilon}\right)\right] \tag{2.21}$$

How about the overall *computational complexity*? That will depend, of course, on specific learning algorithm.

2.4 OCCAM'S RAZOR PRINCIPLE AND OVERFITTING AVOIDANCE

Occam's Razor Principle

The Franciscan monk, Willam of Occam, was born in 1280, much before the invention of modern statistics. His name is linked with machine learning through the basic idea 'The *simpler* explanations are more reasonable, and any unnecessary complexity should be shaved off'. This line of reasoning—Occam's razor principle—is his contribution to the domain of machine learning.

If there are two algorithms and both of them perform equally well on the training set, then according to Occam's razor principle, the *simpler* algorithm can be expected to do better on a test set; 'simpler' may imply needing lesser parameters, lesser training time, fewer attributes for data representation, and lesser computational complexity. Our design methodology itself imposes a bias toward 'simple' solutions in applied machine learning. We usually stop looking for a design when the solution is 'good enough', not necessarily the optimal one.

Occam's razor principle suggests hypothesis functions that avoid *overfitting* of the training data.

Overfitting

In machine learning jargon, a learning machine is said to *overfit* the training examples if certain other learning machine that fits the training examples less well, actually performs better over the total distribution of patterns (i.e., including patterns beyond the training set).

Figure 2.7 depicts the effect of overfitting in a typical application of machine learning. The horizontal axis of the plot shows how *complex* the classifier is (more number of weights in the neural network, large number of nodes in the decision free, more number of rules in a fuzzy logic model, etc.). The vertical axis is an indicator of how *accurate* the forecasts made by the classifier are. The solid line depicts how accurate the classifier is over the training examples, while the broken line shows the accuracy measured over an independent set of test examples (not included in the training set, but class labels known). Predictably, the accuracy of the classifier over training examples increases monotonically as the classifier becomes more complex. But, the accuracy over the independent test examples first grows, then reduces.

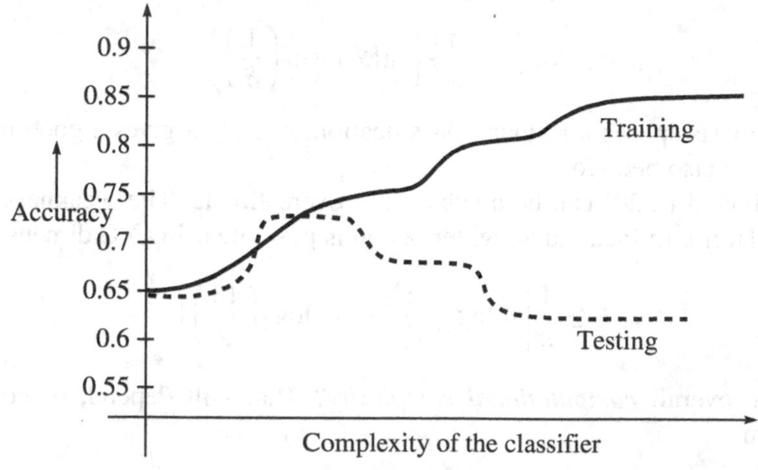

Figure 2.7

Learning is most reliable when the training examples follow a distribution similar to that of future instances unseen during training. If training experience consists of data that lies in a region of state space, then that region must be fully representative of situations over which our learning machine will later be used. Machine learning tasks based on real-world data are unlikely to find this assumption tenable. The real-world data tend to be incomplete, noisy, and inconsistent. Classifier design for such a data results in more complex classifier, if a near-perfect fit on the training data is the goal. To illustrate, consider the effect of a training example with incorrect class label. A classifier that fits the data perfectly will be more complex; complexity in design is simply a consequence of fitting the noisy training example. We expect another classifier that is simpler but does not fit training data perfectly, to outperform this classifier over subsequent data drawn from the same pattern distribution.

Even in noise-free training data, overfitting can occur due to insufficient number of training examples. Coincidental regularities in this dataset lead to a risk of inducing an approximation function with poor generalization performance.

Section 2.1 clearly states that there exist no problem-independent reasons to favor one algorithm over another. Then why is it required to avoid overfitting? Why do we usually promote simpler algorithms which have lesser features and parameters?

In fact, methods that help avoid overfitting are not intrinsically advantageous. They can provide benefit only if they are able to resolve the problems for which they are used. Empirical success depends on the match of the learning algorithm with the problem—not only the imposition of avoiding overfitting. The frequent empirical successes of overfitting avoidance indicate that the classes of real-world problems that have been addressed till now have particular properties that offer a match.

2.5 HEURISTIC SEARCH IN INDUCTIVE LEARNING

Before we move forward, a review of our discussion so far in this chapter, will be helpful.

Review

Machine learning is aimed at building a statistical model of the process, which generates the data, and not to learn an accurate representation of the training data itself. This aim is important for the machine to make good predictions for new inputs, i.e., to demonstrate good generalization.

The success of learning very much depends on hypothesis space complexity and sample complexity. The two are interdependent. The goal is to find a hypothesis (model) simplest in terms of complexity and best in terms of empirical error on the data. Such a choice is expected to give good generalization performance. Basically, *finding a hypothesis function of complexity consistent with the given training data is the search problem in hand.*

Substantial insight into this phenomenon was given in Section 2.2, which introduced the idea of the *bias-variance trade-off*, wherein the mean-square error is decomposed into the sum of the bias-squared plus variance. A function, which is too simple or too inflexible (number of free parameters too less) will have a huge bias, whereas one which is highly complex or too flexible (too many parameters will help fit too much of noise present in the training data) will possess a huge variance. Best generalization is achieved through the best compromise between the contradictory requirements of small bias and small variance.

In order to find the optimum balance between bias and variance, we need to have a way of controlling the effective complexity of the hypothesis functions. Section 2.3 highlighted the *structural-risk minimization principle* to achieve this objective. *VC dimension* was introduced as an important measure of function complexity. Structural-risk minimization uses a set of hypothesis functions ordered in terms of their complexities; hypothesis selection then corresponds to finding the function simplest in terms of complexity and best in terms of empirical error on the data. Relationship between the hypothesis function complexity and sample complexity given by PAC (Probably-Accurately-Correct) model of learning was also highlighted in this section.

What has been the conclusion?

The mathematical models of learning (VC model, PAC model) provide a platform for theoretical characterization of machine learning but lack (at least today) practical tools for real-life problems of varying complexities. Bias-variance trade-off is also a theoretical result because we lack practical tools to obtain optimum values of bias and variance. Machine learning community is, by and large, not using these models to build a learning machine for classification/regression problems; they depend on certain tools, which appear to be *heuristic*, trial-and-error tools. In fact, the motivation for these tools comes from classical statistics and the computational learning theory. Practitioners today, more or less rely on the knowledge of learning domains and their experience with broad range of hypothesis functions. Occam's razor principle and overfitting avoidance, described earlier in Section 2.4, are extensively exploited. Heuristic strategies for overfitting avoidance: regularization, early stopping, pruning (appearing later in this section); and minimum description length principle (detailed in Section 3.9), are some of the heuristic tools explored by practitioners today.

An optimistic viewpoint: the mathematical models give an upper bound on generalization error. We might hope that by our practical heuristic approach (or even mere luck), we would achieve good generalization with fewer training patterns than the number predicted using mathematical learning models developed in classical statistics and computational learning theory.

2.5.1 Search through Hypothesis Space

In this sub-section, we discuss heuristic search strategies. The word 'heuristic' indicates that strategies for the search problem cannot be precisely pre-defined. *Trial-and-error* is the approach of searching for a 'good enough' solution. This approach, however, gets more organized if we exploit the prior knowledge about the learning domains and experience with a broad range of algorithms.

Machine learning entails looking through a space of probable hypotheses to decide on one that fits the data observed, and any previous knowledge held by the learner. Space of probable hypotheses is infinite in theoretical terms. The learning task, therefore, is to explore a widespread space to find the hypothesis that is most consistent with the existing training examples. Luckily, there are merely a few possible candidate hypotheses in use today based on empirical experience (refer to Section 3.5.3).

Applied machine learning organizes the search as per the following two-step procedure (refer to Section 3.5).

(1) The search is first focused on a *class* of the possible hypotheses, chosen for the learning task in hand. Prior knowledge and experience are helpful in this selection. Different classes are appropriate for different kinds of learning tasks.

(2) For each of the classes, the corresponding learning algorithm organizes the search through all possible structures of the learning machine.

In the following paragraphs, we describe principal techniques used in heuristic search to optimize hypothesis complexity for a given training dataset.

Regularization: The regularization model promotes smoother functions by creating a new criterion function that relies not only on the training error, but also on algorithmic intricacy. Particularly, the new criterion function punishes extremely complex hypotheses; looking for the minimum in this criterion is to balance error on the training set with complexity. Formally, it is possible to write the new criterion as a sum of the error on the training set plus a *regularization term*, which depicts constraints or sought after properties of solutions:

$$\bar{E} = E + \lambda\,\Omega \tag{2.22}$$

$$= \text{error on data} + \lambda \times \text{hypothesis complexity where } \lambda \text{ gives the weight of penalty}$$

The second term penalizes complex hypotheses with large variance. When we minimize augmented error function instead of the error on data only, we penalize complex hypotheses and thus decrease variance. When λ is taken too large, only very simple functions are allowed and we risk introducing bias. λ is optimized using cross-validation (discussed later in this chapter).

We consider here the example of neural network hypotheses class (discussion coming up in Chapter 5). The hypothesis complexity may be expressed as,

$$\Omega = \tfrac{1}{2}\sum_{l,j} w_{lj}^2 \tag{2.23}$$

The regularizer encourages smaller weights w_{lj}. For small values of weights, the network mapping is approximately linear. Relatively large values of weights lead to overfitted mapping with regions of large curvature.

Early Stopping: The training of a learning machine corresponds to iterative decrease in the error function defined as per the training data. During a specific training session, this error generally reduces as a function of the number of iterations in the algorithm. Stopping the training before attaining a minimum training error represents a technique of restricting the effective hypothesis complexity.

Pruning: An alternative solution that sometimes is more successful than early stopping the growth (complexity) of the hypothesis is pruning the full-grown hypothesis that is likely to be overfitting the training data. Pruning is the basis of search in many decision-tree algorithms (discussion coming up in Chapter 8); weakest branches of large tree overfitting the training data, which hardly reduce the error rate, are removed.

2.5.2 Ensemble Learning

For the given amount and quality of training data, the output of one hypothesis function may be inappropriate for the problem at hand. The ideal model to make more reliable decisions is to create a combination of outputs of many different hypotheses. Many machine learning algorithms do this by learning an *ensemble* of hypothesis and employing them in a combined form [39–41]. *Bagging* and *boosting* are the most frequently used among these schemes. These general methods may be applied to classification (categorization) and regression (numeric prediction) problems; and, they frequently increase predictive performance over a single hypothesis.

By combining the decisions of various hypotheses, we amalgamate the different outputs into a single prediction. For classification problems, it is done through voting (may be a weighted vote), whereas in case of regression problems, the average is computed (may be a weighted average). Conventionally, various hypotheses in the ensemble possess the same general form—for instance, all neural networks, or all decision trees—simply the final parameter values are different. The difference in parameter values is because of the fact that their training patterns differ—each one handles a certain percentage of data accurately. If the hypotheses complement each other, it would be ideal, as each is a specialist for a part of the domain wherein the other hypotheses fail to do well.

In the **bagging technique**, individual approaches are constructed separately, whereas in **boosting**, each new model is impacted by the performance of those built earlier. In boosting, we first make a model with accuracy on the training set greater than average, and then add new component classifiers to make an ensemble whose joint decision rule possesses a high level of accuracy on the training set. In a scenario such as this, we claim that performance of the learning algorithm has been 'boosted'. The method is capable of training successive classifiers with a subset of 'very informative' training data, considering the present set of component classifiers. **Ada Boost** (an abbreviation for Adaptive Boosting) is a widely used algorithm for boosting.

Bagging is the most straightforward and basic technique of pooling or integrating the outputs of component classifiers.

Yet another ensemble technique is known as **random forests.** Suppose each classifier in the ensemble is a *decision tree* classifier. Therefore, the set of such classifiers gives rise to a 'forest'. The individual decision trees are created with the help of a random choice of attributes at each node to identify the split (Chapter 8). More formally, each tree is dependent on the values of a random vector independently sampled and with the same distribution for all trees in the forest. During classification, each tree votes and the class that is highly popular is returned. You can build random forests with the help of bagging in tandem with the random attribute selection.

Class-Imbalanced Problems: Ensemble methods have been used to solve *class-imbalanced* problems. What are class-imbalanced problems?

Two-class data are *class-imbalanced* if the primary class of interest is represented by just a few samples in the dataset, whereas the majority of the samples represent the other class.

For multiclass-imbalanced data, the data distribution of each class is substantially different, where again the primary class or classes of interest are represented by merely a few samples. The class-imbalanced problem is very similar to *cost-sensitive learning, wherein the costs of errors per class, are unequal*. Metrics for assessment of accuracy of learning machines on the basis of cost-sensitive datasets will be presented in Section 2.8.

The general approaches for improvement of the classification performance of class-imbalanced data are:

(i) Oversampling
(ii) Undersampling
(iii) Threshold moving
(iv) Ensemble methods

Oversampling and undersampling alter sample distribution in the training set; threshold-moving impacts the manner in which the model makes decisions while classifying new data, and ensemble techniques follow the methods discussed earlier.

2.5.3 Evaluation of a Learning System

Before being used, a machine learning system should be evaluated in many aspects, intrinsically linked to what the system is meant for. Some important aspects of evaluation are:

Accuracy: Inductive learning is based on empirical data. The learning system extracts knowledge from the training data. The learned knowledge should be general enough to deal with unknown data. The *generalization* capability of a learning system is an index of *accuracy* of the learning machine.

A machine learning system will not be accepted by its intended users unless it has demonstrated its accuracy in problem solving. This is the single most important objective of learning. We will shortly describe the *accuracy and error measures* and the methods of *error estimation* to quantitatively measure the generalization capability of a learning machine.

Robustness: 'Robustness' means that the machine can perform adequately under all circumstances, including the cases when information is corrupted by noise, is incomplete, and is interfered with irrelevant data. All these conditions seem to be part of the real world (discussed in Chapter 7), and must be considered while evaluating a learning system. Robustness is typically assessed with a series of synthetic datasets representing increasing degrees of inconsistencies in data.

Computational Complexity and Speed: Computational complexity of a learning algorithm and learning speed determine the *efficiency* of a learning system: how fast the systems can arrive at a correct answer, and how much computer memory is required. We know how important speed is in real-time situations.

Online Learning: An online learning system can continue to assimilate new data. This feature is essential for a learning system, which continues to receive inputs from a real-time environment.

Interpretability: This is the level of understanding and insight offered by a learning algorithm. Interpretability is subjective and hence, tougher to evaluate. Interpretation is easy in fuzzy logic systems and decision trees, but still their interpretability may decrease with an increase in their complexity.

Scalability: This is the capability to build the learning machine considering the huge amounts of data. Typically, the assessment of scalability is done with a series of datasets of ascending size.

2.6 ESTIMATING GENERALIZATION ERRORS

The success of learning depends on the hypothesis space complexity and sample complexity. The two are interdependent. The goal is to find a function simplest in terms of complexity and best in terms of empirical error on the data. Such a choice is expected to give good generalization performance. Basically, *finding a hypothesis function of complexity consistent with the given training data* is the problem in hand.

So how is the learning scheme likely to perform in the future on new (unseen) data? We are not interested in the past performance on old (training) data. We are already aware of the output of each instance in the *training set*, which is the reason why it can be used for training. The actual question is: Is the performance on training data likely to be a proper indicator of the performance on unseen data, which is not employed in training? The answer is in the negative. To predict the performance of a learning scheme on new data, the assessment of the success rate (error rate) on a dataset— that had no role in forming the leaning model for classification or numeric prediction (metrics for *error rate* are given in the next two sections)—is required. This independent dataset is known as the *test set*. The assumption is that both the training set and the test set comprise representative samples of the underlying distribution.

Importantly, the test data is in no way employed to build the learning model. Some learning schemes include two stages for constructing a model—one to come up with the elementary structure, and the second to optimize parameters involved in that structure. Separate sets of data may be required in the two stages. In such scenarios, we frequently consider three datasets: the *training data*, the *validation data*, and the *test data*. The training data is employed to create the structure of the learning model. The validation data is employed to optimize parameters of that model, or to choose a specific model if the training data has been made use of to make several learning models. Then the test data is employed to compute the error rate of the final optimized selected model. Each of the three sets has to be selected independently: the validation set should be different from the training set to get perfect performance in the optimization or selection stage, and the test data has to be different from both to get a reliable estimate of the true error rate.

If a lot of data is available, a reasonable large sample is taken up for training, then another independent reasonable large sample is taken up for testing. What to do when vast supply of data is not available. The question of predicting performance on the basis of *limited data* is interesting, yet controversial.

We will take a look at various methods—the probable techniques for success in most practical limited-data situations.

2.6.1 Holdout Method and Random Subsampling

In the *holdout* technique, some amount of data is earmarked for the purpose of testing, while the remainder is employed for training (and a portion of it is kept aside for validation, if needed). Practically speaking, it is common to hold out one-third of the data for testing and use the remainder for training.

In order to divide dataset $\mathcal{D}$ into training and test sets, we can *arbitrarily* sample a set of training examples from $\mathcal{D}$, and keep aside the remaining for testing. If the data is collected over time (*time-series data*), then we can make use of the earlier part to train and the latter part of the data for the purpose of testing. In various applications, this process of dividing time-series data is more suited as the learning machine is made use of in the real world; the unseen data belongs to the future.

The samples used to train and test have to represent the underlying distribution for the problem area. Generally speaking, it is not possible to tell whether a sample is representative or not because we do not know the distribution. However, there is a simple check that may be worthwhile. In case of classification problems, each class in the full dataset needs to be represented in about the right proportion in the training and test sets. The proportion of class-data in training, testing, and full datasets should more or less be same. To make sure this happens, random sampling should be performed in a manner that will guarantee that each class is properly represented in training as well as test sets. This process is known as *stratification*.

Even though *stratified holdout* is generally well worth doing, it offers merely a basic safeguard against irregular representation in training and test sets.

A more general way to alleviate any bias resulting from the specific sample selected for holdout is *random subsampling*, wherein the holdout technique is iterated K times with various arbitrary samples. The accuracy estimate on the whole is considered as the average of the accuracies got from each repetition.

In a single holdout process, we may look at considering swapping the roles of training and test data—that is, train the machine on the test data, and estimate the success rate making use of the training data—and average the two results, thereby decreasing the effect of uneven distribution in training and test sets. This is, however, useful with a 50:50 split of the full data between training and test sets, which is usually not ideal—it is ideal to employ more than 50 per cent of the data for training even at the cost of test data. But, a simple variant becomes the basis of a powerful statistical method, known as *cross-validation*.

2.6.2 Cross-validation

A commonly used technique for forecasting the success rate of a learning method, taking into account a fixed data sample, is the *K-fold cross-validation*. Another estimate prevalent is the *leave-one-out* cross-validation. The description of these two cross-validation techniques follows.

K-Fold Cross-validation

In K-fold cross-validation, the given data $\mathcal{D}$ is randomly divided into K mutually exclusive subsets or 'folds', $\mathcal{D}_k$; $k = 1, \ldots, K$, each of about equal size. Training and testing is done K times. In iteration k, partition $\mathcal{D}_k$ is set aside for testing, and the remainder of the divisions are collectively employed to train the model. That is, in the first iteration, the set $\mathcal{D}_2 \cup \mathcal{D}_3 \cup \cdots \cup \mathcal{D}_K$ serves as the training set to attain the first model, which is tested on $\mathcal{D}_1$; the second iteration is trained on $\mathcal{D}_1 \cup \mathcal{D}_3 \cup \cdots \cup \mathcal{D}_K$ and tested on $\mathcal{D}_2$; and so on.

If stratification is also used, it is known as *stratified K-fold cross-validation* for classification.

Ultimately, the K error estimates received from K iterations are averaged to give rise to an overall error estimate. Out of the 10 machines, the one with lowest error may be deployed.

$K = 10$ folds is the standard number employed to predict the error rate of a learning method. Why 10? Extensive tests on numerous datasets with various learning methods have revealed that

10 is about the right number of folds to achieve the best estimate of error. There is some theoretical proof also that backs up *10-fold cross-validation*. These arguments, although, cannot be said to be conclusive, 10-fold cross-validation is now the standard technique in practical terms.

When you look for a precise estimate, the normal process is to repeat the 10-fold cross-validation procedure 10 times and average the outcomes. This requires applying the learning algorithm 100 times on datasets that are all nine-tenths the size of the original.

Leave-One-Out Cross-Validation

This is an exceptional case of K-fold cross-validation wherein K is set to the number N of initial tuples. In other words, only a single sample is 'left out' for the test set in each iteration. The learning machine is trained on the remainder of the samples. It is judged by its accuracy on the left-out sample. The average of all outcomes of all N judgements in N iterations is taken, and this is the average, which is representative of the error estimate.

The computational expense of this process is quite high as the whole learning process has to be iterated N times, and this is generally not feasible for big datasets. Nevertheless, leave-one-out seems to present an opportunity to squeeze the maximum out of a small dataset and obtain an estimate that is as precise as possible. This process disallows stratification.

2.6.3 Bootstrapping

The *bootstrap technique* is based on the process of *sampling with replacement*. In the earlier techniques, whenever a sample was used from the dataset to form a training or test set, it was never replaced. In other words, the same instance, which was once chosen could not be chosen again. However, most learning techniques can employ an instance several times, and it affects the learning outcome if it is available in the training set more than once. The concept of *bootstrapping* aims to sample the dataset by replacement, so as to form a training set and a test set.

There are many bootstrap techniques. The most popular one is the *0.632 bootstrap*, which works as follows:

A dataset of N instances is sampled N times, with replacements, to give rise to another new dataset of N instances, which is a *bootstrap sample*—a training set of N samples. As certain elements in the bootstrap sample will (almost certainly) be repeated, there will be certain instances in the original dataset $\mathcal{D}$ that have not been selected—these will be used as test instances. If we attempt this many times, on an average, 63.2% of the original data instances will result in the bootstrap sample and the remaining 36.8% will give rise to the test set (therefore, the name, 0.632 bootstrap).

Where does the figure, 63.2%, come from? The probability that a particular instance will be picked is $1/N$; so the probability of not being picked is $\left(1-\frac{1}{N}\right)$. Number of picking opportunities is N, so the probability that an instance will not be picked during the whole sampling cycle is $\left(1-\frac{1}{N}\right)^N$. If N is large, the probability approaches $e^{-1} = 0.368$ (e is the base of natural algorithms that is, e = 2.718; and not the error rate). Thus, for a reasonably large dataset, the test set will contain about 36.8% of the instances, and training set will contain about 63.2% of them. Some instances will be repeated in the training set, bringing it up to a total size of N.

Training a learning system on the training set and calculating the error over the test set will give a pessimistic estimate of the true error because the training set, although its size is N, nevertheless contains only 63.2% of the instances. (In 10-fold cross-validation, 90% of the instances are used for training.) To compensate for this, the bootstrap procedure combines the training error with the test error to give a final error estimate as follows:

$$\begin{aligned}\text{Error estimate} = {} & 0.632 \times \text{Error given by test instances} \\ & + 0.368 \times \text{Error given by training instances}\end{aligned} \tag{2.24}$$

Then, the whole bootstrap procedure is repeated several times, with different replacement samples for the training set, and the results are averaged.

Bootstrapping tends to be overly optimistic. It works best with small datasets.

2.7 METRICS FOR ASSESSING REGRESSION (NUMERIC PREDICTION) ACCURACY

A function,

$$f = \mathbf{X} \rightarrow Y; f(\mathbf{x}) = y$$

maps from $\mathbf{x} \in \mathbf{X}$ to $y \in Y$, where $\mathbf{X}$ is the *input space* and Y is the *output space* of the function. We assume here that $\mathbf{X} \subset \Re^n$, and $Y \subset \Re$. The data is available as samples $(\mathbf{x}, y)$ where the distribution of inputs $\mathbf{x}$ and the function f are both unknown.

The task is to find the model $h(\mathbf{x})$ that explains the underlying data, i.e., $h(\mathbf{x}) \simeq y$ for all samples $(\mathbf{x}, y)$. Equivalently, the task is to approximate function $f(\mathbf{x})$ with unknown properties by $h(\mathbf{x})$.

The term used in statistics for function description of data is *regression*. Also, since $h(\mathbf{x})$ predicts $y \in \Re$ for a given $\mathbf{x}$, the term *numeric prediction* is also in use. Throughout this book, the three terms: function approximation, numeric prediction, and regression, are used interchangeably without any discrimination.

The classic problem of approximation of multivariate function $f(\mathbf{x})$ is the determination of an approximating function $h(\mathbf{x}, \mathbf{w})$, having a fixed finite number of parameters w_j that are entries of the *weight vector* $\mathbf{w}$.

This section presents measures (metrics) for assessing how good or how accurate our regressor (i.e., approximating function $h(\mathbf{x}, \mathbf{w})$) is at predicting the continuous (numeric) response variable. Classical measures of performance are aimed at finding a model $h(\mathbf{x}, \mathbf{w})$ that fits the data well. However, those measures do not tell us much about the ability of the model to predict new cases. Using training data to derive a regressor and then to estimate the accuracy of the resulting learned model can result in misleading overoptimistic estimates due to overspecialization of the learning algorithm to the data. Instead, it is better to measure the regressor's accuracy on a test set/validation set (refer to previous section) consisting of data not used to train the model.

Estimating the *error* in prediction using holdout and random subsampling, cross-validation and bootstrap methods (discussed in the previous section) are common techniques for assessing accuracy of the predictor. Several alternative metrics can be used to assess the accuracy of numeric prediction.

2.7.1 Mean Square Error

Mean square error (MSE) is the principal and most commonly used metric. For calculating MSE, we assume that no statistical information on data is available; the *mean* is obtained from the training data as *arithmative average* (refer to Section 3.2):

$$\text{MSE} = \frac{1}{N}\sum_{i=1}^{N}(y^{(i)} - h(\mathbf{w},\mathbf{x}^{(i)}))^2 \tag{2.25}$$

which measures the average square-deviation of the predicted values from true values. Here, the input data is represented by $\mathbf{x}^{(i)}$; $i = 1, \ldots, N$; with the corresponding output $y^{(i)}$.

Root Mean Square Error (RMSE)

Taking the square root yields,

$$\text{RMSE} = \sqrt{\frac{1}{N}\sum_{i=1}^{N}(y^{(i)} - h(\mathbf{w},\mathbf{x}^{(i)}))^2} \tag{2.26}$$

RMSE more clearly relates to individual errors; it has same dimensions as the predicted value itself (MSE is not as easily interpretable because units are squared).

Sum-of-Error Squares

Sometimes total error, and not the average, is taken for mathematical manipulation by some statistical/machine learning techniques:

$$\textit{Sum-of-Error-Squares} = \sum_{i=1}^{N}(y^{(i)} - h(\mathbf{w},\mathbf{x}^{(i)}))^2 \tag{2.27}$$

(Expected) Mean Square Error

All the metrics given so far pertain to deterministic cases (given only training data, and no statistical information about the data). In a general probabilistic setting, when input and output variables are random variables, the following metric is used:

$$\textit{(Expected) Mean Square Error} = \mathbb{E}\left[\sum_{i=1}^{N}(y^{(i)} - h(\mathbf{w},\mathbf{x}^{(i)}))^2\right] \tag{2.28}$$

where $\mathbb{E}$ is the statistical expectation operator.

2.7.2 Mean Absolute Error

Given the N samples $(\mathbf{x}^{(i)}, y^{(i)})$, an intuitive measure to assess the quality of a model $h(\mathbf{x}, \mathbf{w})$ is the *Mean Absolute Error* (MAE), computed as,

$$\text{MAE} = \frac{1}{N}\sum_{i=1}^{N}|y^{(i)} - h(\mathbf{w}, \mathbf{x}^{(i)})| \tag{2.29}$$

which measures the average deviation of the predicted value from the true value.

In MAE, variance of the model error is not accounted for. Suppose model $h_1(\mathbf{w}, \mathbf{x})$ correctly models 95% of the data but totally differs on the remainder 5%, while model $h_2(\mathbf{w}, \mathbf{x})$ depicts small errors over the full range of the data. They may as well have the same MAE. When each data point possesses the same significance, model $h_2(\mathbf{w}, \mathbf{x})$ may be preferred as its errors possess lower variability. The amount of variability is considered by MSE criterion.

On the contrary, MSE has the tendency to exaggerate the effect of *outliers*—samples when the prediction error is much larger than the others—but MAE does not have this effect; all sizes of errors are treated evenly according to their magnitude.

2.8 METRICS FOR ASSESSING CLASSIFICATION (PATTERN RECOGNITION) ACCURACY

In machine learning systems for pattern recognition, the focus is on recognizing patterns and regularities in data. The term 'pattern recognition' is popular in the context of computer vision wherein there could be more interest in formalizing, explaining, and visualizing the pattern. In machine learning, the focus is traditionally on assigning a label to a given input pattern. In statistics, the introduction of *discriminant analysis* was done for this purpose: referred to as *classification*; it aims to allocate each input pattern to one of the given *classes*. Substantial evolution in all these areas has become progressively similar due to integration of development of ideas with each other. In this book, the terms patterns recognition and classification carry the same meaning.

The evaluation measures described in the previous section are related to numeric prediction situations rather than classification situations. The basic principles—use of an independent test dataset instead of the training set to evaluate performance, the holdout technique and cross-validation—are equally applicable to classification. However, the basic quality measures in terms of error estimates in numeric prediction are not appropriate any more. The errors in numeric prediction arise in various sizes whereas in classification, errors simply exist or are absent.

Several different measures (metrics) can be used to assess the accuracy of a classifier.

2.8.1 Misclassification Error

Traditional classification algorithms aim to minimize the number of errors made during classification. They treat misclassification of all errors equally seriously.

The metric for assessing the accuracy of classification algorithms is: *number of samples misclassified by the model* $h(\mathbf{w}, \mathbf{x})$. For example, for binary classification problems,

$$y^{(i)} \in [0, 1], \quad \text{and} \quad h(\mathbf{w}, \mathbf{x}^{(i)}) = \hat{y}^{(i)} \in [0, 1];\ i = 1, \ldots, N.$$

For 0% error, $(y^{(i)} - \hat{y}^{(i)}) = 0$ for all data points.

$$\textit{Misclassification error} = \frac{\text{Number of data points for which } (y^{(i)} - \hat{y}^{(i)}) \neq 0}{N} \tag{2.30}$$

This accuracy measure works well for the situations where class tuples are more or less *evenly* distributed. However, when the classes are *unbalanced* (e.g., when an important class of interest is rare), decisions made on classifications based on misclassification error lead to poor performance. Misclassification error measure treats misclassification of all classes equally seriously; however, this is often unrealistic. Often certain kinds of misclassification (wrong decisions based on these results) are more serious than other kinds. Let us look at an alternative measure for such situations.

2.8.2 Confusion Matrix

Decisions made on the basis of classifications that are based on misclassification error rate result in poor performance when data is *unbalanced* or highly *skewed*. For instance, in a dataset where financial fraud is detected, the amount of fraud cases is very small (< 1%) in comparison to the normal cases. The data is therefore skewed. In such classification problems, the user is more interested in the minority class. The class, which is of interest to us, is referred to as the *positive class* while the other classes, are the *negative classes* (which may be put together into one *negative class*).

Misclassification error is not a suitable measure in such cases because we may achieve a very high accuracy, but may not identify a single positive class. For example, 99% of the cases are normal in financial fraud dataset. A classifier can achieve 99% accuracy without doing anything by simply classifying every test as 'nonfraud' category: the *costs* of two kinds of error are different—classifying a fraud as nonfraud is for more serious than the reverse. The decisions of doing business with fraud are far more serious than the lost-business because of misclassifying a normal as fraud. We want rules that predict 'fraud' category more accurately than 'nonfraud' category.

Let us look at a clinical example. Considering the parameters of tissue biopsy: there is a possibility of the developed model committing two kinds of error. It can predict the cancerous quality of the tissue sample when it is actually not so. It can even predict that the tissue sample is not cancerous when actually it may be so. When it comes to an actual clinical setting, the second type of error proves more serious than the first as a cancer patient will end up without receiving any treatment, while the first error will lead to more tests being done on the patient. In such scenarios, we may wish to attach *costs* to the various types of misclassification. Instead of sample error rate, we look for a model that will minimize overall loss.

In practice, the costs are rarely known with any degree of accuracy. An alternative strategy is to consider *the ratio of one cost to another*. Consider two-class problems with classes *yes* or *no*. The rarely occurring class (abnormal situation) is marked as *positive* (P) class, and the absence of this class (a normal situation) is marked as *negative* (N) class.

One prediction on the *test set* has four possible results, depicted in Table 2.1. The *true positive* (TP) and the *true negative* (TN) are accurate classifications. A *false positive* (FP) *takes place when the result is inaccurately predicted as positive when it is negative in reality. A false negative* (FN) *is said to occur when the result is inaccurately predicted as negative when in reality it is positive.* Table 2.1 shows the truth in the rows and the decisions of the algorithm in the columns. An ideal algorithm should create a diagonal matrix with FP = FN = 0. *Confusion matrix* in the term commonly used for this form of reporting classification results.

Table 2.1 Confusion matrix

		Hypothesized class (prediction)	
		Classified +ve	**Classified −ve**
Actual class (observation)	Actual +ve	TP	FN
	Actual −ve	FP	TN

Consider a binary classification model $h(\mathbf{x}, \mathbf{w})$ based on the data $\mathcal{D} = \{\mathbf{x}, y\}$. The actual class for a data tuple $\mathbf{x}$ is y, and the hypothesized class is $\hat{y} = h(\mathbf{x}, \mathbf{w})$. The pair of labels $(y, \hat{y})$ specifies the coordinates of each observation within the confusion matrix—the first label specifies the row of the matrix and the second label specifies the column. Therefore, an observation with the label pair $(y, \hat{y})$ will be mapped onto a confusion matrix as follows:

$$(+\text{ve}, +\text{ve}) \rightarrow \text{TP}$$
$$(-\text{ve}, +\text{ve}) \rightarrow \text{FP}$$
$$(+\text{ve}, -\text{ve}) \rightarrow \text{FN}$$
$$(-\text{ve}, -\text{ve}) \rightarrow \text{TN}$$

Example 2.2

A confusion matrix of a model applied to a set of 200 observations is given in Table 2.2.

Table 2.2 A confusion matrix of a model

	Predicted +1	**Predicted −1**
Actual +1	95	7
Actual −1	4	94

The model not only makes 7 false negative errors and 4 false positive errors, but also 95 true positive and 94 true negative predictions. If this were a model for clinical example, the fact that the model makes almost twice as many false negative errors than false positive errors would definitely be worrisome; requiring deeper analysis and, if required, building a new model. Only the confusion matrix is capable of providing insight of this kind.

Metrics for Evaluating Classifier Performance

Now let us look at the evaluation measures, starting with the misclassification rate of the classifier.

Misclassification Error: The overall success rate on a given test set is the number of correct classifications divided by the total number of classifications.

$$\textit{Success rate} = \frac{\text{TP} + \text{TN}}{\text{TP} + \text{TN} + \text{FP} + \text{FN}} \tag{2.31}$$

In the pattern recognition literature, this is also referred to as the overall *recognition rate* of the classifier, that is, it reflects how well the classifier recognizes tuples of various classes.

The *misclassification rate* of a classifier is simply (1 – recognition rate).

$$Misclassificate\ rate = \frac{FP + FN}{TP + TN + FP + FN} \tag{2.32}$$

But is it a fair measure of overall success?

Example 2.3

Suppose that we have trained a classifier to classify medical data tuples, where the class label attribute is '*cancer*' and the possible class values are '*true*' and '*false*'. The confusion matrix of model is as given in Table 2.2.

$$Misclassificate\ rate = \frac{4+7}{95+94+4+7} = 0.055$$

A model that makes 5.5% misclassification errors seems like a reasonable model. But what if only 5.5% of the training samples are actual cancer? Consider, for example, the confusion matrix given in Table 2.3 of a classifier trained on samples with 5.5% of the samples actually cancer (in balanced/skewed data). This model also has misclassification rate of 5.5%, but 7 out of 11 actual 'cancer' patients have been classified as 'noncancer'. Clearly a success rate of 94.5% is not acceptable. Instead, we need other measures which assess how well the classifier can recognize the positive tuples (*cancer* = *true*) and how well it can recognize the negative tuples (*cancer* = *false*). The *sensitivity* (also referred to as *true positive rate*) and *specificity* (also referred to as *true negative rate*) measures can be used, respectively, for this purpose.

Table 2.3 Confusion matrix for clinical example

	Predicted +ve	Predicted –ve
Actual +ve	4	7
Actual –ve	4	185

True Positive Rate: The true positive rate (*tp* rate) of a classifier is estimated as,

$$\begin{aligned} tp\ \text{rate} &\simeq \frac{\text{Positives correctly classified}}{\text{Total positives}} \\ &= \frac{TP}{TP + FN} \end{aligned} \tag{2.33}$$

The *tp* rate tells us how sensitive our decision method is in the detection of the abnormal event. A classification method with high sensitivity will rarely miss the abnormal event when it occurs.

True Negative Rate: The true negative rate (*tn* rate) of a classifier is estimated as,

$$tn \text{ rate} \simeq \frac{\text{Negatives correctly classified}}{\text{Total negatives}} = \frac{\text{TN}}{\text{FP} + \text{TN}} \tag{2.34}$$

This parameter reveals how specific our decision technique is in detecting the abnormal event. A classification technique which is extremely specific will possess an extremly low rate of false alarms resulting from the classification of a normal event as abnormal.

$$\begin{aligned} 1 - \text{specificity} &= 1 - \frac{\text{TN}}{\text{FP} + \text{TN}} = \frac{\text{FP}}{\text{FP} + \text{TN}} \\ &= \frac{\text{Negatives incorrectly classified}}{\text{Total negatives}} \\ &= fp \text{ rate } (\textit{False Positive Rate}) \end{aligned} \tag{2.35}$$

A decision technique is said to be good if it concurrently happens to be highly sensitive (rarely missing the abnormal event as it takes place) and a high specificity (possessing a low rate of false alarm, i.e., low *fp* rate).

Example 2.4

For the confusion matrix given in Table 2.2,

$$\text{Sensitivity} = \frac{\text{TP}}{\text{TP} + \text{FN}} = \frac{95}{95 + 7} = 0.93$$

$$\text{Specificity} = \frac{\text{TN}}{\text{TN} + \text{FP}} = \frac{94}{94 + 4} = 0.96$$

A sensitivity (true positive rate) of 1.0 implies that the model predicts all positive observations in a correct manner; simply put, the approach fails to make any false negative errors. A specificity of 1.0 or false positive rate of 0 indicates that the model predicts all negative observations accurately; that is, the model does not make any false positive predictions. A model is considered good if it has a high *tp* rate and a low *fp* rate at the same time. The model in question, more or less, meets this requirement.

Consider now the confusion matrix of Table 2.3.

$$\text{Sensitivity} = tp \text{ rate} = \frac{\text{TP}}{\text{TP} + \text{FN}} = \frac{4}{4 + 7} = 0.364$$

$$1 - \text{specificity} = fp \text{ rate} = \frac{\text{FP}}{\text{FP} + \text{TN}} = \frac{4}{4 + 185} = 0.021$$

Note that although the classifier has high accuracy (94.5%; Example 2.3), it has low sensitivity, and therefore its ability to accurately recognize positive tuples is poor. Because of low *fp* rate, it can accurately predict negative class.

The model may, therefore, be not an acceptable model for this skewed data case. Techniques for handling class-imbalanced data are given in Section 2.5.

2.8.3 Comparing Classifiers Based on ROC Curves

The true positives, true negatives, false positives and false negatives have different *costs and benefits* (or *risks and gains*) with respect to a classification model.

The costs pertaining to a *false negative* (for instance, wrongly predicting a patient suffering from cancer as not cancerous) are way higher than those related to a *false positive* (wrongly labeling a noncancerous patient as cancerous). Similarly, the benefits associated with a *true positive* (such as correctly predicting that a cancerous patient is cancerous) and *true negative* (correctly predicting that a noncancerous patient is noncancerous) may be different. To evaluate correctness on the basis of *misclassification error*, we make an assumption of equal costs, and divide the sum of false positives and false negatives by the total number of test tuples.

The *ROC Graph* (a two-dimensional graph) plots *tp* rate (sensitivity) on the y-axis and *fp* rate (complement of the specificity) on the x-axis [42]. An ROC graph, hence, shows relative trade-offs between advantages (true positives) and costs (false positives). The origin of the ROC name is in *Receiver Operating Characteristic* curve, which made an appearance in the 1950s as a method of selection of the best voltage threshold discerning pure noise from signal plus noise, in Radar and other similar applications used to detect signals.

Figure 2.8 shows an ROC graph with five classifiers labeled C_1, C_2, C_3, C_4, and C_5. Each classifier produces a single point in ROC space.

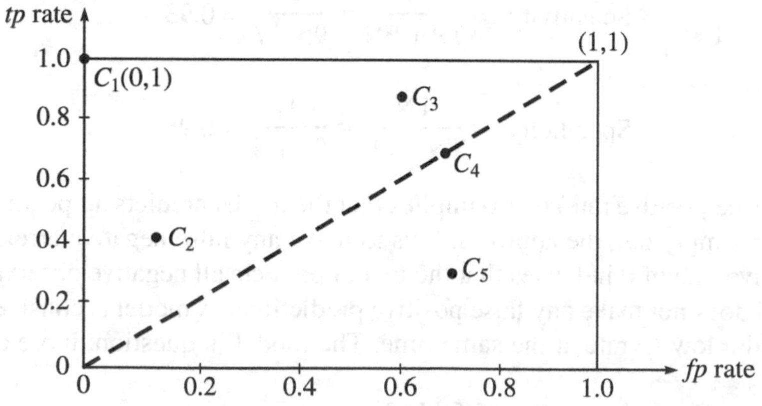

Figure 2.8 A sample ROC graph

Note that the lower left point (0, 0) in ROC space represents a classifier that commits no false positive errors, but it also gains no true positives. The upper right point (1, 1) represents a classifier that unconditionally issues positive classifications. The point (0, 1) represents perfect classifications. Performance of classifier C_1 is thus, perfect.

Points in the ROC space adjacent to the upper left corner (*tp* rate is higher, *fp* rate is lower) are preferred. Classifiers that appear on the left-hand side in the ROC space near the *x*-axis, may be considered as 'conservative': they create positive classifications only on strong evidence. Therefore, they create very few false positive errors, but often have low true positive rates as well (classifier C_2 for example). Classifiers existing on the upper right-hand side of the left-hand triangle in ROC space may be considered 'liberal'; they create positive classifications with feeble evidence; therefore, they classify almost all positives accurately. However, they have high false positive rates (for instance, classifier C_3).

The diagonal $y = x$ represents the strategy of randomly guessing a class. For example, if a classifier randomly guesses the positive class half the time, it can be expected to get half of the positives and half of the negatives correct; this yields the point (0.5, 0.5) in ROC space. If it guesses the positive class 90% of the time, it may be expected to get 90% of the positives correct, but its false positive rate will increase to 90% as well, yielding (0.9, 0.9) in ROC space. Thus, a random classifier (C_4, for example) will produce an ROC point that 'slides' back and forth on the diagonal based on the frequency with which it guesses the positive class.

Any classifier that appears in the lower-right triangle performs worse than random guessing (classifier C_5, for example).

ROC Curves: When a classifier algorithm is applied to a test set, it yields a single confusion matrix, which in turn corresponds to one ROC point. We can create an *ROC curve* by thresholding the classifier with respect to its complexity (for example, different parameter settings for the same learning scheme). Each level of complexity in the space of a hypothesis class produces a different point in ROC space. A sample ROC curve is shown in Fig. 2.9.

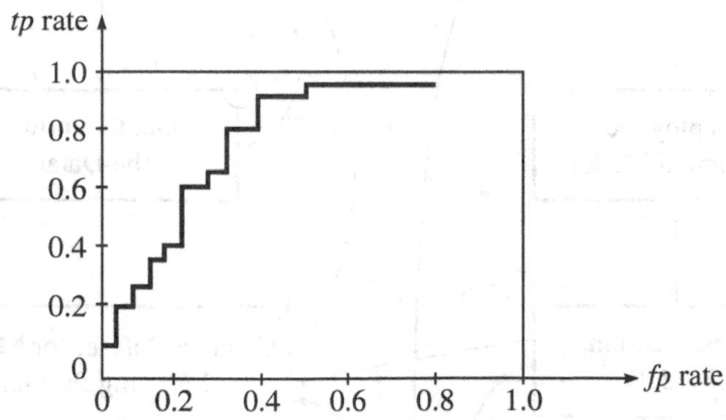

Figure 2.9 A sample ROC curve

Comparison of two different learning schemes on the same domain may be done by analyzing ROC curves in the same ROC space for the learning schemes.

For measuring accuracy of *Information Retrieval* (IR) systems, the IR community has traditionally been using *Precision-Recall Performance Curves*, which are essentially equivalent (except for relabeling of the axes) to the ROC curves. Section 9.5 will provide the details.

2.9 AN OVERVIEW OF THE DESIGN CYCLE AND ISSUES IN MACHINE LEARNING

Machine learning addresses the question of how to build computer programs (*learning systems*) that improve their performance with experience. The process of building a learning system involves a number of design choices. In the following, we present an overview of the design cycle and consider some of the issues that frequently occur [43].

The procedure of building a learning system starts with comprehending the problem domain. We have to first evaluate the problem and establish what data is available and what is required to solve the problem. Once we understand the problem, we can select a suitable technique (support vector machines, neural networks, decision trees, and so on; discussed later in the book) and build the system with this technique. The process, illustrated in Fig. 2.10, is best considered a set of nested loops, instead of a set of sequential steps. While there is a natural order to the steps, it is not really required of them to totally finish with one before going on to the other. Things learned in the later steps will result in earlier ones to be revisited. Various steps in the design cycle usually overlap considerably. The process itself is highly iterative. Let us examine each step in more detail.

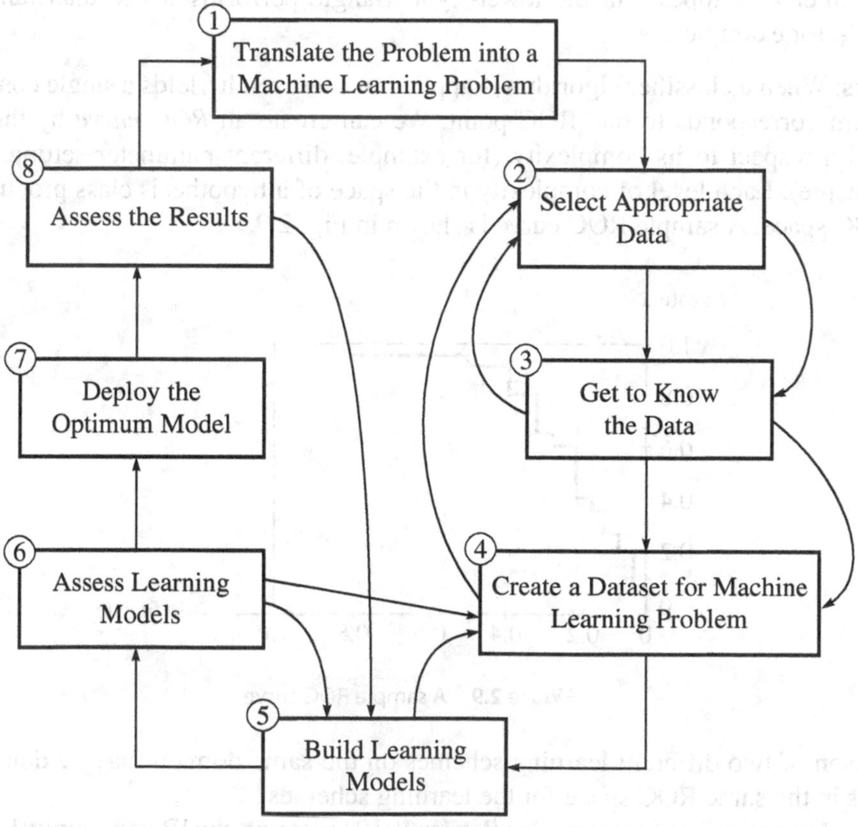

Figure 2.10 An overview of the design cycle

Step 1: **Translate the Problem into a Machine Learning Problem**

During this initial phase of design, we determine the problem's characteristics, specify the objectives and determine what resources are needed for building the learning system.

Typical problems often addressed by machine learning include pattern recognition (classification), numeric prediction (regression), clustering, optimization, and control. The problem type influences our choice of the tool for building a learning system. Of course, choice of building tool also depends on the form and content of the solution.

Often, machine learning is seen as a technical problem of looking for a model that explains the relationship of a target variable with a group of input variables. This requires clear definition of the target variable and identification of appropriate input variables. This task, in turn, relies on a fair understanding of the problem that needs to be looked into. 'How will the outcome be put to use'? and 'In what manner or form will it be delivered'?, are some of the crucial questions for which answers need to be sought. Involving domain experts in finding out how machine learning results will be used; and involving analysts, IT engineers and database administrators in finding out the manner of delivering the outcomes, are helpful. The machine learning expert plays an important role in the team in ensuring that the ultimate statement of the problem is translatable into a well-posed machine learning problem.

Step 2: **Select Appropriate Data**

Once the problem is formulated, we need to create a wish list of data that is needed for machine learning solution. The first source to search for data is the data warehouse. The data existing in the warehouse has already undergone cleaning, verification, and assembled from multiple sources. If this type of data warehouse is not available, or more than one data warehouses are available that do not live up to the needs, the data miners will have to search for data from many other sources.

The misfortune is, however, that there is no easy way of knowing how much data will suffice. The answer is dependent on the specific algorithms used, how complex the data is, and relative frequency of possible outputs.

How do we know when we have collected an adequately large and representative set of data for the learning machine? Very large amount (large N) adds to noise and redundancies, and very small amount (small N) leads to loss of information. Not throwing out large number of variables that seem unlikely to be interesting for the problem, but carefully choosing a few variables that we expect to be important, helps in selection of appropriate set of data. Machine learning approach calls for letting the data reveal what is and what is not important. Domain experts play an important role in the process.

Data collection, and selecting appropriate data can account for, surprisingly, large part of the cost of developing a machine learning system.

Step 3: **Get to Know the Data**

Exploring the unfamiliar dataset before rushing into building models is an important step in developing a learning system. Along the way, we are likely to discover many data quality problems, which could be sorted out early in the design process.

A good first step is to examine a histogram of each variable in the dataset; making note of anything that seems surprising. Do the highest and lowest values seem reasonable for that variable? How many missing values are there?, and so on.

Data visualization tools can be very helpful during the initial exploration of data: scatter plots, bar charts, geographic maps, and other visualization tools are extremely powerful for seeing what is in the data. We can see through correlation analysis, the interdependencies of the variables and dependency of output on input variables.

Data clustering yields information on outliers, and other knowledge elements that are useful in the design process.

In fact, data miners collect lot of important information for the next step of creating a dataset for the machine learning problem, by immersing themselves in the data. For instance, what are the good derived variables to try, what data transformations to try for getting a new reduced dataset which exhibits high 'information packing' properties, etc.

Step 4: **Create a Dataset for Machine Learning Problem**

The choice of the distinguishing features is a critical design requirement, and depends on the characteristics of the problem domain. The goal is to find features that are simple to extract, insensitive to noise, and information-rich. If number of features, n, is small with respect to the learning task, the subsequent design of the machine would lead to poor performance. If n is too large, information carried by one feature may be mutually correlated with that of other. Large n leads to large number of parameters in an algorithm, that leads to loss of generality. Each feature must carry valuable independent information with respect to machine learning task [44, 45].

How do we combine prior knowledge of the problem domain and empirical data to find relevant and effective features? Though the role of domain experts is very significant for this step in design cycle, tools for attribute reduction are very helpful in creating features for the machine learning problem in hand.

For design of learning systems, we collect data from various sources, which is why the data can be of various types. But, a specific technique that builds a learning system needs a specific kind of data. Some techniques deal with continuous variables, whereas others require to have all variables binned into many ranges, or to be normalized to one range, for instance, from 0 to 1. The outcome is that data has to undergo transformation into a form that can be used by a specific technique. Certain issues require to be resolved before data is transformed for a specific tool—issues of *incompatible*, *inconsistent* and *missing* data. The procedure for cleaning data offers a solution to these problems.

If the problem at hand needs a supervised learning solution, the dataset created needs to be divided into three portions. The first portion is the training set, with which the initial model is built. The second part is the validation set, with the help of which the initial model is adjusted to make it more general. The test set is the third part, with which the probable effectiveness of the model is assessed on its application to unseen data.

Step 5: **Build Learning Models**

The details of building learning models vary from technique to technique and are described later in chapters devoted to each machine learning method. In general terms, this is the step where most of the work of creating a model occurs. In supervised learning, the training set is used to generate a hypothesis function (explanation of the dependent or target variable in terms of independent or input variables). Machine learning involves searching through space of possible hypotheses to determine one that fits the observed data and any prior knowledge held by the designer. The hypotheses space is theoretically infinite; the learning task is, thus, to search through this vast space. Applied machine learning organizes the search as per the following two-step procedure;

- First focus on hypothesis class: neural networks, support vector machines, decision trees, fuzzy logic models, Bayesian classification, *k*-Nearest Neighbor (*k-NN*) classifier, ... Initial choice of a class is made on the basis of learning task in hand and prior knowledge held by the designer, based on experience.
- For each of these hypothesis classes, the corresponding learning algorithm organizes the search through all possible underlying functions. The hypothesis class is parameterized and a parametric search is carried out.

How is a class of functions rejected and another one tried? How is a function in a class of functions identified as considerably different from the true model that forms the basis of our patterns, giving rise to the need for a new function? Should designers resort to random and cumbersome trial and error while choosing a hypothesis, without actually ever getting to know whether improved performance can be expected? Or could there be principled techniques to know when to discard a class of functions and summon another?

In general, the process of using data to determine parameters of a learning algorithm is referred to as training the machine. Much of this book is concerned with different procedures for training the machines. How much training data is sufficient? How to establish a relation between the confidence in learned hypothesis and the volume of training experience, and the character of the hypothesis class? When and how can earlier knowledge of a learner, besides data, be of help? These are some of the problems that require to be handled in the model-building step.

Building learning models is the one step of machine learning process that has been truly automated by modern machine learning software. However, the software should not be used as a 'black box' by the designer. Various issues raised here and many other related issues need to be addressed before machine learning software is profitably used.

Step 6: **Assess Learning Models**

Unlike traditional computer programs, learning systems are so designed that they resolve issues that frequently lack adequately defined 'right' and 'wrong' solutions. Assessment of a learning system is done to make sure that it fulfills its purpose to the user's satisfaction.

Assessment of directed models is done for how accurate they are on test data chosen by the designer. The assessment of a model is done on the basis of the context; a model which looks good as per one measure can also appear bad as per another measure. Usually, a measure of assessment is agreed upon for the model-building step, related to the question: How confident can one be about the predictions made by the model?

In addition to accuracy, computational resources necessary and computational complexity of different algorithms (machines) are of considerable practical importance in assessing a model. How are the performance and computational complexity evaluated ? What is the trade-off between computational ease and performance? These are some of the important issues that need to be addressed in the model-assessment step.

Step 7: **Deploy the Optimum Model**

Deployment of a model depends on the purpose of the data mining project. Selecting an optimum model for deployment is application-dependent. Deployment of a model involves integrating the learning system into the environment where it will operate, and establishing an effective maintenance system.

Step 8: **Assess the Results**

Measuring the actual results—both financially and technically—is very important; we can learn and do better next time.

The foregoing steps encompass the methodologies available in most of the available software. A similar methodology, termed CRISP-DM (CRoss-Industry Standard Process for Data Mining) is given in Section 9.2.

We have already seen many issues that arise in the design of learning system. No universal methods have been found for addressing all of these issues. Throughout this book, we will see again and again how methods of learning relate to these central issues.

Chapter

3

STATISTICAL LEARNING

3.1 MACHINE LEARNING AND INFERENTIAL STATISTICAL ANALYSIS

Statistical analysis saw the development of two branches in the 18th century—*Bayesian* and *classical statistics*. In case of the approach that originated from the mathematical works of Thomas Bayes, analysis is based on the concept of *conditional probability*: the probability of an event taking place considering that another event has already taken place. The quantification of the investigator's present state of beliefs, knowledge, and assumptions marks the beginning of the Bayesian analysis. These *subjective priors*, in combination with observed data, are quantified probabilistically using an appropriate objective function.

Regression and correlation were concepts that were developed in the later part of the 19th century, for generic data analysis. A system developed for inference testing in medical sciences, by RA Fisher in the 1920s was based on the concept of *standard deviation*. Bayesian model for inference testing could result in extremely different conclusions by different medical investigators as they used various sets of subjective priors. Fisher developed his system with the objective of providing medical investigators with a common set of tools based on data alone (no subjective priors). To make his system work even for big samples, Fisher had to assume a number of things, including *linear regression*, to define his '*Parametric Model*'.

Mathematical research dominated the 1980s on lines similar to Fisher's statistical inference through the development of nonlinear versions of parametric techniques. Bayesians kept on researching to promote their approach. The line of thinking called *machine learning* emerged from Artificial Intelligence community (1990s) in search of intelligent machines. Machine learning methods allowed the analysis of extremely nonlinear relationships in big datasets which have no known distribution.

Conventional statistical analysis adopts the *deductive* technique to look for relationships in datasets. It employs past knowledge (domain theory) along with training examples to create a model. Machine learning methods, on the other hand, adopt the *inductive* technique to discover feeble patterns of relationships in datasets. In the absence of any more information, it is assumed that the best hypothesis pertaining to unseen patterns is the one *induced* by the training data observed.

However, now, there has been a convergence of the two perspectives. Machine learning methods integrate a lot of statistical thinking. Statistical tests are used by most learning algorithms to construct models, and to prune 'overfitting' in these models. *Statistical learning theory* (PAC framework, VC dimension framework; Section 2.3) tries to provide answers to questions such as what are the conditions necessary for successful learning? What are the conditions under which a specific learning algorithm can be certain of successful learning? This theory provided new directions for complex problems with nonlinear relationships in datasets by 'mapping' data points to high-dimensional spaces with 'kernels'.

Data mining developed much recently, in the 1990s, and grew to become a significant field in the initial years of the 21st century. It is representative of a union of many well-established areas of interest—classical Bayesian statistics, classical parametric statistics, machine learning and database technology. Many off-the-shelf data mining system products and domain-specific data mining application software are on offer in the market, even though several data mining problems still require to be examined in detail.

We do not intend to describe in detail any specific commercial data mining system. Instead, we briefly outline popular data mining techniques for predictive modeling that form part of most of the commercial data mining systems.

We begin this chapter by presenting a handful of key concepts from *descriptive statistics*. Descriptive statistics is only solely concerned with properties of the observed data. It has proven to be useful for data exploration (Section 7.2). It also provides useful tools for inferential statistical analysis/machine learning/data mining. Section 3.2 describes key concepts of descriptive statistics, used later in this book.

Inferential statistics is concerned with making predictions from data. Mathematical methods of inferential statistics employ probability theory for inferring the properties of an underlying distribution from the analysis of properties of a data sample drawn from it. It is concerned also with the testing of the inferences made for precision and reliability (hypotheses testing). A handful of widely used results from the inferential statistical analysis are covered in this chapter.

The widely used technique based on classical Bayesian statistics is *naive Bayes classifier*. The Bayesian learning methods (statistical inference methods) have been found to be competitive with other machine learning algorithms in many cases, and in some cases they outperform. In this chapter, we introduce Bayesian learning, giving detailed coverage of *naive Bayes classifier*, and *k-Nearest Neighbor (k-NN) classifier*.

The inference techniques based on classical parametric statistics: *linear regression, logistic regression, discriminant analysis*; are also covered in this chapter. The other topic based on statistical techniques included in this chapter is the *Minimum Description Length (MDL) principle*.

3.2 DESCRIPTIVE STATISTICS IN LEARNING TECHNIQUES

The standard 'hard-computing' paradigm is based on analytical closed-form models using a reasonable number of equations that can solve the given problem in a reasonable time, at reasonable cost, and with reasonable accuracy. It is a mathematically well-established discipline.

The field of machine learning is also mathematically well-founded; it uses 'soft-models'—modern computer-based applications of standard and novel mathematical and statistical techniques. Each of these techniques is a broad subject by itself with inputs from linear algebra and analytical geometry, vector calculus, unconstrained optimization, constrained optimization, probability theory, and information theory. Our focus in this book is on 'applied' nature of machine learning; in-depth knowledge of these and related topics to understand the content of this book is not essential. However, in-depth knowledge will be essential for advanced study and research in machine learning.

We have assumed that the reader has some knowledge of these techniques. The summary of the properties of the techniques and the notation we use are described as and when a technique appears in our presentation.

Statistics developed as a discipline markedly different from mathematics over the past century and a half, to help scientists derive knowledge from observations, and come up with experiments that give rise to the reproducible and correct outcomes pertaining to the scientific technique. The methods established in the past on small amounts of data in a world of hand calculations have managed to survive and continue to prove how useful they are. These methods have proved how worthwhile they are, not merely in the original spheres but also virtually in all areas wherein data is collected.

This section is aimed at presenting a few primary ideas from descriptive statistics that have confirmed their utility as tools for inferential statistical analysis/machine learning/data mining [46, 47].

3.2.1 Representing Uncertainties in Data: Probability Distributions

One of the common features of existing information for machine learning is the uncertainty associated with it. Real-world data tend to remain incomplete, noisy, and inconsistent. Noise, missing values, and inconsistencies add to the inaccuracy of data. Although *data cleansing* (Chapter 7) procedures try to approximately fill-in the missing values, smooth out noise while identifying outliers, and rectify inconsistencies in the data, inaccuracies do crop up and bring in the factor of uncertainty. Information, in other words, is not always suitable for solving problems. However, machine intelligence can tackle these defects and can usually make the right judgments and decisions. Intelligent systems should possess the capability to deal with uncertainties and derive conclusions.

Most popular uncertainty management paradigms are based on probability theory. Probability can be viewed as a numerical measure of the likelihood of occurrence of an outcome relative to the set of other alternatives. The set of all possible outcomes is the *sample space* and each of the individual outcomes is a *sample point*. Since the outcomes are uncertain, they are termed *random variables*.

For problems of interest to us, the data matrix $\mathcal{D}$ is the sample space; the features/attributes x_j; $j = 1,\ldots, n$, are scalar random variables, and each of N n-dimensional vector random variable $\mathbf{x} \in \Re^n$ is a sample point. Initially, let us focus on features x_j in the data matrix defining random variables; we will add the output column y in our data space later and consider the total sample space.

Probability Mass Function

Assume $n = 1$, and the feature x is a random variable representing a sample point. Random variable x can be *discrete* or *continuous*. When x is discrete, it can possess finite number of discrete values v_{lx}; $l = 1, 2,\ldots, d$. The occurrence of discrete value v_{lx} of the random variable x is expressed by the probability $P(x = v_{lx})$.

From the frequency point of view, $P(x = v_{lx}) \triangleq P(v_{lx})$ can be interpreted as,

$$P(v_{lx}) = \lim_{N\to\infty} \frac{\Delta N}{N} \tag{3.1}$$

where

$$N = \text{number of sample points, and}$$
$$\Delta N = \text{number of times } x = v_{lx}$$

The probabilities of all possible values of x are expressed as a *probability mass function:*

$$\begin{aligned} &P(x) = \langle P(v_{1x}), \ldots, P(v_{dx})\rangle \\ &P(v_{lx}) \geq 0; l = 1, 2, \ldots, d \\ &P(v_{1x}) + P(v_{2x}) + \cdots + P(v_{dx}) = 1 \end{aligned} \tag{3.2}$$

Figure 3.1 graphically displays a probability mass function.

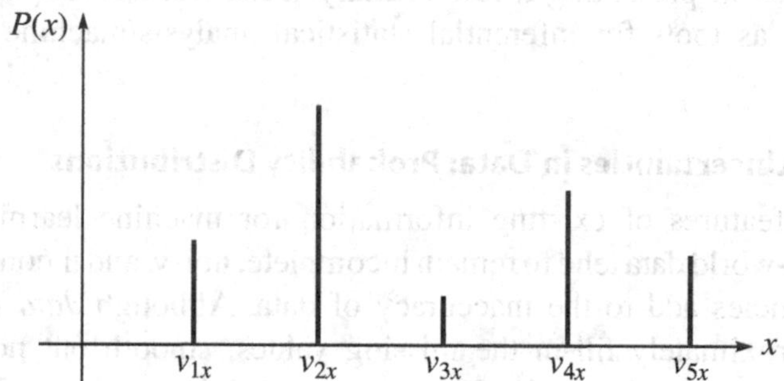

Figure 3.1 A probability mass function

Probability Density Function

A continuous random variable can take infinite values within its domain. In this case, the probability of a particular value within the domain is zero. Thus, we describe a continuous random variable not by the probability of taking on a certain value but by probability of being within a range of values.

For a continuous random variable x, probabilities are associated with the ranges of values of the variable, and consequently, at the specific value of x, only the density of probability is defined. If $p(x)$ is the *probability density function*[1] of x, the probability of x being in the interval (v_{1x}, v_{2x}) is:

$$\begin{aligned} &P(v_{1x} \leq x < v_{2x}) = \int_{v_{1x}}^{v_{2x}} p(x)\,dx \\ &p(x) \geq 0, \text{ and } \int_{-\infty}^{\infty} p(x)\,dx = 1 \end{aligned} \tag{3.3}$$

1 We generally use an uppercase $P(\cdot)$ to denote a probability mass function, and a lowercase $p(\cdot)$ to denote a probability density function.

Figure 3.2 graphically displays a probability density function.

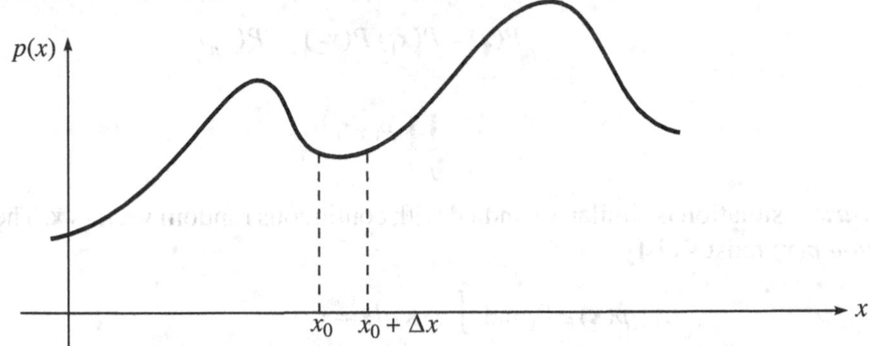

Figure 3.2 A probability density function

Probability that x has a value between x_0 and $x_0 + \Delta x$ is:

$$P(x_0 \le x < x_0 + \Delta x) = \int_{x_0}^{x_0+\Delta x} p(x)\,dx \tag{3.4a}$$

If Δx is incrementally small,

$$P(x_0 \le x < x_0 + \Delta x) = p(x = x_0)\,\Delta x \tag{3.4b}$$

From the frequency point of view,

$$p(x) = \lim_{\Delta x \to 0}\left[\frac{P(x_0 \le x < x_0 + \Delta x)}{\Delta x}\right] = \lim_{\substack{N\to\infty \\ \Delta x \to 0}}\left[\frac{\Delta N/N}{\Delta x}\right] \tag{3.5}$$

where N = number of sample points, and
ΔN = number of times the random variable x lies in the range $x_0 \le x < x_0 + \Delta x$

Vector Random Variables

The results for a scalar random variable x can easily be extended to a vector random variable $\mathbf{x}$, with components x_j: $j = 1,\ldots, n$. The random variable $\mathbf{x}$ can take on values in discrete sample space $\Re_d^n : V_{\mathbf{x}} = \{v_{lx_1}, v_{lx_2}, \ldots, v_{lx_n}\}$; $l = 1, 2, \ldots, d_j$. For each possible value of $\mathbf{x}$ in $\Re_d^n$, we have a *joint probability*,

$$\begin{gathered} P(\mathbf{x}) = P\,(x_1 = v_{lx_1}, x_2 = v_{lx_2}, \ldots, x_n = v_{lx_n}) \\ P(\mathbf{x}) \ge 0, \text{ and } \sum_{\mathbf{x} \in \Re_d^n} P(\mathbf{x}) = 1 \end{gathered} \tag{3.6}$$

Note that $P(\mathbf{x})$ is a function of n variables and can be very complicated multidimensional function. However, if the random components $x_j\,; j = 1, \ldots, n$, are *statistically independent*, then

$$\begin{aligned} P(\mathbf{x}) &= P(x_1)\, P(x_2) \cdots P(x_n) \\ &= \prod_{j=1}^{n} P(x_j) \end{aligned} \tag{3.7}$$

The *multivariate* situation is similarly handled with continuous random vectors **x**. The *probability density function* $p(\mathbf{x})$ must satisfy

$$p(\mathbf{x}) \ge 0, \text{and} \int_{-\infty}^{\infty} p(\mathbf{x})\, d\mathbf{x} = 1 \tag{3.8}$$

where the integral is understood to be an n-fold multiple integral and the element of n-dimensional volume $d\mathbf{x} = dx_1\, dx_2 \cdots dx_n$.

If the components of **x** are statistically independent, then the joint probability density function factors as,

$$p(\mathbf{x}) = \prod_{j=1}^{n} p(x_j) \tag{3.9}$$

Class-conditional Probability Density Function

We now consider total sample space (data matrix) wherein the continuous vector random variable **x** is an n-dimensional vector that corresponds to feature vector for each sample point (each row in data matrix) and y is the discrete scalar random variable that corresponds to the class to which **x** belongs.

$$y \in \{y_1, y_2, \ldots, y_M\} = \{y_q\,; q = 1, \ldots, M\}$$

The variable y has M discrete values: $\{y_1, y_2, \ldots, y_M\} = \{1, 2, \ldots, M\}$. We consider **x** to be a continuous random variable whose density function depends on the class y, expressed as $p(\mathbf{x}|y)$. This is the *class-conditional probability density function*—the probability density function for **x** given that the class is y. The difference between $p(\mathbf{x}|y_q)$ and $p(\mathbf{x}|y_k)$ describes the difference in the feature vector **x** between data of classes y_q and $y_k\,; q, k \in 1, 2, \ldots, M$.

If the attribute values $x_j\,; j = 1, \ldots, n$, are statistically independent, then

$$p(\mathbf{x}|\,y) = \prod_{j=1}^{n} p(x_j|\,y) \tag{3.10}$$

Figure 3.3 illustrates the class-conditional probability density function under the assumption of statistical independence.

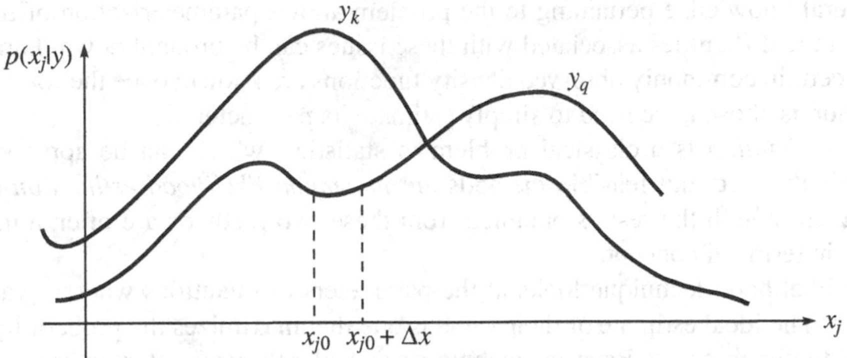

Figure 3.3 Class-conditional probability density function

Probability that x_j is between x_{j0} and $x_{j0} + \Delta x$,

$$P(x_{j0} \le x_j < x_{j0} + \Delta x | y_q) = \int_{x_{j0}}^{x_{j0}+\Delta x} p(x_j | y_q)\, dx_j \tag{3.11a}$$

If Δx is incrementally small,

$$P(x_{j0} \le x_j < x_{j0} + \Delta x\, | y_q) = p(x_j = x_{j0} | y_q)\, \Delta x \tag{3.11b}$$

Estimation of Probability Distributions

The probabilistic structure of a random variable can be characterized completely if its *distribution function* (probability mass function (pmf)/probability density function (pdf)) is known. Therefore pmf/pdf constitutes a mathematical model of the random variable.

Once mathematical models of random variables are specified, it is possible to design optimal classifiers. *Bayesian decision theory* [1, 4] gives a basic statistical model for designing optimal classifiers. This model is based on quantifying the trade-offs between various classification decisions and the risks associated with them. It assumes that the decision problem is posed in probabilistic terms, and that all the relevant pmfs/pdfs; specifically, probability mass function $P(y_q)$ and class-conditional densities $p(\mathbf{x}|y_q)$, $q = 1, \ldots, M$, are known.

Sadly, in pattern recognition applications, we hardly ever have this kind of total knowledge of the probabilistic structure of the problem. Typically, we only have some rough, general knowledge about the situation, along with several *design samples* or *training data*—specific representatives of the patterns we wish to categorize. The problem, therefore, is to discover some method of using this information for designing or training the classifier.

One model for this problem is to make use of the samples to make an estimation of unknown probabilities, and probability densities, and then employ the resulting estimates as if they were the true values. Typically, for supervised pattern classification, the probabilities $P(y_q)$ are estimated quite easily without any problems (refer to Eqn (3.1)). But then, estimation of class-conditional densities $p(\mathbf{x}|y_q)$ is not practical (refer to Eqn (3.5)). The available samples turn out to be small in number for the feasibility of this estimation, and serious issues emerge as a result of the large dimensionality of the vector $\mathbf{x}$.

If our general knowledge pertaining to the problem allows parameterization of the conditional densities, then the difficulties associated with these issues can be brought down. Parametric forms of models of certain commonly observed density functions are known; once the form of underlying density function is chosen, we need to simply estimate its parameters.

Parameter estimation is a classical problem in statistics, which can be approached in many ways. Two widely used and feasible methods are *maximum likelihood estimation* and *Bayesian estimation*. Even though the results obtained from these two methods are often almost same, the models differ in terms of concept.

Maximum-likelihood technique looks at the parameters as quantities wherein values are fixed but not known. The ideal estimate of their value is one that maximizes the probability of obtaining the samples actually observed. Bayesian techniques consider the parameters to be random variables with certain known prior distribution. On observation, the samples convert this to a posterior density, which helps us tune our earlier estimate of the actual values of the parameters. Typically, the effect of observing additional samples is to improve *a posteriori* density function, sharpening it to peak near the true values of the parameters (refer to [4] for details on both the methods).

3.2.2 Descriptive Measures of Probability Distributions

Expected (mean or average) Value

One of the most important descriptive values of a random variable is the point around which distribution is centered, known as measure of central tendency. The *expected (mean* or *average) value* is the best measure for central tendency.

For a scalar random variable x, the expected value, denoted as $\mathbb{E}[x]$ (or μ), is

$$\mathbb{E}[x] = \mu = \int_{-\infty}^{\infty} xp(x)\,dx \tag{3.12a}$$

where $p(x)$ is the probability density function.

For a discrete random variable described by a large random sample S,

$$\mathbb{E}[x] = \mu = \sum_{x \in S} xP(x) \tag{3.12b}$$

where $P(x)$ is probability mass function.

The following results immediately follow:

1. $\mathbb{E}[C] = C;\ C = \text{constant}$ (3.13a)
2. $\mathbb{E}[Cx] = C\,\mathbb{E}[x]$ (3.13b)
3. $\mathbb{E}\left[\sum_{j=1}^{n} x_j\right] = \sum_{j=1}^{n} \mathbb{E}[x_j]$ (3.13c)
4. If there are two random variables x_1 and x_2 defined on a sample space, then

$$\mathbb{E}\,[x_i\,x_2] = \int_{-\infty}^{\infty}\int_{-\infty}^{\infty} x_1\,x_2\,p(x_1, x_2)\,dx_1\,dx_2 \tag{3.13d}$$

where $p(x_1, x_2)$ is *joint probability density function.*

5. If $f(x)$ is any function of x, the expected value of f is defined as,

$$\mathbb{E}\,[f(x)] = \int_{-\infty}^{\infty} f(x)\,p(x)\,dx \tag{3.13e}$$

6. For a vector random variable **x**,

$$\mathbb{E}\,[\mathbf{x}] = \boldsymbol{\mu} = \int_{-\infty}^{\infty} \mathbf{x}\,p(\mathbf{x})\,d\mathbf{x} \tag{3.13f}$$

In discrete case,

$$\mathbb{E}\,[\mathbf{x}] = \boldsymbol{\mu} = \sum_{\mathbf{x}\in S} \mathbf{x}\,P(\mathbf{x}) \tag{3.13g}$$

Variance

The expected value or mean is used more than any other single number to describe a random variable. Usually it is also desired to know the 'spread' of the random variable about the mean. The most convenient measure used for this purpose is variance, denoted as Var[x] or σ^2, defined below:

$$\mathrm{Var}\,[x] = \sigma^2 = \mathbb{E}\,[(x-\mu)^2] = \int_{-\infty}^{\infty} (x-\mu)^2\,p(x)\,dx \tag{3.14a}$$

It may be noted that variance is simply the weighted average of squared deviations from the mean value.

In discrete case,

$$\mathrm{Var}\,[x] = \sigma^2 = \mathbb{E}\,[(x-\mu)^2] = \sum_{x\in S} (x-\mu)^2\,P(x) \tag{3.14b}$$

A more convenient measure of dispersion is the square root of variance, called the *standard deviation*, σ, i.e.,

$$\sigma = \sqrt{\mathrm{Var}\,[x]} \tag{3.14c}$$

The following results immediately follow:

1. $\mathrm{Var}\,[x] = \mathbb{E}[(x - \mathbb{E}[x])^2]$

$$= \mathbb{E}[x^2 - 2x\,\mathbb{E}[x] + (\mathbb{E}\,[x])^2]$$

$$= \mathbb{E}[x^2] - (\mathbb{E}\,[x])^2 \tag{3.14d}$$

$\mathbb{E}[x^2]$ is called the *second moment* ($\mathbb{E}[x]$ is the *first moment*) of the random variable.

2. The moments of difference between a random variable and its mean value are called *central moments*. Variance is thus a *second central moment.*
3. For two random variables x_1 and x_2 defined on a sample space, $\mathbb{E}[x_1\,x_2]$ is the *joint second moment* of x_1 and x_2.

Covariance

The joint second moment of x_1 and x_2 about their respective means μ_1 and μ_2, is the *covariance* of x_1 and x_2, i.e.,

$$\text{Cov}[x_1, x_2] = \sigma_{12} = \mathbb{E}[(x_1 - \mu_1)(x_2 - \mu_2)] \tag{3.15a}$$

$$= \mathbb{E}[x_1 x_2] - \mathbb{E}[x_1]\,\mathbb{E}[x_2] \tag{3.15b}$$

The covariance is an important measure of the degree of statistical dependence between x_1 and x_2. If x_1 and x_2 are statistically independent, $\mathbb{E}[x_1 x_2] = \mathbb{E}[x_1]\,\mathbb{E}[x_2]$, and $\text{Cov}[x_1, x_2] = \sigma_{12} = 0$. If $\text{Cov}[x_1, x_2] = 0$, then x_1 and x_2 are said to be *uncorrelated.*

Sometimes *correlation coefficient*, denoted as ρ, is more convenient to use. It is defined as,

$$\rho = \frac{\text{Cov}[x_1, x_2]}{\text{Var}[x_1]\,\text{Var}[x_2]} = \frac{\sigma_{12}}{\sigma_1^2\,\sigma_2^2} \tag{3.16}$$

ρ is *normalized covariance* and must always be between –1 and +1. If $\rho = +1$, then x_1 and x_2 are maximally positively correlated (the values of x_1 and x_2 tend to be both large or small relative to their respective means); while if $\rho = -1$, they are maximally negatively correlated (the values of x_1 tend to be large when values of x_2 are small and vice versa). If $\rho = 0$, the variables are uncorrelated.

The following results immediately follow:

1. For an n-dimensional random vector $\mathbf{x}$, covariance matrix $\mathbf{\Sigma}$ describes the correlations among its n components $x_1, x_2, \ldots, x_n$; $\boldsymbol{\mu} = [\mu_1\ \mu_2 \cdots \mu_n]^T$ represents the mean vector.

$$\mathbf{\Sigma} = \mathbb{E}[(\mathbf{x} - \boldsymbol{\mu})(\mathbf{x} - \boldsymbol{\mu})^T] \tag{3.17a}$$

$$= \int_{-\infty}^{\infty} (\mathbf{x} - \boldsymbol{\mu})(\mathbf{x} - \boldsymbol{\mu})^T p(\mathbf{x})\,d\mathbf{x} \tag{3.17b}$$

2. The covariance matrix $\mathbf{\Sigma}$ is defined as the (square) matrix whose jk^{th} element σ_{jk} is the covariance of x_j and x_k:

$$\sigma_{jk} = \sigma_{kj} = \mathbb{E}[(x_j - \mu_j)(x_k - \mu_k)];\ j, k = 1, \ldots, n$$

$$\mathbf{\Sigma} = \begin{bmatrix} \mathbb{E}[(x_1-\mu_1)(x_1-\mu_1)] & \mathbb{E}[(x_1-\mu_1)(x_2-\mu_2)] & \cdots & \mathbb{E}[(x_1-\mu_1)(x_n-\mu_n)] \\ \mathbb{E}[(x_2-\mu_2)(x_1-\mu_1)] & \mathbb{E}[(x_2-\mu_2)(x_2-\mu_2)] & \cdots & \mathbb{E}[(x_2-\mu_2)(x_n-\mu_n)] \\ \vdots & \vdots & & \vdots \\ \mathbb{E}[(x_n-\mu_n)(x_1-\mu_1)] & \mathbb{E}[(x_n-\mu_n)(x_2-\mu_2)] & \cdots & \mathbb{E}[(x_n-\mu_n)(x_n-\mu_n)] \end{bmatrix}$$

$$= \begin{bmatrix} \sigma_{11} & \sigma_{12} & \cdots & \sigma_{1n} \\ \sigma_{21} & \sigma_{22} & \cdots & \sigma_{2n} \\ \vdots & \vdots & & \vdots \\ \sigma_{n1} & \sigma_{n2} & \cdots & \sigma_{nn} \end{bmatrix} = \begin{bmatrix} \sigma_1^2 & \sigma_{12} & \cdots & \sigma_{1n} \\ \sigma_{21} & \sigma_2^2 & \cdots & \sigma_{2n} \\ \vdots & \vdots & & \vdots \\ \sigma_{n1} & \sigma_{n2} & \cdots & \sigma_n^2 \end{bmatrix} \tag{3.18}$$

3. $\mathbf{\Sigma}$ is symmetric and its diagonal elements are simply the variances of the individual elements of **x**, which can never be negative; the off-diagonal elements are the covariances, which can be positive or negative. If the variables are statistically independent, the covariances are zero, and the covariance matrix is diagonal.

3.2.3 Descriptive Measures from Data Sample

A *statistic* is a measure of a sample of data. Statistics is the study of these measures and the samples they are measured on. Let us summarize the simplest statistical measures employed in data exploration.

1. **Range:** It is the difference between the smallest and the largest observation in the sample. It is frequently examined along with the minimum and maximum values themselves.
2. **Mean:** It is the arithmetic average value, that is, the sum of all the values divided by the number of values.
3. **Median:** The median value divides the observations into two groups of equal size—one possessing values smaller than the median and another one possessing values bigger than the median
4. **Mode:** The value that occurs most often is called the *mode* of data sample.
5. **Variance and Standard Deviation:** The difference between a given observation and the arithmetic average of the sample is called its *deviation*. The variance is defined as the arithmetic average of the squares of the deviations. It is a measure of dispersion of data values; measures how closely the values cluster around their arithmetic average value. A low variance means that the values stay near the arithmetic average; a high variance means the opposite.
   Standard deviation is the square root of the variance and is commonly employed in measuring dispersion. It is expressed in units similar to the values themselves while variance is expressed in terms of those units squared.
6. **Covariance matrix:** Arithmetic average of the matrix $(\mathbf{x}^{(i)} - \boldsymbol{\mu})(\mathbf{x}^{(i)} - \boldsymbol{\mu})^T$ gives us *sample covariance matrix*; $i = 1,\ldots, N$ are the N observations in the data sample, and $\boldsymbol{\mu}$ is the arithmetic average of the sample. **x** and $\boldsymbol{\mu}$ are both n-dimensional vectors.

$$\boldsymbol{\mu} = \frac{1}{N}\sum_{i=1}^{N} \mathbf{x}^{(i)} \tag{3.19}$$

$$\mathbf{\Sigma} = \frac{1}{N}\sum_{i=1}^{N} (\mathbf{x}^{(i)} - \boldsymbol{\mu})(\mathbf{x}^{(i)} - \boldsymbol{\mu})^T \tag{3.20}$$

Correlation coefficient ρ between two attributes can be calculated from data samples using Eqn (3.16).

3.2.4 Normal Distributions

Of the different density functions explored, the maximum attention has been received by the *multivariate Gaussian (normal) density*. This is primarily because of its analytical tractability. It has a simple parameterized form, and is entirely controlled by numerical values of the two parameters, the *mean* and the *variance/covariance*. Multivariate normal density is, for many applications, an appropriate model for the given data.

According to the *central limit theorem* in statistics:

As random samples are increasingly taken from a population, the distribution of the averages (or any similar statistic) of the sample follows the normal distribution. With an increase in the number of samples, the average of the samples comes closer to the average of the entire population.

The contrapositive is true as well, that is, if the distribution of the values does *not* follow a normal distribution, then it is highly unlikely that the original values were drawn randomly. And, if they were not drawn randomly, some other process is at work.

The *normal* or *Gaussian* probability density function is significant for not only theoretical but also practical reasons. In one dimension, it is defined by (univariate normal probability density function):

$$p(x) = \frac{1}{\sigma\sqrt{2\pi}} \exp\left[-\frac{1}{2}\left(\frac{x-\mu}{\sigma}\right)^2\right] \tag{3.21}$$

The conventional description of normal density is a bell-shaped curve, totally controlled by the numerical values of the two parameters, the mean μ and the variance σ^2. The distribution is symmetrical about the mean; the peak occurring at $x = \mu$, and the width of the bell is proportional to the standard deviation σ (Fig. 3.4).

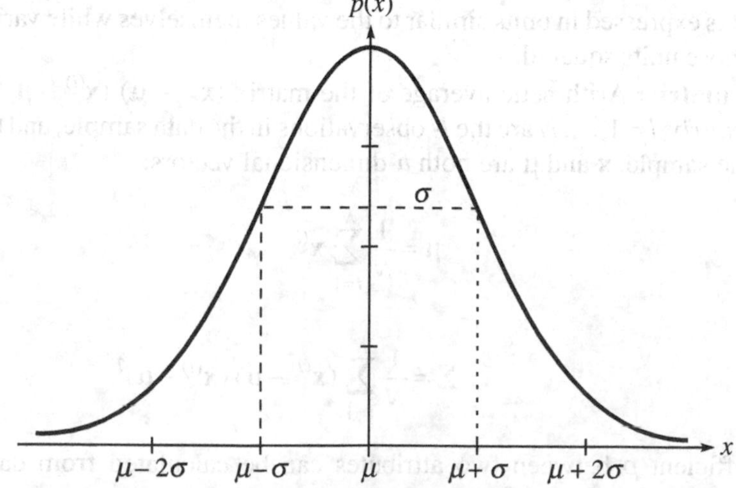

Figure 3.4 A univariate normal distribution

If the random variable is the standardized random variable, $u = (x - \mu)/\sigma$, then the normal distribution is said to be *standardized* having zero mean and unit standard deviation—that is,

$$p(u) = \frac{1}{\sqrt{2\pi}} e^{-u^2/2} \tag{3.22}$$

Consider now a vector random variable **x** with each of the n scalar random variable x_j normally distributed with mean μ_j and variance σ_j^2. If these variables are independent, their joint density has the form

$$p(\mathbf{x}) = \prod_{j=1}^{n} p(x_j)$$

$$= \frac{1}{(2\pi)^{n/2} \prod_{j=1}^{n} \sigma_j} \exp\left[-\frac{1}{2}\sum_{j=1}^{n}\left(\frac{x_j - \mu_j}{\sigma_j}\right)^2\right] \tag{3.23}$$

In this case, covariance matrix $\mathbf{\Sigma}$ is diagonal.

$$\mathbf{\Sigma} = \begin{bmatrix} \sigma_1^2 & 0 & \dots & 0 \\ 0 & \sigma_2^2 & \dots & 0 \\ \vdots & \vdots & & \vdots \\ 0 & 0 & \dots & \sigma_n^2 \end{bmatrix} \tag{3.24a}$$

$$\mathbf{\Sigma}^{-1} = \begin{bmatrix} 1/\sigma_1^2 & 0 & \dots & 0 \\ 0 & 1/\sigma_2^2 & \dots & 0 \\ \vdots & \vdots & & \vdots \\ 0 & 0 & \dots & 1/\sigma_n^2 \end{bmatrix} \tag{3.24b}$$

$$|\mathbf{\Sigma}|^{1/2} = \prod_{j=1}^{n} \sigma_j$$

We can write joint density compactly in terms of quadratic form:

$$p(\mathbf{x}) = \frac{1}{(2\pi)^{n/2} |\mathbf{\Sigma}|^{1/2}} \exp\left[-\frac{1}{2}(\mathbf{x} - \boldsymbol{\mu})^T \mathbf{\Sigma}^{-1} (\mathbf{x} - \boldsymbol{\mu})\right] \tag{3.24c}$$

This, in fact, is the general form of a *multivariate normal density function*, where the covariance matrix $\mathbf{\Sigma}$ is no longer required to be diagonal (normally distributed random variables x_j may be dependent).

3.2.5 Data Similarity

Machine learning, like animal learning, relies heavily on the notion of *similarity*, in its search for valuable knowledge in the database. Machine learning algorithms that interface with numerical form of data representation: patterns are specified by a fixed number (n) of features, where each feature has a numerical value (real) for the pattern, visualize each data as a point in n-dimensional state space. Characterizing the similarity of the patterns in state space can be done through some form of *metric* (distance) measure: distance between two vectors is a measure of similarity between two corresponding patterns. Many measures of 'distance' have been proposed in the literature.

The ease with which humans classify and describe patterns often leads to incorrect assumption that the capability is easy to automate. The choice of *similarity measure* is a deep question that lies at the core of machine learning.

Let us use d_{il} to depict a *distance metric* or *dissimilarity measure*, between patterns i and l. For pattern i, we have the vector of n measurements $(x_1^{(i)}, x_2^{(i)}, \ldots, x_n^{(i)})$, while for pattern l, we have the vector of measurements $(x_1^{(l)}, x_2^{(l)}, \ldots, x_n^{(l)})$.

Distances can be defined in multiple ways, but in general, the following properties are required:

$$\begin{aligned}
&\textit{Nonnegativity: } d_{il} \geq 0 \\
&\textit{Self-proximity: } d_{ii} = 0 \\
&\textit{Symmetry: } d_{il} = d_{li} \\
&\textit{Triangle Inequality: } d_{il} \leq d_{ik} + d_{kl}
\end{aligned} \tag{3.25}$$

Euclidean distance: The most popular distance measure is the *Euclidean distance*, d_{il}, which between two patterns i and l, is defined by,

$$d_{il} = \sqrt{(x_1^{(i)} - x_1^{(l)})^2 + (x_2^{(i)} - x_2^{(l)})^2 + \cdots + (x_n^{(i)} - x_n^{(l)})^2} \tag{3.26}$$

Despite Euclidean distance being the most commonly used similarity measure, there are three primary characteristics that need to be taken into consideration.

(i) It depends highly on scale, and variables that possess bigger scales impact the total distance to a great extent. Thus, we first *normalize* continuous measurements and only then calculate the Euclidean distance. This transforms all measurements to the same scale.

(ii) Euclidean distance entirely ignores the relationship between measurements. If there is a strong correlation between measurements, a *statistical distance* measure seems to be a better candidate.

(iii) Euclidean distance is sensitive to outliers, and if the data comprises outliers, a preferred choice is the use of robust distances like *Manhattan distance*.

Statistical distance: This metric takes into consideration the correlation between measurements. With this metric, measurements extremely correlated with other measurements do not contribute as much as the uncorrelated or less correlated. The *statistical distance*, also referred to as the *Mahalanobis distance*, between patterns i and l is defined as,

$$d_{il} = \sqrt{(\mathbf{x}^{(i)} - \mathbf{x}^{(l)})^T \, \boldsymbol{\Sigma}^{-1} \, (\mathbf{x}^{(i)} - \mathbf{x}^{(l)})} \tag{3.27}$$

where $\mathbf{x}^{(i)}$ and $\mathbf{x}^{(l)}$ are n-dimensional vectors of measurement values of patterns i and l, respectively, and $\boldsymbol{\Sigma}$ is the covariance matrix of these vectors.

Manhattan distance: This distance looks at the absolute differences rather than squared differences, and is defined by,

$$d_{il} = \sum_{j=1}^{n} |x_j^{(i)} - x_j^{(l)}| \tag{3.28}$$

Minkowski metric: One general class of metrics for n-dimensional patterns is the Minkowski metric (also referred to as the L_p norm):

$$L_p(\mathbf{x}^{(i)}, \mathbf{x}^{(l)}) = \left(\sum_{j=1}^{n} |x_j^{(i)} - x_j^{(l)}|^p \right)^{1/p} \tag{3.29}$$

where $p \geq 1$ is a selectable parameter. Setting $p = 2$ gives the familiar Euclidean distance (L_2 norm) and setting $p = 1$ gives the Manhattan distance (L_1 norm). With respect to Eqns (3.26)–(3.29),

$$L_p(\mathbf{x}^{(i)}, \mathbf{x}^{(l)}) = \| \mathbf{x}^{(i)} - \mathbf{x}^{(l)} \|_p = \left(\sum_{j=1}^{n} |x_j^{(i)} - x_j^{(l)}|^p \right)^{1/p} \tag{3.30a}$$

(Minkowski norm)

$$\| \mathbf{x}^{(i)} - \mathbf{x}^{(l)} \|_2 = [(x_1^{(i)} - x_1^{(l)})^2 + (x_2^{(i)} - x_2^{(l)})^2 + \dots + (x_n^{(i)} - x_n^{(l)})^2]^{1/2} \tag{3.30b}$$

(Euclidean norm)

$$\| \mathbf{x}^{(i)} - \mathbf{x}^{(l)} \|_1 = \sum_{j=1}^{n} |x_j^{(i)} - x_j^{(l)}| \tag{3.30c}$$

(Manhattan norm)

In this section, we have presented a handful of key ideas from descriptive statistics that we will be using in the book. Some other concepts/measures from statistics that have proven to be useful in machine learning are: Null Hypothesis; *P*-values; *Z*-scores, Confidence Interval; Chi-Square Distribution; *t*-Distribution; Chi-Square Test; *t*-Test; Analysis of Variance (ANOVA), and other measures [6, 19].

3.3 BAYESIAN REASONING: A PROBABILISTIC APPROACH TO INFERENCE

In the earlier chapters, while defining a prediction problem, the assumption made was that the objective is to maximize the success rate of predictions. In a classification case, for instance, the result of each instance is *correct* (in case the prediction agrees with the actual value for that instance), or *incorrect* (in case it does not agree). Everything is either black or white; there is no question of grey.

The Bayesian model of statistical decision making has a softer edge. It associates a probability each with prediction. The Bayesian classification aims to estimate the probabilities that a pattern to be classified belongs to various possible categories. The Bayesian method assumes that the variables of interest are governed by probability distributions and that *optimal decisions* are made by reasoning about these probabilities along with the observed data [48].

Bayesian learning techniques have relevance in the study of machine learning for two separate reasons.

(i) Bayesian learning algorithms that compute explicit probabilities, for example, the *naive Bayes classifier*, are among the most practical approaches to specific kinds of learning problems. For instance, the naive Bayes classifier is probably among the most effective algorithms for learning tasks to classify text documents. The naive Bayes technique is extremely helpful in case of huge datasets. For example, Google employs naive Bayes classifier to correct the spelling mistakes in the text typed in by users.

    Empirical outcomes reported in the literature show that naive Bayes classifier is competitive with other algorithms, such as decision trees and neural networks; in several cases with huge datasets, it outperforms the other methods.

(ii) The importance of Bayesian techniques to our study of machine learning is also because it gives a meaningful perspective to the comprehension of various learning algorithms that do not explicitly manipulate probabilities. It is essential to have at least a basic familiarity with Bayesian techniques to understand and characterize the operations of several algorithms in machine learning.

Why are other machine learning algorithms required when Bayesian reasoning results in optimal solutions to our classification tasks? It is because Bayesian reasoning offers a probabilistic approach to inference and finds basis in the assumption that variables of interest are governed by probability distributions and that it is possible to make optimal decisions by reasoning about these probabilities, along with observed data. A practical challenge in the application of Bayesian methods is that they need initial knowledge of several probabilities. If there is no prior knowledge of the probabilities, their estimation is often done on the basis of the background knowledge, available data, and assumptions pertaining to the form of underlying distributions. Usually, in pattern recognition problems, the assumption that there was knowledge of forms of underlying density functions is weak; the popular parametric forms rarely fit the densities which are faced in practice. Specifically, all the classical parametric densities possess one local maximum, that is, they are unimodal, while we encounter multimodal densities in many practical problems.

A second practical issue is the high computational cost needed to compute the Bayes optimal hypothesis in the general case. In some special cases, it is possible to significantly reduce this computational cost.

For these reasons, we are forced to look for alternative algorithms requiring tolerance for suboptimal decisions. Naive Bayes classifier provides an approximate solution to the practical problems, particularly when the datasets are very large, but leads to suboptimal decisions.

Bayes theorem is the cornerstone of Bayesian learning methods. Let us begin with the description of this theorem.

3.3.1 Bayes Theorem

Bayes theorem takes its name from Thomas Bayes, for all the initial work he did in probability and decision theory in the 18th century.

We are often interested in determining the best hypothesis h from some space $\mathcal{H}$, considering the observed training data $\mathcal{D}$. We may specify what we imply by the best hypothesis by demanding the *most probable* hypothesis given the observed data $\mathcal{D}$ in addition to any initial knowledge about the prior probabilities of various hypotheses in $\mathcal{H}$. Bayes theorem offers a method of calculating the probability of a hypothesis on the basis of its prior probability, the probabilities of observing different data given the hypothesis, and the observed data itself.

To link it to machine learning problems, we will introduce the Bayes theorem by referring to the data $\mathcal{D}$ as training examples of certain target function, and referring to $\mathcal{H}$ as the space of candidate target functions (in fact, Bayes theorem is much more general; it can be applied equally well to any set $\mathcal{H}$ of mutually exclusive propositions whose probabilities sum to one). We will be concerned with classification problems; the candidate outputs for prediction we consider are, therefore, the classes y_q ; $q = 1, \ldots, M$. We are looking for probability that the data $\mathbf{x}$ belongs to a specified class, given that we know the attribute description of $\mathbf{x}$.

We, therefore, describe and use the Bayes theorem for the following problem setting:

Given the data,

$$\mathcal{D} : \{\mathbf{x}^{(i)}, y^{(i)}\};\ i = 1, 2, \ldots, N \tag{3.31}$$

with patterns

$$\mathbf{x} = [x_1\ x_2 \cdots x_n]^T$$

The n features/attributes are x_j ; $j = 1, \ldots, n$; and the output is y. The question that is often of most significance is: What is the most probable class of new pattern $\mathbf{x}$, given the training data $\mathcal{D}$? The class label is, therefore, the output variable y.

We consider y to be a *random variable* that must be described probabilistically. The domain of random variable y is countable.

$$y : \{y_1, y_2, \ldots, y_q, \ldots, y_M\}$$

The value y_q ; $q = 1, \ldots, M$, corresponds to the class $q \in \{1, \ldots, M\}$. The occurrence of discrete values y_q is expressed by the probability $P(y_q)$. The distribution of all possible values of *discrete random variable* y is expressed as probability distribution,

$$\begin{gathered} P(y) = \langle P(y_1), \ldots, P(y_M) \rangle \\ P(y_1) + \cdots + P(y_M) = 1 \end{gathered} \tag{3.32}$$

We assume that there is some *a priori probability* (or simply *prior*) $P(y_q)$ that the next feature vector belongs to the class q. The prior probabilities reflect our *prior knowledge* of how likely we are to get a class q before the feature vector is actually observed.

The attributes x_j ; $j = 1, \ldots, n$, in the dataset $\mathcal{D}$ may be categorical as well as continuous. We assume first that the continuous attributes are binned (attribute x_j is divided into equal-sized intervals, named *bins*) and converted to categorical variables. Therefore, each attribute x_j is assumed to have value set $V_{x_j} : \{v_{1x_j}, v_{2x_j}, \ldots, v_{d_j x_j}\}$; the values v_{lx_j} ; $l = 1, 2, \ldots, d_j$, are countable.

Bayes theorem provides a way to calculate the probability of a class y_q based on its prior $P(y_q)$, probabilities of observing various patterns given the class, $P(\mathbf{x}|y_q)$, and the probabilities of observing patterns themselves, $P(\mathbf{x})$. Suppose that we have both the prior probabilities $P(y_q)$ and the *conditional probabilities* $P(\mathbf{x}|y_q)$ for $q = 1, \ldots, M$. The probability $P(\mathbf{x})$ can be obtained from the knowledge of $P(y_q)$ and $P(\mathbf{x}|y_q)$. We are interested in the probability $P(y_q|\mathbf{x})$ that class y_q holds given the observed data **x**. $P(y_q|\mathbf{x})$ is called the *posterior probability* (or simply *posterior*) of y_q given **x**, because it reflects our confidence that class y_q holds after we have seen the pattern **x**. Note that the posterior $P(y_q|\mathbf{x})$ reflects the influence of seeing the pattern **x**, in contrast to the prior $P(y_q)$ which is independent of **x**.

Bayes theorem provides a way to calculate posterior $P(y_k|\mathbf{x})$; $k \in \{1, \ldots, M\}$ from the known priors $P(y_q)$, together with known conditional probabilities $P(\mathbf{x}|y_q)$; $q = 1, \ldots, M$.

$$P(y_k|\mathbf{x}) = \frac{P(y_k)\, P(\mathbf{x}|\, y_k)}{P(\mathbf{x})} \tag{3.33a}$$

$$P(\mathbf{x}) = \sum_{q=1}^{M} P(\mathbf{x}|\, y_q)\, P(y_q) \tag{3.33b}$$

$P(\mathbf{x})$ expresses variability of the observed data, independent of the class.

$P(\mathbf{x}|y_k)$ is called the *class likelihood* and is the conditional probability that a pattern belonging to class y_k has the associated observation value **x**. As one might expect, posterior $P(y_k|\mathbf{x})$ increases with prior $P(y_k)$ and with class likelihood $P(\mathbf{x}|y_k)$. It is also reasonable to see that posterior $P(y_k|\mathbf{x})$ decreases as $P(\mathbf{x})$ increases, because the more probable it is that **x** will be observed independent of y_k, the less *evidence* **x** provides in support of y_k.

$$\text{posterior} = \frac{\text{prior} \times \text{likelihood}}{\text{evidence}} \tag{3.34}$$

Because of the normalization by the evidence, the posteriors sum to 1.

Given the priors,

$$P(y_q) \geq 0\,;\ \sum_{q=1}^{M} P(y_q) = 1$$

and likelihoods,

$$P(\mathbf{x}|y_q);\ q = 1, \ldots, M,$$

the posteriors can be calculated as,

$$P(y_k|\mathbf{x}) = \frac{P(y_k)\, P(\mathbf{x}|y_k)}{\sum_{q=1}^{M} P(\mathbf{x}|y_q)\, P(y_q)} \tag{3.35}$$

The practicability of Bayes theorem when the features **x** have discrete values, lies in the fact that conditional probability function $P(y_q|\mathbf{x})$ can be calculated from $P(\mathbf{x}|y_q)$ and $P(y_q)$, which can be estimated from data much more easily than $P(y_q|\mathbf{x})$ itself. The *naive Bayes classifier*, described in the next sub-section, uses data-based probability estimation.

In many learning scenarios, the learner is interested in finding the most probable class k (or at least one of the maximally probable if there are several), given the observed data $\mathbf{x}$. Any such maximally probable class is called a *Maximum A Posteriori* (MAP) class. We can determine the MAP class by choosing:

$$\text{Class } k \text{ if } P(y_k|\mathbf{x}) = \underset{q}{max}\ P(y_q|\mathbf{x}) \tag{3.36}$$

Thus, y_{MAP} corresponds to MAP class provided,

$$\begin{aligned} y_{MAP} &\equiv \underset{q}{arg\ max}\ P(y_q|\mathbf{x}) \\ &\equiv \underset{q}{arg\ max} \frac{P(y_q)\, P(\mathbf{x}|y_q)}{P(\mathbf{x})} \\ &\equiv \underset{q}{arg\ max}\ P(y_q)\, P(\mathbf{x}|y_q) \end{aligned} \tag{3.37}$$

arg max (argument of the maxima) is the point of the domain q (i.e., the class y_q) at which the function is maximized.

Notice that we dropped $P(\mathbf{x})$ because it is a constant independent of y.

In some cases, we will assume that every class is equally probable *a priori* ($P(y_q) = P(y_k)$; $\forall\ q, k$). In this case, we can further simplify Eqn (3.37) and we need only consider the term $P(\mathbf{x}|y_q)$ to find the most probable class. Since $P(\mathbf{x}|y_q)$ represents the likelihood of the data $\mathbf{x}$ given class y_q, any class that maximizes $P(\mathbf{x}|y_q)$ is called *Maximum Likelihood* (ML) class. Thus y_{ML} corresponds to ML class provided,

$$y_{ML} \equiv \underset{q}{arg\ max}\ P(\mathbf{x}|y_q) \tag{3.38}$$

So far, we have used the assumption of binning the continuous data. In fact, Bayes theorem does not need this assumption. It uses *probability density functions* instead of *probability distribution functions* when the data is continuous (refer to Section 3.2). Here, we consider the general Bayes theorem. The version given in Eqn (3.33), based on categorical data assumption, is only a specific application of the theorem.

General Bayes Theorem

Let $\{y_1, \ldots, y_M\}$ be the finite set of M categories (classes) in the observed training data $\mathcal{D}$. We consider y_q ; $q = 1, \ldots, M$, to be a variable that must be described probabilistically. Let feature vector $\mathbf{x}$ in the training data $\mathcal{D}$ be an n-component vector-valued random variable and $p(\mathbf{x})$ be the probability density function that data tuple $\mathbf{x}$ will be observed (i.e., the probability of $\mathbf{x}$ given no knowledge about which class it belongs to). Let $p(\mathbf{x}|y_q)$ be the conditional probability density function for $\mathbf{x}$ conditioned on y_q being its class; $P(y_q)$ describes the prior probability that $\mathbf{x}$ belongs to class y_q.

The posterior probability $P(y_k|\mathbf{x})$ can be computed from $p(\mathbf{x}|y_q)$ and $P(y_q)$ using the Bayes formula:

$$P(y_k|\mathbf{x}) = \frac{p(\mathbf{x}|y_k)\, P(y_k)}{p(\mathbf{x})} \tag{3.39a}$$

where

$$p(\mathbf{x}) = \sum_{q=1}^{M} p(\mathbf{x}|y_q)\, P(y_q) \tag{3.39b}$$

Thus, Bayes theorem provides a way to calculate the probability of a class k based on its prior $P(y_k)$, the probability density function $p(\mathbf{x}|y_k)$—the probabilities of observing various patterns given the class; and the probability density function $p(\mathbf{x})$—the probabilities of observed patterns themselves. The probability density function $p(\mathbf{x})$ for the feature vector in the entire population (independent of the class) can be obtained from the knowledge of $p(\mathbf{x}|y_q)$ and $P(y_q)$; $q = 1, \ldots, M$. We therefore need $P(y_q)$ and $p(\mathbf{x}|y_q)$; $q = 1, \ldots, M$, to compute $P(y_k|\mathbf{x})$; $k \in \{1, \ldots, M\}$.

A practical difficulty in applying Bayes theorem is that it requires initial knowledge of $P(y_q)$ and $p(\mathbf{x}|y_q)$. In real-world applications, these probabilities are not known in advance. Sometimes we have some general knowledge about the probabilistic structure of the problem; this knowledge permits us to parameterize the conditional densities ($P(y_q)$ are relatively easy to determine). Suppose, for instance, we are able to reasonably assume that $p(\mathbf{x}|y_q)$ is a multivariate Gaussian (normal) density with mean $\boldsymbol{\mu}_q$ and covariance matrix $\boldsymbol{\Sigma}_q$, despite not knowing the precise values of these quantities. This information makes the problem simpler, turning it from one of estimating an unknown function $p(\mathbf{x}|y_q)$ to *estimating the parameters* $\boldsymbol{\mu}_q$ and $\boldsymbol{\Sigma}_q$.

The issue of *parameter estimation* is a classical one in statistics, which can be approached in many ways. Two widely used and reasonable methods are *maximum likelihood estimation* and *Bayesian estimation* [1, 4].

Usually in pattern-recognition problems, the assumption that there was knowledge of the forms of probability density functions is weak. The common parametric forms rarely fit the densities truly faced in practice. All the classical parametric densities are unimodal, that is, they have only one local maximum, while we encounter multimodal densities in many practical problems.

Often, our general knowledge pertaining to the problem is not expressed by a parameterized density function. It is, rather, depicted as statistical dependencies (or independencies) or the causal relationships among the features x_j of vector $\mathbf{x}$. There are several cases where we are aware (or it is safe for us to assume) which are the variables that are causally linked, even though it may not be easy to specify accurate probabilistic relationships among these variables. We graphically represent these causal dependencies using *Bayesian belief networks*.

Can we invent efficient algorithms to learn belief nets from training data? If the network structure is provided in advance, it is simple to learn the conditional probabilities. However, learning belief nets where the network structure is unknown in advance is tough.

If the dependency relationships among the features used by a classifier are unknown, we usually proceed by opting for the most basic assumption— the features are conditionally independent given the class. Practically speaking, this so-called *naive Bayes* rule frequently works rather well, even with its manifest simplicity.

In the remainder of this section, we discuss naive Bayes classification in detail, and give a brief exposition of Bayesian belief networks.

3.3.2 Naive Bayes Classifier

The most basic idea we exploit in machine learning is that simple algorithms often work very well. It is recommended that a 'simplicity first' strategy be adopted while analyzing practical datasets in applied machine learning. In the unlimited range of varying possible datasets, there are several different types of structures that can exist. A machine learning tool—irrespective of how advanced—that is searching for one class of structures, may totally miss regularities of a different type.

A comprehensive study has revealed that very simple learning methods perform well on most commonly used datasets. In spite of simplicity, these techniques do well as compared to state-of-the-art learning techniques, only a few percentage points less precise.

A very simple rule for classifying a record into one of the M classes, ignoring all information from $(x_1, x_2, \ldots, x_n)$ attributes that we may have, is to classify the record as a member of the majority class. Take an illustrative example of predicting flight delays [20]. Assume that there are six attributes in the data table (x_1: Day of Week, x_2: Departure Time, x_3: Origin, x_4: Destination, x_5: Carrier, x_6: Weather), and output y gives class labels (Delayed, On Time). Say 82% of the entries in y column record 'On Time'. A *naive rule* for classifying a flight into two classes, ignoring information on $x_1, x_2, \ldots, x_6$ is to classify all flights as being 'On Time'. The naive rule is used as a baseline for evaluating the performance of more complicated classifiers. Clearly, a classifier that uses attribute information should outperform the naive rule.

An easy method is to use all the attributes and permit them to contribute to the decisions, taking into account the features as *equally important* and independent of each other, considering the class. Of course, this is unrealistic; in real-life data, the attributes are definitely not equally important or independent. However, this assumption results in a simple scheme, which, when put in practice, works astonishingly well. The scheme is called *Naive Bayes Classifier*, although there is nothing '*Naïve*' about its use in appropriate circumstances.

Naive Bayes classifier performs well only with categorical attributes. Continuous attributes should be binned (input is divided into equal-sized intervals, named *bins*) and transformed into categorical variables before the naive Bayes classifier can proceed using them.

In order to apply the Bayes theorem to calculate $P(y_k|\mathbf{x})$; $k \in \{1, \ldots, M\}$, we must specify what values are to be used for $P(y_q)$ and $P(\mathbf{x}|y_q)$; $q = 1, \ldots, M$ ($P(\mathbf{x})$ will be determined once we choose the other two). The choice of values of $P(y_q)$ and $P(\mathbf{x}|y_q)$ represents our prior knowledge about the learning task.

In pattern recognition applications, we hardly ever have this type of advance knowledge regarding the probabilistic structure of the problem. Typically, we simply have some rough general knowledge regarding the situation together with training data. The problem is to look for some means of using this information to design or train the classifier.

One approach to this problem is to use the samples to estimate the unknown probabilities $P(y_q)$ and class-conditional probabilities $P(\mathbf{x}|y_q)$; and then use the resulting estimates as if they were the true values. In typical supervised pattern classification problems, the estimation of prior probabilities $P(y_q)$ presents no serious difficulties. Each of the $P(y_q)$ may be estimated simply by counting the frequency with which class y_q occurs in the training data:

$$P(y_q) = \frac{\text{Number of data with class } y_q}{\text{Total number } (N) \text{ of data}} \tag{3.40}$$

If the decision must be made with so little information, it seems logical to use the following rule:

$$\text{Decide } y_k \text{ if } P(y_k) > P(y_l);\ k \neq l$$

How well this rule works depends upon the values of prior probabilities $P(y_q)$; $q = 1, \ldots, M$. If $P(y_k)$ is very much greater than all other $P(y_l)$; $l \neq k$, our decision in favor of y_k will be right most of the time (*unbalanced* data). For *balanced* data, it will not work; we would always make the same decision even though feature vectors belonging to different classes may appear. In most circumstances, we are not asked to make decisions with so little information. We need to estimate class-conditional probabilities $P(\mathbf{x}|y_q)$ as well:

$$P(\mathbf{x}|y_q) = \frac{\text{Number of times pattern } \mathbf{x} \text{ appears with } y_q \text{ class}}{\text{Number of times } y_q \text{ appears in the data}} \tag{3.41}$$

Number of distinct $P(\mathbf{x}|y_q)$ terms that must be estimated from the training data is equal to the number of training patterns (N) times the number of possible classes (M).

In order to obtain a reliable estimate of $P(\mathbf{x}|y_q)$, we need to see every pattern $\mathbf{x}$ in the feature space many times. This is not feasible unless we have a very large set of training data.

The naive Bayes classifier is based on the simplifying assumption that the attribute values are conditionally independent, given the class. In other words, the assumption is that given the class of the pattern, the probability of observing the conjunction $x_1, x_2, \ldots, x_n$ is just the product of the probabilities for the individual attributes:

$$P(x_1, x_2, \ldots, x_n|y_q) = \prod_j P(x_j|y_q) \tag{3.42}$$

Substituting this into Eqn (3.37), we have the naive Bayes algorithm:

$$y_{\text{NB}} = \underset{q}{arg\ max}\ P(y_q) \prod_j P(x_j|y_q) \tag{3.43}$$

where y_{NB} denotes the class output by the naive Bayes classifier. With conditional independence assumption, the MAP classifier becomes the NB classifier.

Notice that in a naive Bayes classifier, the number of distinct $P(x_j|y_q)$ terms that must be estimated from the training data is just the number of distinct attributes (n) times the number of distinct classes (M)—a much smaller number than if we were to estimate the $P(x_1, x_2, \ldots, x_n|y_q) = P(\mathbf{x}|y_q)$ terms.

One interesting difference between the naive Bayes learning and other Bayesian learning methods is that there is no explicit search through the space of possible probability distribution/density functions for the possible values that can be assigned to the various $P(y_q)$ and $P(\mathbf{x}|y_q)$ terms. Instead, the values are generated simply by counting the frequency of various data combinations within the training examples.

Given:

$$\mathcal{D}: \{x_1^{(i)}, x_2^{(i)}, \ldots, x_n^{(i)}; y^{(i)}\} = \{\mathbf{x}^{(i)}, y^{(i)}\}; i = 1, \ldots, N$$

$$\mathbf{x} \in \Re^n; y \in \{y_1, y_2, \ldots, y_M\} = \{y_q; q = 1, \ldots, M\}$$

$$x_j \in \Re; j = 1, \ldots, n$$

Values of $\mathbf{x}$ may be categorical or continuous. If the features are continuous-valued, a discretization process gives us categorical values.

V_{x_j}: value set for attribute x_j.

$v_{1x_j}, v_{2x_j}, \ldots,$ are the values. If d_j are countable values x_j can take, we have,

$$V_{x_j} = \{v_{1x_j}, v_{2x_j}, \ldots, v_{d_j x_j}\} = \{v_{lx_j}; l = 1, 2, \ldots, d_j\}$$

Given the unseen sample $\mathbf{x}: \{x_1, \ldots, x_n\}$, its values belong to the value sets $V_{x_1}, V_{x_2}, \ldots, V_{x_n}$. Let the value of x_j be v_{lx_j}. Then

$$P(x_j | y_q) = N_{q v_{lx_j}} / N_q \tag{3.44}$$

where $N_{q v_{lx_j}}$ is the number of training samples of class y_q having the value v_{lx_j} for attribute x_j, and N_q is the total number of training samples with class y_q.

Class prior probabilities may be calculated as,

$$P(y_q) = N_q / N \tag{3.45}$$

where N is the total number of training samples and N_q is the number of samples of class y_q.

In practice, the computation of $P(\mathbf{x}|y_q)$ can easily prove to be an enormous task. If each component x_j can have d_j values, then $(d_j)^n$ possible values of x_j need to be considered.

Within the framework of error-minimization in machine learning, it is not clear from the training data alone, which hypothesis (model) will give the best generalization; so we resort to partitioning of the dataset to select an appropriate level of complexity through such techniques as cross-validation. The hypothesis function is a *parametric* model and hypothesis complexity is measured in terms of number of parameters.

The naive Bayes classifier is a *non-parametric* method. It uses the Bayes theorem as the model and estimates the priors $P(y_q)$ and likelihoods $P(\mathbf{x}|y_q)$ for an unseen sample $\mathbf{x}$ directly from the given dataset. The data is partitioned into training and test sets, and then naive Bayes classifier is applied to the test samples using training set for estimating probability distributions. Choosing the right training set is perhaps the most important step in the procedure.

Example 3.1

Consider the dataset $\mathcal{D}$ given in Table 3.1.

Table 3.1 Dataset for Example 3.1

	Gender x_1	Height x_2	Class	y
$s^{(1)}$	F	1.6 m	Short	y_1
$s^{(2)}$	M	2 m	Tall	y_3
$s^{(3)}$	F	1.9 m	Medium	y_2
$s^{(4)}$	F	1.88 m	Medium	y_2
$s^{(5)}$	F	1.7 m	Short	y_1
$s^{(6)}$	M	1.85 m	Medium	y_2
$s^{(7)}$	F	1.6 m	Short	y_1
$s^{(8)}$	M	1.7 m	Short	y_1
$s^{(9)}$	M	2.2 m	Tall	y_3
$s^{(10)}$	M	2.1 m	Tall	y_3
$s^{(11)}$	F	1.8 m	Medium	y_2
$s^{(12)}$	M	1.95 m	Medium	y_2
$s^{(13)}$	F	1.9 m	Medium	y_2
$s^{(14)}$	F	1.8 m	Medium	y_2
$s^{(15)}$	F	1.75 m	Medium	y_2

y_1 corresponds to the class 'short', y_2 corresponds to the class 'medium', and y_3 corresponds to the class 'tall'. Therefore,

$$M = 3, N = 15.$$

$$P(y_1) = \frac{N_1}{N} = \frac{4}{15} = 0.267; \quad P(y_2) = \frac{N_2}{N} = \frac{8}{15} = 0.533$$

$$P(y_3) = \frac{N_3}{N} = \frac{3}{15} = 0.2$$

$$V_{x_1} : \{M, F\} = \{v_{1x_1}, v_{2x_1}\}; d_1 = 2$$

$$V_{x_2} = \{v_{1x_2}, v_{2x_2}, v_{3x_2}, v_{4x_2}, v_{5x_2}, v_{6x_2}\}; d_2 = 6$$

$$= \text{bins } \{(0, 1.6], (1.6, 1.7], (1.7, 1.8], (1.8, 1.9], (1.9, 2.0], (2.0, \infty)\}$$

The count table generated from the data is given in Table 3.2.

Table 3.2 Number of training samples, $N_{qv_{lx_j}}$, of class q having value v_{lx_j}

Value v_{lx_j}	Count $N_{qv_{lx_j}}$		
	Short $q = 1$	**Medium** $q = 2$	**Tall** $q = 3$
v_{1x_1}: M	1	2	3
v_{2x_1}: F	3	6	0
v_{1x_2}: (0, 1.6] bin	2	0	0
v_{2x_2}: (1.6, 1.7] bin	2	0	0
v_{3x_2}: (1.7, 1.8] bin	0	3	0
v_{4x_2}: (1.8, 1.9] bin	0	4	0
v_{5x_2}: (1.9, 2.0] bin	0	1	1
v_{6x_2}: (2.0, ∞] bin	0	0	2

We consider an instance from the given dataset (the same procedure applies for a data tuple not in the given dataset (unseen instance)):

$$\mathbf{x} : \{M, 1.95 \text{ m}\} = \{x_1, x_2\}$$

In the discretized domain, 'M' corresponds to v_{1x_1} and '1.95 m' corresponds to v_{5x_2}.

$$P(x_1|y_1) = N_{1v_{1x_1}}/N_1 = 1/4$$

$$P(x_1|y_2) = N_{2v_{1x_1}}/N_2 = 2/8$$

$$P(x_1|y_3) = N_{3v_{1x_1}}/N_3 = 3/3$$

$$P(x_2|y_1) = N_{1v_{5x_2}}/N_1 = 0/4$$

$$P(x_2|y_2) = N_{2v_{5x_2}}/N_2 = 1/8$$

$$P(x_2|y_3) = N_{3v_{5x_2}}/N_3 = 1/3$$

$$P(\mathbf{x}|y_1) = P(x_1|y_1) \times P(x_2|y_1) = \frac{1}{4} \times 0 = 0$$

$$P(\mathbf{x}|y_2) = P(x_1|y_2) \times P(x_2|y_2) = \frac{2}{8} \times \frac{1}{8} = \frac{1}{32}$$

$$P(\mathbf{x}|y_3) = P(x_1|y_3) \times P(x_2|y_3) = \frac{3}{3} \times \frac{1}{3} = \frac{1}{3}$$

$$P(\mathbf{x}|y_1)\, P(y_1) = 0 \times 0.267 = 0$$

$$P(\mathbf{x}|y_2)\, P(y_2) = \frac{1}{32} \times 0.533 = 0.0166$$

$$P(\mathbf{x}|y_3)\, P(y_3) = \frac{1}{3} \times 0.2 = 0.066$$

$$y_{\text{NB}} = \underset{q}{arg\ max}\ P(\mathbf{x}|y_q) \times P(y_q)$$

This gives $q = 3$.

Therefore, for the pattern $\mathbf{x} = \{M\ \ 1.95\text{m}\}$, the predicted class is 'tall'.

The true class in the data table is 'medium'. Note that we are working with an artificial toy dataset. Use of naive Bayes algorithm on real-life datasets will bring out the power of naive Bayes classifier when N is large.

3.3.3 Bayesian Belief Networks

As discussed in the previous sub-section, the *naive Bayes classifier* makes significant use of the assumption that values of attributes $x_1, x_2, \ldots, x_n$ are *conditionally independent* given the target value y_q. This assumption dramatically reduces the complexity of learning the target function. However, in many cases, this conditional independence assumption is clearly overly restrictive. In contrast to the naive Bayes classifier, which assumes that *all* the variables are conditionally independent given the value of the target variable, *Bayesian belief networks* [6, 49] allow stating conditional independence assumptions that apply to *subsets* of variables. Bayesian belief networks, thus, provide an intermediate approach that is less constraining than the global assumption of conditional independence made by naive Bayes classifier, and is more tractable than avoiding conditional independence assumption altogether. (Earlier in this section, a *parametric estimation* procedure was suggested as a solution to problems where conditional independence assumptions were not made, but that solution is not feasible for many applications.)

Generally speaking, a Bayesian belief network is a description of the probability distribution over a set of variables. Suppose there is an arbitrary set of random variables $a_1, a_2, \ldots, a_m$ with each variable a_i capable of taking on the set of possible values V_{a_i}. We define the joint space of variables a_i to be $V_{a_1} \times V_{a_2} \times \cdots \times V_{a_m}$. In other words, each item in the joint space corresponds to one of the possible assignment of values to the tuple of variables $\{a_1, a_2, \ldots, a_m\}$. The probability distribution over this joint space is called the *joint probability distribution* (Section 3.2). It specifies the probability for each of the possible variable bindings for the tuple $\{a_1, a_2, \ldots, a_m\}$. A Bayesian belief network describes the joint probability distribution for a set of variables.

Let a, b, and c be discrete-valued random variables. We say that a is *conditionally independent* of b given c if the probability distribution governing a is independent of the value of b given a value of c; that is, if:

for all values $v_a \in V_a$, for all values $v_b \in V_b$ and for all values $v_c \in V_c$,

$$P(a = v_a | b = v_b, c = v_c) = P(a = v_a | c = v_c) \tag{3.45a}$$

We commonly write this expression in abbreviated form:

$$P(a|b, c) = P(a|c) \tag{3.45b}$$

This definition of conditional independence can be extended to sets of variables as well. Let $\{a_1, \ldots, a_m\}$, $\{b_1, \ldots, b_l\}$, and $\{c_1, \ldots, c_n\}$ be three discrete-valued sets of random variables. We say that the set of variables $\{a_1, \ldots, a_m\}$ is conditionally independent of the set of variables $\{b_1, \ldots, b_l\}$ given the set of variables $\{c_1, \ldots, c_n\}$ if,

$$P(a_1,\ldots, a_m | b_1,\ldots, b_l; c_1,\ldots, c_n) = P(a_1,\ldots, a_m | c_1,\ldots, c_n) \tag{3.46}$$

Note the correspondence between this definition and our use of conditional independence in the naive Bayes classifier, wherein we assume that given the attribute set $\{x_1, x_2, \ldots, x_n\}$, the attribute x_j is conditionally independent of all other attributes x_k given the target value y_q. This allows the naive Bayes classifier to calculate $P(x_1,\ldots, x_n | y_q)$ as follows:

$$P(x_1,\ldots, x_n | y_q) = P(x_1 | y_q)\, P(x_2 | y_q) \cdots P(x_n | y_q)$$

$$= \prod_{j=1}^{n} P(x_j | y_q) \tag{3.47}$$

Let us consider a simple network for illustration [17]. Figure 3.5(a) shows a belief network with six Boolean variables: a_1 (*FamilyHistory*), a_2 (*LungCancer*), a_3 (*PositiveXRay*), a_4 (*Smoker*), a_5 (*Emphysema*), and a_6 (*Dyspnea*). The arcs in Fig. 3.5(a) allow a representation of causal knowledge. For example, having lung cancer is influenced by a person's family history of lung cancer, as well as whether or not the person is a smoker. The variable *PositiveXRay* is independent of whether the patient has a family history of lung cancer or is a smoker, given that we know the patient has lung cancer. In other words, once we know the outcome of the variable *LungCancer*, then the variables *FamilyHistory* and *Smoker* do not provide additional information regarding *PositiveXRay*. The arcs also show that the variable *LungCancer* is conditionally independent of *Emphysema*, given *FamilyHistory* and *Smoker* (we say *LungCancer* is a *descendent* of *FamilyHistory* and *Smoker*). The network arcs, thus, represent that a variable is conditionally independent of other variables in the network, given its immediate predecessors in the network.

Figure 3.5(b) shows the conditional probability table associated with the variable *LungCancer*. The top left entry in the table, for example, expresses the assertion that *P*(*LungCancer* = *True*|*FamilyHistory* = *True*, *Smoker* = *True*) = 0.8. The bottom rightmost entry corresponds to *P*(*LungCancer* = *False*|*FamilyHistory* = *False*, *Smoker* = *False*) = 0.9. The set of local conditional probability tables for all the variables, together with the set of conditional independence assumptions described by the network, give the full joint probability distribution for the network.

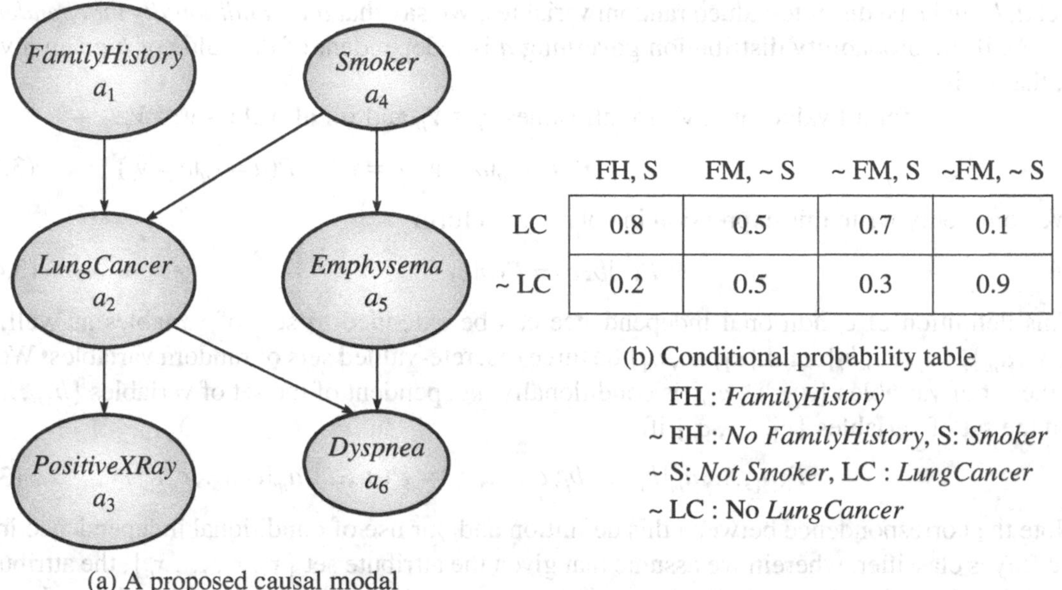

	FH, S	FM, ~ S	~ FM, S	~FM, ~ S
LC	0.8	0.5	0.7	0.1
~ LC	0.2	0.5	0.3	0.9

(a) A proposed causal modal

(b) Conditional probability table

FH : *FamilyHistory*

~ FH : *No FamilyHistory*, S: *Smoker*

~ S: *Not Smoker*, LC : *LungCancer*

~ LC : No *LungCancer*

Figure 3.5

An attractive feature of Bayesian belief networks is that they permit an easy method of representing causal knowledge, such as the fact that *LungCancer* = *True* results in *PositiveXRay* = *True*. In the language of conditional independence, this is expressed through the statement that *PositiveXRay* is conditionally independent of other variables in the network, given the value of *LungCancer*.

The joint probability for any desired assignment of values $(v_{a_1}, \ldots, v_{a_m})$ to the tuple of network variables $(a_1, \ldots, a_m)$ can be computed by the formula,

$$P(v_{a_1}, \ldots, v_{a_m}) = \prod_{i=1}^{m} P(v_{a_i} \mid Parents\ (a_i)) \tag{3.48}$$

where *Parents* (a_i) denotes the set of immediate predecessors of a_i in the network. The values of $P(v_{a_i} \mid Parents\ (a_i))$ are precisely the values stored in the conditional probability table associated with node a_i.

We may like to employ a belief network to infer the value of some target variable (e.g., *Dyspnea*), given the observed values of other variables. It is possible to infer the probability distribution for the target variable, which specifies the probability that it will take on each of its possible values given the observed values of other variables.

Experts who build belief networks for a specific domain at times stand to gain by representing causal effects by directed arcs. Therefore, when we are aware of the causal structure, it is relatively easy to induce models from data. But when machine learning methods are applied to induce models from data whose causal structure is not known, they only build a network on the basis of the correlations observed in the data. Inference of causality from correlations is always tough.

Training Bayesian Belief Networks

Let us first summarize the graphical representation of belief nets. Though these nets are capable of representing continuous multidimensional distributions over their variables, they have enjoyed the greatest application and success for discrete variables. Here, we shall focus on the discrete case.

Each *node* (or *unit*) is representative of one of the data attributes, where it takes on discrete values. The nodes are labeled by *A*, *B*,…, and their values by the corresponding lower case letter. Each *link* in the net (*arc* in the graph) is directional and connects two nodes; the link represents the causal influence of one node on another. For a single node in a net, we demarcate the set of nodes *before* that node—known as its *parents*—and set of those immediately *after* it—known as its *children* (*descendants*).

Let us say there is a belief net, which has causal dependencies depicted by the topology of the links. By applying the Bayes rule directly, it is possible to establish the probability of any configuration of variables in the joint distribution. To progress, we require the *conditional probability tables*, which provide the probability of any variable at a node for each conditioning event—in other words, for the values of the variables in the parent nodes. If the node is without parents, the table simply comprises earlier probabilities of the variable values.

The network and the conditional probability tables consist of all the information of the problem domain. This network model can be used for inference.

A simple belief net is shown in Fig. 3.6 [4].

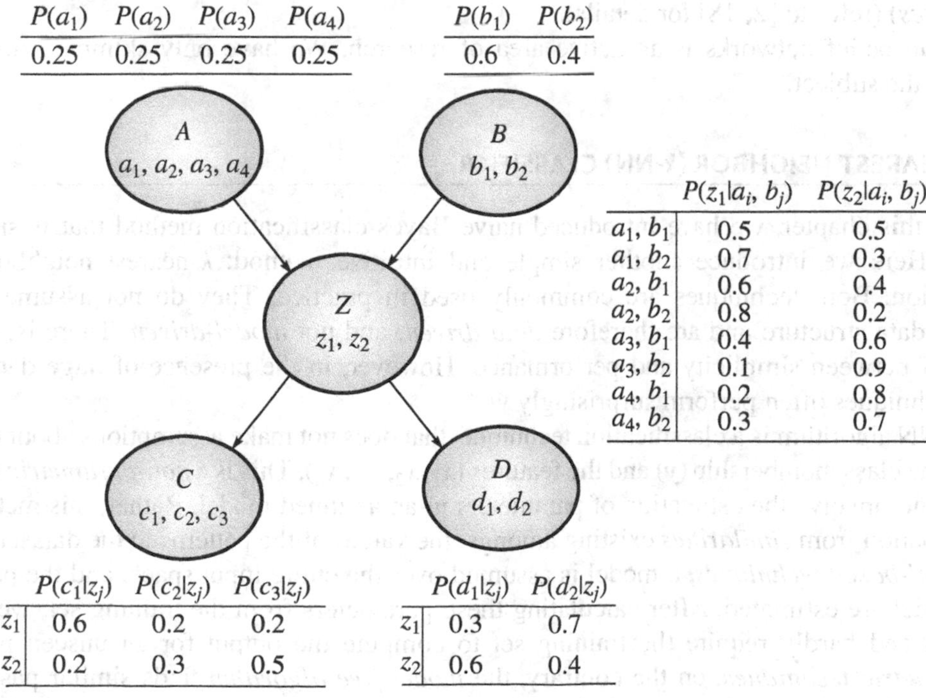

$P(a_1)$	$P(a_2)$	$P(a_3)$	$P(a_4)$
0.25	0.25	0.25	0.25

$P(b_1)$	$P(b_2)$
0.6	0.4

	$P(z_1 \mid a_i, b_j)$	$P(z_2 \mid a_i, b_j)$
a_1, b_1	0.5	0.5
a_1, b_2	0.7	0.3
a_2, b_1	0.6	0.4
a_2, b_2	0.8	0.2
a_3, b_1	0.4	0.6
a_3, b_2	0.1	0.9
a_4, b_1	0.2	0.8
a_4, b_2	0.3	0.7

	$P(c_1 \mid z_j)$	$P(c_2 \mid z_j)$	$P(c_3 \mid z_j)$
z_1	0.6	0.2	0.2
z_2	0.2	0.3	0.5

	$P(d_1 \mid z_j)$	$P(d_2 \mid z_j)$
z_1	0.3	0.7
z_2	0.6	0.4

Figure 3.6 A simple belief net

In situations wherein the network structure is provided in advance and the variables are completely observable (all the variable values are known) in the training examples, it is simple to learn the conditional probability tables. We just estimate the conditional probability table entries as we would for a naive Bayes classifier.

When the network structure is given but only some of the values are observable in the training data, the learning problem is more difficult. There are various methods to choose from for training a belief network (learning the entries in conditional probability tables) in such situations. Two widely used methods are *gradient ascent* procedure and *EM algorithm*.

We represent the set of all possible entries in conditional probability tables by w_{ijk}. The w_{ijk} are viewed as *weights* in the optimization procedure. The gradient ascent optimization procedure searches through the space of weights w_{ijk}. The weights are initialized to random probability values. The gradient ascent strategy performs greedy hill-climbing. At each iteration, weights are updated and will eventually converge to a local optimal solution for maximization of $P(\mathcal{D}|\text{weight space})$ (refer to [1] for details).

A popular alternative to gradient ascent is the *Expectation-Maximization* (*EM*) *Algorithm* for training Bayesian belief networks (refer to [2] for details). *EM algorithm* is also the basis for some clustering strategies (Section 7.6 gives detailed description of EM algorithm).

Learning Bayesian networks when the network structure is not known in advance, is a difficult problem. Many approaches have been proposed for handling such problems. One simple and very fast algorithm is known as *K2 algorithm*. Another good learning algorithm is TAN (*Tree-Augmented Naive Bayes*) (refer to [2, 18] for details).

Bayesian belief networks is an active area of research. We have only skimmed through the surface of the subject.

3.4 *k*-NEAREST NEIGHBOR (*k*-NN) CLASSIFIER

Earlier in this chapter, we have introduced naive Bayes classification method that is simple and intuitive. Here we introduce another simple and intuitive method: *k*-nearest neighbor (*k*-NN) classification. Both techniques are commonly used in practice. They do not assume anything about the data structure and are therefore *data-driven*, and not *model-driven*. There is, naturally, a trade-off between simplicity and performance. However, in the presence of huge datasets, the simple techniques often perform surprisingly well.

The *k*-NN algorithm is a classification technique that does not make assumptions about the model between the class membership (y) and the features ($x_1, x_2, \ldots, x_n$). This is a *non-parametric technique* as it does not involve the estimation of parameters in an assumed model. Rather, this method pulls out information from *similarities* existing amongst the values of the patterns in the dataset.

In *model-based techniques*, a model is assumed over the entire input space, and the parameters of the model are estimated. After calculating these parameters from the training set, we preserve the model and hardly require the training set to compute the output for an unseen pattern. In *non-parametric techniques*, on the contrary, the *model-free algorithm* finds similar past patterns or instances from the training set with the use of a suitable *distance measure* and interpolates from them the correct output. In machine learning, these techniques are also known as *instance-based* or *memory-based* learning algorithms, as they store the training instances and interpolate from these. N is usually quite large, and this increased requirement of memory and computation is the drawback of the non-parametric techniques.

There are several types of non-parametric methods of interest in pattern recognition. One consists of procedures for estimating the density functions $p(\mathbf{x}|y_q)$ from sample patterns (without the assumption that forms of underlying densities are known, as in maximum likelihood or Bayesian estimation methods; and with the assumption that attributes are statistically independent, as in naive Bayes classifier). If these estimates are satisfactory, they can be substituted in the Bayes theorem for the true densities when designing the classifier (*Parzon-Window* approach to estimating densities is quite popular [4]). Another consists of procedures for directly estimating *a posteriori* probabilities $P(y_q|\mathbf{x})$, bypassing probability estimation. Our focus here will be on this method, using *nearest-neighbor rule*.

Let us start with basic concepts of non-parametric estimation of probability density function of a pattern distribution. From these concepts, the *k*-NN approach of classification will emerge.

In density estimation, we assume that the sample $\{\mathbf{x}^{(i)}\}$; $i = 1,\ldots, N$, is drawn independently from some unknown probability density $p(\cdot)$. $\hat{p}(\cdot)$ is our estimate of $p(\cdot)$. We start with the univariate case where $\mathbf{x}^{(i)}$ are scalars ($n = 1$), and later generalize to the multidimensional case.

For the non-parametric estimation, the oldest and the most popular method is the *histogram*, where the input space is divided into equal-sized intervals, named *bins*. Given the origin x° and the bin width h, the bins are the intervals $[x^{\circ} + mh, x^{\circ} + (m + 1)h]$ for positive and negative integers m. The corresponding probability is approximated by the *frequency ratio*:

$$\hat{P}(x) = \frac{\#\{x^{(i)} \text{ in the same bin as } x\}}{N} \tag{3.49}$$

where $\#\{.\}$ denotes the number of training instances $x^{(i)}$ that lie in the same bin as the test instance x. This approximation converges to the true probability as $N \to \infty$. The corresponding probability density function value is assumed constant throughout the bin and is approximated as,

$$\hat{p}(x) \equiv \hat{p}(\bar{x}) = \left[\frac{\#\{x^{(i)} \text{ in the same bin as } x\}}{N}\right]\frac{1}{h} \tag{3.50}$$

where $\bar{x}$ is the midpoint of the bin $\left(|x - \bar{x}| \le \dfrac{h}{2}\right)$, and h is the width of the bin.

In constructing the histogram, we have to choose both an origin and a bin width. With small bins, the estimate is spiky, and with larger bins, the estimate is smoother. There are discontinuities at the boundaries (Fig. 3.7).

The *naive density estimator* frees us from setting the origin: For bins of size $2h$,

$$\hat{p}(x) = \frac{\#\{x - h < x^{(i)} \le x + h\}}{2Nh} \tag{3.51}$$

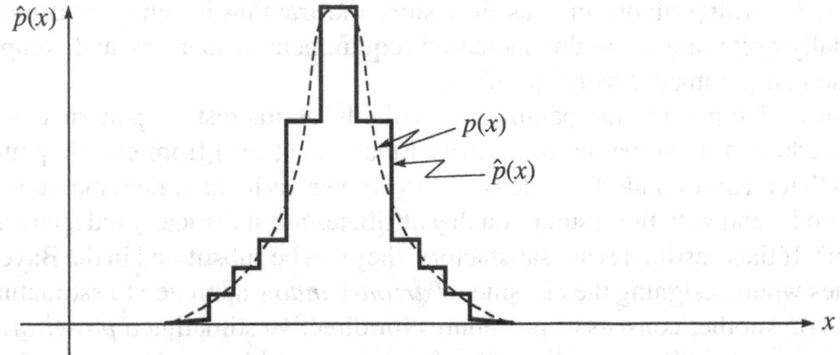

Figure 3.7 Histogram for density estimation.

For fixed N and small h, $p(x)$ is approximated by a finite number of δ-like spiky functions, centered at the training sample points. For a fixed h and $N \to \infty$, the space becomes dense in points and spiky functions are closely located. For a large enough number of samples, the smaller the h, the better the accuracy of the resulting estimate.

Usually, a large N is necessary for acceptable performance. If a one-dimensional interval needs to be filled up with say N equidistant points, the corresponding 2-dimensional square will need N^2 points, the 3-dimensional cube, N^3, and so on. The large number of data points puts a high burden on computational requirements.

The nearest-neighbor class of estimators adopts the reasonable amount of smoothing to the *local* density of the data. The degree of smoothing is controlled by k (the number of neighbors taken into account), which is much smaller than N (the sample size).

For each x ($n = 1$), we define

$$d_1(x) \le d_2(x) \le \cdots \le d_N(x) \tag{3.52}$$

to be distances ($|x - x^{(i)}|$) arranged in ascending order from x to the points in the sample. $d_1(x)$ is the distance to the nearest sample, $d_2(x)$ is the distance to the next nearest, and so on.

$$d_1(x) = \min_i |x - x^{(i)}|$$

If l is the closest sample, namely,

$$l = \arg \min_i |x - x^{(i)}|,$$

then
$$d_2(x) = \min_{i \ne l} |x - x^{(i)}|$$

and so forth.

The k-nearest neighbor (k-NN) density estimate is

$$\hat{p}(x) = \frac{k}{2N\, d_k(x)} \tag{3.53}$$

This is like a naive density estimator with $h = d_k(x)$; the difference being that instead of fixing h and checking how many samples fall in the bin, we fix k, the number of observations to fall in the bin, and compute the bin size. Where density is high, bins are small, and where density is low, bins are larger.

Generalizing to multivariable data,

$$\hat{p}(\mathbf{x}) = \frac{k}{NV_k(\mathbf{x})} \tag{3.54}$$

where $V_k(\mathbf{x})$ is the volume of the n-dimensional hyperspace centered at $\mathbf{x}$ with radius $r = \|\mathbf{x} - \mathbf{x}^{(k)}\|$; $\mathbf{x}^{(k)}$ is the kth nearest observation to $\mathbf{x}$ among the neighbors.

When used for classification, we need the *class-conditional densities*, $p(\mathbf{x}|y_q)$ of the feature-vector distributions. The estimator of the class-conditional density is given as,

$$\hat{p}(\mathbf{x}|y_q) = \frac{k_q}{N_q V_k(\mathbf{x})} \tag{3.55}$$

where k_q is the number of neighbors out of k nearest that belong to class y_q; and N_q is the number of labeled instances belonging to class y_q.

Here, we will not enter into the theoretical details of necessary and sufficient conditions for $\hat{p}(\mathbf{x})$ to coverage to $p(\mathbf{x})$ [4]. We will rather use the practical observation that k-NN estimate is not a good estimate of probability density function, however, k-NN method results in good estimates of *a posteriori* probabilities (particularly when N is large), bypassing probability density estimation.

A reasonable estimate for $P(y_q|\mathbf{x})$ is

$$\hat{P}(y_q|\mathbf{x}) = \frac{\hat{p}(\mathbf{x}|y_q)\,\hat{P}(y_q)}{\hat{p}(\mathbf{x})}$$

$$\hat{p}(\mathbf{x}|\,y_q) = \frac{k_q}{N_q V_k(\mathbf{x})}$$

$$\hat{P}(y_q) = N_q / N$$

$$\hat{p}(\mathbf{x}) = \frac{k}{NV_k(\mathbf{x})}$$

Therefore,

$$\hat{P}(y_q|\mathbf{x}) = \left(\frac{k_q}{N_q V_k(\mathbf{x})} \times \frac{N_q}{N}\right) \Big/ \left(\frac{k}{NV_k(\mathbf{x})}\right)$$

$$= k_q / k \tag{3.56}$$

Thus, the estimate of the *a posteriori* probability that y_q is the class for sample $\mathbf{x}$ is just the fraction of the samples within the cell $V_k(\mathbf{x})$ that are labeled y_q.

We first determine the k points that are closest neighbors of $\mathbf{x}$ with the help of a specific distance metric. The categorization of $\mathbf{x}$ is then given by the class label found in most of the k neighbors. All neighbors have equal vote, and the class with the most number of votes among the k neighbors is selected. Ties are randomly broken or weighted vote is taken. k is usually taken an odd number to reduce ties: confusion exists usually between two neighboring classes.

Note that $k = 1$ is usually not sufficient for determining the class of $\mathbf{x}$ due to noise and outliers in the data. A set of nearest neighbors is needed to accurately decide the class.

The key issue of k-NN algorithm is the *distance/similarity function*, which is selected on the basis of applications and nature of the data. The cautious selection of an appropriate distance function is a crucial step in the use of k-NN. A validation set can be made use of to pick the appropriate distance function for a given dataset by applying all candidates to gauge which one gives better results [50, 51].

Another question is the number of neighbors to select. Again, investigation of varying numbers of neighbors with the help of the validation set can facilitate establishing the optimal number, as the number is dependent on the data distribution and relies heavily on the problem being solved.

Selecting the appropriate training set is probably the most important step in k-NN process. The training set should include enough examples of all probable categories, that is, it should be a *balanced* training set having more or less the same number of instances for all classes. Generally speaking, the size of the training set should have at least thousands, if not hundreds of thousands or millions, of training examples.

The data is partitioned into training, validation, and test sets. Validation set is used to compare error rates for various values of k/various similarity measures, and test set is used to evaluate the performance.

Despite its simplicity, researchers have shown that classification accuracy of k-NN can be as high as those of elaborated methods (to be discussed in later chapters). k-NN is, however, slow at classification task. Due to the fact that there is no model building (in fact, training set itself along with distance measure is the non-parametric model), each test instance is compared with every training example at the classification time, which can be quite time consuming when the training data is large.

In the k-NN procedure, all neighbors have equal vote. In the *Parzon-window* approach [4], the weight of the vote is given by a *kernel function*, typically giving more weight to closer instances.

The computational flow diagram of the Parzon-Window method resembles the connectionist structure of a neural network, hence the name: *Probabilistic Neural Network* (PNN) given to this method [4].

3.5 DISCRIMINANT FUNCTIONS AND REGRESSION FUNCTIONS

In Section 1.7, classification and regression problems were defined as prediction of categorical (class, labels) variables, and prediction of numeric (continuous) variables, respectively. These two kinds of problems are of utmost importance. For example, in the case of speech or character recognition systems, fault detection systems, readers of magnetic-strip codes or credit cards, various alarm systems, and so on, we predict class or category. In control applications, signal processing, and financial markets, we predict (numeric values) various signals and stock prices based on past performance.

3.5.1 Classification and Discriminant Functions

There are many different ways to represent pattern classifiers. One of the most useful ways is in terms of *discriminant functions*. The concepts of discriminant functions are introduced here.

The patterns are feature vectors $\mathbf{x}^{(i)}$; $i = 1, \ldots, N$, and class labels y_q; $q = 1, 2, \ldots, M$. The Bayesian approach to classification assumes that the problem of pattern classification can be expressed in probabilistic terms and that *a priori* probabilities $P(y_q)$ and the conditional probability-density functions $p(\mathbf{x}|y_q)$; $q = 1, \ldots M$, are known. The posterior probability $P(y_q|\mathbf{x})$ is sought for classifying objects into corresponding classes. This probability can be expressed in the form of Bayes rule:

$$P(y_q|\mathbf{x}) = \frac{p(\mathbf{x}|y_q)\, P(y_q)}{\sum_{q=1}^{M} p(\mathbf{x}|y_q)\, P(y_q)}; q = 1, \ldots, M \tag{3.57a}$$

$$\sum_{q=1}^{M} P(y_q) = 1 \tag{3.57b}$$

$$\sum_{q=1}^{M} P(y_q|\mathbf{x}) = 1 \tag{3.57c}$$

The probability density function

$$p(\mathbf{x}) = \sum_{q=1}^{M} p(\mathbf{x}|\, y_q)\, P(y_q) \tag{3.57d}$$

provides necessary scaling, ensuring that sum of posterior probabilities is 1.

Having the posterior probabilities $P(y_q|\mathbf{x})$, one can formulate the following classification decision rule:

Assign an object to a class y_q having the largest value of posterior conditional probability $P(y_q|\mathbf{x})$ for a given feature vector $\mathbf{x}$. In other words, assign a given pattern with an observed feature vector $\mathbf{x}$ to a class y_k when

$$P(y_k|\,\mathbf{x}) > P(y_q|\,\mathbf{x})\ \forall\ q = 1, 2, \ldots, M; q \neq k \tag{3.58a}$$

or for a given feature vector $\mathbf{x}$, decide class y_k if

$$p(\mathbf{x}|\, y_k)\, P(y_k) > p(\mathbf{x}|\, y_q)\, P(y_q)\ \forall\ q = 1, 2, \ldots, M; q \neq k \tag{3.58b}$$

A pattern classifier assigns the feature vector $\mathbf{x}$ to one of the number of possible classes y_q; $q \in \{1, 2, \ldots, M\}$, and in this way partitions feature space into line segments, areas, volumes, and hyper volumes, which are *decision regions* in the case of one-, two-, three-, or higher-dimensional feature space, respectively. All feature vectors belonging to the same class are ideally assigned to the same decision region. The decision regions are often nonoverlapping volumes or hyper volumes, and decision regions of the same class may also be disjoint, consisting of two or more nontouching regions. The boundaries between adjacent regions are *decision boundaries* because classification decisions change across boundaries. These class boundaries are points, straight lines or curves, planes or surfaces, and hyperplanes or hyper surfaces in the case of one-, two-, three-, and higher-dimensional feature space, respectively. In the case of straight lines, planes, and hyperplanes, the decision boundaries are *linear*.

More generally, classification decisions based on feature vector **x** may be stated using a set of explicitly defined *discriminant functions*

$$g_q(\mathbf{x});\ q = 1, 2, \dots, M \tag{3.59}$$

where each discrminant is associated with a particular recognized class y_q; $q = 1, 2, \dots, M$.

The classifier designed using the discriminant functions assigns a pattern with feature vector **x** to a class y_k for which the corresponding discriminant value g_k is the largest:

$$g_k(\mathbf{x}) > g_q(\mathbf{x})\ \forall\ q = 1, 2, \dots, M;\ q \neq k \tag{3.60}$$

Note that the classification is based on the largest discriminant function $g_k(\mathbf{x})$ regardless of how the corresponding discriminant functions are defined. Therefore, any monotonic function of a discriminant function, $f(g(\mathbf{x}))$, will provide identical classification because of the fact that for the monotonic function $f(\cdot)$, the maximal $g_k(\mathbf{x})$ gives rise to the maximal $f(g_k(\mathbf{x}))$. Therefore, if some $g_q(\mathbf{x})$; $q = 1, 2, \dots, M$, are the discriminant functions for a given classifier, so also are the functions $g_q(\mathbf{x}) + C$, or $Cg_q(\mathbf{x})$ for any class-independent constant C; or log functions.

In the case of a Bayes classifier, instead of $g_q(\mathbf{x}) = P(y_q|\mathbf{x})$ or $p(\mathbf{x}|y_q)P(y_q)$, the natural algorithm of $P(y_q|\mathbf{x})$ is used as a discriminant function:

$$\begin{aligned} g_q(\mathbf{x}) &= \ln P(y_q|\mathbf{x}) \text{ or } \ln p(\mathbf{x}|y_q)\, P(y_q) \\ &= \ln p(\mathbf{x}|y_q) + \ln P(y_q);\ q = 1, 2, \dots, M \end{aligned} \tag{3.61}$$

In the case of two-class or binary classification, instead of two discriminants, $g_1(\mathbf{x})$ and $g_2(\mathbf{x})$, applied separately, it is more common to define a single discriminant function

$$g(\mathbf{x}) \equiv g_1(\mathbf{x}) - g_2(\mathbf{x}) \tag{3.62}$$

and to use the following decision rule:

$$\text{Decide } y_1, \text{ if } g(\mathbf{x}) > 0; \text{ otherwise decide } y_2$$

In the case of normally distributed classes (Gaussian classes), which are very common, discriminant functions are *quadratic* [3, 4]. These become *linear* (straight lines, planes, and hyper planes for two-, three-, and *n*-dimensional feature vectors, respectively) when the covariance matrices of corresponding classes are equal [3, 4]. In order to apply the most general Bayes procedure, practically everything about the underlying analyzed process must be known. This includes priors $P(y_q)$ and the class-conditional probability densities $p(\mathbf{x}|y_q)$.

3.5.2 Numeric Prediction and Regression Functions

For simplicity of presentation, we consider a two-dimensional case (whcre only two variables are involved; output *y* and the feature variable *x*). Conceptually, nothing changes in multivariable cases with higher-dimensional inputs and outputs; the regression curves become hyper curves or hyper surfaces.

Let x and y be random variables with a joint probability density function $p(x, y)$. If this function is continuous in y, then the conditional probability density function of y with respect to fixed x can be written as [3]:

$$p(y|x) = \frac{p(x, y)}{p(x)} \tag{3.63}$$

By using this function, the *regression curve* $f(x)$ is defined as the expectation of y for any value of x:

$$\mathbb{E}(y|x) = f(x) = \int_{-\infty}^{\infty} y\, p(y|x)\, dy = \int_{-\infty}^{\infty} y \frac{p(x, y)}{p(x)} dy$$

$$= \frac{\int_{-\infty}^{\infty} y\, p(x, y)\, dy}{p(x)} \tag{3.64}$$

It can easily be shown that the regression curve gives the best estimation of y in the mean-squared-error sense. Depending upon the joint probability density function $p(x, y)$, this function belongs to a certain class, for example, the class of all linear functions or the class of all functions of a given polynomial form.

Regression function for jointly *normally* distributed variables is *linear* [3]. This is an interesting property that was heavily exploited in statistics. Linear regression and correlation analysis, which are closely related, are both highly developed and widely used in diverse fields.

3.5.3 Practical Hypothesis Functions

The classical statistical techniques are based on the fundamental assumption that in most of the real-life problems, the stochastic component of data follows normal distribution. With this assumption, the data can be modeled by linear regresser functions/linear discriminant functions (when covariance matrices of corresponding classes are equal). The Bayes method for the determination of these functions requires prior knowledge on probability distributions.

Linear discriminant functions and linear regressor functions have a variety of pleasant analytical properties. However, the assumptions on which the classical statistical paradigm relied, turned out to be inappropriate for many contemporary real-life problems. The underlying real-life data generation laws may typically be very far from the normal distribution and a model-builder must consider this difference in order to construct an effective algorithm. One obvious way is to look for polynomial regression. But for nonlinear model assumptions, only greedy algorithms, local minima, and heuristic searches were known in the statistical paradigm.

Statistical approach is, thus, very convenient to deal with the design of linear classifiers described by linear discriminant functions, and linear regressors described by linear approximation functions. There is a flood of problems being faced today for which the design of linear classifiers/linear regressors do not lead to a satisfactory solution. The design of nonlinear classifiers/regressors emerges as an inescapable necessity. As we have remarked earlier, statistical approach is not very helpful for such problems. For choosing the appropriate nonlinear function and optimizing the parameters embedded in the function, there is no principled or automatic statistical method.

It should be stressed here that in spite of serious practical limitations of statistical approaches, they still are and will remain very good theoretical and practical tools in the cases where the mentioned assumptions are valid.

What we seek is a way to *learn* nonlinearity. This is the approach of machine learning. As we shall see in later chapters, machine learning can, at least in principle, provide the optimal solution to an arbitrary classification/regression problem.

Machine learning involves searching through a space of possible hypotheses to determine one that best fits the observed data and any prior knowledge held by the learner. Hypotheses space is theoretically infinite. The learning task is, thus, to search through this vast space to locate the hypothesis that is most consistent with the available training examples.

There are many possible approximating functions for a given set of data; unfortunately, there is no theoretical method of determining which out of many possible approximating functions will yield the best approximation. Fortunately, there are only a few possible candidate functions in use today on the basis of empirical experience. The most widely used functions are linear functions, kernel functions, tangent hyperbolic functions, radial basis functions (Gaussians being most popular), and standard membership functions applied in fuzzy logic models (triangle, trapezoidal, singleton, etc.). Our focus in this book will be on these functions only. Through heuristic search, an appropriate function for the problem in hand will be selected. As we shall see in later chapters, an appropriate network of these functional elements can approximate any nonlinearity.

Heuristic search is organized as per the following two-step procedure.

(i) The search is first focused on a *hypothesis class* chosen for the learning task in hand. The different hypotheses classes are appropriate for learning different kinds of functions. The main hypotheses classes are:

1. Linear Models
2. Logistic Models
3. Support Vector Machines
4. Neural Networks
5. Fuzzy Logic Models
6. Decisions Trees
7. k-Nearest Neighbors (k-NN)
8. Naive Bayes

Linear Models (Section 3.6; Chapters 4 and 5), Logistic Models (Section 3.7; Chapter 5), Support Vector Machines (Chapter 4), Neural Networks (Chapter 5), and Fuzzy Logic Models (Chapter 6) are representative of parametric techniques. Parametric techniques make assumptions on the hypotheses classes by means of a fixed number of parameters and optimize these parameters using the learning set. On the other hand, Decision Trees (Chapter 8), and k-NN (Section 3.4) are non-parametric techniques; they can adapt the shape of the modal they produce on the basis of the learning set. Naive Bayes (Section 3.3) is a very simple method of probabilistic framework used for classification, of which Bayesian network is a generalization.

Linear Model: (Section 3.6)

$$h(\mathbf{x}, \mathbf{w}, w_0) = w_0 + w_1 x_1 + w_2 x_2 + \cdots + w_n x_n = w_0 + \mathbf{w}^T\mathbf{x} \tag{3.65}$$

where $\{\mathbf{w}, w_0\}$ is the set of adjustable weights.

The model given by Eqn (3.65) assumes that the given data is approximately linear. We go for nonlinear models when this is not true. The following are the commonly used learning models for creating nonlinear approximating functions.

Logistic Model: (Section 3.7)

$$P(\text{Class } 1|\mathbf{x}) = \frac{1}{1+\exp[-(\mathbf{w}^T\mathbf{x}+w_0)]}$$

where $\{\mathbf{w}, w_0\}$ is the set of adjustable weights in binary classification (nonlinearity is a *sigmoid* function).

Multi-Layer Perceptron (MLP) Model: (Chapter 5)

$$h(\mathbf{x}, \mathbf{w}, \mathbf{v}) = \sum_{l=1}^{m} v_l\, \sigma(\mathbf{x}, \mathbf{w}_l, w_{l0}) + v_0 \tag{3.66}$$

where $\sigma(\cdot)$ is a set of *sigmoidal functions* and $\{v_l, v_0, \mathbf{w}_l, w_{l0}\}$ is the set of adjustable weights.

Radial Basis Function (RBF) Model: (Chapter 5)

$$h(\mathbf{x}, \mathbf{w}, \mathbf{v}) = \sum_{l=1}^{m} w_l\, \phi_l(\mathbf{x}, \mathbf{v}_l) \tag{3.67}$$

where $\phi_l(\cdot)$ is a set of radial basis functions (e.g., Gaussians) and $\{\mathbf{v}_l, \mathbf{w}_l\}$ is the set of adjustable weights.

Fuzzy Logic (FL) Model: (Chapter 6)

$$h(\mathbf{x}, \mathbf{c}, \mathbf{r}) = \frac{\sum_{l=1}^{m} G(\mathbf{x}, \mathbf{c}_l)\, r_l}{\sum_{l=1}^{m} G(\mathbf{x}, \mathbf{c}_l)} \tag{3.68}$$

where m is the number of rules, r_l are the rules, and the basis functions $G(\mathbf{x}, \mathbf{c}_l)$ are the input membership functions centered at $\mathbf{c}_l$.

Kernel Functions: (Chapter 4)

Assume that there exists a mapping from low-dimensional feature space to m-dimensional feature space so that classes can be satisfactorily separated by linear functions. m is generally much higher than the input space dimensionality (n) to make the classes linearly seperable. Kernel functions provide such a mapping.

Commonly used kernels include the following:

Polynomial of degree d: $K(\mathbf{x}^{(i)}, \mathbf{x}^{(k)}) = (\mathbf{x}^{(i)T}\mathbf{x}^{(k)} + 1)^d$ (3.69a)

Gaussian RBF: $$K(\mathbf{x}^{(i)}, \mathbf{x}^{(k)}) = \exp\left(-\frac{1}{2\sigma^2} \| \mathbf{x}^{(i)} - \mathbf{x}^{(k)} \|^2\right) \quad (3.69b)$$

These different hypotheses classes are appropriate for learning different kinds of functions. Initial choice of a class is made on the basis of the learning task in hand, and prior knowledge/experience held by the designer.

(ii) For each of these hypotheses classes, the corresponding learning algorithm organizes the search through all possible underlying functions (structures of the learning machine).

It should be stressed here that the approaches which in pattern recognition problems most often result in linear discriminant functions, and in regression problems result in linear approximating functions, still are and will remain very good theoretical and practical tools if the mentioned assumptions are valid. Very often, in modern real-world applications, many of these postulates are satisfied only approximately. However, even when these assumptions are not totally sound, linear discriminants have shown acceptable performance in classification problems, as linear approximators in regression problems. Because of their simple structure, these techniques do not overfit the training dataset, and for many classification/regression tasks, they may be good starting points or good first estimates of success rate of learning machines for the problems in hand.

A point made earlier in Section 3.1 may be repeated here before concluding this section. The two perspectives: inferential statistical analysis and machine learning, have now converged. We will be using the term 'machine learning' in a broader practical sense—the techniques that include statistical learning, and form the basic tools for data mining applications.

3.6 LINEAR REGRESSION WITH LEAST SQUARE ERROR CRITERION

In Section 3.5, we observed that the practical limitation of statistical approach is the assumed initial knowledge available on the process under investigation. For the Bayes procedure, one should know the underlying probability distributions. If only the forms of underlying distributions were known, we can use the training samples to estimate the values of their parameters.

In this section, we shall instead assume we know the proper forms of the *regressor functions*, and use the samples to estimate the values of parameters of the function. We shall examine both the non-statistical and statistical procedures for determining regressor functions. These procedures, however, do not require the knowledge of the forms of underlying probability distributions.

Linear regressor functions have a variety of good analytical properties. As we observed in Section 3.5, they can be optimal if the underlying distributions are cooperative, such as Gaussians. Even when they are not optimal, we might be willing to sacrifice some performance to be able to gain the benefit of their simplicity. Linear regressor functions are comparatively simple to compute, and if there is no information to suggest otherwise, linear regressors are ideal candidates for early, trial regression.

The problem of finding a linear regressor function will be formulated as a problem of minimizing a criterion function. The widely-used criterion function for regression purposes is the *sum-of-error-squares* (refer to Section 2.7).

In general, regression methods are used to predict the value of *response* (dependent) variable from attribute (independent) variables, where the variables have continuous numeric values. Linear regressor model fits a linear function (relationship) between dependent (output) variable and independent (input) variables—it expresses the output $\hat{y}$ as a linear combination of the attributes $x_1, x_2, \ldots, x_n$; given the N data points $(x_j^{(i)}, y^{(i)}); i = 1, \ldots, N; j = 1, \ldots, n$. The linear regressor model, thus, has the form

$$\hat{y}^{(i)} = w_0 + w_1 x_1^{(i)} + \cdots + w_n x_n^{(i)}; i = 1, \ldots, N \tag{3.70}$$

where $\{w_0, w_1, \ldots, w_n\}$ are the parameters of the model.

Of interest is the difference between the predicted value $\hat{y}$ and the actual value y. The method of linear regression is to choose the $(n + 1)$ coefficients $w_0, w_1, \ldots, w_n$, to minimize the residual sum of squares of these differences over all the N training instances. The performance criterion is thus the *sum-of-error-squares:*

Residual error

$$e^{(i)} = y^{(i)} - \hat{y}^{(i)} = y^{(i)} - \sum_{j=0}^{n} w_j x_j^{(i)}; x_0^{(i)} = 1 \tag{3.71a}$$

Residual sum-of-error-squares

$$E = \sum_{i=1}^{N} (e^{(i)})^2 = \sum_{i=1}^{N} \left(y^{(i)} - \sum_{j=0}^{n} w_j x_j^{(i)} \right)^2 ; x_0^{(i)} = 1 \tag{3.71b}$$

3.6.1 Minimal Sum-of-Error-Squares and the Pseudoinverse

How to minimize sum-of-error-squares E given by Eqn (3.71b)? A matrix $\mathbf{X}$ of input vectors $\mathbf{x}^{(i)}; i = 1, \ldots, N$; vector $\mathbf{y}$ of the desired outputs $y^{(i)}, i = 1, \ldots, N$, and the *weights vector* $\bar{\mathbf{w}}$ of the coefficients of the linear model are introduced as follows:

$$\mathbf{X} = \begin{bmatrix} 1 & 1 & \cdots & 1 \\ x_1^{(1)} & x_1^{(2)} & \cdots & x_1^{(N)} \\ x_2^{(1)} & x_2^{(2)} & \cdots & x_2^{(N)} \\ \vdots & \vdots & & \vdots \\ x_n^{(1)} & x_n^{(2)} & \cdots & x_n^{(N)} \end{bmatrix} = [\bar{\mathbf{x}}^{(1)} \; \bar{\mathbf{x}}^{(2)} \cdots \bar{\mathbf{x}}^{(N)}] \tag{3.72a}$$

$$\bar{\mathbf{x}} = [1 \;\; x_1 \;\; x_2 \cdots x_n]^T \tag{3.72b}$$

$$\mathbf{y} = [y^{(1)} \; y^{(2)} \cdots y^{(N)}]^T \tag{3.72c}$$

$$\bar{\mathbf{w}} = [w_0 \; w_1 \; w_2 \cdots w_n]^T \tag{3.72d}$$

For an optimum solution for $\bar{\mathbf{w}}$, the following equations need to be satisfied:

$$y^{(1)} = \bar{\mathbf{w}}^T \bar{\mathbf{x}}^{(1)} \tag{3.73}$$

$$\vdots$$

$$y^{(N)} = \bar{\mathbf{w}}^T \bar{\mathbf{x}}^{(N)}$$

Therefore,

$$[y^{(1)} \, y^{(2)} \ldots y^{(N)}] = \bar{\mathbf{w}}^T [\bar{\mathbf{x}}^{(1)} \, \bar{\mathbf{x}}^{(2)} \ldots \bar{\mathbf{x}}^{(N)}] = \bar{\mathbf{w}}^T \mathbf{X} \tag{3.74a}$$

or

$$\mathbf{y} = (\bar{\mathbf{w}}^T \mathbf{X})^T \tag{3.74b}$$

The vector of residual errors becomes

$$\mathbf{y} - (\bar{\mathbf{w}}^T \mathbf{X})^T$$

Hence the error function can be written as,

$$\begin{aligned} E(\bar{\mathbf{w}}) &= [\mathbf{y} - (\bar{\mathbf{w}}^T \mathbf{X})^T]^T \, [\mathbf{y} - (\bar{\mathbf{w}}^T \mathbf{X})^T] \\ &= \bar{\mathbf{w}}^T [\mathbf{X}\mathbf{X}^T] \, \bar{\mathbf{w}} - 2\bar{\mathbf{w}}^T \mathbf{X}\mathbf{y} + \mathbf{y}^T \mathbf{y} \end{aligned} \tag{3.75}$$

In this least-squares estimation task, the objective is to find the optimal $\bar{\mathbf{w}}^*$ that minimizes $E(\bar{\mathbf{w}})$. The solution to this classic problem in calculus is found by setting the gradient of $E(\bar{\mathbf{w}})$, with respect to $\bar{\mathbf{w}}$, to zero.

$$\frac{\partial E(\bar{\mathbf{w}})}{\partial \bar{\mathbf{w}}} = 2(\mathbf{X}\mathbf{X}^T) \, \bar{\mathbf{w}} - 2\mathbf{X}\mathbf{y} = \mathbf{0} \tag{3.76a}$$

This gives

$$\bar{\mathbf{w}}^* = (\mathbf{X}\mathbf{X}^T)^{-1} \mathbf{X}\mathbf{y} \tag{3.76b}$$

The fitted output values at the training data are

$$\hat{\mathbf{y}} = \mathbf{X}^T \bar{\mathbf{w}}^* = \mathbf{X}^T (\mathbf{X}\mathbf{X}^T)^{-1} \mathbf{X}\mathbf{y} \tag{3.77}$$

The $(n+1) \times N$ matrix $\mathbf{X}^+ = (\mathbf{X}\mathbf{X}^T)^{-1}\mathbf{X}$ is called the *pseudoinverse* matrix of the matrix $\mathbf{X}^T$. Thus, the optimal solution

$$\bar{\mathbf{w}}^* = \mathbf{X}^+ \mathbf{y} \tag{3.78}$$

It is assumed that the matrix $\mathbf{XX}^T$ is nonsingular. It might happen that the columns of $\mathbf{X}$ are not linearly independent. Then $\mathbf{XX}^T$ is singular and the least squares coefficients $\bar{\mathbf{w}}^*$ are not uniquely defined. The singular case occurs most often when two or more inputs were perfectly correlated. A natural way to resolve the non-unique representation is by dropping redundant columns in $\mathbf{X}$. Most regression software packages detect these redundancies and automatically implement some strategy for removing them.

When the number of training samples is equal to the number of weights to be determined, $\mathbf{X}$ is a square matrix and $\mathbf{X}^+ = (\mathbf{X}^T)^{-1}$; pseudoinverse coincides with regular inverse. This is of little practical interest. Training patterns will almost always be corrupted with noise, and to reduce the influence of these disturbances, the number of training samples must be (much) larger than the adapted weights.

From a computational point of view, the calculation of optimal weights requires the pseudoinverse of $N \times (n+1)$ dimensional matrix. With respect to computing time, the critical part is the inversion of $(n + 1) \times (n + 1)$ matrix $\mathbf{XX}^T$. In the real-life applications, it is quite common for input vectors to be of very high dimension, and in such a situation this part of the calculation may not be easy.

Various methods are available to take care of this problem. For most real-life problems, *Recursive Least Squares Algorithm* [3] might be the best *on-line* (sequential/recursive/iterative/incremental), weight-adapting procedure with high rate of convergence (in terms of iteration steps).

3.6.2 Gradient Descent Optimization Schemes

A classical optimization method that has become widely used in the field of soft computing is the *Gradient Descent Method* (for minimization tasks). Changes of the weights are made according to the following algorithm:

$$\bar{\mathbf{w}}_{k+1} = \bar{\mathbf{w}}_k - \eta \left. \frac{\partial E}{\partial \bar{\mathbf{w}}} \right|_k \tag{3.79}$$

where η denotes the *learning rate*, and k stands for the actual iteration step. There are two approaches for designing the iteration step. The *batch* (off-line, one-shot) *methods* use all the data in one shot. The kth iteration step means the kth presentation of the whole training dataset; the gradient is calculated across the entire set of training patterns. On the other hand, in the *online methods*, k denotes the iteration step after single data pair is presented. Online gradient descent methods share almost all good features of recursive least squares algorithm with reduced computational complexity.

Gradient descent procedure has two advantages over merely computing the pseudoinverse: (1) it avoids the problems that arise when $\mathbf{XX}^T$ is singular (it always yields a solution regardless of whether or not $\mathbf{XX}^T$ is singular); and (2) it avoids the need for working with large matrices. Extended discussion on linear regression will appear in Chapter 5, where we will present the gradient descent procedures.

3.6.3 Least Mean Square (LMS) Algorithm

Till now we have discussed about deterministic algorithms to determine the optimum setting of the weights $\bar{\mathbf{w}}$ that minimizes the criterion function given by Eqn (3.75). Now we explore a digression from this error function. Consider the problem of computing weights vector $\bar{\mathbf{w}}$ so as to minimize *Mean Square Error* (MSE) between desired and true outputs, defined as follows:

$$E(\bar{\mathbf{w}}) = \mathbb{E}\left[\sum_{i=1}^{N} (y^{(i)} - \bar{\mathbf{w}}^T \bar{\mathbf{x}}^{(i)})^2\right] \tag{3.80}$$

where $\mathbb{E}$ is the statistical expectation operator.

The solution to this problem requires the computation of autocorrelation matrix $\mathbb{E}\,[\mathbf{x}\mathbf{x}^T]$ of the set of feature vectors, and cross-correlation matrix $\mathbb{E}[\mathbf{x}y]$ between the desired response and the feature vector. This presupposes knowledge of the underlying distributions, which in general is not known (after all, if it were known, why not use Bayesian method?). Thus, our major goal becomes to see if it is possible to solve the optimization problem without having this statistical information.

The *Least Mean Square* (LMS) algorithm, originally formulated by Widrow and Hoff, is a *stochastic gradient algorithm* that iterates weight vector $\bar{\mathbf{w}}$ in the regressor in the direction of the gradient of the squared amplitude of error signal with respect to that weight vector. LMS is called a stochastic gradient algorithm because the gradient vector is chosen at 'random' and not, as in the steepest descent case—precisely derived from the shape of the total error surface. Random here means the instantaneous value of the gradient. This is then used as the estimator of the true quantity.

In practice, this simply means that LMS uses at each iteration step the actual value of the error after single data pair is presented, not the full sum-of-error-squares function which requires presentation of the entire set of training patterns for calculation of the gradient. Thus, LMS is used in an *online* mode, described earlier in this section, where the weights are updated after each data pair is presented.

The design of the LMS algorithm is very simple, yet a detailed analysis of its convergence behavior is a challenging mathematical task. It turns out that under mild conditions, the solution provided by the LMS algorithm converges in probability to the solution of MSE optimization problem.

Extended discussion on linear regression in Chapter 5 will provide the detailed account of LMS algorithm.

3.7 LOGISTIC REGRESSION FOR CLASSIFICATION TASKS

Linear regression models are well-suited to estimating continuous quantities that can take on a wide range of values (the target variable $y \in \Re$, i.e., it can take on values from $-\infty$ to $+\infty$). Modeling a binary outcome such as yes/no does not seem like a regression task. Because classification problems are extremely common, statisticians have found a way to adopt regression models to this task. The resulting method is called *logistic regression*.

Technically, one can apply linear regression to classification problems treating the dependent variable y in the data matrix as continuous (of course, y must be coded numerically, e.g., 1 for 'yes' and 0 for 'no'). Using the linear regression model to predict y yields predictions that are not necessarily 0 and 1. Therefore, linear regression is inappropriate for categorical response.

The task can be restated as 'what is the probability that an instance belongs to class q?' Because probabilities are continuous numbers, the problem is now a regression task. However, this regression task is also not appropriate for linear regression because linear regression yields predictions that are not necessarily restricted between 0 and 1.

In casual speech, the words *probability* and *odds* are used interchangeably. In statistics, each has a specific meaning. Probability is a number between 0 and 1 indicating the chance of particular outcome occurring. Odds is the ratio of probability of a particular outcome occurring to the probability of it not occurring. The odds is a number between 0 and infinity. This brings us half way to the range of $\Re$. One more step, the log of odds yields a function that goes from $-\infty$ to $+\infty$; log(odds) is called *logit*. The transformation to log(odds), with logit as a dependent variable, converts the problem to a possibly linear regression task. For binary classification problem, it is assumed that

$$\log(\text{odds}) = \log \frac{P(\text{Class 1}|\mathbf{x})}{1 - P(\text{Class 1}|\mathbf{x})} = w_0 + w_1x_1 + \cdots + w_nx_n \tag{3.81}$$

$w_0, w_1, \ldots, w_n$ are $(n + 1)$ parameters of linear logit function.

Solving for probability $P(\text{Class 1}|\mathbf{x})$ from Eqn (3.81) requires a bit of algebra. The result is

$$P(\text{Class 1}|\mathbf{x}) = \frac{1}{1 + e^{-(w_0 + w_1x_1 + \cdots + w_nx_n)}} \tag{3.82a}$$

$$P(\text{Class 2}|\mathbf{x}) = 1 - P(\text{Class 1}|\mathbf{x}) \tag{3.82b}$$

This is the *logistic function* (detailed description of logistic functions given in Chapter 5) that never gets smaller than 0 and never gets larger than 1, but takes on every value in between.

In the following sub-section, we discuss *maximum likelihood* criterion for estimating parameters of logistic regression model.

Maximum Likelihood Estimation of Logistic Regression Models

We are given a set of observed data:

$$\mathcal{D} = \{\mathbf{x}^{(i)}, y^{(i)}\};\ i = 1, \ldots, N$$

$$\mathbf{x} = [x_1 \quad x_2 \cdots x_n]^T \in \Re^n$$

$$y \in \{0, 1\}$$

The proposed model, given by (3.82), is

$$P(y = 1|\mathbf{x}) = \frac{1}{1 - e^{-(w_0 + w_1x_1 + \cdots + w_nx_n)}}$$

$$= \frac{1}{1 - e^{-(\mathbf{w}^T\mathbf{x} + w_0)}}$$

$$P(y=0|\mathbf{x}) = 1 - P(y=1|\mathbf{x})$$

where $$\mathbf{w}^T = [w_1 \; w_2 \cdots w_n]$$

The maximum likelihood approach [4] answers the question: for all possible values of the parameters $\{\mathbf{w}, w_0\}$ that the distribution (3.82) might have, which of these are most likely. We call this 'most likely' estimate, the *maximum likelihood estimate*. It is computed as follows:

- For each of our observed data points, we compute a *likelihood* as a function of the parameters we seek to estimate. This likelihood is just the value of the corresponding probability distribution evaluated at that particular point.
- We then compute the *likelihood function*, which is the combination of likelihoods for all the data points we observed.
- Thereafter, we estimate the set of parameters which will maximize the likelihood function.

Given the input sample $\mathbf{x}^{(i)}$, the probability that it belongs to class $y^{(i)} \in \{0, 1\}$ is $P(y^{(i)}|\mathbf{x}^{(i)})$. We assume that the samples of the distribution $\mathcal{D}$ are *iid* – independent and identically distributed random variables. The likelihood function is, therefore, defined as,

$$\mathcal{L}(\{\mathbf{w}, w_0\}, \mathcal{D}) = \mathcal{L}(\bar{\mathbf{w}}, \mathcal{D}) = \prod_{i=1}^{N} P(y^{(i)}|\mathbf{x}^{(i)})$$

Note that the likelihood function can be viewed as a function with the parameters as function of data rather than the data as function of the parameters. We assume that the data is fixed, but parameters can vary.

In the maximum likelihood problem, our goal is to find $\bar{\mathbf{w}}$ that maximizes $\mathcal{L}$. That is, we wish to find $\bar{\mathbf{w}}^*$ where

$$\bar{\mathbf{w}}^* = \underset{\bar{\mathbf{w}}}{arg\ max}\ \mathcal{L}(\bar{\mathbf{w}}, \mathcal{D})$$

Often we maximize $\log(\mathcal{L}(\bar{\mathbf{w}}, \mathcal{D}))$ instead because it is analytically easier. Because the logarithm is monotonically increasing, the $\bar{\mathbf{w}}^*$ that maximizes log-likelihood, also maximizes the likelihood.

$$\log(\mathcal{L}(\bar{\mathbf{w}}|\mathcal{D})) = \log\left(\prod_{i=1}^{N} P(y^{(i)}|\mathbf{x}^{(i)})\right)$$

$$= \sum_{i=1}^{N} \log P(y^{(i)}|\mathbf{x}^{(i)})$$

If $P(y^{(i)}|\mathbf{x}^{(i)})$ is a well-behaved differentiable function of parameters $\bar{\mathbf{w}}$, $\bar{\mathbf{w}}^*$ can easily be found by standard differential calculus (say, gradient method).

The logistic regression model (3.82) may be expressed in terms of *discriminant function* $g(\mathbf{x}, \bar{\mathbf{w}})$ as,

$$P(y^{(i)}|\mathbf{x}^{(i)}) = g(\mathbf{x}^{(i)}, \bar{\mathbf{w}}) = \frac{1}{1+e^{-(\mathbf{w}^T\mathbf{x}+w_0)}} = \frac{1}{1+e^{-\bar{\mathbf{w}}^T\bar{\mathbf{x}}}} \tag{3.83}$$

where

$$\bar{\mathbf{w}}^T = [w_0 \; w_1 \cdots w_n]; \; \bar{\mathbf{x}} = [1 \; x_1 \cdots x_n]^T$$

Assuming $y^{(i)}$, given $\mathbf{x}^{(i)}$, is Bernoulli[1] with $P(y^{(i)}|\mathbf{x}^{(i)}) = g(\mathbf{x}^{(i)}, \bar{\mathbf{w}})$; we model log-likelihood as,

$$\log(\mathcal{L}(\bar{\mathbf{w}}|\mathcal{D})) = \log\left(\prod_{i=1}^{N} P(y^{(i)}|\mathbf{x}^{(i)})^{y^{(i)}} (P(1-y^{(i)}|\mathbf{x}^{(i)}))^{1-y^{(i)}}\right)$$

$$= \sum_{i=1}^{N} \log\left(g(\mathbf{x}^{(i)}, \bar{\mathbf{w}})^{y^{(i)}} (1-g(\mathbf{x}^{(i)}, \bar{\mathbf{w}}))^{1-y^{(i)}}\right)$$

To maximize log-likelihood with respect to $\bar{\mathbf{w}}$, let us look at each example:

$$\log(\mathcal{L}_i(\bar{\mathbf{w}}|\mathcal{D})) = \log\left(g(\mathbf{x}^{(i)}, \bar{\mathbf{w}})^{y^{(i)}} (1-g(\mathbf{x}^{(i)}, \bar{\mathbf{w}}))^{1-y^{(i)}}\right)$$

$$= y^{(i)} \log g(\mathbf{x}^{(i)}, \bar{\mathbf{w}}) + (1-y^{(i)}) \log(1-g(\mathbf{x}^{(i)}, \bar{\mathbf{w}}))$$

Taking the derivative of $\log(\mathcal{L}_i(\cdot))$ with respect to $\bar{\mathbf{w}}$, we have

$$\nabla_{\bar{\mathbf{w}}}(\log(\mathcal{L}_i(\cdot))) = \frac{y^{(i)}}{g(\mathbf{x}^{(i)}, \bar{\mathbf{w}})} \nabla_{\bar{\mathbf{w}}}(g(\mathbf{x}^{(i)}, \bar{\mathbf{w}})) - \frac{1-y^{(i)}}{1-g(\mathbf{x}^{(i)}, \bar{\mathbf{w}})} \nabla_{\bar{\mathbf{w}}}(g(\mathbf{x}^{(i)}, \bar{\mathbf{w}}))$$

Now

$$g(\mathbf{x}^{(i)}, \bar{\mathbf{w}}) = \frac{1}{1+e^{-a}}; \; a = (\bar{\mathbf{w}}^T \bar{\mathbf{x}}^{(i)})$$

Therefore,

$$\nabla_{\bar{\mathbf{w}}}(g(\mathbf{x}^{(i)}, \bar{\mathbf{w}})) = \frac{\partial}{\partial a}\left(\frac{1}{1+e^{-a}}\right)\frac{\partial a}{\partial \bar{\mathbf{w}}}$$

$$= \left(\frac{1}{1+e^{-a}}\right)\left(1 - \frac{1}{1+e^{-a}}\right)\bar{\mathbf{x}}^{(i)}$$

$$= g(\mathbf{x}^{(i)}, \bar{\mathbf{w}})(1-g(\mathbf{x}^{(i)}, \bar{\mathbf{w}}))\,\bar{\mathbf{x}}^{(i)}$$

This gives

$$\nabla_{\bar{\mathbf{w}}}(\log(\mathcal{L}_i(\bar{\mathbf{w}}|\mathcal{D}))) = (y^{(i)} - g(\mathbf{x}^{(i)}, \bar{\mathbf{w}}))\,\mathbf{x}^{(i)}$$

---

[1] A trial is performed whose outcome is either 'success' or 'failure'. The random variable Y is a 0/1 indicator variable and takes the value 1 for a success and 0 otherwise. $P\{Y=1\} = p$ is the probability that the result of trial is a success.
Then $P\{Y=0\} = 1-p$. This can equivalently be written as,

$P\{Y=y\} = p^y(1-p)^{1-y}; \; y \in [0, 1]$

If Y is Bernoulli, its expected value and variance are [6]

$\mathbb{E}[Y] = p, \; \mathrm{Var}[Y] = p(1-p)$

Considering all training examples, we get,

$$\nabla_{\bar{\mathbf{w}}}(\log(\mathcal{L}(\bar{\mathbf{w}}|\mathcal{D}))) = \sum_{i=1}^{N} (y^{(i)} - g(\mathbf{x}^{(i)}, \bar{\mathbf{w}}))\, \mathbf{x}^{(i)} \tag{3.84a}$$

Note that analytical solution for the optimal $\bar{\mathbf{w}}$ cannot be obtained in this case; so we need to resort to iterative optimization methods, the most commonly employed being that of *gradient descent* (Section 3.6). We get the updated equation (refer to Eqn (3.79)) for *gradient ascent* (maximization problem):

$$\bar{\mathbf{w}} \leftarrow \bar{\mathbf{w}} + \eta \sum_{i=1}^{N} (y^{(i)} - g(\bar{\mathbf{x}}^{(i)}, \bar{\mathbf{w}}))\, \mathbf{x}^{(i)} \tag{3.84b}$$

Batch training and incremental training procedures for gradient methods will be discussed in detail in Chapter 5.

Once training is complete and we have the final $\mathbf{w}$ and w_0, during testing, given $\mathbf{x}^{(k)}$, we calculate $y^{(k)} = g(\mathbf{x}^{(k)}, \mathbf{w}, w_0)$, and we choose Class 1 if $y^{(k)} > 0.5$ and choose Class 2 otherwise. Instead of 0.5, the cut-off can also be chosen to maximize overall accuracy. The overall accuracy is computed for various values of the cut-off value and the one that yields maximum accuracy is chosen.

The procedure given above for a two-class discrimination problem can easily be extended to a multi-class discrimination problem [6].

In Chapter 5, we will discuss the use of neural networks for classification problems. We will use sum-of-error-squares minimization criterion for logistic regression to solve multi-class discrimination problems.

In a general multi-class problem, we have M classes denoted as y_q; $q = 1, \ldots, M$, and input instance belongs to one and exactly one of them. In many cases, the multi-class discrimination problem can be handled using two-class discriminant functions. As we will see in Section 4.9, we can view an M-class classification problem as M two-class problems. Training examples belonging to class y_k are positive instances, and the examples of all other classes y_q; $q \neq k$, are the negative instances for the classification algorithm. An alternative scheme is *pairwise* classification of the M classes. This scheme requires construction of $M(M-1)/2$ two-class classifiers. Section 4.9 gives the details.

3.8 FISHER'S LINEAR DISCRIMINANT AND THRESHOLDING FOR CLASSIFICATION

The primary objective of Fisher's Linear Discriminant (FLD) is to perform dimensionality reduction while preserving as much of the class discriminatory information as possible. As we shall see in this section, FLD provides a specific choice of direction for the projection of n-dimensional data down to one dimension. We shall also see that despite the name, FLD is not a discriminant (classification) function; however, it can be used for classification problems.

3.8.1 Fisher's Linear Discriminant

Consider the problem of projecting the data from n-dimensional space onto a line. Of course, even if the samples formed well-separated compact clusters in n-dimensional space, projection onto an arbitrary line will usually produce a confused mixture of samples from all the classes and thus result in poor recognition performance. However, by moving the line around, we might be able to find an orientation for which the projected samples are well separated. Once we achieve this goal, the problem reduces to classification of one-dimensional data.

Suppose that we have a dataset of N samples $\mathbf{X} = \mathbf{x}^{(i)}$; $i = 1, \ldots, N$; $\mathbf{x} = [x_1\ x_2\ \cdots\ x_n]^T$. N_1 is the subset $\mathbf{X}_1$ labeled Class y_1, and N_2 is the subset $\mathbf{X}_2$ labeled Class y_2: $\mathbf{X}_1 = \mathbf{x}^{(l)}$, $l = 1, \ldots, N_1$; $\mathbf{X}_2 = \mathbf{x}^{(k)}$, $k = 1, \ldots, N_2$; $N_1 + N_2 = N$ and $\mathbf{X}_1 \cup \mathbf{X}_2 = \mathbf{X}$.

If we form a linear combination of the components of $\mathbf{x}$, we obtain

$$z = w_1 x_1 + w_2 x_2 + \cdots + w_n x_n = \mathbf{w}^T \mathbf{x} \tag{3.85}$$

where $\mathbf{w}$ is some $n \times 1$ vector $[w_1\ w_2\ \cdots\ w_n]^T$. This gives us a corresponding set of one-dimensional samples $Z = z^{(i)}$; $i = 1, \ldots, N$; divided into subsets $Z_1 = z^{(l)}$; $l = 1, \ldots, N_1$, and $Z_2 = z^{(k)}$; $k = 1, \ldots, N_2$. Z_1 samples are labeled Class y_1 and Z_2 samples are labeled Class y_2 (Fig. 3.8).

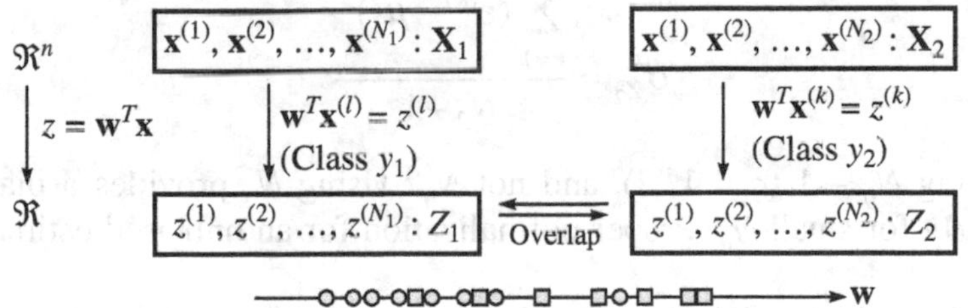

Figure 3.8 Many-to-one mapping

Geometrically, each $z^{(i)}$ is the projection of the corresponding $\mathbf{x}^{(i)}$ onto a line in the direction of $\mathbf{w}$ (Fig. 3.9). The magnitude of $\mathbf{w}$ is of no real significance because it merely scales $z^{(i)}$. The direction of $\mathbf{w}$ is important, however.

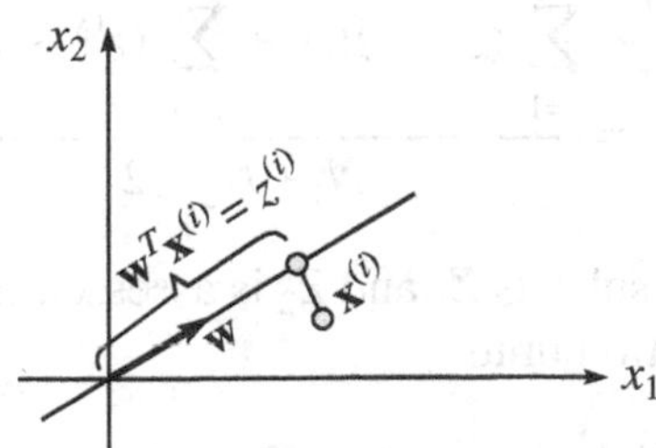

Figure 3.9 Projection of **x** onto a line in two-dimensional space

The goal is to search for the direction $\mathbf{w}$ that results in well-separated univariate subsets Z_1 and Z_2 of the multivariate projected data subsets $\mathbf{X}_1$ and $\mathbf{X}_2$. Of all possible lines, we would like to select the one that maximizes the separability of the scalars $z^{(i)}$; $i = 1, \ldots, N$. In order to find a good projection vector $\mathbf{w}$, we need to define a *measure* of separation between the projections.

The means of data subsets Z_1 and Z_2 are given by

$$\hat{\mu}_1 = \frac{1}{N_1} \sum_{l=1}^{N_1} z^{(l)}; \hat{\mu}_2 = \frac{1}{N_2} \sum_{k=1}^{N_2} z^{(k)} \tag{3.86}$$

We could then choose the distance between the means of data subsets Z_1 and Z_2 as our objective function.

$$\text{Separation} = |\hat{\mu}_1 - \hat{\mu}_2| \tag{3.87}$$

However, the distance between the means $\hat{\mu}_1$ and $\hat{\mu}_2$ is not a very good measure since it does not take into account the variance within the classes. The solution proposed by Fisher is to maximize a function that represents difference between the means, normalized by a measure of the *within-class variance*.

For each class, the variance is given by

$$\sigma_{Z_1}^2 = \frac{\sum_{l=1}^{N_1} (z^{(l)} - \hat{\mu}_1)^2}{N_1 - 1} \tag{3.88a}$$

$$\sigma_{Z_2}^2 = \frac{\sum_{k=1}^{N_2} (z^{(k)} - \hat{\mu}_2)^2}{N_2 - 1} \tag{3.88b}$$

Why are we using $N_q - 1$ ($q = 1, 2$), and not N_q? Using N_q provides a biased estimation of variance, particularly for small N_q. Proper normalization for an unbiased estimator is done using $N_q - 1$ [4].

Within-class variance in data Z is the *pooled* estimate of variance about the means:

$$\sigma_{\text{pooled}}^2 = \frac{(N_1 - 1)\,\sigma_{Z_1}^2 + (N_2 - 1)\,\sigma_{Z_2}^2}{N_1 + N_2 - 2}$$

$$= \frac{\sum_{l=1}^{N_1} (z^{(l)} - \hat{\mu}_1)^2 + \sum_{k=1}^{N_2} (z^{(k)} - \hat{\mu}_2)^2}{N_1 + N_2 - 2} \tag{3.89}$$

The separation of the univariate subsets Z_1 and Z_2 is assessed in terms of difference between the means expressed in standard deviation units:

$$\text{Normalized separation} = \frac{|\hat{\mu}_1 - \hat{\mu}_2|}{\sqrt{\sigma_{\text{pooled}}^2}} \tag{3.90}$$

Therefore, we will be looking for projection where examples from the same class are projected very close to each other (small σ_{pooled}), and at the same time, the means of the projected data belonging to different classes are as farther apart as possible.

In order to define the optimum projection **w**, we need to express normalized separation as an explicit function of **w** and maximize it (straightforward but we need linear algebra and calculus).

$$\hat{\mu}_1 = \frac{1}{N_1}\sum_{l=1}^{N_1} z^{(l)} = \frac{1}{N_1}\sum_{l=1}^{N_1} \mathbf{w}^T\mathbf{x}^{(l)}$$

$$= \mathbf{w}^T\left(\frac{1}{N_1}\sum_{l=1}^{N_1} \mathbf{x}^{(l)}\right) = \mathbf{w}^T\boldsymbol{\mu}_1 \tag{3.91a}$$

where

$$\boldsymbol{\mu}_1 = \frac{1}{N_1}\sum_{l=1}^{N_1} \mathbf{x}^{(l)}$$

= mean of the n-dimensional data in subset $\mathbf{X}_1$ labeled Class y_1.

Similarly,

$$\hat{\mu}_2 = \frac{1}{N_2}\sum_{k=1}^{N_2} z^{(k)} = \frac{1}{N_2}\sum_{k=1}^{N_2} \mathbf{w}^T\mathbf{x}^{(k)}$$

$$= \mathbf{w}^T\left(\frac{1}{N_2}\sum_{k=1}^{N_2} \mathbf{x}^{(k)}\right) = \mathbf{w}^T\boldsymbol{\mu}_2 \tag{3.91b}$$

where

$$\boldsymbol{\mu}_2 = \frac{1}{N_2}\sum_{k=1}^{N_2} \mathbf{x}^{(k)}$$

= mean of the n-dimensional data in subset $\mathbf{X}_2$ labeled Class y_2.

Also,

$$\sum_{l=1}^{N_1} (z^{(l)} - \hat{\mu}_1)^2 = \sum_{l=1}^{N_1} (\mathbf{w}^T\mathbf{x}^{(l)} - \mathbf{w}^T\boldsymbol{\mu}_1)^2$$

$$= \sum_{l=1}^{N_1} (\mathbf{w}^T\mathbf{x}^{(l)} - \mathbf{w}^T\boldsymbol{\mu}_1)(\mathbf{w}^T\mathbf{x}^{(l)} - \mathbf{w}^T\boldsymbol{\mu}_1)^T$$

$$= \sum_{l=1}^{N_1} \mathbf{w}^T(\mathbf{x}^{(l)} - \boldsymbol{\mu}_1)(\mathbf{x}^{(l)} - \boldsymbol{\mu}_1)^T\mathbf{w}$$

$$= \mathbf{w}^T\left[\sum_{l=1}^{N_1} (\mathbf{x}^{(l)} - \boldsymbol{\mu}_1)(\mathbf{x}^{(l)} - \boldsymbol{\mu}_1)^T\right]\mathbf{w} \tag{3.92a}$$

Similarly,

$$\sum_{k=1}^{N_2} (z^{(k)} - \hat{\mu}_2)^2 = \mathbf{w}^T \left[\sum_{k=1}^{N_2} (\mathbf{x}^{(k)} - \boldsymbol{\mu}_2)(\mathbf{x}^{(k)} - \boldsymbol{\mu}_2)^T \right] \mathbf{w} \tag{3.92b}$$

Thus,

$$\sigma^2_{\text{pooled}} = \frac{\sum_{l=1}^{N_1} (z^{(l)} - \hat{\mu}_1)^2 + \sum_{k=1}^{N_2} (z^{(k)} - \hat{\mu}_2)^2}{N_1 + N_2 - 2}$$

$$= \frac{\mathbf{w}^T \left[\sum_{l=1}^{N_1} (\mathbf{x}^{(l)} - \boldsymbol{\mu}_1)(\mathbf{x}^{(l)} - \boldsymbol{\mu}_1)^T \right] \mathbf{w} + \mathbf{w}^T \left[\sum_{k=1}^{N_1} (\mathbf{x}^{(k)} - \boldsymbol{\mu}_2)(\mathbf{x}^{(k)} - \boldsymbol{\mu}_2)^T \right] \mathbf{w}}{N_1 + N_2 - 2}$$

$$= \mathbf{w}^T \left[\frac{\sum_{l=1}^{N_1} (\mathbf{x}^{(l)} - \boldsymbol{\mu}_1)(\mathbf{x}^{(l)} - \boldsymbol{\mu}_1)^T + \sum_{k=1}^{N_2} (\mathbf{x}^{(k)} - \boldsymbol{\mu}_2)(\mathbf{x}^{(k)} - \boldsymbol{\mu}_2)^T}{N_1 + N_2 - 2} \right] \mathbf{w}$$

$$= \mathbf{w}^T \left[\frac{(N_1 - 1)\, \boldsymbol{\Sigma}_{\mathbf{x}_1} + (N_2 - 1)\, \boldsymbol{\Sigma}_{\mathbf{x}_2}}{N_1 + N_2 - 2} \right] \mathbf{w} \tag{3.93a}$$

$$= \mathbf{w}^T \boldsymbol{\Sigma}_{\text{pooled}}\, \mathbf{w};\ \boldsymbol{\Sigma}_{\text{pooled}} \text{ is the covariance matrix} \tag{3.93b}$$

where

$$\boldsymbol{\Sigma}_{\mathbf{x}_1} = \frac{\sum_{l=1}^{N_1} (\mathbf{x}^{(l)} - \boldsymbol{\mu}_1)(\mathbf{x}^{(l)} - \boldsymbol{\mu}_1)^T}{N_1 - 1} \tag{3.93c}$$

$$\boldsymbol{\Sigma}_{\mathbf{x}_2} = \frac{\sum_{k=1}^{N_2} (\mathbf{x}^{(k)} - \boldsymbol{\mu}_2)(\mathbf{x}^{(k)} - \boldsymbol{\mu}_2)^T}{N_2 - 1} \tag{3.93d}$$

Thus, Fisher's linear discriminant is defined as the linear function $\mathbf{w}^T\mathbf{x}$ that maximizes the criterion

$$J(\mathbf{w}) = \frac{|\hat{\mu}_1 - \hat{\mu}_2|}{\sqrt{\sigma^2_{\text{pooled}}}} = \frac{\mathbf{w}^T(\boldsymbol{\mu}_1 - \boldsymbol{\mu}_2)}{\sqrt{\mathbf{w}^T \boldsymbol{\Sigma}_{\text{pooled}}\, \mathbf{w}}} \tag{3.94}$$

$\mathbf{w}$ can be found by solving the equation based on the first derivative of $J(\mathbf{w})$.

$$\frac{\partial}{\partial \mathbf{w}}[J(\mathbf{w})] = \frac{(\boldsymbol{\mu}_1 - \boldsymbol{\mu}_2)}{\sqrt{\mathbf{w}^T \boldsymbol{\Sigma}_{\text{pooled}}\, \mathbf{w}}} - \tfrac{1}{2}\, \mathbf{w}^T (\boldsymbol{\mu}_1 - \boldsymbol{\mu}_2) \frac{2\boldsymbol{\Sigma}_{\text{pooled}}\, \mathbf{w}}{(\mathbf{w}^T \boldsymbol{\Sigma}_{\text{pooled}}\, \mathbf{w})^{3/2}} = \mathbf{0} \tag{3.95}$$

Simplification of this equation gives

$$(\boldsymbol{\mu}_1 - \boldsymbol{\mu}_2) = \left[\frac{\mathbf{w}^T(\boldsymbol{\mu}_1 - \boldsymbol{\mu}_2)}{\mathbf{w}^T \boldsymbol{\Sigma}_{\text{pooled}}\, \mathbf{w}}\right] \boldsymbol{\Sigma}_{\text{pooled}}\, \mathbf{w} \tag{3.96}$$

Multiplication by the inverse of the matrix $\boldsymbol{\Sigma}_{\text{pooled}}$ on the two sides, gives

$$\boldsymbol{\Sigma}_{\text{pooled}}^{-1} (\boldsymbol{\mu}_1 - \boldsymbol{\mu}_2) = \left[\frac{\mathbf{w}^T(\boldsymbol{\mu}_1 - \boldsymbol{\mu}_2)}{\mathbf{w}^T \boldsymbol{\Sigma}_{\text{pooled}}\, \mathbf{w}}\right] \mathbf{w} \tag{3.97}$$

Since $\dfrac{\mathbf{w}^T(\boldsymbol{\mu}_1 - \boldsymbol{\mu}_2)}{\mathbf{w}^T \boldsymbol{\Sigma}_{\text{pooled}}\, \mathbf{w}}$ is a real number, it can be ignored (we are looking for direction of $\mathbf{w}$; the magnitude is not important).

Therefore,

$$\mathbf{w} = \boldsymbol{\Sigma}_{\text{pooled}}^{-1} (\boldsymbol{\mu}_1 - \boldsymbol{\mu}_2) \tag{3.98}$$

The linear function $\mathbf{w}^T\mathbf{x}$ maximizes the distance between the projected means, normalized by the within-class variance.

This is known as Fisher's Linear Discriminant (FLD). Despite the name, FLD is not a discriminant (classification) function, but provides a specific choice of direction for projection of n-dimensional data down to one dimension. FLD is thus a tool that performs dimensionality reduction while preserving as much of the class discriminatory information as possible.

3.8.2 Thresholding

Can Fisher's Linear Discriminant (FLD) also be used for classification? Clearly it can. The classification problem has been converted from an n-dimensional problem to a hopefully more manageable one-dimensional problem. This mapping is many-to-one, and in theory cannot possibly obtain the minimum achievable error rate if we have a very large training set. In general, one is willing to sacrifice some of the theoretically achievable performance for the advantages of working in one dimension.

All that remains is to find the *threshold* τ, that is, the cut-point along one-dimensional subspace of projected data $z^{(i)} = \mathbf{w}^T\mathbf{x}^{(i)}$; $i = 1, \ldots, N$, so that we classify a new data point $\mathbf{x}^{(N+1)}$ as belonging to Class y_1 if $z^{(N+1)} = \mathbf{w}^T\mathbf{x}^{(N+1)} \geq \tau$, and classify it as belonging to Class y_2 otherwise.

$\mathbf{w}$ can be calculated using Eqn (3.98), but Fisher's criterion does not give us the optimum τ. Then, how can we find the threshold if we want to use FLD for classification problems? It makes sense to choose the cut-point that empirically minimizes training error. This is something that has been found to work well in practice.

If the prior probabilities $P(y_1)$ and $P(y_2)$ are equal ($P(y_1) \simeq N_1/N$; $P(y_2) \simeq N_2/N$), then the best choice for the cut-point is the point midway between the projected means. That is (refer to Eqn (3.91) and Eqn (3.98)),

$$\begin{aligned} \tau &= \frac{\hat{\mu}_1 + \hat{\mu}_2}{2} \\ &= \tfrac{1}{2}\, \mathbf{w}^T (\boldsymbol{\mu}_1 + \boldsymbol{\mu}_2) \\ &= \tfrac{1}{2}\, (\boldsymbol{\mu}_1 - \boldsymbol{\mu}_2)^T \, \boldsymbol{\Sigma}_{\text{pooled}}^{-1} \, (\boldsymbol{\mu}_1 + \boldsymbol{\mu}_2) \end{aligned} \tag{3.99}$$

Given a new point $\mathbf{x}^{(N+1)}$. Allocate it to Class y_1 if

$$z^{(N+1)} = \mathbf{w}^T \mathbf{x}^{(N+1)} = (\boldsymbol{\mu}_1 - \boldsymbol{\mu}_2)^T \, \boldsymbol{\Sigma}_{\text{pooled}}^{-1} \, \mathbf{x}^{(N+1)} \geq \tau; \tag{3.100}$$

Otherwise, allocate $\mathbf{x}^{(N+1)}$ to Class y_2.

If the prior probabilities are not equal, moving the cut-point toward the *smaller* class will improve the error rate. By trial-and-error, we then choose the cut-point to minimize misclassification error over the training data. FLD as a classifier has a good track record.

Finally, for the sake of completeness, we underline that FLD can be extended to where there are more than two classes. In this case, the algorithm is called *multiclass* FLD [4].

3.9 MINIMUM DESCRIPTION LENGTH PRINCIPLE

Recall from Chapter 2 the discussion on Occam's razor, a popular inductive bias that can be summarized as 'choose the simplest hypothesis consistent with the data.' In that chapter, we also discussed 'complexity versus accuracy' issue in machine learning using arguments based on over fitting, bias-variance dilema, PAC-theory and VC-theory.

Another framework to discuss hypothesis complexity versus prediction accuracy issue is offered by the *Minimum Description Length* (MDL) principle. Though conceptually quite different, this principle results in a formalism which is identical to the Bayesian one [52].

Suppose there is a 'sender' who desires to transmit output values (y) corresponding to the learning cases (training data $\mathcal{D}$) to a 'receiver' who is already aware of the input values ($\mathbf{x}$), with the use of a message of the shortest possible length (wherein length of the message is measured by the number of bits, for instance). There are many options for this transmission. One approach would be simply to transmit a suitable encoded form of the output values of each case in $\mathcal{D}$ using some fixed coding scheme. In another approach, an accurate model able to recompute the output values from the inputs (the model is expected to be a complex one that makes no errors) is transmitted. Still another approach is to send an approximate (less complex) model together with error corrections. The MDL principle states that the best model for the learning sample is the one which minimizes the length of the message which has to be sent to the receiver:

$$\underbrace{K(h,\mathcal{D})}_{\text{description length}} = \underbrace{K(h)}_{\text{Complexity}} + \underbrace{K(\mathcal{D}\text{ using } h)}_{\text{error}} \tag{3.101}$$

when $K(h)$ is the minimal number of bits necessary to describe the model to the receiver, and $K(\mathcal{D}$ using $h)$ is the minimal number of bits necessary to describe the errors made by the model on the learning sample. Intuitively, the lower is the empirical risk of the model (achieved with increasing model complexity), the lower will be the length $K(\mathcal{D}$ using $h)$; in the limit if the model perfectly fits the data $\mathcal{D}$, this term vanishes. On the other hand, the number of bits used to code a model is certainly an increasing function of the number of parameters used to represent the model and hence of its complexity. Thus the two terms making up the description length are essentially representing a trade-off between the empirical risk and the model complexity.

The two terms in Eqn (3.101) defining $K(h,\mathcal{D})$ are in theory, defined as the *Kolmogorov complexities*. These quantities are essentially non-computable and hence various approximations have been proposed in the literature leading to various practical versions of MDL principle.

3.9.1 Bayesian Perspective

The shortest description length is expected when the hypothesis h results in a precise representation of the statistical process which generated the data, and it is also expected, on an average, this hypothesis will have the ideal generalization properties.

For the discrete hypothesis space and data, we will write $P(h)$ to express the prior probability that hypothesis h holds; it may be based on the background knowledge we possess about the probability that h is the right hypothesis. In the same way, we will write $P(\mathcal{D}|h)$ to imply the probability of observing data $\mathcal{D}$ given a world wherein hypothesis h holds; and $P(\mathcal{D})$ to express the prior probability that training data $\mathcal{D}$ will be observed with no knowledge about which hypothesis holds. We are interested in the posterior probability of h, $P(h|\mathcal{D})$, that h holds given the observed data $\mathcal{D}$. Bayes formula states

$$P(h|\mathcal{D}) = \frac{P(h)\,P(\mathcal{D}|h)}{P(\mathcal{D})} \tag{3.102}$$

The *optimal hypothesis* h^* is the one yielding highest posterior probability, that is,

$$h^* = \underset{h}{arg\ max}[P(h)\,P(\mathcal{D}|h)] \tag{3.103}$$

When h^* in Eqn (3.103) is equivalently expressed in terms of $\log_2$, we have

$$h^* = \underset{h}{arg\ max}[\log_2 P(h) + \log_2 P(\mathcal{D}|h)] \tag{3.104a}$$

or alternatively, minimizing the negative of this quantity:

$$h^* = \underset{h}{arg\ max}[-\log_2 P(h) - \log_2 P(\mathcal{D}|h)] \tag{3.104b}$$

Shannor and Weaver (1949) showed that the optimal code (i.e., the code that minimizes the expected message length) assigns $(-\log_2 P(x))$ bits to encode message x. Equation (3.104b) can thus be interpreted in terms of MDL principle; linking minimum description length to Bayesian approach.

Probability distributions are, however, mostly unknown in real learning problems.

3.9.2 Entropy and Information

A particularly clear application of MDL principle is in the design of decision-tree classifiers. We now turn to a simple naive version of the principle with respect to applications to pattern recognition using *entropy-based error measures*. The concept generalizes to other machine learning applications as well.

To maintain the simplicity of things, suppose that the sequence of instances $\mathbf{x}^{(1)}, \mathbf{x}^{(2)}, \ldots, \mathbf{x}^{(N)}$ is already known to the transmitter as well as the receiver, so that we require to just transmit the categories y_q ($q = 1, \ldots, M$). Now if the y_q given in the training data are exactly the same as the predictions of the hypothesis h, then there is no requirement for transmission of any information pertaining to these instances (the receiver can compute these values once it has received the hypothesis). The description length of the classifications, keeping in mind the hypothesis known is, thus, zero. In situations when some instances are misclassified by h, then for each misclassification, we require to transmit a message that identifies the misclassified instance, as well as its precise classification. The description length then comprises transmission of the approximate hypothesis along with corrections of errors. As per the MDL principle, the most ideal hypothesis for the training sample is one which reduces the length of the message to be sent to the receiver, to the minimum.

As we shall see in Chapter 8, an *entropy-based error criterion* helps design decision trees using MDL principle. The description lengths are equivalently expressed in terms of entropy. We have also used this equivalence in discretization problems in Chapter 7.

The concept of entropy was originally developed by physicists in the context of equilibrium thermodynamics and later extended through development of statistical mechanics. It was introduced into information theory by Shannon. Here we discuss the interpretation of entropy based on information content.

An intuitive notion of information will help understand the notion of entropy. If everything about message or signal is known, there is nothing new to know and hence it carries no information. If the signal is random, i.e., some level of uncertainty is associated, then it carries some information. Thus information content is related to randomness or unpredictability of the message.

Consider a discrete source S of independent messages $s^{(1)}, s^{(2)}, \ldots, s^{(N)}$ with probability of occurrence as $P_1, P_2, \ldots, P_N$. Information carried by the event $s^{(i)}$ is given by $\log_2\left(\frac{1}{P_i}\right)$. It is a measure of randomness or unpredictability of the event. A certain event has probability 1 and hence carries no information. When we use log base 2, the unit of information is bits.

The average information from the discrete source

$$= \sum_{i=1}^{N} P_i \log_2\left(\frac{1}{P_i}\right) = -\sum_{i=1}^{N} P_i \log_2 P_i \tag{3.105}$$

It is called *information entropy*. It represents level of randomness or disorder in S. More the randomness (less the prior knowledge), more the information content, more the information entropy. In case any of the probabilities vanish, we use the fact that

$$\lim_{P \to 0} P \log_2 P = 0 \text{ to define } 0 \log_2 0 = 0 \tag{3.106}$$

Note that entropy does not depend on the messages themselves, just on their probabilities. For a given number N of equally likely messages, the entropy is maximum—we have the maximum uncertainty about the identity of the message that will be chosen. Conversely, if all P_i are 0 except one, we have the minimum entropy—we are certain as to the message that will appear.

For decision-tree classifiers (Chapter 8), minimum description length given by Eqn (3.101) has the following two components:

(i) $K(h)$ representing complexity of the hypothesis. Algorithmic complexity of a decision tree is proportional to the number of nodes and branches.

(ii) $K(\mathcal{D}$ using $h)$ representing errors made by the hypothesis on the training sample, i.e., the amount of randomness or uncertainty created by the hypothesis. For a decision tree, it could be expressed in terms of entropy (in bits) of the data $\mathcal{D}$—the weighted sum of the entropies of the data at the leaf nodes.

Thus, if the tree is grown/pruned based on the entropy criterion, there is an implicit global cost criterion that is equivalent to minimizing a measure of the general form of Eqn (3.101), providing a way of trading off hypothesis complexity for the randomness created by hypothesis (number of errors committed by the hypothesis). It might select a shorter hypothesis that makes a few errors over a larger hypothesis that perfectly classifies the training data.

Chapter

4

LEARNING WITH SUPPORT VECTOR MACHINES (SVM)

4.1 INTRODUCTION

The classical regression and Bayesian classification statistical techniques (overview given in Section 3.5) are based on the very strict assumption that probability distribution models or probability density functions are known. Unfortunately, in real-world practical situations, there is not enough information about the underlying distributions, and *distribution-free regression* or *classification* is needed that does not require knowledge of probability distributions. The only available information is the training dataset.

Under the assumption that the data follows a normal distribution, statistical techniques result in linear regression functions. For the classification problems with normally distributed classes and equal covariance matrices for corresponding classes, we get linear discriminant functions using statistical techniques.

The linear functions are extremely powerful for the regression/classification problems whenever the stated assumptions are true. Unfortunately, the classical statistical paradigm turns out to be inappropriate for many real-life problems because the underlying real-life data generation laws may typically be very far from normal distribution.

Till the 1980s, most of the data analysis and learning methods were confined to linear statistical techniques. Most of the optimal algorithms and theoretical results were available for inference of linear dependencies from data; for nonlinear ones, only greedy algorithms, local minima, and heuristic search were known.

A paradigm shift occurred in the 1980s when researchers, armed with powerful computers of the day, boldly embraced nonlinear methods of learning. The simultaneous introduction of decision trees (Chapter 8) and neural network algorithms (Chapter 5) revolutionalized the practice of pattern recognition and numeric prediction. These methods opened the possibility of efficient learning of nonlinear dependencies.

A second paradigm shift occurred in the 1990s with the introduction of the 'kernel methods'. The differences with the previous approaches are worth mentioning. Most of the earlier learning algorithms had, to a large extent, been based on heuristics or on loose analogies with natural learning systems, e.g., model of nervous systems (neural networks). They were mostly the result of creativity and extensive tuning by the designer, and the underlying reasons for their performance were not fully understood. A large part of the work was devoted to designing heuristics to avoid local minima in hypothesis search process.

With the emergence of computational learning theory (Section 2.3), new efficient representations of nonlinear functions have been discovered and used for the design of learning algorithms. This has led to the creation of powerful algorithms, whose training often amounts to optimization. In other words, they are free from local minima. The use of optimization theory marks a radical departure from the previous greedy search algorithms. In a way, researchers now have the power of nonlinear function learning together with the conceptual and computational convenience, that was, hitherto, a characteristic of linear systems. Support Vector Machine (SVM), probably, represents the best known example of this class of algorithms.

In the present state-of-the-art, no machine learning method is inherently superior to any other; it is the type of problem, prior distribution, and other information that determine which method should provide the best performance. If one algorithm seems to outperform another in a particular situation, it is a consequence of its fit to the particular problem, not the general superiority of the algorithm. Machine learning involves searching through a space of possible hypotheses to determine one that fits the observed data and any prior knowledge held by the learner. We have covered in this book, hypotheses space which is being exploited by the practitioners in data mining problems.

In the current chapter, we present the basic concepts of SVM in an easily digestible way. Applied learning algorithms for support vector classification and regression have been thoroughly explained. The support vector machine is currently considered to be the best off-the-shelf learning algorithm and has been applied successfully in various domains.

Support vector machines were originally designed for binary classification. Initial research attempts were diverted towards making several two-class SVMs to do multiclass classification. Recently, several single-shot multiclass classification algorithms appeared in literature. Our focus in this chapter will be on binary classification; multiclass problems will be reduced to binary classification problems.

Support vector regression is the natural extension of methods used for classification. We will see in this chapter that SVM regression analysis retains all the properties of SVM classifiers. It has already become a powerful technique for predictive data analysis with many applications in varied areas of study.

In Section 3.5, we observed that the practical limitation of statistical approach is the assumed initial knowledge available on the process under investigation. For the Bayes procedure, one should know the underlying probability distributions. If only the forms of underlying distributions were known, we can use the training samples to estimate the values of their parameters.

In this chapter, we shall instead assume that we know the proper forms for the *discriminant functions*, and use the samples to estimate the values of parameters of the classifier. We shall examine various procedures for determining discriminant functions; none of them requiring knowledge of the forms of underlying distributions.

We shall be concerned with linear discriminant functions which have a variety of pleasant analytical properties. As we observed in Section 3.5, they can be optimal if the underlying distributions are cooperative, such as Gaussians having equal covariance. Even when they are not optimal, we might be willing to sacrifice some performance in order to gain the advantage of their simplicity. Linear discriminant functions are relatively easy to compute and in the absence of information suggesting otherwise, linear classifiers are attractive candidates for initial, trial classifiers.

The problem of finding a linear discriminant function will be formulated as a problem of minimizing a criterion function. The obvious criterion function for classification purposes is the *misclassification error* (refer to Section 2.8). Instead of deriving discriminants based on misclassification error, we investigate related criterion function that is analytically more tractable.

4.2 LINEAR DISCRIMINANT FUNCTIONS FOR BINARY CLASSIFICATION

Let us assume two classes of patterns described by two-dimensional feature vectors (coordinates x_1 and x_2) as shown in Fig. 4.1. Each pattern is represented by vector $\mathbf{x} = [x_1\ x_2]^T \in \Re^2$. In Fig. 4.1, we have used circle to denote Class 1 patterns and square to denote Class 2 patterns. In general, patterns of each class will be characterized by random distributions of the corresponding feature vectors.

Figure 4.1 also shows a straight line separating the two classes. We can easily write the equation of the straight line in terms of the coordinates (features) x_1 and x_2 using coefficients or *weights* w_1 and w_2 and a *bias (offset)* or *threshold* term w_0, as given in Eqn (4.1). The weights determine the slope of the straight line, and the bias determines the deviation from the origin of the straight line intersections with the axes:

$$g(\mathbf{x}) = w_1 x_1 + w_2 x_2 + w_0 = 0 \tag{4.1}$$

We say that $g(\mathbf{x})$ is a *linear discriminant function* that divides (categorizes) $\Re^2$ into two *decision regions*.

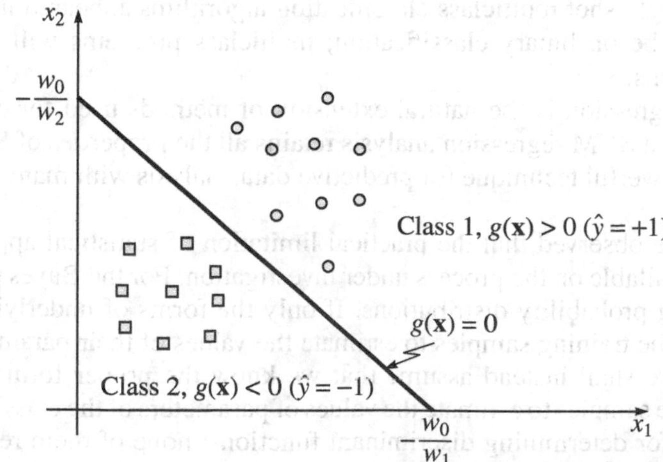

Figure 4.1 Linear discriminant function in two-dimensional space

The generalization of the linear discriminant function for an n-dimensional feature space in $\Re^n$ is straight forward:

$$g(\mathbf{x}) = \mathbf{w}^T\mathbf{x} + w_0 = 0 \tag{4.2}$$

$\mathbf{x} = [x_1\ x_2\ \ldots\ x_n]^T$ is the feature vector

$\mathbf{w} = [w_1\ w_2\ \ldots\ w_n]^T$ is a *weight vector*

w_0 = *bias* parameter

The discriminant function is now a linear n-dimensional surface, called a *hyperplane*; symbolized as $\mathcal{H}$ in the discussion that follows.

For the discriminant function of the form of Eqn (4.2), a two-category classifier implements the following decision rule:

$$\text{Decide Class 1 if } g(\mathbf{x}) > 0 \text{ and Class 2 if } g(\mathbf{x}) < 0. \tag{4.3}$$

Thus, $\mathbf{x}$ is assigned to Class 1 if the inner product $\mathbf{w}^T\mathbf{x}$ exceeds the threshold (bias) $-w_0$, and to Class 2 otherwise. If $g(\mathbf{x}) = 0$, $\mathbf{x}$ can ordinarily be assigned to any class, but in this chapter, we shall leave the assignment undefined.

Figure 4.2 shows the architecture of a typical implementation of the linear classifier. It consists of two computational units: an aggregation unit and an output unit. The aggregation unit collects the n weighted input signals $w_1x_1, w_2x_2, \ldots, w_nx_n$ and sums them together. Note that the summation also has a bias term—a constant input $x_0 = 1$ with a weight of w_0. This sum is then passed on to the output unit, which is a step filter that returns –1 if its input is negative and +1 if its input is positive. In other words, the step filter implements the sign function:

$$\hat{y} = sgn\left(\sum_{j=1}^{n} w_j\, x_j + w_0\right) \tag{4.4a}$$

$$= sgn\,(\mathbf{w}^T\mathbf{x} + w_0) = sgn(g(\mathbf{x})) \tag{4.4b}$$

The *sgn* function extracts the appropriate pattern label from the decision surface $g(\mathbf{x})$. This means that linear classifier implemented in Fig. 4.2 represents a *decision function* of the form,

$$\hat{y} = sgn\,(g(\mathbf{x})) = sgn\,(\mathbf{w}^T\mathbf{x} + w_0) \tag{4.5}$$

The values ±1 in output unit are not unique; only change of sign is important. ±1 is taken for computational convenience.

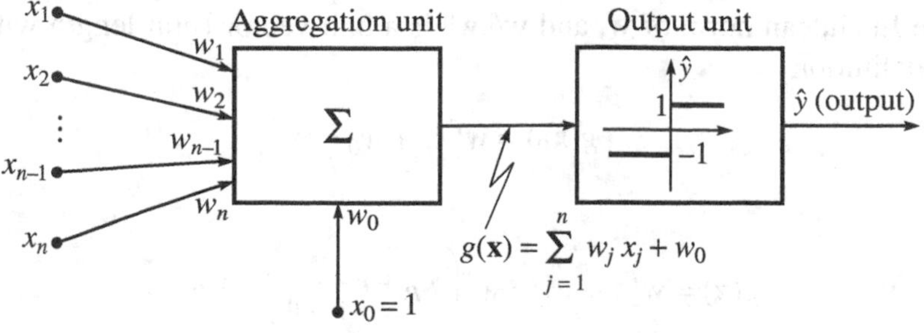

Figure 4.2 A simple linear classifier

If $\mathbf{x}^{(1)}$ and $\mathbf{x}^{(2)}$ are two points on the decision hyperplane, then the following is valid:

$$\mathbf{w}^T\mathbf{x}^{(1)} + w_0 = \mathbf{w}^T\mathbf{x}^{(2)} + w_0 = 0$$

This implies that

$$\mathbf{w}^T(\mathbf{x}^{(1)} - \mathbf{x}^{(2)}) = 0$$

The difference $(\mathbf{x}^{(1)} - \mathbf{x}^{(2)})$ obviously lies on the decision hyperplane for any $\mathbf{x}^{(1)}$ and $\mathbf{x}^{(2)}$. The scalar product is equal to zero, meaning that the weights vector $\mathbf{w}$ is normal (perpendicular) to the decision hyperplane. Without changing the normal vector $\mathbf{w}$, varying w_0 moves the hyperplane parallel to itself. Note also that $\mathbf{w}^T\mathbf{x} + w_0 = 0$ has an inherent degree of freedom. We can rescale the hyperplane to $K\mathbf{w}^T\mathbf{x} + Kw_0 = 0$ for $K \in \Re^+$ (positive real numbers) without changing the hyperplane. Geometry for $n = 2$ with $w_1 > 0$, $w_2 > 0$ and $w_0 < 0$ is shown in Fig. 4.3.

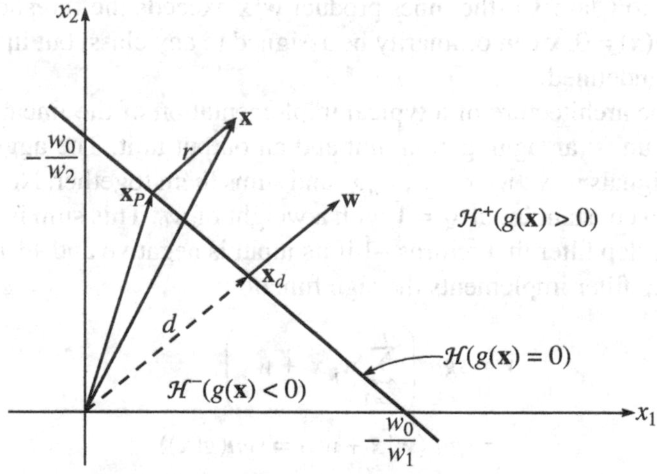

Figure 4.3 Linear decision boundary between two classes

The location of any point $\mathbf{x}$ may be considered relative to the hyperplane $\mathcal{H}$. Defining $\mathbf{x}_P$ as the normal projection of $\mathbf{x}$ onto $\mathcal{H}$ (shown in Fig. 4.3), we may decompose $\mathbf{x}$ as,

$$\mathbf{x} = \mathbf{x}_P + r\frac{\mathbf{w}}{\|\mathbf{w}\|} \tag{4.6}$$

where $\|\mathbf{w}\|$ is the Euclidean norm of $\mathbf{w}$, and $\mathbf{w}/\|\mathbf{w}\|$ is a unit vector (unit length with direction that of $\mathbf{w}$). Since by definition

$$g(\mathbf{x}_P) = \mathbf{w}^T\mathbf{x}_P + w_0 = 0$$

it follows that

$$g(\mathbf{x}) = \mathbf{w}^T\mathbf{x} + w_0 = \mathbf{w}^T\left(\mathbf{x}_P + r\frac{\mathbf{w}}{\|\mathbf{w}\|}\right) + w_0$$

$$= \mathbf{w}^T \mathbf{x}_P + w_0 + \frac{\mathbf{w}^T r\, \mathbf{w}}{\|\mathbf{w}\|}$$

$$= r \frac{\mathbf{w}^T \mathbf{w}}{\|\mathbf{w}\|} = r \frac{\|\mathbf{w}\|^2}{\|\mathbf{w}\|} = r\|\mathbf{w}\|$$

or
$$r = \frac{g(\mathbf{x})}{\|\mathbf{w}\|} \tag{4.7}$$

In other words, $|g(\mathbf{x})|$ is a measure of the Euclidean distance of the point $\mathbf{x}$ from the decision hyperplane $\mathcal{H}$. If $g(\mathbf{x}) > 0$, we say that the point $\mathbf{x}$ is on the *positive side* of the hyperplane, and if $g(\mathbf{x}) < 0$, we say that point $\mathbf{x}$ is on the *negative side* of the hyperplane. When $g(\mathbf{x}) = 0$, the point $\mathbf{x}$ is on the hyperplane $\mathcal{H}$.

In general, the hyperplane $\mathcal{H}$ divides the feature space into two half-spaces: decision region $\mathcal{H}^+$ (positive side of hyperplane $\mathcal{H}$) for Class 1 ($g(\mathbf{x}) > 0$) and region $\mathcal{H}^-$ (negative side of hyperplane $\mathcal{H}$) for Class 2 ($g(\mathbf{x}) < 0$). The assignment of vector $\mathbf{x}$ to $\mathcal{H}^+$ or $\mathcal{H}^-$ can be implemented as,

$$\mathbf{w}^T \mathbf{x} + w_0 \begin{cases} > 0 \text{ if } \mathbf{x} \in \mathcal{H}^+ \\ = 0 \text{ if } \mathbf{x} \in \mathcal{H} \\ < 0 \text{ if } \mathbf{x} \in \mathcal{H}^- \end{cases} \tag{4.8}$$

The perpendicular distance d from the coordinates origin to the hyperplane $\mathcal{H}$ is given by $w_0/\|\mathbf{w}\|$, as is seen below.

$$g(\mathbf{x}_d) = \mathbf{w}^T \mathbf{x}_d + w_0 = 0;\ \mathbf{x}_d = d \frac{\mathbf{w}}{\|\mathbf{w}\|}$$

Therefore,

$$d \frac{\mathbf{w}^T \mathbf{w}}{\|\mathbf{w}\|} + w_0 = 0;\ d \frac{\|\mathbf{w}\|^2}{\|\mathbf{w}\|} = -w_0;\ d = \frac{-w_0}{\|\mathbf{w}\|} \tag{4.9}$$

The origin is on the negative side of $\mathcal{H}$ if $w_0 < 0$, and if $w_0 > 0$, the origin is on the positive side of $\mathcal{H}$. If $w_0 = 0$, the hyperplane passes through the origin.

Geometry for $n = 3$ is shown in Fig. 4.4.

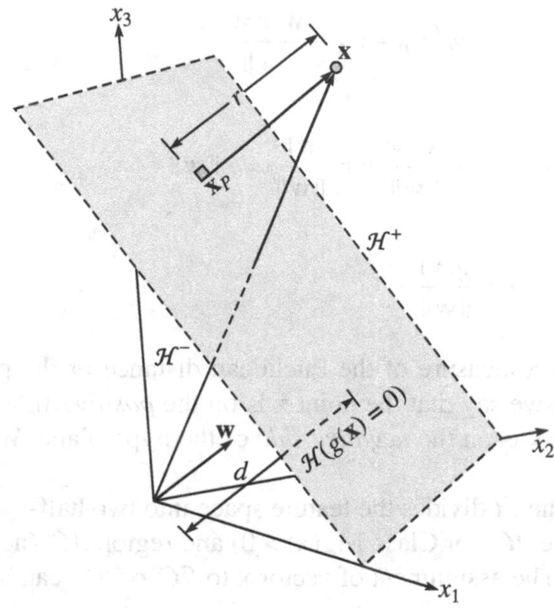

Figure 4.4 Hyperplane $\mathcal{H}$ separates the feature space into two half-space $\mathcal{H}^+(g(x) > 0)$ and $\mathcal{H}^-(g(x) < 0)$

4.3 PERCEPTRON ALGORITHM

Let us assume that we have a set of N samples $\mathbf{x}^{(1)}, \mathbf{x}^{(2)}, \ldots, \mathbf{x}^{(N)}$; some labeled Class 1 and some labeled Class 2:

$$\mathcal{D}: \{(\mathbf{x}^{(1)}, y^{(1)}), \ldots, (\mathbf{x}^{(N)}, y^{(N)})\} \tag{4.10}$$

with $\mathbf{x}^{(i)} \in \Re^n$, and $y^{(i)} \in \{+1, -1\}$.

We want to use these samples to determine the weights $\mathbf{w}$ and w_0 in a linear discriminant function

$$g(\mathbf{x}) = \mathbf{w}^T\mathbf{x} + w_0 \tag{4.11}$$

Let us say there is a reason behind the belief that there is a solution for which the likelihood of error is quite low. This leads to the desire to seek a weight vector that correctly classifies all the samples. If there does exist such a weight vector, the samples are said to be *linearly separable* (Fig. 4.5a).

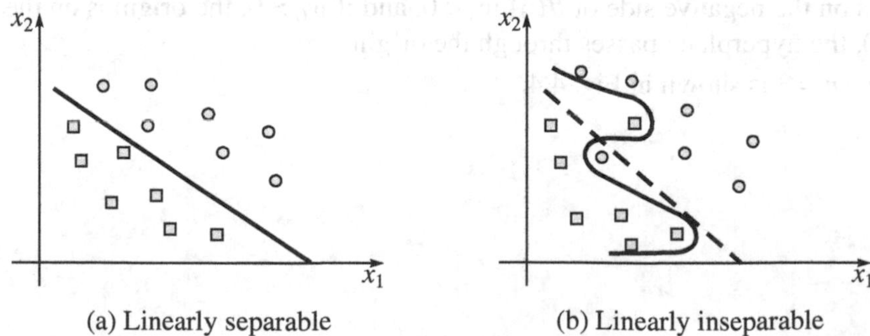

Figure 4.5

Mostly, the classes are *overlapped* and the genuine separability is given by *nonlinear* decision boundaries (Fig. 4.5b). The samples in these cases are said to *linearly inseparable*.

Consider now the problem of constructing a *criterion function* for solving the linear classification problem to determine the weights **w** and w_0. The criterion function, which is the most obvious for the purpose of classification, is the *number of samples misclassified* by the weight vector. But since this function is stepwise constant, it is naturally a weak candidate for gradient search. The *Perceptron Algorithm* seems to be a better alternative for this criterion function.

Rosenblatt (1950s) proposed the machine—the perceptron—whose architecture encodes the structure of a linear discriminant function (Fig. 4.3). Although it seemed initially promising, it was quickly proved that perceptrons could not be trained to recognize many classes of problems. Inspite of limitations, the perceptron is an interesting machine because, even in such a simple system, we can find most of the central concepts that we will need for the theory of Neural Networks (discussed in the next chapter) and Support Vector Machines (discussed in the present chapter).

In the following, we describe *perceptron training algorithm* and its limitations for classification problems.

A perceptron takes a vector of real-valued inputs x_j ; $j = 1, \ldots, n$, calculates a linear combination of these inputs $\left(\sum_{j=1}^{n} w_j x_j = \mathbf{w}^T \mathbf{x} \right)$, then outputs a +1 if the result is greater than the threshold $(-w_0)$ and –1 if the result is less than the threshold (Fig. 4.3). The perceptron algorithm tests the decision function $g(\mathbf{x})$ on each element in the training set, and if the test fails, it adjusts the free parameters **w** and w_0 incrementally. This process continues until all the elements of the training set are perfectly classified.

At the heart of the algorithm are two *update rules* (Here $y^{(i)}$ is the target output for the current training sample $\mathbf{x}^{(i)}(i = 1, \ldots, N)$, and $\hat{y}^{(i)}$ is the output generated by the perceptron):

$$\mathbf{w} \leftarrow \mathbf{w} + \Delta\mathbf{w} = \mathbf{w} + \eta\, y^{(i)}\, \mathbf{x}^{(i)} \tag{4.12a}$$

$$w_0 \leftarrow w_0 + \Delta w_0 = w_0 + \eta\, y^{(i)}\, R^2 \tag{4.12b}$$

The quantity R is called the *radius* of the training data and can be considered the radius of the hypersphere centered at the origin of our coordinate system that encloses all the points of the dataset. In our case where the data universe is $\Re^n$, this is simply the position vector length of the training set point located farthest from the origin [53]:

$$R \leftarrow \max_{1 \le i \le N} \| \mathbf{x}^{(i)} \| \tag{4.13}$$

where $\| \cdot \|$ stands for the Euclidean norm.

The quantity η is a positive scale factor $(0 < \eta \le 1)$ that sets the step size. Called the *learning rate*, it controls the convergence speed of the search heuristic. Note that if η is too small, convergence is needlessly slow, whereas if η is too large, the correction process will overshoot, and can even diverge. So choice of η is crucial.

The intuition behind the update rules is that if $\hat{y}^{(i)} = y^{(i)}$ (the point is correctly classified), there is no weights update ($\Delta\mathbf{w} = 0$, $\Delta w_0 = 0$). In case of a misclassified point ($\hat{y}^{(i)} \neq y^{(i)}$), the rules attempt to correct the position of the decision surface in such a way that the point is no longer misclassified.

One may come across different expressions of the weight changes $\Delta\mathbf{w}$, Δw_0 in the literature compared to the ones given in (4.12). However, the conceptual framework of the weight updates remains the same. For example, the following update rules are probably most commonly employed:

$$\begin{aligned} \mathbf{w} &\leftarrow \mathbf{w} + \Delta\mathbf{w} = \mathbf{w} + \eta\,(y^{(i)} - \hat{y}^{(i)})\,\mathbf{x}^{(i)} \\ w_0 &\leftarrow w_0 + \Delta w_0 = w_0 + \eta\,(y^{(i)} - \hat{y}^{(i)}) \end{aligned} \tag{4.14}$$

Here, we describe the conceptual framework with respect to the rules given in (4.12).

Consider a training dataset point $(\mathbf{x}^{(i)}, y^{(i)})$ with $y^{(i)} = +1$. If this point is correctly classified by the decision surface ($\hat{y}^{(i)} = +1$), $\Delta\mathbf{w}$ and Δw_0 are zero and no weights are updated. Suppose the perceptron outputs a –1, when the target output is +1. The update rule (4.12a) attempts to correct this misclassification by *rotating* the decision surface in the direction of $\mathbf{x}^{(i)}$. The rotation is accomplished by *adding* a scaled version of $\mathbf{x}^{(i)}$ to the normal vector $\mathbf{w}$ (refer to Fig. 4.6). An analogous computation can be performed for a misclassified point with a target value of –1. In this case, the adjustment term will be *subtracted* from the normal vector, causing the rotation in the opposite direction.

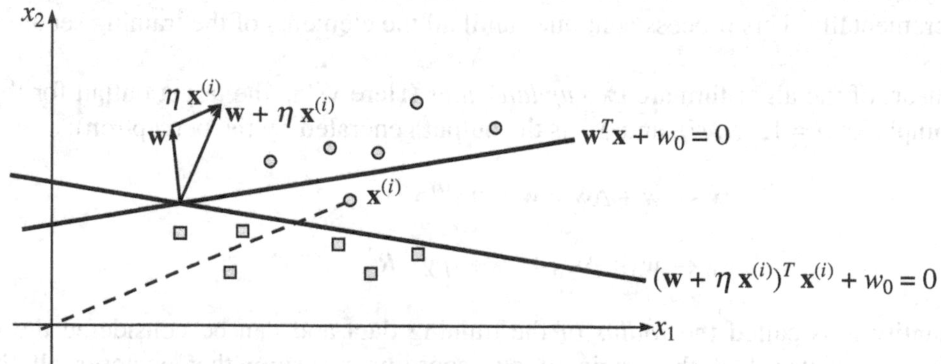

Figure 4.6 The point $\mathbf{x}^{(i)}$ is no more misclassified after the rotation of hyperplane

The second update rule (4.12b) attempts to correct a misclassification by *translating* the decision surface. We may rewrite the rule as,

$$-w_0 \leftarrow -w_0 - \eta\, y^{(i)}\, R^2$$

or

$$b \leftarrow b - \eta\, y^{(i)}\, R^2 \tag{4.15}$$

For a misclassified point with $y^{(i)} = +1$, b is reduced and we need to translate the decision surface in the direction opposite to the normal vector (refer to Fig. 4.7).

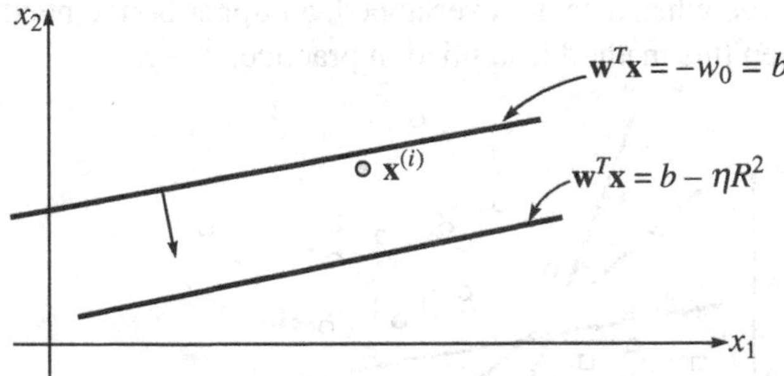

Figure 4.7 Translation of hyperplane

The overall effect of the two update rules is demonstrated in Fig. 4.8. A square point is misclassified at time step *t*. Decision surface is rotated and translated at time step t+1. The square point is now correctly classified but overcompensation results in misclassification of a circle point. This misclassification in turn forces the perceptron algorithm to apply the update rules in the opposite direction during the next iteration, leading to a decision surface at time step t+2. This process continues. If one *episode* is completed (all data used), another episode starts from the first sample. The algorithm terminates when all the points are classified correctly.

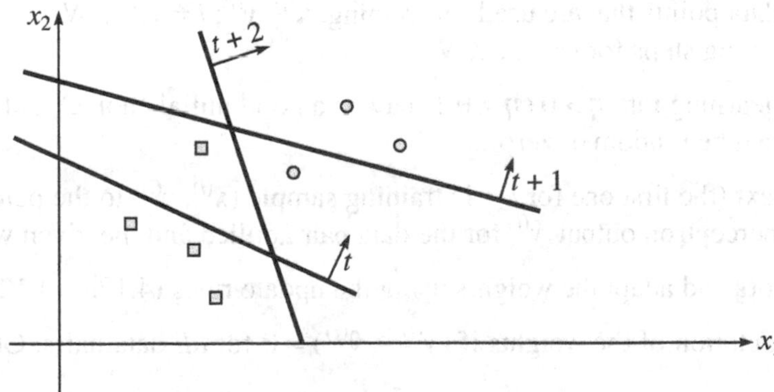

Figure 4.8 Rotation and translation of decision surface

The solution is *nonunique* because there are more than one hyperplanes separating two linearly separable classes (refer to Fig. 4.9). The decision surface search stops as soon as some surface is found that separates the training set. This can lead to decision surfaces that are positioned close to training set points. Considering that the training dataset is only an approximate representation of the rest of the data universe, such solutions can lead to misclassifications of unseen points.

Attractiveness of the perceptron algorithm lies in its simplicity. There is, however, a major problem associated with this algorithm for real-world solutions: datasets are almost certainly *not* linearly separable, while the algorithm finds a separating hyperplane only for linearly separable data. When the dataset is linearly inseparable, the test of the decision surface will always fail for some subset of training points, regardless of the adjustments we make to the free parameters, and the algorithm will loop forever. There may, however, be situations when linear separating hyperplane can be a good solution even when data are overlapped; an upper bound needs to be imposed on the number of iterations when this method is applied in practice.

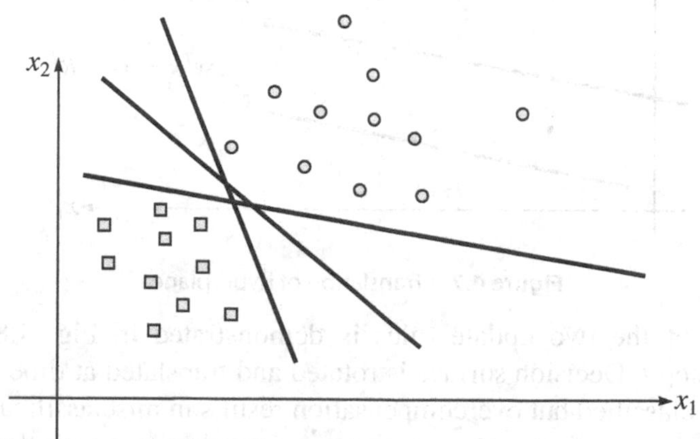

Figure 4.9 Nonunique solution

Summary of the Perceptron Learning

Given the set of N data points that are used for training: $\mathbf{x}^{(i)}, y^{(i)}; i = 1, \ldots, N$.

Perform the following steps for $i = 1, \ldots, N$.

Step 1: Choose the learning rate $\eta > 0$ ($\eta = 0.1$ may be a good initial choice) and initial weights $\mathbf{w}$, w_0 (initial weights can be random or zero).

Step 2: Apply the next (the first one for $i = 1$) training sample ($\mathbf{x}^{(i)}, y^{(i)}$) to the perceptron and using Eqn (4.5), find the perceptron output $\hat{y}^{(i)}$ for the data pair applied and the given weights $\mathbf{w}$ and w_0.

Step 3: Find the errors and adapt the weights using the update rules (4.12a)/(4.12b).

Step 4: Stop the adaptation of the weights if $(y^{(i)} - \hat{y}^{(i)}) = 0$ for *all* data pairs. Otherwise, go back to Step 2.

How do we cope with problems which are not linearly separable? The perceptron gives us a simple method when the samples are linearly separable. The minimization of classification error is the perceptron criterion function. It is an error-correcting process, as it requires that the weights be modified, only when an error crops up. Since no weight vector can accurately classify each sample in a nonseparable group, it is quite clear that the error-correcting process in perceptron algorithm can never stop. Support Vector Machines (SVM), as we shall see in this chapter, seek a weight vector that *maximizes the margin* (the minimum distance from the samples to the separating hyperplane), and employ an optimization procedure that works well for both the linearly separable and inseparable samples. The *SVM criterion function* of *largest margin* provides a unique solution and promises a good classification with previously unseen data.

Another strong alternative is available in *Neural Networks* (NN). History has proved that limitations of Rosenblatt's perceptron can be overcome by Neural Networks (discussed in the next chapter). The perceptron criterion function considers misclassified samples, and the gradient procedures for minimization are not applicable. The neural networks primarily solve the regression problems considering all the samples and *minimum squared-error criterion*, and employ gradient procedures for minimization. The algorithms for separable—as well as inseparable—data classification are first developed in the context of regression problems and then adapted for classification problems.

In the present chapter, our interest is in SVM-based solutions to real-life (nonlinear) classification and regression problems. To explain how a support vector machine works for these problems, it is perhaps easiest to start with the case of linearly separable patterns in the context of binary pattern classification. In this context, the main idea of a support vector machine is to construct a hyperplane as the decision surface in such a way that the margin of separation between Class 1 and Class 2 examples is maximized. We will then take up the more difficult case of linearly nonseparable patterns. With the material on how to find the optimal hypersurface for linearly nonseparable patterns at hand, we will formally describe the construction of a support vector machine for real-life (nonlinear) pattern recognition task. As we shall see shortly, basically the idea of a support vector machine hinges on the following two mathematical operations:

(i) Nonlinear mapping of input patterns into a high-dimensional feature space.
(ii) Construction of optimal hyperplane for linearly separating the feature vectors discovered in Step (i).

The final stage of our presentation will be to extend these results for application to multiclass classification problems, and nonlinear regression problems.

4.4 LINEAR MAXIMAL MARGIN CLASSIFIER FOR LINEARLY SEPARABLE DATA

Let the set of training (data) examples $\mathcal{D}$ be

$$\mathcal{D} = \{\mathbf{x}^{(1)}, y^{(1)}), (\mathbf{x}^{(2)}, y^{(2)}), \ldots, (\mathbf{x}^{(N)}, y^{(N)})\} \tag{4.16}$$

where $\mathbf{x} = [x_1\ x_2\ \ldots\ x_n]^T$ is an n-dimensional *input vector* (pattern with n-features) for the ith example in a real-valued space $\mathbf{X} \subseteq \Re^n$; y is its *class label* (output value), and $y \in \{+1, -1\}$. +1 denotes Class 1 and –1 denotes Class 2.

To build a classifier, SVM finds a linear function of the form

$$g(\mathbf{x}) = \mathbf{w}^T\mathbf{x} + w_0 \tag{4.17}$$

so that the input vector $\mathbf{x}^{(i)}$ is assigned to Class 1 if $g(\mathbf{x}^{(i)}) > 0$, and to Class 2 if $g(\mathbf{x}^{(i)}) < 0$, i.e.,

$$y^{(i)} = \begin{cases} +1 \text{ if } \mathbf{w}^T\mathbf{x}^{(i)} + w_0 > 0 \\ -1 \text{ if } \mathbf{w}^T\mathbf{x}^{(i)} + w_0 < 0 \end{cases} \tag{4.18}$$

Hence, $g(\mathbf{x})$ is a real-valued function; $g\colon \mathbf{X} \subseteq \Re^n \to \Re$.

$\mathbf{w} = [w_1\ w_2\ \ldots\ w_n]^T \in \Re^n$ is called the *weight vector* and $w_0 \in \Re$ is called the *bias*.

In essence, SVM finds a hyperplane

$$\mathbf{w}^T\mathbf{x} + w_0 = 0 \tag{4.19}$$

that separates Class 1 and Class 2 training examples. This hyperplane is called the *decision boundary* or *decision surface*. Geometrically, the hyperplane (4.19) divides the input space into two half spaces: one half for Class 1 examples and the other half for Class 2 examples. Note that hyperplane (4.19) is a line in a two-dimensional space and a plane in a three-dimensional space.

For linearly separable data, there are many hyperplanes (lines in two-dimensional feature space; Fig. 4.9) that can perform separation. How can one find the best one? The SVM framework provides a good answer to this question. Among all the hyperplanes that minimize the training error, find the one with the largest *margin*—the gap between the data points of the two classes. This is an intuitively acceptable approach: select the decision boundary that is far away from both the classes (Fig. 4.10). Large-margin separation is expected to yield good classification on previously unseen data, i.e., good generalization.

From Section 4.2, we know that in $\mathbf{w}^T\mathbf{x} + w_0 = 0$, $\mathbf{w}$ defines a direction perpendicular to the hyperplane. $\mathbf{w}$ is called the *normal vector* (or simply *normal*) of the hyperplane. Without changing the normal vector $\mathbf{w}$, varying w_0 moves the hyperplane parallel to itself. Note also that $\mathbf{w}^T\mathbf{x} + w_0 = 0$ has an inherent degree of freedom. We can rescale the hyperplane to $K\mathbf{w}^T\mathbf{x} + Kw_0 = 0$ for $K \in \Re^+$ (positive real numbers), without changing the hyperplane.

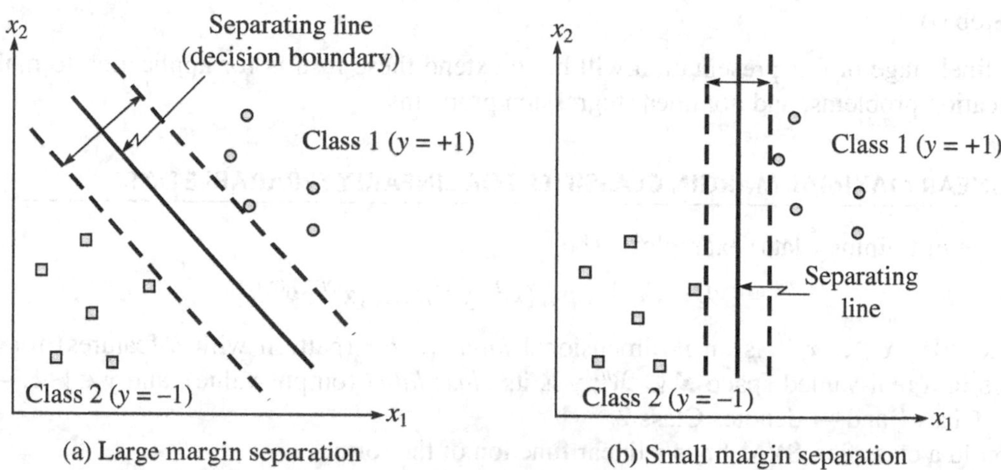

Figure 4.10

Since SVM maximizes the margin between Class 1 and Class 2 data points, let us find the margin. The linear function $g(\mathbf{x}) = \mathbf{w}^T\mathbf{x} + w_0$ gives an algebraic measure of the distance r from $\mathbf{x}$ to the hyperplane $\mathbf{w}^T\mathbf{x} + w_0 = 0$. We have seen earlier in Section 4.2 that this distance is given by (Eqn (4.7))

$$r = \frac{g(\mathbf{x})}{\|\mathbf{w}\|} \tag{4.20}$$

Now consider a Class 1 data point $(\mathbf{x}^{(i)}, +1)$ that is closest to the hyperplane $\mathbf{w}^T\mathbf{x} + w_0 = 0$ (Fig. 4.11).

The distance d_1 of this data point from the hyperplane is

$$d_1 = \frac{g(\mathbf{x}^{(i)})}{\|\mathbf{w}\|} = \frac{\mathbf{w}^T\mathbf{x}^{(i)} + w_0}{\|\mathbf{w}\|} \tag{4.21a}$$

Similarly,

$$d_2 = \frac{g(\mathbf{x}^{(k)})}{\|\mathbf{w}\|} = \frac{\mathbf{w}^T\mathbf{x}^{(k)} + w_0}{\|\mathbf{w}\|} \tag{4.21b}$$

where $(\mathbf{x}^{(k)}, -1)$ is a Class 2 data point closest to the hyperplane $\mathbf{w}^T\mathbf{x} + w_0 = 0$.

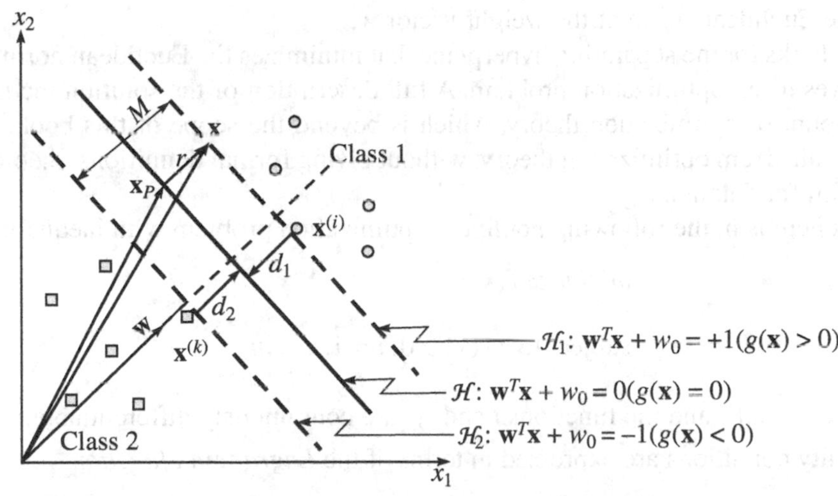

Figure 4.11 Geometric interpretation of algebraic distances of points to a hyperplane for two-dimensional case

We define two parallel hyperplanes $\mathcal{H}_1$ and $\mathcal{H}_2$ that pass through $\mathbf{x}^{(i)}$ and $\mathbf{x}^{(k)}$, respectively. $\mathcal{H}_1$ and $\mathcal{H}_2$ are also parallel to the hyperplane $\mathbf{w}^T\mathbf{x} + w_0 = 0$. We can rescale $\mathbf{w}$ and w_0 to obtain (this rescaling, as we shall see later, simplifies the quest for significant patterns, called *support vectors*)

$$\begin{aligned} \mathcal{H}_1 &: \mathbf{w}^T\mathbf{x} + w_0 = +1 \\ \mathcal{H}_2 &: \mathbf{w}^T\mathbf{x} + w_0 = -1 \end{aligned} \tag{4.22}$$

such that

$$\begin{aligned} &\mathbf{w}^T\mathbf{x}^{(i)} + w_0 \geq 1 \text{ if } y^{(i)} = +1 \\ &\mathbf{w}^T\mathbf{x}^{(i)} + w_0 \leq -1 \text{ if } y^{(i)} = -1 \end{aligned} \tag{4.23a}$$

or equivalently

$$y^{(i)}(\mathbf{w}^T\mathbf{x}^{(i)} + w_0) \geq 1 \tag{4.23b}$$

which indicates that no training data fall between hyperplanes $\mathcal{H}_1$ and $\mathcal{H}_2$. The distance between the two hyperplanes is the margin M. In the light of rescaling given by (4.22),

$$d_1 = \frac{1}{\|\mathbf{w}\|}; d_2 = \frac{-1}{\|\mathbf{w}\|} \tag{4.24}$$

where the '–' sign indicates that $\mathbf{x}^{(k)}$ lies on the side of the hyperplane $\mathbf{w}^T\mathbf{x} + w_0 = 0$ opposite to that where $\mathbf{x}^{(i)}$ lies. From Fig. 4.11, it follows that

$$M = \frac{2}{\|\mathbf{w}\|} \tag{4.25}$$

Equation (4.25) states that maximizing the margin of separation between classes is equivalent to minimizing the Euclidean norm of the weight vector $\mathbf{w}$.

Since SVM looks for the separating hyperplane that minimizes the Euclidean norm of the weight vector, this gives us an optimization problem. A full description of the solution method requires a significant amount of optimization theory, which is beyond the scope of this book. We will only use relevant results from optimization theory, without giving formal definitions, theorems or proofs (refer to [54, 55] for details).

Our interest here is in the following nonlinear optimization problem with inequality constraints:

$$\begin{aligned} &\textit{minimize } f(\mathbf{x}) \\ &\text{subject to } g_i(\mathbf{x}) \ge 0;\ i = 1, \ldots, m \end{aligned} \tag{4.26}$$

where $\mathbf{x} = [x_1\ x_2\ \cdots\ x_n]^T$, and the functions f and g_i are continuously differentiable.

The optimality conditions are expressed in terms of the *Lagrangian function*

$$L(\mathbf{x}, \boldsymbol{\lambda}) = f(\mathbf{x}) - \sum_{i=1}^{m} \lambda_i g_i(\mathbf{x}) \tag{4.27}$$

where $\boldsymbol{\lambda} = [\lambda_1\ \cdots\ \lambda_m]^T$ is a vector of Lagrange multipliers.

An optimal solution to the problem (4.26) must satisfy the following necessary conditions, called *Karush-Kuhn-Tucker* (KKT) *conditions:*

$$\begin{aligned} &\text{(i)}\ \frac{\partial L(\mathbf{x}, \boldsymbol{\lambda})}{\partial x_j} = 0;\ j = 1, \ldots, n \\ &\text{(ii)}\ g_i(\mathbf{x}) \ge 0;\ i = 1, \ldots, m \\ &\text{(iii)}\ \lambda_i \ge 0;\ i = 1, \ldots, m \\ &\text{(iv)}\ \lambda_i\, g_i(\mathbf{x}) = 0;\ i = 1, \ldots, m \end{aligned} \tag{4.28}$$

In view of condition (iii), the vector of Lagrange multipliers belongs to the set $\{\boldsymbol{\lambda} \in \Re^m, \boldsymbol{\lambda} \ge \mathbf{0}\}$. Also note that condition (ii) is the original set of constraints.

Our interest, as we will see shortly, is in convex functions f and linear functions g_i. For this class of optimization problems, when there exist vectors $\mathbf{x}^0$ and $\boldsymbol{\lambda}^0$ such that the point $(\mathbf{x}^0, \boldsymbol{\lambda}^0)$ satisfies the

KKT conditions (4.28), then $\mathbf{x}^0$ gives the global minimum of the function $f(\mathbf{x})$, with the constraint given in (4.26).

Let

$$L^*(\mathbf{x}) = \underset{\lambda \in \Re^m}{max}\ L(\mathbf{x}, \lambda), \text{ and } L_*(\lambda) = \underset{x \in \Re^n}{\min}\ L(\mathbf{x}, \lambda)$$

It is clear from these equations that for any $\mathbf{x} \in \Re^n$ and $\lambda \in \Re^m$,

$$L_*(\lambda) \le L(\mathbf{x}, \lambda) \le L^*(\mathbf{x})$$

and thus, in particular

$$L_*(\lambda) \le L^*(\mathbf{x})$$

This holds for any $\mathbf{x} \in \Re^n$ and $\lambda \in \Re^m$; so it holds for the λ that maximizes the left-hand side, and the $\mathbf{x}$ that minimizes the right-hand side. Thus,

$$\underset{\lambda \in \Re^m}{max}\ \underset{\mathbf{x} \in \Re^n}{min}\ L(x, \lambda) \le \underset{\mathbf{x} \in \Re^n}{min}\ \underset{\lambda \in \Re^m}{max}\ L(\mathbf{x}, \lambda)$$

The two problems, *min-max* and *max-min*, are said to be *dual* to each other. We refer to the min-max problem as the *primal problem*. The objective to be minimized, $L^*(\mathbf{x})$, is referred to as the *primal function*. The max-min problem is referred to as the *dual problem*, and $L_*(\lambda)$ as the *dual function*. The optimal primal and dual function values are equal when f is a convex function and g_i are linear functions. The concept of duality is widely used in the optimization literature. The aim is to provide an alternative formulation of the problem which is more convenient to solve computationally and/or has some theoretical significance. In the context of SVM, the dual problem is not only easy to solve computationally, but also crucial for using *kernel functions* to deal with nonlinear decision boundaries. This will be clear later.

The nonlinear optimization problem defined in (4.26) can be represented as min-max problem, as follows:

For the Lagrangian (4.27), we have

$$L^*(\mathbf{x}) = \underset{\lambda \in \Re^m}{max} \left[f(\mathbf{x}) - \sum_{i=1}^{m} \lambda_i g_i(\mathbf{x}) \right]$$

Since $g_i(\mathbf{x}) \ge 0$ for all i, $\lambda_i = 0$ $(i = 1, \ldots, m)$ would maximize the Lagrangian; thus,

$$L^*(\mathbf{x}) = f(\mathbf{x})$$

Therefore, our original constrained problem (4.26) becomes the min-max primal problem:

$$\underset{\mathbf{x} \in \Re^n}{minimize}\ L^*(\mathbf{x})$$

$$\text{subject to } g_i(\mathbf{x}) \ge 0;\ i = 1, \ldots, m$$

The concept of duality gives the following formulation for max-min dual problem:

$$\underset{\lambda \in \Re^m,\, \lambda \ge 0}{maximize}\ L_*(\lambda)$$

More explicitly, this nonlinear optimization problem with *dual variables* $\boldsymbol{\lambda}$, can be written in the form:

$$\underset{\boldsymbol{\lambda} \geq \mathbf{0}}{maximize} \; \underset{\mathbf{x} \in \Re^n}{min} \left[f(\mathbf{x}) - \sum_{i=1}^{m} \lambda_i g_i(\mathbf{x}) \right] \tag{4.29}$$

Let us now state the learning problem in SVM.

Given a set of linearly separable training examples,

$$\mathcal{D} = \{(\mathbf{x}^{(1)}, y^{(1)}), (\mathbf{x}^{(2)}, y^{(2)}), \ldots, (\mathbf{x}^{(N)}, y^{(N)})\},$$

the learning problem is to solve the following constrained minimization problem:

$$\begin{aligned} &minimize \quad f(\mathbf{w}) = \tfrac{1}{2}\mathbf{w}^T\mathbf{w} \\ &\text{subject to} \quad y^{(i)}(\mathbf{w}^T\mathbf{x}^{(i)} + w_0) \geq 1; \, i = 1, \ldots, N \end{aligned} \tag{4.30}$$

This formulation is called the primal formulation of *hard-margin* SVM. Solving this problem will produce the solutions for $\mathbf{w}$ and w_0 which in turn, give us the maximal margin hyperplane $\mathbf{w}^T\mathbf{x} + w_0 = 0$ with the margin $2/\|\mathbf{w}\|$.

The objective function is quadratic and convex in parameters $\mathbf{w}$, and the constraints are linear in parameters $\mathbf{w}$ and w_0. The dual formulation of this constrained optimization problem is obtained as follows.

First we construct the Lagrangian:

$$L(\mathbf{w}, w_0, \boldsymbol{\lambda}) = \tfrac{1}{2}\mathbf{w}^T\mathbf{w} - \sum_{i=1}^{N} \lambda_i [y^{(i)}(\mathbf{w}^T\mathbf{x}^{(i)} + w_0) - 1] \tag{4.31}$$

The KKT conditions are as follows:

$$\begin{aligned} &\text{(i)} \quad \frac{\partial L}{\partial \mathbf{w}} = \mathbf{0}; \text{ which gives } \mathbf{w} = \sum_{i=1}^{N} \lambda_i y^{(i)}\mathbf{x}^{(i)} \\ &\qquad \frac{\partial L}{\partial w_0} = 0; \text{ which gives } \sum_{i=1}^{N} \lambda_i y^{(i)} = 0 \\ &\text{(ii)} \quad y^{(i)}(\mathbf{w}^T\mathbf{x}^{(i)} + w_0) - 1 \geq 0; \, i = 1, \ldots, N \\ &\text{(iii)} \quad \lambda_i \geq 0; \, i = 1, \ldots, N \\ &\text{(iv)} \quad \lambda_i[y^{(i)}(\mathbf{w}^T\mathbf{x}^{(i)} + w_0) - 1] = 0; \, i = 1, \ldots, N \end{aligned} \tag{4.32}$$

From condition (i) of KKT conditions (4.32), we observe that the solution vector has an expansion in terms of training examples. Note that although the solution $\mathbf{w}$ is unique (due to the strict convexity of the function $f(\mathbf{w})$), the dual variables λ_i need not be. There is a dual variable λ_i for each training data point. Condition (iv) of KKT conditions (4.32) shows that for data points not on the margin hyperplanes (i.e., $\mathcal{H}_1$ and $\mathcal{H}_2$), $\lambda_i = 0$:

$$y^{(i)}(\mathbf{w}^T\mathbf{x}^{(i)} + w_0) - 1 > 0 \Rightarrow \lambda_i = 0$$

For data points on the margin hyperplanes, $\lambda_i \geq 0$:

$$y^{(i)}(\mathbf{w}^T\mathbf{x}^{(i)} + w_0) - 1 = 0 \Rightarrow \lambda_i \geq 0$$

However, the data points on the margin hyperplanes with $\lambda_i = 0$ do not contribute to the solution **w**, as is seen from condition (i) of KKT conditions (4.32). The data points on the margin hyperplanes with associated dual variables $\lambda_i > 0$ are called *support vectors*, which give the name to the algorithm, *support vector machines*.

To postulate the dual problem, we first expand Eqn (4.31), term by term, as follows:

$$L(\mathbf{w}, w_0, \boldsymbol{\lambda}) = \tfrac{1}{2}\,\mathbf{w}^T\,\mathbf{w} - \sum_{i=1}^{N} \lambda_i y^{(i)}\mathbf{w}^T\mathbf{x}^{(i)} - w_0 \sum_{i=1}^{N} \lambda_i y^{(i)} + \sum_{i=1}^{N} \lambda_i \tag{4.33}$$

Transformation from the primal to the corresponding dual is carried out by setting the partial derivatives of the Lagrangian (4.33) with respect to the *primal variables* (i.e., **w** and w_0) to zero, and substituting the resulting relations back into the Lagrangian. The objective is to merely substitute condition (i) of KKT conditions (4.32) into the Lagrangian (4.33) to remove the primal variables; which gives us the dual objective function.

The third term on the right-hand side of Eqn (4.33) is zero by virtue of condition (i) of KKT conditions (4.32). Furthermore, from this condition we have,

$$\mathbf{w}^T\,\mathbf{w} = \sum_{i=1}^{N} \lambda_i y^{(i)}\mathbf{w}^T\mathbf{x}^{(i)} = \sum_{i=1}^{N}\sum_{k=1}^{N} \lambda_i\lambda_k y^{(i)}y^{(k)}\mathbf{x}^{(i)T}\,\mathbf{x}^{(k)}$$

Accordingly, minimization of function L in Eqn (4.33) with respect to primal variables **w** and w_0, gives us the following dual objective function:

$$L_*(\boldsymbol{\lambda}) = \sum_{i=1}^{N} \lambda_i - \tfrac{1}{2}\sum_{i=1}^{N}\sum_{k=1}^{N} \lambda_i\lambda_k y^{(i)}y^{(k)}\mathbf{x}^{(i)T}\,\mathbf{x}^{(k)} \tag{4.34}$$

We may now state the dual optimization problem.

Given a set of linearly separable training examples $\{(\mathbf{x}^{(i)}, y^{(i)})\}_{i=1}^{N}$, find the dual variables $\{\lambda_i\}_{i=1}^{N}$, that maximize the objective function (4.34) subject to the constraints

- $\sum_{i=1}^{N} \lambda_i y^{(i)} = 0$ (4.35)
- $\lambda_i \geq 0;\ i = 1, \ldots, N$

This formulation is dual formulation of the *hard-margin SVM*.

Having solved the dual problem numerically (using MATLAB's **quadprog** function, for example), the resulting optimum λ_i values are then used to compute **w** and w_0. **w** is computed using condition (i) of KKT conditions (4.32):

$$\mathbf{w} = \sum_{i=1}^{N} \lambda_i y^{(i)}\mathbf{x}^{(i)} \tag{4.36}$$

and w_0 is computed using condition (iv) of KKT conditions (4.32):

$$\lambda_i [y^{(i)}(\mathbf{w}^T\mathbf{x}^{(i)} + w_0) - 1] = 0; i = 1, \ldots, N \tag{4.37}$$

Note that though there are N values of λ_i in Eqn (4.36), most vanish with $\lambda_i = 0$ and only a small percentage have $\lambda_i > 0$. The set of $\mathbf{x}^{(i)}$ whose $\lambda_i > 0$ are the *support vectors*, and as we see in Eqn (4.36), $\mathbf{w}$ is the weighted sum of these training instances that are selected as the support vectors:

$$\mathbf{w} = \sum_{i \in svindex} \lambda_i y^{(i)} \mathbf{x}^{(i)} \tag{4.38}$$

where *svindex* denotes the set of indices of support vectors.

From Eqn (4.38), we see that the support vectors $\mathbf{x}^{(i)}$; $i \in svindex$, satisfy

$$y^{(i)}(\mathbf{w}^T\mathbf{x}^{(i)} + w_0) = 1$$

and lie on the margin. We can use this fact to calculate w_0 from any support vector as,

$$w_0 = \frac{1}{y^{(i)}} - \mathbf{w}^T\mathbf{x}^{(i)}$$

For $y^{(i)} \in [+1, -1]$, we can equivalently express this equation as,

$$w_0 = y^{(i)} - \mathbf{w}^T\mathbf{x}^{(i)} \tag{4.39}$$

Instead of depending on one support vector to compute w_0, in practice, all support vectors are used to compute w_0, and then their average is taken for the final value of w_0. This is because the values of λ_i are computed numerically and can have numerical errors.

$$w_0 = \frac{1}{|svindex|} \sum_{i \in svindex} (y^{(i)} - \mathbf{w}^T\mathbf{x}^{(i)}) \tag{4.40}$$

where $|svindex|$ corresponds to total number of indices in the set *svindex*, i.e., total number of support vectors.

The majority of λ_i are 0, for which $y^{(i)}(\mathbf{w}^T\mathbf{x}^{(i)} + w_0) > 1$. These are the $\mathbf{x}^{(i)}$ points that exist more than adequately away from the discriminant, and have zero effect on the hyperplane. The instances that are not support vectors have no information; the same solution will be obtained on removing any subset from them. From this viewpoint, the SVM algorithm can be said to be similar to the k-NN algorithm (Section 3.4) which stores only the instances neighboring the class discriminant.

During testing, we do not enforce a margin. We calculate

$$g(\mathbf{x}) = \mathbf{w}^T\mathbf{x} + w_0 \tag{4.41a}$$

and choose the class according to the sign of $g(\mathbf{x})$: $sgn\ (g(\mathbf{x}))$ which we call the *indicator function* i_F,

$$i_F = \hat{y} = sgn\ (\mathbf{w}^T\mathbf{x} + w_0) \tag{4.41b}$$

Choose Class 1 ($\hat{y} = +1$) if $\mathbf{w}^T\mathbf{x} + w_0 > 0$, and Class 2 ($\hat{y} = -1$) otherwise.

Example 4.1

In this example, we visualize SVM (hard-margin) formulation in two variables. Consider the toy dataset given in Table 4.1.

SVM finds a hyperplane

$$\mathcal{H}: w_1 x_1 + w_2 x_2 + w_0 = 0$$

and two bounding planes

$$\mathcal{H}_1: w_1 x_1 + w_2 x_2 + w_0 = +1$$

$$\mathcal{H}_2: w_1 x_1 + w_2 x_2 + w_0 = -1$$

such that

$$w_1 x_1 + w_2 x_2 + w_0 \geq +1 \quad \text{if } y^{(i)} = +1$$

$$w_1 x_1 + w_2 x_2 + w_0 \leq -1 \quad \text{if } y^{(i)} = -1$$

or equivalently

$$y^{(i)}(w_1 x_1 + w_2 x_2 + w_0) \geq 1$$

We write these constraints explicitly as (refer to Table 4.1),

$(-1)\ [w_1 + w_2 + w_0] \geq 1$

$(-1)\ [2w_1 + w_2 + w_0] \geq 1$

$(-1)\ [w_1 + 2w_2 + w_0] \geq 1$

$(-1)\ [2w_1 + 2w_2 + w_0] \geq 1$

$(+1)\ [4w_1 + 4w_2 + w_0] \geq 1$

$(+1)\ [4w_1 + 5w_2 + w_0] \geq 1$

$(+1)\ [5w_1 + 4w_2 + w_0] \geq 1$

$(+1)\ [5w_1 + 5w_2 + w_0] \geq 1$

Table 4.1 Data for classification

Sample i	$x_1^{(i)}$	$x_2^{(i)}$	$y^{(i)}$
1	1	1	–1
2	2	1	–1
3	1	2	–1
4	2	2	–1
5	4	4	+1
6	4	5	+1
7	5	4	+1
8	5	5	+1

The constraint equations in matrix form:

$$\underbrace{\begin{bmatrix} -1 & 0 & 0 & 0 & 0 & 0 & 0 & 0 & 0 & 0 \\ 0 & -1 & 0 & 0 & 0 & 0 & 0 & 0 & 0 & 0 \\ 0 & 0 & -1 & 0 & 0 & 0 & 0 & 0 & 0 & 0 \\ 0 & 0 & 0 & -1 & 0 & 0 & 0 & 0 & 0 & 0 \\ 0 & 0 & 0 & 0 & -1 & 0 & 0 & 0 & 0 & 0 \\ 0 & 0 & 0 & 0 & 0 & +1 & 0 & 0 & 0 & 0 \\ 0 & 0 & 0 & 0 & 0 & 0 & +1 & 0 & 0 & 0 \\ 0 & 0 & 0 & 0 & 0 & 0 & 0 & +1 & 0 & 0 \\ 0 & 0 & 0 & 0 & 0 & 0 & 0 & 0 & +1 & 0 \\ 0 & 0 & 0 & 0 & 0 & 0 & 0 & 0 & 0 & +1 \end{bmatrix}}_{\mathbf{Y}} \left[\underbrace{\begin{bmatrix} 1 & 1 \\ 2 & 1 \\ 1 & 2 \\ 2 & 2 \\ 1.5 & 1.5 \\ 4 & 4 \\ 4 & 5 \\ 5 & 4 \\ 5 & 5 \\ 4.5 & 4.5 \end{bmatrix}}_{\mathbf{X}} \underbrace{\begin{bmatrix} w_1 \\ w_2 \end{bmatrix}}_{\mathbf{w}} + w_0 \underbrace{\begin{bmatrix} 1 \\ 1 \\ 1 \\ 1 \\ 1 \\ 1 \\ 1 \\ 1 \\ 1 \\ 1 \end{bmatrix}}_{\mathbf{e}} \right] \geq \underbrace{\begin{bmatrix} 1 \\ 1 \\ 1 \\ 1 \\ 1 \\ 1 \\ 1 \\ 1 \\ 1 \\ 1 \end{bmatrix}}_{\mathbf{e}}$$

or

$$\mathbf{Y}(\mathbf{X}\mathbf{w} + w_0\mathbf{e}) \geq \mathbf{e} \tag{4.42a}$$

For a dataset with N samples, and n features in vector $\mathbf{x}$,

$$\underset{(N\times N)}{\mathbf{Y}} = \begin{bmatrix} y^{(1)} & 0 & \dots & 0 \\ 0 & y^{(2)} & \dots & 0 \\ 0 & 0 & \dots & 0 \\ \vdots & \vdots & & \vdots \\ 0 & 0 & \dots & y^{(N)} \end{bmatrix}; \underset{(N\times n)}{\mathbf{X}} = \begin{bmatrix} \mathbf{x}^{(1)T} \\ \mathbf{x}^{(2)T} \\ \vdots \\ \mathbf{x}^{(N)T} \end{bmatrix}; \underset{n\times 1}{\mathbf{w}} = \begin{bmatrix} w_1 \\ w_2 \\ \vdots \\ w_n \end{bmatrix}; \underset{(N\times 1)}{\mathbf{e}} = \begin{bmatrix} 1 \\ 1 \\ \vdots \\ 1 \end{bmatrix} \tag{4.42b}$$

$$\mathbf{Y} = diag(\mathbf{y});\ \mathbf{y} = [y^{(1)}\ y^{(2)}\ \dots\ y^{(N)}]^T \tag{4.42c}$$

Our aim is to find the weight matrix $\mathbf{w}$ and the bias term w_0 that maximize the margin of separation between the hyperplanes $\mathcal{H}_1$ and $\mathcal{H}_2$, and at the same time satisfy the constraint equations (4.42). It gives us an optimization problem

$$\begin{aligned} &\underset{\mathbf{w},\, w_0}{maximize} \left(\tfrac{1}{2}\, \mathbf{w}^T\mathbf{w}\right) \\ &\text{subject to } \mathbf{Y}(\mathbf{X}\mathbf{w} + w_0\mathbf{e}) \geq \mathbf{e} \end{aligned} \tag{4.43}$$

Once we obtain $\mathbf{w}$ and w_0, we have our decision boundary:

$$\mathbf{w}^T\mathbf{x} + w_0 = 0$$

and for a new unseen data point $\mathbf{x}$, we assign $sgn\ (\mathbf{w}^T\mathbf{x} + w_0)$ as the class value.

For solving the above problem in *primal*, we need to rewrite the problem in the standard QP (Quadratic Programming) format.

Another way to solve the above primal problem is the Lagrangian dual. The problem is more easily solved in terms of its Lagrangian dual variables:

$$\underset{\boldsymbol{\lambda} \geq \mathbf{0}}{maximize} \left[\underset{\mathbf{w}, w_0}{min}\ L(\mathbf{w}, w_0, \boldsymbol{\lambda}) \right]$$

$$\text{where (Eqn (4.31))} \tag{4.44}$$

$$L(\mathbf{w}, w_0, \boldsymbol{\lambda}) = \tfrac{1}{2}\mathbf{w}^T\mathbf{w} - \boldsymbol{\lambda}^T [\mathbf{Y}(\mathbf{X}\mathbf{w} + w_0\mathbf{e}) - \mathbf{e}]$$

$$\underset{(N\times 1)}{\boldsymbol{\lambda}} = [\lambda_1\ \lambda_2\ \dots\ \lambda_N]^T \text{ is a vector of Lagrange multipliers.}$$

It leads to the *dual optimization problem* (Eqns (4.34–4.35))

$$\begin{aligned} &\underset{\boldsymbol{\lambda}}{maximize} \left[\boldsymbol{\lambda}^T\mathbf{e} - \tfrac{1}{2}\boldsymbol{\lambda}^T \mathbf{Y}\mathbf{X}\mathbf{X}^T\mathbf{Y}\boldsymbol{\lambda} \right] \\ &\text{subject to } \mathbf{e}^T\mathbf{Y}\boldsymbol{\lambda} = 0;\ \boldsymbol{\lambda} \geq \mathbf{0} \end{aligned} \tag{4.45}$$

Some standard quadratic optimization programs typically *minimize* the given objective function:

$$\begin{aligned} &\underset{\boldsymbol{\lambda}}{minimize} \left[\tfrac{1}{2}\boldsymbol{\lambda}^T\mathbf{Q}\boldsymbol{\lambda} - \mathbf{e}^T\boldsymbol{\lambda} \right] \\ &\text{subject to } \mathbf{y}^T\boldsymbol{\lambda} = 0;\ \boldsymbol{\lambda} \geq \mathbf{0} \\ &\text{where } \mathbf{Q} = \mathbf{Y}\,\mathbf{X}\,\mathbf{X}^T\,\mathbf{Y} \end{aligned} \tag{4.46}$$

The input required for the above program is only $\mathbf{X}$ and $\mathbf{y}$. It returns the Lagrange multiplier vector $\boldsymbol{\lambda}$.

Having solved the dual problem numerically (using a standard optimization program), optimum λ_i values are then used to compute $\mathbf{w}$ and w_0 (Eqns (4.38, 4.40)):

$$\mathbf{w} = \sum_{i \in svindex} \lambda_i\, y^{(i)}\mathbf{x}^{(i)} \tag{4.47a}$$

$$w_0 = \frac{1}{|svindex|}\left[\sum_{i \in svindex} [y^{(i)} - \mathbf{w}^T\mathbf{x}^{(i)}] \right] \tag{4.47b}$$

Using the MATLAB **quadprog** routine for the dataset of Table 4.1, we obtain [56]

$$\boldsymbol{\lambda}^T = [0\ \ 0\ \ 0\ \ 0.25\ \ 0\ \ 0.25\ \ 0\ \ 0\ \ 0\ \ 0]$$

This means that data points with index 4 and 6 are *support vectors*; i.e., support vectors are:

$$\begin{bmatrix} 2 \\ 2 \end{bmatrix};\ \begin{bmatrix} 4 \\ 4 \end{bmatrix}$$

and *svindex* is {4, 6}; |*svindex*| = 2.

$$\mathbf{w} = \sum_{i \in svindex} \lambda_i y^{(i)} \mathbf{x}^{(i)} = 0.25\,[-1]\begin{bmatrix}2\\2\end{bmatrix} + 0.25\,[1]\begin{bmatrix}4\\4\end{bmatrix} = \begin{bmatrix}0.5\\0.5\end{bmatrix}$$

$$\begin{aligned} w_0 &= \tfrac{1}{2}\,[y^{(4)} - \mathbf{w}^T\mathbf{x}^{(4)} + y^{(6)} - \mathbf{w}^T\mathbf{x}^{(6)}] \\ &= \tfrac{1}{2}\left[-1 - [w_1\ w_2]\begin{bmatrix}2\\2\end{bmatrix} + 1 - [w_1\ w_2]\begin{bmatrix}4\\4\end{bmatrix}\right] \\ &= -3 \end{aligned}$$

Therefore, the decision hyperplane $g(\mathbf{x})$ is

$$g(\mathbf{x}) = 0.5x_1 + 0.5x_2 - 3$$

and the *indicator function*

$$i_F = \hat{y} = sgn(g(\mathbf{x})) = sgn(0.5x_1 + 0.5x_2 - 3)$$

4.5 LINEAR SOFT MARGIN CLASSIFIER FOR OVERLAPPING CLASSES

The linear hard-margin classifier gives a simple SVM when the samples are linearly separable. In practice, however, the training data is almost always linearly nonseparable because of random errors owing to different reasons. For instance, certain instances may be wrongly labeled. The labels could be different even for two input vectors that are identical.

If SVM has to be of some use, it should permit noise in the training data. But, with noisy data, the linear SVM algorithm described in earlier section, will not obtain a solution as the constraints cannot be satisfied. For instance, in Fig. 4.12, there exists a Class 2 point (square) in the Class 1 region, and a Class 1 point (circle) in the Class 2 area. However, in spite of the couple of mistakes, the decision boundary seems to be good. But the *hard-margin* classifier presented previously cannot be used, because all the constraints

$$y^{(i)}(\mathbf{w}^T\mathbf{x}^{(i)} + w_0) \geq 1;\ i = 1, \ldots, N$$

cannot be satisfied.

So the constraints have to be modified to permit mistakes. To allow errors in data, we can relax the margin constraints by introducing *slack* variables, $\zeta_i (\geq 0)$, as follows:

$$\begin{aligned} \mathbf{w}^T\mathbf{x}^{(i)} + w_0 &\geq 1 - \zeta_i \quad &\text{for} \quad y^{(i)} = +1 \\ \mathbf{w}^T\mathbf{x}^{(i)} + w_0 &\leq -1 + \zeta_i \quad &\text{for} \quad y^{(i)} = -1 \end{aligned}$$

Thus, we have the new constraints

$$\begin{gathered} y^{(i)}(\mathbf{w}^T\mathbf{x}^{(i)} + w_0) \geq 1 - \zeta_i;\ i = 1, \ldots, N \\ \zeta_i \geq 0 \end{gathered} \tag{4.48}$$

The geometric interpretation is shown in Fig. 4.12.

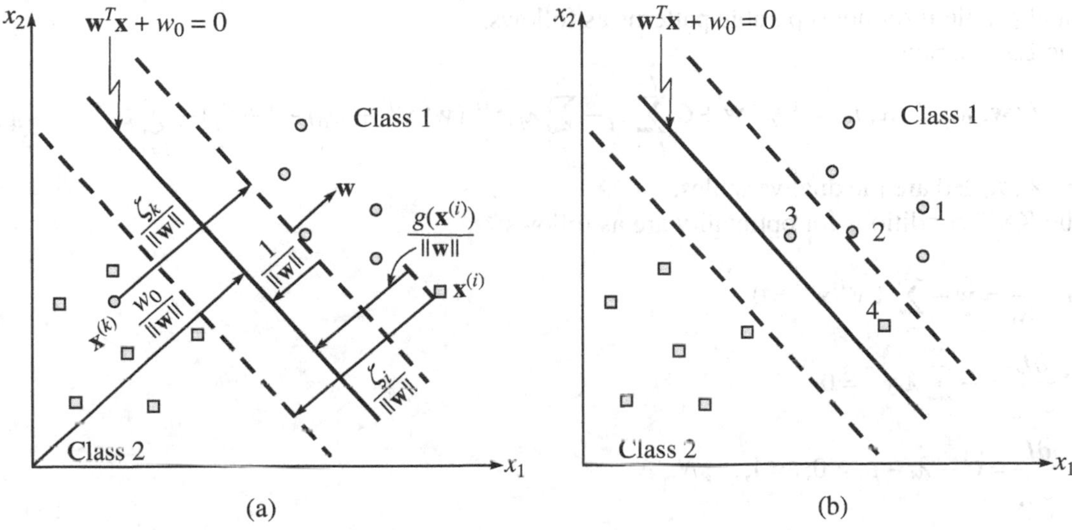

Figure 4.12 Soft decision boundary

In classifying an instance, there are four possible cases (see Fig. 4.12(b)). Instance 1 is on the correct side and far away from the origin; $y^{(i)}g(\mathbf{x}^{(i)}) > 1$, $\zeta_i = 0$. Instance 2 is on the correct side and on the margin; $\zeta_i = 0$. For instance 3, $\zeta_i = 1 - g(\mathbf{x}^{(i)})$, $0 < \zeta_i < 1$, the instance is on the correct side but in the margin and not sufficiently away. For instance 4, $\zeta_i = 1 + g(\mathbf{x}^{(i)}) > 1$, the instance is on the wrong side—this is a misclassification.

We also need to penalize the errors in the objective function. A natural way is to assign an extra cost for errors to change the objective function to

$$\tfrac{1}{2}\mathbf{w}^T\mathbf{w} + C\left(\sum_{i=1}^{N}\zeta_i\right); C \ge 0$$

where C is a user specified penalty parameter. This parameter is a trade-off parameter between margin and mistakes.

The parameter C trades off complexity, as measured by norm of weight vector, and data misfit, as measured by the number of nonseperable points. Note that we are penalizing not only the misclassified points but also the ones in the margin for better generalization. Increasing C corresponds to assigning a high penalty to errors, simultaneously resulting in larger weights. The width of the soft-margin can be controlled by penalty parameter C.

The new optimization problem becomes

$$\begin{aligned} &\textit{minimize}\ \tfrac{1}{2}\mathbf{w}^T\mathbf{w} + C\sum_{i=1}^{N}\zeta_i \\ &\text{subject to } y^{(i)}(\mathbf{w}^T\mathbf{x}^{(i)} + w_0) \ge 1 - \zeta_i;\ i = 1, \ldots, N \\ &\qquad\qquad \zeta_i \ge 0;\ i = 1, \ldots, N \end{aligned} \tag{4.48a}$$

This formulation is called the *soft-margin SVM*.

Proceeding in the manner similar to that described earlier for separable case, we may formulate the dual problem for nonseparable patterns as follows.

The Lagrangian

$$L(\mathbf{w}, w_0, \zeta, \lambda, \mu) = \tfrac{1}{2}\mathbf{w}^T\mathbf{w} + C\sum_{i=1}^{N}\zeta_i - \sum_{i=1}^{N}\lambda_i[y^{(i)}(\mathbf{w}^T\mathbf{x}^{(i)} + w_0) - 1 + \zeta_i] - \sum_{i=1}^{N}\mu_i\zeta_i \tag{4.49}$$

where λ_i, $\mu_i \geq 0$ are the dual variables.

The KKT conditions for optimality are as follows:

$$\begin{aligned}
&\text{(i)}\quad \frac{\partial L}{\partial \mathbf{w}} = \mathbf{w} - \sum_{i=1}^{N}\lambda_i y^{(i)}\mathbf{x}^{(i)} = 0\\
&\qquad \frac{\partial L}{\partial w_0} = -\sum_{i=1}^{N}\lambda_i y^{(i)} = 0\\
&\qquad \frac{\partial L}{\partial \zeta_i} = C - \lambda_i - \mu_i = 0;\ i = 1, \ldots, N\\
&\text{(ii)}\quad y^{(i)}(\mathbf{w}^T\mathbf{x}^{(i)} + w_0) - 1 + \zeta_i \geq 0;\ i = 1, \ldots, N\\
&\qquad \zeta_i \geq 0;\ i = 1, \ldots, N\\
&\text{(iii)}\quad \lambda_i \geq 0;\ i = 1, \ldots, N\\
&\qquad \mu_i \geq 0;\ i = 1, \ldots, N\\
&\text{(iv)}\quad \lambda_i(y^{(i)}(\mathbf{w}^T\mathbf{x}^{(i)} + w_0) - 1 + \zeta_i) = 0;\ i = 1, \ldots, N\\
&\qquad \mu_i\zeta_i = 0;\ i = 1, \ldots, N
\end{aligned} \tag{4.50}$$

We substitute the relations in condition (i) of KKT conditions (4.50) into the Lagrangian (4.49) to obtain dual objective function. From the relation $C - \lambda_i - \mu_i = 0$, we can deduce that $\lambda_i \leq C$ because $\mu_i \geq 0$. Thus, the dual formulation of the *soft-margin SVM* is

$$\begin{aligned}
&\textit{maximize } L_*(\boldsymbol{\lambda}) = \sum_{i=1}^{N}\lambda_i - \tfrac{1}{2}\sum_{i=1}^{N}\sum_{k=1}^{N}\lambda_i\lambda_k y^{(i)}y^{(k)}\mathbf{x}^{(i)T}\mathbf{x}^{(k)}\\
&\text{subject to } \sum_{i=1}^{N}\lambda_i y^{(i)} = 0\\
&\qquad 0 \leq \lambda_i \leq C;\ i = 1, \ldots, N
\end{aligned} \tag{4.51}$$

Interestingly, ζ_i and μ_i are not in the dual objective function; the objective function is identical to that for the separable case. The only difference is the constraint $\lambda_i \leq C$ (inferred from $C - \lambda_i - \mu_i = 0$ and $\mu_i \geq 0$). The dual problem (4.51) can also be solved numerically, and the resulting λ_i values are then used to compute $\mathbf{w}$ and w_0. The weight vector $\mathbf{w}$ is computed using Eqn (4.36).

The bias parameter w_0 is computed using condition (iv) of KKT conditions (4.50):

$$\lambda_i(y^{(i)}(\mathbf{w}^T\mathbf{x}^{(i)} + w_0) - 1 + \zeta_i) = 0 \tag{4.52a}$$

$$\mu_i\zeta_i = 0 \tag{4.52b}$$

Since we do not have values for ζ_i, we have to get around it. λ_i can have values in the interval $0 \le \lambda_i \le C$. We will separate it into the following three cases:

Case 1: $\lambda_i = 0$

We know that $C - \lambda_i - \mu_i = 0$. With $\lambda_i = 0$, we get $\mu_i = C$. Since $\mu_i\zeta_i = 0$ (Eqn (4.52b)), this implies that $\zeta_i = 0$; which means that the corresponding ith pattern is correctly classified without any error (as it would have been with hard-margin SVM). Such patterns may lie on margin hyperplanes or outside the margin. However, they do not contribute to the optimum value of **w**, as is seen from Eqn (4.36).

Case 2: $0 < \lambda_i < C$

We know that $C - \lambda_i - \mu_i = 0$. Therefore, $\mu_i = C - \lambda_i$, which means $\mu_i > 0$. Since $\mu_i\zeta_i = 0$ (Eqn (4.52b), this implies that $\zeta_i = 0$. Again the corresponding ith pattern is correctly classified. Also, from Eqn (4.52a), we see that for $\zeta_i = 0$ and $0 < \lambda_i < C$, $y^{(i)}(\mathbf{w}^T\mathbf{x}^{(i)} + w_0) = 1$; so the corresponding patterns are on the margin.

Case 3: $\lambda_i = C$

With $\lambda_i = C$, $y^{(i)}(\mathbf{w}^T\mathbf{x}^{(i)} + w_0) + \zeta_i = 1$, and $\zeta_i > 0$. But $\zeta_i \ge 0$ is a constraint of the problem. So $\zeta_i > 0$; which means that the corresponding pattern is mis-classified or lies inside the margin.

Note that support vectors have their $\lambda_i > 0$, and they define **w** as given by Eqn (4.36). We can compute **w** from the following equation (refer Eqn (4.38)):

$$\mathbf{w} = \sum_{i \in svindex} \lambda_i y^{(i)}\mathbf{x}^{(i)} \tag{4.53a}$$

where *svindex* denotes the set of indices of support vectors (patterns with $\lambda_i > 0$).

Of all the support vectors, those whose $\lambda_i < C$ (Case 2), are the ones that are on the margin, and we can use them to calculate w_0; they satisfy

$$y^{(i)}(\mathbf{w}^T\mathbf{x}^{(i)} + w_0) = 1$$

We can compute w_0 from the following equation (refer Eqn (4.40)):

$$w_0 = \frac{1}{|svmindex|} \sum_{i \in svmindex} (y^{(i)} - \mathbf{w}^T\mathbf{x}^{(i)}) \tag{4.53b}$$

where *svmindex* are the set of support vectors that fall on the margin.

Finally, expressions for both the decision function $g(\mathbf{x})$ and an indicator function $i_F = sgn\ (g(\mathbf{x}))$ for a soft-margin classifier are the same as for linearly separable classes (refer Eqns (4.41)):

$$g(\mathbf{x}) = \mathbf{w}^T\mathbf{x} + w_0 \tag{4.54a}$$

$$i_F = \hat{y} = sgn\ (g(\mathbf{x})) = sgn\ (\mathbf{w}^T\mathbf{x} + w_0) \tag{4.54b}$$

The following points need attention of the reader:

- A significant property of SVM is that the solution is sparse in λ_i. Majority of the training data points exist outside the margin area and their λ_i's in the solution are 0. The data points on the margin with $\lambda_i = 0$, do not contribute to the solution either. Only those data points that are on the margin hyperplanes with $0 < \lambda_i < C$, and those mis-classified or inside the margin ($\lambda_i = C$) make contribution to the solution. In the absence of this sparsity property, SVM would prove to be impractical for huge datasets.
- Parameter C in the optimization formulation (4.51) is the regularization parameter, fine-tuned with the help of cross-validation. It defines the trade-off between margin maximization and error minimization. In case it is too big, there is high penalty for nonseparable points, and we may store several support vectors and overfit. In case it is too small, we may come across very simple solutions that underfit.

  The tuning process can be rather time-consuming for huge datasets. Many heuristic rules for selection of C have been recommended in the literature. Refer to [57] for a heuristic formula for the selection for the parameter, which has proven to be close to optimal in many practical situations. More about the difficulty associated with choice of C will appear in a later section.
- The final decision boundary is

$$\mathbf{w}^T\mathbf{x} + w_0 = 0$$

Substituting for $\mathbf{w}$ and w_0 from Eqns (4.53), we obtain

$$\left(\sum_{i \in svindex} \lambda_i y^{(i)} \mathbf{x}^{(i)}\right)^T \mathbf{x} + \frac{1}{|svmindex|}\left[\sum_{k \in svmindex}\left[y^{(k)} - \left(\sum_{i \in svindex} \lambda_i y^{(i)} \mathbf{x}^{(i)}\right)^T \mathbf{x}^{(k)}\right]\right] = 0$$

or

$$\sum_{i \in svindex} (\lambda_i y^{(i)} \mathbf{x}^{(i)T} \mathbf{x})$$

$$+ \frac{1}{|svmindex|}\left[\sum_{k \in svmindex}\left[y^{(k)} - \left(\sum_{i \in svindex} \lambda_i y^{(i)} \mathbf{x}^{(i)T} \mathbf{x}^{(k)}\right)\right]\right] = 0 \tag{4.55}$$

We notice that $\mathbf{w}$ and w_0 do not need to be explicitly computed. As we will see in a later section, this is crucial for using kernel functions to handle nonlinear decision boundaries.

Example 4.2

In Example 4.1, SVM (hard margin) formulation was developed in matrix form. In the following, we give matrix form of SVM (soft margin) formulation [56].

If all the data points are not linearly separable, we allow training error or in other words, allow points to lie between bounding hyperplanes and beyond. When a point, say $\mathbf{x}^{(i)} = [x_1^{(i)}\ x_2^{(i)}\ \cdots\ x_n^{(i)}]^T$

with $y^{(i)} = +1$ lies either between the bounding hyperplanes or beyond (into the region where $\mathbf{w}^T\mathbf{x}^{(i)} + w_0 \leq -1$), we add a positive quantity ζ_i to the left of inequality (refer to (4.23a)) to satisfy the constraint $\mathbf{w}^T\mathbf{x}^{(i)} + w_0 + \zeta_i \geq +1$. Similarly, when a point with $y^{(i)} = -1$ lies either between the bounding hyperplanes or beyond (into the region where $\mathbf{w}^T\mathbf{x}^{(i)} + w_0 \geq +1$), we subtract a positive quantity ζ_i from the left of the inequality (refer to (4.23a)) to satisfy the constraint $\mathbf{w}^T\mathbf{x}^{(i)} + w_0 - \zeta_i \leq -1$. For all other points, let us assume that we are adding ζ_i terms with zero values. Thus, we have the constraints (Eqn (4.47))

$$y^{(i)}(\mathbf{w}^T\mathbf{x}^{(i)} + w_0) \geq 1 - \zeta_i\,;\ \zeta_i \geq 0$$

The constraint equation in matrix form can be written as,

$$\mathbf{Y}(\mathbf{X}\mathbf{w} + w_0\mathbf{e}) + \boldsymbol{\zeta} \geq \mathbf{e} \tag{4.56}$$

where $$\underset{(N\times 1)}{\boldsymbol{\zeta}} = \begin{bmatrix} \zeta_1 \\ \zeta_2 \\ \vdots \\ \zeta_N \end{bmatrix} \geq \begin{bmatrix} 0 \\ 0 \\ \vdots \\ 0 \end{bmatrix}$$

and matrices/vectors **Y**, **X**, **w**, and **e** have already been defined in Eqns (4.42).

The optimization problem becomes (Eqn (4.48a))

$$\underset{\mathbf{w},\boldsymbol{\zeta}}{Minimize}\left[\tfrac{1}{2}\mathbf{w}^T\mathbf{w} + C\,\mathbf{e}^T\boldsymbol{\zeta}\right]$$

$$\text{subject to } \mathbf{Y}(\mathbf{X}\mathbf{w} + w_0\mathbf{e}) + \boldsymbol{\zeta} \geq \mathbf{e}, \text{ and } \boldsymbol{\zeta} \geq \mathbf{0} \tag{4.57}$$

Here, C is a scalar value (≥ 0) that controls the trade-off between margin and errors. This C is to be supplied at the time of training. Proper choice of C is crucial for good generalization performance of the classifier. Usually, the value of C is obtained by trial-and-error with cross-validation.

Minimization of the quantity $\frac{1}{2}\mathbf{w}^T\mathbf{w} + C\,\mathbf{e}^T\boldsymbol{\zeta}$ with respect to **w** and $\boldsymbol{\zeta}$ causes maximum separation between the bounding planes with minimum number of points crossing their respective bounding planes.

Proceeding in the manner similar to that described in Example 4.1 for hard-margin SVM, we may formulate the dual problem for soft-margin SVM as follows (Eqn (4.51)):

$$\underset{\boldsymbol{\lambda}}{Minimize}\left\{-\mathbf{e}^T\boldsymbol{\lambda} + \tfrac{1}{2}\boldsymbol{\lambda}^T\mathbf{Q}\,\boldsymbol{\lambda}\right\}$$

$$\text{subject to } \mathbf{y}^T\boldsymbol{\lambda} = 0,\ \mathbf{0} \leq \boldsymbol{\lambda} \leq C\mathbf{e} \tag{4.58}$$

where $\boldsymbol{\lambda} = [\lambda_1\ \lambda_2\ \ldots\ \lambda_N]^T$ are dual variables, and all other matrices/vectors have been defined earlier in Example 4.1.

The main difference between soft-margin and hard-margin SVM classifiers is that in case of soft margin, the Lagrange multipliers λ_i are bounded; $0 \le \lambda_i \le C$. We will separate the bounds on λ_i into three cases:

1. $\lambda_i = 0$. This leads to $\zeta_i = 0$, implying that the corresponding *i*th pattern is correctly classified.
2. $0 < \lambda_i < C$. This also leads to $\zeta_i = 0$, implying that corresponding *i*th pattern is correctly classified. The points with $\zeta_i = 0$ and $0 < \lambda_i < C$, fall on the bounding planes of the class to which the points belong. These are the support vectors that lie on the margin.
3. $\lambda_i = C$. In this case, $\zeta_i > 0$; this implies that the corresponding *i*th pattern is mis-classified or lies inside the margin. Since $\lambda_i \neq 0$, these are also support vectors

Once we obtain Lagrange multipliers, $\boldsymbol{\lambda}$, using quadratic programming, we can compute $\mathbf{w}$ and w_0 using Eqns (4.53).

As we will see in Section 4.7, SVM formulation for nonlinear classifiers is similar to the one given in this example. There, we will consider a toy dataset to illustrate numerical solution of SVM (soft-margin) problem.

4.6 KERNEL-INDUCED FEATURE SPACES

So far we have considered parametric models for classification (and regression) in which the form of mapping from input vector $\mathbf{x}$ to class label y (or continuous real-valued output y) is linear. One common strategy in machine learning is to learn nonlinear functions with a linear machine. For this, we need to change the representation of the data:

$$\mathbf{x} = \{x_1, \ldots, x_n\} \Rightarrow \boldsymbol{\phi}(\mathbf{x}) = \{\phi_1(\mathbf{x}), \ldots, \phi_m(\mathbf{x})\} \tag{4.59}$$

where $\boldsymbol{\phi}(.)$ is a nonlinear map from input feature space to some other feature space. The selection of $\boldsymbol{\phi}$ is constrained to yield a new feature space in which the linear machine can be used. Hence, the set of hypotheses we consider will be functions of the type

$$g(\boldsymbol{\phi}(\mathbf{x}), \mathbf{w}) = \sum_{l=1}^{m} w_l\, \phi_l(\mathbf{x}) + w_0 = \mathbf{w}^T \boldsymbol{\phi} + w_0 \tag{4.60}$$

where $\mathbf{w}$ is now the m-dimensional weight parameter.

This means that we will build nonlinear machines in two steps:

(i) first a fixed nonlinear mapping transforms the data into a new feature space, and
(ii) then a linear machine is used to classify the data in the new feature space.

By selecting m functions $\phi_l(\mathbf{x})$ judiciously, one can approximate any nonlinear discriminant function in $\mathbf{x}$ by such a linear expansion. The resulting discriminant function is not linear in $\mathbf{x}$, but it is linear in $\boldsymbol{\phi}(\mathbf{x})$. The m functions merely map points in the n-dimensional $\mathbf{x}$-space to points in the m-dimensional $\boldsymbol{\phi}$-space. The homogeneous discriminant function $\mathbf{w}^T\boldsymbol{\phi} + w_0$ separates points in the transformed space by a hyperplane. Thus, the mapping from $\mathbf{x}$ to $\boldsymbol{\phi}$ reduces the problem of finding nonlinear discriminant function in $\mathbf{x}$ to one of finding linear discriminant function in $\boldsymbol{\phi}(\mathbf{x})$. With a clever choice of nonlinear $\boldsymbol{\phi}$-functions, we can obtain arbitrary nonlinear decision regions in $\mathbf{x}$-space, in particular those leading to minimum errors.

The central difficulty is naturally choosing the appropriate m-dimensional mappings $\boldsymbol{\phi}(\mathbf{x})$ of the original input feature vectors $\mathbf{x}$. This approach of the design of nonlinear machines is linked to our expectation that the patterns which are not linearly separable in $\mathbf{x}$-space become linearly separable in $\boldsymbol{\phi}$-space. This expectation is based on the observations that by selecting $\phi_l(\mathbf{x})$; $l = 1, \ldots, m$, functions judiciously and letting m sufficiently large, one can approximate any nonlinear discriminant function in $\mathbf{x}$ by linear expansion (4.60).

In the sequel, we will first try to justify our expectations that by going to a higher dimensional space, the classification task may be transformed into a linear one, and then study popular alternatives for the choice of functions $\phi_l(\cdot)$.

Let us first attempt to intuitively understand why going to a higher dimensional space increases the chances of a linear separation. For linear function expansions such as,

$$g(\mathbf{x}, \mathbf{w}) = \sum_{j=1}^{n} w_j x_j + w_0$$

or

$$g(\boldsymbol{\phi}(\mathbf{x}), \mathbf{w}) = \sum_{l=1}^{m} w_l \, \phi_l(\mathbf{x}) + w_0$$

the VC dimension increases as the number of weight parameters increases. Using a transformation $\boldsymbol{\phi}(\mathbf{x})$ to a higher dimensional $\boldsymbol{\phi}$-space ($m > n$), typically amounts to increasing the *capacity* (refer to Section 2.3) of the learning machine, and rendering problems separable that are not linearly separable to start with.

Cover's theorem [58] formalizes the intuition that the *number of linear separations increases with the dimensionality*. The number of possible linear separations of *well distributed* N points in an n-dimensional space ($N > n + 1$), as per this theorem, equals

$$2\sum_{j=0}^{n} \binom{N-1}{j} = 2\sum_{j=0}^{n} \frac{(N-1)!}{(N-1-j)!\, j!}$$

The more we increase n, the more terms there are in the sum, and thus the larger is the resulting number.

The linear separability advantage of high-dimensional feature spaces comes with a cost: the *computational complexity*. The approach of selecting $\phi_l(\mathbf{x})$; $l = 1, \ldots, m$, with $m \to \infty$ may not work; such a classifier would have too many free parameters to be determined from a limited number of training data. Also, if the number of parameters is too large relative to the number of training examples, the resulting model will overfit the data, affecting the generalization performance.

So there are problems. First how do we choose the nonlinear mapping to a higher dimensional space? Second, the computations involved will be costly. It so happens that in solving the quadratic optimization problem of the linear SVM (i.e., when searching for a linear SVM in the new higher dimensional space), the training tuples appear only in the form of dot products (refer to Eqn (4.51)):

$$\langle \boldsymbol{\phi}(\mathbf{x}^{(i)}), \boldsymbol{\phi}(\mathbf{x}^{(k)}) \rangle = [\boldsymbol{\phi}(\mathbf{x}^{(i)})]^T \, [\boldsymbol{\phi}(\mathbf{x}^{(k)})]$$

Note that the dot product requires one multiplication and one addition for each of the m-dimensions. Also, we have to compute the dot product with every one of the support vectors. In training, we have to compute a similar dot product several times. The dot product computation required is very heavy and costly. We need a *trick* to avoid the dot product computations.

Luckily, we can use a math trick. Instead of computing the dot product on the transformed data tuples, it turns out that it is mathematically equivalent to instead apply a *kernel function*, $K(\mathbf{x}^{(i)}, \mathbf{x}^{k})$, to the original input data. That is,

$$K(\mathbf{x}^{(i)}, \mathbf{x}^{(k)}) = \langle \boldsymbol{\phi}(\mathbf{x}^{(i)}), \boldsymbol{\phi}(\mathbf{x}^{(k)}) \rangle$$

In other words, everywhere that $\langle \boldsymbol{\phi}(\mathbf{x}^{(i)}), \boldsymbol{\phi}(\mathbf{x}^{(k)}) \rangle$ appears in the training algorithm, we can replace it with $K(\mathbf{x}^{(i)}, \mathbf{x}^{(k)})$. In this way, all calculations are made in the original input space, which is of potentially much lower dimensionality.

Another feature of the *kernel trick* addresses the problem of choosing nonlinear mapping $\boldsymbol{\phi}(\mathbf{x})$. It turns out that we don't even have to know what the mapping is. That is, admissible kernel substitutions $K(\mathbf{x}^{(i)}, \mathbf{x}^{(k)})$ can be determined without the need of first selecting a mapping function $\boldsymbol{\phi}(\mathbf{x})$.

Let us study the kernel trick in a little more detail for appreciating this important property of SVMs to solve nonlinear classification problems.

Example 4.3

Given a database with feature vectors $\mathbf{x} = [x_1 \; x_2]^T$. Consider the nonlinear mapping

$$(x_1, x_2) \xrightarrow{\boldsymbol{\phi}} (\phi_1(\mathbf{x}), \phi_2(\mathbf{x}), \phi_3(\mathbf{x}), \phi_4(\mathbf{x}), \phi_5(\mathbf{x}), \phi_6(\mathbf{x}))$$

$$\boldsymbol{\phi}(\mathbf{x}) = [1 \;\; \sqrt{2}x_1 \;\; \sqrt{2}x_2 \;\; \sqrt{2}x_1x_2 \;\; x_1^2 \;\; x_2^2]^T$$

For this mapping, the dot product is

$$[\boldsymbol{\phi}(\mathbf{x}^{(i)})]^T \boldsymbol{\phi}(\mathbf{x}^{(k)}) = [1 \;\; \sqrt{2}x_1^{(i)} \;\; \sqrt{2}x_2^{(i)} \;\; \sqrt{2}x_1^{(i)}x_2^{(i)} \;\; (x_1^{(i)})^2 \;\; (x_2^{(i)})^2] \begin{bmatrix} 1 \\ \sqrt{2}x_1^{(k)} \\ \sqrt{2}x_2^{(k)} \\ \sqrt{2}x_1^{(k)}x_2^{(k)} \\ (x_1^{(k)})^2 \\ (x_2^{(k)})^2 \end{bmatrix}$$

$$= 1 + 2x_1^{(i)}\, x_1^{(k)} + 2x_2^{(i)}\, x_2^{(k)} + 2x_1^{(i)}\, x_2^{(i)}\, x_1^{(k)}\, x_2^{(k)}$$

$$+ (x_1^{(i)})^2\, (x_1^{(k)})^2 + (x_2^{(i)})^2\, (x_2^{(k)})^2$$

$$= [(\mathbf{x}^{(i)})^T\, \mathbf{x}^{(k)} + 1]^2$$

The inner product of vectors in new (higher dimensional) space has been expressed as a function of the inner product of the corresponding vectors in the original (lower dimensional) space. The *kernel function* is

$$K(\mathbf{x}^{(i)}, \mathbf{x}^{(k)}) = [\mathbf{x}^{(i)T}\,\mathbf{x}^{(k)} + 1]^2$$

Constructing Kernels

In order to exploit *kernel trick*, we need to be able to construct valid kernel functions. A straightforward way of computing kernel $K(\cdot)$ from a map $\phi(\cdot)$ is to choose a feature mapping $\phi(\mathbf{x})$ and then use it to find the corresponding kernel:

$$K(\mathbf{x}^{(i)}, \mathbf{x}^{(k)}) = [\phi(\mathbf{x}^{(i)})]^T\phi(\mathbf{x}^{(k)}) \tag{4.61}$$

The kernel trick makes it possible to map the data *implicitly* into a higher dimensional feature space, and to train a linear machine in such a space, potentially side-stepping the computational problems inherent in this high-dimensional space. One of the curious facts about using a kernel is that we do not need to know the underlying feature map $\phi(\cdot)$ in order to be able to learn in the new feature space. Kernel trick shows a way of computing dot products in these high-dimensional spaces without *explicitly* mapping into the spaces.

The straightforward way of computing kernel $K(\cdot)$ from the map $\phi(\cdot)$ given in Eqn (4.61), can be inverted, i.e., we can choose a kernel rather than the mapping and apply it to the learning algorithm directly.

We can, of course, first propose a kernel function and then expand it to identify $\phi(\mathbf{x})$. Identification of ϕ is not needed if we can show whether the proposed function is a kernel or not without the need of corresponding mapping function.

Mercer's Theorem: In the following, we introduce Mercer's theorem, which provides a test whether a function $K(\mathbf{x}^{(i)},\mathbf{x}^{(k)})$ constitutes a valid kernel without having to construct the function $\phi(\mathbf{x})$ explicitly.

Let $K(\mathbf{x}^{(i)}, \mathbf{x}^{(k)})$ be a *symmetric* function on the finite input space. Then $K(\mathbf{x}^{(i)}, \mathbf{x}^{(k)})$ is a kernel function if and only if the matrix

$$\mathbf{K} = \begin{bmatrix} K(\mathbf{x}^{(1)},\mathbf{x}^{(1)}) & K(\mathbf{x}^{(1)},\mathbf{x}^{(2)}) & \dots & K(\mathbf{x}^{(1)},\mathbf{x}^{(N)}) \\ \vdots & \vdots & K(\mathbf{x}^{(i)},\mathbf{x}^{(k)}) & \vdots \\ K(\mathbf{x}^{(N)},\mathbf{x}^{(1)}) & K(\mathbf{x}^{(N)},\mathbf{x}^{(2)}) & \dots & K(\mathbf{x}^{(N)},\mathbf{x}^{(N)}) \end{bmatrix} \tag{4.62}$$

is *positive semidefinite*.

For any function $K(\mathbf{x}^{(i)}, \mathbf{x}^{(k)})$ satisfying Mercer's theorem, there exists a space in which $K(\mathbf{x}^{(i)}, \mathbf{x}^{(k)})$ defines an inner product. What, however, Mercer's theorem does not disclose to us is how to find this space. That is, we do not have a general tool to construct the mapping function $\phi(\cdot)$ once we know the kernel function (in simple cases, we can expand $K(\mathbf{x}^{(i)}, \mathbf{x}^{(k)})$ and rearrange it to give $[\phi(\mathbf{x}^{(i)})]^T\phi(\mathbf{x}^{(k)})$). Furthermore, we lack the means to know the dimensionality of the space, which can even be infinite. For further details, see [53].

What are some of the kernel functions that could be used? Properties of the kinds of kernel functions that could be used to replace the dot product scenario just described have been studied. The most popular general-purpose kernel functions are:

Polynomial kernel of degree d

$$K(\mathbf{x}^{(i)}, \mathbf{x}^{(k)}) = (\mathbf{x}^{(i)T}\,\mathbf{x}^{(k)} + c)^d;\ c > 0,\ d \geq 2 \tag{4.63}$$

Gaussian radial basis function kernel

$$K(\mathbf{x}^{(i)}, \mathbf{x}^{(k)}) = \exp\left(-\frac{\|\mathbf{x}^{(i)} - \mathbf{x}^{(k)}\|^2}{2\sigma^2}\right);\ \sigma > 0 \tag{4.64}$$

The feature vector that corresponds to the Gaussian kernel has infinite dimensionality.

Sigmoidal kernel

$$K(\mathbf{x}^{(i)}, \mathbf{x}^{(k)}) = \tanh\,(\beta\,\mathbf{x}^{(i)T}\,\mathbf{x}^{(k)} + \gamma) \tag{4.65}$$

for appropriate values of β and γ so that Mercer's conditions are satisfied. One possibility is $\beta = 2$, $\gamma = 1$.

Each of these results in a different nonlinear classifier in (the original) input space.

There are no golden rules for determining which admissible kernel will result in the most accurate SVM. In practice, the kernel chosen does not generally make a large difference in resulting accuracy. SVM training always finds a global solution, unlike neural networks (discussed in the next chapter) where many local minima usually exist.

4.7 NONLINEAR CLASSIFIER

The SVM formulations debated till now, need Class 1 and Class 2 examples to be capable of linear representation, that is, with the decision boundary being a hyperplane. But, for several real-life datasets, the decision boundaries are nonlinear. To deal with nonlinear case, the formulation and solution methods employed for the linear case are still applicable. Only input data is transformed from its original space into another space (generally, a much higher dimensional space) so that a linear decision boundary can separate Class 1 examples from Class 2 in the transformed space, called the *feature space*. The original data space is known as the *input space*.

Let the set of training (data) examples be

$$\mathcal{D} = \{\mathbf{x}^{(1)}, y^{(1)}), (\mathbf{x}^{(2)}, y^{(2)}), \ldots, (\mathbf{x}^{(N)}, y^{(N)})\} \tag{4.66}$$

where $\mathbf{x} = [x_1\ x_2\ \ldots\ x_n]^T$.

Figure 4.13 illustrates the process. In the input space, the training examples cannot be linearly separated; in the feature space, they can be separated linearly.

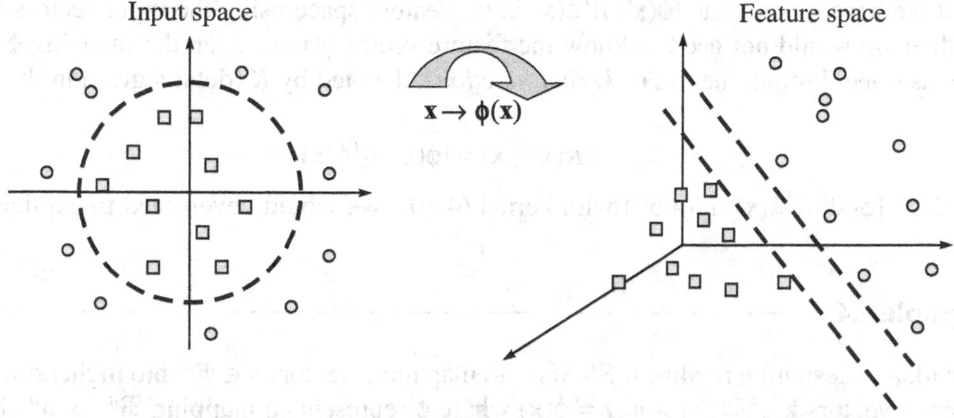

Figure 4.13 Transformation from the input space to feature space

With the transformation, the optimization problem in (4.48a) becomes

$$minimize\ \tfrac{1}{2}\mathbf{w}^T\mathbf{w} + C\sum_{i=1}^{N}\zeta_i \tag{4.67}$$

$$\text{subject to } y^{(i)}(\mathbf{w}^T\phi(\mathbf{x}^{(i)}) + w_0) \geq 1 - \zeta_i;\ i = 1, \ldots, N$$

$$\zeta_i \geq 0;\ i = 1, \ldots, N$$

The corresponding dual is (refer to (4.51))

$$minimize\ L_*(\boldsymbol{\lambda}) = \sum_{i=1}^{N}\lambda_i - \tfrac{1}{2}\sum_{i=1}^{N}\sum_{k=1}^{N}\lambda_i\,\lambda_k y^{(i)}y^{(k)}\,[\phi(\mathbf{x}^{(i)})]^T\,\phi(\mathbf{x}^{(k)})$$

$$\text{subject to } \sum_{i=1}^{N}\lambda_i y^{(i)} = 0 \tag{4.68}$$

$$0 \leq \lambda_i \leq C;\ i = 1, \ldots, N$$

The potential issue with this strategy is that there are chances of it suffering from the curse of dimensionality. The number of dimensions in the feature space may be very large with certain useful transformations, even with a reasonable number of attributes in the input space. Luckily, explicit transformations are not required as we see that for the dual problem (4.68), the building of the decision boundary only requires the assessment of $[\phi(\mathbf{x}^{(i)})]^T\phi(\mathbf{x})$ in the feature space. With reference to (4.55), we have the following decision boundary in the feature space:

$$\sum_{i=1}^{N}\lambda_i y^{(i)}[\phi(\mathbf{x}^{(i)})]^T\,\phi(\mathbf{x}) + w_0 = 0 \tag{4.69}$$

Thus, if we have to compute $[\boldsymbol{\phi}(\mathbf{x}^{(i)})]^T\boldsymbol{\phi}(\mathbf{x})$ in the feature space using the input vectors $\mathbf{x}^{(i)}$ and $\mathbf{x}$ directly, then we would not need to know the feature vector $\boldsymbol{\phi}(\mathbf{x})$ or even the mapping $\boldsymbol{\phi}$ itself. In SVM, this is done through the use of *kernel functions*, denoted by K (details given in the previous section):

$$K(\mathbf{x}^{(i)}, \mathbf{x}) = [\boldsymbol{\phi}(\mathbf{x}^{(i)})]^T\boldsymbol{\phi}(\mathbf{x}) \tag{4.70}$$

We replace $[\boldsymbol{\phi}(\mathbf{x}^{(i)})]^T\boldsymbol{\phi}(\mathbf{x})$ in (4.69) with kernel (4.70). We would never need to explicitly know what $\boldsymbol{\phi}$ is.

Example 4.4

The basic idea in designing nonlinear SVMs is to map input vectors $\mathbf{x} \in \Re^n$ into higher dimensional feature-space vectors $\mathbf{z} \in \Re^m$; $m > n$. $\mathbf{z} = \boldsymbol{\phi}(\mathbf{x})$ where $\boldsymbol{\phi}$ represents a mapping $\Re^n \to \Re^m$. Note that input space is spanned by components x_j; $j = 1, \ldots, n$, of an input vector $\mathbf{x}$, and feature space is spanned by components $\mathbf{z}_l$; $l = 1, \ldots, m$, of vector $\mathbf{z}$. By performing such a mapping, we expect that in feature space, the learning algorithm will be able to linearly separate the mapped data by applying the linear SVM formulation. This approach leads to a solution of a quadratic optimization problem with inequality constraints in $\mathbf{z}$-space. The solution for an *indicator function*, *sgn* $(\mathbf{w}^T\mathbf{z} + w_0)$, which is a linear classifier in feature space, creates a nonlinear separating hypersurface in the original input space.

There are two basic problems in taking this approach when mapping an input $\mathbf{x}$-space into higher-order $\mathbf{z}$-space:

1. Choice of $\boldsymbol{\phi}(\mathbf{x})$, which should result in a rich class of decision hypersurfaces.
2. Calculation of the scalar product $\mathbf{z}^T\mathbf{z}$, which can be computationally very discouraging if the feature-space dimension m is very large.

The explosion in dimensionality from n to m can be avoided in calculations by noticing that in the quadratic optimization problem (Eqn (4.51)), training data only appear in the form of scalar products $\mathbf{x}^T\mathbf{x}$. These products are replaced by scalar products $\mathbf{z}^T\mathbf{z}$ in feature space, and the latter are expressed by using a *symmetric* kernel function $K(\mathbf{x}^{(i)}, \mathbf{x}^{(k)})$ that results in *positive semidefinite* kernel matrix (Eqn (4.62))

$$\mathbf{K}(\mathbf{x}^{(i)}, \mathbf{x}^{(k)}) = \begin{bmatrix} K(\mathbf{x}^{(1)}, \mathbf{x}^{(1)}) & K(\mathbf{x}^{(1)}, \mathbf{x}^{(2)}) & \cdots & K(\mathbf{x}^{(1)}, \mathbf{x}^{(N)}) \\ \vdots & \vdots & K(\mathbf{x}^{(i)}, \mathbf{x}^{(k)}) & \vdots \\ K(\mathbf{x}^{(N)}, \mathbf{x}^{(1)}) & K(\mathbf{x}^{(N)}, \mathbf{x}^{(2)}) & \cdots & K(\mathbf{x}^{(N)}, \mathbf{x}^{(N)}) \end{bmatrix} \tag{4.71}$$

The required scalar products $\mathbf{z}^T\mathbf{z}$ in feature space are calculated directly by computing kernels $\mathbf{K}(\mathbf{x}^{(i)}, \mathbf{x}^{(k)})$ for given training data vectors in an input space. In this way, we bypass the computational complexity of an extremely high dimensionality of feature space. By applying kernels, we do not even have to know what the actual mapping $\boldsymbol{\phi}(\mathbf{x})$ is. Thus, using the chosen kernel $\mathbf{K}(\mathbf{x}^{(i)}, \mathbf{x}^{(k)})$, an SVM can be constructed that operates in an infinite dimensional space.

Another problem with kernel-induced nonlinear classification approach is regarding the choice of a particular type of kernel function. There is no clear-cut answer. No theoretical proofs yet exist supporting applications for any particular type of kernel function. For the time being, one can only suggest that various models be tried on a given dataset and that the one with the best generalization capacity be chosen.

The learning algorithm for nonlinear soft-margin SVM classifier has already been given. Let us give the matrix formulation here, which is helpful in using standard quadratic optimization software.

In the matrix form, nonlinear SVM (soft margin) formulation is (Eqn (4.58))

$$\underset{\boldsymbol{\lambda}}{Minimize}\left\{-\mathbf{e}^T\boldsymbol{\lambda}+\tfrac{1}{2}\boldsymbol{\lambda}^T\mathbf{Q}\,\boldsymbol{\lambda}\right\}$$
$$\text{subject to } \mathbf{y}^T\boldsymbol{\lambda}=0,\ \mathbf{0}\le\boldsymbol{\lambda}\le C\mathbf{e} \tag{4.72}$$

Here, $\mathbf{Q}=\mathbf{YKY}$, and $\mathbf{K}$ is kernel matrix (Eqn (4.71). This formulation follows from Eqn (4.68). As an illustration, with consider toy dataset with $n = 1$.

$$\mathbf{X}=\begin{bmatrix}1\\2\\5\\6\end{bmatrix};\ \mathbf{y}=\begin{bmatrix}-1\\-1\\1\\-1\end{bmatrix}$$

For the choice of the kernel function $K(\mathbf{x}^{(i)},\mathbf{x}^{(k)})=(\mathbf{x}^{(i)T}\mathbf{x}^{(k)}+1)^2$, the matrix $\mathbf{Q}=\mathbf{YKY}$ is given by:

$$\mathbf{Q}=\begin{bmatrix}4 & 9 & -36 & 49\\ 9 & 25 & -121 & 169\\ -36 & -121 & 676 & -961\\ 49 & 169 & -961 & 1369\end{bmatrix}$$

The SVM formulation with $C = 50$, yields [56]

$$\boldsymbol{\lambda}=[0\quad 2.5\quad 7.333\quad 4.833]^T=[\lambda_1\quad \lambda_2\quad \lambda_3\quad \lambda_4]^T.$$

From Eqn (4.55), we have,

$$\mathbf{w}=\sum_{i\in svindex}\lambda_i y^{(i)}\mathbf{z}^{(i)}\ ;\ svindex=\{2,3,4\}$$

This gives

$$\mathbf{w}^T\mathbf{z}=\sum_{i\in svindex}\lambda_i y^{(i)}\mathbf{z}^{(i)T}\mathbf{z}$$

$$= \sum_{i \in svindex} \lambda_i y^{(i)} K(\mathbf{x}^{(i)}, \mathbf{x}) \tag{4.73}$$

$$= \sum_{i \in svindex} \lambda_i y^{(i)} (\mathbf{x}^{(i)T} \mathbf{x} + 1)^2$$

For the given data,

$$\mathbf{w}^T \mathbf{z} = (2.5)(-1)(2x+1)^2 + (7.333)(1)(5x+1)^2 + (4.833)(-1)(6x+1)^2$$
$$= -0.667x^2 + 5.333x$$

The bias w_0 is determined from the requirement that at the support-vector points $x = 2, 5$, and 6, the outputs must be -1, $+1$, and -1, respectively. From Eqn (4.55), we have

$$w_0 = \frac{1}{|svmindex|} \left[\sum_{k \in svmindex} \left[y^{(k)} - \left(\sum_{i \in svindex} \lambda_i y^{(i)} \mathbf{z}^{(i)T} \mathbf{z}^{(k)} \right) \right] \right]$$

$$= \frac{1}{|svmindex|} \left[\sum_{k \in svmindex} \left[y^{(k)} - \left(\sum_{i \in svindex} \lambda_i y^{(i)} K(\mathbf{x}^{(i)}, \mathbf{x}^{(k)}) \right) \right] \right] \tag{4.74}$$

$$= \frac{1}{|svmindex|} \left[\sum_{k \in svmindex} \left[y^{(k)} - \left(\sum_{i \in svindex} \lambda_i y^{(i)} (\mathbf{x}^{(i)T} \mathbf{x}^{(k)} + 1)^2 \right) \right] \right]$$

$$= \tfrac{1}{3}[-1 - (-0.667(2)^2 + 5.333(2)) + 1 - (-0.667(5)^2$$
$$+ 5.333(5)) - 1 - (-0.667(6)^2 + 5.333(6))]$$
$$= -9$$

Therefore, the nonlinear decision function in the input space:

$$g(\mathbf{x}) = -0.667x^2 + 5.333x - 9$$

and indicator function

$$i_F = \hat{y} = sgn(g(\mathbf{x}))$$
$$= sgn(-0.667x^2 + 5.333x - 9)$$

The nonlinear decision function and the indicator function for one-dimensional data under consideration, are shown in Fig. 4.14.

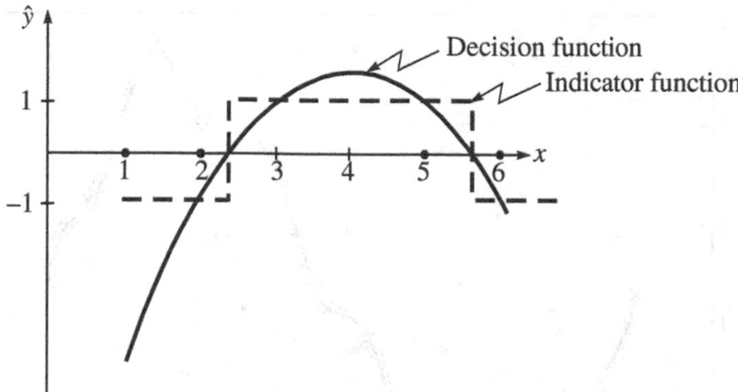

Figure 4.14 Nonlinear SV classification for Example 4.4

4.8 REGRESSION BY SUPPORT VECTOR MACHINES

Initially developed for solving classification problems, SV techniques can also be successfully applied in regression (numeric prediction) problems. Unlike classification (pattern recognition) problems where the desired outputs are discrete values: $y \in [+1, -1]$, here the system responses $y \in \Re$ are continuous values. The general regression learning problem is set as follows:

The learning machine is given N training data

$$\mathcal{D}: \{(\mathbf{x}^{(1)}, y^{(1)}), \ldots, (\mathbf{x}^{(N)}, y^{(N)})\};\ \mathbf{x} \in \Re^n,\ y \in \Re \tag{4.75}$$

where inputs $\mathbf{x}$ are n-dimensional vectors and scalar output y has continuous values. The objective is to learn the input-output relationship $\hat{y} = f(\mathbf{x})$: a nonlinear regression model.

In regression, typically some measure for *error of approximation* is used instead of margin between an optimal separating hyperplane and support vectors, which was used in the design of SV classifiers. In our regression formulations described in earlier chapters, we have used *sum-of-error-squares* criterion (Section 2.7). Here, in SV regression, our goal is to find a function $\hat{y} = f(\mathbf{x})$ that has at most ε deviation (when ε is a prescribed parameter) from the actually obtained targets y for all the training data. In other words, we do not care about errors as long as they are less than ε, but any deviation larger than this will be treated as *regression error*.

To account for the regression error in our SV formulation, we use *ε-insensitive loss function*:

$$|y - f(\mathbf{x})|_\varepsilon \triangleq \begin{cases} 0 & \text{if } |y - f(\mathbf{x})| \le \varepsilon \\ |y - f(\mathbf{x})| - \varepsilon & \text{otherwise} \end{cases} \tag{4.76}$$

This loss (error) function defines an *ε-insensitivity zone* (*ε-tube*). We tolerate errors up to ε (data point $(\mathbf{x}^{(i)}, y^{(i)})$ within the ε-insensitivity zone or ε-tube), and the errors beyond (above/below the ε-tube) have a *linear* effect (unlike sum-of-error-squares criterion). The error function is, therefore, more tolerant to noise and is thus more *robust*. There is a region of no error, which results in sparseness.

The parameter ε defines the requirements on the accuracy of approximation. An increase in ε means a reduction in accuracy requirements; it results in smoothing effects on modeling highly noisy polluted data. On the other hand, a decrease in ε may result in complex model that overfits the data (refer to Fig. 4.15).

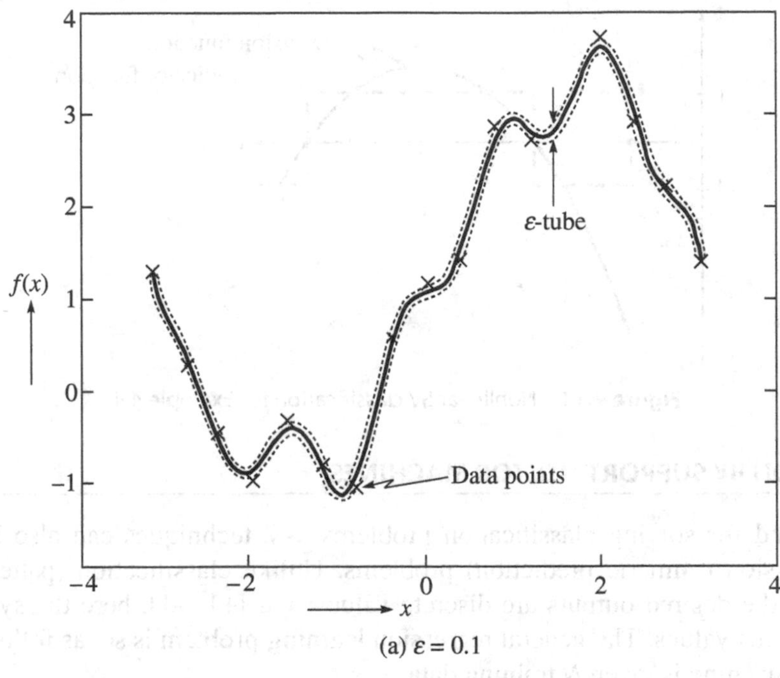

(a) $\varepsilon = 0.1$

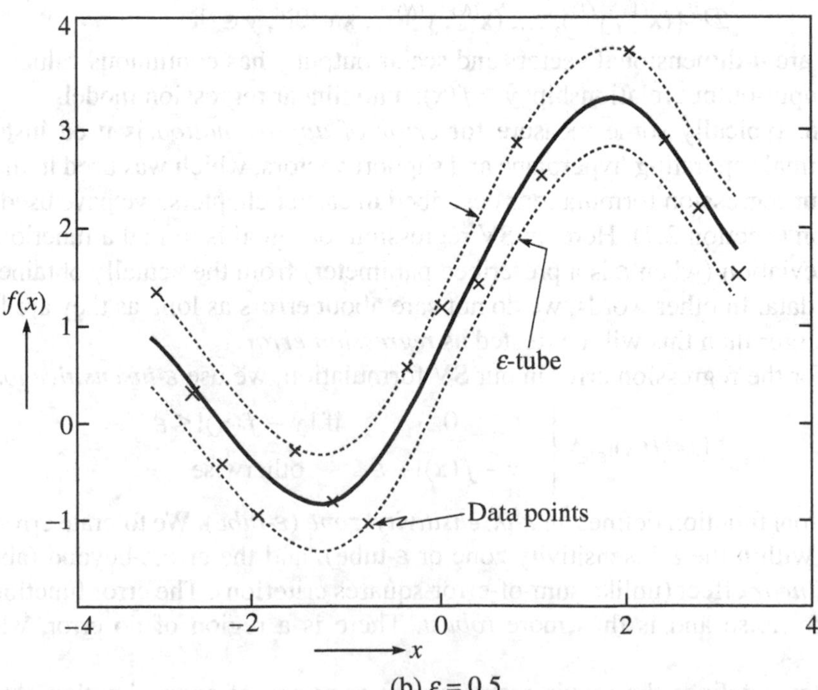

(b) $\varepsilon = 0.5$

Figure 4.15 One-dimensional SV regression

In formulating an SV algorithm for regression, the objective is to minimize the error (loss) and $\|\mathbf{w}\|^2$ simultaneously. The role of $\|\mathbf{w}\|^2$ in the objective function is to reduce model complexity; thereby preventing the problem of overfitting, thus improving generalization.

4.8.1 Linear Regression

For pedagogical reasons, we begin by describing SV formulation for *linear regression*; functions $f(\cdot)$ taking the form

$$f(\mathbf{x}) = \mathbf{w}^T\mathbf{x} + w_0;\ \mathbf{w} \in \Re^n,\ w_0 \in \Re \tag{4.77}$$

Analogous to the 'soft margin' classifier described earlier, we introduce (non-negative) slack variables ζ_i, ζ_i^*; $i = 1, \ldots, N$; to measure the deviation of training examples outside the ε-insensitivity zone. Figure 4.16 shows how the ε-insensitivity zone looks like when the regression is linear.

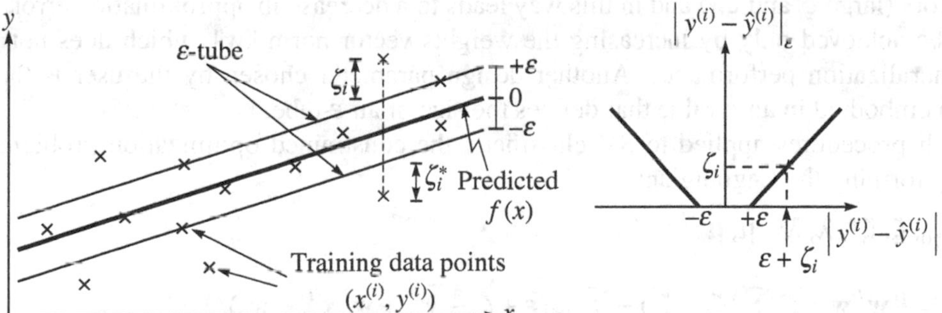

(a) One-dimensional support vector linear regression (b) ε-insensitive loss function

Figure 4.16

If a point $(\mathbf{x}^{(i)}, y^{(i)})$ falls within the ε-tube, the associated ζ_i, ζ_i^* is zero. If it is above the tube, $\zeta_i > 0$, $\zeta_i^* = 0$ and $\zeta_i = 0$, $\zeta_i^* > 0$ if the point is below it.

$$|y^{(i)} - f(\mathbf{x}^{(i)})| - \varepsilon = \zeta_i \text{ for data 'above' the } \varepsilon\text{-insensitivity zone} \tag{4.78a}$$

$$|y^{(i)} - f(\mathbf{x}^{(i)})| - \varepsilon = \zeta_i^* \text{ for data 'below' the } \varepsilon\text{-insensitivity zone} \tag{4.78b}$$

The loss (error) is equal to zero for training data points inside the tube ($|y^{(i)} - \hat{y}^{(i)}| \le \varepsilon$); the loss is ζ_i for data 'above' the tube ($y_i - \hat{y}^{(i)} - \varepsilon = \zeta_i$), and ζ_i^* for data 'below' the tube ($\hat{y}^{(i)} - y^{(i)} - \varepsilon = \zeta_i^*$). Only the data points outside the tube contribute to the loss (error) with deviations penalized in a linear fashion.

Minimizing $\|\mathbf{w}\|^2$ simultaneously with minimizing the loss, results in small values for $\mathbf{w}$ and thereby a *flat* function $f(\mathbf{x})$ given by (4.77).

Analogous to the SV formulation for soft margin linear classifiers, we arrive at the following formulation of linear regression:

$$\begin{aligned} &\textit{minimize } \tfrac{1}{2}\mathbf{w}^T\mathbf{w} + C\sum_{i=1}^{N}(\zeta_i + \zeta_i^*) \\ &\text{subject to } y^{(i)} - \mathbf{w}^T\mathbf{x}^{(i)} - w_0 \le \varepsilon + \zeta_i;\ i = 1, \ldots, N \\ &\qquad \mathbf{w}^T\mathbf{x}^{(i)} + w_0 - y^{(i)} \le \varepsilon + \zeta_i^*;\ i = 1, \ldots, N \\ &\qquad \zeta_i, \zeta_i^* \ge 0;\ i = 1, \ldots, N \end{aligned} \tag{4.79}$$

Note that the constant $C > 0$, which influences a trade-off between an approximation error and the weights vector norm $\|\mathbf{w}\|$, is a design parameter chosen by the user. An increase in C penalizes larger errors (large ζ_i and ζ_i^*) and in this way leads to a decrease in approximation error. However, this can be achieved only by increasing the weights vector norm $\|\mathbf{w}\|$, which does not guarantee good generalization performance. Another design parameter chosen by the user is the required precision embodied in an ε value that defines the size of an ε-tube.

As with procedures applied to SV classifiers, the constrained optimization problem (4.79) is solved by forming the Lagrangian:

$$\begin{aligned} &L(\mathbf{w}, w_0, \boldsymbol{\zeta}, \boldsymbol{\zeta}^*, \boldsymbol{\lambda}, \boldsymbol{\lambda}^*, \boldsymbol{\mu}, \boldsymbol{\mu}^*) \\ &= \tfrac{1}{2}\mathbf{w}^T\mathbf{w} + C\sum_{i=1}^{N}(\zeta_i + \zeta_i^*) - \sum_{i=1}^{N}\lambda_i(\varepsilon + \zeta_i - y^{(i)} + \mathbf{w}^T\mathbf{x}^{(i)} + w_0) \\ &\quad - \sum_{i=1}^{N}\lambda_i^*(\varepsilon + \zeta_i^* + y^{(i)} - \mathbf{w}^T\mathbf{x}^{(i)} - w_0) - \sum_{i=1}^{N}(\mu_i\zeta_i + \mu_i^*\zeta_i^*) \end{aligned} \tag{4.80}$$

where $\mathbf{w}$, w_0, ζ_i and ζ_i^* are the primal variables, and λ_i, λ_i^*, μ_i, $\mu_i^* \ge 0$ are the dual variables.

The KKT conditions are as follows:

$$\begin{aligned} &\text{(i)}\quad \frac{\partial L}{\partial \mathbf{w}} = \mathbf{w} - \sum_{i=1}^{N}(\lambda_i - \lambda_i^*)\,\mathbf{x}^{(i)} = \mathbf{0} \\ &\qquad \frac{\partial L}{\partial w_0} = \sum_{i=1}^{N}(\lambda_i^* - \lambda_i) = 0 \\ &\qquad \frac{\partial L}{\partial \zeta_i} = C - \lambda_i - \mu_i = 0;\ i = 1, \ldots, N \\ &\qquad \frac{\partial L}{\partial \zeta_i^*} = C - \lambda_i^* - \mu_i^* = 0;\ i = 1, \ldots, N \\ &\text{(ii)}\quad \varepsilon + \zeta_i - y^{(i)} + \mathbf{w}^T\mathbf{x}^{(i)} + w_0 \ge 0;\ i = 1, \ldots, N \\ &\qquad \varepsilon + \zeta_i^* + y^{(i)} - \mathbf{w}^T\mathbf{x}^{(i)} - w_0 \ge 0;\ i = 1, \ldots, N \\ &\qquad \zeta_i, \zeta_i^* \ge 0;\ i = 1, \ldots, N \end{aligned} \tag{4.81}$$

(iii) $\lambda_i, \lambda_i^*, \mu_i, \mu_i^* \geq 0; i = 1, \ldots, N$

(iv) $\lambda_i(\varepsilon + \zeta_i - y^{(i)} + \mathbf{w}^T\mathbf{x}^{(i)} + w_0) = 0; i = 1, \ldots, N$

$$\lambda_i^*(\varepsilon + \zeta_i^* + y^{(i)} - \mathbf{w}^T\mathbf{x}^{(i)} - w_0) = 0; i = 1, \ldots, N$$

$$\mu_i \zeta_i = 0; i = 1, \ldots, N$$

$$\mu_i^* \zeta_i^* = 0; i = 1, \ldots, N$$

Substituting the relations in condition (i) of KKT conditions (4.81) into Lagrangian (4.80), yields the dual objective function. The procedure is parallel to what has been followed earlier. The resulting dual optimization problem is

maximize $L_*(\boldsymbol{\lambda}, \boldsymbol{\lambda}^*) =$

$$-\varepsilon \sum_{i=1}^{N} (\lambda_i + \lambda_i^*) + \sum_{i=1}^{N} (\lambda_i - \lambda_i^*) y^{(i)} - \tfrac{1}{2} \sum_{i=1}^{N} \sum_{k=1}^{N} (\lambda_i - \lambda_i^*)(\lambda_k - \lambda_k^*) \mathbf{x}^{(i)T} \mathbf{x}^{(k)} \tag{4.82}$$

subject to $\sum_{i=1}^{N} (\lambda_i - \lambda_i^*) = 0; \lambda_i, \lambda_i^* \in [0, C]$

From condition (i) of KKT conditions (4.81), we have,

$$\mathbf{w} = \sum_{i=1}^{N} (\lambda_i - \lambda_i^*) \mathbf{x}^{(i)} \tag{4.83}$$

Thus, the weight vector **w** is completely described as a linear combination of the training patterns $\mathbf{x}^{(i)}$. One of the most important properties of SVM is that the solution is sparse in λ_i, λ_i^*. For $|\hat{y}^{(i)} - y^{(i)}| < \varepsilon$, the second factor in the following KKT conditions (conditions (iv) in (4.81)):

$$\begin{aligned} \lambda_i(\varepsilon + \zeta_i - y^{(i)} + \mathbf{w}^T\mathbf{x}^{(i)} + w_0) &= \lambda_i(\varepsilon + \zeta_i - y^{(i)} + \hat{y}^{(i)}) = 0 \\ \lambda_i^*(\varepsilon + \zeta_i^* + y^{(i)} - \mathbf{w}^T\mathbf{x}^{(i)} - w_0) &= \lambda_i^*(\varepsilon + \zeta_i^* + y^{(i)} - \hat{y}^{(i)}) = 0 \end{aligned} \tag{4.84}$$

are nonzero; hence λ_i, λ_i^* have to be zero. This equivalently means that all thc data points inside the ε-insensitive tube (a large number of training examples belong to this category) have corresponding λ_i, λ_i^* equal to zero. Further, from Eqn (4.84), it follows that only for $|\hat{y}^{(i)} - y^{(i)}| \geq \varepsilon$, the dual variables λ_i, λ_i^* may be nonzero. Since there can never be a set of dual variables λ_i, λ_i^* which are both simultaneously nonzero, as this would require slacks in both directions ('above' the tube and 'below' the tube), we have $\lambda_i \times \lambda_i^* = 0$.

From conditions (i) and (iv) of KKT conditions (4.81), it follows that

$$\begin{aligned} (C - \lambda_i)\zeta_i &= 0 \\ (C - \lambda_i^*)\zeta_i^* &= 0 \end{aligned} \tag{4.85}$$

Thus, the only samples $(\mathbf{x}^{(i)}, y^{(i)})$ with corresponding $\lambda_i, \lambda_i^* = C$ lie outside the ε-insensitive tube around f. For $\lambda_i, \lambda_i^* \in (0, C)$, we have $\zeta_i, \zeta_i^* = 0$ and moreover the second factor in Eqn (4.84) has to vanish. Hence, w_0 can be computed as follows:

$$\begin{aligned} w_0 &= y^{(i)} - \mathbf{w}^T\mathbf{x}^{(i)} - \varepsilon \quad \text{for} \quad \lambda_i \in (0, C) \\ w_0 &= y^{(i)} - \mathbf{w}^T\mathbf{x}^{(i)} + \varepsilon \quad \text{for} \quad \lambda_i^* \in (0, C) \end{aligned} \tag{4.86}$$

All the data points with $\lambda_i, \lambda_i^* \in (0, C)$ may be used to compute w_0, and then their average taken as the final value for w_0.

Once we solve the quadratic optimization problem for $\boldsymbol{\lambda}$ and $\boldsymbol{\lambda}^*$, we see that all instances that fall in the ε-tube have $\lambda_i = \lambda_i^* = 0$; these are the instances that are fitted with enough precision. The support vectors satisfy either $\lambda_i > 0$ or $\lambda_i^* > 0$, and are of two types. They may be instances that are on the boundary of the tube (either λ_i or λ_i^* is between 0 and C), and we use these to calculate w_0. Instances that fall outside the ε-tube ($\lambda_i = C$) are of second type of support vectors. For these instances, we do not have a good fit.

Using condition (i) of KKT conditions (4.81), we can write the fitted line as a weighted sum of the support vectors;

$$f(\mathbf{x}) = \mathbf{w}^T\mathbf{x} + w_0 = \sum_{i \in svindex} (\lambda_i - \lambda_i^*)\mathbf{x}^{(i)T}\mathbf{x} + w_0 \tag{4.87a}$$

where *svindex* denotes the set of indices of support vectors. Note that for each $i \in svindex$, one element of the pair (λ_i, λ_i^*) is zero.

The parameter w_0 may be obtained from either of the equations in (4.86). If the former one is used for which $i \in$ set of instances that correspond to support vectors on the upper boundary ($\lambda_i \in (0, C)$) of the ε-tube (let us denote these points as belonging to the set *svm1index* of indices of support vectors that fall on the upper boundary), then we have (refer to Eqn (4.55)),

$$w_0 = \frac{1}{|svm1index|}\left[\sum_{k \in svm1index}\left[y^{(k)} - \varepsilon - \left(\sum_{i \in svindex} (\lambda_i - \lambda_i^*)\mathbf{x}^{(i)T}\mathbf{x}^{(k)}\right)\right]\right] \tag{4.87b}$$

4.8.2 Nonlinear Regression

For nonlinear regression, the quadratic optimization problem follows from Eqn (4.82):

$$\begin{aligned} \textit{maximize } L_*(\boldsymbol{\lambda}, \boldsymbol{\lambda}^*) = &-\varepsilon\sum_{i=1}^{N}(\lambda_i + \lambda_i^*) + \sum_{i=1}^{N}(\lambda_i - \lambda_i^*)\, y^{(i)} \\ &-\tfrac{1}{2}\sum_{i=1}^{N}\sum_{k=1}^{N}(\lambda_i - \lambda_i^*)(\lambda_k - \lambda_k^*)\,\mathbf{z}^{(i)T}\mathbf{z}^{(k)} \end{aligned}$$

$$\text{subject to } \sum_{i=1}^{N}(\lambda_i - \lambda_i^*) = 0;\ \lambda_i\ \lambda_i^* \in [0, C] \tag{4.88}$$

where $\mathbf{z} = \boldsymbol{\phi}(\mathbf{x})$

The dot product $\mathbf{z}^{(i)T}\mathbf{z}^{(k)} = [\boldsymbol{\phi}(x^{(i)})]^T\boldsymbol{\phi}(\mathbf{x}^{(k)})$ in Eqn (4.88) can be replaced with a kernel $K(\mathbf{x}^{(i)}, \mathbf{x}^{(k)})$.

The nonlinear regression function (refer to Eqn (4.87a))

$$f(\mathbf{x}) = \mathbf{w}^T \boldsymbol{\phi}(\mathbf{x}) + w_0 = \sum_{i \in svindex} (\lambda_i - \lambda_i^*)[\boldsymbol{\phi}(\mathbf{x}^{(i)})]^T \boldsymbol{\phi}(\mathbf{x}) + w_0$$

$$= \sum_{i \in svindex} (\lambda_i - \lambda_i^*)\, K(\mathbf{x}^{(i)}, \mathbf{x}) + w_0 \tag{4.89}$$

The parameter w_0 for this nonlinear regression solution may be obtained from either of the equations in (4.86). If the former one is used for which $i \in$ set of instances that correspond to support vectors on the upper boundary ($\lambda_i \in (0, C)$) of the ε-tube (let us denote these points as belonging to the set *svm1index*), then we have (refer to Eqn (4.87b))

$$w_0 = \frac{1}{|svm\,1index|}\left[\sum_{k \in svm1index}\left[y^{(k)} - \varepsilon - \left(\sum_{i \in svindex} (\lambda_i - \lambda_i^*)\, K(\mathbf{x}^{(i)}, \mathbf{x}^{(k)})\right)\right]\right] \tag{4.90}$$

Regression problems when solved using SVM algorithm presented in this section, are basically *quadratic optimization problems*. Attempting standard quadratic programming routines (e.g., MATLAB) will be a rich learning experience for the readers. To help the readers use the standard routine, we give matrix formulation of SVM regression in the example that follows.

Example 4.5

In this example, we illustrate the SVM nonlinear regressor formulation; the formulation will be described in matrix form so as to help use of a standard quadratic optimization software.

As with nonlinear classification, input vector $\mathbf{x} \in \Re^n$ are mapped into vectors $\mathbf{z}$ of a higher-dimensional feature space: $\mathbf{z} = \boldsymbol{\phi}(\mathbf{x})$, where $\boldsymbol{\phi}$ represents a mapping $\Re^n \rightarrow \Re^m$. The linear regression problem is then solved in this feature space. The solution for a regression hypersurface, which is linear in a feature space, will create a nonlinear regressing hypersurface in the original input space.

It can easily be shown that $\hat{y} = \mathbf{w}^T\mathbf{z} + w_0$ is a regression expression, and with the ε-insensitive loss function, the formulation leads to the solution equations of the form (refer to (4.82))

$$\text{Minimize } \tfrac{1}{2}\bar{\boldsymbol{\lambda}}^T \mathbf{Q}\, \bar{\boldsymbol{\lambda}} + \mathbf{g}^T \bar{\boldsymbol{\lambda}}$$

$$\text{subject to } [\mathbf{e}^T \;\; -\mathbf{e}^T]\, \bar{\boldsymbol{\lambda}} = 0, \text{ and } 0 \le \lambda_i, \lambda_i^* \le C;\; i = 1, \ldots, N \tag{4.91}$$

where

$$\mathbf{Q} = \begin{bmatrix} \mathbf{K} & -\mathbf{K} \\ -\mathbf{K} & \mathbf{K} \end{bmatrix};\; \bar{\boldsymbol{\lambda}} = \begin{bmatrix} \boldsymbol{\lambda} \\ \boldsymbol{\lambda}^* \end{bmatrix}, \text{ and } \mathbf{g} = \begin{bmatrix} \varepsilon - \mathbf{y} \\ \varepsilon + \mathbf{y} \end{bmatrix}$$

K as given in Eqn (4.71),

$$\boldsymbol{\lambda} = [\lambda_1\; \lambda_2 \ldots \lambda_N]^T$$

$$\varepsilon - \mathbf{y} = [\varepsilon - y^{(1)} \;\; \varepsilon - y^{(2)} \ldots \varepsilon - y^{(N)}]^T$$

After computing Lagrange multipliers λ_i and λ_i^* using a quadratic optimization routine, we find optimal desired nonlinear regression function as (Eqns (4.89–4.90))

$$f(\mathbf{x}) = \sum_{i \in svindex} (\lambda_i - \lambda_i^*)\, \mathbf{z}^{(i)T}\mathbf{z} + w_0$$

$$= \sum_{i \in svindex} (\lambda_i - \lambda_i^*)\, [\boldsymbol{\phi}(\mathbf{x}^{(i)})]^T \boldsymbol{\phi}(\mathbf{x}) + w_0$$

$$= \sum_{i \in svindex} (\lambda_i - \lambda_i^*)\, K(\mathbf{x}^{(i)}, \mathbf{x})$$

$$+ \frac{1}{|svm\,1\,index|}\left[\sum_{k \in svm1index}\left[y^{(k)} - \varepsilon - \left(\sum_{i \in svindex} (\lambda_i - \lambda_i^*)\, K(\mathbf{x}^{(i)}, \mathbf{x}^{(k)})\right)\right]\right] \tag{4.92}$$

There are a number of learning parameters that can be utilized for constructing SV machines for regression. The two most relevant are insensitivity parameter ε and penalty parameter C. Both the parameters are chosen by the user. An increase in ε has smoothing effects on modeling highly noisy polluted data. An increase in ε means a reduction in requirements for the accuracy of approximation. We have already commented on the selection of parameter C. More on it will appear later.

The SV training works almost perfectly for not too large datasets. However, when the number of data points is large, the quadratic programming problem becomes extremely difficult to solve with standard methods. Some approaches to resolve the quadratic programming problem for large datasets have been developed. We will talk about this aspect of SV training in a later section.

4.9 DECOMPOSING MULTICLASS CLASSIFICATION PROBLEM INTO BINARY CLASSIFICATION TASKS

Support vector machines were originally designed for binary classification. Initial research attempts were directed towards making several two-class SVMs to do multiclass classification. Recently, several single-shot multiclass classification algorithms appeared in the literature.

At present, there are two types of approaches for multiclass classification. In the first approach called 'indirect methods', we construct several binary SVMs and combine the outputs for predicting the class. In the second approach called 'direct methods', we consider all in a single optimization problem. Because of computational complexity of training in the direct methods, indirect methods so far are most widely used as they do not pose any numerical difficulties while training. We limit our discussion to indirect methods.

There are two popular methods in the category of indirect methods: One-Against-All (OAA), and One-Against-One (OAO). In the general case, both OAA and OAO are special cases of *error-correcting-output codes* that decompose a multiclass problem to a set of two-class problems [59].

4.9.1 One-Against-All (OAA)

Consider the training set

$$\mathcal{D}: \{(\mathbf{x}^{(1)}, y^{(1)}), (\mathbf{x}^{(2)}, y^{(2)}), \ldots, (\mathbf{x}^{(N)}, y^{(N)})\}$$

where the label $y^{(i)}$ for each observation can take on any value $\{y_q\}$; $q = 1, \ldots, M$. The precise nature of the label set is not important as long as there exists a unique label for each class in the classification problem.

The task is to design M linear discriminant functions:

$$g_q(\mathbf{x}) = \mathbf{w}_q^T\mathbf{x} + w_{q0};\ q = 1, 2, \ldots, M \tag{4.93a}$$

The decision function is given by $g_k(\mathbf{x})$, where

$$\begin{aligned} k &= arg \max_{1 \le q \le M}(g_q(\mathbf{x})) \\ &= arg \max_{1 \le q \le M}(\mathbf{w}_q^T\mathbf{x} + w_{q0}) \end{aligned} \tag{4.93b}$$

New point $\mathbf{x}$ is assigned to class k.

Geometrically, this is equivalent to associating a hyperplane to each class and to assigning a new point $\mathbf{x}$ to the class whose hyperplane is farthest from it. The design of M linear discriminants (4.93a) follows the following procedure.

In the OAA technique [53, 60], given M classes, we construct M linear decision surfaces g_1, g_2, …, g_M. Each decision surface is trained to separate one class from the others. In other words, the decision surface g_1 is trained to separate Class 1 from all other classes; the decision surface g_2 is trained to separate Class 2 from all other classes, and so on. For the classification of an unknown point, a *voting scheme* based on which of the M decision surfaces return the largest value for this unknown point is used. We then use the decision surface that returns the largest value for the unknown point to assign this point to the class:

$$\text{Class} = arg \max_{q \in \{1, \ldots, M\}} g_q(\mathbf{x}) \tag{4.94}$$

This approach is called the *winner-takes-all* approach.

Let us examine this construction in more detail. To train M decision surfaces, we construct M binary training sets:

$$\mathcal{D}^q = \mathcal{D}_+^q \cup \mathcal{D}_-^q;\ q = 1, \ldots, M$$

where

$\mathcal{D}_+^q$: the set of all observations in $\mathcal{D}$ that are members of the class q

and

$\mathcal{D}_-^q$: the set of all remaining observations in $\mathcal{D}$, i.e., the set of all observations in $\mathcal{D}$ that are *not* members of the class q.

For convenience, we label the training set $\mathcal{D}^q$ with the class labels $\{+1, -1\}$; the label $+1$ is used for observations in class q, and the label -1 is used for observations that are not in class q.

We train decision surface g_q on the corresponding dataset $\mathcal{D}^q$, which gives rise to

$$g_q(\mathbf{x}) = \mathbf{w}_q^T\mathbf{x} + w_{q0}$$

The decision surface $g_q: \Re^n \rightarrow \Re$ returns a signed real value, which can be interpreted as the distance of the point $\mathbf{x} \in \Re^n$ to the decision surface. If the value returned is +ve, the point $\mathbf{x}$ is above the decision surface and is taken as a member of class +1 with respect to the decision surface, and if the value returned is –ve, the point is below the decision surface and is considered to be a member of the class –1 with respect to the decision surface. The value returned can also be interpreted as a *confidence* value: If our decision surface returns a large +ve value for a specific point, we are quite confident that the point belongs to class +1; on the contrary, if our decision surface returns a large –ve value for a point, we can confidently say that the point does not belong to Class +1. Since our training set $\mathcal{D}^q$ for the decision surface g_q was laid out in such a way that all observations in class q are considered +ve examples, it follows that a decision surface g_k that returns the largest positive value for some point $\mathbf{x}$ among all other decision surfaces $g_1, \ldots, g_M$, assigns the point to class k with $k \in \{1, 2, \ldots, M\}$

Although the OAA classification technique has shown to be robust in real-world applications, the fact that the individual training sets for each decision surface are highly *unbalanced* could be a potential source of problems. The *pairwise classification* technique (One-Against-One (OAO) classification technique) avoids this situation by constructing decision surface for each pair of classes [53, 60].

4.9.2 One-Against-One (OAO)

In pairwise classification, we train a classifier for each pair of classes. For M classes, this results in $M(M-1)/2$ binary classifiers.

We let $g_{q,k}$ denote the decision surface that separates the pair of classes $q \in \{1, \ldots, M\}$ and $k \in \{1, \ldots, M\}$ with $q \neq k$. We label class q samples by +1 and class k samples by –1, and train each decision surface

$$g_{q,k}(\mathbf{x}) = \mathbf{w}_{q,k}^T \mathbf{x} + w_{q,k0} \tag{4.95}$$

on the data set

$$\mathcal{D}^{q,k} = \mathcal{D}^q \cup \mathcal{D}^k$$

where

$\mathcal{D}^q$: the set of all observations in $\mathcal{D}$ with the label q

and

$\mathcal{D}^k$: the set of all observations in $\mathcal{D}$ with the label k

Once we have constructed all the pairwise decision surfaces $g_{q,k}$ using the corresponding training sets $\mathcal{D}^{q,k}$, we can classify an unseen point by applying each of the $M(M-1)/2$ decision surfaces to this point, keeping track of how many times the point was assigned to what class label. The class label with the highest count is then considered the label for the unseen point.

Voting Scheme

c_q = the frequency of 'wins' for class q computed by applying $g_{q,k}$ for all $k \neq q$.

This results in a vector

$$\mathbf{c} = [c_1, \ldots, c_M]$$

of frequencies of 'wins' of each class. The final decision is made by the most frequent class:

$$\text{Class } q = arg \max_{q=1,\ldots M} c_q \tag{4.96}$$

Frequencies of 'wins' = number of votes.

There is a likelihood of a tie in the voting scheme of OAO classification. We can breakdown the tie through the interpretation of the actual values returned by decision surfaces as confidence values. On adding up absolute values of the confidence values assigned to each of the tied labels, we take the winner to be the tied label possessing the maximum sum of confidence values.

It seems that OAO classification solves our problem of unbalanced datasets. However, it solves the problem at the expense of introducing a new complication: the fact that for M classes, we have to construct $M(M-1)/2$ decision surfaces. For small M, the difference between the number of decision surfaces we have to build for the OAA and OAO techniques is not that drastic. (For $M = 4$, OAA requires 4 binary classifiers and OAO requires 6). However, for large M, the difference can be quite drastic (For $M = 10$, OAA requires 10 binary classifiers and OAO requires 45).

The individual classifiers in OAO technique, however, are usually smaller in size (they have fewer support vectors) then they would be in the OAA approach. This is for two reasons: first, the training sets are smaller, and second, the problems to be learned are usually easier, since the classes have less overlap. Since the size of QP in each classifier is smaller, it is possible to train fast.

Nevertheless, if M is large, then the resulting OAO system may be slower than the corresponding OAA. Platt et al. [61] improved the OAO approach and proposed a method called *Directed Acyclic Graph SVM* (DAGSVM) that forms a tree-like structure to facilitate the testing phase.

4.10 VARIANTS OF BASIC SVM TECHNIQUES

Basically, support vector machines form a learning algorithm family rather than a single algorithm. The basic concept of the SVM-based learning algorithm is pretty simple: find a good learning boundary while maximizing the margin (i.e., the distance between the closest learning samples that correspond to different classes). Each algorithm that optimizes an objective function in which the maximal margin heuristic is encoded, can be considered a *variant* of basic SVM.

In the variants, improvements are proposed by researchers to gain speed, accuracy, low computer-memory requirement and ability to handle multiple classes. Every variant holds good in a particular field under particular circumstances.

Since the introduction of SVM, numerous variants have been developed. In this section, we highlight some of these which have earned popularity in their usage or are being actively researched.

Changing the Metric of Margin from L_2-norm to L_1-norm

The standard L_2-norm SVM is a widely used tool in machine learning. The L_1-norm SVM is a variant of the standard L_2-norm SVM.

The L_1-norm formulation is given by (Eqns (4.48a)):

$$\text{minimize } \tfrac{1}{2}\mathbf{w}^T\mathbf{w} + C\sum_{i=1}^{N}\zeta_i$$

$$\text{subject to } y^{(i)}(\mathbf{w}^T\mathbf{x}^{(i)} + w_0) \ge 1 - \zeta_i;\ i = 1, \ldots, N$$

$$\zeta_i \ge 0; \qquad i = 1, \ldots, N$$

In L_2-norm SVM, the sum of squares of error (slack) variables are minimized along with the reciprocal of the square of the margin between the boundary planes. The formulation of the problem is given by:

$$\begin{aligned} &\text{minimize } \tfrac{1}{2}\mathbf{w}^T\mathbf{w} + \frac{C}{2}\sum_{i=1}^{N}(\zeta_i)^2 \\ &\text{subject to } y^{(i)}(\mathbf{w}^T\mathbf{x}^{(i)} + w_0) \ge 1 - \zeta_i;\ i = 1, \ldots, N \\ &\zeta_i \ge 0;\ i = 1, \ldots, N \end{aligned} \tag{4.97}$$

It has been argued that the L_1-norm penalty has advantages over the L_2-norm penalty under certain scenarios, especially when there are redundant noise variables. L_1-norm SVM is able to delete many noise features by estimating their coefficients by zero, while L_2-norm SVM will use all the features. When there are many noise variables, the L_2-norm SVM suffers severe damage caused by the noise features. Thus, L_1-norm SVM has inherent variable selection property, while this is not the case for L_2-norm SVM. In this book, our focus has been on L_1-norm SVM formulations. Refer to [56] for L_2-norm SVM formulations.

Replacing Control Parameter *C* in Basic SVM (*C*-SVM); $C \ge 0$, by Parameter ν (ν-SVM); $\nu \in [0, 1]$

As we have seen in the formulations of basic SVM (*C*-SVM) presented in earlier sections, *C* is a user-specified penalty parameter. It is a trade-off between margin and mistakes. Proper choice of *C* is crucial for good generalization power of the classifier. Usually, the parameter is selected by trial-and-error with cross-validation.

Tuning of parameter *C* can be quite time-consuming for large datasets. In the scheme proposed by Schölkopf et al. [62], the parameter $C \ge 0$ in the basic SVM is replaced by a parameter $\nu \in [0, 1]$. ν has been shown to be a lower bound on the fraction of support vectors and an upper bound on the fraction of instances having margin errors (instances that lie on the wrong side of the hyperplane). By playing with ν, we can control the fraction of support vectors, and this is advocated to be more intuitive than playing with *C*. However, as compared to *C*-SVM, its formulations are more complicated.

A formulation of ν-SVM is given by:

$$\text{minimize } \tfrac{1}{2}\mathbf{w}^T\mathbf{w} - \nu\rho + \frac{1}{N}\sum_{i=1}^{N}\zeta_i$$

$$\text{subject to } y^{(i)}(\mathbf{w}^T\mathbf{x}^{(i)} + w_0) \geq \rho - \zeta_i;\ i = 1, \ldots, N \quad (4.98)$$
$$\zeta_i \geq 0;\ i = 1, \ldots, N;\ \rho \geq 0$$

Note that parameter C does not appear in this formulation; instead there is parameter ν. An additional parameter ρ also appears that is a variable of the optimization problem and scales the margin: the margin is now $2\rho/\|\mathbf{w}\|$ (refer to Eqn (4.25)).

The formulation given by (4.98) represents modification of the basic SVM classification (C-SVM) given by (4.48a) to obtain ν-SVM classification. With analogous modifications of the basic SVM regression (ε-SVM regression), we obtain ν-SVM regression.

Sequential Minimization Algorithms

Support vector machines have attracted many researchers in the last two decades due to many interesting properties they enjoy: immunity to overfitting by means of regularization, guarantees on the generalization error, robust training algorithms that are based on well established mathematical programming techniques, and above all, their success in many real-world classification problems. Despite the many advantages, basic SVM suffers from a serious drawback; it requires Quadratic Programming (QP) solver to solve the problem. The amount of computer memory needed for a standard QP solver increases exponentially with the size of the data. Therefore, the issue is whether it is possible for us to scale up the SVM algorithm for huge datasets comprising thousands and millions of instances. Many decomposition techniques have been developed to scale up the SVM algorithm.

Techniques based on decomposition, break down a large optimization problem into smaller problems, with each one involving merely some cautiously selected variables so that efficient optimization is possible. Platt's SMO algorithm (Sequential Minimal Optimization) [63] is an extreme case of the decomposition techniques developed, which works on a set of two data points at a time. Owing to the fact that the solution for a working set of two data points can be arrived at analytically, the SMO algorithm does not invoke standard QP solvers. Due to its analytical foundation, the SMO and its improved versions [64–66] are particularly simple and at the moment in the widest use. Many free software packages are available, and the ones that are most popular are *SVM light* [67] and *LIBSVM* [68].

Variants based on Trade-off between Complexity and Accuracy

Decomposition techniques handle memory issue alone by dividing a problem into a series of smaller ones. However, these smaller problems are rather time consuming for big datasets. Number of techniques for reduction in the training time have been suggested at the cost of accuracy.

Many variants have been reported in the literature. Some of the popular ones are on follows.

LS-SVM (Least Squares Support Vector Machine): It is a *Least Squares* version of the classical SVM. LS-SVM classification formulation implicitly corresponds to a regression interpretation with binary targets $y^{(i)} = \pm 1$. Proposed by Suykens and Vandewalla [69], its formulation has equality constraints; a set of linear equations has to be solved instead of a quadratic programming problem for classical SVMs.

PSVM (Proximal Support Vector Machine): Developed by Fung and Mangasarian [70], *Proximal* SVM leads to an extremely fast and simple algorithm for generating a classifier that is obtained by solving a set of linear equations. Proximal SVM is comparable with standard SVM in performance.

The key idea of PSVM is that it classifies points by assigning them to the closer of the two parallel planes that are pushed apart as far as possible. These planes are pushed apart by introducing the term $\mathbf{w}^T\mathbf{w} + w_0^2$ in the objective function of classical L_2-norm optimization problem.

LSVM (Lagrangian Support Vector Machine): A fast and simple algorithm, based on an implicit *Lagrangian* formulation of the dual of a simple reformulation of the standard quadratic program of a support vector machine, was proposed by Mangasarian [71]. This algorithm minimizes unconstrained differentiable convex function for classifying N points in a given n-dimensional input space. An iterative Lagrangian Support Vector Machine (LSVM) algorithm is given for solving the modified SVM. This algorithm can resolve problems accurately with millions of points, at a pace greater than SMO (if n is less than 100) without any optimization tools, like linear or quadratic programming solvers.

Multiclass-based SVM Algorithms

Originally, the SVM was developed for binary classification; the basic idea to apply SVM technique to multiclass problems is to decompose the multiclass problem into several two-class problems that can be addressed directly using several SVMs (refer to Section 4.9). This decomposition approach gives 'indirect methods' for multiclass classification problems.

Instead of creating several binary classifiers, a more natural way is to distinguish all classes in one single optimization processing. This approach gives 'direct methods' for multiclass classification problems [72].

Weston and Watkins' Multiclass SVM: In the method (the idea is similar to OAA approach) proposed by Weston and Watkins [73], for an M-class problem, a single objective function is designed for training all M-binary classifiers simultaneously and maximizing the margin from each class to the remaining classes. The main disadvantage of this approach is that the computational time may be very high due to the enormous size of the resulting QP. The OAA approach is generally preferred over this method.

Crammer and Singer's Multiclass SVM: Crammer and Singer [74] presented another 'all-together' approach. This approach gives a compact set of constraints; however, the number of variables in its dual problem are high. This value may explode even for small datasets.

Simplified Multiclass SVM (SimMSVM): A simplified method, named SimMSVM [75], relaxes the constraints of Crammer and Singer's approach.

The support vector machine is currently considered to be the best off-the-shelf learning algorithm and has been applied successfully in various domains. Scholkopf and Smola [53] is a classic book on the subject.

Chapter

5

LEARNING WITH NEURAL NETWORKS (NN)

5.1 TOWARDS COGNITIVE MACHINE

Human intelligence possesses robust attributes with complex sensory, control, affective (emotional processes), and cognitive (thought processes) aspects of information processing and decision making. There are over a hundred billion biological neurons in our central nervous system (CNS), playing a key role in these functions. CNS obtains information from the external environment via numerous natural sensory mechanisms—vision, hearing, touch, taste, and smell. With the help of cognitive computing, it assimilates the information and offers the right interpretation. The cognitive process then progresses towards some attributes, such as learning, recollection, and reasoning, which results in proper actions via muscular control.

The progress in technology based on information in recent times has widened the capabilities and applications of computers. If we wish a machine (computer) to exhibit certain cognitive functions, such as learning, remembering, reasoning, and perceiving, which humans are known to exhibit, we need to define 'information' in a general manner and develop new mathematical techniques and hardware with the ability to handle the simulation and processing of cognitive information. Mathematics, in its present form, was developed to comprehend physical processes, but cognition, as a process, does not essentially follow these mathematical laws. So what exactly is *cognitive mathematics* then? The question is rather difficult. However, scientists have converged to the understanding that if certain 'mathematical aspects' of our process of thinking are re-examined along with the 'hardware aspects' of 'the neurons'—which is the primary component of the brain—we may, to a certain level, be able to successfully emulate the process.

Biological neuronal procedures are rather complex [76], and the advancement made in terms of understanding the field with the help of experiments is raw and inadequate. However, with the help of this limited understanding of the biological processes, it has been possible to emulate some human learning behaviors, via the fields of mathematics and systems science. Neuronal information processing involves a range of complex mathematical processes and mapping functions. And they serve as a parallel-cascade computing structure in synergism. The aim of system scientists is to create an intelligent cognitive system on the basis of this limited understanding of the brain—a system that can help human beings to perform all kinds of tasks requiring decision making. Various new computing theories of the *neural networks* field have been developing, which it is hoped will be capable of providing a *thinking machine*. Given that they are based on neural networks architecture, they should hopefully be able to create a low-level cognitive machine, which scientists have been trying to build for so long.

The subject of cognitive machines is in an exciting state of research and we believe that we are slowly progressing towards the development of truly cognitive machines.

5.1.1 From Perceptrons to Deep Networks

Historically, research in neural networks was inspired by the desire to produce artificial systems capable of sophisticated 'intelligent' processing similar to the human brain. The *perceptron* is the earliest of the artificial neural networks paradigms. Frank Rosenblatt built this learning machine device in hardware in 1958. In 1959, Bernard Widrow and Marcian Hoff developed a learning rule, sometimes known as *Widrow-Haff rule*, for ADALINE (ADAptive LINear Elements). Their learning rule was simple and yet elegant.

Affected by the predominantly rosy outlook of the time, some people exaggerated the potential of neural networks. Biological comparisons were blown out of proportion. In 1969, significant limitations of perceptrons, a fundamental block for more powerful models, were highlighted by Marvin Minsky. It brought to a halt much of the activity in neural network research.

Nevertheless, a few dedicated scientists, such as Teuvo Kohonen and Stephen Grossberg, continued their efforts. In 1982, John Hopfield introduced a *recurrent*-type neural network that was based on the interaction of neurons through a feedback mechanism. The *backpropagation learning rule* arrived on the neural-network scene at approximately the same time from several independent sources (Werbos; Parker; Rumelhart, Hinton, and Williams). Essentially a refinement of the Widrow-Hoff learning rule, the backpropagation learning rule provided a systematic means for training *multilayer feedforward* networks, thereby overcoming the limitations presented by Minsky. Research in the 1980s triggered a boom in the scientific community. New and better models have been proposed. A number of today's technological problems are in the areas where neural-network technology has demonstrated potential.

As the research in neural networks is evolving, more and more types of networks are being introduced. For reasonably complex problems, neural networks with backpropagation learning have serious limitations. The learning speed of these feedforward neural networks is, in general, far slower than required and it has been a major bottleneck in their applications. Two reasons behind this limitation may be: (i) the slow gradient-based learning algorithms extensively used to train neural networks, and (ii) all the parameters of the network are tuned iteratively by using learning algorithms. A new learning algorithm was proposed in 2006 by Huang, Guang-Bin et. al. [77], called *Extreme Learning Machine* (ELM), for single hidden layer feedforward neural networks, in which the weights connecting input to hidden nodes are randomly chosen and never updated and weights connecting hidden nodes to output are analytically determined. Experimental results based on real-world benchmarking function approximation (regression) and classification problems show that ELM can produce best generalization in some cases and can be thousands of times faster than traditional popular learning algorithms for feedforward neural networks.

Novel research investigations in ELM and related areas have produced a suite of machine learning techniques for (single and multi-) hidden layer feedforward networks in which hidden neurons need not be tuned. ELM theories argue that random hidden neurons capture the essence of some brain learning mechanisms. ELM has a great potential as a viable alternative technique for large-scale computing and AI (*Artificial Intelligence*).

The 'traditional' neural networks are based on what we might interpret as 'shallow' learning; in fact, this learning methodology has very little resemblance to the brain, and one might argue that it would be more fair to regard them simply as a discipline under statistics.

The subject of cognitive machines is in an exciting state of research and we believe that we are slowly progressing towards the development of truly intelligent systems. A step towards realizing strong AI has been taken through the recent research in 'deep learning'. Considering the far-reaching applications of AI, coupled with the awareness that deep learning is evolving as one of its most powerful methods, today it is not possible for one to enter the machine learning community without any knowledge of deep networks.

Deep learning algorithms are in sharp contrast to shallow learning algorithms in terms of the number of parameterized transformations a signal comes across as it spreads from input layer to the output layer. A parameterized transformation refers to a processing unit containing trainable parameters—weights and thresholds. A chain of transformations from input to output is a *Credit Assignment Path* (CAP), which describes potentially causal connections from input to output and may have varied lengths. In case of a feedforward neural network, the depth of the CAPs, and therefore, the *depth of the network*, is a number of hidden layers plus one (the output layer is also parameterized).

Today, deep learning, based on *learning representations of data*, is a significant member of family of machine learning techniques. It is possible to represent an observation, for instance an image, in various ways, such as a vector of intensity values per pixel or in a more abstract way as a set of edges, regions of particular shape, and so on. Some representations make learning tasks simpler. Deep learning aims to replace hand-crafted *features* with efficient algorithms for *supervised* or *unsupervised feature learning*, and hierarchical *feature extraction*.

The field of deep learning has been characterized in several ways. These definitions have in common:

(i) multiple layers of nonlinear processing units.
(ii) the supervised/unsupervised learning of feature representations in each layer, with the layers giving rise to a hierarchy from low-level to high-level characteristics.

What a layer of nonlinear processing unit, employed in a deep learning algorithm, consists of is dependent on the problem that needs to be solved.

Deep learning is linked closely to a category of brain-development theories published by cognitive neuroscientists in the early 1990s. Some of the deep-learning representations are inspired by progress in neuroscience and are roughly based on interpretation of information processing and communication patterns in a nervous system, such as neural coding which tries to describe the relationship between a range of stimuli and related neuronal responses in the brain.

The term 'deep learning' gained traction in the mid 2000s, after a publication by Geoffrey Hinton and Ruslan Salakhutdinov. They showed how many-layered feedforward network could be effectively pre-trained one layer at a time. Since its resurgence, deep learning has become part of the many-of-the-art systems in various disciplines, particularly computer vision and speech recognition. The real impact of deep learning in industry began in large-scale speech recognition around 2010. Recent useful references on the subject are [78–82].

Our focus in this chapter will be on traditional neural networks [83, 84]. These networks are being used today for many real-world regression and classification problems. Also, a sound understanding of these networks is a prerequisite to learn ELM and deep learning algorithms. The two recent research developments are outside the scope of this book.

The terms 'Neural Networks (NN)' and 'Artificial Neural Networks (ANN)' are both commonly used in the literature for the same field of study. We will use the term 'Neural Networks' in this book.

Broadly speaking, AI (*Artificial Intelligence*) is any computer program that does something smart [2, 5]. *Machine learning* is a subfield of AI. That is, all machine learning counts as AI, but not all AI counts as machine learning. For example, rule-based expert systems, frame-based expert systems, knowledge graphs, evolutionary algorithms could be described by AI but none of them is in machine learning. *Deep learning* may be considered as subfield of machine learning. Deep neural networks are a set of algorithms setting new records in accuracy for many important problems. *Deep* is a technical term; it refers to number of layers in a neural network. Multiple hidden layers allow deep neural networks to learn features of the data in a hierarchy.

Deep learning may share elements of traditional machine learning, but some researchers feel that it will emerge as a *class* by itself, as a subfield of AI.

5.2 NEURON MODELS

A discussion of anthropomorphism to introduce neural network technology may be worthwhile—as it helps explain the terminology of neural networks. However, anthropomorphism can lead to misunderstanding when the metaphor is carried too far. We give here a brief description of how the brain works; a lot of details of the complex electrical and chemical processes that go on in the brain, have been ignored. A pragmatic justification for such a simplification is that by starting with a simple model of the brain, scientists have been able to achieve very useful results.

5.2.1 Biological Neuron

To the extent a human brain is understood today, it seems to operate as follows: bundles of neurons, or nerve fibers, form nerve structures. There are many different types of neurons in the nerve structure, each having a particular shape, size, and length depending upon its function and utility in the nervous system. While each type of neuron has its own unique features needed for specific purposes, all neurons have two important structural components in common. These may be seen in the typical biological neuron shown in Fig. 5.1. At one end of the neuron are a multitude of tiny, filament-like appendages called *dendrites*, which come together to form larger branches and trunks where they attach to *soma*, the body of the nerve cell. At the other end of the neuron is a single filament leading out of the soma, called an *axon*, which has extensive branching on its far end. These two structures have special electrophysiological properties which are basic to the function of neurons as *information processors*, as we shall see next.

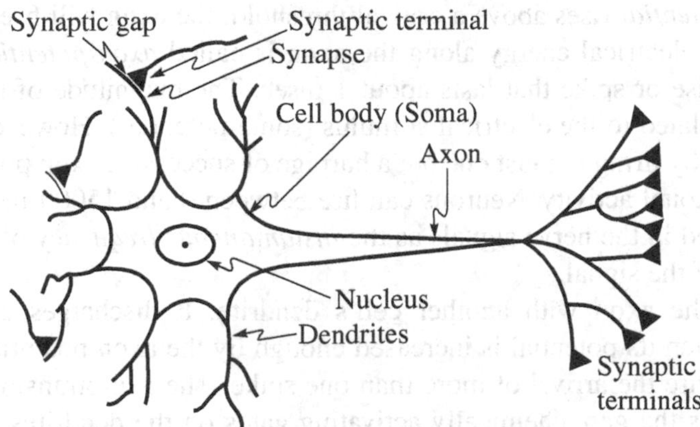

Figure 5.1 A typical biological neuron

Neurons are connected to each other via their axons and dendrites. Signals are sent through the axon of one neuron to the dendrites of other neurons. Hence, dendrites may be represented as the inputs to the neuron, and the axon as its output. Note that each neuron has many inputs through its multiple dendrites, whereas it has only one output through its single axon. The axon of each neuron forms connections with the dendrites of many other neurons, with each branch of the axon meeting exactly one dendrite of another cell at what is called a *synapse*. Actually, the axon terminals do not quite touch the dendrites of the other neurons, but are separated by a very small distance of between 50 and 200 angstroms. This separation is called the *synaptic gap*.

A conventional computer is typically a single processor acting on explicitly programmed instructions. Programmers break tasks into tiny components, to be performed in sequence rapidly. On the other hand, the brain is composed of ten billion or so neurons. Each nerve cell can interact directly with up to 200,000 other neurons (though 1000 to 10,000 is typical). In place of explicit rules that are used by a conventional computer, it is the pattern of connections between the neurons in the human brain, which seems to embody the 'knowledge' required for carrying out various information-processing tasks. In human brain, there is no equivalent of a CPU that is in overall control of the actions of all the neurons.

The brain is organized into different regions, each responsible for different functions. The largest parts of the brain are the *cerebral hemispheres*, which occupy most of the interior of the skull. They are layered structures; the most complex being the outer layer, known as the *cerebral cortex*, where the nerve cells are extremely densely packed to allow greater interconnectivity. Interaction with the environment is through the visual, auditory, and motion control (muscles and glands) parts of the cortex.

In essence, neurons are tiny electrophysiological information-processing units which communicate with each other through electrical signals. The synaptic activity produces a voltage pulse on the dendrite which is then conducted into the soma. Each dendrite may have many synapses acting on it, allowing massive interconnectivity to be achieved. In the soma, the *dendrite potentials* are added. Note that neurons are able to perform more complex functions than simple addition on the inputs they receive, but considering a simple summation is a reasonable approximation.

When the *soma potential* rises above a critical threshold, the axon will fire an electrical signal. This sudden burst of electrical energy along the axon is called *axon potential* and has the form of an electrical impulse or spike that lasts about 1 msec. The magnitude of the axon potential is constant and is not related to the electrical stimulus (soma potential). However, neurons typically respond to a stimulus by firing not just one but a barrage of successive axon potentials. What varies is the frequency of axonal activity. Neurons can fire between 0 and 1500 times per second. Thus, information is encoded in the nerve signals as the *instantaneous frequency* of axon potentials and the *mean frequency* of the signal.

A synapse pairs the axon with another cell's dendrite. It discharges chemicals known as *neurotransmitters*, when its potential is increased enough by the axon potential. The triggering of the synapse may require the arrival of more than one spike. The neurotransmitters emitted by the synapse diffuse across the gap, chemically activating gates on the dendrites, which, on opening, permit the flow of charged ions. This flow of ions, changes the dendritic potential and generates voltage pulse on the dendrite, which is then conducted into the neuron body. At the synaptic junction, the number of gates that open on the dendrite is dependent on the number of neurotransmitters emitted. It seems that some synapses *excite* the dendrites they impact, while others act in a way that *inhibits* them. This results in changing the local potential of the dendrite in a positive or negative direction.

Synaptic junctions alter the effectiveness with which the signal is transmitted; some synapses are good junctions and pass a large signal across, whilst others are very poor, and allow very little through.

Essentially, each neuron receives signals from a large number of other neurons. These are the inputs to the neuron which are 'weighted'. That is, some signals are stronger than others. Some signals excite (are positive), and others inhibit (are negative). The effects of all weighted inputs are summed. If the sum is equal to or greater than the *threshold* for the neuron, the neuron *fires* (gives output). This is an 'all-or-nothing' situation. Because the neuron either fires or does not fire, the *rate of firing*, not the amplitude, conveys the magnitude of information.

The ease of transmission of signals is altered by activity in the nervous system. The neural pathway between two neurons is susceptible to fatigue, oxygen deficiency, and agents like anesthetics. These events create a resistance to the passage of impulses. Other events may increase the rate of firing. This ability to adjust signals is a mechanism for *learning*.

After carrying a pulse, an axon fiber is in a condition of complete non-excitability for a specific time period known as the *refractory period*. During this interval, the nerve conducts no signals, irrespective of how intense the excitation is. Therefore, we could segregate the time scale into successive intervals, each equal to the length of the refractory period. This will permit a discrete-time description of the neurons' performance in terms of their states at discrete-time instances.

5.2.2 Artificial Neuron

Artificial neurons bear only a modest resemblance to real things. They model approximately three of the processes that biological neurons perform (there are at least 150 processes performed by neurons in the human brain).

An artificial neuron

(i) evaluates the input signals, determining the strength of each one;
(ii) calculates a total for the combined input signals and compares that total to some threshold level; and
(iii) determines what the output should be.

Input and Outputs

Just as there are many inputs (stimulation levels) to a biological neuron, there should be many input signals to our artificial neuron (AN). All of them should come to our AN simultaneously. In response, a biological neuron either 'fires' or 'doesn't fire' depending upon some *threshold* level. Our AN will be allowed a single output signal, just as is present in a biological neuron: many inputs, one output (Fig. 5.2).

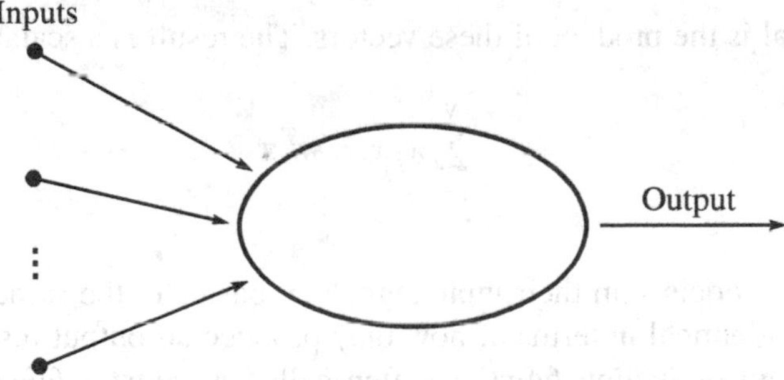

Figure 5.2 Many inputs, one output model of a neuron

Weighting Factors

Each input will be given a relative weighting, which will affect the impact of that input (Fig. 5.3). This is something like varying synaptic strengths of the biological neurons—some inputs are more important than others in the way they combine to produce an impulse. Weights are adaptive coefficients within the network, that determine the intensity of the input signal. In fact, this adaptability of connection strength is precisely what provides neural networks their ability to learn and store information, and, consequently, is an essential element of all neuron models.

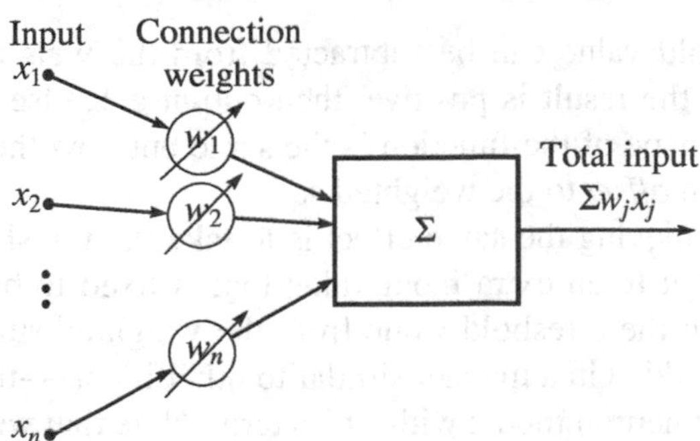

Figure 5.3 A neuron with weighted inputs

Excitatory and inhibitory inputs are represented simply by positive or negative connection weights, respectively. Positive inputs promote the firing of the neuron, while negative inputs tend to keep the neuron from firing.

Mathematically, we could look at the inputs and the weights on the inputs as vectors.

The *input vector*

$$\mathbf{x} = \begin{bmatrix} x_1 \\ x_2 \\ \vdots \\ x_n \end{bmatrix} \tag{5.1a}$$

and the connection *weight vector*

$$\mathbf{w}^T = [w_1 \; w_2 \; \ldots \; w_n] \tag{5.1b}$$

The total input signal is the product of these vectors. The result is a scalar

$$\sum_{j=1}^{N} w_j x_j = \mathbf{w}^T \mathbf{x} \tag{5.1c}$$

Activation Functions

Although most neuron models sum their input signals in basically the same manner, as described above, they are not all identical in terms of how they produce an output response from this input. Artificial neurons use an *activation function*, often called a *transfer function*, to compute their activation as a function of total input stimulus. Several different functions may be used as activation functions, and, in fact, the most distinguishing feature between existing neuron models is precisely which function they employ.

We will, shortly, take a closer look at the activation functions. We first build a neuron model, assuming that the transfer function has a threshold behavior, which is, in fact, the type of response exhibited by biological neurons: when the total stimulus exceeds a certain threshold value θ, a constant output is produced, while no output is generated for input levels below the threshold. Figure 5.4a shows this neuron model. In this diagram, the neuron has been represented in such a way that the correspondence of each element with its biological counterpart may be easily seen.

Equivalently, the threshold value can be subtracted from the weighted sum and the resulting value compared to zero; if the result is positive, then output a 1, else output a 0. This is shown in Fig. 5.4b; note that the shape of the function is the same but now the jump occurs at zero. The threshold effectively adds an *offset* to the weighted sum.

An alternative way of achieving the same effect is to take the threshold out of the body of the model neuron, and connect it to an extra input value that is fixed to be 'on' all the time. In this case, rather than subtracting the threshold value from the weighted sum, the extra input of +1 is multiplied by a weight and added in a manner similar to other inputs—this is known as *biasing* the neuron. Figure 5.4c shows a neuron model with a bias term. Note that we have taken constant input '1' with an adaptive weight 'w_0' in our model.

The first formal definition of a synthetic neuron model, based on the highly simplified considerations of the biological neuron, was formulated by McCulloch and Pitts (1943). The two-port model (inputs—activation value—output mapping) of Fig. 5.4 is essentially the *MP neuron model*. It is important to look at the features of this unit—which is an important and popular neural network building block.

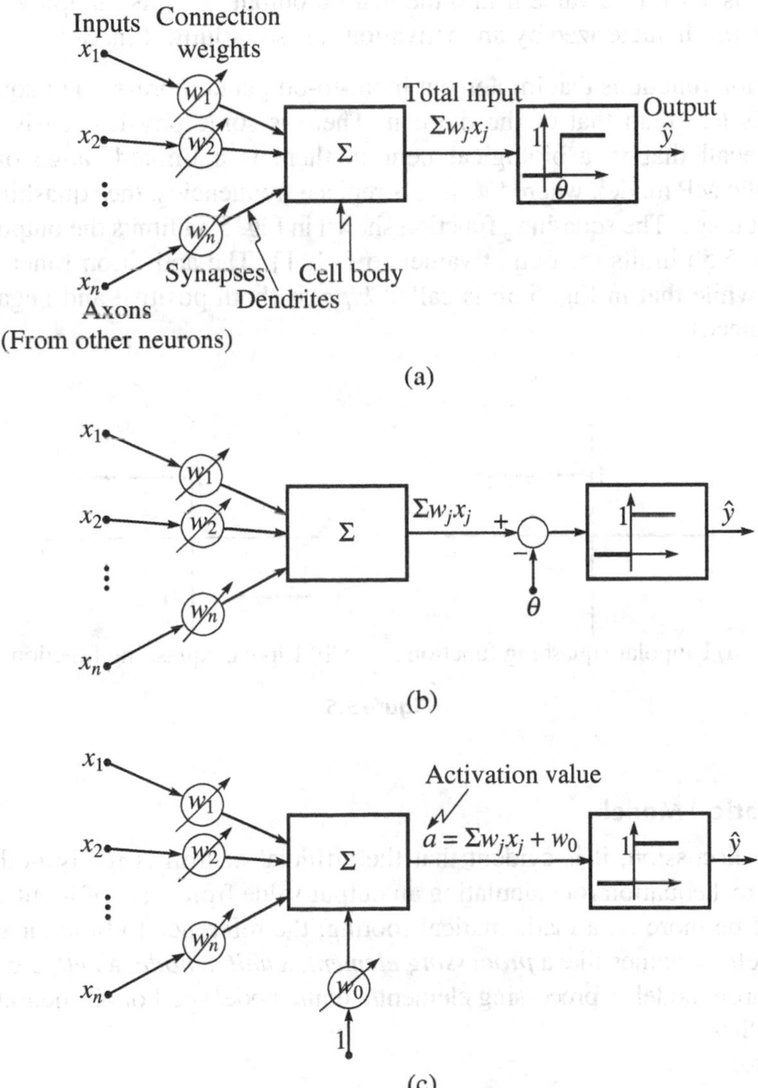

Figure 5.4 The MP neuron model

It is a basic unit, thresholding a weighted sum of its inputs to get an output. It does not particularly consider the complex patterns and timings of the real nervous activity in real neural systems.

It does not have any of the complex characteristics existing in the body of biological neurons. It is, therefore, a *model*, and not a *copy* of a real neuron.

The MP artificial neuron model involves two important processes:

(i) Forming net activation by combining inputs. The input values are amalgamated by a weighted additive process to achieve the neuron activation value a (refer to Fig. 5.4c).

(ii) Mapping this activation value a into the neuron output $\hat{y}$. This mapping from activation to output may be characterized by an 'activation' or 'squashing' function.

For the activation functions that implement input-to-output compression or squashing, the range of the function is less than that of the domain. There is some physical basis for this desirable characteristic. Recall that in a biological neuron, there is a limited range of output (spiking frequencies). In the MP model, where DC levels replace frequencies, the squashing function serves to limit the output range. The squashing function shown in Fig. 5.5a limits the output values to $\{0, 1\}$, while that in Fig. 5.5b limits the output values to $\{-1, 1\}$. The activation function of Fig. 5.5a is called *unipolar*, while that in Fig. 5.5b is called *bipolar* (both positive and negative responses of neurons are produced).

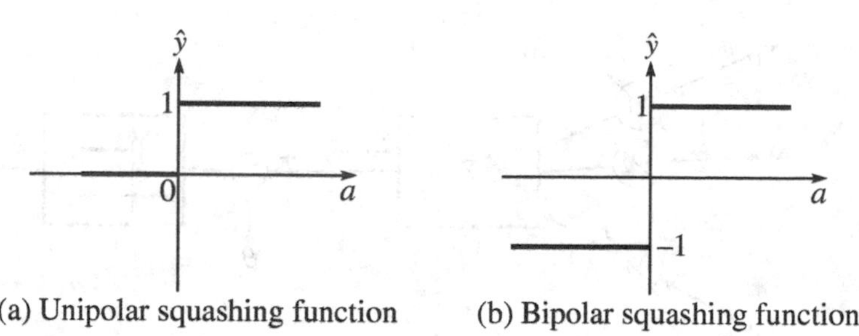

(a) Unipolar squashing function (b) Bipolar squashing function

Figure 5.5

5.2.3 Mathematical Model

From the earlier discussion, it is evident that the artificial neuron is really nothing more than a simple mathematical equation for calculating an output value from a set of input values. From now onwards, we will be more on a mathematical footing; the reference to biological similarities will be reduced. Therefore, names like a ***processing element***, a *unit*, a *node*, a *cell*, etc., may be used for the neuron. A neuron model (a processing element/a unit/a node/a cell of our neural network) will be represented as follows:

The input vector

$$\mathbf{x} = [x_1\ x_2\ \ldots\ x_n]^T;$$

the connection weight vector

$$\mathbf{w}^T = [w_1\ w_2\ \ldots\ w_n];$$

the unity-input weight w_0 (bias term), and the output $\hat{y}$ of the neuron are related by the following equation:

$$\hat{y} = \sigma(\mathbf{w}^T\mathbf{x} + w_0) = \sigma\left(\sum_{j=1}^{n} w_j x_j + w_0\right) \tag{5.2}$$

where $\sigma(\cdot)$ is the activation function (transfer function) of the neuron.

The weights are always adaptive. We can simplify our diagram as in Fig. 5.6a; adaptation need not be specifically shown in the diagram.

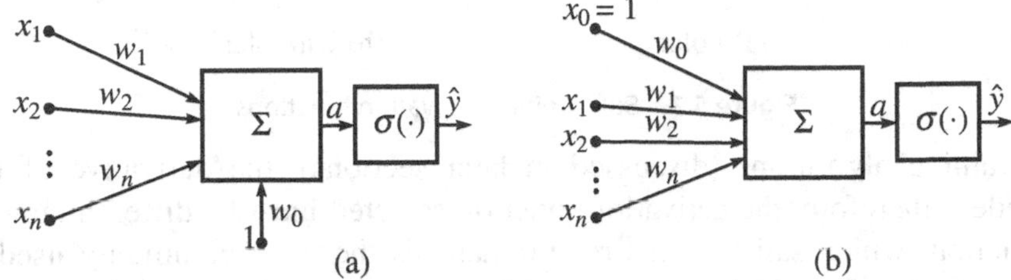

Figure 5.6 Mathematical model of a neuron (perceptron)

The bias term may be absorbed in the input vector itself as shown in Fig. 5.6b.

$$\hat{y} = \sigma(a)$$

$$= \sigma\left(\sum_{j=0}^{n} w_j x_j\right); x_0 = 1 \tag{5.3a}$$

$$= \sigma\left(\sum_{j=1}^{n} w_j x_j + w_0\right) = \sigma(\mathbf{w}^T\mathbf{x} + w_0) \tag{5.3b}$$

In the literature, this model of an artificial neuron is also referred to as a *perceptron* (the name was given by Rosenblatt in 1958).

The expressions for the neuron output $\hat{y}$ are referred to as the *cell recall mechanism.* They describe how the output is reconstructed from the input signals and the values of the cell parameters.

The artificial neural systems under investigation and experimentation today, employ a variety of activation functions that have more diversified features than the one presented in Fig. 5.5. Below, we introduce the main activation functions that will be used later in this chapter.

The MP neuron model shown in Fig. 5.4 used the *hard-limiting activation function.* When artificial neurons are cascaded together in layers (discussed in the next section), it is more common to use a *soft-limiting activation function.* Figure 5.7a shows a possible bipolar soft-limiting semilinear activation function. This function is, more or less, the ON-OFF type, as before, but has a sloping region in the middle. With this smooth thresholding function, the value of the output will be practically 1 if the weighted sum exceeds the threshold by a huge margin and, conversely, it will be practically –1 if the weighted sum is much less than the threshold value. However, if the threshold and the weighted sum are almost the same, the output from the neuron will have a value somewhere between the two extremes. This means that the output from the neuron can be related to its inputs in a more useful and informative way. Figure 5.7b shows a unipolar soft-limiting function.

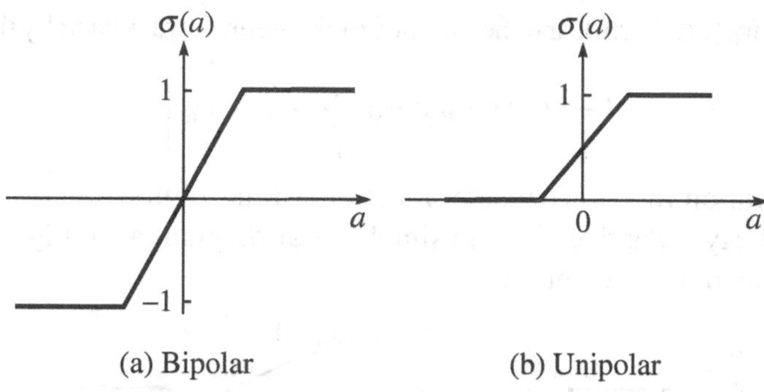

(a) Bipolar (b) Unipolar

Figure 5.7 Soft-limiting activation functions.

For many training algorithms (discussed in later sections), the derivative of the activation function is needed; therefore, the activation function selected must be differentiable. The *logistic* or *sigmoid* function, which satisfies this requirement, is the most commonly used soft-limiting activation function. The sigmoid function (Fig. 5.8a):

$$\sigma(a) = \frac{1}{1 + e^{-\lambda a}} \tag{5.4}$$

is continuous and varies monotonically from 0 to 1 as a varies from $-\infty$ to ∞. The gain of the sigmoid, λ, determines the steepness of the transition region. Note that as the gain approaches infinity, the sigmoid approaches a hard-limiting nonlinearity. One of the advantages of the sigmoid is that it is *differentiable*. This property had a significant impact historically, because it made it possible to derive a gradient search learning algorithm for networks with multiple layers (discussed in later sections).

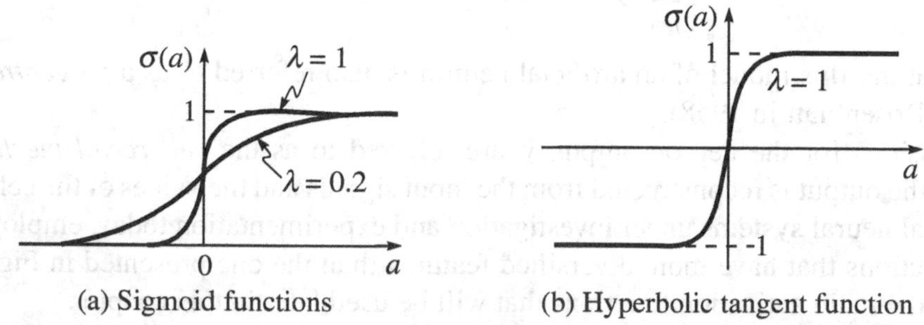

(a) Sigmoid functions (b) Hyperbolic tangent function

Figure 5.8

The sigmoid function is unipolar. A bipolar function with similar characteristics is a *hyperbolic tangent* (Fig. 5.8b):

$$\sigma(a) = \frac{1 - e^{-\lambda a}}{1 + e^{-\lambda a}} = \tanh\left(\tfrac{1}{2}\lambda a\right) \tag{5.5}$$

The biological basis of these activation functions can easily be established. It is known that neurons located in different parts of the nervous system have different characteristics. The neurons of the ocular motor system have a sigmoid characteristic, while those located in the visual area have a Gaussian characteristic. As we said earlier, anthropomorphism can lead to misunderstanding when the metaphor is carried too far. It is now a well-known result in neural network theory that a two-layer neural network is capable of solving any classification problem. It has also been shown that a two-layer network is capable of solving any nonlinear function approximation problem [3, 83]. This result does not require the use of sigmoid nonlinearity. The proof assumes only that nonlinearity is a continuous, smooth, monotonically increasing function that is bounded above and below. Thus, numerous alternatives to sigmoid could be used, without a biological justification. In addition, the above result does not require that the nonlinearity be present in the second (output) layer. It is quite common to use linear output nodes since this tends to make learning easier. In other words,

$$\sigma(a) = \lambda a; \lambda > 0 \tag{5.6}$$

is used as an activation function in the output layer. Note that this function does not 'squash' (compress) the range of output.

Our focus in this chapter will be on two-layer *perceptron networks* with the first (hidden) layer having *log-sigmoid*

$$\sigma(a) = \frac{1}{1+e^{-a}} \tag{5.7a}$$

or *tan-sigmoid*

$$\sigma(a) = \frac{1-e^{-a}}{1+e^{-a}} \tag{5.7b}$$

activation function, and the second (output) layer having *linear* activation function

$$\sigma(a) = a \tag{5.8}$$

The log-sigmoid function has historically been a very popular choice, but since it is related to the tan-sigmoid by the simple transformation

$$\sigma_{\text{log-sigmoid}} = (\sigma_{\text{tan-sigmoid}} + 1)/2 \tag{5.9}$$

both of these functions are in use in neural network models.

We have so far described two classical neuron models:

- perceptron—a neuron with sigmoidal activation function (sigmoidal function is a softer version of the original perceptron's hard limiting or threshold activation function); and
- linear neuron—a neuron with linear activation function.

5.3 NETWORK ARCHITECTURES

In the biological brain, a huge number of neurons are interconnected to form the network and perform advanced intelligent activities. The artificial neural network is built by neuron models. Many different types of artificial neural networks have been proposed, just as there are many theories on how biological neural processing works. We may classify the organization of the neural networks largely into two types: a feedforward net and a recurrent net. The feedforward net has a hierarchical structure that consists of several layers, without interconnection between neurons in each layer, and signals flow from input to output layer in one direction. In the recurrent net, multiple neurons in a layer are interconnected to organize the network. In the following, we give typical characteristics of the feedforward net and the recurrent net, respectively.

5.3.1 Feedforward Networks

A feedforward network consists of a set of *input terminals* which feed the input patterns to a layer or subgroup of neurons. The layer of neurons makes independent computations on data that it receives, and passes the results to another layer. The next layer may, in turn, make its independent computations and pass on the results to yet another layer. Finally, a subgroup of one or more neurons determines the output from the network. This last layer of the network is the *output layer*. The layers that are placed between the input terminals and the output layer are called *hidden layers*.

Some authors refer to the input terminals as the input layer of the network. We do not use that convention since we wish to avoid ambiguity. Note that each neuron in a network makes its computation based on the weighted sum of its inputs. There is one exception to this rule: the role of the 'input layer' is somewhat different as units in this layer are used only to hold input data, and to distribute the data to units in the next layer. Thus, the 'input layer' units perform no function—other than serving as a buffer, fanning out the inputs to the next layer. These units do not perform any computation on the input data, and their weights, strictly speaking, do not exist.

The network outputs are generated from the output layer units. The output layer makes the network information available to the outside world. The hidden layers are internal to the network and have no direct contact with the external environment. There may be from zero to several hidden layers. The network is said to be *fully connected* if every output from a single node is channeled to every node in the next layer.

The number of input and output nodes needed for a network will depend on the nature of the data presented to the network, and the type of the output desired from it, respectively. The number of neurons to use in a hidden layer, and the number of hidden layers required for processing a task, is less obvious. Further comments on this question will appear later.

A Layer of Neurons

A one-layer network with n inputs and M neurons is shown in Fig. 5.9. In the network, each input x_j; $j = 1, 2, \ldots, n$ is connected to the qth neuron input through the weight w_{qj}; $q = 1, 2, \ldots, M$. The qth neuron has a *summer* that gathers its weighted inputs to form its own scalar output

$$\sum_{j=1}^{n} w_{qj}\, x_j + w_{q0};\ q = 1, 2, \ldots, M$$

Finally, the qth neuron outputs $\hat{y}_q$ through its activation function $\sigma(\cdot)$:

$$\hat{y}_q = \sigma\left(\sum_{j=1}^{n} w_{qj}\, x_j + w_{q0}\right);\ q = 1, 2, \ldots, M \tag{5.10a}$$

$$= \sigma(\mathbf{w}_q^T \mathbf{x} + w_{q0});\ q = 1, 2, \ldots, M \tag{5.10b}$$

where weight vector $\mathbf{w}_q$ is defined as,

$$\mathbf{w}_q^T = [w_{q1}\ w_{q2}\ \ldots\ w_{qn}] \tag{5.10c}$$

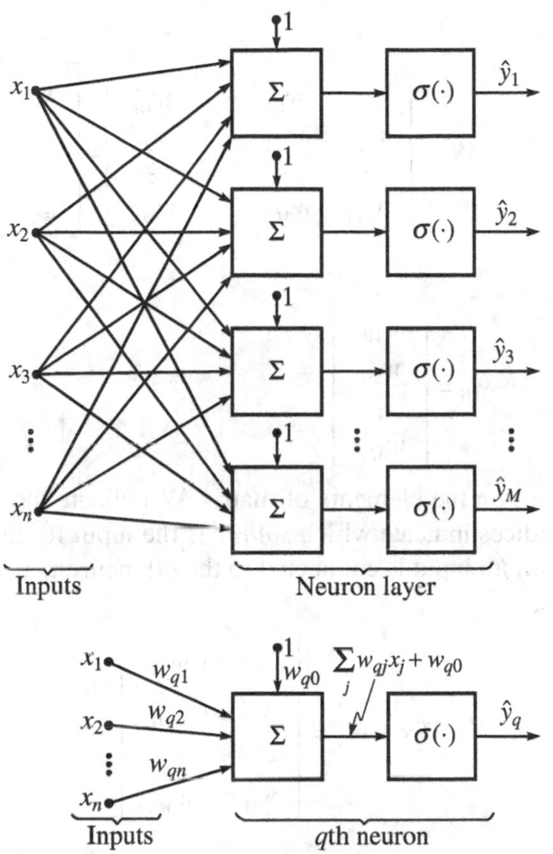

Figure 5.9 A one-layer network

Note that it is common for the number of inputs to be different from the number of neurons (i.e., $n \neq M$). A layer is not constrained to have the number of its inputs equal to the number of its neurons.

The layer shown in Fig. 5.9 has $M \times 1$ output vector

$$\hat{\mathbf{y}} = \begin{bmatrix} \hat{y}_1 \\ \hat{y}_2 \\ \vdots \\ \hat{y}_M \end{bmatrix}, \tag{5.11a}$$

$n \times 1$ input vector

$$\mathbf{x} = \begin{bmatrix} x_1 \\ x_2 \\ \vdots \\ x_n \end{bmatrix} \tag{5.11b}$$

$M \times n$ weight matrix

$$\mathbf{W} = \begin{bmatrix} w_{11} & w_{12} & \cdots & w_{1n} \\ w_{21} & w_{22} & \cdots & w_{2n} \\ \vdots & \vdots & & \vdots \\ w_{M1} & w_{M2} & \cdots & w_{Mn} \end{bmatrix} = \begin{bmatrix} \mathbf{w}_1^T \\ \mathbf{w}_2^T \\ \vdots \\ \mathbf{w}_M^T \end{bmatrix} \tag{5.11c}$$

and $M \times 1$ bias vector

$$\mathbf{w}_0 = \begin{bmatrix} w_{10} \\ w_{20} \\ \vdots \\ w_{M0} \end{bmatrix} \tag{5.11d}$$

Note that the row indices on the elements of matrix **W** indicate the *destination* neuron for the weight, and the column indices indicate which *source* is the input for that weight. Thus, the index w_{qj} says that the signal from jth input is connected to the qth neuron.

The activation vector is,

$$\mathbf{W}\mathbf{x} + \mathbf{w}_0 = \begin{bmatrix} \mathbf{w}_1^T\mathbf{x} + w_{10} \\ \mathbf{w}_2^T\mathbf{x} + w_{20} \\ \vdots \\ \mathbf{w}_M^T\mathbf{x} + w_{M0} \end{bmatrix} \tag{5.11e}$$

The outputs are,

$$\begin{aligned} \hat{y}_1 &= \sigma(\mathbf{w}_1^T\mathbf{x} + w_{10}) \\ \hat{y}_2 &= \sigma(\mathbf{w}_2^T\mathbf{x} + w_{20}) \\ &\vdots \\ \hat{y}_M &= \sigma(\mathbf{w}_M^T\mathbf{x} + w_{M0}) \end{aligned} \tag{5.11f}$$

The input-output mapping is of the feedforward and instantaneous type since it involves no time delay between the input **x** and the output $\hat{\mathbf{y}}$.

Consider a neural network with a single output node. For a dataset with n attributes, the output node receives $x_1, x_2, \ldots, x_n$, takes a weighted sum of these and applies the $\sigma(\cdot)$ function. The output of the neural network is therefore $\sigma\left(\sum_{j=1}^{n} w_j x_j + w_0\right)$.

First consider a numerical output y (i.e., $y \in \Re$). If $\sigma(\cdot)$ is a linear activation function (Eqn (5.8)), the output is simply

$$\hat{y} = \sum_{j=1}^{n} w_j x_j + w_0$$

This is exactly equivalent to the formulation of linear regression given earlier in Section 3.6 (refer to Eqn (3.70)).

Now consider binary output variable y. If $\sigma(\cdot)$ is log-sigmoid function (Eqn (5.7a)), the output is simply

$$\hat{y} = \frac{1}{1+\exp\left(-\sum_{j=1}^{n} w_j x_j + w_0\right)}$$

$$= \frac{1}{1+e^{-(w_1x_1+w_2x_2+\cdots+w_nx_n+w_0)}}$$

which is equivalent to logistic regression formulation given in Section 3.7 (refer to Eqn (3.84)). Note that here $\hat{y}$ takes continuous values in the interval $\{0, 1\}$ and represents the probability of belonging to Class q, i.e., $\hat{y} = P(\text{Class } 1|\mathbf{x})$, and $P(\text{Class } 2|\mathbf{x}) = 1 - \hat{y}$.

In both cases, although the neural network models are equivalent to the linear and logistic regression models, the resulting estimates for the weights in neural network models will be different from those in linear and logistic regression. This is because the estimation methods are different. As we will shortly see, the neural network estimation method is different from maximum likelihood method used in logistic regression, and may be different from least-squares method used in linear regression.

We will use multiple output nodes $\hat{y}_q$; $q = 1, \ldots, M$, for multiclass discrimination problems (detailed in Section 5.8). For regression (function approximation) problems, multiple output nodes correspond to multiple response variables we are interested in for numeric prediction. In this case, a number of regression problems are learned at the same time. An alternative is to train separate networks for separate regression problems (with one output node). In this chapter, we will focus on this alternative approach. Our focus is justified on the ground that in many real-life applications, we are interested in only one response variable, i.e., scalar output variable.

Multi-Layer Perceptrons

Neural networks normally have at least two layers of neurons, with the first layer neurons having nonlinear and differentiable activation functions. Such networks, as we will see, can approximate any nonlinear function. In real life, we are faced with nonlinear problems, and multilayer neural network structures have the capability of providing solutions to these problems.

Figure 5.10 shows a two-layer NN, with n inputs and two layers of neurons. The first of these layers has m neurons feeding into the second layer possessing M neurons. The first layer or the *hidden layer*, has m *hidden-layer neurons*; the second or the output layer, has M *output-layer neurons*. It is not uncommon for different layers to have different numbers of neurons. The outputs of the hidden layer are inputs to the following layer (output layer); and the network is fully connected. Neural networks possessing several layers are known as *Multi-Layer Perceptrons* (MLP); their computing power is meaningfully improved over the one-layer NN.

All continuous functions, which display certain smoothness, can be approximated to any desired accuracy with a network of one hidden layer of sigmoidal hidden units, and a layer of linear output units [83]. Does this mean that it is not required to employ more than one hidden layer and/or mix different kinds of activation functions? In fact, the accuracy may be enhanced with the help of network architectures with more hidden layers/mixing activation functions. Especially when the mapping to be learned is highly complicated, there is a likelihood of performance improvement. However, as the implementation and training of the network become increasingly complex with sophisticated network architectures, it is normal to apply only a single hidden layer of similar activation functions, and an output layer of linear units. We will focus on two-layer feedforward neural networks with sigmoidal/hyperbolic tangent hidden units and linear output units for function approximation problems. For classification problems, the linear output units will be replaced with sigmoidal units. These are widely used network architectures, and work very well in many practical applications.

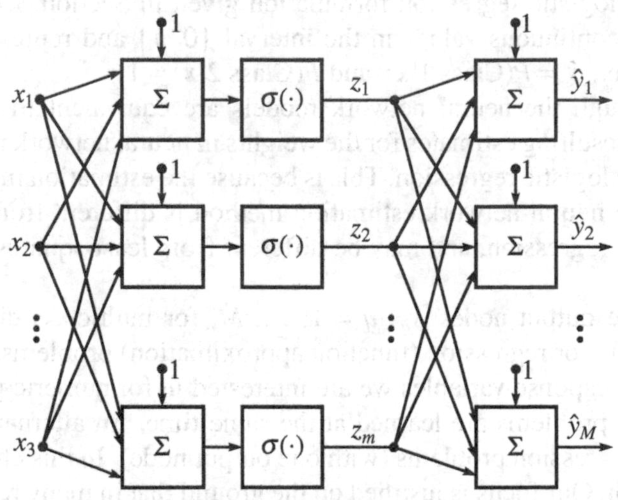

Figure 5.10 A two-layer network

Defining the input terminals as x_j; $j = 1, \ldots, n$; and the hidden-layer outputs as z_l, allows one to write

$$\begin{aligned} z_l &= \sigma\left(\sum_{j=1}^{n} w_{lj}x_j + w_{l0}\right); \; l = 1, 2, \ldots, m \\ &= \sigma(\mathbf{w}_l^T\mathbf{x} + w_{l0}) \end{aligned} \tag{5.12a}$$

where

$$\mathbf{w}_l^T \triangleq [w_{l1} \; w_{l2} \; \ldots \; w_{ln}]$$

are the weights connecting input terminals to hidden layer.

Defining the output-layer nodes as $\hat{y}_q$, one may write the *NN* output as,

$$\hat{y}_q = \left(\sum_{l=1}^{m} v_{ql}\, z_l + v_{q0} \right); q = 1, \ldots, M$$
$$= \mathbf{v}_q^T \mathbf{z} + v_{q0} \tag{5.12b}$$

where

$$\mathbf{v}_q^T \triangleq [v_{q1}\; v_{q2}\; \ldots\; v_{qm}]$$

are the weights connecting hidden layer to output layer.

For the multiclass discrimination problems, our focus will be on two-layer feedforward neural networks with sigmoidal/hyperbolic tangent hidden units (outputs of hidden units given by (5.12a)), and sigmoidal output units. The *NN* output of this multilayer structure may be written as,

$$\hat{y}_q = \sigma\left(\sum_{l=1}^{m} v_{ql} z_l + v_{q0} \right); q = 1, \ldots, M \tag{5.12c}$$

The inputs to the output-layer units (refer to Eqns (5.12b)-(5.12c)) are the nonlinear basis function values z_l; $l = 1, \ldots, m$, computed by the hidden units. It can be said that the hidden units make a nonlinear transformation from the n-dimensional input space to the m-dimensional space spanned by the hidden units and in this space, the output layer implements a linear/logistic function.

5.3.2 Recurrent Networks

The feedforward networks (Figs 5.9–5.10) implement fixed-weight mappings from the input space to the output space. Because the networks have fixed weights, the *state* of any neuron is solely determined by the input to the unit, and not the initial and past states of the neurons. This independence of initial and past states of the network neurons limits the use of such networks because no *dynamics* are involved. The maps implemented by the feedforward networks of the type shown in Figs 5.9–5.10, are *static* maps.

To allow initial and past state involvement along with serial processing, *recurrent neural networks* utilize *feedback*. Recurrent neural networks are also characterized by use of nonlinear processing units; thus, such networks are nonlinear dynamic systems (networks of the form shown in Figs 5.9–5.10 are nonlinear static systems).

The architectural layout of a recurrent network takes diverse forms. Feedback may come from the output neurons of a feedforward network to the input terminals. Feedback may also come from the hidden neurons of the network to the input terminals. In case the feedforward network possesses two or more hidden layers, the likely forms of feedback expand further. Recurrent networks possess a rich collection of architectural layouts.

It often turns out that several real-world problems, which are thought to be solvable only through recurrent architectures, are solvable with feedforward architectures also. A multilayer feedforward network, which realizes a static map, is capable of representing the input/output behavior of a dynamic system. To make this possible, the neural network has to be provided with information regarding the system history—delayed inputs and outputs (refer to Section 1.4.1). The amount of history required is dependent on the level of accuracy sought, and the resulting computational complexity. Large number of inputs increase the number of weights in the network that may result in higher accuracy, but then it may significantly increase the training time. Trial-and-error on the number of inputs, as well as the network structures, is the search process as in other machine learning systems. (Later sections will give more details.)

From several practical applications published over the past decade, there seems to be considerable evidence that multilayer feedforward networks have an extraordinary capability to do quite well in most cases.

We will focus on two-layer feedforward neural networks with sigmoidal or hyperbolic tangent hidden units and linear/sigmoidal output units. This, in all likelihood, is the most popular network architecture as it works well in many practical applications.

The rest of this chapter is organized as follows: We first consider principles of design for the primitive units that make up artificial neural networks (perceptrons, linear units, and sigmoid units), along with learning algorithms for training single units. We then present the BACKPROPAGATION algorithm for training multilayer networks of such units, and several general issues related to the algorithm. We conclude the chapter with our discussion on RBF networks.

5.4 PERCEPTRONS

Classical NN systems are based on units called PERCEPTRON and ADALINE (ADAptive Linear Element). Perceptron was developed in 1958 by Frank Rosenblatt, a researcher in neurophysiology, to perform a kind of pattern recognition tasks. In mathematical terms, it resulted from the solution of classification problem. ADALINE was developed by Bernard Widrow and Marcian Hoff; it originated from the field of signal processing, or more specifically from the adaptive noise cancellation problem. It resulted from the solution of the regression problem; the regressor having the properties of noise canceller (linear filter).

The perceptron takes a vector of real-valued inputs, calculates a linear combination of these inputs; then outputs +1 if the result is greater than the threshold and –1 otherwise (refer to Fig. 4.2).

The ADALINE in its early stage consisted of a neuron with a linear activation function (Eqn 5.8), a hard limiter (a thresholding device with a signum activation function) and the Least Mean Square (LMS) learning algorithm. We focus on the two most important parts of ADALINE—its linear activation function and the LMS learning rule. The hard limiter is omitted, not because it is irrelevant, but for being of lesser importance to the problems to be solved. The words ADALINE and *linear neuron* are both used here for a neural processing unit with a linear activation function and a corresponding learning rule (not necessarily LMS).

The roots of both the perceptron and the ADALINE were in the linear domain. However, in real life, we are faced with nonlinear problems, and the perceptron was superseded by more sophisticated and powerful neuron and neural network structures (multilayer neural networks). What is the type of unit to be used to construct multilayer networks? Firstly, we may be encouraged to select the linear units. But, multiple layers of cascaded linear units still produce only linear functions, and we prefer networks capable of representing highly nonlinear functions. Another likely selection could be perceptron unit. However, because of its discontinuous threshold, it is not differentiable, and therefore, not suited to the gradient descent approach for optimizing the performance criterion. What is required is a unit with output , which is a nonlinear function of its inputs—an output which is also a differentiable function of its inputs. One solution is the *sigmoid unit*, a unit similar to perceptron, but based on a smoothened, differentiable threshold function (Fig. 5.8; Eqns (5.7)). These activation functions are nothing but *softer* versions of original perceptron's hard-limiting threshold functions. In literature, these softer versions are also referred to as perceptrons, and the multilayer neural networks are also referred to as *Multi-Layer Perceptron (MLP) Networks*.

In the following, we discuss principles of perceptron learning for classification tasks. The next section gives the principles for linear-neuron learning. There after principles of 'soft' perceptron (sigmoid unit) learning will be presented.

5.4.1 Limitations of Perceptron Algorithm for Linear Classification Tasks

The roots of the Rosenblatt's perceptron were in the linear domain. It was developed as the simplest yet powerful classifier providing the *linear separability* of class patterns or examples. In Section 4.3, we have presented a detailed account of *perceptron algorithm*. It was observed that there is a major problem associated with this algorithm for real-world solutions: datasets are almost certainly *not* linearly separable, while the algorithm finds a separating hyperplane only for linearly separable data. When the dataset is linearly inseparable, the test of the decision surface will always fail for some subset of training points regardless of the adjustments we make to the free parameters, and the algorithm will loop forever. So, an upperbound needs to be imposed on the number of iterations. Thus, when perceptron algorithm is applied in practice, we have to live with the errors—true outputs will not always be equal to the desired ones.

History has proved that limitations of Rosenblatt's perceptron can be overcome by neural networks. The perceptron criterion function is based on *misclassification error* (number of samples misclassified) and the gradient procedures for minimization are not applicable. The neural networks primarily solve the regression problems, are based on *minimum squared-error criterion* (Eqn (3.71)) and employ gradient procedures for minimization. The algorithms for separable—as well as inseparable—data classification are first developed in the context of regression problems and then adapted for classification problems. Some methods for minimization of squared-error criterion were discussed in Section 3.6; the gradient procedures will be discussed in the present chapter.

5.4.2 Linear Classification using Regression Techniques

In the following, we present basic concepts of *linear* classification using regression techniques employing classical hard-limiting perceptrons.

In real-life, we are faced with *nonlinear* classification problems. The perceptron was superseded by more sophisticated and powerful neuron and neural network structures. A popular network used today—the *multilayer network*, has hidden layers of neurons with sigmoidal activations (discussed in later sections). These activation functions are nothing but *softer* versions of original perceptron's hard-limiting threshold functions. In literature, these softer versions are also referred to as perceptrons. Using a Multi-Layer Perceptron (MLP) network for nonlinear classification is not radically different; it directly follows from the concepts for linear classification discussed below.

The regression techniques discussed in Section 3.6, and also later in this chapter, can be used for linear classification with a careful choice of the target values associated with classes. Let the set of training (data) examples $\mathcal{D}$ be

$$\begin{aligned}\mathcal{D} &= \{x_j^{(i)}, y^{(i)}\}; i = 1, \ldots, N; j = 1, \ldots, n \\ &= \{\mathbf{x}^{(i)}, y^{(i)}\}\end{aligned} \tag{5.13}$$

where $\mathbf{x}^{(i)} = [x_1^{(i)}\ x_2^{(i)} \ldots x_n^{(i)}]^T$ is an n-dimensional *input vector* (pattern with n-features) for the ith example in a real-valued space; $y^{(i)}$ is its *class label* (output value), and $y^{(i)} \in [+1, -1]$, +1 denotes Class 1 and –1 denotes Class 2. To build a linear classifier, we need a linear function of the form

$$g(\mathbf{x}) = \mathbf{w}^T\mathbf{x} + w_0 \tag{5.14}$$

so that the input vector $\mathbf{x}^{(i)}$ is assigned to Class 1 if $g(\mathbf{x}^{(i)}) > 0$, and to Class 2 if $g(\mathbf{x}^{(i)}) < 0$, i.e.,

$$\hat{y}^{(i)} = \begin{cases} +1 \text{ if } \mathbf{w}^T\mathbf{x}^{(i)} + w_0 > 0 \\ -1 \text{ if } \mathbf{w}^T\mathbf{x}^{(i)} + w_0 < 0 \end{cases} \tag{5.15}$$

$\mathbf{w} = [w_1\ w_2 \ldots w_n]^T$ is the *weight vector* and w_0 is the *bias*. In terms of regression, we can view this classification problem as follows.

Given a vector $\mathbf{x}^{(i)}$, the output of the *summing unit* (linear combiner) will be $\mathbf{w}^T\mathbf{x}^{(i)} + w_0$ (decision hyperplane) and thresholding the output through a *sgn* function gives us perceptron output $\hat{y}^{(i)} = \pm 1$ (refer to Fig. 5.11).

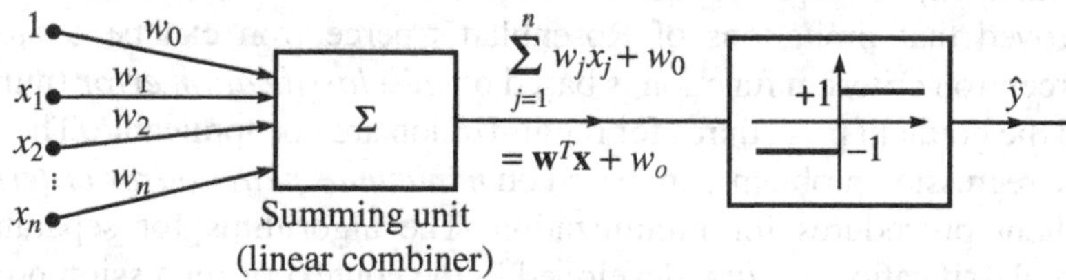

Figure 5.11 Linear classification using regression technique

The sum of error squares for the classifier becomes (Eqns (3.71))

$$E = \sum_{i=1}^{N} (e^{(i)})^2 = \sum_{i=1}^{N} (y^{(i)} - \hat{y}^{(i)})^2 \tag{5.16}$$

We require E to be a function of $(\mathbf{w}, w_0)$ to design the linear function $g(\mathbf{x}) = \mathbf{w}^T\mathbf{x} + w_0$ that minimizes E. To obtain $E(\mathbf{w}, w_0)$, we replace the perceptron outputs $\hat{y}^{(i)}$ by the linear combiner outputs $\mathbf{w}^T\mathbf{x}^{(i)} + w_0$; this gives us the error function

$$E(\mathbf{w}, w_0) = \sum_{i=1}^{N} (y^{(i)} - (\mathbf{w}^T\mathbf{x}^{(i)} + w_0))^2 \tag{5.17}$$

Consider pattern i. If $\mathbf{x}^{(i)} \in$ Class 1, the desired output $y^{(i)} = +1$ with summing unit output $\mathbf{w}^T\mathbf{x}^{(i)} + w_0 > 0$ ($\hat{y}^{(i)} = +1$, refer to Fig. 5.11), the contribution of correctly classified pattern i to $E(\mathbf{w}, w_0)$ is small when compared with wrongly classified pattern ($\mathbf{w}^T\mathbf{x}^{(i)} + w_0 < 0$; $\hat{y}^{(i)} = -1$).

The error function $E(\mathbf{w}, w_0)$ in Eqn (5.17), is a continuous, differentiable function; therefore, gradient descent approach (discussed later in this sub-section) for minimization of $E(\mathbf{w}, w_0)$ will be applicable. The training algorithm based on this $E(\mathbf{w}, w_0)$ can be seen as the training algorithm of a *linear neuron* without the nonlinear (signum) activation function. Nonlinearity is ignored during training; after training and once the weights have been fixed, the model is the perceptron model with the hard limiter following the linear combiner.

- If the unthresholded output $\mathbf{w}^T\mathbf{x}^{(i)} + w_0$ can be trained to fit the desired values $y^{(i)} = \pm 1$ in a perfect way, then the thresholded output will fit them as well (because sgn (1) = 1 and sgn (– 1) = –1). Even when the target values cannot fit perfectly in the unthresholded case, the thresholded value will correctly fit the ±1 target value whenever the unthresholded output has the correct sign. Note, however, that while gradient descent procedure will learn weights that minimize the error in the unthresholded output, these weights will not necessarily minimize the number of training examples misclassified by the thresholded output.
- The perceptron training rule (Section 4.3) converges after a finite number of iterations to a weight vector that perfectly classifies the training data provided the training examples are linearly separable. The gradient descent rule converges only asymptotically toward the minimum-error weight vector, possibly requiring unbounded time, but converges regardless of whether the training data is linearly separable or not.

5.4.3 Standard Gradient Descent Optimization Scheme: Steepest Descent

Gradient descent serves as the basis for learning algorithms that search the hypothesis space of possible weight vectors to find the weights that best fit the training examples. The gradient descent training rule for a single neuron is important because it provides the basis for the BACKPROPAGATION algorithm, which can learn networks with many interconnected units.

For linear classification using regression techniques, the task is to train unthresholded perceptron (it corresponds to the first stage of perceptron, without the threshold; Fig. 5.11) for which the output is given by

$$g(\mathbf{x}) = \sum_{j=1}^{n} w_j x_j + w_0 = \mathbf{w}^T\mathbf{x} + w_0 \tag{5.18}$$

Let us define a single weight vector $\bar{\mathbf{w}}$ for the weights $(\mathbf{w}, w_0)$:

$$\bar{\mathbf{w}}^T = [w_0\ w_1\ w_2\ \ldots\ w_n]^T \tag{5.19}$$

In terms of the weight vector $\bar{\mathbf{w}}$, the output

$$g(\mathbf{x}) = \bar{\mathbf{w}}^T\bar{\mathbf{x}} \tag{5.20a}$$

where

$$\bar{\mathbf{x}}^T = [x_0\ x_1\ x_2\ \ldots\ x_n]^T;\ x_0 = 1 \tag{5.20b}$$

The unthresholded output $\overline{\mathbf{w}}^T\overline{\mathbf{x}}^{(i)}$ is to be trained to fit the desired values $y^{(i)}$ minimizing the error (Eqn (5.17))

$$E(\overline{\mathbf{w}}) = \tfrac{1}{2}\sum_{i=1}^{N}(y^{(i)} - \overline{\mathbf{w}}^T\overline{\mathbf{x}}^{(i)})^2 \tag{5.21}$$

This error function is a continuous, differentiable function; therefore, gradient descent approach for minimization of $E(\overline{\mathbf{w}})$ will be applicable (the constant $\frac{1}{2}$ is used for computational convenience only; it gets cancelled out by the differentiation required in the error minimization process).

To understand the gradient descent algorithm, it is helpful to visualize the error space of possible weight vectors and the associated values of the *performance criterion* (*cost function E*). For the unthresholded unit (a *linear* weighted combination of inputs), the error surface is parabolic with a single global minimum. The specific parabola will depend, of course, on the particular set of training examples.

How can we calculate the direction of steepest descent along the error surface? This direction can be found by computing the derivative of E with respect to each component of the vector $\overline{\mathbf{w}}$. This vector-derivative is called the *gradient* of E with respect to $\overline{\mathbf{w}}$, written $\nabla E(\overline{\mathbf{w}})$.

$$\nabla E(\overline{\mathbf{w}}) = \left[\frac{\partial E}{\partial w_0}\;\frac{\partial E}{\partial w_1}\cdots\frac{\partial E}{\partial w_n}\right]^T \tag{5.22}$$

When interpreted as a vector in weight space, the gradient specifies the direction that produces the steepest increase in E. The negative of this vector, therefore, gives the direction of *steepest decrease*. Therefore, the training rule for gradient descent is,

$$\overline{\mathbf{w}} \leftarrow \overline{\mathbf{w}} + \Delta\overline{\mathbf{w}} \tag{5.23a}$$

where

$$\Delta\overline{\mathbf{w}} = -\eta\,\nabla E(\overline{\mathbf{w}}) \tag{5.23b}$$

Here η is a positive constant (less than one), called the *learning rate* which determines the step size in the gradient descent search. This training rule can also be written in its component form:

$$w_j \leftarrow w_j + \Delta w_j;\, j = 0, 1, 2, \ldots, n \tag{5.24a}$$

where

$$\Delta w_j = -\eta\,\frac{\partial E}{\partial w_j} \tag{5.24b}$$

which shows that steepest descent is achieved by altering each component w_j of $\overline{\mathbf{w}}$ in proportion to $\dfrac{\partial E}{\partial w_j}$.

Gradient descent search helps determine a weight vector that minimizes E by starting with an arbitrary initial weight vector and then altering it again and again in small steps. At each step, the weight vector is changed in the direction producing the steepest descent along the error surface. The process goes on till the global minimum error is attained.

To build a practical algorithm for repeated updation of weight according to (5.24), we require an effective technique to calculate the gradient at each step. Luckily, this is quite easy. The gradient with respect to weight w_j; $j = 1, \ldots, n$, can be obtained by differentiating E from Eqn (5.21) as,

$$\frac{\partial E}{\partial w_j} = \frac{\partial}{\partial w_j}\left[\frac{1}{2}\sum_{i=1}^{N}\left(y^{(i)} - \left(\sum_{j=1}^{n} w_j x_j^{(i)} + w_0\right)\right)^2\right]$$

The error $e^{(i)}$ for the ith sample of data is given by $e^{(i)} = y^{(i)} - \left(\sum_{j=1}^{n} w_j x_j^{(i)} + w_0\right)$. It follows that

$$\frac{1}{2}\sum_{i=1}^{N}\left[\frac{\partial}{\partial w_j}(e^{(i)})^2\right] = \sum_{i=1}^{N} e^{(i)}\frac{\partial e^{(i)}}{\partial w_j}$$

$$\frac{\partial e^{(i)}}{\partial w_j} = -x_j^{(i)}$$

$$\frac{\partial E}{\partial w_j} = \frac{\partial}{\partial w_j}\left[\frac{1}{2}\sum_{i=1}^{N}(e^{(i)})^2\right]$$

$$= -\sum_{i=1}^{N} e^{(i)} x_j^{(i)}$$

$$= -\sum_{i=1}^{N}\left(y^{(i)} - \left(\sum_{j=1}^{n} w_j x_j^{(i)} + w_0\right)\right)x_j^{(i)}$$

The gradient with respect to bias,

$$\frac{\partial E}{\partial w_0} = -\sum_{i=1}^{N} e^{(i)} = -\sum_{i=1}^{N}\left(y^{(i)} - \left(\sum_{j=1}^{n} w_j x_j^{(i)} + w_0\right)\right)$$

Therefore, the weight update rule for gradient descent becomes

$$w_j \leftarrow w_j + \eta\sum_{i=1}^{N}\left(y^{(i)} - \left(\sum_{j=1}^{n} w_j x_j^{(i)} + w_0\right)\right)x_j^{(i)} \tag{5.25a}$$

$$w_0 \leftarrow w_0 + \eta\sum_{i=1}^{N}\left(y^{(i)} - \left(\sum_{j=1}^{n} w_j x_j^{(i)} + w_0\right)\right) \tag{5.25b}$$

An *epoch* is a complete run through all the N associated pairs. Once an epoch is completed, the pair $(\mathbf{x}^{(1)}, y^{(1)})$ is presented again and a run is performed through all the pairs again. After several epochs, the output error is expected to be sufficiently small.

The iteration index k corresponds to the number of times the set of N pairs is presented and cumulative error is compounded. That is, k corresponds to the epoch number.

In terms of iteration index k, the weight update equations are

$$\mathbf{w}(k+1) = \mathbf{w}(k) + \eta \sum_{i=1}^{N} \left(y^{(i)} - \left(\sum_{j=1}^{n} w_j(k) x_j^{(i)} + w_0(k) \right) \right) \mathbf{x}^{(i)} \tag{5.26a}$$

$$w_0(k+1) = w_0(k) + \eta \sum_{i=1}^{N} \left(y^{(i)} - \left(\sum_{j=1}^{n} w_j(k) x_j^{(i)} + w_0(k) \right) \right) \tag{5.26b}$$

5.5 LINEAR NEURON AND THE WIDROW-HOFF LEARNING RULE

The perceptron ('softer' version; sigmoid unit) has been a fundamental building block in the present-day neural models. Another important building block has been ADALINE (ADAptive LINear Element), developed by Bernard Widrow and Marcian Hoff in 1959. It originated from the field of signal processing, or more specifically, from the adaptive noise cancellation problem. It resulted from the solution of regression problem; the regressor having the properties of noise canceller (linear filter). All its power in linear domain is still in full service, and despite being a simple neuron, it is present (without a thresholding device) in almost all the neural models for regression functions. The words ADALINE and *linear neuron* are both used here for a neural processing unit with a linear activation function shown in Fig. 5.12a. The neuron labeled with summation sign only (Fig. 5.12b) is equivalent to linear neuron of Fig. 5.12a.

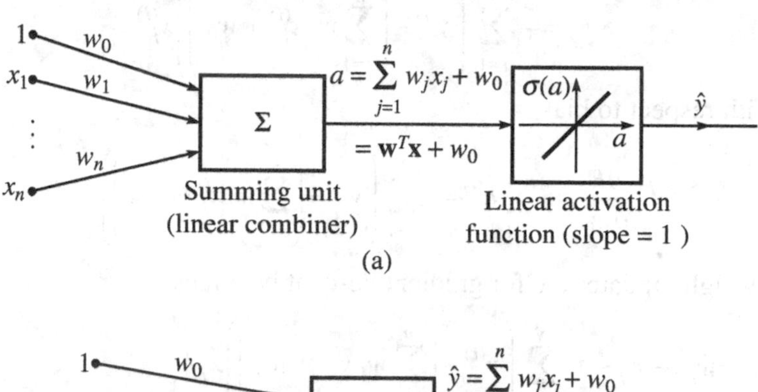

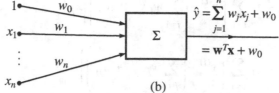

Figure 5.12 Neural processing unit with a linear activation function

In the last section, we have discussed gradient descent optimization scheme to determine the optimum setting of the weights ($\mathbf{w}$, w_0) that minimize the criterion function given by Eqn (5.21). Note that this 'sum of error squares' criterion function is deterministic and the gradient descent scheme gives a deterministic algorithm for minimization of this function. Now we explore a digress from this criterion function. Consider the problem of computing weights ($\mathbf{w}$, w_0) so as to minimize *Mean Square Error* (MSE) between desired and true outputs, defined as follows.

$$E(\mathbf{w}, w_0) = \mathbb{E}\left[\frac{1}{2}\sum_{i=1}^{N}\left(y^{(i)} - (\mathbf{w}^T\mathbf{x}^{(i)} + w_0)\right)^2\right] \tag{5.27}$$

where $\mathbb{E}$ is the statistical expectation operator.

The solution to this problem requires the computation of autocorrelation matrix $\mathbb{E}[\mathbf{x}\mathbf{x}^T]$ of the set of feature vectors, and cross-correlation matrix $\mathbb{E}[\mathbf{x}y]$ between the desired response and the feature vector. This presupposes knowledge of the underlying distributions, which, in general, is not known. Thus, our major goal becomes to see if it is possible to solve this optimization problem without having this statistical information.

The *Least Mean Square* (LMS) algorithm, originally formulated by Widrow and Hoff, is a *stochastic gradient algorithm* that iterates weights ($\mathbf{w}$, w_0) in the regressor after each presentation of data sample, unlike the standard gradient descent that iterates weights after presentation of the whole training dataset. That is, the kth iteration in standard gradient descent means the kth *epoch*, or the kth presentation of the whole training dataset, while k^{th} iteration in stochastic gradient descent means the presentation of kth single training data pair (drawn in sequence or randomly). Thus, the calculation of the weight change $\Delta\bar{\mathbf{w}}$ or the gradient needed for this, is *pattern-based*, not *epoch-based* ($\Delta\bar{\mathbf{w}} = -\eta\,\nabla E(\bar{\mathbf{w}})$; Eqn (5.23b)).

LMS is called a stochastic gradient algorithm because the gradient vector is chosen at 'random' and not, as in steepest descent case, precisely derived from the shape of the total error surface. Random means here the instantaneous value of the gradient. This is then used as the estimator of the true quantity.

The design of the LMS algorithm is very simple, yet a detailed analysis of its convergence behavior is a challenging mathematical task. It turns out that under mild conditions, the solution provided by the LMS algorithm converges in probability to the solution of the sum-of-error-squares optimization problem.

5.5.1 Stochastic Gradient Descent

While the standard gradient descent training rule of Eqn (5.26) calculates weight updates after summing errors over *all* the training examples in the given dataset $\mathcal{D}$; the concept behind stochastic gradient descent is to approximate this gradient descent search by updating weights *incrementally*, following the calculation of the error for *each* individual example. This modified training rule is like the training rule given by Eqns (5.26) except that as we iterate through each training example, we update the weights according to the gradient with respect to the distinct error function,

$$E(k) = \tfrac{1}{2}[y^{(i)} - \hat{y}(k)]^2 = \tfrac{1}{2}[e(k)]^2 \tag{5.28a}$$

$$\hat{y}(k) = \sum_{j=1}^{n} w_j(k)\, x_j^{(i)} + w_0(k) \tag{5.28b}$$

where k is the iteration index. Note that the input components $x_j^{(i)}$ and the desired output $y^{(i)}$ are not functions of the iteration index. Training pairs $(\mathbf{x}^{(i)}, y^{(i)})$, drawn in sequence or randomly, are presented to the network at each iteration. The gradients with respect to weights and bias are computed as follows:

$$\frac{\partial E(k)}{\partial w_j(k)} = e(k)\frac{\partial e(k)}{\partial w_j(k)} = -\,e(k)\frac{\partial \hat{y}(k)}{\partial w_j(k)}$$

$$= -\,e(k)\,x_j^{(i)}$$

$$\frac{\partial E(k)}{\partial w_0(k)} = -\,e(k)$$

The stochastic gradient descent algorithm becomes,

$$w_j(k+1) = w_j(k) + \eta\; e(k)\; x_j^{(i)} \tag{5.29a}$$

$$w_0(k+1) = w_0(k) + \eta\; e(k) \tag{5.29b}$$

In terms of vectors, this algorithm may be expressed as,

$$\mathbf{w}(k+1) = \mathbf{w}(k) + \eta\; e(k)\,\mathbf{x}^{(i)} \tag{5.30a}$$

$$w_0(k+1) = w_0(k) + \eta\; e(k) \tag{5.30b}$$

Stochastic gradient training algorithm iterates over the training examples $i = 1, 2, \ldots, N$ (drawn in sequence or randomly); at each iteration, altering weights as per the above equations. The sequence of these weight updates, iterated over all the training examples, gives rise to reasonable approximation to the gradient with respect to the entire set of training data. *By making the values of η small enough, stochastic gradient descent can be made to approximate standard gradient descent (steepest descent) arbitrarily closely.*

At each presentation of data $(\mathbf{x}^{(i)}, y^{(i)})$, one step of training algorithm is performed which updates both the weights and the bias. Note that teaching the network one fact at a time from one data pair, does not work. All the weights and the bias set so meticulously for one fact, could be drastically altered in learning the next fact. The network has to learn everything together, finding the best weights and bias settings for the total set of facts. Therefore, with incremental learning, the training should stop only after an epoch has been completed.

5.6 THE ERROR-CORRECTION DELTA RULE

In this section, gradient descent strategy for adapting weights for a single neuron having differentiable activation function is demonstrated. This will just be a small (nonlinear) deviation from the derivation of adaptive rule for the linear activation function, given in the previous section. Including this small deviation will be a natural step for deriving gradient-descent based algorithm for multilayer neural networks (in the next section, we will derive this algorithm).

A neural unit with any differentiable function $\sigma(a)$ is shown in Fig. 5.13. It first computes a linear combination of its inputs (activation value a); then applies nonlinear activation function $\sigma(a)$

to the result. The output $\hat{y}$ of nonlinear unit is a continuous function of its input a. More precisely, the nonlinear unit computes its output as,

$$\hat{y} = \sigma(a) \tag{5.31a}$$

$$a = \sum_{j=1}^{n} w_j x_j + w_0 \tag{5.31b}$$

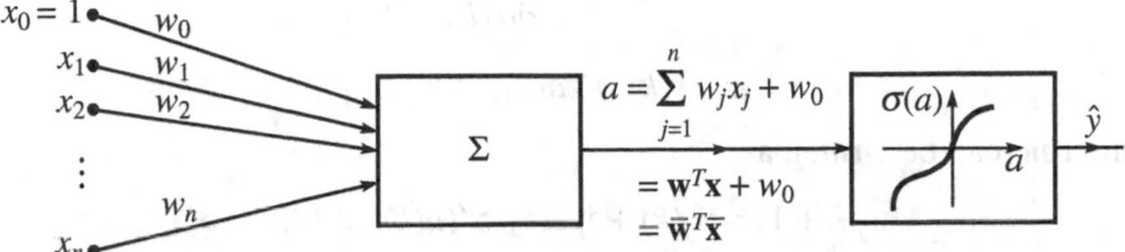

Figure 5.13 Neural unit with any differentiable activation function

The problem is to find the expression for the learning rule for adapting weights using a training set of pairs of input and output patterns; the learning is in stochastic gradient descent mode, as in the last section. We begin by defining error function $E(k)$:

$$E(k) = \tfrac{1}{2}(y^{(i)} - \hat{y}(k))^2 = \tfrac{1}{2}[e(k)]^2 \tag{5.32a}$$

$$e(k) = y^{(i)} - \hat{y}(k) \tag{5.32b}$$

$$\hat{y}(k) = \sigma\left(\sum_{j=1}^{n} w_j(k)\, x_j^{(i)} + w_0(k)\right) \tag{5.32c}$$

For each training example i, weights w_j; $j = 1, \ldots, n$ (and bias w_0) are updated by adding to it Δw_j (and Δw_0).

$$\Delta w_j(k) = -\eta \frac{\partial E(k)}{\partial w_j(k)} \tag{5.33a}$$

$$w_j(k+1) = w_j(k) - \eta \frac{\partial E(k)}{\partial w_j(k)} \tag{5.33b}$$

$$w_0(k+1) = w_0(k) - \eta \frac{\partial E(k)}{\partial w_0(k)} \tag{5.33c}$$

Note that $E(k)$ is a nonlinear function of the weights now, and the gradient cannot be calculated following the equations derived in the last section for a linear neuron. Fortunately, the calculation of the gradient is straight forward in the nonlinear case as well. For this purpose, the chain rule is,

$$\frac{\partial E(k)}{\partial w_j(k)} = \frac{\partial E(k)}{\partial a(k)} \frac{\partial a(k)}{\partial w_j(k)} \tag{5.34}$$

where the first term on the right-hand side is a measure of an *error change* due to the activation value $a(k)$ at the kth iteration, and the second term shows the influence of the weights on that particular activation value $a(k)$. Applying the chain rule again, we get,

$$\frac{\partial E(k)}{\partial w_j(k)} = \frac{\partial E(k)}{\partial e(k)} \frac{\partial e(k)}{\partial \hat{y}(k)} \frac{\partial \hat{y}(k)}{\partial a(k)} \frac{\partial a(k)}{\partial w_j(k)}$$

$$= e(k)\,[-1]\,\frac{\partial \sigma(a(k))}{\partial a(k)}\,x_j^{(i)}$$

$$= -\,e(k)\,\sigma'(a(k))\,x_j^{(i)} \tag{5.35}$$

The learning rule can be written as,

$$w_j(k+1) = w_j(k) + \eta\, e(k)\, \sigma'(a(k))\, x_j^{(i)} \tag{5.36a}$$

$$w_0(k+1) = w_0(k) + \eta\, e(k)\, \sigma'(a(k)) \tag{5.36b}$$

This is the most general learning rule that is valid for a single neuron having any nonlinear and differentiable activation function and whose input is formed as a product of the pattern and weight vectors. It follows the LMS algorithm for a linear neuron presented in the last section, which was an early powerful strategy for adapting weights using data pairs only.

This rule is also known as *delta learning rule* with delta defined as,

$$\delta(k) = e(k)\,\sigma'(a(k))$$

$$= (y^{(i)} - \hat{y}(k))\,\sigma'(a(k)) \tag{5.37}$$

In terms of $\delta(k)$, the weights-update equations become

$$w_j(k+1) = w_j(k) + \eta\,\delta(k)\,x_j^{(i)} \tag{5.38a}$$

$$w_0(k+1) = w_0(k) + \eta\,\delta(k) \tag{5.38b}$$

It should be carefully noted that the $\delta(k)$ in these equations is not the *error* but the *error change* $-\dfrac{\partial E(k)}{\partial a(k)}$ due to the input $a(k)$ to the nonlinear activation function at the kth iteration:

$$-\frac{\partial E(k)}{\partial a(k)} = -\frac{\partial E(k)}{\partial e(k)} \frac{\partial e(k)}{\partial \hat{y}(k)} \frac{\partial \hat{y}(k)}{\partial a(k)}$$

$$= -\,e(k)\,[-1]\,\sigma'(a(k))$$

$$= e(k)\,\sigma'(a(k)) = \delta(k) \tag{5.39}$$

Thus, $\delta(k)$ will generally not be equal to the *error* $e(k)$. We will use the term *error signal* for $\delta(k)$, keeping in mind that, in fact, it represents the error change.

In the world of neural computing, the error signal $\delta(k)$ is of highest importance. After a hiatus in the development of learning rules for multilayer networks for about 20 years, the adaptation rule based on delta rule made a breakthrough in 1986 and was named the *generalized delta learning rule*. Today, the rule is also known as the *error backpropagation learning rule* (discussed in the next section).

Interestingly, for a linear activation function (Fig. 5.12),

$$\sigma(a(k)) = a(k)$$

Therefore,

$$\sigma'(a(k)) = 1$$

and

$$\delta(k) = e(k)\,\sigma'(a(k)) = e(k) \tag{5.40}$$

That is, delta represents the error itself. Therefore, the delta rule for a linear neuron is same as the LMS learning rule presented in the previous section.

5.6.1 Sigmoid Unit: Soft-Limiting Perceptron

In the neural unit of Fig. 5.13, any nonlinear, smooth, differentiable, and preferably nondecreasing function can be used. The requirement for the activation function to be differentiable is basic for the error backpropagation algorithm. On the other hand, the requirement that a nonlinear activation function should monotonically increase is not so strong, and it is connected with the desirable property that its derivative does not change the sign.

The activation functions that are most commonly used in multilayer neural networks are the squashing sigmoidal functions. The sigmoidal unit is very much like a perceptron, but is based on smoothed differentiable threshold function.

The neural unit shown in Fig. 5.13 becomes a *sigmoidal unit* when $\sigma(\cdot)$ represents the sigmoidal nonlinearity illustrated in Fig. 5.8. The sigmoidal unit first computes a linear combination of its inputs (activation value a), and then applies a threshold to the result. The thresholded output $\hat{y} = \sigma(a)$ is a continuous function of its input. More precisely, the sigmoid unit computes its output as,

$$\hat{y} = \sigma(a) \tag{5.41a}$$

$$= \sigma\left(\sum_{j=1}^{n} w_j x_j + w_0\right) \tag{5.41b}$$

Because a sigmoid unit maps a very large input down to a small range outputs, it is often referred to as *squashing function.*

The most common squashing sigmoidal functions are *unipolar logistic function* (Fig. 5.8a, Eqn (5.7a)) and the *bipolar sigmoidal function* (related to a tangent hyperbolic; Fig. 5.8b, Eqn (5.7b)). The unipolar logistic function, henceforth referred to as *log-sigmoid,* squashes the inputs to outputs between 0 and 1, while the bipolar function, henceforth referred to as *tan-sigmoid,* squashes the inputs to the outputs between –1 and +1.

Log-sigmoid

$$\sigma(a) = \frac{1}{1+e^{-a}} \tag{5.42a}$$

Tan-sigmoid

$$\sigma(a) = \frac{1-e^{-a}}{1+e^{-a}} \tag{5.42b}$$

Sigmoidal unit has the useful property that its derivative is easily expressed in terms of its output:

Log-sigmoid

$$\frac{d\sigma(a)}{da} = \frac{d}{da}\left[\frac{1}{1+e^{-a}}\right] = \frac{e^{-a}}{(1+e^{-a})^2} = \frac{1}{1+e^{-a}}\left[1 - \frac{1}{1+e^{-a}}\right]$$

$$= \sigma(a)\,[1-\sigma(a)] \tag{5.43a}$$

$$= \hat{y}\,(1-\hat{y}) \tag{5.43b}$$

Tan-sigmoid

$$\frac{d\sigma(a)}{da} = \frac{d}{da}\left[\frac{1-e^{-a}}{1+e^{-a}}\right] = \frac{2e^{-a}}{(1+e^{-a})^2} = \frac{1}{2}\left[1-\left(\frac{1-e^{-a}}{1+e^{-a}}\right)\right]\left[1+\left(\frac{1-e^{-a}}{1+e^{-a}}\right)\right]$$

$$= \tfrac{1}{2}(1-\sigma(a))\,(1+\sigma(a)) \tag{5.44a}$$

$$= \tfrac{1}{2}(1-\hat{y})\,(1+\hat{y}) \tag{5.44b}$$

As we shall see, the gradient descent learning makes use of these derivatives.

The most general learning rule that is valid for a single neuron having any nonlinear and differentiable activation function is given by Eqns (5.37–5.38). For the specific case of sigmoidal (log-sigmoid) nonlinearity, we have,

$$\sigma'(a(k)) = \frac{d}{da(k)}\,\sigma(a(k)) = \sigma(a(k))\,[1-\sigma(a(k))]$$

$$= \hat{y}(k)[1-\hat{y}(k)]$$

Therefore,

$$\delta(k) = e(k)\,\sigma'(a(k)) = (y^{(i)} - \hat{y}(k))\,\hat{y}(k)\,[1-\hat{y}(k)]$$

The weight-update equations become,

$$w_j(k+1) = w_j(k) + \eta\,\delta(k)\,x_j^{(i)} \tag{5.45a}$$

$$w_0(k+1) = w_0(k) + \eta\,\delta(k) \tag{5.45b}$$

$$\delta(k) = [y^{(i)} - \hat{y}(k)]\,\hat{y}(k)\,[1-\hat{y}(k)] \tag{5.45c}$$

We construct multilayer networks using sigmoid units (next section will describe commonly used structures). Initially we may be tempted to select the linear units discussed earlier. But, multiple layers of cascaded linear units continue to produce only linear functions and we favor networks possessing the ability to represent highly nonlinear functions. The (hard-limiting) perceptron unit is another likely selection, but its discontinuous threshold makes it undifferentiable and therefore, not suited for gradient descent. We require a unit whose output is a nonlinear function of its inputs, but whose output is also a differentiable function of its inputs. There are many possible choices satisfying these requirements; sigmoid unit is the most popular choice.

5.7 MULTI-LAYER PERCEPTRON (MLP) NETWORKS AND THE ERROR-BACKPROPAGATION ALGORITHM

As noted in previous sections, single (hard-limiting) perceptron can only express linear decision surfaces, and single linear neuron can only approximate linear functions. In contrast, MLP networks trained by the backpropagation algorithm are capable of expressing a rich variety of nonlinear decision surfaces/approximating nonlinear functions. This section discusses how to learn such MLP networks using gradient descent algorithms similar to the ones described in previous sections. The backpropagation algorithm learns the weights for an MLP network, given a network with a fixed set of units and interconnections. It employs gradient descent to attempt to minimize the squared error between the network outputs and the target values for these outputs.

A typical feedforward neural network is made up of a hierarchy of layers, and the neurons in the network are arranged along these layers. The external environment is connected to the network through *input terminals*, and the *output-layer neurons*. To build an artificial neural network, we must first decide how many layers of neurons are to be used and how many neurons in each layer. In other words, we must first choose the network architecture. The number of input terminals, and the number of output nodes in output layer depend on the nature of the data presented to the network, and the type of the output desired from it, respectively. For scalar-output applications, the network has a single output unit, while for vector-output applications, it has multiple output units.

A multi-layer perceptron (MLP) network is a feedforward neural network with one or more hidden layers. Each hidden layer has its own specific function. Input terminals accept input signals from the outside world and redistribute these signals to all neurons in a hidden layer. The output layer accepts a stimulus pattern from a hidden layer and establishes the output pattern of the entire network. Neurons in the hidden layers perform transformation of input attributes; the weights of the neurons represent the features in the transformed domain. These features are then used by the output layer in determining the output pattern. A hidden layer 'hides' its desired output. The training data provides the desired output of the network; that is, the desired outputs of output layer. There is no obvious way to know what the desired outputs of the hidden layers should be.

The derivation of error-backpropagation algorithm will be given here for two MLP structures shown in Figs 5.14 and 5.15. The functions $\sigma_o(\cdot)/\sigma_{oq}(\cdot)$ are the linear/log-sigmoid activation functions of the output layer, and the functions $\sigma_{hl}(\cdot)$ are the activation functions of the hidden layer (log-sigmoid or tan-sigmoid). These structures have one hidden layer only. Though more than one hidden layer in the network may provide some advantage for approximating some complex nonlinear functions (more on this in a later section), most of the practical MLP networks use one hidden layer of sigmoidal units.

We begin by considering network structure of Fig. 5.15 with log-sigmoid activation functions in the output layer. The results for structure of Fig. 5.14 with one linear output node will then easily follow.

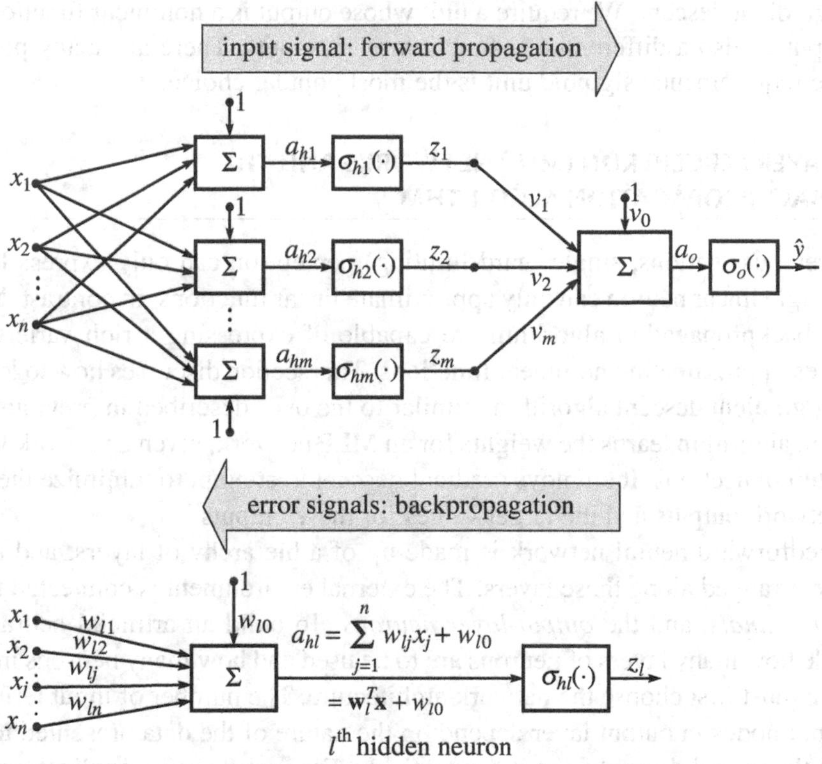

Figure 5.14 Scalar-output MLP network

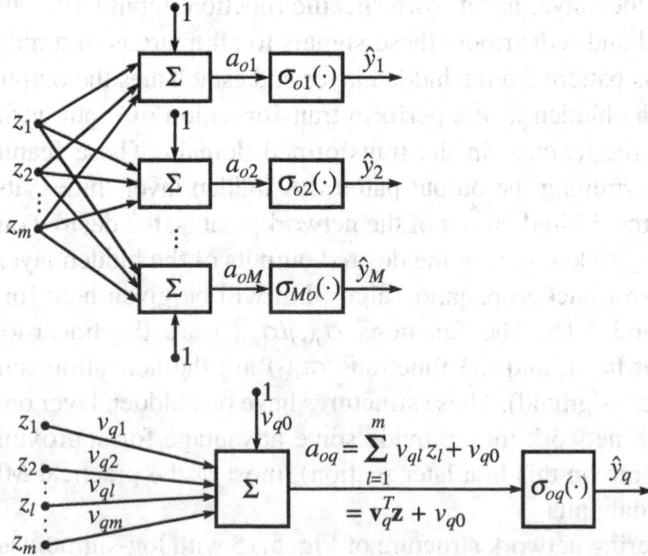

Figure 5.15 Output layer of vector-output MLP network

When **x** is fed to the input terminals (including the bias), the activation spreads in the feedforward direction, and the values z_l of the hidden units are computed. (Each hidden unit is a perceptron on its own and applies the nonlinear sigmoid function to its weighted sum of inputs.) The outputs $\hat{y}_q$ in the output layer are computed taking the hidden-layer outputs z_l as their inputs. Each output unit applies the log-sigmoid function to its weighted sum of inputs. It can be said that the hidden units make a nonlinear transformation from the n-dimensional input space to the m-dimensional space spanned by the hidden units, and in this space the output layer implements a log-sigmoid function.

Training Protocols

In the previous sections, we have presented two useful gradient descent optimization schemes for a single neuron: the standard gradient descent (steepest descent), and the stochastic gradient descent. In the steepest descent mode, the error function E is given by,

$$E(k) = \tfrac{1}{2}\sum_{i=1}^{N}(y^{(i)} - \hat{y}(k))^2 = \tfrac{1}{2}\sum_{i=1}^{N}[e(k)]^2 \tag{5.46a}$$

$$\hat{y}(k) = \sum_{j=1}^{n} w_j(k)\, x_j^{(i)} + w_0(k) \tag{5.46b}$$

All the patterns are presented to the network before a step of weight-update takes place.

In the stochastic gradient descent, the error function is given by,

$$E(k) = \tfrac{1}{2}(y^{(i)} - \hat{y}(k))^2 = \tfrac{1}{2}[e(k)]^2 \tag{5.47a}$$

$$\hat{y}(k) = \sum_{j=1}^{n} w_j(k)\, x_j^{(i)} + w_0(k) \tag{5.47b}$$

The weights are updated for each of the training pairs.

From now onwards, the two schemes will be referred to as two protocols of training: batch training, and incremental training[1]. Whereas the *batch training rule* computes the weight updates after summing errors over *all* the training examples (*batch of training data*), the idea behind *incremental training rule* is to update weights incrementally, following the calculation of error for *each* individual example (incremental training rule gives stochastic approximation of batch training which implements standard gradient descent (steepest descent)).

The gradient descent weight-update rules for MLP networks are similar to the *delta training rule* described in the previous section for a single neuron. The training data supplies target values $y_q^{(i)}$; $q = 1, \ldots, M$, of the outputs. Given the initial values of the weights w_{lj}; $l = 1, \ldots, m$; $j = 1, \ldots, n$, with bias weights w_{l0} between input terminals and hidden layer, and the weights v_{ql} with bias weights v_{q0} between hidden layer and output layer, the network outputs $\hat{y}_q^{(i)}$ are calculated by propagating input signals $x_j^{(i)}$ in the forward direction, computing hidden-layer outputs $z_l^{(i)}$, and therefrom the outputs $\hat{y}_q^{(i)}$. Using delta training rule, the weights v_{ql} and v_{q0} are updated with $z_l^{(i)}$ as inputs to the output neurons. The variable (or signal) δ designates an *error signal*, but not the error itself as defined in Eqn (5.45), that is, δ will generally not be equal to the error $e_q^{(i)} = y_q^{(i)} - \hat{y}_q^{(i)}$. (Interestingly, the equality does hold for linear activation function.)

[1] In *on-line* training, each pattern is presented once and only once; there is no use of memory for storing the patterns. This explains the difference between incremental training and on-line training.

There is, however, a problem in updating the weights w_{lj} and w_{l0} between input terminals and the hidden layer. The weight updates of w_{lj} and w_{l0} should ideally reduce the network output error $\sum_{q=1}^{M}(y_q^{(i)} - \hat{y}_q^{(i)})$. The delta training rule requires the target values for the outputs of hidden-layer neurons, which are unknown. Hidden layer affects the network output error *indirectly* through its weights.

Since we do not know the desired values of hidden-layer outputs, the weights w_{lj} and w_{l0} could not be updated using delta training rule. After a hiatus in the development of learning rules for multilayer networks for about 20 years, the delta training rule made a breakthrough in 1986 and was named the *generalized delta rule* (today, it is more popularly known as *error-backpropagation learning rule*). The error is calculated at the output nodes, and then *propagated backwards* through the network from the output layer to the input terminals. This *error backpropagation* gives *error terms* (and not the errors themselves) for hidden units outputs. Using these error terms, the weights w_{lj} and w_{l0} between input terminals and hidden units are updated in a usual incremental/batch training mode.

Thus, in a backpropagation network, the learning algorithm has two phases: *the forward propagation to compute MLP outputs, and the backpropagation to compute backpropagated errors*. In the forward propagation phase, the input signals **x** are propagated through the network from input terminals to output layer, while in the backpropagation phase, the errors at the output nodes are propagated backwards from output layer to input terminals (see Fig. 5.14). This is an indirect way in which the weights w_{lj} and w_{l0} can influence the network outputs and hence the cost function E.

The error-backpropagation algorithm (as we shall see shortly) begins by constructing a network with the desired number of hidden units and initializing the network weights to small random values. Typically, in a backpropagation network, the layers are *fully connected*, that is, every input terminal and every neuron in each layer is connected to every other neuron in the adjacent forward layer. Given this fixed network structure, the main loop of the algorithm repeatedly iterates over the training examples. For each training example, it applies the network to the example, computes the gradient with respect to the error on this example, then updates all weights in the network to implement incremental training protocol. Similar procedure is followed for implementing batch training rule.

The weight-update rule for an MLP network may be repeated thousands of times in a typical application. A range of termination conditions can be employed to stop the process. One may opt to stop after a fixed number of iterations through the loop, or once the error on the training examples drops below some threshold, or once error on a separate validation set of examples fulfills a specific criterion. The selection of proper termination criterion is important as very few iterations can be unsuccessful in reducing error sufficiently, and too many can cause overfitting of data.

5.7.1 The Generalized Delta Rule

We develop the learning rule for the case when there are multiple neurons in the output layer (Fig. 5.15). The derivation here is of learning rule for the adaptation of weights in an incremental mode.

Update Rule for Output-Units Weights

As before, we begin by defining cost function (sum-of-error-squares) for this neural network having M output-layer neurons. At each iteration step k,

$$E(k) = \frac{1}{2}\sum_{q=1}^{M}\left(y_q^{(i)} - \hat{y}_q(k)\right)^2 = \frac{1}{2}\sum_{q=1}^{M}[e_q(k)]^2 \tag{5.48a}$$

$$\hat{y}_q(k) = \sigma_{oq}\left(\sum_{l=1}^{m}[v_{ql}(k)\, z_l(k)] + v_{q0}(k)\right) \tag{5.48b}$$

$$= \sigma_{oq}(a_{oq}(k)) \tag{5.48c}$$

From forward propagation phase, we have,

$$z_l(k) = \sigma_{hl}(u_{hl}(k)) \tag{5.49a}$$

$$= \sigma_{hl}\left(\sum_{j=1}^{n}[w_{lj}(k)\, x_j^{(i)}] + w_{l0}(k)\right) \tag{5.49b}$$

We have used subscript 'o' for the output layer and subscript 'h' for the hidden layer. This is necessary because the *error signal terms* (the delta values) for output layer neurons must be distinguished from those for hidden layer processing units.

The input a_{oq} to the qth output unit (Eqn (5.48b)) is given as,

$$a_{oq}(k) = \sum_{l=1}^{m}(v_{ql}(k)\, z_l(k)) + v_{q0}(k)$$

The *error signal term* for the qth neuron is defined as (Eqn (5.39)),

$$\delta_{oq}(k) = -\frac{\partial E(k)}{\partial a_{oq}(k)}$$

Applying the chain rule, the gradient of the cost function with respect to the weight v_{ql} is,

$$\frac{\partial E(k)}{\partial v_{ql}(k)} = \frac{\partial E(k)}{\partial a_{oq}(k)}\frac{\partial a_{oq}(k)}{\partial v_{ql}(k)}$$

$$= -\,\delta_{oq}(k)\, z_l(k)$$

The weight change from Eqns (5.33) can now be written as,

$$\Delta v_{ql}(k) = -\,\eta\,\frac{\partial E(k)}{\partial v_{ql}(k)} = \eta\,\delta_{oq}(k)\, z_l(k)$$

Applying the chain rule, the expression for error signal is,

$$\delta_{oq}(k) = -\frac{\partial E(k)}{\partial a_{oq}(k)} = -\frac{\partial E(k)}{\partial \hat{y}_q(k)} \frac{\partial \hat{y}_q(k)}{\partial a_{oq}(k)}$$

$$= e_q(k)\ \sigma'_{oq}(a_{oq}(k))$$

where the term $\sigma'_{oq}(a_{oq}(k))$ represents the slope $\partial \hat{y}_q(k) / \partial a_{oq}(k)$ of the qth output neuron's activation function, assumed in general, to be any nonlinear differentiable function. For our specific case of log-sigmoid output units (Eqn (5.43)),

$$\sigma'_{oq}(a_{oq}(k)) = \frac{\partial \sigma_{oq}(a_{oq}(k))}{\partial a_{oq}(k)} = \hat{y}_q(k)\,[1 - \hat{y}_q(k)]$$

resulting in a simple expression of the error signal term:

$$\delta_{oq}(k) = e_q(k)\ \hat{y}_q(k)\,[1 - \hat{y}_q(k)]$$

Finally, the weight adjustments can be calculated from

$$v_{ql}(k+1) = v_{ql}(k) + \eta\ \delta_{oq}(k)\ z_l(k) \tag{5.50a}$$

$$v_{q0}(k+1) = v_{q0}(k) + \eta\ \delta_{oq}(k) \tag{5.50b}$$

Update-Rule for Hidden-Units Weights

The problem at this point is to calculate the error signal terms δ_{hl} for the hidden-layer neurons, to update the weights w_{lj} between input terminals and hidden layer. The derivation of the expression for δ_{hl} was a major breakthrough in the learning procedure for multilayer neural networks. Unlike the output nodes, the desired outputs of the hidden nodes (and hence the errors at the hidden nodes) are unknown. If the 'target' outputs for hidden nodes were known for any input, the input terminals to hidden layer weights could be adjusted by a procedure similar to the one used for output nodes. However, there is no explicit 'supervisor' to state what the hidden units' outputs should be.

Training examples provide only the target values for the network outputs. The error signal terms δ_{oq} for the output nodes are easily calculated as we have seen.

The *generalized delta rule* provides a way out for the computation of error signal terms for hidden units. An *intuitive* understanding of the procedure will be helpful in the derivation that follows.

To begin, notice that input terminals to hidden layer weights w_{lj} can influence the rest of the network and hence the output error only through a_{hl} (refer to Figs 5.14–5.15). The generalized delta rule computes the error signal terms δ_{hl} for hidden layer units by summing the output error signal terms δ_{oq} for each output unit influenced by a_{hl} (from Fig. 5.15, we see that a_{hl} contributes to errors at all output layer neurons), weighting each of the δ_{oq}'s by v_{ql}—the weights from the hidden unit to output units. Thus, by *error backpropagation* from output layer to the hidden layer, we take into account the *indirect* ways in which w_{lj} can influence the network outputs, and hence the error E. The power of backpropagation is that it allows us to calculate an 'effective' error for each hidden node and, thus, derive the learning rule for input terminals to hidden layer weights.

The derivation of the learning rule or of the equations for the weight change Δw_{lj} of any hidden-layer neuron follows the gradient procedure used earlier for output-layer neurons:

$$\Delta w_{lj}(k) = -\eta \frac{\partial E(k)}{\partial w_{lj}(k)} \tag{5.51a}$$

$$w_{lj}(k+1) = w_{lj}(k) - \eta \frac{\partial E(k)}{\partial w_{lj}(k)} \tag{5.51b}$$

Similar equations hold for bias weights w_{l0}.

Applying the chain rule, the gradient of the cost function with respect to the weight w_{lj} is,

$$\frac{\partial E(k)}{\partial w_{lj}(k)} = \frac{\partial E(k)}{\partial a_{hl}(k)} \frac{\partial a_{hl}(k)}{\partial w_{lj}(k)}$$

where input a_{hl} to each hidden layer activation function is given as,

$$a_{hl}(k) = \sum_{j=1}^{n} w_{lj}(k)\, x_j^{(i)} + w_{l0}(k)$$

The error signal term for l^{th} neuron is given as,

$$\delta_{hl}(k) = -\frac{\partial E(k)}{\partial a_{hl}(k)}$$

Therefore, the gradient of the cost function becomes,

$$\frac{\partial E(k)}{\partial w_{lj}(k)} = \frac{\partial E(k)}{\partial a_{hl}(k)} \frac{\partial a_{hl}(k)}{\partial w_{lj}(k)}$$

$$= -\,\delta_{hl}(k)\, x_j^{(i)}$$

The weight-update equation (Eqn (5.51b)) takes the form:

$$w_{lj}(k+1) = w_{lj}(k) + \eta\, \delta_{hl}(k)\, x_j^{(i)} \tag{5.52a}$$

For the bias weights,

$$w_{l0}(k+1) = w_{l0}(k) + \eta\, \delta_{hl}(k) \tag{5.52b}$$

Now the problem in hand is to calculate the error signal term δ_{hl} for hidden-layer neuron in terms of error signal terms δ_{oq} of the output-layer neurons employing error backpropagation.

The activation a_{hl} of lth hidden-layer neuron is given as,

$$a_{hl}(k) = \sum_{j=1}^{n} w_{lj}(k)\, x_j^{(i)} + w_{l0}(k)$$

The error signal term for l^{th} neuron,

$$\delta_{hl}(k) = -\frac{\partial E(k)}{\partial a_{hl}(k)}$$

Since a_{hl} contributes to errors at all output-layer neurons, we have the chain rule

$$\frac{\partial E(k)}{\partial a_{hl}(k)} = \sum_{q=1}^{M} \frac{\partial E(k)}{\partial a_{oq}(k)} \frac{\partial a_{oq}(k)}{\partial a_{hl}(k)}$$

$$= \sum_{q=1}^{M} -\delta_{oq}(k) \frac{\partial a_{oq}(k)}{\partial a_{hl}(k)}$$

$$= \sum_{q=1}^{M} -\delta_{oq}(k) \frac{\partial a_{oq}(k)}{\partial z_l(k)} \frac{\partial z_l(k)}{\partial a_{hl}(k)}$$

Since,

$$a_{oq}(k) = \sum_{l=1}^{m} v_{ql}(k)\, z_l(k) + v_{q0}$$

we have,

$$-\delta_{hl}(k) = \frac{\partial E(k)}{\partial a_{hl}(k)} = \sum_{q=1}^{M} -\delta_{oq}(k)\, v_{ql}(k)\, \sigma'_{hl}(a_{hl})$$

Assuming unipolar sigmoid activation functions in hidden-layer neurons, we have,

$$\sigma'_{hl}(a_{hl}(k)) = z_l(k)\,[1 - z_l(k)]$$

This gives,

$$\delta_{hl}(k) = z_l(k)\,[1 - z_l(k)] \sum_{q=1}^{M} \delta_{oq}(k)\, v_{ql}(k)$$

We now use the equations derived above and give the final form (to be used in algorithms) of weights update rules for the two-layer MLP networks shown in Figs (5.14 and 5.15) with unipolar sigmoidal units in the hidden layer and unipolar sigmoid/linear units in the output layer.

Weights-update equations employing incremental training for the multi-output structure (log-sigmoid output units): Our actual aim lies in learning from all the data pairs known to us. We teach the neural network with the help of one data pair at a time; weights and biases are updated after each presentation of data pair. The iteration index k corresponds to presentation of each data pair. But since the network is supposed to be learning from all the data pairs together, the stopping criterion is applied after completion of each epoch (presentation of all the N data pairs).

Forward Recursion to Compute MLP Output

Present the input $\mathbf{x}^{(i)}$ to the MLP network, and compute the output using

$$z_l(k) = \sigma_{hl}\left(\sum_{j=1}^{n} (w_{lj}(k)\, x_j^{(i)}) + w_{l0}(k)\right);\ l = 1, \ldots, m \tag{5.53a}$$

$$\hat{y}_q(k) = \sigma_{oq}\left(\sum_{l=1}^{m} (v_{ql}(k)\, z_l(k)) + v_{q0}(k))\right);\ q = 1, \ldots, M \tag{5.53b}$$

with initial weights w_{lj}, w_{l0}, v_{ql}, v_{q0}, randomly chosen.

Backward Recursion for Backpropagated Errors

$$\delta_{oq}(k) = [y_q^{(i)} - \hat{y}_q(k)]\ \hat{y}_q(k)\,[1 - \hat{y}_q(k)] \tag{5.53c}$$

$$\delta_{hl}(k) = z_l(k)\ [1 - z_l(k)]\ \sum_{q=1}^{M} \delta_{oq}(k)\ v_{ql}(k) \tag{5.53d}$$

Computation of MLP Weights and Bias Updates

$$v_{ql}(k+1) = v_{ql}(k) + \eta\ \delta_{oq}(k)\ z_l(k) \tag{5.53e}$$

$$v_{q0}(k+1) = v_{q0}(k) + \eta\ \delta_{oq}(k) \tag{5.53f}$$

$$w_{lj}(k+1) = w_{lj}(k) + \eta\ \delta_{hl}(k)\ x_j^{(i)} \tag{5.53g}$$

$$w_{l0}(k+1) = w_{l0}(k) + \eta\ \delta_{hl}(k) \tag{5.53h}$$

Weights-update equations employing batch training for the multi-output structure (log-sigmoid output units): Our interest lies in learning to minimize the total error over the entire batch of training examples. All N pairs are presented to the network (one at a time) and a cumulative error is computed after all pairs have been presented. At the end of this procedure, the neuron weights and biases are updated once. In batch training, the iteration index corresponds to the number of times the set of N pairs is presented and cumulative error is compounded. That is, k corresponds to epoch number.

The initial weights are randomly chosen, and iteration index k is set to 0. All the training data pairs; $i = 1, \ldots, N$, are presented to the network before the algorithm moves to iteration index $k + 1$ to update the weights. The weights-update is done using the following equations:

$$z_l^{(i)} = \sigma_{hl}\left(\sum_{l=1}^{m} (w_{lj}(k)\, x_j^{(i)}) + w_{l0}(k)\right) \tag{5.54a}$$

$$\hat{y}_q^{(i)} = \sigma_{oq}\left(\sum_{l=1}^{m} (v_{ql}(k)\, z_l^{(i)}) + v_{q0}(k)\right) \tag{5.54b}$$

$$\delta_{oq}^{(i)} = (y_q^{(i)} - \hat{y}_q^{(i)})\ \hat{y}_q^{(i)}\,(1 - \hat{y}_q^{(i)}) \tag{5.54c}$$

$$\Delta v_{ql}(k) = \eta \sum_{i=1}^{N} \delta_{oq}^{(i)}\ z_l^{(i)} \tag{5.54d}$$

$$\delta_{hl}^{(i)} = z_l^{(i)}(1 - z_l^{(i)}) \sum_{q=1}^{M} \delta_{oq}^{(i)}\, v_{ql}(k) \tag{5.54e}$$

$$\Delta w_{lj}(k) = \eta \sum_{i=1}^{N} \delta_{hl}^{(i)}\, x_j^{(i)} \tag{5.54f}$$

$$v_{ql}(k+1) = v_{ql}(k) + \Delta v_{ql}(k) \tag{5.54g}$$

$$w_{lj}(k+1) = w_{lj}(k) + \Delta w_{lj}(k) \tag{5.54h}$$

Weights-update equations for the multi-output structure (linear output units): Equations for multi-output structure with linear output nodes directly follow from the above equations with appropriate changes in $\sigma_{oq}(\cdot)$ and δ_{oq}.

Incremental Training

$$z_l(k) = \sigma_{hl}\left(\sum_{j=1}^{n} (w_{lj}(k)\, x_j^{(i)}) + w_{l0}(k)\right);\ l = 1, \ldots, m \tag{5.55a}$$

$$\hat{y}_q(k) = \sum_{l=1}^{m} (v_{ql}(k)\, z_l(k)) + v_{q0}(k)\ ;\ q = 1, \ldots, M \tag{5.55b}$$

$$\delta_{oq}(k) = (y_q^{(i)} - \hat{y}_q(k)) \tag{5.55c}$$

$$\delta_{hl}(k) = z_l(k)\,[1 - z_l(k)] \sum_{q=1}^{M} \delta_{oq}(k)\, v_{ql}(k) \tag{5.55d}$$

$$v_{ql}(k+1) = v_{ql}(k) + \eta\, \delta_{oq}(k)\, z_l(k) \tag{5.55e}$$

$$v_{q0}(k+1) = v_{q0}(k) + \eta\, \delta_{oq}(k) \tag{5.55f}$$

$$w_{lj}(k+1) = w_{lj}(k) + \eta\, \delta_{hl}(k)\, x_j^{(i)} \tag{5.55g}$$

$$w_{l0}(k+1) = w_{l0}(k) + \eta\, \delta_{hl}(k) \tag{5.55h}$$

Batch Training

$$z_l^{(i)} = \sigma_{hl}\left(\sum_{j=1}^{n} (w_{lj}(k)\, x_j^{(i)}) + w_{l0}(k)\right) \tag{5.56a}$$

$$\hat{y}_q^{(i)} = \sum_{l=1}^{m} (v_{ql}(k)\, z_l^{(i)}) + v_{q0}(k) \tag{5.56b}$$

$$\delta_{oq}^{(i)} = (y_q^{(i)} - \hat{y}_q^{(i)}) \tag{5.56c}$$

$$\Delta v_{ql}(k) = \eta \sum_{i=1}^{N} \delta_{oq}^{(i)}\, z_l^{(i)} \tag{5.56d}$$

$$\delta_{hl}^{(i)} = z_l^{(i)}(1 - z_l^{(i)}) \sum_{q=1}^{M} \delta_{oq}^{(i)} \, v_{ql}(k) \tag{5.56e}$$

$$\Delta w_{lj}(k) = \eta \sum_{i=1}^{N} \delta_{hl}^{(i)} \, x_j^{(i)} \tag{5.56f}$$

Weights-update equations for single-output structure: Equations for single-output structure directly follow from the above equations with $M = 1$.

A summary of the error-propagation algorithm is given in Table 5.1 for the vector-output network structure, employing incremental learning.

Table 5.1 Summary of error backpropagation algorithm

Given a set of N data pairs $\{\mathbf{x}^{(i)}, \mathbf{y}^{(i)}\}$; $i = 1, 2, \ldots, N$, that are used for training:

$$\mathbf{x} = [x_1 \; x_2 \; \ldots \; x_n]^T = \{x_j\}; j = 1, 2, \ldots, n$$

$$\mathbf{y} = [y_1 \; y_2 \; \ldots \; y_M]^T = \{y_q\}; q = 1, 2, \ldots, M$$

Forward Recursion

Step 1: Choose the number of units m in the hidden layer, the learning rate η, and predefine the maximally allowed (desired) error E_{des}.

Step 2: Initialize weights w_{lj}, w_{l0}, v_{ql}, v_{q0}; $l = 1, \ldots, m$.

Step 3: Present the input $\mathbf{x}^{(i)}$ (drawn in sequence or randomly) to the network.

Step 4: Consequently, compute the output from the hidden and output layer neurons using Eqns (5.53a–5.53b).

Step 5: Find the value of the sum of error-squares cost function $E(k)$ at the kth iteration for the data pair applied and given weights (in the first step of an epoch, initialize $E(k) = 0$):

$$E(k) \leftarrow \tfrac{1}{2} \sum_{q=1}^{M} (y_q^{(i)} - \hat{y}_q(k))^2 + E(k)$$

Note that the value of the cost function is accumulated over all the data pairs.

Backward Recursion

Step 6: Calculate the error signals δ_{oq} and δ_{hl} for the output layer neurons and hidden layer neurons, respectively, using Eqns (5.53c–5.53d).

Update Weights

Step 7: Calculate the updated output layer weights $v_{ql}(k+1)$, $v_{q0}(k+1)$ and updated hidden layer weights $w_{lj}(k+1)$, $w_{l0}(k+1)$ using Eqns (5.53e–5.53h).

Contd.

Step 8: If $i < N$, go to step 3; otherwise go to step 9.

Step 9: The learning epoch (sweep through all data pairs) completed: $i = N$. For $E_N < E_{des}$, terminate training. Otherwise go to step 3 and start a new learning epoch.

Example 5.1

For sample calculations on learning by backpropagation algorithm, we consider a multilayer feedforward network shown in Fig. 5.16. Let the learning rate η be 0.9. The initial weight and bias values are given in Table 5.2. We will update the weight and bias values for one step with input vector $\mathbf{x} = [1\ 0\ 1]^T$ and desired output $y = 1$. Activation function for hidden nodes is a unipolar sigmoidal function, and for output-layer node is the linear function:

$$\sigma_{hl}(a_{hl}) = \frac{1}{1+e^{-a_{hl}}}$$

$$\sigma_o(a_o) = a_o$$

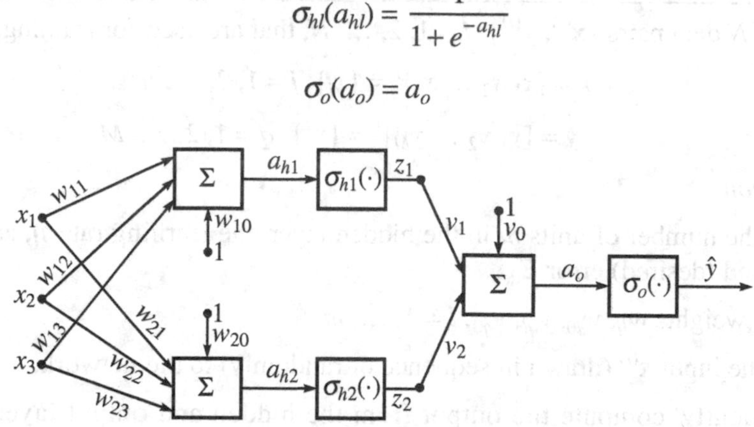

Figure 5.16 Example of a multilayer feedforward network

Table 5.2 Initial weight and bias values

Weight/Bias	w_{11}	w_{21}	w_{12}	w_{22}	w_{13}	w_{23}	w_{10}	w_{20}	v_1	v_2	v_0
Initial value	0.2	–0.3	0.4	0.1	–0.5	0.2	–0.4	0.2	–0.3	–0.2	0.1

Initial tuple $\mathbf{x} = [1\ 0\ 1]^T = [x_1\ x_2\ x_3]^T$ is presented to the network, and outputs from each hidden and output layer neurons are computed using Eqns (5.53a–5.53b):

$$a_{hl}(k) = \sum_{j=1}^{3} (w_{lj}(k)\, x_j^{(i)}) + w_{l0}(k);\ l = 1, 2$$

$$z_l(k) = \sigma_{hl}(a_{hl})$$

$$a_o(k) = v_1(k)\, z_1(k) + v_2(k)\, z_2(k) + v_0(k)$$

$$\hat{y}(k) = \sum_{l=1}^{2} (v_l(k)\, z_l(k)) + v_0(k) = a_o(k)$$

Outputs of hidden and output layer neurons are shown in Table 5.3.

The error of output unit is computed and propagated backward. The error terms are shown in Table 5.3 (Eqns (5.53c–5.53d)).

$$\delta_o(k) = y^{(i)} - \hat{y}(k)$$
$$\delta_{hl}(k) = z_l(k)\,[1 - z_l(k)]\,\delta_o(k)\,v_l(k)$$

The weight and bias updates are shown in Table 5.4 (Eqns (5.53e–5.53h)).

$$v_l(k+1) = v_l(k) + \eta(y^{(i)} - \hat{y}(k))\,z_l(k)$$
$$v_0(k+1) = v_0(k) + \eta(y^{(i)} - \hat{y}(k))$$
$$w_{lj}(k+1) = w_{lj}(k) + \eta\,z_l(k)\,[1 - z_l(k)]\,x_j^{(i)}(y^{(i)} - \hat{y}(k))\,v_l(k)$$
$$w_{l0}(k+1) = w_{l0}(k) + \eta\,z_l(k)\,[1 - z_l(k)]\,(y^{(i)} - \hat{y}(k))\,v_l(k)$$

Table 5.3 Input, output and error-term calculations for hidden and output nodes

Node	Input	Output	Error-term
Hidden 1	$a_{h1} = -0.7$	$z_1 = 0.3348$	$\delta_{h1} = -0.0735$
Hidden 2	$a_{h2} = 0.1$	$z_2 = 0.5250$	$\delta_{h2} = -0.0511$
Output	$a_o = -0.1045$	$\hat{y} = -0.1045$	$\delta_o = 1.1045$

Table 5.4 Calculations for weight and bias updates

Weight/Bias	w_{11}	w_{21}	w_{12}	w_{22}	w_{13}	w_{23}	w_{10}	w_{20}
New value	0.1339	–0.3496	0.4	0.1	–0.5661	0.1504	–0.46661	0.1504
Weight/Bias	v_1	v_2	v_0					
New value	0.0298	0.3219	1.0941					

Consider now that the activation function for hidden nodes is a unipolar sigmoid function, and for output node also, it is unipolar sigmoid. In this case,

$$a_{hl}(k) = \sum_{j=1}^{3}(w_{lj}(k)\,x_j^{(i)}) + w_{l0}(k);\ l = 1, 2$$
$$z_l(k) = \sigma_{hl}(a_{hl})$$
$$a_o(k) = v_1(k)\,z_1(k) + v_2(k)\,z_2(k) + v_0(k)$$
$$\hat{y}(k) = \sigma_o(a_o)$$
$$\delta_o(k) = (y^{(i)} - \hat{y}(k))\;\hat{y}(k)\,(1 - \hat{y}(k))$$

$$\delta_{hl}(k) = z_l(k)\,[1 - z_l(k)]\;\delta_o(k)\;v_l(k)$$

$$v_l(k+1) = v_l(k) + \eta\;\delta_o(k)\;z_l(k)$$

$$v_0(k+1) = v_0(k) + \eta\;\delta_o(k)$$

$$w_{lj}(k+1) = w_{lj}(k) + \eta\;\delta_{hl}(k)\;x_j^{(i)}$$

$$w_{l0}(k+1) = w_{l0}(k) + \eta\;\delta_{hl}(k)$$

Sample calculations are left as an exercise for the reader.

5.7.2 Convergence and Local Minima

As discussed earlier, the backpropagation algorithm implements a gradient descent search through the space of possible network weights, reducing the error E between the target values and the network outputs. As the error surface for MLP networks may comprise various local minima, gradient descent could get trapped in any of these. Due to this, error backpropagation over MLP networks is only assured to converge toward a local minima in E and not essentially to the global minimum error.

Let us consider, as an illustrative example, minimization of the following nonlinear function [85]:

$$E(w) = 0.5\,w^2 - 8\sin w + 7$$

where w is a *scalar* weight parameter. This function has two local minima and a global minimum (Fig. 5.17).

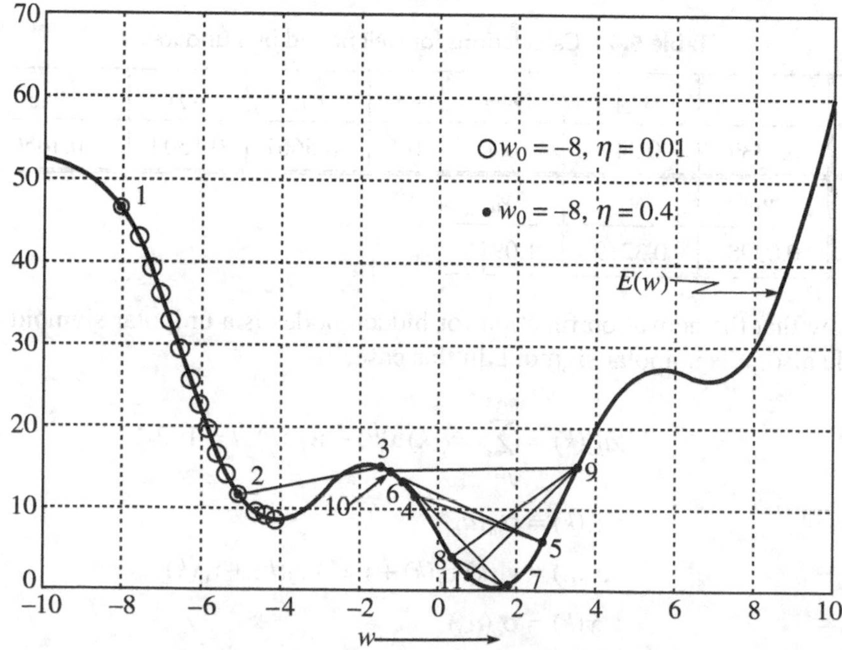

Figure 5.17 Convergence of gradient-descent algorithm

The gradient-descent update law for parameter w is obtained as,

$$\Delta w = -\eta \frac{\partial E}{\partial w} = -\eta (w - 8 \cos w)$$

Figure 5.17 shows the convergence of cost function $E(w)$ with respect to the weight w. It can be seen from the figure that when $\eta = 0.01$ and $w_0 = -8$, the final weight $w = -4$ corresponds to a local minimum (descent indicated by 'O' in the figure). When $\eta = 0.4$ and $w_0 = -8$, the local minimum can be avoided and the final weight ultimately settles down at the global minimum (descent indicated by '•' in the figure). It shows that for a small step size, error backpropagation algorithm may get stuck at a local minimum. For a larger step size, it may come out of local minima and get into global minimum. (However, the algorithm may not converge for a large step size.) Moreover, when the step size is large, it zigzags its way about the true direction to the global minimum, thus leading to a slow convergence. There is no comprehensive method to select the initial weight w_0 and learning rate η for a given problem so that the global minimum can be reached. When trapped in a local minima, the whole learning procedure must be repeated starting from some other initial weights vector; a costly exercise in terms of computing time.

Despite the lack of assured convergence to the global minimum, error-backpropagation algorithm is a highly effective function approximation method in practice. In many practical applications, the problem of local minima has not been found to be as severe as one might fear. Intuitively, more weights in the network correspond to high dimensional error surfaces (one dimension per weight) that might provide 'escape routes' for gradient descent to fall away from the local minimum.

Despite these comments, gradient descent over the complex error surfaces represented by MLP networks is still poorly understood, and no methods are known to predict with certainty when local minima will cause difficulties. Common heuristics to attempt to alleviate the problem of local minima include:

- Add a momentum term to the weight-update rule (described in the next sub-section).
- Use incremental training rather than batch training. Incremental training effectively descends a different error surface for each training example, relying on the average of these to approximate the gradient with respect to the full training set.

5.7.3 Adding Momentum to Gradient Descent

Despite the fact that error-backpropagation algorithm triggered a revival of the whole neural networks field, it was clear from the beginning that the standard error-backpropagation algorithm is not a serious candidate for finding the optimal weights (the global minimum of the cost function) for large-scale nonlinear problems. Many improved algorithms have been proposed in order to find a reliable and fast strategy for optimizing the learning rate in a reasonable amount of computing time. One of the first, simple yet powerful, improvements of the standard error-backpropagation algorithm is given here—the *momentum gradient descent*.

A general weight-update equation for gradient descent algorithms is given by,

$$w_j(k+1) = w_j(k) - \eta \frac{\partial E(k)}{\partial w_j(k)} \tag{5.57a}$$

Weights w_j are updated at each iteration number k so that the prescribed cost function $E(k)$ decreases; η is the learning rate with $0 < \eta < 1$.

In momentum gradient descent rule, the weight-update equation is altered by making the update on k^{th} iteration depend partially on the update that occurred during the $(k-1)^{th}$ iteration:

$$w_j(k+1) = w_j(k) - \eta \frac{\partial E(k)}{\partial w_j(k)} + \alpha[w_j(k) - w_j(k-1)] \tag{5.57b}$$

Here, $0 \le \alpha < 1$ is a constant called the *momentum rate*.

Momentum term $\alpha[w_j(k) - w_j(k-1)]$ allows a network to respond, not only to the local gradient, but also to recent trends in error surface. Without momentum, a network may get stuck in a shallow local minimum; adding momentum may help the NN 'ride through' local minima. Also, with momentum term, the flat regions in the error surface can be crossed at a much faster rate.

The choice of both the learning rate η and momentum rate α is highly problem-dependent and usually a trial-and-error procedure. There have been many proposals on how to improve learning by calculating and using *adaptive learning rate* and *adaptive momentum rate*, which follow and adjust to changes in the nonlinear error surface.

5.7.4 Heuristic Aspects of the Error-backpropagation Algorithm

Multi-Layer Perceptron (MLP) networks are of great interest because they have a sound theoretical basis, meaning that they have *universal function approximation* property with atleast two layers (one layer NNs do not generally have universal function approximation property). According to the basic universal approximation result [86], it is possible for any smooth function $\mathbf{f}(\mathbf{x})$; $\mathbf{x} = [x_1\ x_2 \ldots x_n]^T$, $\mathbf{f}(\cdot) = [f_1(\cdot)\ f_2(\cdot) \ldots f_q(\cdot)]^T$, to be approximated randomly closely on a compact set in n-dimensional state space for certain (large enough) number m of hidden-layer neurons. In this result, the activation functions are not required on the NN output layer (i.e., the output layer activation functions may be linear). Also, the bias terms on the output layer are not required, even though the hidden layer bias terms are needed.

Despite this sound theoretical foundation concerning the representational capabilities of NNs, and the success of error-backpropagation learning algorithm, there are many practical problems. The most troublesome is the usually long training process, which does not ensure that the best performance of the network (the absolute minimum of the cost function) will be achieved. The algorithm may get stuck at some local minimum, and such a termination with a suboptimal solution will require repetition of the whole training process by changing the structure or some of the learning parameters that influence the iteration scheme.

Further note that, though the universal function approximation property says 'there exists an NN that approximates $\mathbf{f}(\mathbf{x})$', it does not show how to determine such a network. Many practical questions still remain open, and for a broad range of applications, the design of neural networks, their learning procedures and the corresponding training parameters is still an empirical art.

The discussion that follows in this section regarding the structure of a network and learning parameters does not yield conclusive answers, but it does represent a useful aggregate of experience of extensive application of error-backpropagation algorithm and many related learning techniques [3].

Number of Hidden Layers

Theoretical results and many simulations in different fields support the result that there is no need to have more than one hidden layer. But, at times, the performance of the network can be enhanced with the help of a more complex architecture, particularly, when the mapping to be learned is highly complicated—there is a likelihood of the performance being made better by placing more hidden layers with their own weights, after the first layer, with sigmoid hidden units. When the number of hidden layer units in a single hidden-layer network is too large, it may be practical to employ multiple hidden layers, choosing 'long and narrow' networks over 'short and flat' networks. Experimental evidence tends to show that using a two hidden-layer network for continuous functions has sometimes advantages over a one hidden-layer network, as the former requires shorter training times.

A rule of thumb that might be useful is to try solving the problem at hand using an NN with one hidden layer first.

Type of Activation Functions in a Hidden Layer

It has been shown that a two-layer network is capable of solving any nonlinear function approximation problem. This theoretical result does not require the use of sigmoidal nonlinearity. The proof assumes only that nonlinearity is a continuous, smooth, monotonically increasing function that is bounded above and below. Thus, numerous alternatives to sigmoid could be used without a biological justification.

The most serious competitor to MLP networks are the networks that use radial basis functions (RBFs) in hidden layer neurons. The most representative and popular RBF is a Guassian function. Whether a sigmoidal or Gaussian activation function is preferable is difficult to say. Both types have certain advantages and disadvantages, and final choice depends mostly on the problem (data set).

A fundamental difference between these two types of NNs is that feedforward MLP NNs are representative of *global approximation* schemes, whereas NNs with RBFs (typically the Gaussian activation functions) are representative of *local approximation* schemes. The adjectives *global* and *local* are connected with the region of input space of the network for which the NN has nonzero output (RBF networks will be discussed in a later section).

Is it possible to use a mix of different types of activation functions in a layer of a network? The answer is 'theoretically yes'. It is possible to improve the accuracy with the help of a more sophisticated network architecture with mixed activation functions in a layer. When the mapping to be learned is highly complicated, there is a likelihood that the performance may be improved. But, because the implementation and training of the network becomes more complex, it is the practice to apply only one hidden layer of sigmoidal activation functions and an output layer of linear/ sigmoidal units.

Number of Neurons in a Hidden Layer

The number of neurons in a hidden layer is the most important design parameter with respect to approximation capabilities of a neural network. Recall that both the number of input components (attributes/features) and the number of output neurons is, in general, determined by the nature of the problem. Thus, the real representational power of an NN and its generalization capability are primarily determined by the number of hidden-layer neurons. In the case of general nonlinear regression, the main task is to model the underlying function between the given inputs and outputs by filtering out the disturbances embedded in the noisy training dataset. Similar statements can be made for pattern recognition (classification) problems. By changing the number of hidden-layer nodes, two extreme solutions should be avoided: filtering out the underlying function (not enough hidden-layer neurons) and modeling of noise or overfitting the data (too many hidden-layer neurons). These problems have been discussed earlier in Sections 2.4, and 2.2.

The guidelines to select the appropriate number of hidden neurons are rather empirical at the moment. To avoid large number of neurons and the corresponding inhibitively large training times, the smaller number of hidden layer units are used in the first trial. One increases accuracy by adding in steps more hidden neurons.

Variants of Gradient-Descent Procedure

Two main issues in selecting a training algorithm are fast convergence and accuracy of function approximation. The widely used technique to train a feedforward network has been the backpropagation algorithm with gradient descent optimization. The primary drawbacks of this algorithm are its slow rate of convergence, and its inability to ensure global convergence. Several variants have been proposed to improve the speed of response, and also to account for generalization ability and avoidance of local minimum. Standard optimization techniques that use quasi-Newton techniques have been suggested. The issues with quasi-Newton techniques are that the storage and memory need increase as the square of the size of the network. The nonlinear optimization techniques such as Newton-Raphson method or the conjugate gradient method may be adopted for training the feedforward networks. Though these algorithms converge in fewer iterations than the backpropagation algorithm with gradient descent, too much computations per pattern are required. These variants may converge faster in some cases and slower in others. Other algorithms for faster convergence include extended Kalman filtering, and Levenberg-Marquardt. However, all these variants do not come closer to backpropagation algorithm with gradient descent as far as simplicity and ease of implementation is concerned.

The most popular variant used in backpropagation algorithm is probably altering the weight-update rule in the algorithm by ensuring that at least partly the weight update on the kth iteration relies on the update that took place during the $(k-1)^{\text{th}}$ iteration. This augmented learning process has been presented in the earlier sub-section. To handle the overfitting issue for backpropagation learning, the weight decay approach may be used. That is, bring down each weight by a certain small factor during each iteration. This can be achieved by altering the definition of E so that it includes a penalty term which corresponds to the total magnitude of the network weights. The approach keeps weight values small to bias learning against complex decision surfaces/functions (Eqn (2.22)).

Genetic Algorithm (Appendix A) based optimization in neural networks is capable of global search and is not easily fooled by local minima. Genetic algorithms do not use the derivative for the fitness function. Therefore, they are possibly the best tool when the activation functions are not differentiable. Note that genetic algorithm can take an unrealistically huge processing time to find the best solution. Details are given in Section 5.10 on Genetic-Neural systems.

Learning Rate

In the weight-update rule of gradient descent procedures, the term η, called the *learning rate*, appears. It determines the magnitude of the change $\Delta\bar{\mathbf{w}}$ but not its direction. The learning rate controls the stability and rate of adaptation. The general approach to choosing η is to decrease the learning rate as soon as it is observed that adaptation does not converge. The smaller η is, the smoother the convergence of search but the higher the number of iteration steps needed. Descending by small η will lead to the nearest minimum when the error $E(\bar{\mathbf{w}})$ is a nonlinear function of the weights. If $E_{\min}(\bar{\mathbf{w}})$ is larger than the predefined maximally allowed error, the whole learning procedure must be repeated starting from some other initial weights vector. Therefore, working with small η may be rather costly in terms of computing time.

A typical rule of thumb is to start with some larger learning rate and reduce it during optimization. Clearly, what is considered a small or large learning rate is highly problem-dependent, and proper η should be established in the first few runs for a given problem.

In a recent work [87], two novel algorithms have been proposed on *adaptive learning rate* using the Lyapunov stability theory. It is observed that this adaptive learning rate increases the speed of convergence.

Weights Initialization

The learning procedure using the error-backpropagation algorithm begins with some initial set of weights which is usually randomly chosen. However, the initialization is a controlled process. This first step in choosing weights is important because with 'bad' initial choice, the training may get lost forever without any significant learning, or it may stop soon at some local minima. (Note that the problems with local minima are not related only to initialization.) Initialization by using random numbers is very important in avoiding the effects of symmetry in the network. In other words, all the hidden-layer neurons should start with guaranteed different weights.

A first guess, and a good one, is to start learning with small initial weights. How small the weights must be depends on the training dataset (very small hidden-layer weights must also be avoided). Learning is often a kind of empirical art, and there are many rules of thumb about how small the weights should be. One is that the practical range of hidden-layer initial weights should be $[-2/n, 2/n]$. Another similar rule of thumb is that the weights should be uniformly distributed inside the range $[-2.4/n, 2.4/n]$. (MATLAB uses the techniques proposed by Nguyan and Widrow.)

The initialization of output-layer weights should not result in small weights. If the output-layer weights are small, then so is the contribution of hidden-layer neurons to the output error, and consequently the effect of the hidden-layer weights is not visible enough. Also recall that the output-layer weights are used in calculating the error signal terms for hidden-layer neurons. If the output-layer weights are too small, these terms also become very small, which in turn leads to small initial changes in hidden-layer weights; learning in the initial phase will, therefore, be too slow.

Stopping Criterion at Training

The error-backpropagation algorithm resulted from a combination of sum-of-error-squares (Eqn (5.21)) as a cost function to be optimized and the gradient descent method for weights adaptation.

If the training patterns are colored by noise, the minimization of sum-of-error-squares criterion by incremental learning is equivalent to LMS criterion (Eqn (5.27)). The learning process is controlled by the prescribed maximally allowed or desired error E_{des}. More precisely, one should have an expectation about the magnitude of the error at the end of the learning process.

After a learning epoch (the sweep through all the training pairs) is completed in incremental learning process, the sum of error-squares over all the training pairs is accumulated and the total error E_N is compared with the acceptable (desired) value E_{des}. Learning is terminated if $E_N < E_{des}$; otherwise a new learning epoch is started. In the batch mode, weight updating is performed after the presentation of all the training examples that constitute an epoch. The error E_N is compared with E_{des} after each iteration of a learning epoch.

The sum-of-error-squares is not good as *stopping criterion* because E_N increases with the increase of number of data pairs. The more data, the larger is E_N. Scaling of error function gives a better stopping criterion. The *Root Mean Square Error* (RMSE) is widely used scalar function (Eqn (2.26)):

$$E_{RMS} = \sqrt{\frac{1}{N}\sum_{i=1}^{N}(e^{(i)})^2} \tag{5.58}$$

There will be no need to change the learning algorithm derived earlier. Training is performed using sum-of-error-squares as the cost function (performance criterion), and RMSE is used as a stopping criterion at training. However, if desired, for the batch mode, the learning algorithm with *Mean Square Error* (MSE) (refer to Eqn (2.25))

$$E_{av} = \frac{1}{N}\sum_{i=1}^{N}(e^{(i)})^2 \tag{5.59}$$

may be used as the cost function for training the network.

The condition for termination on the basis of $E_N < E_{des}$ is, in fact, a weak strategy as backpropagation is vulnerable to overfitting the training examples at the cost of reducing generalization accuracy over other unseen examples. One of the most successful techniques for overcoming the overfitting issue is to merely provide a validation dataset. The algorithm supervises the error with regard to this validation set, while making use of the training set to drive the gradient descent search (refer to Section 2.6).

5.8 MULTI-CLASS DISCRIMINATION WITH MLP NETWORKS

In Section 5.2, we discussed the *linear classification* problem using regression techniques. We found that a single perceptron can be used to realize binary classification.

Multi-Layer Perceptron (MLP) networks can be used to realize *nonlinear classifier*. Most of the real-world problems require nonlinear classification.

Consider the network of Fig. 5.14 with a single log-sigmoid neuron in the output layer. For a binary classification problem, the network is trained to fit the desired values of $y^{(i)} \in [0, 1]$

representing Classes 1 and 2, respectively. The network output $\hat{y}$ approximates $P(\text{Class } 1|\mathbf{x})$; $P(\text{Class } 2|\mathbf{x}) = 1 - \hat{y}$.

When training results in perfect fitting, then the classification outcome is obvious. Even when target values cannot fit perfectly, we put some thresholds on the network output $\hat{y}$. For [0, 1] target values:

$$\text{Class} = \begin{cases} 1 & \text{if } 0 < \hat{y} \le 0.5 \\ 2 & \text{if } \hat{y} > 0.5 \end{cases} \tag{5.60}$$

What about an example that is very close to the boundary $\hat{y} = 0.5$? Ambiguous classification may result in near or exact ties.

Instead of '0' and '1' values for classification, we may use values of 0.1 and 0.9. 0 and 1 is avoided because it is not possible for the sigmoidal units to produce these values, considering finite weights. If we try to train the network to fit target values of precisely 0 and 1, gradient descent will make the weights grow without limits. On the other hand, values of 0.1 and 0.9 are attainable with the use of the sigmoid unit with finite weights.

Consider now a multi-class discrimination problem. The inputs to the MLP network are just the values of the feature measurements (suitably normalized). We may use a single log-sigmoid node for the output. For the *M*-class discrimination problem, the network is trained to fit the desired values of $y^{(i)} = \{0.1, 0.2, \ldots, 0.9\}$ for Classes 1, 2, …, 9 (for $M = 9$), respectively. We put some thresholds on the network output $\hat{y}$. For {0.1, …, 0.9} desired values:

$$\text{Class} = \begin{cases} 1 & \text{if } 0 < \hat{y} \le 0.1 \\ 2 & \text{if } 0.1\,\hat{y} \le 0.2 \\ \vdots & \\ 9 & \text{if } \quad \hat{y} > 0.8 \end{cases} \tag{5.61}$$

Note that the difference between the desired network output for various classes is small; ambiguous classification may result in near or exact ties. The approach may work satisfactorily if the classes are well separated. This gets impractical as the number of classes gets large and the boundaries are artificial.

A more realistic approach is *1–of–M encoding*. A separate node is used to represent each possible class and the target vectors consist of 0s everywhere except for the element that corresponds to the correct class. For a four-class ($M = 4$) discrimination problem, the target vector

$$\mathbf{y}^{(i)} = [y_1^{(i)}\ y_2^{(i)}\ y_3^{(i)}\ y_4^{(i)}]^T$$

To encode four possible classes, we use

$$\mathbf{y}^{(i)} = [1\ 0\ 0\ 0]^T \text{ to encode Class 1}$$

$$\mathbf{y}^{(i)} = [0\ 1\ 0\ 0]^T \text{ to encode Class 2}$$

$$\mathbf{y}^{(i)} = [0\ 0\ 1\ 0]^T \text{ to encode Class 3}$$

and

$$\mathbf{y}^{(i)} = [0\ 0\ 0\ 1]^T \text{ to encode Class 4}$$

We are therefore using four log-sigmoid nodes in the output layer of the MLP network (Fig. 5.15), and binary output values (we want each output to be 0 or 1; instead of 0 and 1 values, we use values of 0.1 and 0.9).

Once the network has been trained, performing the classification is easy: simply choose the element $\hat{y}_k$ of the output vector that is the largest element of $\mathbf{y}$, i.e., pick the $\hat{y}_k$ for which $\hat{y}_l > \hat{y}_q \forall q \neq k; q = 1, \ldots, M$.

This generates an unambiguous decision, since it is very unlikely that two output neurons will have identical largest output values.

The 1-of-*M* encoding approach is, in fact, equivalent to designing *M* binary classifiers.

Neural networks perform best when the features and response variables are on a scale of [0, 1]. For this reason, all variables should be scaled to a {0, 1} interval before feeding them into the network. For a numerical variable x that takes values in the range $\{x_{\min}, x_{\max}\}$, we normalize the measurements as follows:

$$x_{\text{norm}} = \frac{x - x_{\min}}{x_{\max} - x_{\min}}$$

Even if new data exceed this range by a small amount, this will not affect the results much.

For more details on multiclass classification using neural networks, refer to [59, 88–90].

Example 5.2

Consider the toy dataset shown in Table 5.5, for a certain processed cheese [20]. The two attributes x_1 and x_2 are scores for fat and salt, indicating the relative presence of fat and salt in the particular cheese sample. The output variable y is the cheese sample's acceptance by a taste-test panel.

Table 5.5 A toy dataset

Sample i	x_1 (Fat score)	x_2 (Salt score)	y (Acceptance)	$\mathbf{y} = [y_1\ y_2]^T$ (Target vector)
1	0.2	0.9	yes	$[0.9\ \ 0.1]^T$
2	0.1	0.1	no	$[0.1\ \ 0.9]^T$
3	0.2	0.4	no	$[0.1\ \ 0.9]^T$
4	0.2	0.5	no	$[0.1\ \ 0.9]^T$
5	0.4	0.5	yes	$[0.9\ \ 0.1]^T$
6	0.3	0.8	yes	$[0.9\ \ 0.1]^T$

Figure 5.18 describes an example of a typical neural net that could be used for predicting the acceptance for the given scores on fat and salt. 1-of-*M* encoding scheme has been used for classification. Two separate nodes have been used to represent two possible classes: 'yes' and 'no'.

The target vector $\mathbf{y} = [0.9\ 0.1]^T$ encodes Class 'yes'; and $\mathbf{y} = [0.1\ 0.9]^T$ encodes Class 'no'. The encoded desired output for each sample is shown in Table 5.5.

We choose the following initial weight and bias values: $w_{11} = 0.05$, $w_{21} = -0.01$, $w_{31} = 0.02$, $w_{12} = 0.01$, $w_{22} = 0.03$, $w_{32} = -0.01$, $w_{10} = -0.3$, $w_{20} = 0.2$, $w_{30} = 0.05$, $v_{11} = 0.01$, $v_{21} = -0.01$, $v_{12} = 0.05$, $v_{22} = 0.03$, $v_{13} = 0.015$, $v_{23} = 0.02$, $v_{10} = -0.015$, $v_{20} = 0.05$.

We use a learning rate $\eta = 0.5$ to update weight and bias values. The activation function for hidden and output nodes is unipolar sigmoid.

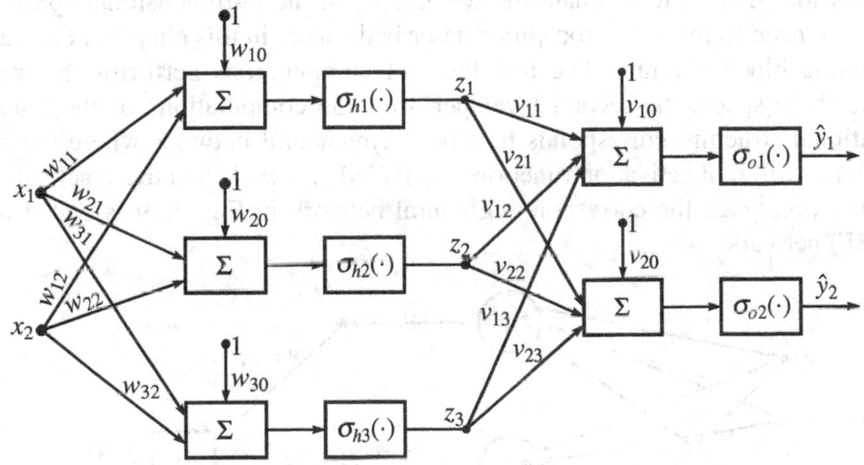

Figure 5.18 A typical neural net for Example 5.2

Sample calculations may be carried out in a way similar to that used in Example 5.1. The reader is encouraged to do this practice on the first sample of the training data.

5.9 RADIAL BASIS FUNCTIONS (RBF) NETWORKS

Let us consider our feature vectors $\mathbf{x}$ to be in the n-dimensional space $\Re^n$. Let $\phi_1(\cdot), \phi_2(\cdot), \ldots, \phi_m(\cdot)$ be m nonlinear functions

$$\phi_l: \Re^n \to \Re;\ l = 1, \ldots, m \tag{5.62a}$$

which define the mapping

$$\mathbf{x} \in \Re^n \to \mathbf{z} \in \Re^m \tag{5.62b}$$

$$\mathbf{z} = \begin{bmatrix} \phi_1(\mathbf{x}) \\ \phi_2(\mathbf{x}) \\ \vdots \\ \phi_m(\mathbf{x}) \end{bmatrix}$$

Our goal now is to investigate whether there is an appropriate value for m and functions $\phi_l(\cdot)$ so that a nonlinear function $f(\mathbf{x})$ can be approximated as a linear combination of $\phi_l(\mathbf{x})$, that is,

$$\hat{f}(\mathbf{x}) = \sum_{l=1}^{m} w_l \phi_l(\mathbf{x}) \tag{5.63}$$

where the *weights* w_l; $l = 1, \ldots, m$, are the parameters of the linear approximation. This is equivalent to representing nonlinear function $f(\mathbf{x}, \mathbf{w})$ in terms of *interpolation functions*, also called *basis functions*, $\phi_l(\cdot)$.

Once the basis functions $\phi_l(\cdot)$ have been selected, the problem becomes a typical design problem of linear regression, that is, to estimate the weights w_l in the m-dimensional space. (In case of classification, we need to resort to procedures described earlier in this chapter.) Figure 5.19 shows the corresponding block diagram. The first layer of computations performs the mapping from the **x**-space to the **z**-space; the second layer performs the computations of the linear regressor. The computational structure corresponds to a two-layer neural network where the nodes of the hidden layer have different activation functions $\phi_l(\cdot)$; $l = 1, \ldots, m$. When these activation functions are *radial basis functions*, the corresponding neural network in Fig. 5.19 is called *Radial Basis Function* (RBF) network.

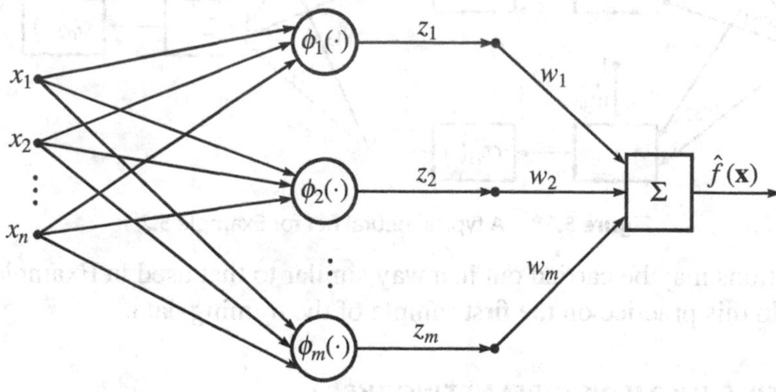

Figure 5.19 Computational structure for function approximation using interpolation (basis) functions $\phi_l(\cdot)$

RBF neuron uses *radially symmetric* activation function $\phi_l(\| \mathbf{x} - \mathbf{c}_l \|)$; the argument of the function is the Euclidean distance of the input vector **x** from a center $\mathbf{c}_l$. This justifies the name *Radial Basis Function* (RBF). The RBF function $\phi_l(\cdot)$ can take various forms; the Gaussian form is more widely used. (This is not a Gaussian density, but we use the same name anyway.) A Gaussian basis function is typically parameterized by two parameters: the *center* $\mathbf{c}_l$ which defines its position, and a *spread* parameter that determines its shape. The spread parameter is equal to the *standard deviation* σ_l in case of a one-dimensional Gaussian function (do not confuse the standard deviation parameter σ of Gaussian function with the sigmoidal activation function $\sigma(\cdot)$). A one-dimensional Gaussian function is depicted in Fig. 5.20.

In the case of multivariate input vector **x**, the parameters that define the shape of the hyper-Gaussian function, are elements of a covariance matrix $\mathbf{\Sigma}$ (refer to Section 3.2). With the selection of the same spread parameter σ for all components of the input vector, the covariance matrix $\mathbf{\Sigma} = \text{diag}(\sigma^2)$.

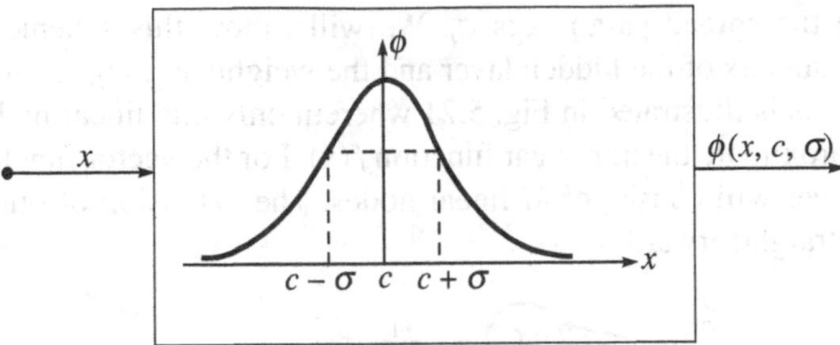

Figure 5.20 Gaussian function in RBF neuron model

With the assumption of same spread parameter for the components of the input vector, a Gaussian function for the lth neuron in hidden layer; $l = 1, \ldots, m$, is

$$\phi_l(\mathbf{x}, \mathbf{c}_l, \sigma_l) = \exp\left(-\frac{\|\mathbf{x}-\mathbf{c}_l\|^2}{2\sigma_l^2}\right) \tag{5.64a}$$

$$= \exp\left(-\frac{(\mathbf{x}-\mathbf{c}_l)^T(\mathbf{x}-\mathbf{c}_l)}{2\sigma_l^2}\right) \tag{5.64b}$$

where $\mathbf{c}_l$ is the center and σ_l is the spread parameter of the function $\phi_l(\cdot)$. Dimension of each center $\mathbf{c}_l$ for an $\mathbf{x}$-input network is $n \times 1$, the same as the dimension of $\mathbf{x}$. Each σ_l is a scalar.

The idea behind using such local basis functions is that in the input data, there are groups or clusters of instances and for each such cluster, we define a basis function $\phi_l(\cdot)$, which becomes nonzero if instance $\mathbf{x}^{(i)}$ belongs to cluster l. One can use *K*-means/Self-Organizing Maps (discussed in Chapter 7) to find the centers $\mathbf{c}_l$. Once we have the centers, there is a simple and effective heuristic to find the spreads: For each center, we find the most distant instant covered by that center and set σ_l to half or one-third of that value. We can also use the *EM* algorithm (discussed in Chapter 7) on Gaussian mixtures to find the cluster parameters. The parameter l is the complexity parameter (similar to K in *K*-means) like the number of hidden units in MLP network; it trades-off simplicity with accuracy; with more units, we approximate the training data better but we get a complex model and risk overfitting; too few may underfit. Again the optimal value is determined by cross-validation.

The hidden-layer neuron receives the Euclidean distance $\| \mathbf{x} - \mathbf{c}_l \|$ and computes the *scalar* value of the basis function $\phi_l(\mathbf{x}, \mathbf{c}_l, \sigma_l)$. Each hidden unit has its own receptive field in the input space, i.e., each center is representative of some of the input patterns. Consider, for instance, an input vector $\mathbf{x}$ which lies in the receptive field for center $\mathbf{c}_l$. This would activate the hidden unit with center $\mathbf{c}_l$. Suppose an input vector lies between two receptive field centers, then both of these hidden units will be appropriately activated. The output will be a weighted average of the basis function values.

There are two sets of parameters governing the mapping properties of the RBF network: Weights w_l; $l = 1, \ldots, m$, in the output layer, and the parameters $\{\mathbf{c}_l, \sigma_l\}$ of the radial basis functions. A popular scheme to learn the parameters of the RBF functions is to learn only the centers $\mathbf{c}_l$ and therefrom determine the spread parameters σ_l. We will follow this scheme wherein centers $\mathbf{c}_1$, $\mathbf{c}_2$, ..., $\mathbf{c}_m$ are the parameters of the hidden layer and the weights $w_1, w_2, \ldots, w_m$ are the parameters of the output layer. This is illustrated in Fig. 5.21 wherein only one linear node in the output layer has been taken to approximate the nonlinear function $f(\mathbf{x})$. For the vector function, $\mathbf{f}(\mathbf{x}) = [f_1(\mathbf{x}), \ldots, f_M(\mathbf{x})]^T$; the output layer will consist of M linear nodes. The extension of our presentation to this multivariate case is straightforward.

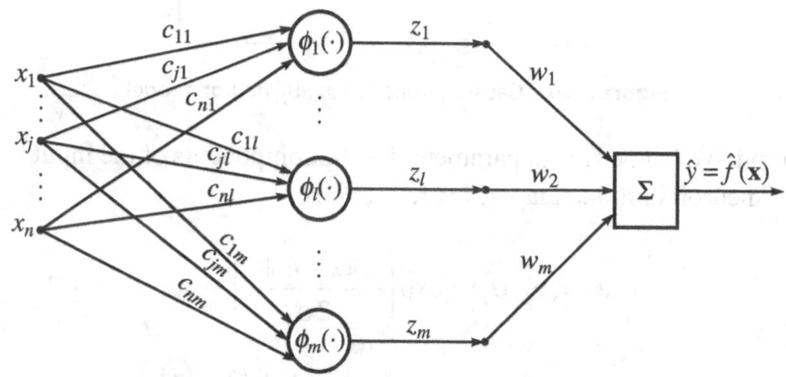

Figure 5.21 Architecture of an RBF network for $\Re^n \to \Re$ mapping

RBF networks have gained considerable attention as an alternative to MLP networks trained by the backpropagation algorithm [3]. Both the MLP and RBF networks are the basic constituents of the feedforward neural networks. They are structurally equivalent. MLP network of Fig. 5.14 has one hidden layer with sigmoidal activation functions (more than one hidden layer may be taken, in general) and an output layer containing one or more neurons with linear activation functions (sigmoidal units may be used in the output layer for classification tasks). RBF network of Fig. 5.21 has RBF functions in the hidden layer's neurons and linear units in the output layer. Unlike MLP networks, there is only one hidden layer in all RBF networks, and output layer units are strictly linear. Also unlike MLP networks, one does not augment the n-dimensional input vector $\mathbf{x}$ with a bias term +1; and usually hidden layer outgoing vector $\mathbf{z}$ is also not augmented with the bias term.

In the MLP networks, the input signal to the neuron's activation functions is equal to $\mathbf{w}^T\mathbf{x}$; $\mathbf{w}$ is the weight vector and $\mathbf{x}$ is the n-dimensional input vector. The input signal to the radial basis functions is equal to the distance between the input vector $\mathbf{x}$ and a center $\mathbf{c}_l$ of a specific function. In an MLP network, the input to the sigmoidal activation functions of the (first) hidden layer are linear combinations of input features $\left(\sum_j w_{lj} x_j\right)$, and the output of each sigmoidal unit saturates to the same value with increasing $\sum_j w_{lj} x_j$. In contrast, in the RBF networks, output of each RBF node is the same for all input points $\mathbf{x}$ having the same Euclidean distance from the respective center $\mathbf{c}_l$, and decreases exponentially with the distance. In other words, the activation responses of the nodes are of a local nature in the RBF, and of a global nature in the MLP networks. This intrinsic difference has important repercussions for both the convergence speed and the generalization performance. In general, multilayer perceptrons learn slower than RBF counterparts. In contrast, MLP exhibit improved generalization performance, especially for regions that are not sufficiently represented in the training set.

To attain performance similar to MLP networks, an RBF network must have much higher order (big value of m). This is because of the local nature of the RBF activation functions, which make it mandatory to employ a huge number of centers to fill the space where $f(\mathbf{x})$ is defined, and this number displays an exponential dependence on the dimension n of the input space.

The advantages of RBF networks, such as linearity in the parameters, and the availability of fast and efficient training methods, have been observed in many publications. Like an MLP network, an RBF network has *universal function approximation* ability.

5.9.1 Training the RBF Network

Training of an RBF network requires optimal selection of the centers $\mathbf{c}_l$, and weights w_l; $l = 1, \ldots, m$. This is a two-fold problem, unlike in an MLP network. As the two layers of the network perform different tasks, both the layers are optimized using different techniques and in different time scales. Trial-and-error nature of training an RBF network follows the heuristic guidelines given earlier for MLP networks. We will present here those features of RBF-network training that are different from training MLP networks.

Several strategies are applicable for training RBF networks; some of these strategies are discussed below.

Fixed Centers

What we can do for training RBF networks is a two-stage process; we use an unsupervised technique to determine the centers, and then construct a supervised layer. This is referred to as *hybrid learning*. Because Gaussians are local, ideally several more units will be required than what would be required in case an MLP network were to be used.

Once the centers $\mathbf{c}_{jl}$; $j = 1, \ldots, n$; $l = 1, \ldots, m$ (and the spread parameters σ_l) of the RBFs are fixed employing unsupervised learning (for example, K-means, self-organizing map, Gaussian mixture model (Chapter 7 will give the details)), the training problem reduces to determination of output-layer weights w_l in the RBF model of Fig. 5.21. This is a linear regression problem and the best solution is the *least-squares* solution (a supervised learning problem (Sections 5.4–5.5)).

Another method, known as the *anchor* technique, sets the centers to the arbitrarily selected patterns from the training set without any further update. Provided that the training set is distributed in a representative manner over the entire feature-vector space, this seems to be reasonable way to choose the m centers.

Moving Centers

In most real-life problems, during the learning phase, it may be practical to simultaneously adopt both the centers (and the shape parameters) and the output-layer weights in a supervised manner. The RBF function of Eqn (5.64) is differentiable and we can backpropagate, just as we back propagated in MLP networks to update the first-layer weights. The structure is similar to an MLP network with $\phi_l(\cdot)$ as the hidden units and $\mathbf{c}_l$ (and σ_l) as the first-layer parameters; the Gaussian as the activation function in the hidden layer and w_l as the second-layer weights (Fig. 5.21). Using gradient-descent update rules for the centers (if desired, for spread parameters as well; however, the spread parameters may be computed from the centers using some appropriate rule of thumb) in *incremental mode* (weight update after each input pattern is presented to the network), and the LMS algorithm (stochastic gradient descent (Section 5.5)) for computing output-layer weights (incremental mode of learning), gives a simple and effective approach for simultaneously adopting all the parameters of an RBF network.

The update rule for centers-learning is given below:

$$c_{jl} \leftarrow c_{jl} - \eta \frac{\partial E}{\partial c_{jl}} \tag{5.65}$$

for $j = 1, \ldots, n$; $l = 1, \ldots, m$; and the cost function

$$E = \tfrac{1}{2}(y - \hat{y})^2 \tag{5.66}$$

where y is the desired output corresponding to input $\mathbf{x}$, and $\hat{y} = \hat{f}(\mathbf{x})$ is the actual output of the network.

The actual response of the RBF network shown in Fig. 5.21, is computed as,

$$\hat{y} = \sum_{l=1}^{m} w_l z_l$$

$$= \sum_{l=1}^{m} w_l \phi_l(\mathbf{x}, \mathbf{c}_l, \sigma_l) \tag{5.67a}$$

For the Gaussian radial basis function,

$$\phi_l(\mathbf{x}, \mathbf{c}_l, \sigma_l) = \exp(-r_l^2/2\sigma_l^2) \tag{5.67b}$$

where $r_l = \| \mathbf{x} - \mathbf{c}_l \|$ is the Euclidean distance of input vector $\mathbf{x}$ from the center $\mathbf{c}_l$.

Differentiating E with respect to c_{jl} gives:

$$\frac{\partial E}{\partial c_{jl}} = \frac{\partial E}{\partial \hat{y}} \times \frac{\partial \hat{y}}{\partial \phi_l} \times \frac{\partial \phi_l}{\partial c_{jl}}$$

$$= -(y - \hat{y}) \times w_l \times \frac{\partial \phi_l}{\partial r_l} \times \frac{\partial r_l}{\partial c_{jl}}$$

$$\frac{\partial \phi_l}{\partial r_l} = -\frac{r_l}{\sigma_l^2} \phi_l$$

$$\frac{\partial r_l}{\partial c_{jl}} = \frac{\partial}{\partial c_{jl}} \left(\sum_{j=1}^{n} (x_{jl} - c_{jl})^2 \right)^{1/2}$$

$$= -(x_{jl} - c_{jl})/r_l$$

After simplication, the update rule for the centers becomes,

$$c_{jl} \leftarrow c_{jl} + \eta\,(y - \hat{y})\, w_l(x_{jl} - c_{jl})\, \phi_l/\sigma_l^2 \tag{5.68}$$

The gradient $\partial E/\partial c_{jl}$ exhibits a *clustering* effect.

Despite the fact that it is simple to implement gradient descent for hidden-layer weights, it suffers from more difficulties in RBF networks than in MLP networks. The derivative of RBF changes sign; this is not the case for sigmoidal functions. Also, unlike MLP networks with sigmoidal activation functions, obtaining gradient information in RBF networks is difficult or expensive (we may use *genetic algorithm* for the nonlinear optimization problem (refer to Appendix A)).

One usually obtains faster convergence with the completely supervised scheme. Although computational requirement increases by adjusting the centers using gradient descent update rule, the number of centers can be substantially reduced by this approach. The generalization performance of such a network is much better as compared to hybrid learning with fixed centers.

Training using completely supervised method is, however, slow. Hybrid learning trains one layer at a time, and is faster.

5.10 GENETIC-NEURAL SYSTEMS

Neural-network learning is a search procedure to minimize a performance criterion (error function). To utilize existing learning algorithms, one requires to choose several parameters—the number of layers, the number of units in each layer, the activation functions, as well as learning paramenters. Learning procedure is generally performed using the error backpropagation for connection weights, and the trial-and-error model for the other parameters. These design steps often require much time and experience, but here genetic algorithms can be of help. The genetic algorithms are presented in Appendix A.

Genetic algorithms can be introduced into neural networks at various levels [91]:

- learning connection weights including biases;
- determining optimal architecture; or
- simultaneously determining architecture and weights.

We will limit our presentation to the first task, i.e., the use of genetic algorithms to the problems of optimization of neural network weights.

Optimization of Neural Network Weights

The gradient-based algorithms for learning weights of neural networks usually run multiple times to avoid local minima, and also gradient information must be available. Two of the important arguments for the use of genetic algorithms to the problems of optimization of neural network weights, are

- a global search of space of weights, avoiding local minima; and
- useful where getting gradient information is expensive and not easy.

It is to be noted that once gradient information is easily available, the techniques based on gradient may be more effective in terms of computation speed than the GA for weight optimization of neural networks. In fact, there is no best training algorithm, because the most suited technique is always dependent on problem.

With a fixed topology, the weights of a neural network are coded in a chromosome. Each individual of the population is determined by an entire group of neural network weights. The sequence of placement of the weights in the chromosome is random, but can't be altered once the procedure of learning starts.

The fitness of individuals will be assessed on the basis of the fitness function, defined as the sum of squares of errors, being the differences between the signal sought by the network and the network output signal for various input data. The genetic algorithm works on the population of individuals (chromosomes representative of neural networks with the same architecture but with varying weights values) as per the typical genetic cycle, which consists of the following steps:

1. Decoding each individual of the present population to the set of weights and building the corresponding neural network with this set of weights; while the network architecture and the learning rule are predefined.
2. Computing the total mean squared error of the difference between the desired signals and output signals for all the input data. This error determines the fitness of the individual (built network).
3. Choosing and applying genetic operators—for instance, crossover and mutation—and obtaining a new generation.

Example 5.3

We are given the two-layer backpropagation network shown in Fig. 5.22. The activation function for all the nodes in the hidden layer is,

$$\sigma(a_l) = \frac{1}{1 + e^{-a_l}}$$

and the activation function for the node in the output layer is,

$$\bar{\sigma}(a) = \frac{e^a - e^{-a}}{e^a + e^{-a}}$$

The learning constant $\eta = 0.2$. The desired output is y.

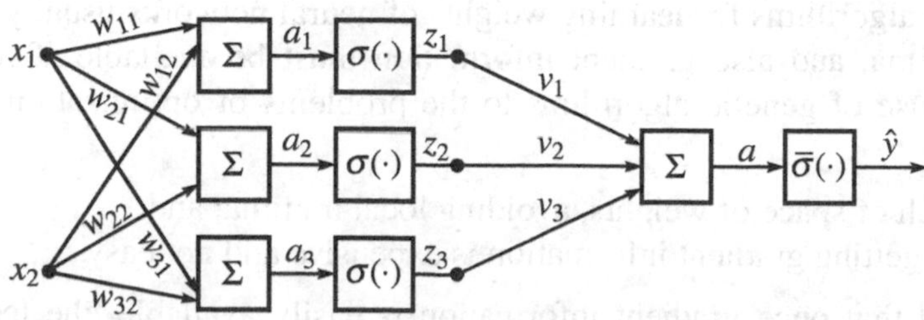

Figure 5.22

The problem is concerned with optimization of the connection weights of the neural network shown in Fig. 5.22. In the following, we describe how a binary-coded GA could be used (instead of gradient algorithm) to update the connection weights of this network.

One of the GA-strings is given below, in which five bits ($L = 5$) are used to represent each connection weight (all the weights are assumed to vary in the range 0.0 to 1.0):

$$\{w_{11}\ w_{12}\ w_{21}\ w_{22}\ w_{31}\ w_{32}\ v_1\ v_2\ v_3\} =$$
$$10110\ \ 01011\ \ 01101\ \ 11011\ \ 10001\ \ 00011\ \ 11001\ \ 11110\ \ 11101 \quad (5.69)$$

The parameter w_{11} is represented by the binary substring 10110. Its decoded value is $b = 22$. It varies in the range of $\{w_{11}^{min}, w_{11}^{max}\} = \{0.0, 1.0\}$. Using the mapping rule (refer to Eqn (A.10) in Appendix A), its real value can be determined as follows:

$$w_{11} = w_{11}^{min} + \frac{b}{2^L - 1}(w_{11}^{max} - w_{11}^{min}) = 0.0 + \frac{22}{2^5 - 1}(1.0 - 0.0) = 0.709677$$

Similarly, the real values of all the parameters represented by the GA-string (5.69) can be calculated. The real values are:

$$\{w_{11}, w_{12}, w_{21}, w_{22}, w_{31}, w_{32}, v_1, v_2, v_3\} =$$
$$\{0.709677,\ 0.354839,\ 0.419355,\ 0.870968,\ 0.548387,\ 0.096774,\ 0.806452,\ 0.967742,\ 0.935484\} \quad (5.70)$$

The first training pattern of the data $\{\mathbf{x}^{(i)}, y^{(i)}; i = 1, 2, \ldots, N\}$ is assumed to be $\{x_1 = 0.6, x_2 = 0.7, y = 0.9\}$. The outputs of the hidden units for an input ($x_1 = 0.6, x_2 = 0.7$) and the connection weights given by (5.70), are found from Fig. 5.22 as follows:

$$a_1 = 0.674194;\ a_2 = 0.861291;\ a_3 = 0.396774;\ z_1 = 0.662442;\ z_2 = 0.702930;\ z_3 = 0.597912$$

The activation value a of the neuron in the output layer is obtained as follows:

$$a = v_1 z_1 + v_2 z_2 + v_3 z_3 = 1.773820$$

and the predicted output of the network is,

$$\hat{y} = \frac{e^a - e^{-a}}{e^a + e^{-a}} = 0.9440$$

Since the target output for this training pattern is equal to 0.9, the error in prediction is found to be equal to 0.0440.

A population of GA-strings represents a number of candidate neural networks. When the batch mode of training is adopted, the whole training data is passed through the neural network represented by a GA string. This gives Mean Square Error (MSE):

$$MSE = \frac{1}{N}\sum_{i=1}^{N}(y^{(i)} - \hat{y}^{(i)})^2 \quad (5.71)$$

Since GA is a maximization algorithm, we may choose the fitness function

$$J = \frac{1}{MSE + \varepsilon} \tag{5.72}$$

where ε is a small positive number.

The population of GA-strings is then modified using the selection, crossover, and mutation operators. The GA, through its search, is expected to evolve an optimal neural network (refer to Appendix A).

Chapter

6

FUZZY INFERENCE SYSTEMS

6.1 INTRODUCTION

So far, we were mostly concerned with learning from experimental data (examples, samples, measurements, records, patterns or observations), expressed in the form of numerical data. In neural networks, for example, we used the following machine learning problem setting:

There is some unknown dependency (mapping, function) $y = f(\mathbf{x})$ between some high-dimensional input vector $\mathbf{x}$ and a scalar output y (or vector output $\mathbf{y}$). The only information available about the underlying dependency is a training dataset $\{\mathbf{x}^{(i)}, y^{(i)}; i = 1, 2, \ldots, N\}$. The number of neurons, their link structure and the corresponding synaptic weights $\mathbf{w}$ were the subjects of learning procedure: $\hat{y} = h(\mathbf{x}, \mathbf{w})$ represents the learning function.

It may be noted that it is hard to find any physical meaning of the neural-network weights $\mathbf{w}$. One cannot select a single synaptic weight as a discrete piece of knowledge. Here, knowledge is embedded in the entire network, and any change in a synaptic weight may lead to unpredictable results. Neural network learning is thus a 'black box' design situation in which the process is entirely unknown. Knowledge (information) is available only in the form of data pairs $\{\mathbf{x}^{(i)}, y^{(i)}\}$; $i = 1, 2, \ldots, N$. The network is required to be trained using this knowledge before the machine could be used for prediction.

Generally speaking, learning is most highly dependable when the training instances correspond to the probability distribution of a whole set of examples over which the final system performance has to be measured. The present theory of machine learning has its foundation in the critical assumption that the probability distribution of training examples is the same as (though not known) the distribution of novel examples—the data unseen by the machine during its training stage. In spite of our necessity to assume this in order to achieve theoretical results, it is essential to take into consideration that the assumption is frequently violated in practice. One of the common features of available information for machine learning is that the real-world data has the tendency to be incomplete, noisy and inconsistent, which contribute to data inaccuracy. Though procedures for cleaning data try to put in the missing values, smooth out noise while detecting outliers and data inconsistencies, inaccuracies do bring in an element of uncertainty.

Since uncertainty models (probability distributions) for the available data are unknown, the uncertainty management in our applied machine learning procedures is approximately carried out by heuristic approach of finding the hypothesis function simplest in terms of complexity and best in terms of empirical error on the data.

Our focus in this chapter is on another machine learning problem setting where structured human knowledge expressed in the form of IF-THEN rules (and not data) serves as a model for knowledge representation. In this *rules-based setting* for machine learning, a set of IF-THEN rules constitutes the knowledge base.

How can humans reason about complex systems, when the complete description of such a system often requires more detailed information than a human could ever hope to recognize simultaneously and assimilate with understanding? The answer is that humans have the capacity to reason approximately about a complex system, thereby maintaining only a generic understanding about the problem.

Structured human knowledge is based on the basic premise—human solution to the problem exists and it is acceptable (it is a function of our tolerance for imprecision). Also, the mere existence of acceptable human solution in some linguistic form is not sufficient. One must be able to articulate to structure the human solution in the language of learning machine, for example, in the form of IF-THEN rules. The key idea is that the structured human knowledge describes the operation of the process of interest from the standpoint of some (human expert) operator/s of the process, and captures the empirical knowledge of the operation of that process that has been acquired through direct experience with the actual operation of the process. Today, after several thousand successful applications, there is more or less convergence on the trustworthiness of the premise.

The neural network and rule-based systems lie at the two extreme poles of system modeling. At the neural-network pole, there is a *black box design* situation in which the process is entirely unknown; there are only input-output data-pair examples (measurements, records, observations, samples). Also, no physical meaning can be attached to the individual parameters of the trained network. At the other pole, a solution to the problem is known, that is, structured human knowledge (expertise, experience, heuristics) about the process exists. Also, knowledge in the rule-based model can be divided into individual rules and the user can see and understand the piece of knowledge applied by the system. Rule-based system is, thus, a *white box design* approach; its *interpretability* feature is able to explain its reasoning and justify the conclusion.

In neural-network design, uncertainty creeps in through incomplete, noisy and inconsistent data. In rule-based systems, uncertainty creeps in through weak implications, imprecise language, combining views of different domain experts; in addition to other unknown sources. One of the common characteristics of the information available from domain experts is the use of imprecise natural language. We use different terms to describe the same linkage requirement. Knowledge engineers have the painful job of structuring domain knowledge. Also domain experts have, usually, contradictory opinions that lead to conflicting rules. Care has to be taken by the knowledge engineers to create a knowledge base from the available domain knowledge, establishing concrete correlations between IF (condition) and THEN (action) parts of the rules. Naturally, conflict resolutions at the level of design of knowledge base always give rise to uncertainties in the created model.

The most serious problem in applying rule-based models is a *rule-explosion* phenomenon. The number of rules increases exponentially with the dimensions of the input and output spaces. An exhaustive search through all the rules in knowledge base is applied during each cycle of the inference mechanism. Systems based on knowledge embedded in a huge rule-base tend to be slow, and therefore, such systems are best avoided for real-time applications.

In this chapter, we will be concerned with the following three technologies for the design of intelligent machines.

1. Construction of *fuzzy rule-based models* that have been successful in the development of intelligent machines. The word 'fuzzy' may sound to mean intrinsically imprecise, but there is nothing 'fuzzy' about fuzzy rule-based models. They are firmly based on multivalued logic (*fuzzy logic*) and do not violate any well-proven laws of logic. They are aimed at handling imprecise and approximate concepts of human-like solutions, which cannot be processed by any other modeling tool. Knowledge base for these models is created from the information available from domain experts, which is based on their experience and expertise.
2. Can the fuzzy rule-based systems be used when the knowledge about the process is available only in the form of input-output data pairs or knowledge is available in the form of data with additional source of information from domain experts? The answer is 'yes'; we will deal with the design of such systems.
3. The goals of intelligent technologies—fuzzy rule-based systems and neural networks—are common. They both try to mimic human intelligence and ultimately build an intelligent machine. But, they use different means to achieve their goals (refer to Section 1.1). The theory of fuzzy logic offers mathematical power to emulate cognitive functions of thought and perception, and neural networks, through limited understanding of biological neuronal processes of human brain, provide emulation of certain human learning behaviors through mathematics and systems science. We will consider integration of the two technologies.

In neural information processing, there are a variety of complex mathematical operations and mapping functions involved, that, in synergism, act as a parallel computing machine that emulates the biological neuronal processes. Therefore, while a fuzzy rule-based system is dependent on logical inferences, and has its focus on modeling human reasoning, a neural network is dependent on parallel data processing, wherein the focus is on modeling a human brain, examining its structure and functions, in particular, its learning capability.

In fuzzy rule-based systems, knowledge is depicted in the form of IF-THEN rules suggested by domain experts. Once the storage of the rules in the knowledge base of the fuzzy system has taken place, no modification is possible. It is not possible for fuzzy models to learn from experience or take to new/unfamiliar environments. It is only possible for a human expert to alter the knowledge base manually through the addition, change or deletion of certain rules. In neural networks, knowledge is stored in synaptic weights between neurons. This knowledge is procured while learning, that is, when a set of training data is offered to the network.

Therefore, fuzzy rule-based systems are not capable of learning, but possess the ability to explain the manner in which a specific solution has been arrived at. A neural network is capable of learning, but serves as a 'black box' for the user. Through integration of the two technologies, we can combine the advantages of each and create a more powerful and effective intelligent machine.

We will see in this chapter that integrated *neuro-fuzzy systems* bring together the parallel computing and learning capabilities of neural networks along with human-like knowledge representation and explanation capabilities of fuzzy systems.

When pure fuzzy methods have unacceptable performance for classification (pattern recognition) tasks, hybridization with decision trees may result in acceptable performance. *Fuzzy decision trees* have received considerable attention of the researchers. We will discuss this hybrid method in Chapter 8.

6.2 COGNITIVE UNCERTAINTY AND FUZZY RULE-BASE

The basic syntax of an IF-THEN rule is:

IF < antecedent > THEN < consequent > (6.1a)

The 'IF' part of a rule is a proposition known as the *rule antecedent* (*premise* or *condition*). The 'THEN' part is called the *rule consequent* (*conclusion* or *action*). A rule with a single antecedent is said to have an *atomic* proposition.

Atomic propositions do not, usually, make a knowledge base in real-life situations. Many propositions connected by logical connectives may be needed. A set of such *compound propositions*, connected by IF-THEN rules, makes a knowledge base. A rule with compound propositions takes the form:

IF < antecedent 1 >
AND < antecedent 2 > (6.1b)
⋮
THEN < consequent >

The rules are representative of the mapping of the inputs to the outputs for a rule-based system. It is possible to organize the IF-THEN rules in many different forms. Multi-input-multi-output (MIMO) and multi-input-single-output (MISO) are the two common forms. The MIMO rules have multiple antecedents and multiple consequents, while MISO rules have multiple antecedents and single consequent.

Keywords link several different antecedents in a rule— AND (conjunction), OR (disjunction) or both in combination. But, it is better to not mix conjunctions and disjunctions (defined later in Section 6.3) in the same rule.

All parts of the antecedent are calculated simultaneously and resolved in a single number, using fuzzy set operations (described later in Section 6.3). All parts of a consequent are affected equally by the antecedent. It is usually possible to implement a MIMO rule with q consequent parts by specifying q MISO rules.

The antecedent of a rule incorporates two parts: a feature/attribute (*linguistic*) and its *value*. The feature and its value are linked by operators, such as *is*, *are*, *is not*, *are not*. Mathematical operators can also be used.

Rule-based systems based on two-valued Boolean logic have the characteristic that they involve sharp cut-offs for continuous attributes. For example, consider the following rule for customer credit application approval.

IF (*years employed* ≥ 2) AND (*income* ≥ 50K) THEN *credit* = *approved*

Basically, the rule states that applications of customers who have held a job for two or more years and earn a minimum of \$ 50,000, are approved. According to this rule, customers who have held a job for a minimum of two years will obtain credit if they earn, say, \$ 50,000, but will not be eligible for credit if their income is \$ 49,900. Rigid cut-offs such as this may appear unjust. Instead, dividing *income* into classes, such as low-income, medium-income or high-income, and then applying 'multi-valued logic' to permit 'soft' boundaries for each class, may appear to be more sensible for such an application. Our intelligent machine (computer) has to know *computing with words*, and also the multi-valued logic. The categories: low_income, medium_income, high_income, are *vague* and *ambiguous* terms for the machine; it has to know interpretation of such terms. Can we make the machines learn how to solve such problems?

Problems involving vagueness and ambiguity have been successfully addressed consciously or subconsciously by humans. In process industry, for example, a process operator employs a set of IF-THEN rules to control a process. The operator estimates the important process variables, and based on this information, he/she manipulates the process through control variable. The estimation of the process variables is *not* done in numerical form; it is rather done in linguistic form. For example, he/she may select 'error' and 'error-rate' as important process variables, and manipulate the process through incremental 'change-in-control'.

$$\text{error} = e = \text{desired output} - \text{actual (current) output} \tag{6.2a}$$

$$\text{error-rate} = \dot{e} = \frac{de}{dt} = \frac{\text{error at action time } t - \text{error at previous action}}{\text{Time duration between two actions}}$$

$$= \frac{e(t) - e(t-T)}{T} \tag{6.2b}$$

$$\text{Change-in-control} = \Delta u = u(t) - u(t-T) \tag{6.2c}$$

The process operator may categorize the variables: e, $\dot{e}$, Δu, into following labels over their entire operating range.

$$[\text{Negative } (N), \text{ Zero } (Z), \text{ Positive } (P)] \tag{6.2d}$$

The categories (linguistic labels) of the process variables and control variable are, in general, vague and ambiguous. It is easy for process operators to understand and interpret these classes as they possess the background to deal with the problems and solutions so described.

A commonly used way of expressing the knowledge based on human experience and understanding is through IF-THEN rules. A typical rule for the selected variables and their categories, may be of the form:

$$\text{IF } e \text{ is positive AND } \dot{e} \text{ is positive THEN } \Delta u \text{ is positive} \tag{6.3}$$

This 'linguistic rule' is formed exclusively using linguistic variables and values. As linguistic values do not precisely represent the underlying quantities described by them, the linguistic rules also lack precision. They remain merely abstract concepts pertaining to the manner in which good control can be achieved and could mean different things to various people. But, they are at an abstraction level that process operators are familiar with in terms of the manner of controlling a process.

There is just a limited number of probable rules for a limited number of linguistic variables and values. With two inputs of three linguistic values for each input, there are maximum $3^2 = 9$ likely rules (all possible combinations of linguistic values for two inputs). Thus, the set of rules (called the *rule-base*) employed by the process operator over the entire range of operation of the process consists of nine rules.

The following set of rules represents abstract knowledge that the domain expert has about how to control the process given the error and error-rate as inputs (Note that this rule-base is commonly employed in process industry; it, in fact, realizes the well-known PI control scheme [33]).

$$\begin{aligned}
&\text{Rule 1 : IF } e \text{ is positive AND } \dot{e} \text{ is positive THEN } \Delta u \text{ is Positive}\\
&\text{Rule 2 : IF } e \text{ is positive AND } \dot{e} \text{ is Negative THEN } \Delta u \text{ is Zero}\\
&\text{Rule 3 : IF } e \text{ is positive AND } \dot{e} \text{ is Zero THEN } \Delta u \text{ is positive}\\
&\text{Rule 4 : IF } e \text{ is Negative AND } \dot{e} \text{ is Positive THEN } \Delta u \text{ is Zero}\\
&\text{Rule 5 : IF } e \text{ is Negative AND } \dot{e} \text{ is Negative THEN } \Delta u \text{ is Negative}\\
&\text{Rule 6 : If } e \text{ is Negative AND } \dot{e} \text{ is Zero THEN } \Delta u \text{ is Negative}\\
&\text{Rule 7 : IF } e \text{ is Zero AND } \dot{e} \text{ is Positive THEN } \Delta u \text{ is Positive}\\
&\text{Rule 8 : IF } e \text{ is Zero AND } \dot{e} \text{ is Negative THEN } \Delta u \text{ is Negative}\\
&\text{Rule 9 : IF } e \text{ is Zero AND } \dot{e} \text{ is Zero THEN } \Delta u \text{ is Zero}
\end{aligned} \tag{6.4}$$

Rule-base for a specific application is created by domain experts with deep knowledge of application and strong experience.

The rule-base can be represented in a table format for the cases where there are two or three inputs. Table 6.1 shows the rule-base given above in a tabular representation. Note that the body of the table lists the linguistic consequents of the rules, and the left column and top row of the table hold the linguistic values of the premise terms. For instance, the cell defined by the crossing of the third row and the third column depicts Rule 1 in the given rule set, and the cell demarcated by the third row and first column intersection depicts Rule 2.

Table 6.1 Rule table for process control

'Change-in-control Δu'		'error-rate' $\dot{e}$		
		N	*Z*	*P*
'error' e	*N*	*N*	*N*	*Z*
	Z	*N*	*Z*	*P*
	P	*Z*	*P*	*P*

Providing an intelligent machine (computer) with the level of understanding the expert operator possesses, is a difficult task. How can we represent 'expert knowledge' that uses vague and ambiguous terms, in a computer? Can it be done at all? Now we know that the answer is 'yes' and the solution has been provided by *fuzzy logic Theory*.

As we will see later in this chapter that when an acceptable human solution to a problem exists (i.e., there is usable human knowledge), *fuzzy logic* provides the tools to transform it into an efficient algorithm. Fuzzy logic controllers are being extensively used (replacing human operators) for automation and control in process industry.

Let us consider another simple illustration of cognitive uncertainty—a Car-Following System. Assume that there are two vehicles on the highway; the driver of the following vehicle will regulate the intervehicle spacing. In this man-machine control system, a driver employs consciously or subconsciously a set of IF-THEN rules to regulate the intervehicle spacing.

Important variables describing the car-following system are intervehicle spacing (distance), the vehicle velocity (speed), and the braking force applied to the vehicle (braking-force). Based on the estimation of input variables *distance* and *speed*, the driver manipulates the *braking-force*. The estimation of the system variables is *not* done in numerical form; it is rather done in linguistic form. For example, the driver may categorize the variables 'distance' into the following labels.

[Very Small, Small, Moderate, Large, Very Large] (6.5a)

Analogously, categories of 'speed':

[Very Low, Low, Moderate, High, Very High] (6.5b)

and that of 'braking-force':

[Zero, One-Fourth, One-Half, Three-Fourths, Full] (6.5c)

We use our common knowledge in controlling the distance between our car and the vehicle in front of us while driving. Rule-Base comprises 25 rules of the following type:

IF *distance* is Very Small AND *speed* Very Low THEN *braking-force* One-Half (6.6)

This knowledge is imprecise. We all drive a car differently. Despite many involved uncertainty factors, the resulting knowledge model with regard to solution of the given problem is usually an acceptable one.

If there is a usable knowledge, fuzzy logic provides the tools to transfer it into an efficient algorithm. Using fuzzy logic, one tries to model structured human knowledge.

Fuzzy logic may play an important role in intelligent vehicle highway systems. Vehicles will be automatically driven with on-board lateral and longitudinal controllers. The lateral controllers help steer the vehicles around corners, change lanes, and also perform other steering tasks. If a vehicle is moving alone, it is possible to maintain a stable velocity with the help of the longitudinal controllers. These controllers are also used to follow a lead vehicle at a harmless distance, or execute other speed/tracking functions. On the present-day highways, human beings do this kind of driving and handle these tasks.

Fuzzy or *multi-valued* logic was introduced by Jan Lukasiewicz in the 1920s, scrutinized by Max Black in the 1930s, and rediscovered, extended to formal system of mathematical logic by Dr Lotfi Zadeh, a professor from University of California in Berkeley, in the 1960s. The seminal work: *Fuzzy Sets* (published in 1965) by Zadeh is considered to be the foundation of current theory in fuzzy logic.

But, the technical community accepted the fuzzy set theory a little slowly and with some difficulty, partly because of the name 'fuzzy', which lacked seriousness. Ultimately, the fuzzy theory which was not given importance in the West, was seriously looked at in the East. The Japanese have used it for over four decades, with great success.

Since then, the subject has been the focus of many independent research investigations by mathematicians, scientists and engineers around the world. Fuzzy logic has advanced at a fast pace. It has grown to be one of the most successful technologies, used to develop sophisticated control systems.

Although most fuzzy technology applications continue to be reported in control systems, fuzzy *decision-making systems* are applied widely in several areas. For example, in engineering some potential application areas include *robotics* (path planning, task scheduling, navigation); *computers* (memory allocation, task scheduling); *process industries* (failure diagnosis performance, assessment monitoring); *manufacturing* (scheduling and planning material flow, resource allocation); *power industry* (power distribution, load estimation.); *traffic systems* (routing and signal switching).

In business and health sectors, fuzzy logic has been used for the following applications: *finance, credit evaluation, stock market analysis*; *medical diagnostic systems, health monitoring*.

The list is only representative of the range of possible applications for fuzzy logic. Others have been studied, while still others are yet to be identified.

Let us summarize the philosophical ideas behind fuzzy logic. Fuzzy logic does not mean logic which is fuzzy but refers to logic used to describe fuzziness. According to Zadeh, the term fuzzy is concrete, immediate and descriptive. Fuzzy logic adds a range of logical values between 0 (completely false) and 1 (completely true) to Boolean logic to address issues of vagueness and ambiguity in our thinking. Classical binary logic can be considered as a special case of multi-valued fuzzy logic.

Fuzzy logic has often been perceived incorrectly as a direct competitor to probability theory, whereas, in fact, it addresses a different set of issues. Probability theory handles *statistical uncertainty* dealing with nature's random behavior. All the methods described in earlier chapters address the uncertainty resulting from the natural world from statistical variations or randomness. While these methods may work well when used to measure the likelihood of a hypothesis; they do not reveal anything about the meaning of the hypothesis. *Cognitive uncertainty*, on the contrary, concerns human cognition. Cognitive uncertainty may be further categorized into two subclasses: *vagueness* and *ambiguity*. Ambiguity results from situations that have two or more options, such that the choice between them is left unspecified. Vagueness results from a challenge in precisely defining the concepts.

Our focus in this chapter is on the essential ideas and tools necessary for the construction of fuzzy-rules based systems. We start by introducing certain important ideas, terminology, notations, and arithmetic of fuzzy sets and fuzzy logic. We include just a minimum, though adequate, amount of fuzzy mathematics. To help easy reading, the background material is given in a very informal fashion, with basic and clear notation and explanation. Wherever possible, extremely rigorous mathematics is not used.

Books on Computational Intelligence usually give a detailed account of Fuzzy Systems. We have listed some books of this kind in the references [26–29]. The books [3, 5, 92] give a good account of fuzzy logic, while [85, 93] are more oriented towards control applications. Integrated neural-fuzzy systems are dealt with in detail in [94].

6.3 FUZZY QUANTIFICATION OF KNOWLEDGE

Till this point, we have only quantified in an abstract fashion, the knowledge that the human expert has about how to control the process. Next, we will show how to use fuzzy logic to fully quantify the meaning of linguistic descriptions so that we may automate in the fuzzy system, the rules specified by the expert.

6.3.1 Fuzzy Logic

Knowledge is structured information and knowledge acquisition is done through learning and experience, which are forms of high-level processing of information. Representing and processing knowledge hold the keys to all intelligent systems. When it comes to *logic*, knowledge representation is done by propositions and its processing is performed using reasoning, by applying various laws of logic, including a suitable *rule of inference*.

The focus of *fuzzy logic* is on linguistic variables in natural language, wherein the objective is to provide foundations for *approximate reasoning* with imprecise propositions.

In *classical logic*, a proposition is either TRUE, denoted by 1, or FALSE, denoted by 0. Consider the following proposition p:

'Team member is female'

Let X be a collection of 10 people: $x_1, x_2, ..., x_{10}$, who form a project team. The entire object of discussion is,

$$X = \{x_1, x_2, \ldots, x_{10}\}$$

In general, the entire object of discussion is called a 'universe of discourse', and each constituent member x is called an 'element' of the universe (the fact that x is an element of X, is written as $x \in X$).

If x_1, x_2, x_3 and x_4 are female members in the project team, then the proposition p on the universe of discourse X is equally well represented by the *crisp* (nonfuzzy) set A defined below:

$$A = \{x_1, x_2, x_3, x_4\}$$

The fact that A is a subset of X is denoted as $A \subset X$.

The proposition can also be expressed by a mapping μ_A from X into the binary space $\{0, 1\}$.

$$\mu_A : X \rightarrow \{0, 1\} \tag{6.7a}$$

such that,

$$\mu_A = \begin{cases} 0; x = x_5, x_6, x_7, x_8, x_9, x_{10} \\ 1; x = x_1, x_2, x_3, x_4 \end{cases}$$

That is to say, the value $\mu_A(x) = 1$ when the element x satisfies the attributes of set A; 0 when it does not. μ_A is called the *characteristic function* of A.

Next, supposing that, within A, only x_1 and x_2 are below age 20; we may call them 'minors'. Then,

$$B = \{x_1, x_2\}$$

consists of minor team members. In this case

$$\mu_B(x) = \begin{cases} 1; x = x_1, x_2 \\ 0; \text{otherwise} \end{cases}$$

B is obviously a subset of A; we write $B \subset A$.

We have considered the 'set of females A', and the 'set of minor females B' in X. Is it also possible to consider a 'set of young females C'? If, for convenience, we consider the attribute 'young' to be same as 'minor', then $C = B$; but, in this case, we have created a sharp boundary under which x_2 who is 19 is still young ($\mu_C(x_2) = 1$), but x_3 who just turned 20 today is no longer young ($\mu_C(x_3) = 0$); in just one day, the value changed from yes (1) to no (0), and x_3 is now an old maid.

However, is it not possible that a young woman becomes an old maid over a period of 10 to 15 years, so that we ought to be patient with her? Prof Zadeh admitted values such as 0.8 and 0.9 that are intermediate between 0 and 1, thus creating the concept of a 'fuzzy set'. Whereas, a crisp set is defined by the characteristic function that can assume only the two values {0, 1}, a *fuzzy set* is defined by a 'membership function' that can assume an infinite number of values; any real number in the closed interval [0, 1].

With this definition, the concept of 'young woman' in X can be expressed flexibly in terms of *membership function* (Fuzzy sets are denoted in this book by a set symbol with a tilde understrike):

$$\mu_{\underset{\sim}{C}} : X \rightarrow [0, 1] \tag{6.7b}$$

such that,

$$\mu_{\underset{\sim}{C}} = \begin{cases} 1; & x = x_1, x_2 \\ 0.9; & x = x_3 \\ 0.2; & x = x_4 \\ 0; & \text{otherwise} \end{cases}$$

The significance of such terms as 'patient' and 'flexibly' in the above description may be explained as follows. For example, we have taken $\mu_{\underset{\sim}{C}}(x_3) = 0.9$, but suppose that x_3 objects that 'you are being unfair; I really ought to be a 1 but if you insist we can compromise on 0.95'. There is a good amount of subjectivity in the choice of membership values. A great deal of research is being done on the question of assignment of *membership values*. However, even with this restriction, it has become possible to deal with many problems that could not be handled with only crisp sets.

Since [0, 1] incorporates {0, 1}, the concept of fuzzy set can be considered as an extended concept, which incorporates the concept of crisp set. For example, the crisp set B of 'minors' can be regarded as a fuzzy set $\underset{\sim}{B}$ with the membership function:

$$\mu_{\underset{\sim}{B}}(x) = \begin{cases} 1; & x = x_1, x_2 \\ 0; & \text{otherwise} \end{cases}$$

Example 6.1

One of the widely used instances of a fuzzy set is the one that consists of tall people. In this case, the universe of discourse is potential heights (the real line), suppose, 3 feet to 9 feet tall. If the set of tall people is defined properly as a distinct set, we may consider people with height beyond six feet as officially tall. The characteristic function of the set A = {tall men} then, is

$$\mu_A(x) = \begin{cases} 1 & \text{for } 6 \le x \\ 0 & \text{for } 3 \le x < 6 \end{cases}$$

Such a condition is expressed by a Venn diagram shown in Fig 6.1(a) and a characteristic function shown in Fig 6.2(a).

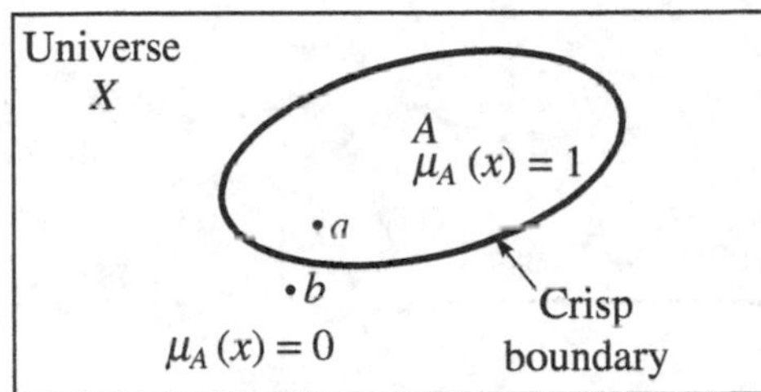

(a) The crisp set A and the universe of discourse

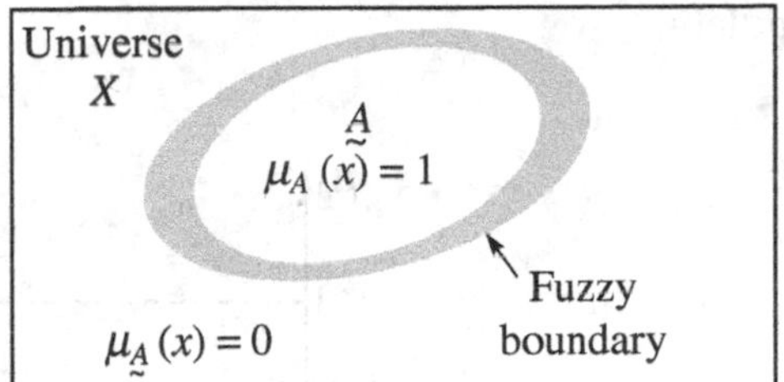

(b) The fuzzy set $\underset{\sim}{A}$ and the universe of discourse

Figure 6.1

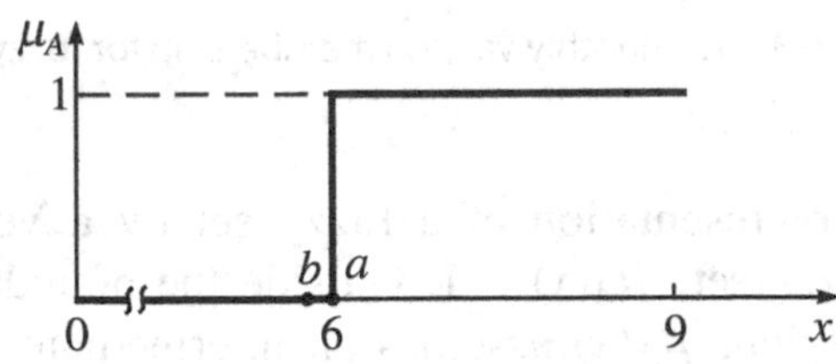

(a) Characteristic function of crisp set A

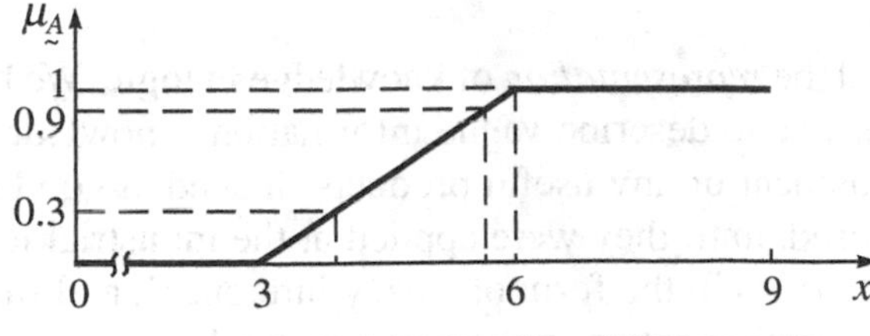

(b) Membership function of fuzzy set $\underset{\sim}{A}$

Figure 6.2

For our example of universe X of heights of people, the crisp set A of all people with $x \ge 6$ has a sharp boundary: individual, 'a' corresponding to $x = 6$ is a member of the crisp set A, and individual 'b' corresponding to $x = 5.9$ is unambiguously *not* a member of set A. Is it not an absurd statement for the situation under consideration? *A* 0.1″ reduction in the height of a person has changed μ_A from 1 to 0, and the person is no more tall.

A crisp set of all real numbers greater than 6 may make sense as the numbers are part of an abstract plane, but when it comes to real people, it is not logical to consider one person short and another tall, when the difference in their height is only the width of a hair. However, if this type of distinction is not sensible, then how can the set of all tall men be rightly defined? Just like our example of 'set of young women', the word 'tall' is like a curve describing the level to which any person is tall. Figure 6.2(b) depicts a possible membership function of this fuzzy set.

Note that there is inherent subjectivity in fuzzy set description. Figure 6.3 shows a smoothly varying curve (*S*-shaped) for transition from not tall to tall. Compared to Fig. 6.2(b), the membership values are lower for heights close to 3′ and are higher for heights close to 6′. This looks more reasonable; however, the price paid is in terms of a more complex function, which is more difficult to handle.

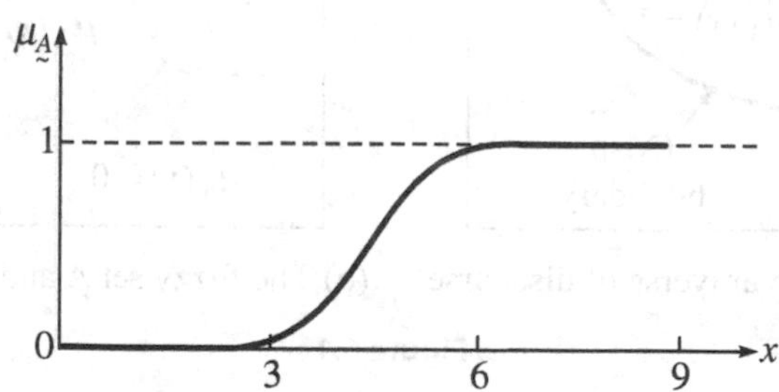

Figure 6.3 A smoothly varying membership for fuzzy set $\underset{\sim}{A}$

Figure 6.1(b) shows the representation of a fuzzy set by a Venn diagram. In the central (unshaded) region of the fuzzy set, $\mu_{\underset{\sim}{A}}(x) = 1$. Outside the boundary region of fuzzy set, $\mu_{\underset{\sim}{A}}(x) = 0$. On the boundary region, $\mu_{\underset{\sim}{A}}(x)$ assumes an intermediate value in the interval (0, 1). Presumably, the membership value of an x in fuzzy set $\underset{\sim}{A}$ approaches a value of 1 as it moves closer to the central (unshaded) region; it approaches a value of 0 as it moves closer to leaving the boundary region of $\underset{\sim}{A}$.

So far we have discussed the *representation* of knowledge in logic. We have seen that the concept of fuzzy sets makes it possible to describe vague information (knowledge). But description alone will not lead to the development of any useful products. Indeed, a good deal of time passed after fuzzy sets were first proposed, until they were applied at the industrial level. However, eventually it became possible to apply them in the form of 'fuzzy inference', and fuzzy logic theory has now become legitimized as one component of applied high technology.

In fuzzy logic theory, nothing is done at random or haphazardly. Information containing a certain amount of vagueness is expressed as faithfully as possible, without the distortion produced by forcing it into a 'crisp' mould, and it is then processed by applying an appropriate rule of inference.

'Approximate reasoning' is the best known form of fuzzy logic processing and covers a variety of inference rules.

Fuzziness is often confused with probability. The basic difference between them is that fuzziness handles deterministic plausibility, whereas probability deals with the possibility of nondeterministic (stochastic) events. Fuzziness is one facet of uncertainty—it is the vagueness found in definition of a concept (weak implications in rule-based systems, for example), and/or ambiguity in meaning of linguistic terms. However, the uncertainty of probability usually pertains to the occurrence of phenomena, not their ambiguity. For instance, 'There is a 50 percent chance that he will live' is a very uncertain statement because of inherent randomness. On the other hand, 'she is a young woman' is a statement with uncertainty in definition of 'young woman'. Therefore, fuzziness is the description of the vagueness and ambiguity of an event, while randomness describes the uncertainty in occurrence of an event.

We can now give a formal definition to fuzzy sets.

6.3.2 Fuzzy Sets

A *universe of discourse*, X, is a collection of objects all having the same characteristics. The individual elements in the universe X will be denoted as x.

A universe of discourse and a membership function that spans the universe, completely define a *fuzzy set*. Consider a universe of discourse X with x representing its generic element. A fuzzy set $\underset{\sim}{A}$ in X has the membership function $\mu_{\underset{\sim}{A}}(x)$ which maps the elements of the universe onto numerical values in the interval [0, 1]:

$$\mu_{\underset{\sim}{A}}(x) : X \to [0, 1] \tag{6.8a}$$

Every element x in X has a membership function $\mu_{\underset{\sim}{A}}(x) \in [0, 1]$. $\underset{\sim}{A}$ is then defined by the set of ordered pairs:

$$\underset{\sim}{A} = \{(x, \mu_{\underset{\sim}{A}}(x)) \mid x \in X, \mu_{\underset{\sim}{A}}(x) \in [0, 1]\} \tag{6.8b}$$

A membership value of zero implies that the corresponding element is definitely *not* an element of the fuzzy set $\underset{\sim}{A}$. A membership function of unity means that the corresponding element is definitely an element of fuzzy set $\underset{\sim}{A}$. A grade of membership greater than zero and less than unity, corresponds to a noncrisp (or fuzzy) membership of the fuzzy set $\underset{\sim}{A}$. Classical sets can be considered as special case of fuzzy sets with all membership grades equal to unity or zero.

A fuzzy set $\underset{\sim}{A}$ is formally given by its membership function $\mu_{\underset{\sim}{A}}(x)$. We will identify any fuzzy set with its membership function, and use these two terms interchangeably.

Membership functions characterize the fuzziness in a fuzzy set. However, the shape of the membership functions, used to describe the fuzziness, has very few restrictions indeed. It might be claimed that the rules used to describe fuzziness are also fuzzy. Just as there are an infinite number of ways to characterize fuzziness, there are an infinite number of ways to graphically depict the membership functions that describe fuzziness. Although the selection of membership functions is *subjective, it cannot be arbitrary*, it should be *plausible*.

To avoid unjustified complications, $\mu_{\underset{\sim}{A}}(x)$ is usually constructed without a high degree of precision. It is advantageous to deal with membership functions involving a small number of parameters. Indeed, one of the key issues in the theory and practice of fuzzy sets is how to define the proper membership functions. Primary approaches include (1) asking the human expert to define them; (2) using data from the system to generate them; and (3) making them in a trial-and-error manner. In more than 25 years of practice, it has been found that the third approach, though ad hoc, works effectively and efficiently in many real-world applications.

Numerous applications have shown that only four types of membership functions are needed in most circumstances: trapezoidal, triangular (a special case of trapezoidal), Gaussian, and bell-shaped. Figure 6.4 shows an example of each type. Among the four, the first two are more widely used. All these fuzzy sets are continuous, normal and convex.

A fuzzy set is said to be *continuous* if its membership function is continuous.

A fuzzy set is said to be *normal* if its *height* is one (the largest membership value of a fuzzy set is called the height of the fuzzy set).

The *convexity* property of fuzzy sets is viewed as a generalization of the classical concept of convexity of crisp sets. Consider the universe X to be a set of real numbers $\Re$. A subset A of $\Re$ is said to be *convex* if, and only if, for all $x_1, x_2, \in A$ and for every real number λ satisfying $0 \le \lambda \le 1$, we have

$$\lambda x_1 + (1 - \lambda)x_2 \in A \tag{6.9}$$

It can easily be established that any set defined by a single interval of real numbers is convex; any set defined by more than one interval, which does not contain some points between the intervals, is not convex.

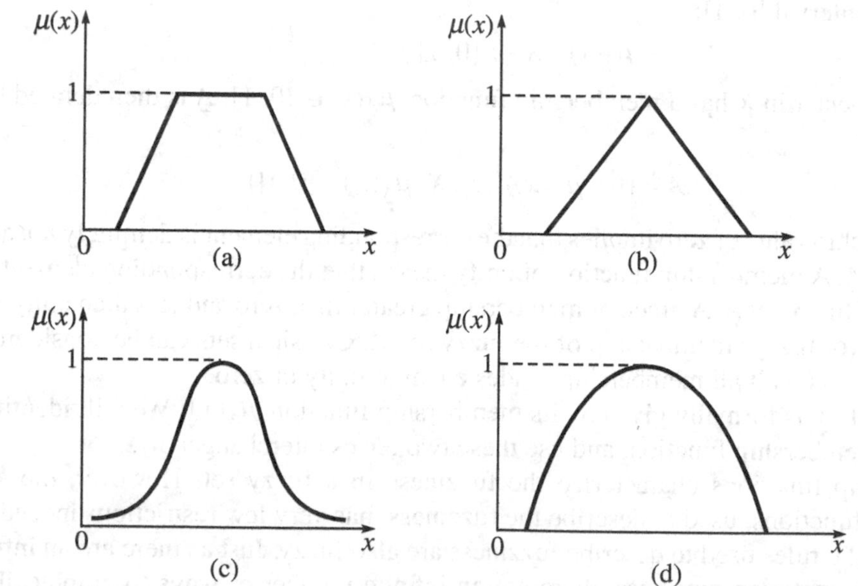

Figure 6.4 Examples of fuzzy sets: (a) Trapezoidal (b) Triangular (c) Gaussian (d) Bell-shaped

An *alpha-cut* of a fuzzy set $\underset{\sim}{A}$ is a crisp set A_α that contains all the elements of the universal set X that have a membership grade in $\underset{\sim}{A}$ greater than or equal to α (refer to Fig. 6.5). The convexity property of fuzzy sets, as said earlier, is viewed as a generalization of the classical concept of convexity of crisp sets. In order to make the generalized convexity consistent with the classical definition of convexity, it is required that α-cuts of a *convex fuzzy set* be convex for all $\alpha \in (0, 1]$

in the classical sense (0-cut is excluded here since it is always equal to $\Re$ in this sense and thus includes $-\infty$ to $+\infty$). Figure 6.5(a) shows a fuzzy set that is convex. Two of the α-cuts shown in this figure are clearly convex in the classical sense, and it is easy to see that any other α-cuts for $\alpha > 0$ are convex as well. Figure 6.5(b) illustrates a fuzzy set that is not convex. The lack of convexity of this fuzzy set can be demonstrated by identifying some of its α-cuts ($\alpha > 0$) that are not convex.

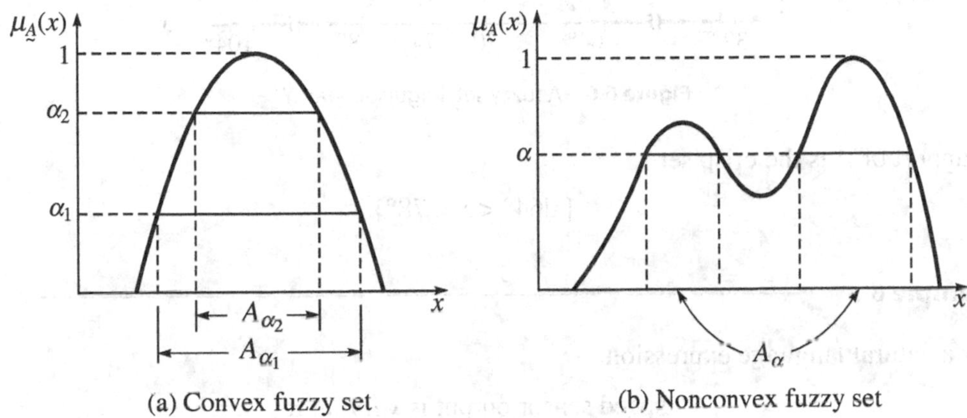

(a) Convex fuzzy set (b) Nonconvex fuzzy set

Figure 6.5

The *support* of a fuzzy set $\underset{\sim}{A}$ is the crisp set of all $x \in X$ such that $\mu_{\underset{\sim}{A}}(x) > 0$. That is,

$$\text{supp}(\underset{\sim}{A}) = \{x \in X \mid \mu_{\underset{\sim}{A}}(x) > 0\} \tag{6.10}$$

The element $x \in X$ at which $\mu_{\underset{\sim}{A}}(x) = 0.5$, is called the *crosspoint*.

A fuzzy set $\underset{\sim}{A}$ whose support is a single point in X with $\mu_{\underset{\sim}{A}}(x) = 1$, is referred to as a *fuzzy singleton*.

Example 6.2

Consider the fuzzy set described by membership function depicted in Fig 6.6, where the universe of discourse is

$$X = [32°\text{F}, 104°\text{F}]$$

This fuzzy set $\underset{\sim}{A}$ is linguistic 'warm', with membership function

$$\mu_{\underset{\sim}{A}}(x) = \begin{cases} 0\,;\, x < 64° \\ (x - 64°)/6\,;\, 64° \le x < 70° \\ 1\,;\, 70° < x \le 74° \\ (78° - x)/4\,;\, 74 < x \le 78° \\ 0\,;\, x > 78° \end{cases}$$

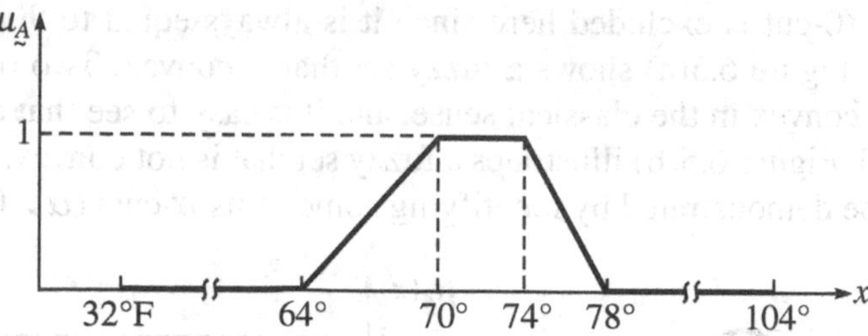

Figure 6.6 A fuzzy set: linguistic 'warm'

The support of $\tilde{A}$ is the crisp set

$$\{x|64° < x < 78°\}$$

Example 6.3

Consider a natural language expression:

'speed sensor output is very large'

The formal, symbolic translation of this natural language expression, in terms of linguistic variables, proceeds as follows:

(i) An abbreviation 'Speed' may be chosen to denote the physical variable 'speed sensor output'.
(ii) An abbreviation 'Xfast' (i.e., extra fast) may be chosen to denote the particular value 'very large' of speed.
(iii) The above natural language expression is rewritten as 'Speed is Xfast'.

Such an expression is an *atomic fuzzy proposition*. The 'meaning' of the atomic proposition is then defined by a fuzzy set $X\tilde{F}ast$, or a membership function $\mu_{X\tilde{F}ast}(x)$, defined on the physical domain X = [0 mph, 100 mph] of the physical variable 'speed'.

Many atomic propositions may be associated with a linguistic variable, e.g.,

'Speed is Fast'
'Speed is Moderate'
'Speed is Slow'
'Speed is XSlow'

Thus, the set of *linguistic values* that the linguistic variable 'Speed' may take is

$$\{XFast, Fast, Moderate, Slow, XSlow\}$$

These linguistic values are called *terms* of the linguistic variable. Each term is defined by an appropriate membership function.

It is usual in approximate reasoning to have the following *frame* associated with the notion of a linguistic variable:

$$\langle \tilde{A}, \mathcal{L}\tilde{A}, X, \mu_{\mathcal{L}\tilde{A}} \rangle \tag{6.11}$$

where $\underset{\sim}{A}$ denotes the symbolic name of a linguistic variable, e.g., speed, temperature, level, error, change-of-error, etc. $\mathcal{L}\underset{\sim}{A}$ is the set of linguistic values that $\underset{\sim}{A}$ can take, i.e., $\mathcal{L}\underset{\sim}{A}$ is the term set of $\underset{\sim}{A}$. X is the actual physical domain over which linguistic variable $\underset{\sim}{A}$ takes its quantitative (crisp) values, and $\mu_{\mathcal{L}\underset{\sim}{A}}$ is a set of membership functions which gives a meaning to the linguistic values in terms of the quantitative elements of X.

Example 6.4

Consider speed interpreted as a linguistic variable with $X = [0 \text{ mph}, 100 \text{ mph}]$, i.e., x = 'Speed'. Its term set could be

$$\{Slow, Moderate, Fast\}$$

$\underset{\sim}{Slow}$ = the fuzzy set for 'a speed below about 40 miles per hour (mph)', with membership function $\mu_{\underset{\sim}{Slow}}$,

$\underset{\sim}{Moderate}$ = the fuzzy set for 'a speed close to 55 mph', with membership function $\mu_{\underset{\sim}{Moderate}}$,

$\underset{\sim}{Fast}$ = the fuzzy set for 'a speed above about 70 mph', with membership function $\mu_{\underset{\sim}{Fast}}$.

The frame of speed is

$$\langle \underset{\sim}{Speed}, \mathcal{L}\underset{\sim}{Speed}, X, \mu_{\mathcal{L}\underset{\sim}{Speed}} \rangle$$

where

$$\mathcal{L}\underset{\sim}{Speed} = \{\underset{\sim}{Slow}, \underset{\sim}{Moderate}, \underset{\sim}{Fast}\}$$

$$X = [0, 100] \text{ mph}$$

$\mu_{\underset{\sim}{Slow}}, \mu_{\underset{\sim}{Moderate}}, \mu_{\underset{\sim}{Fast}}$ are given in Fig. 6.7.

The frame of speed helps us to decide the degree to which an atomic proposition associated with 'speed' is satisfied, given a specific physical value of speed. For example, for crisp input

$$Speed = 50 \text{ mph}$$

$$\mu_{\underset{\sim}{Slow}}(50) = 1/3$$

$$\mu_{\underset{\sim}{Moderate}}(50) = 2/3$$

$$\mu_{\underset{\sim}{Fast}}(50) = 0$$

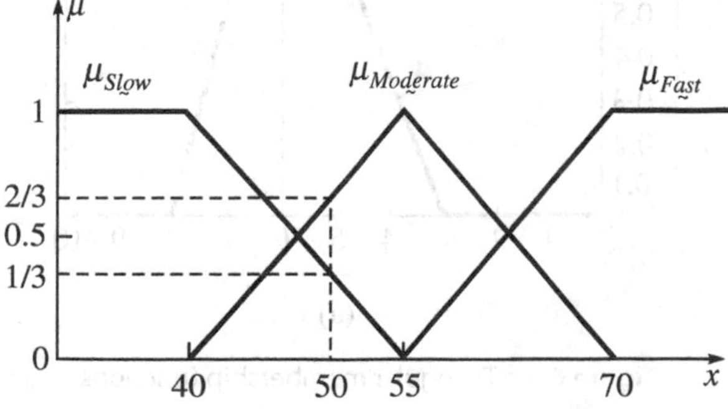

Figure 6.7 Terms of linguistic variable 'Speed'

Therefore, the proposition 'Speed is Slow' is satisfied to a degree of 1/3, the proposition 'Speed is Moderate' is satisfied to a degree of 2/3, and the proposition 'Speed is Fast' is not satisfied.

An extension of ordinary fuzzy sets is to allow the membership values to be a fuzzy set—instead of a crisply defined degree. A fuzzy set whose membership function is itself a fuzzy set, is called a *Type*-2 *fuzzy set* [94]. A *Type*-1 *fuzzy set* is an ordinary fuzzy set. We will limit our discussion to *Type*-1 fuzzy sets. The reference to a fuzzy set in this chapter, implies a Type-1 fuzzy set.

We have observed that a fuzzy set is completely characterized by its membership function. Widely used *continuous membership functions* have been illustrated in Fig 6.4. In the following, we give most general form of these membership functions. Note that all membership functions given in Fig. 6.4 are specific cases of these analytical forms.

We also define here *discrete* and *singleton* functions that we will be using later in this chapter.

Continuous Membership Functions

Triangular: A triangular membership function is specified by three parameters (a, b, c) as follows:

$$\mu(x) = \begin{cases} 0 & \text{if} \quad x \le a \\ \dfrac{x-a}{b-a} & \text{if} \quad a \le x \le b \\ \dfrac{c-x}{c-b} & \text{if} \quad b \le x \le c \\ 0 & \text{if} \quad c \le x \end{cases} \tag{6.12}$$

The parameters (a, b, c) with $a < b < c$ determine the x-coordinates of the three corners of the triangle (Fig. 6.8(a)).

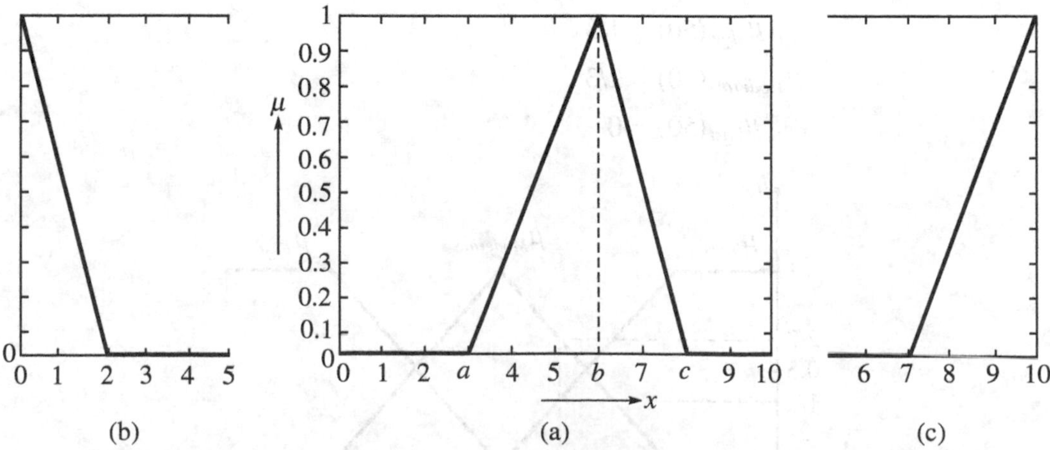

Figure 6.8 Triangular membership functions

The two membership functions shown in Figs 6.8(b) and 6.8(c), defined at the corners of universe of discourse, are special cases of Eqn (6.12).

Trapezoidal: A trapezoidal membership function is specified by four parameters (a, b, c, d) as follows:

$$\mu(x) = \begin{cases} 0 & \text{if} \quad x \le a \\ \dfrac{x-a}{b-a} & \text{if} \quad a \le x \le b \\ \dfrac{d-x}{d-c} & \text{if} \quad c \le x \le d \\ 0 & \text{if} \quad d \le x \end{cases} \tag{6.13}$$

The parameters (a, b, c, d) with $a < b \le c < d$ determine the x-coordinates of the four corners of the trapezoid (Fig. 6.9(a)). Note that trapezoidal membership function with parameters (a, b, c, d) reduces to a triangular membership function when b is equal to c.

The two membership functions shown in Figs 6.9(b) and 6.9(c), are special cases of Eqn (6.13). These functions are inherently open right or left and thus are appropriate for representing concepts such as 'very large' or 'very small'.

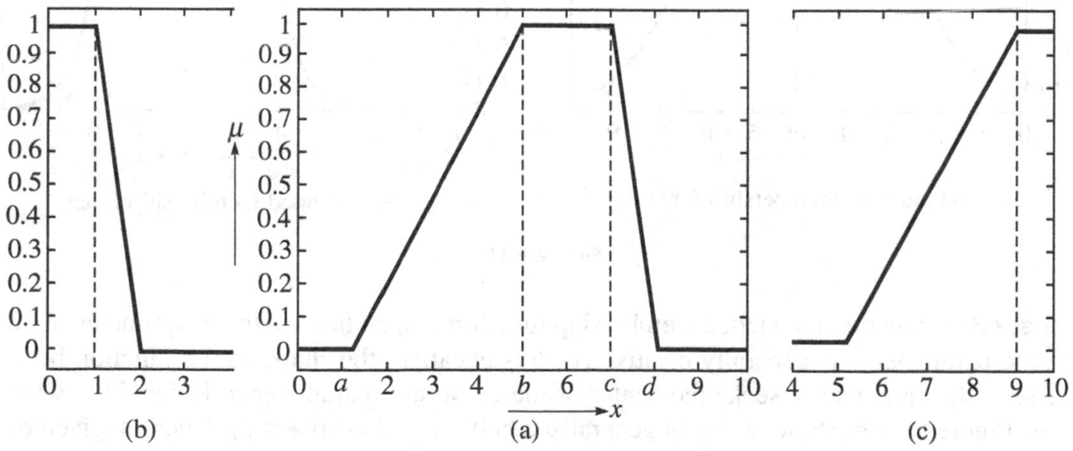

Figure 6.9 Trapezoidal membership functions

Piecewise linear functions (generally either triangular or trapezoidal) constitute the simplest type of membership functions. As we have observed, these may be *symmetrical* or *asymmetrical* in shape.

Gaussian and *bell-shaped* membership functions display smoothness, due to which their use for specifying fuzzy sets is increasing.

Even though the Gaussian membership functions and bell-shaped membership functions are smooth, they do not have the ability to specify asymmetric membership functions, which are significant in some applications. *Sigmoidal* membership functions are either open left or right, and are used for synthesizing asymmetric and closed membership functions.

In the following, we describe Gaussian, bell-shaped and sigmoidal membership functions.

Gaussian: A Gaussian membership function is specified by two parameters c and σ; c represents the center of the membership function and σ determines the width. Figure 6.10(a) shows a plot of general Gaussian membership function, given by the expression:

$$\mu(x) = e^{-\frac{1}{2}\left(\frac{x-c}{\sigma}\right)^2} \tag{6.14}$$

Gaussian membership functions have the features that they are symmetrical, smooth, and non-zero at all points.

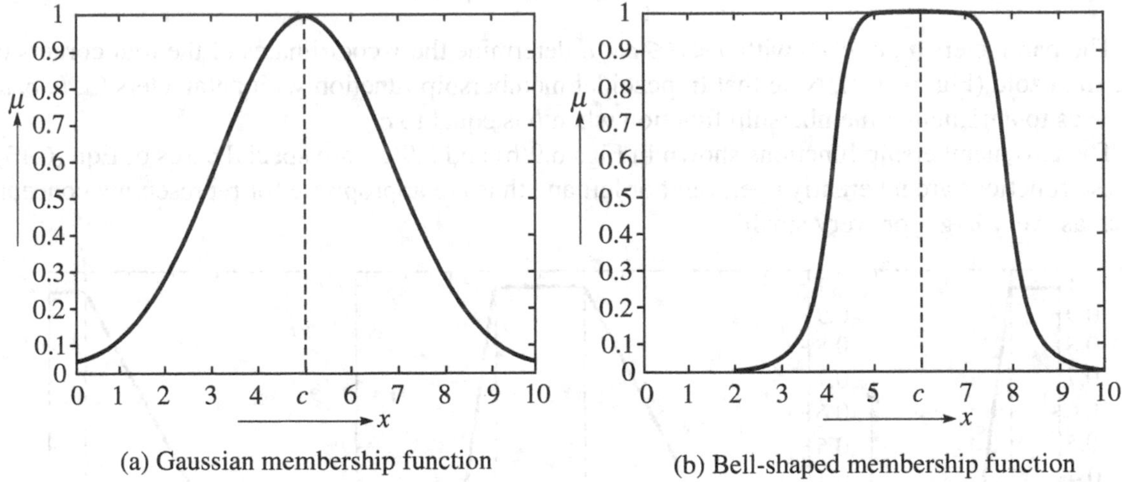

(a) Gaussian membership function

(b) Bell-shaped membership function

Figure 6.10

Bell-shaped: A general bell-shaped membership function is specified via three parameters (a, b, c), where the parameter b is generally positive (IF b is negative, the shape of the function becomes a reverse bell). Parameter c seeks the center of the curve and parameter a depicts how wide the curve is. Figure 6.10(b) shows a plot of generalized bell-shaped membership function, given by the expression:

$$\mu(x) = \frac{1}{1+\left|\frac{x-c}{a}\right|^{2b}} \tag{6.15}$$

Bell-shaped function given by Eqn (6.15) is symmetrical, smooth, and non-zero at all points.

Sigmoidal: A sigmoidal type membership function can generally be open to the right or to the left. The former type is commonly used as activation function in neural networks. The general expression of sigmoidal membership function is given as,

$$\mu(x) = \frac{1}{1 + e^{-a(x-c)}} \tag{6.16}$$

where c determines distance from the origin and a determines steepness of the function. If a is positive, the membership function is open to the right whereas if it is negative, it is open to the left. Figure 6.11(a) illustrates sigmoidal function open to the right.

Symmetrical or asymmetrical but closed (not open to the right or left) membership functions can be constructed by using either the difference or product of the two sigmoidal membership functions described above. The membership function formed by the difference between two sigmoidal membership functions is defined as *difference-sigmoidal*, and the one formed by the product of two sigmoidal functions is defined as *product-sigmoid*. These are shown in Figs 6.11(b) and 6.11(c), respectively.

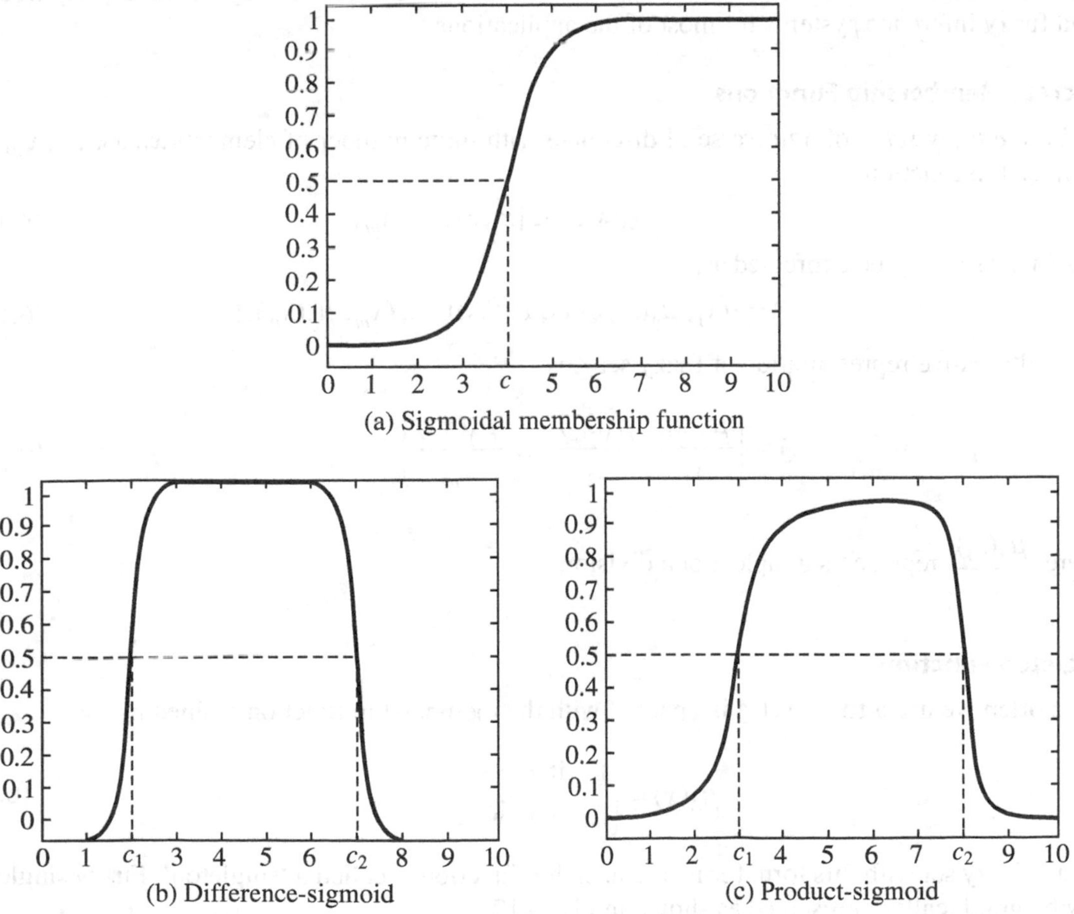

Figure 6.11

A sigmoidal function depends upon two parameters a and c:

$$f(x; a, c) = \frac{1}{1 + e^{-a(x-c)}}$$

Difference-sigmoid depends upon four parameters a_1, c_1, a_2 and c_2, and is the difference between two of these sigmoidal functions (Fig. 6.11(b)): $f_1(x; a_1, c_1,) - f_2(x; a_2, c_2)$.

Product-sigmoid again depends upon four parameters a_1, c_1, a_2 and c_2, and is the product of two of these sigmoidal functions (Fig. 6.11(c)): $f_1(x; a_1, c_1) \times f_2(x; a_2, c_2)$.

All the membership functions in sigmoid family are smooth and non-zero at all points.

The list of membership functions introduced above is by no means exhaustive; other specialized membership functions can be created for specific applications. Available software tools provide a wide selection to choose from when we are selecting a membership function for the application in hand. It may, however, be noted that exotic membership functions are by no means required for good fuzzy inference systems for most of the applications.

Discrete Membership Functions

A *discrete fuzzy set* $\underset{\sim}{A}$ of a universe of discourse with finite number of elements $x_1, x_2, \ldots, x_m$, has membership function

$$\mu_{\underset{\sim}{A}}(x);\ x \in \{x_1, x_2, \ldots, x_m\} \tag{6.17a}$$

This fuzzy set may be expressed as,

$$\underset{\sim}{A} = \{(x_1, \mu_{\underset{\sim}{A}}(x_1)), (x_2, \mu_{\underset{\sim}{A}}(x_2)), \ldots, (x_m, \mu_{\underset{\sim}{A}}(x_m))\} \tag{6.17b}$$

An alternative representation of fuzzy set $\underset{\sim}{A}$ is,

$$\underset{\sim}{A} = \left\{ \frac{\mu_{\underset{\sim}{A}}(x_1)}{x_1}, \frac{\mu_{\underset{\sim}{A}}(x_2)}{x_2}, \ldots, \frac{\mu_{\underset{\sim}{A}}(x_m)}{x_m} \right\} \tag{6.17c}$$

where $\dfrac{\mu_{\underset{\sim}{A}}(x_i)}{x_i}$ represents a tuple, not a division.

Singleton Function

Quite often we use a fuzzy set $\underset{\sim}{A}$ in space X with the membership function defined by

$$\mu_{\underset{\sim}{A}}(x) = \begin{cases} 1 & \text{if } x = x_0 \\ 0 & \text{otherwise} \end{cases} \tag{6.18}$$

Any fuzzy set with this form for its membership function is called a 'singleton'. Fuzzy singleton may be graphically represented as shown in Fig. 6.12.

Note that the discrete impulse function can be used to represent the singleton membership function. Basically, singleton is a fuzzy set representation of a crisp number x_0.

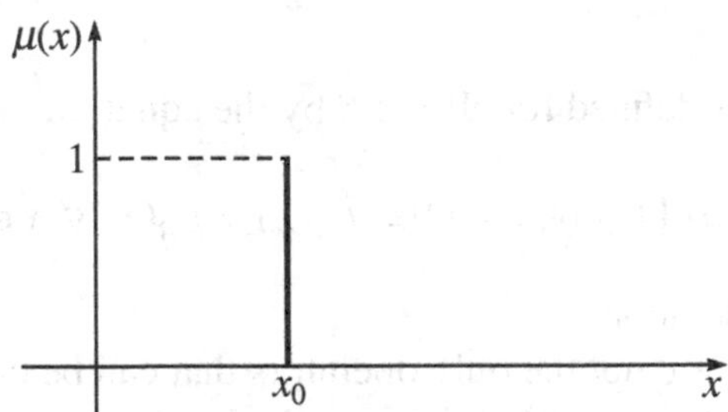

Figure 6.12 Fuzzy singleton

6.3.3 Fuzzy Set Operations

There are different kinds of fuzzy set theories differing from one another on the basis of the set operations (complement, intersection, union) they use. The fuzzy complement, intersection and union are not unique operations, unlike their crisp counterparts; different functions may be suitable to represent these operations in various contexts. Not only membership functions of fuzzy sets, but also, operations on fuzzy sets rely on the context. The ability to establish suitable membership functions and meaningful fuzzy operations in the context of each specific application, is essential to make fuzzy set theory really useful.

The intersection and union operations on fuzzy sets are often called *triangular norms* (*t*-norms), and *triangular conorms* (*t*-conorms; also called *s*-norms), respectively. The reader is advised to refer to [94] for the axioms which *t*-norms, *t*-conorms, and the complements of fuzzy sets are required to satisfy.

In the following, we define *standard fuzzy operations*, which are generalizations of the corresponding crisp set operations.

Consider the fuzzy sets $\underset{\sim}{A}$ and $\underset{\sim}{B}$ in the universe X.

$$\underset{\sim}{A} = \{(x, \mu_{\underset{\sim}{A}}(x)) \mid x \in X;\ \mu_{\underset{\sim}{A}}(x) \in [0,1]\} \tag{6.19a}$$

$$\underset{\sim}{B} = \{(x, \mu_{\underset{\sim}{B}}(x)) \mid x \in X;\ \mu_{\underset{\sim}{B}}(x) \in [0,1]\} \tag{6.19b}$$

The operations with $\underset{\sim}{A}$ and $\underset{\sim}{B}$ are introduced *via* operations on their membership functions $\mu_{\underset{\sim}{A}}(x)$ and $\mu_{\underset{\sim}{B}}(x)$ correspondingly.

Complement

The standard complement, $\bar{\underset{\sim}{A}}$, of fuzzy set $\underset{\sim}{A}$ with respect to the universal set X, is defined for all $x \in X$ by the equation

$$\mu_{\bar{\underset{\sim}{A}}}(x) \triangleq 1 - \mu_{\underset{\sim}{A}}(x)\ \forall\ x \in X \tag{6.20}$$

Intersection

The standard intersection, $\underset{\sim}{A} \cap \underset{\sim}{B}$, is defined for all $x \in X$ by the equation

$$\mu_{\underset{\sim}{A} \cap \underset{\sim}{B}}(x) \triangleq min\,[\mu_{\underset{\sim}{A}}(x), \mu_{\underset{\sim}{B}}(x)] \equiv \mu_{\underset{\sim}{A}}(x) \wedge \mu_{\underset{\sim}{B}}(x)\ \forall\ x \in X \tag{6.21}$$

where $\wedge$ indicates the **min** operation.

Union

The standard union, $\underset{\sim}{A} \cup \underset{\sim}{B}$, is defined for all $x \in X$ by the equation

$$\mu_{\underset{\sim}{A} \cup \underset{\sim}{B}}(x) \triangleq max\,[\mu_{\underset{\sim}{A}}(x), \mu_{\underset{\sim}{B}}(x)] \equiv \mu_{\underset{\sim}{A}}(x) \vee \mu_{\underset{\sim}{B}}(x)\ \forall\ x \in X \tag{6.22}$$

where $\vee$ indicates the **max** operation.

The *min* and *max* operators are not the only operators that can be chosen to model, respectively, the intersection and union of fuzzy sets, but they are the most commonly used ones in engineering applications. Popular alternatives to *min* and *max* operators are *algebraic product* and *algebraic sum*:

Algebraic Product

$$\mu_{\underset{\sim}{A}}(x) \wedge \mu_{\underset{\sim}{B}}(x) = \mu_{\underset{\sim}{A}}(x) \cdot \mu_{\underset{\sim}{B}}(x)\ \forall\ x \in X \tag{6.23}$$

Algebraic Sum

$$\mu_{\underset{\sim}{A}}(x) \vee \mu_{\underset{\sim}{B}}(x) = \mu_{\underset{\sim}{A}}(x) + \mu_{\underset{\sim}{B}}(x) - \mu_{\underset{\sim}{A}}(x)\ \mu_{\underset{\sim}{B}}(x)\ \forall\ x \in X \tag{6.24}$$

There are many more classes of *t*-norms for intersection and *t*-conorms for union which have not been listed in this Chapter [3, 94].

6.3.4 Fuzzy Relations

Consider two universes (crisp sets) X and Y. The *Cartesian product* of two sets X and Y (in this order) is a set of all ordered pairs such that the first element in each pair is a member of X and the second element is a member of Y. Formally,

$$X \times Y = \{(x, y);\ x \in X, y \in Y\} \tag{6.25}$$

where $X \times Y$ denotes the Cartesian product.

A *fuzzy relation* on $X \times Y$, denoted by $\underset{\sim}{R}$, or $\underset{\sim}{R}(X, Y)$, is defined as the set

$$\underset{\sim}{R} = \{((x, y), \mu_{\underset{\sim}{R}}(x, y)) \mid (x, y) \in X \times Y, \mu_{\underset{\sim}{R}}(x, y) \in [0, 1]\} \tag{6.26}$$

where $\mu_{\underset{\sim}{R}}(x, y)$ is a function of two variables, called membership function of the fuzzy relation. The degree of membership of the ordered pair (x, y) in $\underset{\sim}{R}$, associated with each pair (x, y) in $X \times Y$, is a real number in the interval [0, 1]. The degree of membership indicates the degree to which x is in relation with y. It is clear that a fuzzy relation is basically a fuzzy set.

If X and Y are discrete universes of discourse, a relation $\underset{\sim}{R}$ on $X \times Y$ can be presented in the form of a *relational matrix*, or graphically as a discrete set of points in a three-dimensional space $(x, y, \mu_{\underset{\sim}{R}}(x, y))$. When the universes of discourse are continuous sets comprising an infinite number of elements, the membership function $\mu_{\underset{\sim}{R}}(x, y)$ is a *surface* over the Cartesian product $X \times Y$, not a *curve* as in the case of one-dimensional fuzzy sets. A relation defined on the space $X \times Y$ is a *binary relation*.

A Cartesian product can be generalized for m universes of discourse $X_1, X_2, \ldots, X_m$. A relation defined on this space is given by

$$\underset{\sim}{R} = \{((x_1, x_2, \ldots, x_m), \mu_{\underset{\sim}{R}}(x_1, x_2, \ldots, x_m)) \mid (x_1, x_2, \ldots, x_m) \in X_1 \times X_2 \times \cdots \times X_m, \\ \mu_{\underset{\sim}{R}}(x_1, x_2, \ldots, x_m) \in [0, 1]\} \qquad (6.27)$$

Example 6.5

Let X and Y be two given sets

$$X = \{1, 2, 3\}, Y = \{2, 3, 4\}$$

The relation $\underset{\sim}{R}$: 'x is smaller than y', may be presented in the form of a relational matrix as follows:

$\underset{\sim}{R}$		y		
		2	3	4
x	1	1	1	1
	2	0	1	1
	3	0	0	1

The elements of the relational matrix are the degrees of membership $\mu_{\underset{\sim}{R}}(x, y)$, i.e., degrees of belonging of a specific pair (x, y) to the given relation $\underset{\sim}{R}$. For example, the pair (3, 2) belongs with a degree 0 to the relation 'x is smaller than y' or the possibility that 3 is smaller than 2 is zero. This relation is a typical example of a *crisp relation*. The condition involved is this relation is precise one that is either fulfilled or not fulfilled.

Example 6.6

Let us now consider a typical example of an imprecise relation $\underset{\sim}{R}$: 'x is approximately equal to y'. The degree of belonging $\mu_{\underset{\sim}{R}}(x, y)$ of pairs (x, y) from the Cartesian product $X \times Y$ to this relation $\underset{\sim}{R}$ can be any number between 0 and 1.

For continuous universes X and Y, given by [3]

$$X = \{x \in \Re \mid 1 \le x \le 3\}, Y = \{y \in \Re \mid 2 \le y \le 4\},$$

the membership function $\mu_{\underset{\sim}{R}}(x, y)$ of the relation '$x \simeq y$', is a surface over the Cartesian product $X \times Y$, shown in Fig. 6.13.

The relational matrix after discretization by a step of 0.5, is as follows:

$\underset{\sim}{R}$		y				
		2	2.5	3	3.5	4
x	1	0.67	0.5	0.33	0.17	0
	1.5	0.83	0.67	0.5	0.33	0.17
	2	1	0.83	0.67	0.5	0.33
	2.5	0.83	1	0.83	0.67	0.5
	3	0.67	0.83	1	0.83	0.67

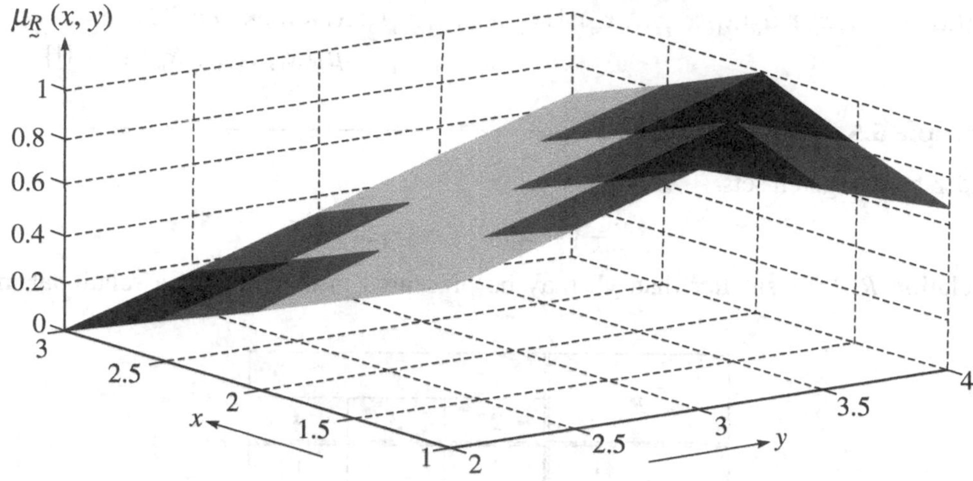

Figure 6.13 Membership function of the relation: "*x* is approximately equal to *y*".

Example 6.7

Consider an example of fuzzy set: the set of people with normal weight. In this case, the universe of discourse appears to be all potential weights (the real line). However, the knowledge representation in terms of this universe, is not useful. The normal weight of a person is a function of his/her height.

$$\text{Body Mass Index (BMI)} = \frac{\text{Weight, kg}}{(\text{Height, m})^2} \tag{6.28}$$

Males with BMI values that range from 20 to 25 are members of the *healthy set*. Males with BMI values greater than 27 or less than 18 are not members of the healthy set. BMI values in the range 18–20 and 25–27 represent a fuzzy situation.

Suppose we wish to express fuzzy relation $\underset{\sim}{R}$: 'healthy male adult', in terms of height and weight of a person. We consider the range of height (cm): {140, 145, 150, ..., 180}, and range of weight (kg): {45, 50, ..., 105}; denoted by x and y, respectively.

$$\begin{aligned} x \in X &= \{140, 145, \ldots, 180\} \\ y \in Y &= \{45, 50, \ldots, 105\} \end{aligned} \tag{6.29a}$$

Membership values for the set 'healthy' may be calculated from the following function:

$$\text{healthy}(x) = \begin{cases} 0 \;; & x < 18 \\ (x-18)/2 \;; & 18 \le x \le 20 \\ 1 \;; & 20 < x < 25 \\ (27-x)/2 \;; & 25 \le x \le 27 \\ 0 \;; & x > 27 \end{cases} \tag{6.29b}$$

A membership value of 1 is healthy; a membership value of 0 is not healthy; a membership value between 0 and 1 is the degree of membership in the healthy set.

A relational matrix $\underset{\sim}{R}(X, Y)$, obtained using Eqns (6.28–6.29), is given below (relational matrix (6.30)).

$\underset{\sim}{R}$		y (kg)												
		45	**50**	**55**	**60**	**65**	**70**	**75**	**80**	**85**	**90**	**95**	**100**	**105**
x (cm)	140	1.00	0.74	0.00	0.00	0.00	0.00	0.00	0.00	0.00	0.00	0.00	0.00	0.00
	145	1.00	1.00	0.42	0.00	0.00	0.00	0.00	0.00	0.00	0.00	0.00	0.00	0.00
	150	1.00	1.00	1.00	0.17	0.00	0.00	0.00	0.00	0.00	0.00	0.00	0.00	0.00
	155	0.37	1.00	1.00	1.00	0.00	0.00	0.00	0.00	0.00	0.00	0.00	0.00	0.00
	160	0.00	0.77	1.00	1.00	0.80	0.00	0.00	0.00	0.00	0.00	0.00	0.00	0.00
	165	0.00	0.18	1.00	1.00	1.00	0.64	0.00	0.00	0.00	0.00	0.00	0.00	0.00
	170	0.00	0.00	0.52	1.00	1.00	1.00	0.52	0.00	0.00	0.00	0.00	0.00	0.00
	175	0.00	0.00	0.00	0.80	1.00	1.00	1.00	0.44	0.00	0.00	0.00	0.00	0.00
	180	0.00	0.00	0.00	0.26	1.00	1.00	1.00	1.00	0.38	0.00	0.00	0.00	0.00

(6.30)

Each entry in the matrix, $\mu_{\underset{\sim}{R}}(x, y)$, indicates the degree a person with a corresponding height and weight is considered to be healthy. For instance, the entry corresponding to a height of 160 cm and a weight of 50 kg has a value 0.77, which is the degree to which a person will be considered a healthy person, i.e., $\mu_{\underset{\sim}{R}}(x, y) = 0.77$ for $x = 160$ cm and $y = 50$ kg.

The relational matrix in this example is actually a surface over the Cartesian product $X \times Y$ (Fig. 6.14).

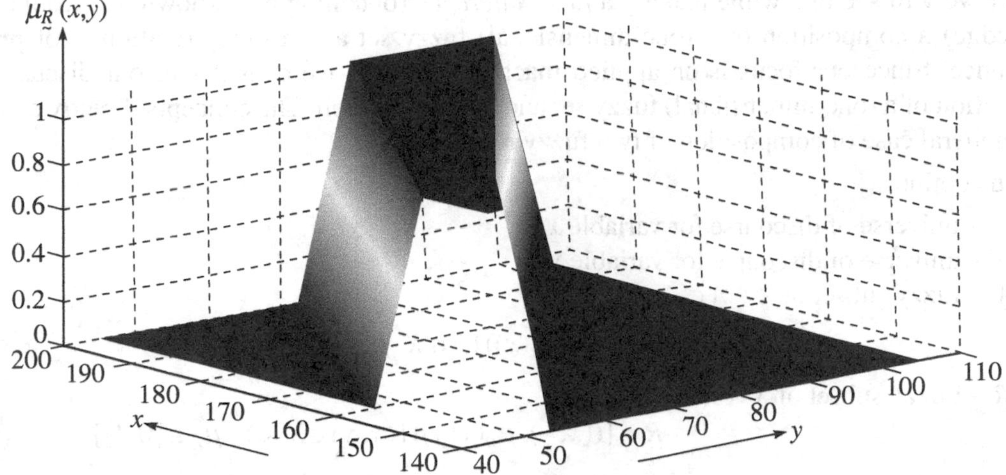

Figure 6.14 Membership functions $\mu_{\underset{\sim}{R}}(x, y)$ of the relation $\underset{\sim}{R}$: 'healthy male adult'; x is height in cm and y is weight in kg.

Composition of Fuzzy Relations

Fuzzy relations in different product spaces can be combined with each other by *composition*. Note that fuzzy sets can also be combined with fuzzy relations in the same way as relations are also fuzzy sets. A composition is also a fuzzy set because relations are fuzzy sets.

Let $\underset{\sim}{R}$ and $\underset{\sim}{S}$ be two binary fuzzy relations in product spaces $X \times Y$ and $Y \times Z$, respectively. The composition of the two relations $\underset{\sim}{R}$ and $\underset{\sim}{S}$ is a relation in product space $X \times Z$, denoted as $\underset{\sim}{R} \circ \underset{\sim}{S}$ (the description that follows can easily be extended to relations defined on Cartesian product of m universes; $m > 2$).

Composition of $\underset{\sim}{R}$ and $\underset{\sim}{S}$ is the result of three operations:

- Extending each relation so that their dimensions are identical
- Intersection of the two extended relations
- Projecting the intersection to the dimensions not shared by the two original relations

The *extension* operation extends the dimensions of $\underset{\sim}{R}$ and $\underset{\sim}{S}$ to $X \times Y \times Z$. The *projection* operation projects the *intersection* of the two extended relations in product space $X \times Y \times Z$ to $X \times Z$. Extension and projection are dual operations. The former extends the dimension of a fuzzy relation while the latter reduces the dimension of a fuzzy relation.

Typically, we apply *cylindrical extension* to the two fuzzy relations $\underset{\sim}{R}$ and $\underset{\sim}{S}$ so that they have the same dimensionality in order to apply set operation: intersection (in sub-section 6.3.3, we described set operations on fuzzy sets defined in the same universe of discourse). The projection operation, as the name implies, projects a fuzzy relation to a subset of selected dimensions. This operation is often used to extract the possibility distribution (membership functions) of a few selected variables from a given fuzzy relation.

A mathematically rigorous treatment of extension and projection requires *extension principle* of fuzzy set theory, which is beyond the scope of this book. We will instead give basic description of characterization of the principle in terms of cylindrical extension and projection, with respect to our focus on applied machine learning.

Later we will see that while making a *fuzzy inference* (obtaining new knowledge from existing knowledge) a composition of a (one-dimensional) fuzzy set and a fuzzy relation is of practical importance. Since our focus is on applied machine learning, we consider in our discussion the composition of a (one-dimensional) fuzzy set and a fuzzy relation. The concepts directly carry over to the general case of composition of two fuzzy relations.

Let us define

- X – universe of discourse for variable x
- Y – universe of discourse for variable y
- $\underset{\sim}{A}$ – fuzzy subset of X ; $\underset{\sim}{A} \subseteq X$

$$\underset{\sim}{A} = \{(x, \mu_{\underset{\sim}{A}}(x)) \mid x \in X, \mu_{\underset{\sim}{A}} \in [0, 1]\} \tag{6.31a}$$

- $\underset{\sim}{R}$ – Fuzzy subset of $X \times Y$

$$\underset{\sim}{R} = \{((x, y), \mu_{\underset{\sim}{R}}(x, y)) \mid (x, y) \in X \times Y, \mu_{\underset{\sim}{R}} \in [0, 1]\} \tag{6.31b}$$

$\underset{\sim}{A}$ is a fuzzy set in space X, and $\underset{\sim}{R}$ is a fuzzy relation in product space $X \times Y$. The composition of the fuzzy set $\underset{\sim}{A}(X)$ and relation $\underset{\sim}{R}(X, Y)$ is a fuzzy set in space Y. The composition operation can be viewed as a combination of cylindrical extension, intersection, and projection of fuzzy sets.

- Cylindrical extension is a step for extending fuzzy set $\underset{\sim}{A}(X)$ to the product space $X \times Y$. This step enables us to take the intersection of $\underset{\sim}{A}$ and $\underset{\sim}{R}$ (intersection operation requires both $\underset{\sim}{A}$ and $\underset{\sim}{R}$ to be defined on the same space; Section 6.3.3).
- Intersection of the extended relation (fuzzy set) with $\underset{\sim}{R}$ in the product space $X \times Y$, for matching $\underset{\sim}{A}$ with $\underset{\sim}{R}$.
- Projection of the intersection on space Y; this results in a fuzzy set $\underset{\sim}{B}$ in space Y, *induced* by the composition.

Cylindrical Extension: The cylindrical extension of $\underset{\sim}{A}$ (a fuzzy set defined on X) in $X \times Y$, denoted $ce\underset{\sim}{A}$, is the set of all tuples $(x, y) \in X \times Y$, with membership function equal to $\mu_{\underset{\sim}{A}}(x)$, i.e.,

$$\mu_{ce\underset{\sim}{A}}(x, y) = \mu_{\underset{\sim}{A}}(x) \text{ for every } y \in Y \tag{6.32}$$

Figure 6.15 shows graphically cylindrical extension of a fuzzy set $\underset{\sim}{A}$ in X to a fuzzy relation $ce\underset{\sim}{A}$ in space $X \times Y$.

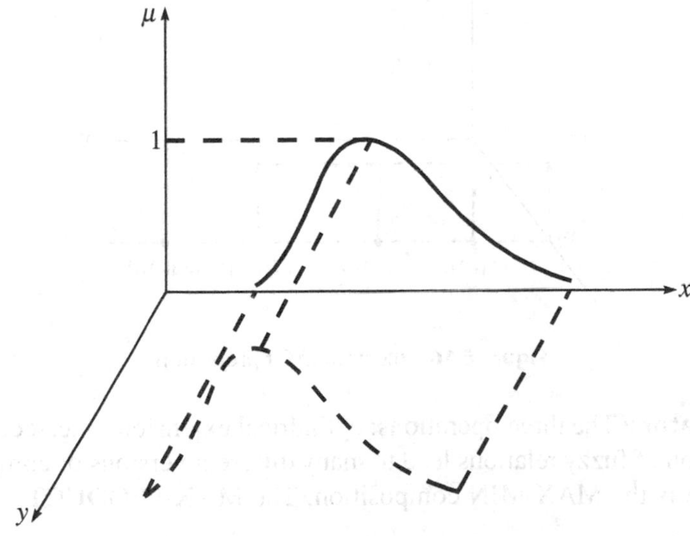

Figure 6.15 Illustration for cylindrical extension

Intersection: The intersection of the fuzzy sets (relations)

$ce\underset{\sim}{A}$ and $\underset{\sim}{R}$ in $X \times Y$ is given by the following equations.

Min-Operator

$$\begin{aligned}\mu_{ce\underset{\sim}{A} \cap \underset{\sim}{R}}(x, y) &= \mu_{\underset{\sim}{T}}(x, y) = \mu_{ce\underset{\sim}{A}}(x, y) \wedge \mu_{\underset{\sim}{R}}(x, y)\\ &= min\{\mu_{ce\underset{\sim}{A}}(x, y), \mu_{\underset{\sim}{R}}(x, y)\}\\ &= min\{\mu_{\underset{\sim}{A}}(x), \mu_{\underset{\sim}{R}}(x, y)\}\ \forall\, x \in X, \forall\, y \in Y\end{aligned} \tag{6.33a}$$

Product operator

$$\mu_{ce\underset{\sim}{A}\cap\underset{\sim}{R}}(x,y) = \mu_{\underset{\sim}{T}}(x,y) = \mu_{ce\underset{\sim}{A}}(x,y) \wedge \mu_{\underset{\sim}{R}}(x,y)$$
$$= \{\mu_{ce\underset{\sim}{A}}(x,y)\, \mu_{\underset{\sim}{R}}(x,y)\}$$
$$= \{\mu_{\underset{\sim}{A}}(x)\,[\mu_{\underset{\sim}{R}}(x,y)]\}\ \forall\, x \in X, \forall\, y \in Y \tag{6.33b}$$

Projection: The projection of $\underset{\sim}{T}$ (a fuzzy set defined on $(X \times Y)$) on Y, denoted *proj* $\underset{\sim}{T}$, is a set of all $y \in Y$ with membership grade equal to $\underset{x}{max}\{\mu_{\underset{\sim}{T}}(x,y)\}$; '$\underset{x}{max}$' represents *maximum* with respect to x while y is considered fixed, i.e.,

$$\mu_{proj\,\underset{\sim}{T}}(y) = \underset{x}{max}\{\mu_{\underset{\sim}{T}}(x,y)\} \tag{6.34}$$

Projection on Y means y_0 is assigned the highest membership degree from the tuples (x_1, y_0), (x_2, y_0), …; $x_1, x_2, \ldots \in X$, and $y_0 \in Y$. The rational for using the *max* operator on the membership functions of $\underset{\sim}{T}$ comes from the extension principle (many-to-one mapping).

Figure 6.16 graphically illustrates the projection step.

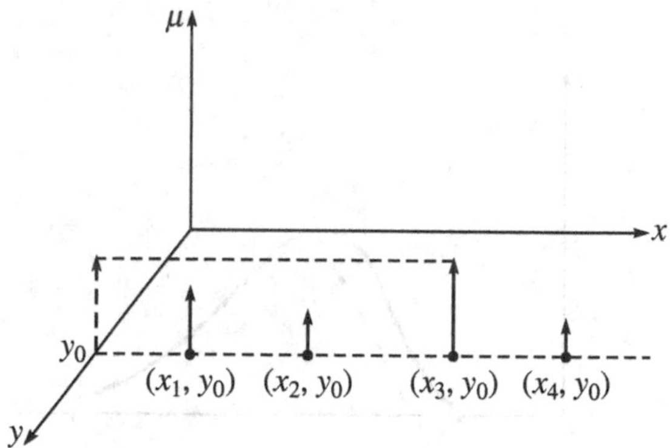

Figure 6.16 Illustration for projection

Composition Operators: The three operations: cylindrical extension, intersection, and projection, involved in imposition of fuzzy relations lead to many different versions of compositional operator. The best known one is the MAX-MIN composition. The MAX-PRODUCT composition is often the best alternative.

Let $\underset{\sim}{A}$ be a fuzzy set in space X, and $\underset{\sim}{R}$ be a fuzzy relation in product space $X \times Y$.

$$\underset{\sim}{A} = \{(x, \mu_{\underset{\sim}{A}}(x)) \mid x \in X, \mu_{\underset{\sim}{A}}(x) \in [0,1]\} \tag{6.35a}$$

$$\underset{\sim}{R} = \{((x,y), \mu_{\underset{\sim}{R}}(x,y)) \mid (x,y) \in X \times Y, \mu_{\underset{\sim}{R}}(x,y) \in [0,1]\} \tag{6.35b}$$

Max-Min Composition

The MAX-MIN composition $\underset{\sim}{A} \circ \underset{\sim}{R}$ is given by the following equation.

$$\underset{\sim}{A} \circ \underset{\sim}{R} = \{y, \underset{x}{max}[min(\mu_{\underset{\sim}{A}}(x), \mu_{\underset{\sim}{R}}(x,y))] \mid x \in X, y \in Y\} \tag{6.36a}$$

Max-Product Composition

The MAX-PRODUCT composition is given by the following equation.

$$\underset{\sim}{A} \circ \underset{\sim}{R} = \{(y, \max_{x} [\mu_{\underset{\sim}{A}}(x)\mu_{\underset{\sim}{R}}(x, y)]) \mid x \in X, y \in Y\} \tag{6.36b}$$

Example 6.8

This example will facilitate the understanding of three steps: cylindrical extension, intersection, and projection, involved in composition operation.

We wish to know the possible weight of a healthy person who is about 160 cm tall. In Example 6.7, we obtained fuzzy relation $\underset{\sim}{R}$: 'healthy male adult', in terms of height x and weight y of a person. For the ranges, X and Y, of height and weight given in Eqn (6.29a), a relational matrix $\underset{\sim}{R}(X, Y)$ was obtained, giving $\mu_{\underset{\sim}{R}}(x, y)\ \forall\, x \in X$ and $\forall\, y \in Y$. This relational matrix is given in Eqn (6.30).

Assume 'about 160 cm' is a discrete fuzzy set $\underset{\sim}{A}$ in X, defined as,

$$\left[\frac{0}{140}, \frac{0}{145}, \frac{0.4}{150}, \frac{0.8}{155}, \frac{1}{160}, \frac{0.8}{165}, \frac{0.4}{170}, \frac{0}{175}, \frac{0}{180}\right] \tag{6.37a}$$

The 'healthy male adult' fuzzy relation $\underset{\sim}{R}$ on $X \times Y$ product space is given by the matrix in (6.30). Cylindrical extension $ce\underset{\sim}{A}$ of $\underset{\sim}{A}$ in X to the product space $X \times Y$ (relational matrix (6.37b)):

x \ y	45	50	55	60	65	70	75	80	85	90	95	100	105
140	0	0	0	0	0	0	0	0	0	0	0	0	0
145	0	0	0	0	0	0	0	0	0	0	0	0	0
150	0.4	0.4	0.4	0.4	0.4	0.4	0.4	0.4	0.4	0.4	0.4	0.4	0.4
155	0.8	0.8	0.8	0.8	0.8	0.8	0.8	0.8	0.8	0.8	0.8	0.8	0.8
160	1	1	1	1	1	1	1	1	1	1	1	1	1
165	0.8	0.8	0.8	0.8	0.8	0.8	0.8	0.8	0.8	0.8	0.8	0.8	0.8
170	0.4	0.4	0.4	0.4	0.4	0.4	0.4	0.4	0.4	0.4	0.4	0.4	0.4
175	0	0	0	0	0	0	0	0	0	0	0	0	0
180	0	0	0	0	0	0	0	0	0	0	0	0	0

(6.37b)

The two fuzzy relations $ce\underset{\sim}{A}(x, y)$ and $\underset{\sim}{R}(x, y)$ are defined on the same product space $X \times Y$. The intersection of the two relations gives

$$ce\underset{\sim}{A} \cap \underset{\sim}{R} = \mu_{ce\underset{\sim}{A}}(x, y) \wedge \mu_{\underset{\sim}{R}}(x, y)$$
$$= min\,[\mu_{ce\underset{\sim}{A}}(x, y), \mu_{\underset{\sim}{R}}(x, y)]$$

The result is given in the matrix in (6.38a). We have used 'min' operator for the intersection of two fuzzy sets. As said earlier, a popular alternative is 'product' operator.

x \ y	45	50	55	60	65	70	75	80	85	90	95	100	105
140	0	0	0	0	0	0	0	0	0	0	0	0	0
145	0	0	0	0	0	0	0	0	0	0	0	0	0
150	0.4	0.4	0.4	0.17	0	0	0	0	0	0	0	0	0
155	0.37	0.8	0.8	0.8	0	0	0	0	0	0	0	0	0
160	0	0.77	1	1	0.8	0	0	0	0	0	0	0	0
165	0	0.18	0.8	0.8	0.8	0.64	0	0	0	0	0	0	0
170	0	0	0.4	0.4	0.4	0.4	0.4	0	0	0	0	0	0
175	0	0	0	0	0	0	0	0	0	0	0	0	0
180	0	0	0	0	0	0	0	0	0	0	0	0	0

(6.38a)

The projected fuzzy set gives the weight possibility distribution of a healthy male adult who is about 160 cm tall. The result is given in (6.38b).

$$\left[\frac{0.4}{45}, \frac{0.8}{50}, \frac{1}{55}, \frac{1}{60}, \frac{0.8}{65}, \frac{0.64}{70}, \frac{0.4}{75}, \frac{0}{80}, \frac{0}{85}, \frac{0}{90}, \frac{0}{95}, \frac{0}{100}, \frac{0}{105}\right] \tag{6.38b}$$

If we know the exact measured height of a male adult, say, it is 160 cm, then we should first transform the 'crisp' measurement of height into a discrete fuzzy set. It is a *fuzzification* process that converts a crisp value into a fuzzy set.

Height x is a singleton at $x_0 = 160$; therefore (Eqn (6.18), Fig. 6.12),

$$\mu_{\underset{\sim}{A}}(x) = \begin{cases} 1 & \text{if } x = 160 \\ 0 & \text{otherwise} \end{cases} \tag{6.39a}$$

This gives

$$\underset{\sim}{A} = \left[\frac{0}{140}, \frac{0}{145}, \frac{0}{150}, \frac{0}{155}, \frac{1}{160}, \frac{0}{165}, \frac{0}{170}, \frac{0}{175}, \frac{0}{180}\right] \tag{6.39b}$$

The possibility weight distribution for the male adult to be a 'healthy male adult', can now be calculated.

6.4 FUZZY RULE-BASE AND APPROXIMATE REASONING

Problems featuring uncertainty and ambiguity have been successfully addressed subconsciously by humans. Humans can adapt to unfamiliar situations and they are able to gather information in an efficient manner and discard irrelevant details. The information gathered need not be complete and precise, and could be general, qualitative and vague, because humans can reason, infer and deduce new information and knowledge. They can learn, perceive and improve their skills through experience.

Is there a way for humans to reason about complex systems when the entire description of a system of this type frequently demands more thorough information than is possible for a human to recognize simultaneously, and assimilate with understanding? The answer is 'yes'. Humans are capable of *reasoning approximately*. While reasoning about a complicated system, humans reason approximately about its behavior, thus preserving just a generic understanding of the problem.

The fuzzy set theory provides a mathematical power to capture the uncertainties related to human cognitive processes, for instance, thinking and reasoning. Fuzzy logic offers an *inference* morphology that enables approximate human reasoning capabilities to knowledge-based systems.

Knowledge representation and processing are the keys to any intelligent system. In *logic*, knowledge is represented by *propositions* and is processed through *reasoning* by the application of various laws of logic. *Fuzzy logic* focuses on linguistic variables in natural language, and aims to provide foundations for *approximate reasoning* with *imprecise propositions*.

In the previous sections, we provided an intuitive explanation of many of the concepts related to fuzzy logic. The objective set for this section is to provide a complete exposition on the details of the operation of fuzzy systems. We will see that the knowledge-base of a fuzzy system consists of *fuzzy rule-base*, and *database*—membership functions of the fuzzy sets used in fuzzy rules. *Approximate reasoning*, which can be viewed as a process by which a possible imprecise conclusion is deduced from the collection of imprecise rule-base, will be performed through an *inference mechanism*.

In this section, we will first describe the concepts in terms of general fuzzy systems, followed by examples as tutorial illustrations.

Consider a fuzzy system with n inputs x_j; $j = 1, 2, \ldots, n$, and output y. The ordinary ('crisp') sets X_j and Y are universes of discourse for the crisp inputs x_j and crisp output y, respectively. To specify rules for the rule-base, the expert uses 'linguistic' variables for inputs and output, and their characteristics. For our fuzzy systems, linguistic variables denoted by $\underset{\sim}{x}_j$ are used to describe the inputs x_j. Similarly, linguistic variable denoted by $\underset{\sim}{y}$ is used to describe output y.

Just as x_j and y take on values over each universe of discourse X_j and Y, respectively, linguistic variables $\underset{\sim}{x}_j$ and $\underset{\sim}{y}$ take on 'linguistic values' that are used to describe characteristics of the variables. We assume that there exist k_j linguistic values for the linguistic variable $\underset{\sim}{x}_j$; $k_j = 1, 2, \ldots, K_j$. Then the linguistic variable $\underset{\sim}{x}_j$ takes on the elements from the set of linguistic values denoted by

$$\underset{\sim}{A}_j = \{\underset{\sim}{A}_{jk_j} \,;\, j = 1, 2, \ldots, n;\, k_j = 1, 2, \ldots, K_j\} \tag{6.40a}$$

Similarly, let $\underset{\sim}{B}_k$ denote the kth linguistic value of the linguistic variable $\underset{\sim}{y}$; then $\underset{\sim}{y}$ takes on elements from the set of linguistic values

$$\underset{\sim}{B} = \{ \underset{\sim}{B}_k;\ k = 1, 2, ..., K\} \tag{6.40b}$$

The mapping of the inputs to the output in a fuzzy system is characterized by fuzzy IF-THEN rules. The inputs $\underset{\sim}{x}_j$ of the fuzzy system are associated with the premise, and the output $\underset{\sim}{y}$ is associated with the consequent. The form of a linguistic rule is

$$\text{IF } \underset{\sim}{x}_1 \text{ is } \underset{\sim}{A}_{1k_1}, \text{ AND } \underset{\sim}{x}_2 \text{ is } \underset{\sim}{A}_{2k_2} \text{ AND ... AND } \underset{\sim}{x}_n \text{ is } \underset{\sim}{A}_{nk_n} \text{ THEN } \underset{\sim}{y} \text{ is } \underset{\sim}{B}_k; \tag{6.41}$$

$k_1 = 1, ..., K_1;\ k_2 = 1, ..., K_2, ..., k_n = 1, ..., K_n;\ k = 1, ..., K.$

This rule has n propositions connected by logical connectives. Quite often, in engineering applications, we come across logical connective AND (conjunction operation). Our rule example has been taken with AND connective; however, the underlying concepts are valid for OR connectives (disjunction operation) as well.

Note that we allow for the case where the expert does not use all the linguistic terms (and hence, the fuzzy sets that characterize them) to state some rules. Stated equivalently, some premise terms in (6.41) may be missing; there does not need to be a premise term for each input in each rule, although often there is.

Consider now a fuzzy system with one input $\underset{\sim}{x}$ and one output $\underset{\sim}{y}$; X and Y are universes of discourse for x and y, respectively. We assume that there are K_A linguistic values (fuzzy subsets) for the linguistic variable $\underset{\sim}{x}$, and $\underset{\sim}{A}$ represents an element from this set. Similarly, we assume that there are K_B linguistic values (fuzzy subsets) for the linguistic variable $\underset{\sim}{y}$, and $\underset{\sim}{B}$ represents an element from this set. The form of the linguistic rule is

$$\text{IF } \underset{\sim}{x} \text{ is } \underset{\sim}{A} \text{ THEN } \underset{\sim}{y} \text{ is } \underset{\sim}{B} \tag{6.42a}$$

A fuzzy rule can be interpreted as a *fuzzy implication*

$$\underset{\sim}{A} \rightarrow \underset{\sim}{B} \tag{6.42b}$$

In rule (6.42a), the premise is an *atomic proposition* Atomic fuzzy propositions do not, usually make a knowledge-base in real-life situations. Many propositions connected by logical connectives may be needed. A set of such *compound propositions* connected by IF-THEN rules, makes a knowledge-base.

Rule (6.41) is a general form for compound propositions connected by logical AND connectives. Compound propositions connected by logical OR connectives, may be dealt with on similar lines.

In Section 6.3, we defined fuzzy operations; conjunction (AND), disjunction (OR) on fuzzy sets that lie in the same universe of discourse. In fuzzy rules of the form (6.41), these operations are to be performed on fuzzy sets that lie in different universes of discourse. The *fuzzy Cartesian product* is used to quantify operations on many universes of discourse. If $\underset{\sim}{A}_{1k_1}, \underset{\sim}{A}_{2k_2},..., \underset{\sim}{A}_{nk_n}$ are fuzzy sets defined on the universes of discourse $X_1, X_2, ..., X_n$, respectively, their Cartesian product, denoted by $\underset{\sim}{A}_{1k_1} \times \underset{\sim}{A}_{2k_2} \times \cdots \times \underset{\sim}{A}_{nk_n}$ is a fuzzy set specified by relation $\underset{\sim}{R}(X_1, X_2, ..., X_n)$. The fuzzy implication is then

$$\underset{\sim}{A}_{1k_1} \times \underset{\sim}{A}_{2k_2} \times \cdots \times \underset{\sim}{A}_{nk_n} \rightarrow \underset{\sim}{B}_k \tag{6.43}$$

This represents a mapping from n-dimensional product space $X_1 \times X_2 \times \cdots \times X_n$ to a single universe Y.

It is an entire set of rules of the form (6.41) that the expert specifies for a *fuzzy inference system*. We assume that there are a total of R rules in the *rule-base* numbered $r = 1, 2, \ldots, R$; and we naturally assume that the rules in the rule-base are distinct (i.e., there are no two rules exactly with the same premise and consequent).

We note that if all the premise terms are present in every rule and a rule is formed for each possible combination of premise elements, then there are

$$\prod_{j=1}^{n} K_j = K_1 \times K_2 \times \cdots \times K_n \tag{6.44}$$

rules in the rule-base. Clearly, the number of rules increases exponentially with an increase in the number of fuzzy system inputs or membership functions.

Given a linguistic variable $\underset{\sim}{x}_j$ with a linguistic value $\underset{\sim}{A}_{jk_j}$ defined on the universe of discourse X_j. The function $\mu(\underset{\sim}{x}_j)$ associated with $\underset{\sim}{A}_{jk_j}$ that maps X_j to [0, 1] is the membership function. This membership function describes the certainty that an element of X_j, denoted x_j, with a linguistic variable $\underset{\sim}{x}_j$ may be classified linguistically as $\underset{\sim}{A}_{jk_j}$.

In humans reasoning, our rules express cause-effect relations; fuzzy logic is a tool for transferring such structured knowledge into workable algorithms. Figure 6.17 expresses the cause-effect relationship for fuzzy logic algorithms.

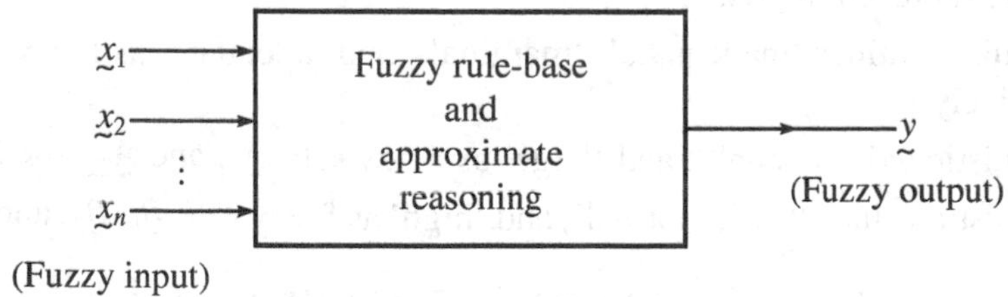

Figure 6.17 Cause-effect relationship

A fuzzy logic algorithm works on the existing imprecise knowledge available in the form of fuzzy rule-base. The cause (input) is fuzzy, and matching of the fuzzy rule-base with fuzzy input yields the effect (output), which is also fuzzy.

In real-world applications, we are usually given the crisp inputs and we require crisp outputs for decision-making. Therefore, the given crisp inputs are *fuzzified* to create knowledge in appropriate form for the fuzzy logic algorithm. Also the fuzzy outputs of the algorithm are *defuzzified* to create crisp output for the user.

From our discussion in Section 6.3, it follows that for the fuzzy logic algorithm to perform matching operation, the knowledge expressed by IF-THEN rules should first be quantified via fuzzy relations. Only after the appropriate fuzzy relation of the rules is calculated, can some *inference* (based on approximate reasoning) to obtain new knowledge (output) from the existing knowledge (rule-base and input), take place.

In this section, we will describe the required operations: quantification of fuzzy rules via fuzzy relations, fuzzification of input, inference mechanism, and defuzzification of inferred fuzzy sets.

Example 6.9

We examine here the rule-base of a simple (toy) two-input, one-output problem that has three rules. The objective is to use this problem in this section for the purpose of illustrating the concepts involved in design and basic mechanics of fuzzy decision-making systems [5].

Rule-Base

Rule 1: IF *project_funding* is *adequate* OR *project_staffing* is *small* THEN *risk* is *low*

Rule 2: IF *project_funding* is *marginal* AND *project_staffing* is *large* THEN *risk* is *normal*

Rule 3: IF *project_funding* is *inadequate* THEN *risk* is *high* (6.45a)

Let us define

- X_1 : Universe of discourse for linguistic variable 'project funding'
- X_2 : Universe of discourse for linguistic variable 'project staffing'
- Y : Universe of discourse for linguistic variable 'risk'
- $x_1 \in X_1$, variable representing project funding
- $x_2 \in X_2$, variable representing project staffing
- $y \in Y$ variable representing risk
- Denote linguistic values 'inadequate', 'marginal', and 'adequate' as fuzzy sets $\underset{\sim}{A}_{11}$, $\underset{\sim}{A}_{12}$, and $\underset{\sim}{A}_{13}$, respectively
- Denote linguistic values 'small', and 'large' as fuzzy sets $\underset{\sim}{A}_{21}$ and $\underset{\sim}{A}_{22}$, respectively
- Denote linguistic values 'low', 'normal', and 'high' as fuzzy sets $\underset{\sim}{B}_1$, $\underset{\sim}{B}_2$, and $\underset{\sim}{B}_3$, respectively

$$\begin{aligned}
\underset{\sim}{A}_{1k_1} &= \{(x_1, \mu_{\underset{\sim}{A}_{1k_1}}(x_1)) \mid x_1 \in X_1, \mu_{\underset{\sim}{A}_{1k_1}} \in [0,1]\};\ k_1 = 1, 2, 3 \\
\underset{\sim}{A}_{2k_2} &= \{(x_2, \mu_{\underset{\sim}{A}_{2k_2}}(x_2)) \mid x_2 \in X_2, \mu_{\underset{\sim}{A}_{2k_2}} \in [0,1]\};\ k_2 = 1, 2 \\
\underset{\sim}{B}_{k} &= \{(y, \mu_{\underset{\sim}{B}_{k}}(y)) \mid y \in Y, \mu_{\underset{\sim}{B}_{k}} \in [0,1]\};\ k = 1, 2, 3
\end{aligned} \tag{6.45b}$$

The rule-base for the problem in terms of the defined variables and fuzzy sets, becomes:

$$\begin{aligned}
&\text{Rule 1: IF } \underset{\sim}{x}_1 \text{ is } \underset{\sim}{A}_{13} \text{ OR } \underset{\sim}{x}_2 \text{ is } \underset{\sim}{A}_{21} \text{ THEN } \underset{\sim}{y} \text{ is } \underset{\sim}{B}_1 \\
&\text{Rule 2: IF } \underset{\sim}{x}_1 \text{ is } \underset{\sim}{A}_{12} \text{ AND } \underset{\sim}{x}_2 \text{ is } \underset{\sim}{A}_{22} \text{ THEN } \underset{\sim}{y} \text{ is } \underset{\sim}{B}_2 \\
&\text{Rule 3: IF } \underset{\sim}{x}_1 \text{ is } \underset{\sim}{A}_{11} \text{ THEN } \underset{\sim}{y} \text{ is } \underset{\sim}{B}_3
\end{aligned} \tag{6.45c}$$

The membership functions for all the 8 linguistic values (three for input $\underset{\sim}{x}_1$, two for input $\underset{\sim}{x}_2$, and three for the output $\underset{\sim}{y}$) are shown in Fig. 6.18. The ranges of the universes of discourse X_1, X_2, and Y have been taken as 0% to 100%.

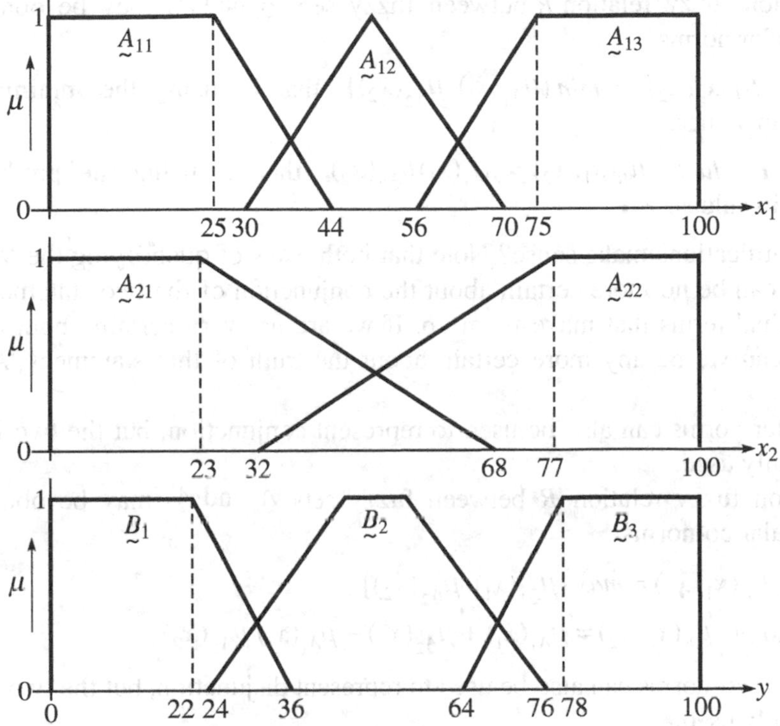

Figure 6.18 Membership functions for the rule base (6.45c)

6.4.1 Quantification of Rules via Fuzzy Relations

A fuzzy logic algorithm based on approximate reasoning works on the existing imprecise knowledge available in the form of fuzzy rule-base. The algorithm obtains new knowledge (imprecise) from the existing knowledge through some inference mechanism. To perform inference, it requires quantification of fuzzy rules. To do this, we first quantify the meaning of the premises of the rules that are composed of several terms. This is followed by quantification of the fuzzy implications.

Premise Quantification: Here, we examine the ways of quantifying the logical 'AND' and 'OR' operations that combine the meaning of linguistic terms in the premise of a rule, i.e., we examine *fuzzy conjunction* and *fuzzy disjunction* functions defined on fuzzy sets that lie in different universes of discourse. The *fuzzy Cartesian product* is used to quantify operations on many universes of discourse.

The Cartesian product of two crisp sets X_1 and X_2 is denoted as,

$$X_1 \times X_2 = \{(x_1, x_2) | \; x_1 \in X_1 \text{ and } x_2 \in X_2\} \tag{6.46}$$

The Cartesian product of two fuzzy sets $\underset{\sim}{A}_1$ and $\underset{\sim}{A}_2$ that lie in universes X_1 and X_2, respectively, with membership functions, $\mu_{\underset{\sim}{A}_1}(x_1)$ and $\mu_{\underset{\sim}{A}_2}(x_2)$, is denoted by $\underset{\sim}{A}_1 \times \underset{\sim}{A}_2$. The conjunction or disjunction operation on these fuzzy sets results in a *fuzzy relation* $\underset{\sim}{R}$, which itself is a fuzzy set with membership function $\mu_{\underset{\sim}{R}}(x_1, x_2)$.

The conjunction fuzzy relation $\underset{\sim}{R}$ between fuzzy sets $\underset{\sim}{A}_1$ and $\underset{\sim}{A}_2$ may be obtained using the following triangular norms.

- *Minimum*: $\mu_{\underset{\sim}{R}}(x_1, x_2) = min\{\mu_{\underset{\sim}{A}_1}(x_1), \mu_{\underset{\sim}{A}_2}(x_2)\}$, that is, using the minimum of the two membership values. (6.47a)
- *Algebraic Product*: $\mu_{\underset{\sim}{R}}(x_1, x_2) = \mu_{\underset{\sim}{A}_1}(x_1)\mu_{\underset{\sim}{A}_2}(x_2)$, that is, using the product of the two membership values. (6.47b)

Do these quantifications make sense? Note that both ways of quantifying the 'AND' operation indicate that we can be no more certain about the conjunction of the two statements than we are about the individual terms that make them up. If we are not very certain about the truth of one statement, how can we be any more certain about the truth of that statement 'AND' the other statement?

Other triangular norms can also be used to represent conjunction, but the two listed above are the most commonly used.

The disjunction fuzzy relation $\underset{\sim}{R}$ between fuzzy sets $\underset{\sim}{A}_1$ and $\underset{\sim}{A}_2$ may be obtained using the following triangular co-norms.

- *Maximum*: $\mu_{\underset{\sim}{R}}(x_1, x_2) = max\{\mu_{\underset{\sim}{A}_1}(x_1), \mu_{\underset{\sim}{A}_2}(x_2)\}$ (6.48a)
- *Algebraic sum*: $\mu_{\underset{\sim}{R}}(x_1, x_2) = \mu_{\underset{\sim}{A}_1}(x_1) + \mu_{\underset{\sim}{A}_2}(x_2) - \mu_{\underset{\sim}{A}_1}(x_1)\,\mu_{\underset{\sim}{A}_2}(x_2)$ (6.48b)

Other triangular co-norms can also be used to represent disjunction, but the two listed above are the most commonly used.

We have earlier listed the commonly used triangular norms and triangular co-norms in Section 6.3.3 for the *intersection* and *union* functions defined on fuzzy sets that lie in the same universe of discourse. Triangular norms are employed for intersection while triangular co-norms for union.

The set-theoretic operations such as intersection or union applied to fuzzy sets defined in different domains result in a multi-dimensional fuzzy set in the Cartesian product of these domains. The operation is, in fact, performed by first extending (cylindrical extension (Section 6.3.4)) the original fuzzy sets into Cartesian product domain and then computing the operation on these multi-dimensional fuzzy sets. Consider for example *Cartesian-product intersection*. Fuzzy sets $\underset{\sim}{A}_1$ and $\underset{\sim}{A}_2$ are subsets of universes of discourse X_1 and X_2, respectively. The conjunction fuzzy set $\underset{\sim}{A}_1 \times \underset{\sim}{A}_2$ is given by

$$\underset{\sim}{A}_1 \times \underset{\sim}{A}_2 = ce\underset{\sim}{A}_1 \cap ce\underset{\sim}{A}_2 \tag{6.49}$$

where both $ce\underset{\sim}{A}_1$ and $ce\underset{\sim}{A}_2$ (cylindrical extensions of $\underset{\sim}{A}_1$ and $\underset{\sim}{A}_2$, respectively) are defined on the product space $X_1 \times X_2$.

$$\mu_{\underset{\sim}{A}_1 \times \underset{\sim}{A}_2}(x_1, x_2) = \mu_{ce\underset{\sim}{A}_1}(x_1, x_2) \wedge \mu_{ce\underset{\sim}{A}_2}(x_1, x_2) \tag{6.50a}$$

The cylindrical extension is usually considered *implicitly* and it is not stated in the notation:

$$\mu_{\underset{\sim}{A}_1 \times \underset{\sim}{A}_2}(x_1, x_2) = \mu_{\underset{\sim}{A}_1}(x_1) \wedge \mu_{\underset{\sim}{A}_2}(x_2) \tag{6.50b}$$

In fuzzy rule-base, intersection is used to represent the 'AND' operation, while union is used to represent the 'OR' operation. Conjunction and disjunction functions defined on fuzzy sets that lie in different universes of discourse represent intersection and union, respectively.

The fuzzy relation given by conjunction/disjunction function is actually a surface over the Cartesian product $X_1 \times X_2$, a three-dimensional plot $(x_1, x_2, \mu_{\underset{\sim}{A}_1 \times \underset{\sim}{A}_2}(x_1, x_2))$.

Fuzzy Implication Relations: In everyday human reasoning, implications are used to combine somehow related statements. Human reasoning is based on cause-effect relations; there has to be a causality between the antecedent (IF part) and the consequent (THEN part). Fuzzy logic is a tool for transferring such structured knowledge into workable algorithms.

For fuzzy implication, there can be no effect (output) without a cause (input). The quantification of fuzzy implication may be based on the logic that truth value of the conclusion must not be larger than that of the premise, i.e., *we can be no more certain about our conclusions than we are about our premises*.

Consider the rule 'IF $\underset{\sim}{x}$ is $\underset{\sim}{A}$ THEN $\underset{\sim}{y}$ is $\underset{\sim}{B}$', written as $\underset{\sim}{A} \to \underset{\sim}{B}$, and defined on universes X and Y, respectively. The *minimum* and *product* fuzzy implications are the two most commonly used functions to quantify the implication operator.

- *Minimum fuzzy implication*:

$$\mu_{\underset{\sim}{A} \to \underset{\sim}{B}}(x, y) = min\,\{\mu_{\underset{\sim}{A}}(x), \mu_{\underset{\sim}{B}}(y)\} \tag{6.51a}$$

- *Product fuzzy implication*:

$$\mu_{\underset{\sim}{A} \to \underset{\sim}{B}}(x, y) = \mu_{\underset{\sim}{A}}(x)\,\mu_{\underset{\sim}{B}}(y) \tag{6.51b}$$

Note that $\mu_{\underset{\sim}{A} \to \underset{\sim}{B}}(x, y)$ defines a fuzzy set implied by the rule. The implication $\underset{\sim}{A} \to \underset{\sim}{B}$ is, thus, a fuzzy relation defined on $\{(x, y) | x \in X \text{ and } y \in Y\}$.

Many fuzzy implications have been proposed in the literature. The two listed above use the definition of fuzzy conjunction (intersection operator).

6.4.2 Fuzzification of Input

In the previous section, we have seen that fuzzy sets are used to quantify the knowledge in the rule-base. Next section will describe *inference mechanism* that operates on fuzzy sets to produce fuzzy sets. As we shall see, the mechanism works on the principle of matching input fuzzy sets with rule-base fuzzy sets, requiring input to the fuzzy system being specified in the form of fuzzy sets.

In the real-world applications, we are usually given the crisp inputs. A fuzzy logic system works on fuzzy inputs; therefore, an additional process of *fuzzification* is required.

Quite often, *singleton fuzzification* is used for crisp inputs. Singleton fuzzification, given by Eqn (6.18) and graphically shown in Fig. 6.12, is simply a different representation for the number x_0. This fuzzification is usually employed in implementations because in the absence of noise, we are undeniably sure that x_0 takes on its measured value (and no other value), and it offers some savings in calculations required for implementation of a fuzzy system (relative to, for instance, Gaussian fuzzification). Most practical implementations of fuzzy systems use singleton fuzzification. Our focus in this book on applied machine learning, will therefore be on singleton inputs to fuzzy systems.

6.4.3 Inference Mechanism

In Section 6.3.4, we discussed *relational inference*. A summary will be helpful for ongoing discussion.

Suppose there exists a fuzzy relation $\underset{\sim}{R}$ in product space $X \times Y$, and $\underset{\sim}{A}$ is a fuzzy set in X. Then fuzzy set $\underset{\sim}{B}$ in Y can be induced by $\underset{\sim}{A}$ through the *composition* of $\underset{\sim}{A}$ and $\underset{\sim}{R}$.

$$\underset{\sim}{B} = \underset{\sim}{A} \circ \underset{\sim}{R} \tag{6.52a}$$

This composition is defined by

$$\underset{\sim}{B} = proj(\underset{\sim}{R} \cap ce\underset{\sim}{A}) \text{ on } Y \tag{6.52b}$$

where $\underset{\sim}{R}$ and $ce\underset{\sim}{A}$ (cylindrical extension of $\underset{\sim}{A}$) are fuzzy sets in space $X \times Y$, and $proj(\underset{\sim}{R} \cap ce\underset{\sim}{A})$ is the projection of $\underset{\sim}{R} \cap ce\underset{\sim}{A}$ (intersection (combination) of $\underset{\sim}{R}$ and $ce\underset{\sim}{A}$) on Y.

We considered two popular composition operators:

Max-Min composition

$$\mu_{\underset{\sim}{B}}(y) = \underset{x}{max}\,[min(\mu_{\underset{\sim}{A}}(x),\ \mu_{\underset{\sim}{R}}(x, y))] \tag{6.53a}$$

Max-Product composition

$$\mu_{\underset{\sim}{B}}(y) = \underset{x}{max}\,[\mu_{\underset{\sim}{A}}(x)\,\mu_{\underset{\sim}{R}}(x, y)] \tag{6.53b}$$

Note that cylindrical extension of $\underset{\sim}{A}$ into $X \times Y$ is implicit, and Max-Min and Max-Product composition operators show projection and combination (intersection) phases.

Equations (6.53a) and (6.53b) may be represented by a single general equation.

$$\mu_{\underset{\sim}{B}}(y) = \underset{x}{max}\,[\mu_{\underset{\sim}{A}}(x) \wedge \mu_{\underset{\sim}{R}}(y)];\ \wedge \text{ is a } t\text{-norm.} \tag{6.54}$$

Inference in fuzzy rule-based systems is the process of deriving an output fuzzy set given the rules and the inputs. A rule-base in a fuzzy system can be regarded as a fuzzy relation. Consider a fuzzy rule-base consisting of rules with atomic propositions. General r^{th} rule; $r = 1, \ldots, R$, in this rule-base, is of the form:

$$\text{Rule } r : \text{IF } x \text{ is } \underset{\sim}{A}^{(r)} \text{ THEN } y \text{ is } \underset{\sim}{B}^{(r)};\ r = 1, \ldots, R;\ x \in X,\ y \in Y \tag{6.55}$$

Each rule in the rule-base can be represented by a fuzzy relation $\underset{\sim}{R}^{(r)}: X \times Y \rightarrow [0, 1]$, computed by *implication operator*. Two popular implication operations have been listed earlier in this section:

Minimum fuzzy implication

$$\mu_{\underset{\sim}{R}^{(r)}}(x, y) = min\,(\mu_{\underset{\sim}{A}^{(r)}}(x),\ \mu_{\underset{\sim}{B}^{(r)}}(y)) = \mu_{\underset{\sim}{A}^{(r)} \rightarrow \underset{\sim}{B}^{(r)}}(x, y) \tag{6.56a}$$

Product fuzzy implication

$$\mu_{\underset{\sim}{R}^{(r)}}(x, y) = [\mu_{\underset{\sim}{A}^{(r)}}(x)\,\mu_{\underset{\sim}{B}^{(r)}}(y)] = \mu_{\underset{\sim}{A}^{(r)} \rightarrow \underset{\sim}{B}^{(r)}}(x, y) \tag{6.56b}$$

The entire rule-base (6.55) is represented by aggregating the relations $\underset{\sim}{R}^{(r)}$ of the individual rules into a single fuzzy relation $\underset{\sim}{R}$. The relations $\underset{\sim}{R}^{(r)}$ represent implications; the aggregated relation is obtained by a *union* operator. In other words, the rules are implicitly connected by 'OR' operator.

$$\underset{\sim}{R} = \bigcup_{r=1}^{R} \underset{\sim}{R}^{(r)}, \text{ that is, } \mu_{\underset{\sim}{R}}(x, y) = \max_{1 \le r \le R} [\mu_{\underset{\sim}{R}^{(r)}}(x, y)] \tag{6.57}$$

Substituting from Eqn (6.56a)/Eqn (6.56b), the rule-base relation $\underset{\sim}{R}$ may be expressed as,

$$\underset{\sim}{R} = \max_{1 \le r \le R} [\mu_{\underset{\sim}{A}^{(r)}}(x) \wedge \mu_{\underset{\sim}{B}^{(r)}}(y)]; \wedge \text{ is a } t\text{-norm} \tag{6.58}$$

Once the rule-base is represented as a fuzzy relation, the output of the rule-based model can then be computed by the relational composition operator given by Eqn (6.52).

For the rule-based models, the reasoning scheme can be simplified bypassing the relational calculus. This is advantageous as discretization of domains and storing of relation $\underset{\sim}{R}$ can be avoided. In the following, we present a reasoning scheme for rule-based systems which does not require any discretization; it works directly with continuous membership functions.

Suppose an input fuzzy set is $x = \underset{\sim}{A}'$. Given the fuzzy rule-base (6.55) and the fuzzy input, the inference mechanism yields output fuzzy set $\underset{\sim}{B}'$ (refer to Eqn (6.54)):

$$\mu_{\underset{\sim}{B}'}(y) = \max_{x} \{\mu_{\underset{\sim}{A}'}(x) \wedge \mu_{\underset{\sim}{R}}(x, y)\} \tag{6.59a}$$

Substituting for $\mu_{\underset{\sim}{R}}(x, y)$ from Eqn (6.58),

$$\mu_{\underset{\sim}{B}'}(y) = \max_{x} \{\mu_{\underset{\sim}{A}'}(x) \wedge [\max_{1 \le r \le R} (\mu_{\underset{\sim}{A}^{(r)}}(x) \wedge \mu_{\underset{\sim}{B}^{(r)}}(y))]\} \tag{6.59b}$$

Since the *max* and $\wedge$ operations are taken over different domains, their order can be changed as follows:

$$\begin{aligned}\mu_{\underset{\sim}{B}'}(y) &= \max_{1 \le r \le R} \{\max_{x} [\mu_{\underset{\sim}{A}'}(x) \wedge (\mu_{\underset{\sim}{A}^{(r)}}(x) \wedge \mu_{\underset{\sim}{B}^{(r)}}(y))]\} \\ &= \max_{1 \le r \le R} \{\max_{x} [\mu_{\underset{\sim}{A}'}(x) \wedge \mu_{\underset{\sim}{A}^{(r)}}(x) \wedge \mu_{\underset{\sim}{B}^{(r)}}(y)]\}\end{aligned} \tag{6.60}$$

Denote

$$\alpha^{(r)} = \max_{x} [\mu_{\underset{\sim}{A}'}(x) \wedge \mu_{\underset{\sim}{A}^{(r)}}(x)] \tag{6.61}$$

= the degree of fulfillment of the rth-rule antecedent

It is also called *matching degree*, representing the compatibility of the antecedent condition of the rule and the input.

The output fuzzy set of the linguistic model is thus,

$$\mu_{\underset{\sim}{B}'}(y) = \max_{1 \le r \le R} [\alpha^{(r)} \wedge \mu_{\underset{\sim}{B}^{(r)}}(y)], \; y \in Y \tag{6.62}$$

The union (*max* operator in Eqn (6.62)) of the fuzzy sets representing conclusions $[\alpha^{(r)} \wedge \mu_{\underset{\sim}{B}^{(r)}}(y)]$ from various rules results in an overall implied fuzzy set $\underset{\sim}{B}'$.

The *fuzzy inference algorithm* is summarized below.

1. Compute the degree of fulfillment for each rule:

$$\alpha^{(r)} = \max_{x} [\mu_{\underset{\sim}{A}'}(x) \wedge \mu_{\underset{\sim}{A}^{(r)}}(x)]; \; 1 \le r \le R \tag{6.63a}$$

Note that for singleton fuzzy set $\underset{\sim}{A}'$,

$$\mu_{\underset{\sim}{A}'}(x) = \begin{cases} 1 & x = x_0 \\ 0 & \text{otherwise} \end{cases} \tag{6.63b}$$

The equation for $\alpha^{(r)}$ simplifies to

$$\alpha^{(r)} = \mu_{\underset{\sim}{A}^{(r)}}(x_0) \tag{6.63c}$$

2. Derive the output fuzzy sets $\mu_{\underset{\sim}{B}'^{(r)}}$ for each rule:

$$\mu_{\underset{\sim}{B}'^{(r)}}(y) = \alpha^{(r)} \wedge \mu_{\underset{\sim}{B}^{(r)}}(y);\ y \in Y,\ 1 \le r \le R \tag{6.64}$$

3. Aggregate the output fuzzy sets:

$$\mu_{\underset{\sim}{B}'}(y) = \max_{1 \le r \le R} [\mu_{\underset{\sim}{B}'^{(r)}}(y)];\ y \in Y \tag{6.65}$$

Rules with Several Inputs

We have so far described the inference mechanism for the rule-based models having antecedents with atomic (single) propositions (Eqns (6.42)). In real-life applications, we come across rule-based models with rules having compound (several) propositions. Fuzzy logic operators (connectives) such as conjunction or disjunction, can be used to combine the propositions. The connectives AND and OR are implemented by t-norms, and t-conorms, respectively.

Consider the following antecedent of a rule:

$$\text{IF } x_1 \text{ is } \underset{\sim}{A}_1 \text{ AND } x_2 \text{ is } \underset{\sim}{A}_2 \text{ THEN } y \text{ is } \underset{\sim}{B} \tag{6.66a}$$

where $\underset{\sim}{A}_1$ and $\underset{\sim}{A}_2$ have membership functions $\mu_{\underset{\sim}{A}_1}(x)$ and $\mu_{\underset{\sim}{A}_2}(x)$, respectively. The premise of the rule can be represented by a fuzzy set with the membership function

$$\mu_{\text{premise}}(x_1, x_2) = \mu_{\underset{\sim}{A}_1}(x) \wedge \mu_{\underset{\sim}{A}_2}(x) \tag{6.66b}$$

where '$\wedge$' stands for the selected t-norm to model the AND connective.

The relation $\underset{\sim}{R} = (\underset{\sim}{A}_1 \text{ AND } \underset{\sim}{A}_2) \rightarrow \underset{\sim}{B}$ is defined by

$$\mu_{\underset{\sim}{R}}(x_1, x_2, y) = \mu_{(\underset{\sim}{A}_1 \text{ AND } \underset{\sim}{A}_2) \rightarrow \underset{\sim}{B}}(x_1, x_2, y) \tag{6.66c}$$

When the inputs are crisp and treated as fuzzy singletons: $x_1 = x_{10}$, and $x_2 = x_{20}$, then output

$$\underbrace{\mu_{\underset{\sim}{B}'}(y)}_{\substack{\text{Inference}\\ \text{Output}}} = \underbrace{[\mu_{\underset{\sim}{A}_1}(x_{10}) \text{ AND } \mu_{\underset{\sim}{A}_2}(x_{20})]}_{\text{IF}} \rightarrow \underbrace{\mu_{\underset{\sim}{B}}(y)}_{\text{THEN}}$$

Therefore,

$$\mu_{\underset{\sim}{B}'}(y) = \alpha \wedge \mu_{\underset{\sim}{B}}(y) \tag{6.66d}$$

where

$$\alpha = \mu_{\underset{\sim}{A}_1}(x_{10}) \wedge \mu_{\underset{\sim}{A}_2}(x_{20}) \tag{6.66e}$$

is the degree of fulfillment (matching degree) of the inputs with rule antecedent.

When inputs are fuzzy sets $\underset{\sim}{A}_1'$ and $\underset{\sim}{A}_2'$ (this may correspond to situations where inputs are based on expert estimate) then

$$\mu_{\underset{\sim}{B}'}(y) = [\alpha \wedge \mu_{\underset{\sim}{B}}(y)];\ y \in Y \tag{6.67a}$$

where

$$\alpha = [\max_x (\mu_{\underset{\sim}{A}_1'}(x) \wedge \mu_{\underset{\sim}{A}_1}(x))] \wedge [\max_x (\mu_{\underset{\sim}{A}_2'}(x) \wedge \mu_{\underset{\sim}{A}_2}(x))] \tag{6.67b}$$

For systems with several inputs, the inference mechanism given by Eqns (6.63–6.65) is applicable with the only change in computation of degree of fulfillment $\alpha^{(r)}$ for rule r. The degree of fulfillment $\alpha^{(r)}$ now is obtained by applying logical connectives AND/OR to the antecedents of the fuzzy rules.

Fuzzy logic researchers have proposed and applied several t-norms/t-conorms to execute '$\wedge$' operations in Eqn (6.66), and of course different methods may lead to different results. Most of the fuzzy-system software tools allow us to customize the t-norms for computation of degree of fulfillment (Eqn (6.66e)), AND/OR logical connectives t-norms/t-conorms (Eqn (6.66b)), and t-norm for fuzzy implication (Eqn (6.66d)). The user is required to make the choice.

The most popular choice is *min* operator for computation of degree of fulfillment, *min*/*max* for AND/OR, and *min* for fuzzy implication. Our focus in application examples will be on this widely used choice.

Singleton Model for Consequents

A special case of the linguistic fuzzy model is obtained when the consequent fuzzy set $\underset{\sim}{B}$ is a singleton set. This set can be represented by real number b, yielding the following type of fuzzy rule for rule r in the rule-base; $r = 1, \ldots, R$.

$$\text{IF } x_1 \text{ is } \underset{\sim}{A}_1^{(r)} \text{ AND } x_2 \text{ is } \underset{\sim}{A}_2^{(r)} \text{ THEN } y = b^{(r)} \tag{6.68}$$

We will see in Section 6.6 that the singleton model can also be seen as a special case of Takagi-Sugeno fuzzy model. The advantage of singleton model for consequents is that the consequent parameters $b^{(r)}$ can easily be estimated from data using least-squares techniques.

Graphical Visualization of the Fuzzy-Inference Algorithm

For graphical visualization of the fuzzy inference algorithm, let us consider different kinds of rule-base conditions.

Inferred Fuzzy Set of a Single Atomic Fuzzy Rule

Fuzzy logic allows inference when the input is close to the premise of a rule, but not equal. It produces an output which is close to the consequent of the rule, but not equal. A typical problem in fuzzy approximate reasoning is as follows:

$$\begin{aligned} &\text{Implication : IF } x \text{ is } \underset{\sim}{A} \text{ THEN } y \text{ is } \underset{\sim}{B} \\ &\text{Given input : } x \text{ is } \underset{\sim}{A}' \\ &\text{Output} \quad\ \ : ? \end{aligned} \tag{6.69a}$$

Given the fuzzy rule and the fuzzy input, the inference mechanism yields

$$\text{Output}: y = \underset{\sim}{B}' \tag{6.69b}$$

A graphical representation of fuzzy inference algorithm is shown in Fig. 6.19.

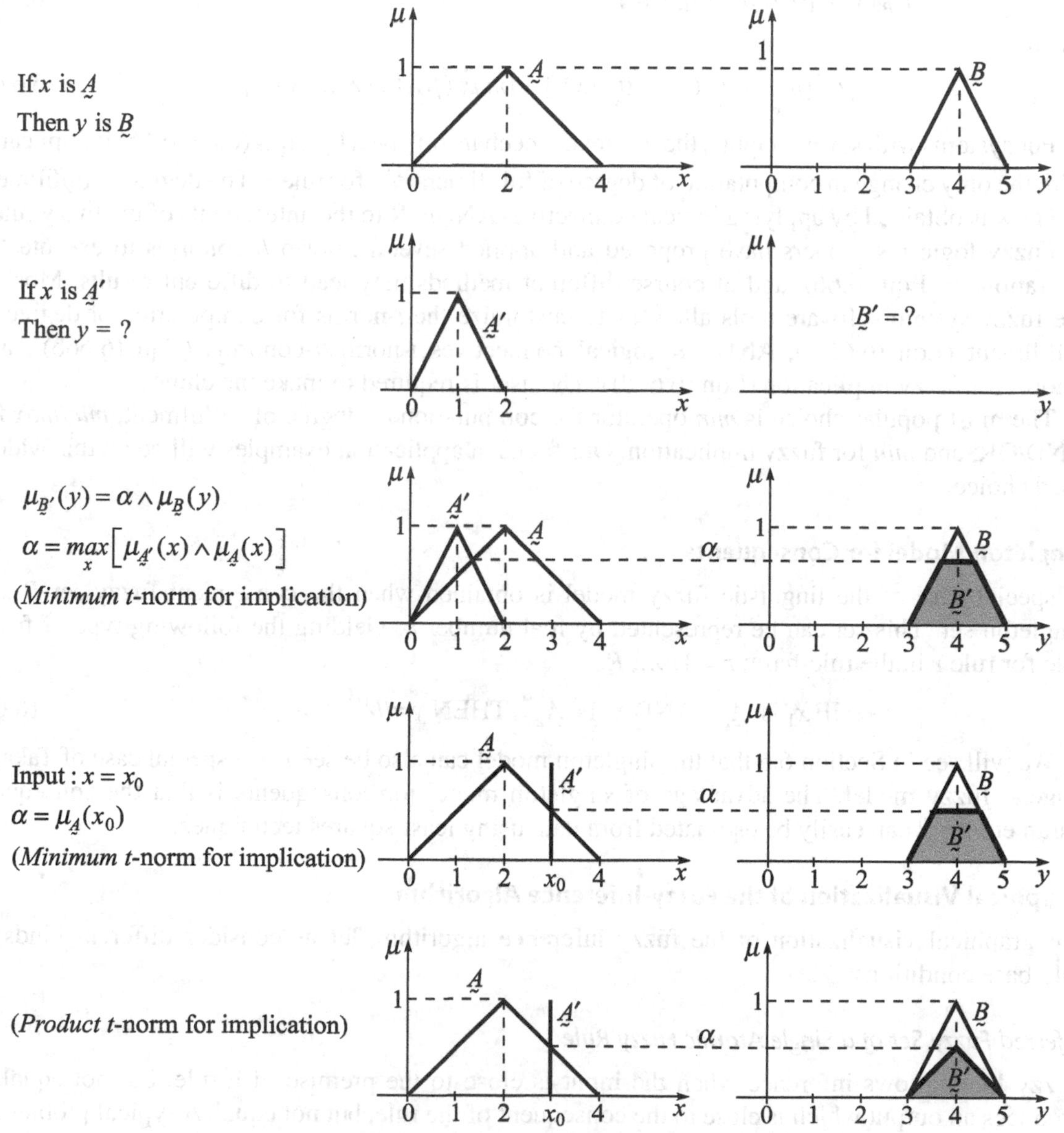

Figure 6.19 Inference of a single atomic fuzzy rule

It may be noted that the consequent membership function is *clipped* or *scaled* to the level of degree of fulfillment of the rule antecedent, leading to *not-normal* fuzzy sets.

The commonly used technique of correlating the rule consequent with the degree of fulfillment of the rule antecedent is to merely cut the consequent membership function at the level of the degree of fulfillment (minimum fuzzy implication). This technique is known as *clipping*. With the top of the membership function being sliced, the clipped fuzzy set suffers some loss of information. But clipping is often favored as it results in algorithms that are simpler and faster.

Scaling is useful in preserving the original shape of the fuzzy set: the original membership function of the rule consequent is adjusted through the multiplication of all the membership levels by the degree of fulfillment of the rule antecedent (*product* fuzzy implication).

Inferred Fuzzy Set of a Single Compound Fuzzy Rule

Implication: IF x_1 is $\underset{\sim}{A}_1$ AND x_2 is $\underset{\sim}{A}_2$ THEN y is $\underset{\sim}{B}$

Given input: $x_1 = \underset{\sim}{A}_1'$ and $x_2 = \underset{\sim}{A}_2'$

Output: $y = \underset{\sim}{B}'$?

Figure 6.20 shows the graphical representation of fuzzy inference algorithm for

(i) crisp input: x_{10}, x_{20} (ii) fuzzy input: $\underset{\sim}{A}_1', \underset{\sim}{A}_2'$

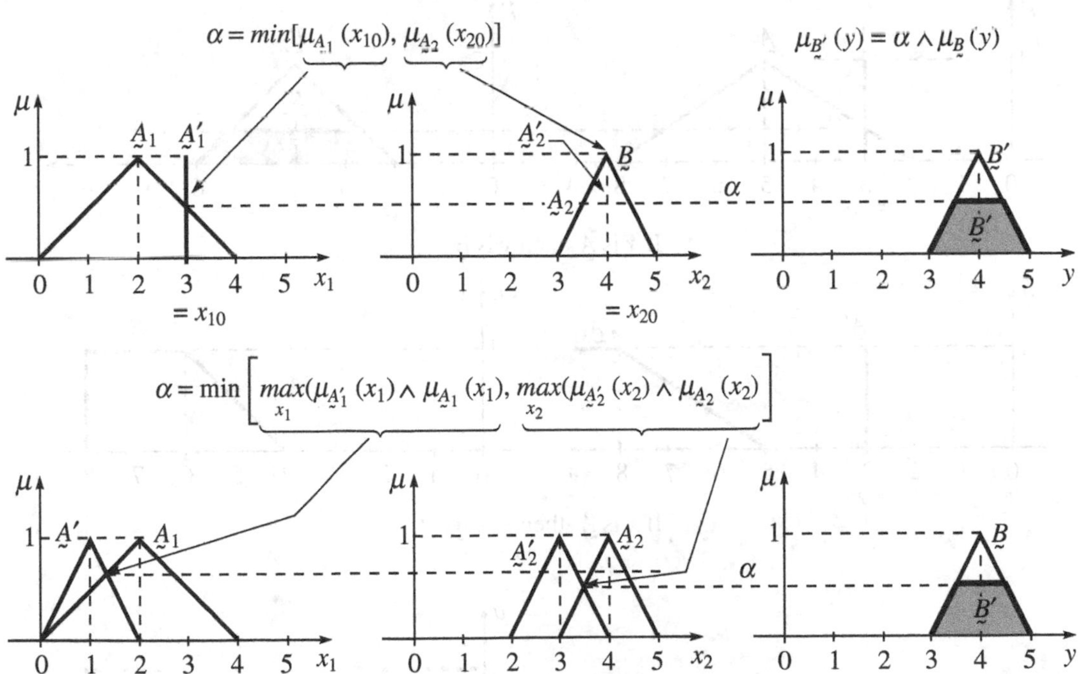

Figure 6.20 Inference of a single compound fuzzy rule

Inferred Fuzzy Set of a Fuzzy Rule-Base with Atomic Propositions

$$\text{Rule-base:} \begin{cases} \text{IF } x \text{ is } \underset{\sim}{A}_1 \text{ THEN } y \text{ is } \underset{\sim}{B}_1 \\ \text{IF } x \text{ is } \underset{\sim}{A}_2 \text{ THEN } y \text{ is } \underset{\sim}{B}_2 \\ \text{IF } x \text{ is } \underset{\sim}{A}_3 \text{ THEN } y \text{ is } \underset{\sim}{B}_3 \end{cases}$$

Given input: x is $\underset{\sim}{A}'$

Output: y is $\underset{\sim}{B}'$?

Figure 6.21 shows the graphical representation of fuzzy inference algorithm for crisp input x_0: a fuzzy singleton.

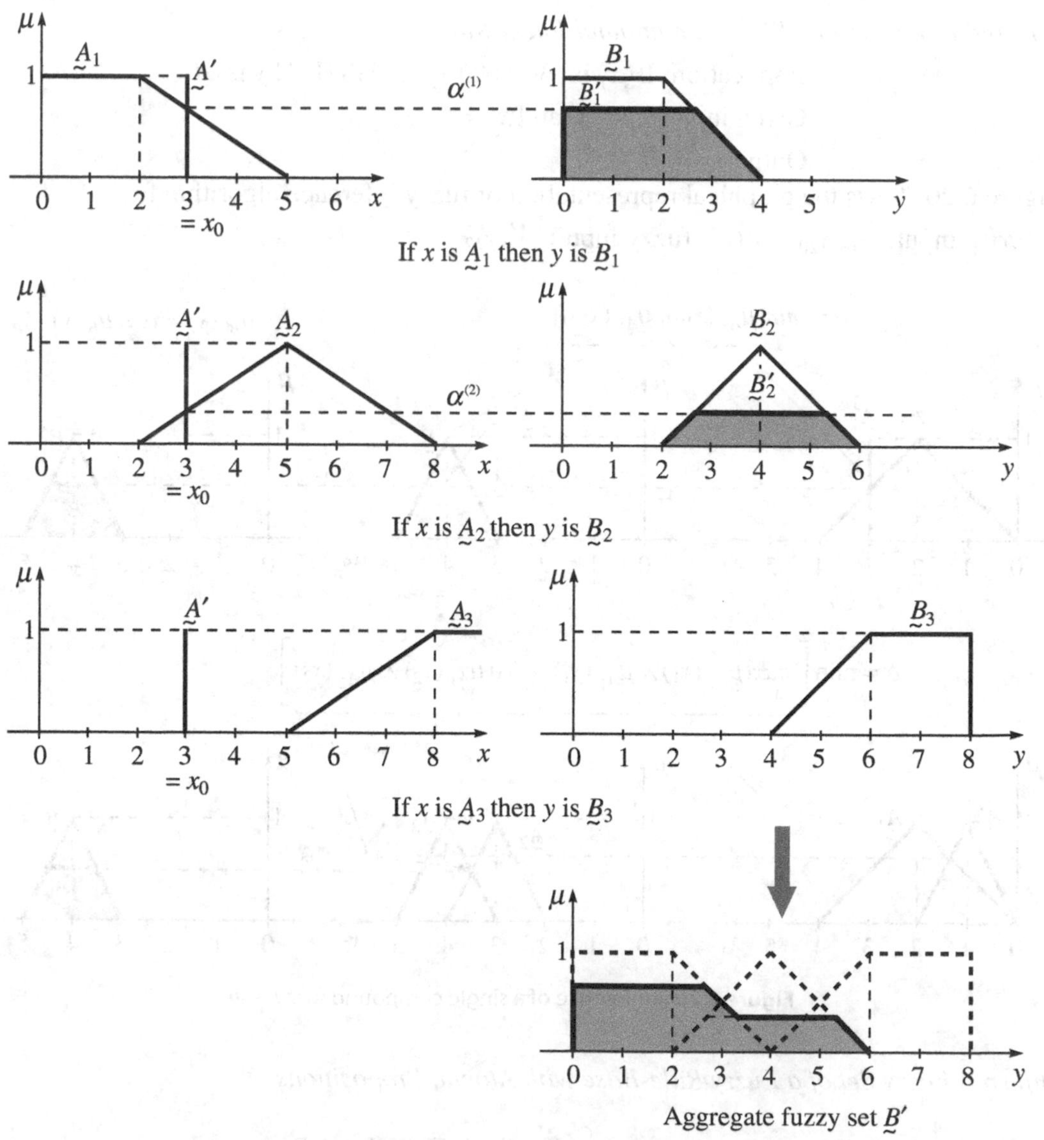

Figure 6.21 Inference of rule-base with atomic propositions

Note that for the given input x_0, the degree of fulfillment of the antecedent of rule 1 is $\alpha^{(1)} = \mu_{\underset{\sim}{A}_1}(x_0)$, and degree of fulfillment of rule 2 is $\alpha^{(2)} = \mu_{\underset{\sim}{A}_2}(x_0)$. Since $\mu_{\underset{\sim}{A}_3}(x_0) = 0$, the degree of fulfillment $\alpha^{(3)} = 0$. This equivalently means that for the given input, rule 3 is not 'ON', i.e., given input does not 'trigger' rule 3.

Determining applicability of a rule is called 'firing'. We say that a rule fires if its degree of fulfillment is greater than zero. The degree of fulfillment is also called *firing strength*. The inference mechanism seeks to determine which rules fire to find out which rules are relevant to the current situation. The inference mechanism then combines the recommendations of all the rules that fire, to come up with the single conclusion.

The firing strength of the antecedent of a rule determines the certainty with which that rule applies, and typically we will more strongly take into account the recommendations of the rules that we are more certain apply to the current situation. For example, from Fig. 6.21 we observe that $\mu_{\underset{\sim}{A}_1}(x_0) > \mu_{\underset{\sim}{A}_2}(x_0)$. Therefore, certainty of rule 1 is higher than that of rule 2; recommendation of rule 1 gets greater weightage in the overall conclusion derived from the rule-base (This will become clear later when we discuss defuzzification process to obtain crisp value of the overall conclusion). However, all the rules that fire, contribute to the overall conclusion.

Figure 6.21 shows two types of output of the inference process:

- Recommendations represented by *implied fuzzy sets* from all the rules that fire.
- Recommendation represented by *aggregated implied fuzzy set*. Aggregation is the process of unification (union operator) of the outputs of all the rules. In other words, we take the membership functions of all the rule consequents previously clipped (or scaled) and combine them into a single fuzzy set (obtained by drawing all the inferred fuzzy sets on one axis, as shown in Fig. 6.21).

In the next section, we will describe the *defuzzification* process which operates on the implied fuzzy sets produced by the rules that fire, and combines their effects to provide the 'most certain' crisp value of the output y. We will also describe an alternative defuzzification scheme of obtaining the crisp output from aggregated implied fuzzy set.

One observation can, however, be made here itself. The crisp value of the output is certainly a function of the area under the inferred fuzzy sets. For the output to be finite, the area must be finite. Therefore for the output y, the membership functions at the outermost edges ($\underset{\sim}{B}_1$ and $\underset{\sim}{B}_3$ in Fig. 6.21) cannot be 'saturated'. This is essential for the fuzzy system to be properly defined. For the input, there is no such restriction; the membership $\underset{\sim}{A}_3$ in Fig. 6.21 saturates at a value of one.

Inferred Fuzzy Set of Fuzzy Rule-Base with Compound Propositions

Consider a fuzzy inference system characterized by three fuzzy variables:

$$\text{Input variables: } x_1 \in X_1 = [0, 125]; x_2 \in X_2 = [0, 10]$$

$$\text{Output variable: } y \in Y = [0, 10]$$

The linguistic values of x_1 are described by fuzzy subsets $\underset{\sim}{A}_{11}, \underset{\sim}{A}_{12}, \underset{\sim}{A}_{13}, \underset{\sim}{A}_{14}$ and $\underset{\sim}{A}_{15}$. The linguistic values of x_2 are described by fuzzy subsets $\underset{\sim}{A}_{21}, \underset{\sim}{A}_{22}, \underset{\sim}{A}_{23}, \underset{\sim}{A}_{24}$, and $\underset{\sim}{A}_{25}$, and those of y by $\underset{\sim}{B}_1$, $\underset{\sim}{B}_2$, $\underset{\sim}{B}_3$, $\underset{\sim}{B}_4$ and $\underset{\sim}{B}_5$.

With the membership function encompassing all fuzziness for a specific fuzzy set, the way it is described reflects the essence of a fuzzy operation. Due to the significance of the 'shape' of the membership function, several shapes for assigning membership values to fuzzy variables have been proposed (refer to Section 6.3.2). Optimization of these assignments is often done through trial-and-error. Figures 6.22–6.24 show an assignment of ranges and fuzzy membership functions for x_1, x_2, and y.

Now that we have inputs and output in terms of fuzzy variables, we need to construct a set of rules for the operation of the fuzzy inference system. The rule-base matrix for our example is given in Table 6.2. Note that we do not need to specify all the cells in the matrix. No entry signifies no action (inference) for that situation.

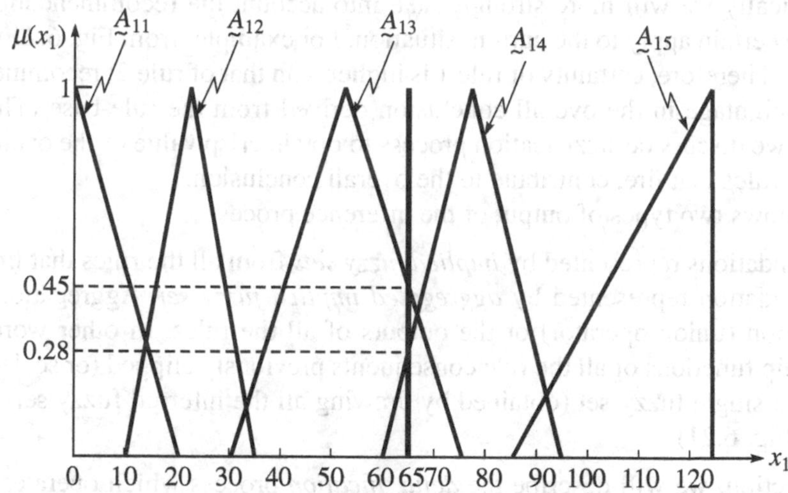

Figure 6.22 Fuzzy membership functions for x_1

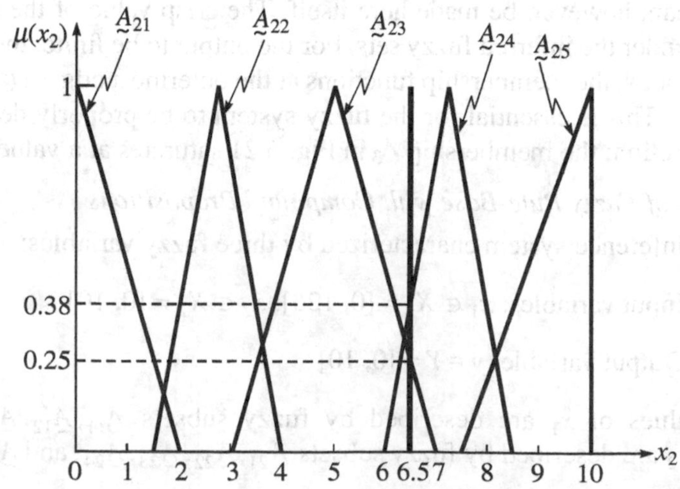

Figure 6.23 Fuzzy membership functions for x_2

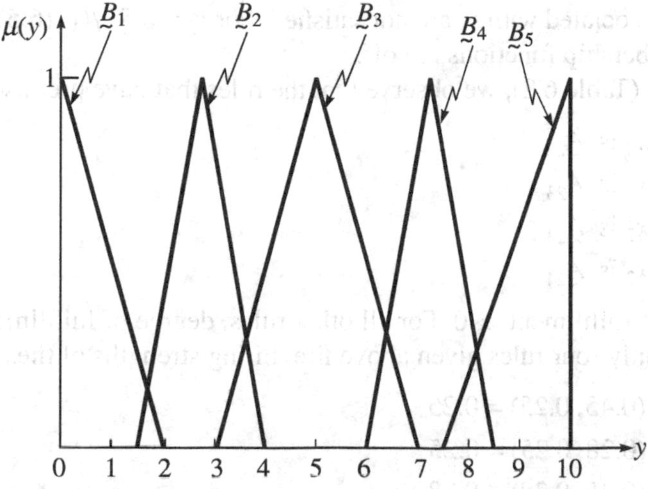

Figure 6.24 Fuzzy membership functions for the output y

We can translate the table entries into IF-THEN rules. We give here a couple of examples.

- IF x_1 is $\underset{\sim}{A}_{12}$ AND x_2 is $\underset{\sim}{A}_{22}$ THEN y is $\underset{\sim}{B}_3$
- IF x_1 is $\underset{\sim}{A}_{11}$ AND x_2 is $\underset{\sim}{A}_{24}$ THEN y is $\underset{\sim}{B}_5$

Table 6.2 Rule Table

Output y		**Input x_1**				
		$\underset{\sim}{A}_{11}$	$\underset{\sim}{A}_{12}$	$\underset{\sim}{A}_{13}$	$\underset{\sim}{A}_{14}$	$\underset{\sim}{A}_{15}$
Input x_2	$\underset{\sim}{A}_{21}$	$\underset{\sim}{B}_3$	$\underset{\sim}{B}_2$	$\underset{\sim}{B}_1$		
	$\underset{\sim}{A}_{22}$	$\underset{\sim}{B}_4$	$\underset{\sim}{B}_3$	$\underset{\sim}{B}_1$	$\underset{\sim}{B}_1$	
	$\underset{\sim}{A}_{23}$	$\underset{\sim}{B}_5$	$\underset{\sim}{B}_4$	$\underset{\sim}{B}_3$	$\underset{\sim}{B}_1$	
	$\underset{\sim}{A}_{24}$	$\underset{\sim}{B}_5$	$\underset{\sim}{B}_4$	$\underset{\sim}{B}_4$	$\underset{\sim}{B}_2$	
	$\underset{\sim}{A}_{25}$	$\underset{\sim}{B}_5$	$\underset{\sim}{B}_4$	$\underset{\sim}{B}_4$	$\underset{\sim}{B}_3$	

We assume that input is defined as $x_{10} = 65$, $x_{20} = 6.5$. Figures 6.22–6.23 show the fuzzy singletons corresponding to the given input. For $x_1 = 65$, $\mu_{\underset{\sim}{A}_{13}}(65) = 0.45$ and $\mu_{\underset{\sim}{A}_{14}}(65) = 0.28$, and all other membership functions are off (i.e., their values are zero). Therefore, the proposition 'x_1 is $\underset{\sim}{A}_{13}$' is satisfied to a degree of 0.45 and the proposition 'x_1 is $\underset{\sim}{A}_{14}$' is satisfied to a degree of 0.28; all other atomic propositions associated with x_1 are not satisfied. For $x_2 = 6.5$, $\mu_{\underset{\sim}{A}_{23}}(6.5) = 0.25$ and $\mu_{\underset{\sim}{A}_{24}}(6.5) = 0.38$; all other membership functions are off.

From the rule table (Table 6.2), we observe that the rules that have premise terms

(i) x_1 is $\underset{\sim}{A}_{13}$ AND x_2 is $\underset{\sim}{A}_{23}$

(ii) x_1 is $\underset{\sim}{A}_{14}$ AND x_2 is $\underset{\sim}{A}_{23}$

(iii) x_1 is $\underset{\sim}{A}_{13}$ AND x_2 is $\underset{\sim}{A}_{24}$

(iv) x_1 is $\underset{\sim}{A}_{14}$ AND x_2 is $\underset{\sim}{A}_{24}$

have $\mu_{premise}$(degree of fulfillment) > 0. For all other rules, degree of fulfillment is zero. Therefore, for the given inputs, only four rules given above fire. Firing strengths of these rules are:

(i) $\mu_{\underset{\sim}{A}_{13} \times \underset{\sim}{A}_{23}} = min(0.45, 0.25) = 0.25$

(ii) $\mu_{\underset{\sim}{A}_{14} \times \underset{\sim}{A}_{23}} = min(0.28, 0.25) = 0.25$

(iii) $\mu_{\underset{\sim}{A}_{13} \times \underset{\sim}{A}_{24}} = min(0.45, 0.38) = 0.38$

(iv) $\mu_{\underset{\sim}{A}_{14} \times \underset{\sim}{A}_{24}} = min(0.28, 0.38) = 0.28$

Output recommendation of each rule is shown in Fig. 6.25. Figure 6.26 shows the aggregated output $\underset{\sim}{B}'$.

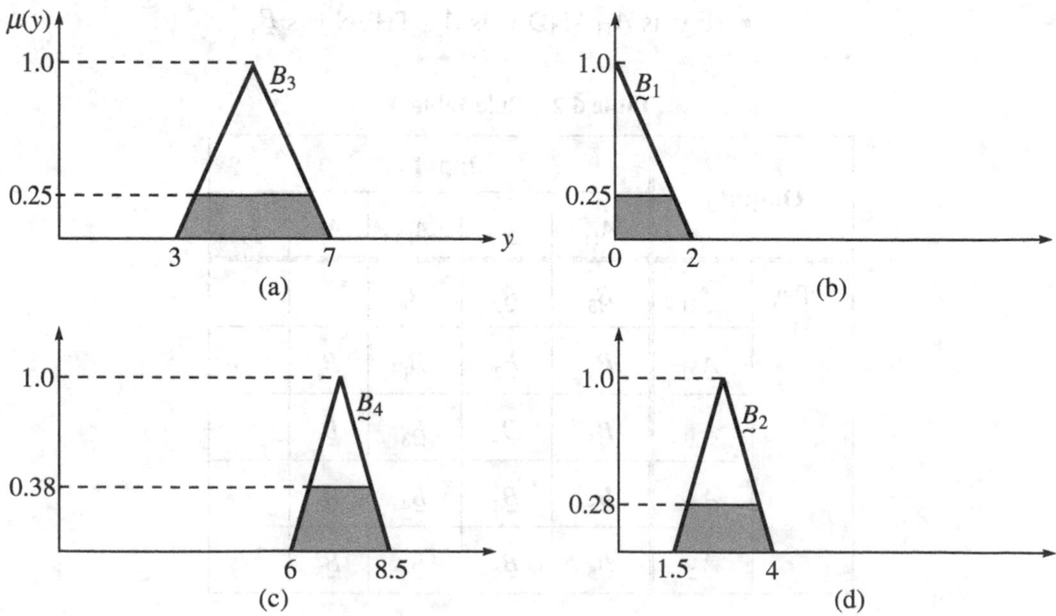

Figure 6.25 Inference for each rule

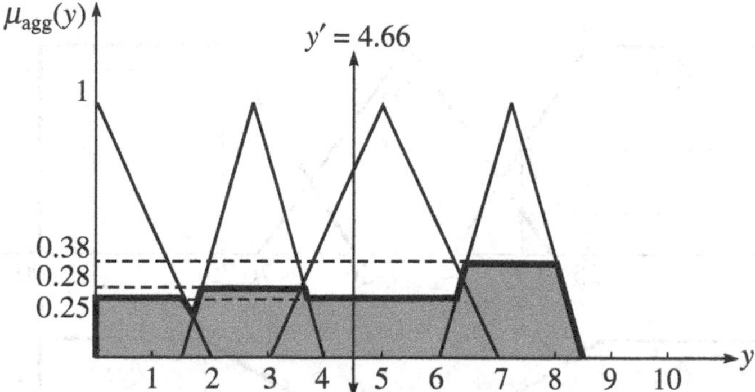

Figure 6.26 Aggregated fuzzy set $\underset{\sim}{B}'$

There are several ways to select the parameters of the fuzzy inference system that seem sensible. Which of the membership functions are the best? What is the number of linguistic values and rules that should exist? Should *minimum* or *product* be representing 'AND' in the premise and the implication? Which defuzzification technique should one select? Such questions must necessarily be answered while designing a fuzzy inference system.

Generally, the selection of all the parts of the fuzzy system is done using a kind of trial-and-error approach. But eventually, the aforementioned questions can be answered through the experience of designing fuzzy inference systems for a variety of applications presenting challenging characteristics. This is what we do in this chapter.

Example 6.10

Let us revisit Example 6.9. The rule-base of the fuzzy system considered in the example is given by (6.45a). Fuzzy variables and fuzzy sets defined for this system are given in (6.45b), and the rule-base in terms of these variables and fuzzy sets is given in (6.45c). The membership functions for all the linguistic values are shown in Fig. 6.18.

We consider the crisp input: x_1 (project_funding) = 35%, and x_2 (project_staffing) = 65%. As seen in Fig. 6.27, the crisp input x_{10} corresponds to the membership functions of $\underset{\sim}{A}_{11}$ (*inadequate*) and $\underset{\sim}{A}_{12}$ (*marginal*) to the degrees of 0.5 and 0.2, respectively. The crisp input x_{20} corresponds to the membership functions of $\underset{\sim}{A}_{21}$ (*small*) and $\underset{\sim}{A}_{22}$ (*large*) to the degrees of 0.1 and 0.7, respectively. Firing strengths (degrees of fulfillment) of the three rules are obtained as follows:

$$\alpha^{(1)} = max[\mu_{\underset{\sim}{A}_{13}}(x_{10}), \mu_{\underset{\sim}{A}_{21}}(x_{20})] = max[0.0, 0.1] = 0.1$$
$$\alpha^{(2)} = min[\mu_{\underset{\sim}{A}_{12}}(x_{10}), \mu_{\underset{\sim}{A}_{22}}(x_{20})] = min[0.2, 0.7] = 0.2$$
$$\alpha^{(3)} = \mu_{\underset{\sim}{A}_{11}}(x_{10}) = 0.5$$

Now for each rule, the result of the antecedent evaluation can be applied to the membership functions of the consequent; the consequent membership function is *clipped* to the level of firing strength of the rule. The implied fuzzy sets of the rules are then aggregated. Figure 6.28 shows the implied fuzzy sets of the rules and the aggregated fuzzy set.

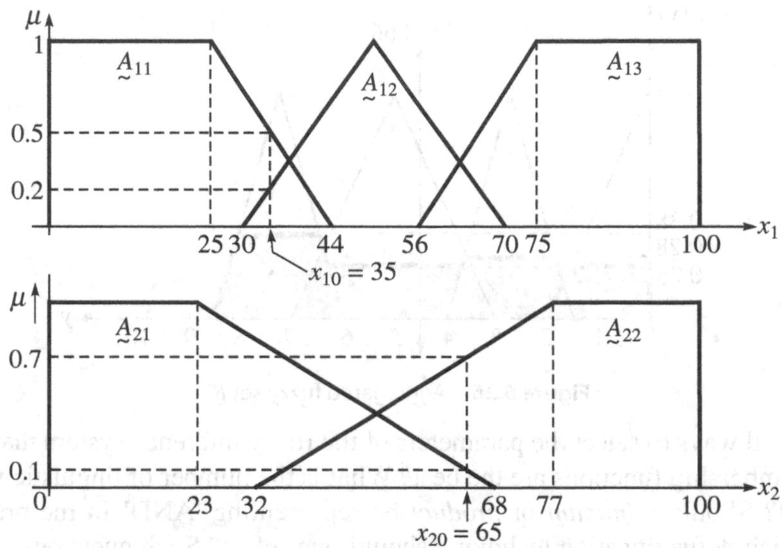

Figure 6.27 Fuzzy membership functions

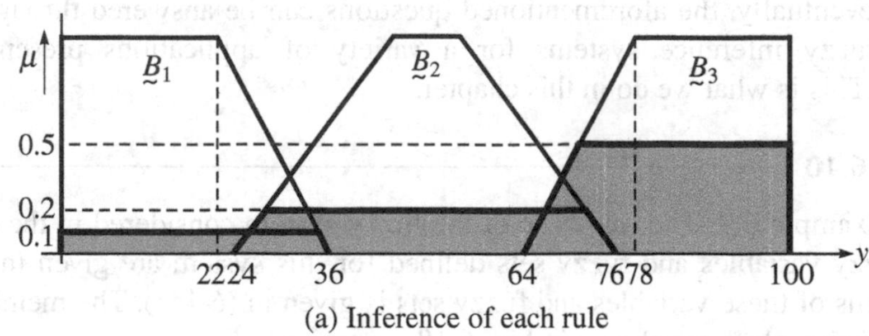

(a) Inference of each rule

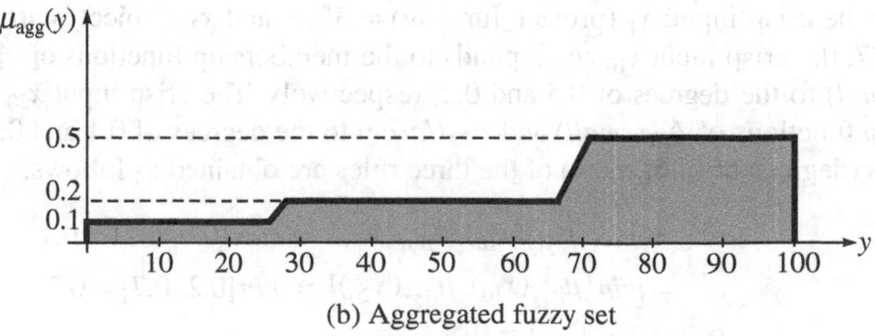

(b) Aggregated fuzzy set

Figure 6.28

Rule-Base with Singleton Consequents

It is possible to make use of one spike, a *singleton*, as the membership function of the rule consequent. A fuzzy singleton is a fuzzy set possessing a membership that is unity at a single particular point on the universe of discourse and zero everywhere else.

Fuzzy inference mechanism for singleton consequents is similar to the one described earlier for fuzzy-set consequents in Figs 6.22–6.26; instead of triangular fuzzy sets $\underset{\sim}{B}_1$, $\underset{\sim}{B}_2$, $\underset{\sim}{B}_3$, $\underset{\sim}{B}_4$, and $\underset{\sim}{B}_5$, we use fuzzy singletons.

Example 6.11

Let us reconsider the fuzzy system described earlier in Examples 6.9 and 6.10; now with singleton consequents.

The output of each fuzzy rule is now constant. The rule-base is now given as,

$$\text{Rule-base:} \begin{cases} \text{IF } x_1 \text{ is } \underset{\sim}{A}_{13} \text{ OR } x_2 \text{ is } \underset{\sim}{A}_{21} \text{ THEN } y \text{ is } b^{(1)} \\ \text{IF } x_1 \text{ is } \underset{\sim}{A}_{12} \text{ AND } x_2 \text{ is } \underset{\sim}{A}_{22} \text{ THEN } y \text{ is } b^{(2)} \\ \text{IF } x_1 \text{ is } \underset{\sim}{A}_{11} \text{ THEN } y \text{ is } b^{(3)} \end{cases}$$

Given input: x_{10}, x_{20}

For the given input (x_{10}, x_{20}), the degree of fulfillment of rule 1 is $\alpha^{(1)} = max[\mu_{\underset{\sim}{A}_{13}}(x_{10}), \mu_{\underset{\sim}{A}_{21}}(x_{20})]$; for rule 2, $\alpha^{(2)} = min[\mu_{\underset{\sim}{A}_{12}}(x_{10}), \mu_{\underset{\sim}{A}_{22}}(x_{20})]$; and for rule 3, $\alpha^{(3)} = \mu_{\underset{\sim}{A}_{11}}(x_{10})$.

Figure 6.29 shows graphical representation of fuzzy inference process. It shows the implied singleton fuzzy set of each rule, as well as the aggregated output fuzzy set.

The similarity of fuzzy inference represented in Figs 6.27–6.28 and Fig. 6.29, is quite noticeable. The only distinction is that rule consequents are singletons in Fig. 6.29. From Fig. 6.27, we have $\alpha^{(1)} = 0.1$, $\alpha^{(2)} = 0.2$, and $\alpha^{(3)} = 0.5$. The fuzzy sets of Fig. 6.28 are now replaced by singletons $b^{(1)} = 20$, $b^{(2)} = 50$, and $b^{(3)} = 80$.

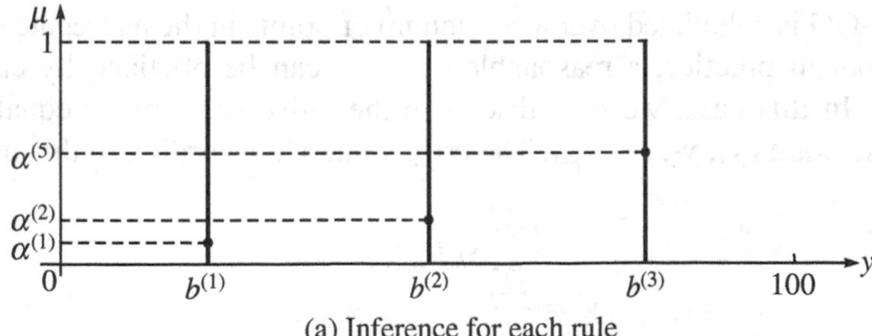

(a) Inference for each rule
$\alpha^{(1)} = 0.1$, $\alpha^{(2)} = 0.2$, $\alpha^{(3)} = 0.5$
$b^{(1)} = 20$, $b^{(2)} = 50$, $b^{(3)} = 80$

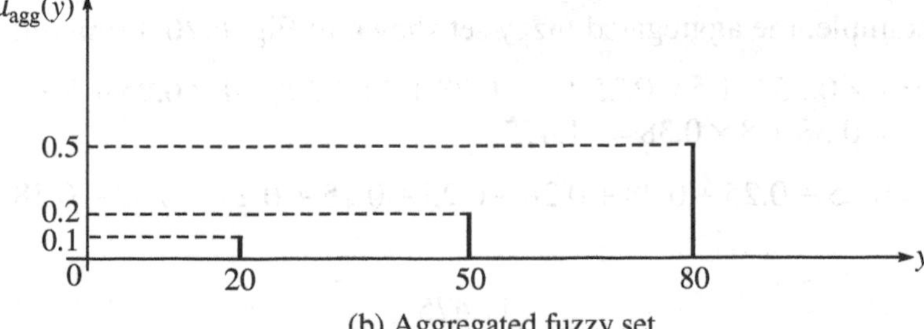

(b) Aggregated fuzzy set

Figure 6.29

6.4.4 Defuzzification of Inferred Fuzzy Set

Next, we consider the defuzzification operation, which is the final component of the fuzzy inference system. The operation of defuzzification is carried out on the implied fuzzy sets created by the inference mechanism and the results of this process are combined to provide the 'most certain' output. Defuzzification may be thought of as 'decoding' the fuzzy set information produced by implied fuzzy sets of the rules that fire, into numeric output.

As an alternative, the inference mechanism could, in addition, calculate the aggregated implied fuzzy set representing the conclusion arrived at, taking into account all the rules firing simultaneously. Defuzzification then operates on this aggregated implied fuzzy set to provide the 'most certain' output.

There are several defuzzification strategies, and it is easy to create/devise many more. Each one offers a means to select one crisp output, on the basis of either aggregated implied fuzzy set or implied fuzzy sets of each rule individually.

In the following, the two most popular defuzzification methods are presented: the *center-of-gravity* (COG) and the *Mean-of-maxima* (MOM).

COG method: It finds the point where a vertical line would slice the aggregate fuzzy set $\underset{\sim}{B}'$ into two equal masses (areas). The method is also called *centroid defuzzification* or *center-of-area* (COA) defuzzification.

Mathematically, the COG can be expressed as,

$$\text{COG} = \frac{\int_y \mu_{\underset{\sim}{B}'}(y)\, y\, dy}{\int_y \mu_{\underset{\sim}{B}'}(y)\, dy} \tag{6.70}$$

In theory, the COG is calculated over a continuum of points in the aggregate output membership function $\mu_{\underset{\sim}{B}'}(y)$, but in practice, a reasonable estimate can be obtained by calculating it over a sample of points. In this case, we may discretize the universe Y into q equal (or almost equal) subintervals by the points $y_1, y_2, \ldots, y_{q-1}$. The crisp value y', according to this method, is

$$y' = \frac{\sum_{k=1}^{q-1} y_k\, \mu_{\underset{\sim}{B}'}(y_k)}{\sum_{k=1}^{q-1} \mu_{\underset{\sim}{B}'}(y_k)} \tag{6.71a}$$

Consider, for example, the aggregated fuzzy set shown in Fig. 6.26. From this figure, we obtain

$$\sum y_k\, \mu_{\underset{\sim}{B}'}(y_k) = 1 \times 0.25 + 1.5 \times 0.25 + 2 \times 0.28 + 3 \times 0.28 + 4 \times 0.25 + 5 \times 0.25 + 6 \times 0.25 + 7 \times 0.38 + 8 \times 0.38 = 11.475$$

$$\sum \mu_{\underset{\sim}{B}'}(y_k) = 0.25 + 0.25 + 0.28 + 0.28 + 0.25 + 0.25 + 0.25 + 0.38 + 0.38 = 2.57$$

Therefore,

$$y' = \frac{11.475}{2.57} = 4.46$$

The physical interpretation of y' is that, if the area is cut of a thin piece of metal or wood, the center of the area will be the center of gravity.

In fact, there is hardly any need of discretization of the universe for situations like the one shown in Fig. 6.26; we can split up geometry into pieces and place a straight edge (centroid) through the area of each piece to have it perfectly balanced with an equal area of the figure on either side. The crisp output can be calculated using the expression given below:

$$y' = \frac{\sum_{p=1}^{P} (\text{area of sub-region } p) \times (\text{center of area of } p^{\text{th}} \text{ sub-region})}{\sum_{p=1}^{P} (\text{area of sub-region } p)} \tag{6.71b}$$

where P indicates the number of small pieces of area or sub-regions.

Using the aggregated implied fuzzy set for defuzzification has two associated problems: (1) The aggregated implied fuzzy set itself is difficult to compute; (2) The defuzzification of aggregated fuzzy set is also difficult to compute.

It may turn out to be a computationally simple and acceptable solution if individual implied fuzzy sets from each fired rule are used in defuzzification process. Area and center of area of each implied fuzzy set is calculated, and then COG is computed.

Let $b^{(r)}$ denote the center of membership function $\mu^{(r)}(y)$ of the implied fuzzy set of rule r, and $\int_y \mu^{(r)}(y)dy$ denote the area under the membership function $\mu^{(r)}(y)$. The COG method computes crisp output y' to be

$$y' = \frac{\sum_{r=1}^{R} b^{(r)} \int_y \mu^{(r)}(y)dy}{\sum_{r=1}^{R} \int_y \mu^{(r)}(y)dy} \tag{6.72a}$$

$$= \frac{\sum_{r=1}^{R} (\text{Area of the implied fuzzy set of rule } r) \times (\text{Center of area of this fuzzy set})}{\sum_{r=1}^{R} \text{Area of the implied fuzzy set of rule } r} \tag{6.72b}$$

Center-Average Method: In the center-average method of defuzzification, we let

$$y' = \frac{\sum_{r=1}^{R} b^{(r)} \alpha^{(r)}}{\sum_{r=1}^{R} \alpha^{(r)}} \tag{6.73}$$

where $b^{(r)}$ denotes the center of membership function $\mu^{(r)}(y)$ of the implied fuzzy set of rule r, and $\alpha^{(r)}$ is the degree of fulfillment (firing strength) of rule r. We call it the 'center-average' technique as Eqn (6.73) is a *weighted average* of the center values of the output membership functions of the rules. The center-average technique replaces the areas of the implied fuzzy sets employed in COG with the values of firing strengths $\alpha^{(r)}$. This replacement is a valid one because the area of the implied fuzzy set is usually proportional to $\alpha^{(r)}$.

Weighted-Average of Implied Singletons

For the singleton model, the weighted-average of the implied singletons results in defuzzification of the model to the crisp output y':

$$y' = \frac{\sum_{r=1}^{R} \alpha^{(r)} b^{(r)}}{\sum_{r=1}^{R} \alpha^{(r)}} \tag{6.74}$$

Note that in this equation, $b^{(r)}$ are singletons.

Example 6.12

Let us revisit the fuzzy system considered in Examples 6.9–6.11.

The COG defuzzification of the aggregated fuzzy set in Fig. 6.28:

$$\text{COG} = \frac{(0+10+20)\times 0.1 + (30+40+50+60)\times 0.2 + (70+80+90+100)\times 0.5}{0.1+0.1+0.1+0.2+0.2+0.2+0.2+0.5+0.5+0.5+0.5}$$

$$= 67.4$$

It means that the risk involved in our 'fuzzy' project is 67.4%.

Weighted average of the singletons in Fig. 6.29:

$$\text{Weighted-Average} = \frac{0.1\times 20 + 0.2\times 50 + 0.5\times 80}{0.1+0.2+0.5}$$

$$= 65$$

MOM Method: A crisp output y' may be chosen as the point on the output universe of discourse for which the aggregated implied fuzzy set $\underset{\sim}{B}'$ achieves a maximum.

Sometimes, the same maximum value can occur at more than one point in $\underset{\sim}{B}'$. In this case, we need to specify a strategy on how to pick only one point. In the *mean-of-maximum* (MOM) strategy, we choose crisp output y' that represents the mean of all elements $y \in Y$ whose membership in $\underset{\sim}{B}'$ is the maximum.

MOM method can also be applied on individual implied fuzzy sets of rules r. The implied fuzzy sets $\underset{\sim}{B}^{(r)}$ are first defuzzified in order to obtain crisp values $b^{(r)}$—representatives of fuzzy sets $\underset{\sim}{B}^{(r)}$:

$$b^{(r)} = mom(\underset{\sim}{B}^{(r)}) \tag{6.75a}$$

A crisp output value is then computed by taking the weighted average of $b^{(r)}$'s:

$$y' = \frac{\sum_{r=1}^{R} \alpha^{(r)} b^{(r)}}{\sum_{r=1}^{R} \alpha^{(r)}} \tag{6.75b}$$

where $\alpha^{(r)}$ is the degree of fulfillment (firing strength) of fired rule r.

Choice of a Defuzzification Method: While various values calculated by various defuzzification techniques may offer reasonable outputs y', it is not easy to say which is the most ideal. As for other components of fuzzy inference system, trial-and-error approach is generally used to select defuzzification method. Experience of designing fuzzy inference systems for a variety of applications presenting challenging characteristics may yield some guidelines for choosing the inference strategy and defuzzification technique.

6.5 MAMDANI MODEL FOR FUZZY INFERENCE SYSTEMS

A fuzzy inference system (FIS) is a static nonlinear mapping between its inputs and outputs. It is assumed that the fuzzy system has inputs x_j, $j = 1, \ldots, n$; and outputs y_q; $q = 1, 2, \ldots, M$. For simplicity of discussion, we will consider a system with only one output, as shown in Fig. 6.30. (It is usually possible to consider a multi-output system in terms of q single-output systems). The inputs and output are "crisp"—that is, they are real numbers, not fuzzy sets.

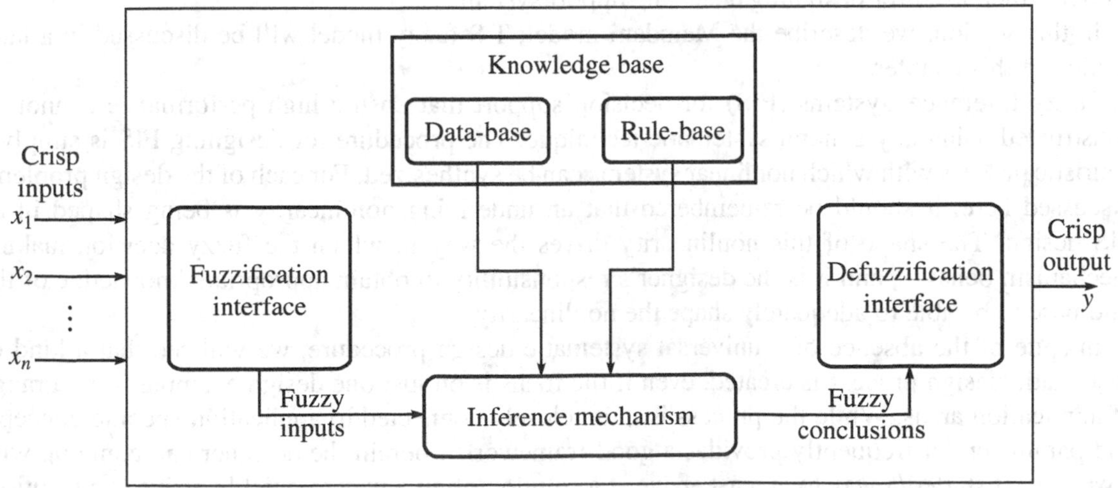

Figure 6.30 Structure of fuzzy inference systems

An FIS comprises five traditional blocks: a *rule-base* of IF-THEN rules comprising a fuzzy logic quantification of the expert's linguistic description of the way to arrive at appropriate decisions;

a *database* which defines the membership functions of the fuzzy sets used in the fuzzy rules; an *inference system* (also known an 'inference engine", or 'fuzzy inference module') which emulates the decision making of the expert; a *fuzzification interface*, which converts the crisp inputs into information (fuzzy sets) that the inference mechanism can conveniently employ to activate and apply rules; and a *defuzzification interface*, which converts the fuzzy output of the inference mechanism into a crisp output (refer to Fig. 6.30).

The fuzzy inference technique most widely in use is the Mamdani technique. In 1975, Professor Mamdani of London University established what is probably the first fuzzy mechanism for controlling a steam engine and boiler combination [95, 96]. A set of fuzzy rules were applied, which were supplied by process operators with experience.

The fuzzy inference technique suggested by Takagi, Sugeno, and Kang is a significant topic in theoretical studies as well as in practical applications of fuzzy modeling and control. The fundamental idea of this technique is to decompose the input space into fuzzy areas and to approximate the system in each area using a simple model. The overall fuzzy model, therefore, comprises of interlinked subsystems with less complex models. In fuzzy systems literature, this approach is being referred to by various titles: Sugeno [97], Takagi-Sugeno [98], Takagi-Sugeno-Kang [98, 99] fuzzy model. We will use the title 'Takagi-Sugeno fuzzy model', popularly known as T-S fuzzy model.

Thus, two typical fuzzy inference systems are based on *Mamdani model* and *T-S fuzzy model*. Mamdani's method is a fuzzy inference mechanism that is designed directly based on the set of fuzzy rules supplied by experienced human operators. The underlying nonlinearity of the decision-making system is shaped heuristically in the design process. Explicit nonlinear system identification is not done in this case. On the other hand, the Takagi-Sugeno method follows an indirect approach: identification of the nonlinear system in terms of a T-S fuzzy model is first carried out; the T-S fuzzy model is then used for designing decision-support system.

In this section, we describe the Mamdani model; T-S fuzzy model will be discussed in a later section of this chapter.

Fuzzy Inference Systems (FIS) for decision support that ensure high performance cannot be constructed using any general systematic technique. The procedure for designing FIS is simply a heuristic process with which nonlinear systems can be synthesized. For each of the design problems discussed here, it should be remembered that an underlying nonlinearity is being shaped in an FIS design. The shape of this nonlinearity drives the way in which the fuzzy decision-making mechanism behaves, and it is the designer's responsibility to obtain appropriate knowledge of the rule-base to be able to adequately shape the nonlinearity.

In spite of the absence of a universal systematic design procedure, we will see that a kind of systematic design process is created, even if the focus is on just one design example from a range of application areas. While the process is quite closely connected to application-specific concepts and parameters, it frequently provides a good framework wherein the designer can come up with a way to get started/a way to at least arrive at a solution/often a way to quickly arrive at a solution.

In the following sub-sections, we discuss two design examples for different kinds of challenging applications: mobile robot navigation among moving obstacles, and mortgage loan assessment. We begin with the problem of mobile robot navigation. Through this example, we provide a detailed view of the general methodology for design of fuzzy inference systems. Our discussion on the other example is relatively brief.

6.5.1 Mobile Robot Navigation Among Moving Obstacles

In robotics, one of the significant areas of research is to create intelligent robots capable of planning their own motion during navigation via two-dimensional or three-dimensional terrains. A significant amount of work has been carried out for motion planning of a mobile robot among moving obstacles. Here, our objective is limited: illustration of design procedure for fuzzy inference systems. We therefore consider here a simple fuzzy-logic based solution to the motion planning problem.

A typical process for developing FIS incorporates the following steps.

Step 1: **Specify the problem and define linguistic variables**

The first and the most significant step in building an FIS is problem specification. We need to determine the problem input and output variables as well as their ranges.

For our problem, a mobile robot is required to navigate from point S (starting point) to point T (target point). A schematic representation of our problem is shown in Fig. 6.31. There are three main fuzzy variables which the FIS for our problem should consider: distance of the *nearest obstacle forward* from the robot (distance $|PO_2|$ in Fig. 6.31, assuming O_2 to be the nearest obstacle forward); relative angle between the path joining the robot and the target point and the path to the nearest obstacle forward ($\angle TPO_2$ in Fig. 6.31, assuming O_2 to be the nearest obstacle forward); and the relative velocity vector of the nearest obstacle forward with respect to the robot.

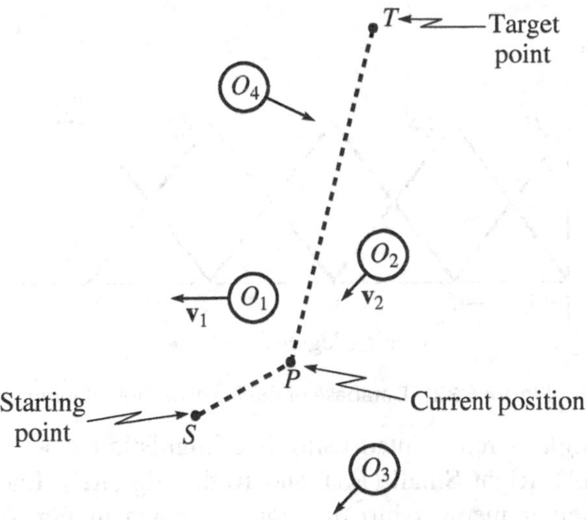

Figure 6.31 A schematic of input (*distance* and *angle*) and output (*deviation*) variables

Pratihar et. al. [100, 101] have proposed a simple practical procedure for creating a rule-base for the problem. It is possible for the robot to locate and find the velocity of each obstacle at a consistent interval of time with the help of sensors. The robot, thus, is aware of the position and velocity of each hurdle at the end of each time step. Therefore, the nearest obstacle forward can be defined and altered with the help of the relative velocity information of the obstacles. In such cases, (Fig. 6.31), even if the obstacle O_1 is closer in comparison to obstacle O_2, it is assumed that the obstacle O_2 is the closest obstacle forward as the relative velocity $\mathbf{v}_1$ of O_1 directs away from the robot's way in the direction of the target point T, while the relative velocity $\mathbf{v}_2$ of O_2 directs towards the robot. This practical consideration permits us to get rid of the third variable (relative velocity) from the rule base. Therefore, we have only two input variables: $x_1 = distance$, and $x_2 = angle$. The action (consequent) variable is the *deviation* of the robot from its path towards the target (Fig. 6.31).

Step 2: **Determine fuzzy sets**

In practice, all linguistic variables, linguistic values and their ranges are usually chosen by domain experts. For our problem, the fuzzy variable *distance* is represented using four linguistic terms: Very Near ($\tilde{VN}$), Near ($\tilde{N}$), Far ($\tilde{F}$), and Very Far ($\tilde{VF}$). Each of these terms is assumed to take a triangular membership function as shown in Fig. 6.32.

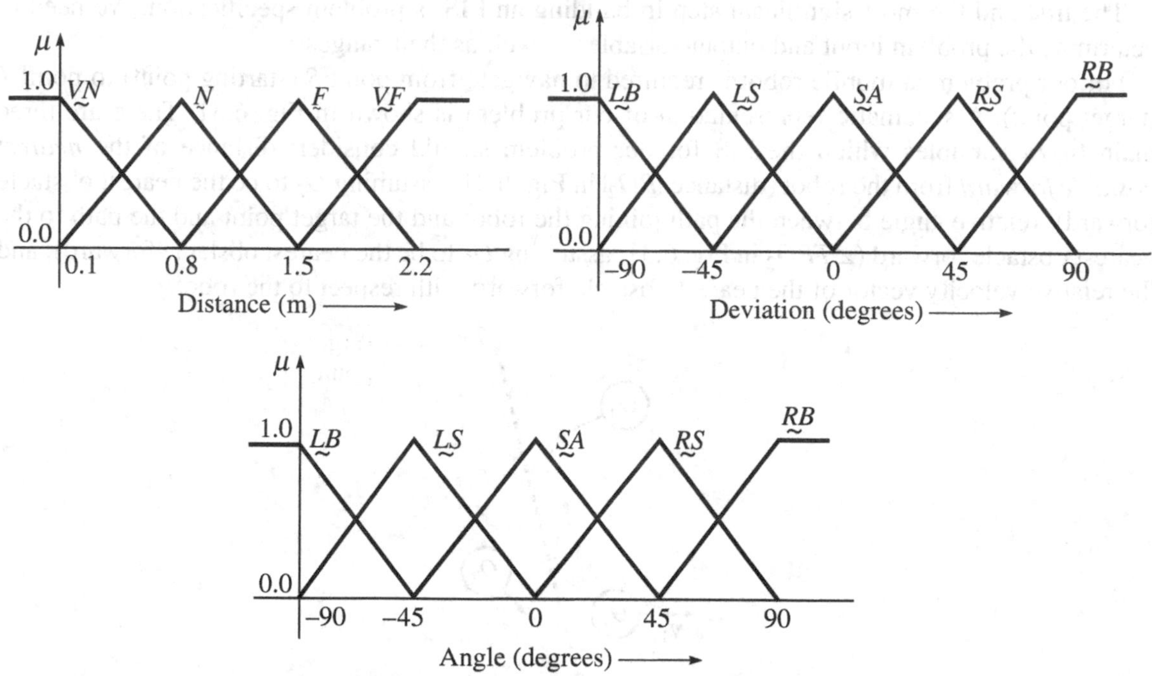

Figure 6.32 Database of the fuzzy motion planner

The fuzzy variable *angle* is represented using five linguistic terms: Left Big ($\tilde{LB}$), Left Small ($\tilde{LS}$), Straight Ahead ($\tilde{SA}$), Right Small ($\tilde{RS}$), and Right Big ($\tilde{RB}$). Each of these terms is also assumed to take a triangular membership function as shown in Fig. 6.32. The output (action) variable *deviation* is considered to have five fuzzy values: $\tilde{LB}, \tilde{LS}, \tilde{SA}, \tilde{RS}$, and $\tilde{RB}$. The triangular membership functions for *deviation* are also shown in Fig. 6.32.

Step 3: **Construct fuzzy rules**

Next we need to obtain fuzzy rules. There are two input variables and one output variable in our example. It is often convenient to represent fuzzy rules in a matrix form. For a two input-one output system, the rules are depicted as $K_1 \times K_2$ matrix, where K_1 corresponds to the number of linguistic terms of one input variable, and K_2 corresponds to the number of linguistic terms of the other. The linguistic values of one input variable form the rows of the matrix and the linguistic values of the other input variable form the columns. At the intersection of a row and a column lies the linguistic value of the output variable (Note that for a three input-one output system, the representation takes the form of a cube, and so on).

A typical rule will look like the following:

IF *distance* is $\underset{\sim}{VN}$ AND *angle* is $\underset{\sim}{SA}$ THEN *deviation* is $\underset{\sim}{LS}$

When an obstacle is very near and straight ahead, the robot deviates towards left by a small amount. However, when the obstacle is very near but on the left (big) of the robot, the robot goes straight ahead.

If *distance* is $\underset{\sim}{VN}$ AND *angle* is $\underset{\sim}{LB}$ THEN *deviation* is $\underset{\sim}{SA}$

With four choices of *distance* and five choices for *angle*, there could be a total of 4 × 5 or 20 combinations of the two antecedent variables possible. For each of these 20 combinations, there will be one value of the consequent variable. Thus, the maximum number of rules that may be present in the rule-base is 20. All the 20 rules which are used in this study are given in Table 6.3 [91].

Table 6.3 Rule-base of the fuzzy motion planner

Deviation		*Angle*				
		$\underset{\sim}{LB}$	$\underset{\sim}{LS}$	$\underset{\sim}{SA}$	$\underset{\sim}{RS}$	$\underset{\sim}{RB}$
Distance	$\underset{\sim}{VN}$	$\underset{\sim}{SA}$	$\underset{\sim}{RS}$	$\underset{\sim}{LS}$	$\underset{\sim}{LS}$	$\underset{\sim}{SA}$
	$\underset{\sim}{N}$	$\underset{\sim}{SA}$	$\underset{\sim}{SA}$	$\underset{\sim}{RB}$	$\underset{\sim}{SA}$	$\underset{\sim}{SA}$
	$\underset{\sim}{F}$	$\underset{\sim}{SA}$	$\underset{\sim}{SA}$	$\underset{\sim}{RS}$	$\underset{\sim}{SA}$	$\underset{\sim}{SA}$
	$\underset{\sim}{VF}$	$\underset{\sim}{SA}$	$\underset{\sim}{SA}$	$\underset{\sim}{SA}$	$\underset{\sim}{SA}$	$\underset{\sim}{SA}$

Step 4: **Encode the fuzzy sets and fuzzy rules, and set the design-procedure to perform fuzzy inference.**

The structure of the fuzzy inference system (FIS), shown in Fig. 6.30, comprises four principal components: knowledge base, inference mechanism, fuzzification interface, and defuzzification interface. The knowledge base (data-base and rule-base) for the problem is given in Fig. 6.32 and Table 6.3. The input fuzzifier takes the crisp numeric inputs and converts them into fuzzy form needed by the inference mechanism. At the output, the defuzzification interface combines the conclusions reached by the inference mechanism and converts them into crisp numeric values for robot steering action.

After defining fuzzy sets and fuzzy rules, the next step is to encode them along with the procedure for fuzzy inference. To accomplish this task, we may choose one of the two options: to build our system using a programming language such as C/C++ or to apply a fuzzy logic development tool such as 'MATLAB Fuzzy Logic Tool box'.

In the following paragraphs, we describe the inference procedure through a hand-calculation exercise.

Step 4(a): **Fuzzification**

The first step in the FIS design procedure is to take the crisp inputs x_1 = *distance*, and x_2 = *angle*, and determine the degree to which these inputs belong to each of the appropriate fuzzy sets.

Assume the given set of inputs to be $x_{10}(distance) = 1.04$ m; $x_{20}(angle) = 30$ degrees. Figure 6.33a shows fuzzy singletons for the given crisp inputs. From this figure, we find that *distance* of 1.04 m corresponds to membership functions $\underset{\sim}{N}$(Near) and $\underset{\sim}{F}$(Far) to the degrees of 0.66 and 0.34 respectively. In the same way, *angle* of 30 degrees corresponds to membership functions $\underset{\sim}{SA}$ (Straight Ahead) and $\underset{\sim}{RS}$ (Right Small) with values 0.33 and 0.67, respectively.

Step 4(b): **Rule evaluation**

The second step in the FIS design procedure is to take the fuzzified inputs $\mu_{\underset{\sim}{N}}(x_{10}) = 0.66$, $\mu_{\underset{\sim}{F}}(x_{10}) = 0.34$, $\mu_{\underset{\sim}{SA}}(x_{20}) = 0.33$, and $\mu_{\underset{\sim}{RS}}(x_{20}) = 0.67$, and apply them to the antecedents of the fuzzy rules. To evaluate the conjunction of the rule antecedents, we apply the *min t*-norm and to evaluate the disjunction, we apply the *max t*-conorm. The antecedent evaluation results in a single number that represents the degree of fulfillment (firing strength) of a rule.

For the given inputs, the following four rules are being fired from a total of 20:

IF *distance* is $\underset{\sim}{N}$ AND *angle* is $\underset{\sim}{SA}$ THEN *deviation* is $\underset{\sim}{RB}$

IF *distance* is $\underset{\sim}{N}$ AND *angle* is $\underset{\sim}{RS}$ THEN *deviation* is $\underset{\sim}{SA}$

IF *distance* is $\underset{\sim}{F}$ AND *angle* is $\underset{\sim}{SA}$ THEN *deviation* is $\underset{\sim}{RS}$

IF *distance* is $\underset{\sim}{F}$ AND *angle* is $\underset{\sim}{RS}$ THEN *deviation* is $\underset{\sim}{SA}$

The degree of fulfillment (firing strength) of the fired rules are calculated as follows:

$$\alpha^{(1)} = \min(\mu_{\underset{\sim}{N}}(x_{10}), \mu_{\underset{\sim}{SA}}(x_{20})) = \min(0.66, 0.33) = 0.33$$

$$\alpha^{(2)} = \min(\mu_{\underset{\sim}{N}}(x_{10}), \mu_{\underset{\sim}{RS}}(x_{20})) = \min(0.66, 0.67) = 0.66$$

$$\alpha^{(3)} = \min(\mu_{\underset{\sim}{F}}(x_{10}), \mu_{\underset{\sim}{SA}}(x_{20})) = \min(0.34, 0.33) = 0.33$$

$$\alpha^{(4)} = \min(\mu_{\underset{\sim}{F}}(x_{10}), \mu_{\underset{\sim}{RS}}(x_{20})) = \min(0.34, 0.67) = 0.34$$

With *min fuzzy implication*, the fuzzified outputs (implied fuzzy sets) corresponding to the above four rules are shown in Fig. 6.33b. Figure 6.33c shows union of these fuzzified outputs (aggregated implied fuzzy set).

Step 4(c): **Defuzzification**

The COG method is a well-balanced method, sensitive to the height and width of the total fuzzy region. Therefore, this technique may be used for defuzzification unless we have a strong reason to believe that our fuzzy system will behave better under other defuzzification methods.

As said earlier, a reasonable estimate of COG can be obtained by calculating it over a sample of points (refer to Eqn (6.71a)). This can be achieved by discretizing the universe. Alternatively, we can split up geometry into pieces and place a straight edge (centroid) through area of each piece to have it perfectly balanced with equal area on either side. Equation (6.71b) is then used to compute COG. For example, the shaded region in Fig. 6.33c representing combined output (aggregated fuzzy set) corresponding to the four fired rules can be divided into four regular sub-regions; two triangles and two rectangles. Area and center of area of each of these regions can easily be determined. Crisp output can then be obtained using Eqn (6.71b). We leave this as an exercise for the reader; and instead, illustrate the method given by Eqn (6.72) that uses implied fuzzy sets of the rules.

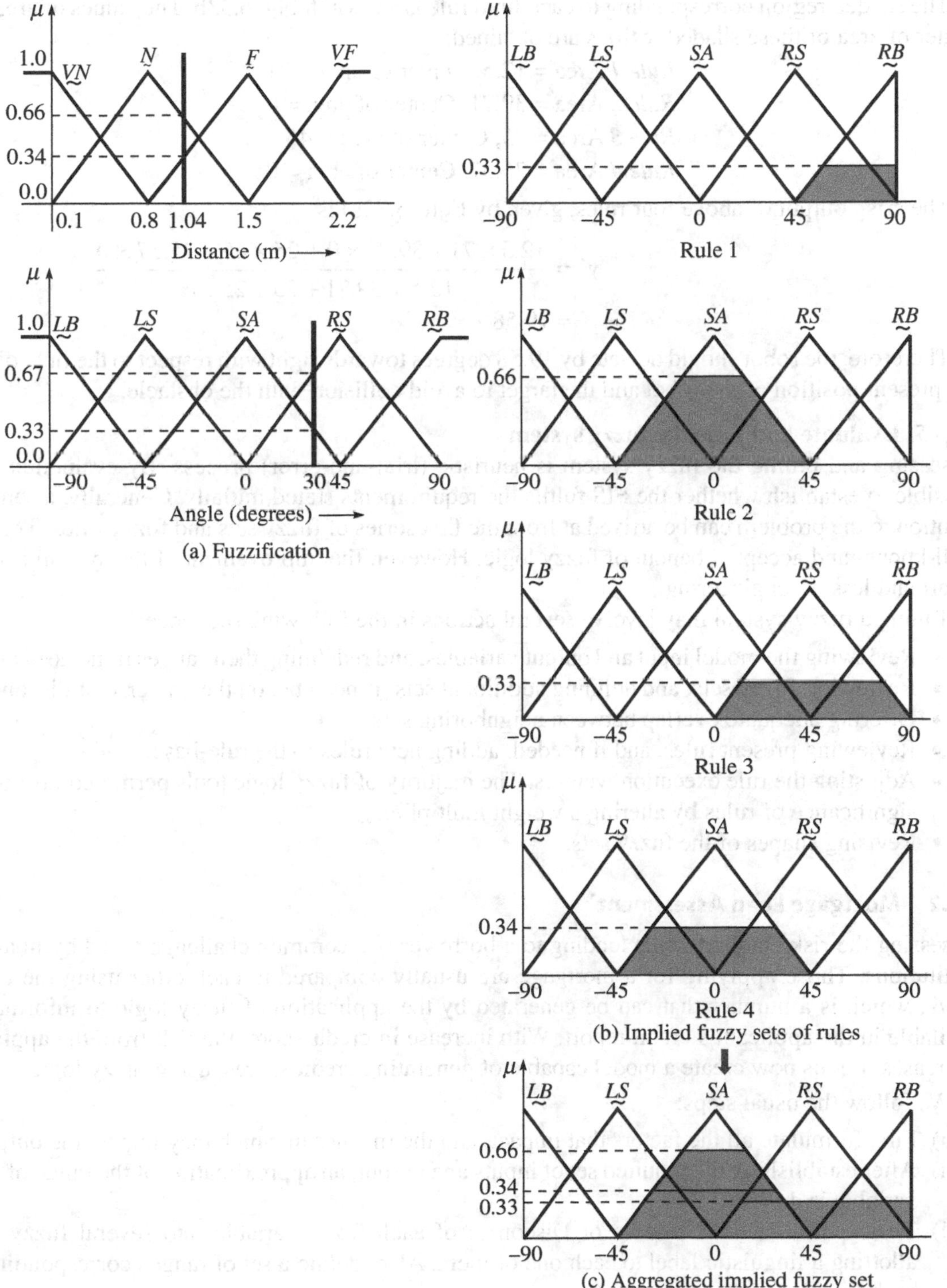
μ
1.0
0.66
0.34
0.0
VN
N
F
VF
0.1
0.8
1.04
1.5
2.2
Distance (m)
μ
1.0
0.67
0.33
0.0
LB
LS
SA
RS
RB
-90
-45
0
30
45
90
Angle (degrees)
(a) Fuzzification
μ
LB
LS
SA
RS
RB
0.33
-90
-45
0
45
90
Rule 1
0.66
Rule 2
0.33
Rule 3
0.34
Rule 4
(b) Implied fuzzy sets of rules
0.66
0.34
0.33
(c) Aggregated implied fuzzy set

Figure 6.33

The shaded region corresponding to each fired rule is shown in Fig. 6.33b. The values of area and center of area of these shaded regions are obtained:

Rule 1 Area = 12.5, Center of area = 71
Rule 2 Area = 39.71, Center of area = 0
Rule 3 Area = 25, Center of area = 45
Rule 4 Area = 25.57, Center of area = 0

The crisp output of above four rules, given by Eqn (6.72b) is

$$y' = \frac{12.5 \times 71 + 39.71 \times 0 + 25 \times 45 + 25.57 \times 0}{12.5 + 39.71 + 25 + 25.57}$$
$$= 19.58$$

Therefore, the robot should deviate by 19.58 degrees towards right with respect to the line joining the present position of the robot and the target to avoid collision with the obstacle.

Step 5: **Evaluate and tune the fuzzy system**

Assessing and tuning the fuzzy system is heuristic (trial-and-error) process. By evaluation, it is possible to establish whether the FIS fulfils the requirements stated initially. Generally, a sensible solution to the problem can be arrived at from the first series of fuzzy sets and fuzzy rules. This is a well-known and accepted benefit of fuzzy logic. However, the improvement of the system is quite an art and less of engineering.

Tuning a fuzzy system may involve several actions in the following sequence:

- Reviewing the model input and output variables, and redefining their ranges if the need arises.
- Reviewing fuzzy sets, and defining additional sets, if need be, on the universe of discourse.
- Offering adequate overlap between neighboring sets.
- Reviewing present rules, and if needed, adding new rules to the rule-base.
- Adjusting the rule execution weights. The majority of fuzzy logic tools permit control of the significance of rules by altering a weight multiplier.
- Revising shapes of the fuzzy sets.

6.5.2 Mortgage Loan Assessment

Assessing the risk entailed while lending to a borrower is a common challenge faced by financial institutions. Those applying for a mortgage are usually compared to each other using the *credit score*, which is a number that can be generated by the application of fuzzy logic to information available in the applicant's credit report. With increase in credit score, the risk from the applicant decreases. Let us now create a model capable of generating credit scores, using fuzzy logic.

We follow the usual steps:

(i) First formulate all the factors that impact, and the manner in which they impact the output.
(ii) After establishing the required set of inputs and output, an approximation of the range of each variable is done.
(iii) Then, partition the Universe of Discourse of each fuzzy variable into several fuzzy sets, allotting a linguistic label to each one of them. Also, define a set of ranges corresponding to each linguistic label.

(iv) The next step is to create a fuzzy rule-base to establish the manner in which a variable influences the result.
(v) Then the fuzzy system can be modeled with the help of a fuzzy logic development tool.
(vi) Finally, evaluate and tune the fuzzy inference system.

Membership functions can be defined and fuzzy rules constructed based on expert advice from mortgage advisors who have experience. The help of bank managers is also sought for developing the mortgage granting policies. For our problem, the input and output variables are:

- *Input Variables*
  1. Consumer Evaluation [0–10]
  2. Market Value of House [0–1000 × 10^3] $
  3. Income [0–100, 000] $
  4. Interest on Loan [0–10] %
- *Output Variable*
  Credit Score [0–10]

Figure 6.34 depicts fuzzy sets for linguistic variables employed in our problem. Triangular and trapezoidal membership functions can appropriately represent the knowledge of the mortgage expert.

Decision-support systems may include hundreds of fuzzy rules. For example, a fuzzy system for *Credit Score* evaluation developed by BMW Bank used 413 fuzzy rules [102]. We should examine whether the knowledge base allows dividing the problem into several modules deployed in a hierarchical structure. For our problem, the range of *Consumer Evaluation* is 0–10: a normalized value. The value of this variable follows from fuzzy evaluation of *Demographics*, *Finance*, and *Financial Security*. A module with inputs *Demographics*, *Finance*, and *Financial Security* gives the output *Consumer Evaluation*, which in turn, becomes an input for *Credit Score* evaluation module. Further module division deals with dependence of *Demographics* on *Age*, *Education*, *Marital Status*, and *Number of Children*; *Finance* on *Income*, *Length of Employment*, and *Type of Employment*, and *Financial Security* on current *Living Arrangement*, *Value of Car*, and *Value of Assets*.

For the purpose of illustration, we give here an oversimplified rule-base for *Credit Score* evaluation module [5].

Let

x_1 = Market Value of House, with linguistic values: *Very Low*, *Low*, *Medium*, *High*, *Very High*.

x_2 = Income, with linguistic values: *Low*, *Medium*, *High*, *Very High*

x_3 = Interest on Loan, with linguistic values: *Low*, *Medium*, *High*

x_4 = Consumer Evaluation, with linguistic values: *Low*, *Medium*, *High*

y = Credit Score, with linguistic values: *Very Low*, *Low*, *Medium*, *High*, *Very High*

A rule-base for *Credit Score* evaluation follows.

Rule 1: IF x_2 is *Low* AND x_3 is *Medium* THEN y is *Very Low*

Rule 2: IF x_2 is *Low* AND x_3 is *High* THEN y is *Very Low*

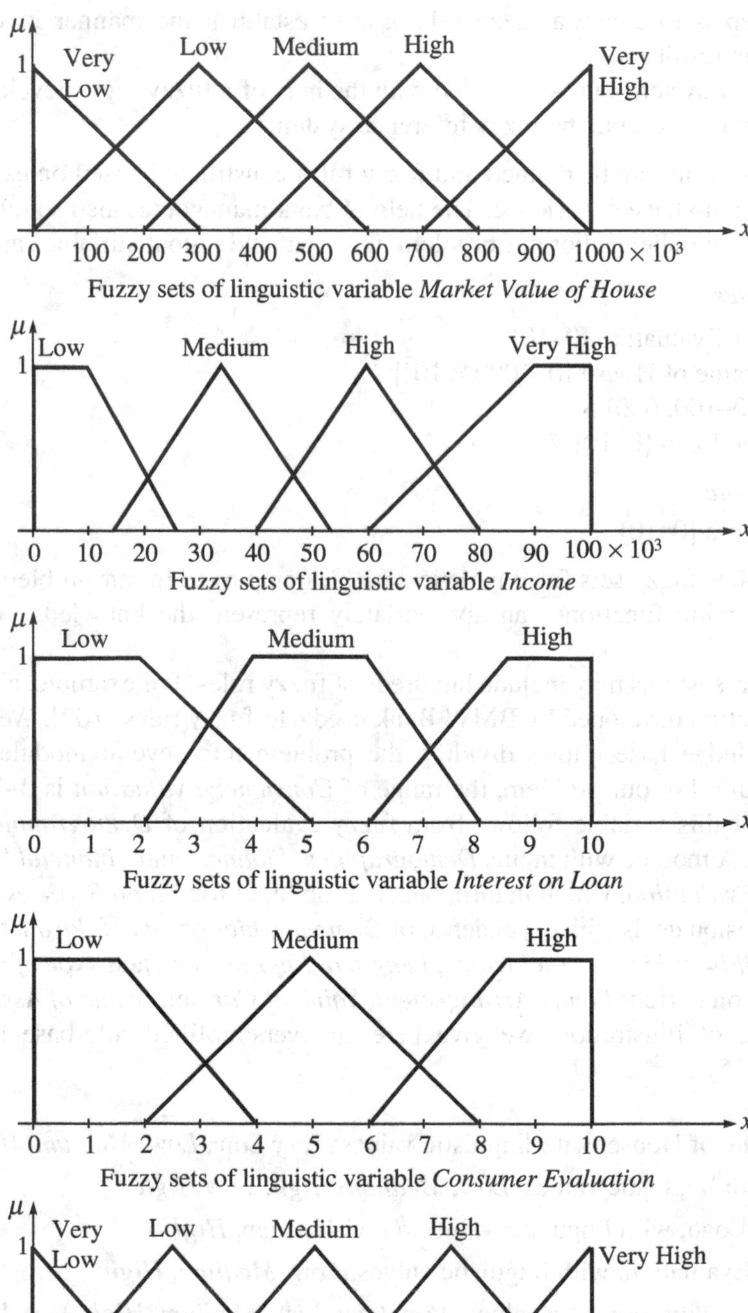
μ
1
Very Low
Low
Medium
High
Very High
x_1
0 100 200 300 400 500 600 700 800 900 1000 × 10³
Fuzzy sets of linguistic variable *Market Value of House*
μ
1
Low
Medium
High
Very High
x_2
0 10 20 30 40 50 60 70 80 90 100 × 10³
Fuzzy sets of linguistic variable *Income*
μ
1
Low
Medium
High
x_3
0 1 2 3 4 5 6 7 8 9 10
Fuzzy sets of linguistic variable *Interest on Loan*
μ
1
Low
Medium
High
x_4
0 1 2 3 4 5 6 7 8 9 10
Fuzzy sets of linguistic variable *Consumer Evaluation*
μ
1
Very Low
Low
Medium
High
Very High
y
0 1 2 3 4 5 6 7 8 9 10
Fuzzy sets of linguistic variable *Credit Score*

Figure 6.34

Rule 3: IF x_2 is *Medium* AND x_3 is *High* THEN y is *Low*
Rule 4: IF x_4 is *Low* THEN y is *Very Low*
Rule 5: IF x_1 is *Very Low* THEN y is *Very Low*
Rule 6: IF x_4 is *Medium* AND x_1 is *Very Low* THEN y is *Low*
Rule 7: IF x_4 is *Medium* AND x_1 is *Low* THEN y is *Low*
Rule 8: IF x_4 is *Medium* AND x_1 is *Medium* THEN y is *Medium*
Rule 9: IF x_4 is *Medium* AND x_1 is *High* THEN y is *High*
Rule 10: IF x_4 is *Medium* AND x_1 is *Very High* THEN y is *High*
Rule 11: IF x_4 is *High* AND x_1 is *Very Low* THEN y is *Low*
Rule 12: IF x_4 is *High* AND x_1 is *Low* THEN y is *Medium*
Rule 13: IF x_4 is *High* AND x_1 is *Medium* THEN y is *High*
Rule 14: IF x_4 is *High* AND x_1 is *High* THEN y is *High*
Rule 15: IF x_4 is *High* AND x_1 is *Very High* THEN y is *Very High*

To build our system, we may use the MATLAB fuzzy Logic Tool box, or any other fuzzy system software tool. We leave this as an exercise for the reader. Hand calculations will help the reader understand the design procedure.

6.6 TAKAGI-SUGENO FUZZY MODEL

As said earlier in Section 6.5, two typical fuzzy inference systems are based on Mamdani model and Takagi-Sugeno (T-S) fuzzy model. The Mamdani model results in a 'standard' fuzzy system, and is widely used for capturing expert knowledge. It allows us to describe the expertise in more intuitive, more human-like form. The underlying nonlinearity of the decision-making system is shaped heuristically in the design process. Explicit nonlinear system identification is not done in this case.

On the other hand, the Takagi-Sugeno method follows an indirect approach: identification of the nonlinear system in terms of a T-S fuzzy model is first carried out; the T-S fuzzy model is then used for designing decision-support system.

If a nonlinear system is effectively represented as a fuzzy cluster of linear systems defined locally, then understanding of the complex nonlinear system can become more profound as linear systems are well understood. Such a powerful representation has been introduced by Takagi, Sugeno, and Kang [97–99], which we will refer to as T-S fuzzy model. Numerous research works have been carried out using the T-S fuzzy model, as such a representation can provide a better understanding.

The T-S fuzzy modeling is thus a *multi-model* approach for modeling complex nonlinear systems, in which simple sub-models (typically linear models) are combined to describe the global behavior of the system. The idea of a multi-model approach is not new, but the idea of building multi-models using fuzzy set theory offers a new technique.

The rule base of a fuzzy inference system consists of a set of R fuzzy IF-THEN rules of the form:

'If a set of conditions is satisfied THEN a set of consequences can be inferred'

Different types of consequent parts have been used in fuzzy rules. In the earlier section, we have studied fuzzy rules based on Mamdani's approach, in which linguistic term associated with a membership function is used as consequent. The T-S fuzzy model is based on a different approach: instead of a linguistic term with an associated membership function, we use a *function* in the consequent of a rule, which does not have an associated membership function. Typically linear functions are used as consequent of the R rules; the overall fuzzy model of a nonlinear system is achieved by fuzzy 'blending' of the R linear system models.

One mapping that has proven to be particularly useful is to have linear dynamic system as the output function, so that r^{th} rule has the form ($r = 1, \ldots, R$):

Rule r: IF $\underset{\sim}{x}_1$ is $\underset{\sim}{A}_{1k_1}$ AND $\underset{\sim}{x}_2$ is $\underset{\sim}{A}_{2k_2}$ AND $\cdots$ AND $\underset{\sim}{x}_n$ is $\underset{\sim}{A}_{nk_n}$

$$\text{THEN } \dot{\mathbf{x}}^{(r)}(t) = \mathbf{A}^{(r)}\mathbf{x}(t) + \mathbf{B}^{(r)}\mathbf{u}(t) \tag{6.76}$$

where $\mathbf{x} = [x_1\ x_2\ \cdots\ x_n]^T$ is n-dimensional state vector and $\mathbf{u} = [u_1\ u_2\ \cdots\ u_m]^T$ is m-dimensional input vector. Each rule represents a fuzzy zone in the state space. For an n-dimensional state-space system, the number of such fuzzy rules (and hence fuzzy zones) is

$$K_1 \times K_2 \times \cdots \times K_n \tag{6.77a}$$

That is, the n-dimensional input space is divided into $K_1 \times K_2 \times \cdots \times K_n$ fuzzy partition spaces:

$$(\underset{\sim}{A}_{1k_1}, \underset{\sim}{A}_{2k_2}, \ldots, \underset{\sim}{A}_{nk_n});\ k_1 = 1, \ldots, K_1, \ldots, k_n = 1, \ldots, K_n$$

or
$$\{\underset{\sim}{A}_{jk_j}\};\ j = 1, \ldots, n,\ k_j = 1, 2, \ldots, K_j \tag{6.77b}$$

Thus, the nonlinear dynamic system is composed of R such rules where each rule is associated with a local linear model, with parameters $\mathbf{A}^{(r)}$ and $\mathbf{B}^{(r)}$.

The decision-making logic employs fuzzy IF-THEN rules from the rule-base to infer the output by a fuzzy reasoning method. The contribution of each local linear model (i.e., each rule) in the estimated output of the FIS is dictated by the *firing strength* of the rule. We use *product* strategy to assign firing strength $\mu^{(r)}$ to each rule $r = 1, \ldots, R$:

$$\mu^{(r)} = \mu_{1k_1}(x_1) \times \mu_{2k_2}(x_2) \times \cdots \times \mu_{nk_n}(x_n) \tag{6.78a}$$

$$= \prod_{(j,\,k_j) \in I_r} \mu_{jk_j}(x_j) \tag{6.78b}$$

where I_r is the set of all $\underset{\sim}{A}_{jk_j}$ associated with the premise part of rule r.

Given a current state vector $\mathbf{x}$, and an input vector $\mathbf{u}$, the T-S fuzzy model infers the system dynamics as,

$$\dot{\mathbf{x}} = \frac{1}{\sum_{r=1}^{R} \mu^{(r)}} \sum_{r=1}^{R} \mu^{(r)}(\mathbf{A}^{(r)}\mathbf{x} + \mathbf{B}^{(r)}\mathbf{u}) \tag{6.79}$$

By defining the normalized membership grade associated with the rth rule as,

$$\bar{\mu}^{(r)} = \frac{\mu^{(r)}}{\sum_{r=1}^{R} \mu^{(r)}} \tag{6.80}$$

the T-S fuzzy model representation of any nonlinear continuous-time system is expressed as,

$$\dot{\mathbf{x}} = \sum_{r=1}^{R} \bar{\mu}^{(r)} (\mathbf{A}^{(r)}\mathbf{x} + \mathbf{B}^{(r)}\mathbf{u}) \tag{6.81}$$

The linear model parameters $\mathbf{A}^{(r)}$ and $\mathbf{B}^{(r)}$ can be found from the input-output dataset using a fuzzy neural network [103, 104].

If the nonlinear model of a system is exactly known, then the T-S fuzzy representation can be derived using linearization techniques [85], i.e., the parameters $\mathbf{A}^{(r)}$ and $\mathbf{B}^{(r)}$ can be directly obtained from the system dynamics by linearizing the nonlinear dynamics at different operating points.

Given a nonlinear system in terms of T-S fuzzy model, various control schemes can be designed using traditional control techniques [85].

T-S Fuzzy Model with Linear Static Mapping as Rule Consequents

An alternative setting in which T-S fuzzy models have been used is to use a linear static mapping as rule consequents. For example,

Rule r: IF $\underset{\sim}{x}_1$ is $\underset{\sim}{A}_{1k_1}$ AND $\cdots$ AND $\underset{\sim}{x}_n$ is $\underset{\sim}{A}_{nk_n}$

$$\text{THEN } \hat{y}^{(r)} = a_0^{(r)} + a_1^{(r)}x_1 + \cdots + a_n^{(r)}x_n;\ r = 1, 2, \ldots, R \tag{6.82}$$

The consequent part is a linear function of the input variables x_j; $j = 1, \ldots, n$; and $a_0, a_1, \ldots, a_n$ are the $(n + 1)$ parameters that determine the real consequent value.

We limit our discussion to linear mappings of this form, and describe a procedure for finding the parameters $a_0, a_1, \ldots, a_n$ from the input-output data.

T-S Fuzzy Model from Input-output Data: We consider here, a single-output FIS in the n-dimensional input space. Let us assume that the following N input-output pairs are given as training data for constructing FIS model:

$$\{\mathbf{x}^{(i)}, y^{(i)} |\ i = 1, 2, \ldots, N\} \tag{6.83}$$

where $\mathbf{x}^{(i)} = [x_1^{(i)}\ x_2^{(i)} \cdots x_n^{(i)}]^T$ is the input vector of ith input-output pair and $y^{(i)}$ is the corresponding output.

The *fuzzification* interface transforms crisp values of input variables into fuzzy singletons. A singleton is a fuzzy set, possessing a membership function that is unity at a single particular point on the universe of discourse (the numerical-data value), and zero everywhere else. Basically, a fuzzy singleton is an accurate value, and therefore, fuzzification does not introduce any fuzziness in this case. This strategy has been employed widely in fuzzy modeling applications, as it is implemented with ease. Two aspects define a *database* (i) a fuzzy partition of an input space, and (ii) membership functions of antecedent fuzzy sets. Assume that the domain interval of the *j*th input variable x_j is equally divided into K_j fuzzy sets labeled $\underset{\sim}{A}_{j1}, \underset{\sim}{A}_{j2}, \ldots, \underset{\sim}{A}_{jK_j}$, for $j = 1, 2, \ldots, n$. Then the n-dimensional input space is divided into $K_1 \times K_2 \times \cdots \times K_n$ fuzzy partition spaces:

$$(\underset{\sim}{A}_{1k_1}, \underset{\sim}{A}_{2k_2}, \ldots, \underset{\sim}{A}_{nk_n});\ k_1 = 1, 2, \ldots, K_1; \ldots; k_n = 1, \ldots, K_n \tag{6.84}$$

Though any type of membership functions (e.g., triangle-shaped, trapezoid-shaped, bell-shaped, etc.) can be used for fuzzy sets, we employ the symmetric triangle-shaped fuzzy sets, $\underset{\sim}{A}_{jk_j}$, with the following membership functions:

$$\mu_{\underset{\sim}{A}_{jk_j}}(x_j) \triangleq \mu_{jk_j}(x_j) = 1 - \frac{2\left|x_j - c_{(j,k_j)}\right|}{w_{(j,k_j)}};\ k_j = 1, 2, \ldots, K_j \tag{6.85}$$

$c_{(j,k_j)}$ is the center of the membership function, where the membership grade is equal to 1, and $w_{(j,k_j)}$ denotes the width of the membership function (Fig. 6.35).

By means of the input-output data, the range $[x_j^{\min}, x_j^{\max}]$ of the *j*th input variable is determined, where

$$x_j^{\min} = \min_{i=\{1,\ldots,N\}} x_j^{(i)},\ x_j^{\max} = \max_{i=\{1,\ldots,N\}} x_j^{(i)} \tag{6.86a}$$

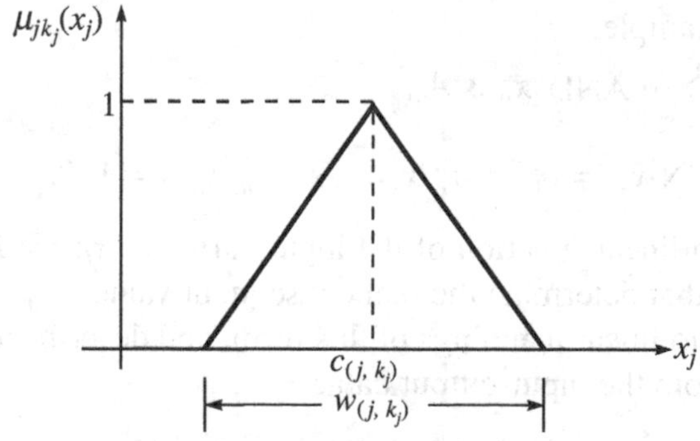

Figure 6.35 Parameters of a membership function

The position of center of each membership function with respect to the *j*th variable is determined by,

$$c_{(j,k_j)} = x_j^{\min} + (k_j - 1)\,[x_j^{\max} - x_j^{\min})/(K_j - 1];\ c_{(j,1)} = x_j^{\min};\ c_{(j,K_j)} = x_j^{\max} \tag{6.86b}$$

To achieve sufficient overlap from one linguistic label to another, we take,

$$w_{(j,k_j)} = 2(c_{(j,k_j+1)} - c_{(j,k_j)}) \tag{6.86c}$$

Figure 6.36 shows an example where the domain interval of x_1 is divided into $K_1 = 5$ fuzzy sets.

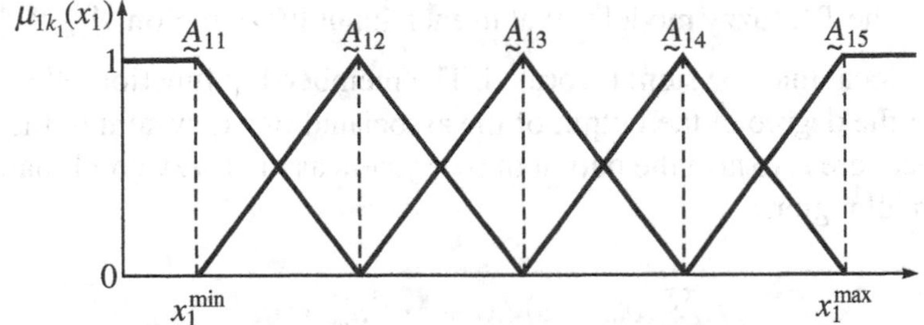

Figure 6.36 Fuzzy partition of an input space and membership functions of fuzzy sets

The rule base consists of a set of fuzzy IF-THEN rules in the form 'IF a set of conditions is satisfied THEN a set of consequences can be inferred'. Different types of consequent parts have been used in fuzzy rules; here we focus on linear static mapping in T-S fuzzy architecture: the domain interval of y is represented by R linear functions, giving rise to R fuzzy rules. All the rules corresponding to the possible combinations of the inputs are implemented. The total number of rules R for an n-input system is: $K_1 \times K_2 \times \cdots \times K_n$.

The format of fuzzy rules is,

Rule r: IF x_1 is $\underset{\sim}{A}_{1k_1}$ AND $\cdots$ AND x_n is $\underset{\sim}{A}_{nk_n}$ THEN

$$\hat{y}^{(r)} = a_0^{(r)} + a_1^{(r)}x_1 + \cdots + a_n^{(r)}x_n;\; r = 1, 2, \ldots, R \tag{6.87}$$

The consequent part is a linear function of the input variables x_j; $a_0, a_1, \ldots, a_n$ are the $(n + 1)$ parameters that determine the real consequent value. The aim of the linear function is to describe the local linear behavior of the system. Each rule r gives rise to a local linear model. The selected R rules are required to approximate the function that theoretically underlines the system behavior most consistently, with the given sample of input-output data (6.83) (when $\hat{y}$ is a constant in (6.87), we get a standard fuzzy model in which the consequent of a rule is specified by a singleton).

The decision-making logic employs fuzzy IF-THEN rules from the rule base to infer the output by a fuzzy reasoning method. The contribution of each local linear model (i.e., each rule) in the estimated output of the FIS is dictated by the *firing strength* of the rule. We use *product strategy* to assign firing strength $\mu^{(r)}$ to each rule $r = 1, 2, \ldots, R$.

Given an input vector, $\mathbf{x}^{(i)} = [x_1^{(i)}\; x_2^{(i)} \cdots x_n^{(i)}]^T$, the degree of compatibility of $\mathbf{x}^{(i)}$ to the rth fuzzy IF-THEN rule is the firing strength $\mu^{(r)}$ of the rule, and is given by (note that we have used **product** *t*-norm operator on the premise part of the rule),

$$\begin{aligned}\mu^{(r)}(\mathbf{x}^{(i)}) &= \mu_{1k_1}(x_1^{(i)}) \times \mu_{2k_2}(x_2^{(i)}) \times \cdots \times \mu_{nk_n}(x_n^{(i)}) \\ &= \prod_{(j,\, k_j) \in I_r} \mu_{jk_j}(x_j^{(i)})\end{aligned} \tag{6.88}$$

where I_r is the set of all $\underset{\sim}{A}_{jk_j}$ associated with the premise part of rule r.

The main idea of the T-S fuzzy model is that in each input fuzzy region $\underset{\sim}{A}_{1k_1} \times \underset{\sim}{A}_{2k_2} \times \cdots \times \underset{\sim}{A}_{nk_n}$ of the input domain, a local linear system is formed. The membership function $\mu^{(r)}(\mathbf{x}^{(i)})$ of each region is a map indicating the degree of the output of the associated linear system to the region. A simple *defuzzification* procedure is to take the output of the system as the fuzzy combination of the outputs of local systems in all regions:

$$\hat{y} = \frac{\sum_{r=1}^{R} (a_0^{(r)} + a_1^{(r)} x_1 + \cdots + a_n^{(r)} x_n)\mu^{(r)}}{\sum_{r=1}^{R} \mu^{(r)}} \tag{6.89a}$$

$$= \sum_{r=1}^{R} (a_0^{(r)} + a_1^{(r)} x_1 + \cdots + a_n^{(r)} x_n)\bar{\mu}^{(r)} \tag{6.89b}$$

where

$$\bar{\mu}^{(r)} = \frac{\mu^{(r)}}{\sum_{r=1}^{R} \mu^{(r)}} \tag{6.89c}$$

is the *normalized firing strength* of rule r; a ratio of firing strength of rule r to the sum of the firing strengths of all the rules.

Note that the output of the fuzzy model can be determined only if the parameters in rule consequents are known. However, it is often difficult or even impossible to specify a rule consequent in a polynomial form. Fortunately, it is not necessary to have any prior knowledge of rule consequent parameters for the T-S fuzzy modeling approach to deal with a problem. These parameters can be determined using *least squares estimation method* as follows.

Given the values of the membership parameters and a training set of N input-output patterns $\{\mathbf{x}^{(i)}, y^{(i)}; i = 1, 2, \ldots, N\}$, we can form N linear equations in terms of the consequent parameters.

$$\begin{aligned} y^{(i)} = {} & \bar{\mu}^{(1)}(\mathbf{x}^{(i)})[a_0^{(1)} + a_1^{(1)} x_1^{(i)} + \cdots + a_n^{(1)} x_n^{(i)}] + \bar{\mu}^{(2)}(\mathbf{x}^{(i)})[a_0^{(2)} + a_1^{(2)} x_1^{(i)} + \cdots + a_n^{(2)} x_n^{(i)}] + \cdots \\ & + \bar{\mu}^{(R)}(\mathbf{x}^{(i)})[a_0^{(R)} + a_1^{(R)} x_1^{(i)} + \cdots + a_n^{(R)} x_n^{(i)}]; i = 1, 2, \cdots N \end{aligned} \tag{6.90}$$

where $\bar{\mu}^{(r)}(\mathbf{x}^{(i)})$ is the normalized firing strength of rule r, fired by the input pattern $\mathbf{x}^{(i)}$.

In terms of vectors

$$\begin{aligned} \bar{\mathbf{x}}^{(i)} &= [1 \;\; x_1^{(i)} \;\; x_2^{(i)} \cdots x_n^{(i)}]^T \\ \boldsymbol{\theta}^{(r)} &= [a_0^{(r)} \;\; a_1^{(r)} \cdots a_n^{(r)}] \\ \boldsymbol{\Theta} &= [a_0^{(1)} \;\; a_1^{(1)} \cdots a_n^{(1)} \;\; a_0^{(2)} \cdots a_n^{(2)} \cdots a_0^{(R)} \cdots a_n^{(R)}] \end{aligned} \tag{6.91}$$

we can write the N linear equations as follows:

$$
\begin{aligned}
y^{(1)} &= \bar{\mu}^{(1)}(\mathbf{x}^{(1)})\,[\boldsymbol{\theta}^{(1)}\bar{\mathbf{x}}^{(1)}] + \bar{\mu}^{(2)}(\mathbf{x}^{(1)})\,[\boldsymbol{\theta}^{(2)}\bar{\mathbf{x}}^{(1)}] + \cdots + \bar{\mu}^{(R)}(\mathbf{x}^{(1)})\,[\boldsymbol{\theta}^{(R)}\bar{\mathbf{x}}^{(1)}] \\
y^{(2)} &= \bar{\mu}^{(1)}(\mathbf{x}^{(2)})\,[\boldsymbol{\theta}^{(1)}\bar{\mathbf{x}}^{(2)}] + \bar{\mu}^{(2)}(\mathbf{x}^{(2)})\,[\boldsymbol{\theta}^{(2)}\bar{\mathbf{x}}^{(2)}] + \cdots + \bar{\mu}^{(R)}(\mathbf{x}^{(2)})\,[\boldsymbol{\theta}^{(R)}\bar{\mathbf{x}}^{(2)}] \\
&\ \vdots \\
y^{(N)} &= \bar{\mu}^{(1)}(\mathbf{x}^{(N)})\,[\boldsymbol{\theta}^{(1)}\bar{\mathbf{x}}^{(N)}] + \bar{\mu}^{(2)}(\mathbf{x}^{(N)})\,[\boldsymbol{\theta}^{(2)}\bar{\mathbf{x}}^{(N)}] + \cdots + \bar{\mu}^{(R)}(\mathbf{x}^{(N)})\,[\boldsymbol{\theta}^{(R)}\bar{\mathbf{x}}^{(N)}]
\end{aligned}
\tag{6.92}
$$

These N equations can be rearranged into a single vector-matrix equation:

$$
\begin{bmatrix} y^{(1)} \\ y^{(2)} \\ \vdots \\ y^{(N)} \end{bmatrix} =
\begin{bmatrix}
[\bar{\mathbf{x}}^{(1)}]^T \bar{\mu}^{(1)}(\mathbf{x}^{(1)}) & [\bar{\mathbf{x}}^{(1)}]^T \bar{\mu}^{(2)}(\mathbf{x}^{(1)}) \cdots [\bar{\mathbf{x}}^{(1)}]^T \bar{\mu}^{(R)}(\mathbf{x}^{(1)}) \\
[\bar{\mathbf{x}}^{(2)}]^T \bar{\mu}^{(1)}(\mathbf{x}^{(2)}) & [\bar{\mathbf{x}}^{(2)}]^T \bar{\mu}^{(2)}(\mathbf{x}^{(2)}) \cdots [\bar{\mathbf{x}}^{(2)}]^T \bar{\mu}^{(R)}(\mathbf{x}^{(2)}) \\
\vdots & \\
[\bar{\mathbf{x}}^{(N)}]^T \bar{\mu}^{(1)}(\mathbf{x}^{(N)}) & [\bar{\mathbf{x}}^{(N)}]^T \bar{\mu}^{(2)}(\mathbf{x}^{(N)}) \cdots [\bar{\mathbf{x}}^{(N)}]^T \bar{\mu}^{(R)}(\mathbf{x}^{(N)})
\end{bmatrix}
\begin{bmatrix} [\boldsymbol{\theta}^{(1)}]^T \\ [\boldsymbol{\theta}^{(2)}]^T \\ \vdots \\ [\boldsymbol{\theta}^{(R)}]^T \end{bmatrix}
\tag{6.93a}
$$

or

$$\mathbf{y} = \mathbf{X}^T\,\boldsymbol{\Theta}^T \tag{6.93b}$$

Solution methods for least squares estimation problem have earlier been described in Section 3.8. The optimal solution to this classic problem in calculus is given by (refer to Eqn (3.78)),

$$\boldsymbol{\Theta}^{*T} = (\mathbf{X}\mathbf{X}^T)^{-1}\mathbf{X}\mathbf{Y} = \mathbf{X}^{+}\mathbf{y} \tag{6.93c}$$

where $\mathbf{X}^{+}$ is the *pseudoinverse* of matrix $\mathbf{X}^T$.

In the T-S fuzzy model given above, we have used the most intuitive approach of implementing all possible combinations of the given fuzzy sets as rules. In fact, if data is not uniformly distributed, some rules may never be fired. This and other drawbacks are handled by many variants of the basic ANFIS model, described in the next section.

6.7 NEURO-FUZZY INFERENCE SYSTEMS

The intelligent technologies—fuzzy inference systems (FIS) and neural networks (NN)—both have same objectives. Both try to mimic human intelligence and ultimately generate a machine with intelligence. But, the means they use to attain their objectives differ. The theory of fuzzy logic, which finds basis in the idea of graded membership, offers mathematical power to emulate cognitive functions—the processes of thought and perception. In neural information processing, there are a variety of complex mathematical operations and mapping functions involved, that, in synergism, act as a parallel computing machine that emulates the biological neuronal processes. Therefore, while an FIS is dependent on logical inferences and concentrates on modeling human reasoning, an NN depends on parallel data processing and concentrates on modeling a human brain, by taking a look at its structure and functions, in particular, at its learning ability. The knowledge representation and data processing techniques employed in FIS and NN reflect these basic differences.

The IF-THEN rules suggested by domain experts represent the knowledge in FIS. Once the rules are stored in knowledge base, they cannot be modified. FIS cannot learn from experience or adapt to new environments. It is only possible for a human expert to manually alter the knowledge base through addition, change or deletion of certain rules.

Knowledge in NN is stored in synaptic weights between neurons. This knowledge is procured while learning, when the network is presented with a training set of data. Unlike FIS, NN learns without human involvement.

In FIS, however, knowledge can be classified into individual rules and the user can see and comprehend the knowledge the system applies. On the contrary, in NN, it is not possible to choose a single synaptic weight as a discrete piece of knowledge. Knowledge is entrenched in the whole network; it is not possible to break it into individual pieces. Any alteration in a synaptic weight may result in random outcomes.

An FIS cannot learn, but can explain how it arrives at a particular solution. It is thus, a 'white-box' for the user. An NN can learn but acts as a 'black-box' for the user. Through marriage between the two technologies, we can combine the advantages of each and create a more powerful and effective intelligent machine. Integrated neuro-fuzzy systems are a combination of the parallel computation and learning capabilities of neural networks with the human-like knowledge representation and explanation capabilities of fuzzy systems. The outcome is that neural networks become more transparent, whereas fuzzy systems gain the ability to learn.

Neural networks have been integrated with fuzzy logic techniques in two different ways. In one approach, the neurons of an NN have been designed using the concept of fuzzy set theory with the following combinations of input signals and connection weights:

- Real input signals but fuzzy weights.
- Fuzzy input signals but real weights.
- Fuzzy input signals and fuzzy weights.

The developed network is generally called *Fuzzy Neural Network* [105].

In another approach, an FIS is represented using the structure of an NN, and trained using backpropagation algorithm. Thus, the performance of FIS can be tuned using NN techniques. Such an integrated system is popularly known as *Neuro-Fuzzy Inference System* [94].

Neuro-fuzzy inference systems have been developed by various investigators to solve a variety of problems. Attempts have been made to model Mamdani approach of FIS using the structure of feedforward network [5, 91]. The Takagi-Sugeno approach is by far the most popular candidate for data-based fuzzy modeling. Roger Jang from the Tsing Hua University (Taiwan) proposed a neural network that is functionally equivalent to a T-S fuzzy model [106]. He called it an *Adaptive Neuro-Fuzzy Inference System* (ANFIS). We limit our discussion in this section to ANFIS only.

6.7.1 ANFIS Architecture

Figure 6.37 shows the ANFIS architecture. For simplicity, we assume that the ANFIS has two inputs, x_1 and x_2, and one output $\hat{y}$. Each input is represented by two fuzzy sets, and the output by a first-order polynomial. The ANFIS implements the following four rules:

$$\begin{aligned}
&\textit{Rule 1}: \text{ IF } x_1 \text{ is } \underset{\sim}{A}_{11} \text{ and } x_2 \text{ is } \underset{\sim}{A}_{21} \text{ THEN } y^{(1)} = a_0^{(1)} + a_1^{(1)}x_1 + a_2^{(1)}x_2 \\
&\textit{Rule 2}: \text{ IF } x_1 \text{ is } \underset{\sim}{A}_{12} \text{ and } x_2 \text{ is } \underset{\sim}{A}_{22} \text{ THEN } y^{(2)} = a_0^{(2)} + a_1^{(2)}x_1 + a_2^{(2)}x_2 \\
&\textit{Rule 3}: \text{ IF } x_1 \text{ is } \underset{\sim}{A}_{12} \text{ and } x_2 \text{ is } \underset{\sim}{A}_{21} \text{ THEN } y^{(3)} = a_0^{(3)} + a_1^{(3)}x_1 + a_2^{(3)}x_2 \\
&\textit{Rule 4}: \text{ IF } x_1 \text{ is } \underset{\sim}{A}_{11} \text{ and } x_2 \text{ is } \underset{\sim}{A}_{22} \text{ THEN } y^{(4)} = a_0^{(4)} + a_1^{(4)}x_1 + a_2^{(4)}x_2
\end{aligned} \tag{6.94}$$

where $\underset{\sim}{A}_{11}$ and $\underset{\sim}{A}_{12}$ are fuzzy sets on the universe of discourse of input variable x_1, $\underset{\sim}{A}_{21}$ and $\underset{\sim}{A}_{22}$ are fuzzy sets on the universe of discourse of input variable x_2; $a_0^{(r)}$, $a_1^{(r)}$ and $a_2^{(r)}$ is a set of parameters specified for rule r.

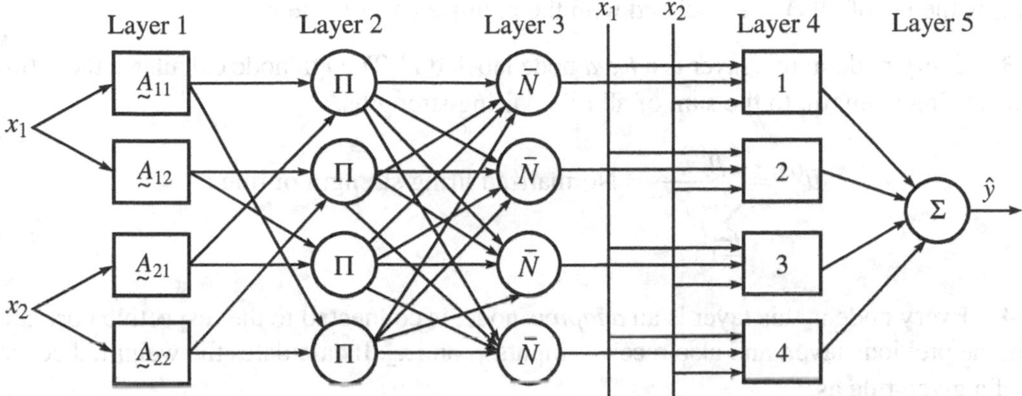

Figure 6.37 An Adaptive Neuro-Fuzzy Inference System (ANFIS)

Let us now discuss the purpose of each layer in ANFIS of Fig. 6.37.

Layer 1 The inputs to the nodes in the first layer are the input fuzzy sets of the ANFIS. Since input fuzzy sets are fuzzy singletons, numerical inputs are directly transmitted to the first-layer nodes.

Nodes in this layer represent the membership functions associated with each linguistic term of input variables. Every node here is an *adaptive* node. Links in this layer are fully connected between input terminals and their corresponding membership function nodes. Membership functions can be any appropriate parameterized function; we use Gaussian function.

$$\mu_{\underset{\sim}{A}_{jk_j}}(x_j) \triangleq \mu_{jk_j}(x_j) = \exp\left[-\left(\frac{x_j - c_{(j,\,k_j)}}{w_{(j,\,k_j)}}\right)^2\right] \tag{6.95}$$

The nodes are labeled $\underset{\sim}{A}_{jk_j}$; $j = 1, 2$; $k = 1, 2$. Total number of nodes in this layer is, therefore, four. $c_{(j,\,k_j)}$ is the center (mean) and $w_{(j,\,k_j)}$ is the width (variance), respectively, of the membership function corresponding to the node $\underset{\sim}{A}_{jk_j}$; x_j is the input and μ_{jk_j} is the output of the node. The adjusted weights in Layer 1 are $c_{(j,\,k_j)}$'s and $w_{(j,\,k_j)}$'s. As the values of these parameters change, the Gaussian function varies accordingly; thus exhibiting various forms of membership functions of fuzzy set $\underset{\sim}{A}_{jk_j}$. Parameters in this layer are referred to as *premise parameters*.

Layer 2 Every node in this layer is a fixed node labeled $\prod$, whose output is the product of all the incoming signals. Each node output represents firing strength of a rule. In fact, other *t*-norm operators could also be used as node functions.

Each node, representing a single T-S fuzzy rule, has the output

$$\mu^{(r)}(\mathbf{x}) \triangleq \prod_{(j,\, k_j) \in I_r} \mu_{jk_j}(x_j) \tag{6.96}$$

where I_r is the set of all $\underset{\sim}{A}_{jk_j}$ associated with the premise part of rule r.

Layer 3 Every node in this layer is a *fixed* node labeled $\overline{N}$. The rth node calculates the ratio of the rth rule's firing strength, to the sum of all rules' firing strengths:

$$\overline{\mu}^{(r)} = \frac{\mu^{(r)}}{\sum_{r=1}^{R} \mu^{(r)}} = \text{Normalized firing strength of rule } r \tag{6.97}$$

Layer 4 Every node is this layer is an *adaptive* node, is connected to the respective normalization node in the previous layer, and also receives inputs x_1 and x_2. It calculates the weighted consequent value of a given rule as,

$$\hat{y}^{(r)} = \overline{\mu}^{(r)}\,[a_0^{(r)} + a_1^{(r)} x_1 + a_2^{(r)} x_2] \tag{6.98}$$

where $\overline{\mu}^{(r)}$ is the normalized firing strength from layer 3, and $a_0^{(r)}$, $a_1^{(r)}$ and $a_2^{(r)}$ are the parameters of this node. Parameters in this layer are referred to as *consequent parameters.*

Each node in Layer 4 is a local linear model of the T-S fuzzy system; integration of outputs of all local linear models yields global output.

Layer 5 The single node in this layer is a *fixed* node labeled Σ, which computes the overall output as the summation of all incoming signals:

$$\hat{y} = \sum_{r=1}^{R} (a_0^{(r)} + a_1^{(r)} x_1 + a_2^{(r)} x_2)\overline{\mu}^{(r)};\ R = 4 \tag{6.99}$$

6.7.2 How Does an ANFIS Learn?

An ANFIS employs a hybrid learning algorithm—a combination of the least squares estimator and the gradient descent technique. To begin with, each membership neuron is allocated initial activation functions. The function centers of the neurons linked to input x_j, are set in such a manner that the domain of x_j is equally divided, and the widths are set to permit enough overlapping of the respective functions.

A forward and a backward pass are involved in each epoch of an ANFIS training algorithm. In the forward pass, the ANFIS is presented with a training set of input patterns (input vector **x**), neurons outputs are computed on layer-by-layer basis, and the rules consequent parameters are identified by the least squares estimator. In the T-S fuzzy inference, an output $\hat{y}$ is a linear function. Thus, given the values of the membership parameters and a training set of N input-output patterns, we can form N linear equations in terms of the consequent parameters (refer to Eqn (6.93)). Least-squares solution of these equations yields the consequent parameters.

As soon as the rule consequent parameters are established, we can compute actual network output, $\hat{y}$, and determine the error

$$e = y - \hat{y} \tag{6.100}$$

The backpropagation algorithm is applied in the backward pass. The error signals are transmitted back, and the premise parameters are updated as per the chain rule.

The goal is to minimize the error function

$$E = \tfrac{1}{2}(y - \hat{y})^2 \tag{6.101}$$

The error at Layer 5:

$$\frac{\partial E}{\partial \hat{y}} = (\hat{y} - y) \tag{6.102}$$

Backpropagating to Layer 3 via Layer 4 (refer to Eqn (6.99)),

$$\frac{\partial E}{\partial \bar{\mu}^{(r)}} = \frac{\partial E}{\partial \hat{y}}\frac{\partial \hat{y}}{\partial \bar{\mu}^{(r)}} = \frac{\partial E}{\partial \hat{y}}[(a_0^{(r)} + a_1^{(r)}x_1 + a_2^{(r)}x_2)] \tag{6.103}$$

Backpropagating to Layer 2 (refer to Eqn (6.97),

$$\frac{\partial E}{\partial \mu^{(r)}} = \frac{\partial E}{\partial \bar{\mu}^{(r)}}\frac{\partial \bar{\mu}^{(r)}}{\partial \mu^{(r)}} = \frac{\partial E}{\partial \bar{\mu}^{(r)}}\left[\frac{\bar{\mu}^{(r)}(1 - \bar{\mu}^{(r)})}{\mu^{(r)}}\right] \tag{6.104}$$

The error at Layer 1:

I_r is the set of all $\underset{\sim}{A}_{jk_j}$ associated with the premise part of rule r. Reverse pass: $I_{(j,k_j)}$ is the set of all rule nods in Layer 2 connected to $(j, k_j)^{\text{th}}$ node (corresponding to $\underset{\sim}{A}_{jk_j}$) of Layer 1.

Backpropagating error to Layer 1 (refer to Eqn (6.96)),

$$\begin{aligned}\frac{\partial E}{\partial \mu_{jk_j}} &= \sum_{r \in I_{(j,\,k_j)}} \frac{\partial E}{\partial \mu^{(r)}}\frac{\partial \mu^{(r)}}{\partial \mu_{jk_j}} \\ &= \sum_{r \in I_{(j,\,k_j)}} \frac{\partial E}{\partial \mu^{(r)}}\left[\frac{\mu^{(r)}}{\mu_{jk_j}}\right]\end{aligned} \tag{6.105a}$$

From Eqn (6.95), we obtain,

$$\frac{\partial \mu_{jk_j}}{\partial c_{(j,\,k_j)}} = 2\mu_{jk_j}(x_j - c_{(j,\,k_j)})/w^2_{(j,k_j)} \tag{6.105b}$$

$$\frac{\partial \mu_{jk_j}}{\partial w_{(j,\,k_j)}} = 2\mu_{jk_j}(x_j - c_{(j,\,k_j)})^2/w^3_{(j,k_j)} \tag{6.105c}$$

Denoting the iteration index by t (refer to Eqn (5.24)),

$$c_{(j,\,k_j)}(t+1) = c_{(j,\,k_j)}(t) - \eta \frac{\partial E(t)}{\partial c_{(j,\,k_j)}(t)} \tag{6.106a}$$

$$w_{(j,\,k_j)}(t+1) = w_{(j,\,k_j)}(t) - \eta \frac{\partial E(t)}{\partial w_{(j,\,k_j)}(t)} \tag{6.106b}$$

where η is the learning rate.

For given input-output pairs $(\mathbf{x}^{(i)}, y^{(i)};\ i = 1, 2, \ldots, N)$, the batch-updating algorithm back propagates the cumulative error resulting from the difference between $y^{(i)};\ i = 1, \ldots, N$ and $\hat{y}^{(i)};\ i = 1, \ldots, N$, from output layer to the previous layers to update weights of the network.

Example 6.13

Figure 6.37 shows the schematic diagram of an ANFIS, used to model a process with two inputs, x_1 and x_2, and one output y. Two fuzzy sets $\underset{\sim}{A}_{11}$ and $\underset{\sim}{A}_{12}$ have been utilized to represent x_1; and x_2 has been expressed using two other fuzzy sets $\underset{\sim}{A}_{21}$ and $\underset{\sim}{A}_{22}$. The membership function distributions of x_1 and x_2 are shown in Fig. 6.38.

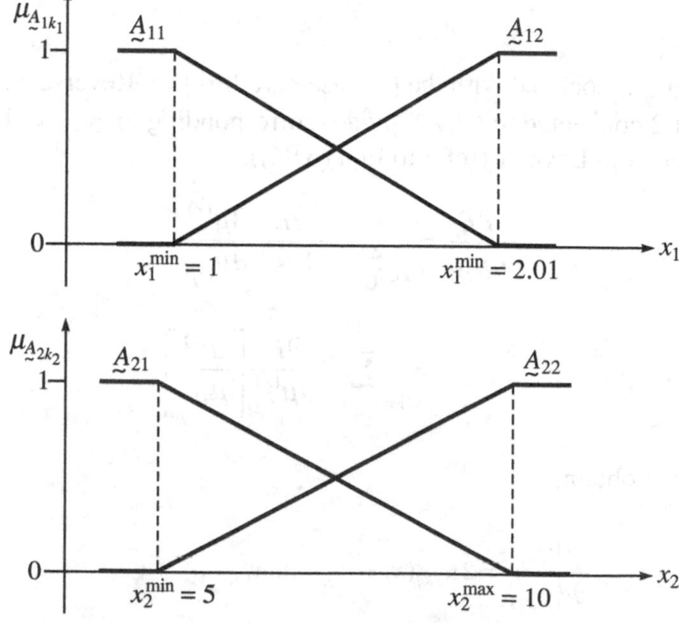

Figure 6.38 Membership function distributions

There are a maximum of 2 × 2 possible rules (refer to (6.94)); the values of the coefficients of the consequent part of the rules are as follows:

$$a_0^{(1)} = 0.10, a_0^{(2)} = 0.11, a_0^{(3)} = 0.13, a_0^{(4)} = 0.14, a_1^{(1)} = 0.2, a_1^{(2)} = 0.2, a_1^{(3)} = 0.3,$$

$$a_1^{(4)} = 0.3, a_2^{(1)} = 0.3, a_2^{(2)} = 0.4, a_2^{(3)} = 0.3, a_2^{(4)} = 0.4.$$

The objective is to determine the predicted output $\hat{y}$ of ANFIS when $x_1 = 1.1$ and $x_2 = 6.0$.

For given values of x_1 and x_2, we find, using the principle of similar triangles, from Fig. 6.38 (Layer 1 in Fig. 6.37):

$$\mu_{A_{11}}(x_1) = \left(\frac{2.01 - 1.1}{2.01 - 1}\right) \times 1 = 0.900990$$

$$\mu_{A_{12}}(x_1) = \left(\frac{1.1 - 1}{2.01 - 1}\right) \times 1 = 0.099010$$

$$\mu_{A_{21}}(x_2) = \left(\frac{10 - 6}{10 - 5}\right) \times 1 = 0.8$$

$$\mu_{A_{22}}(x_2) = \left(\frac{6 - 5}{10 - 5}\right) \times 1 = 0.2$$

All the possible four rules, given in Eqn (6.94), will be fired. Firing strengths of the rules are (Layer 2 in Fig. 6.37; Eqn (6.96)):

$$\mu^{(1)}(\mathbf{x}) = 0.900990 \times 0.8 = 0.720792$$
$$\mu^{(2)}(\mathbf{x}) = 0.099010 \times 0.2 = 0.019802$$
$$\mu^{(3)}(\mathbf{x}) = 0.099009 \times 0.8 = 0.079208$$
$$\mu^{(4)}(\mathbf{x}) = 0.900990 \times 0.2 = 0.180198$$

The normalized firing strengths of the rules are (Layer 3 in Fig. 6.37; Eqn (6.97)):

$$\bar{\mu}^{(1)} = \mu^{(1)} \bigg/ \sum_{r=1}^{4} \mu^{(r)} = \mu^{(1)} = 0.720792$$

$$\bar{\mu}^{(2)} = \mu^{(2)}, \bar{\mu}^{(3)} = \mu^{(3)}, \bar{\mu}^{(4)} = \mu^{(4)}$$

Weighted consequent values of the rules are (Layer 4 in Fig. 6.37; Eqn (6.98)):

$$\hat{y}^{(1)} = 0.720792(0.10 + 0.2 \times 1.1 + 0.3 \times 6.0) = 1.528079;$$

$$\hat{y}^{(2)} = 0.054059;\ \hat{y}^{(3)} = 0.179010;\ \hat{y}^{(4)} = 0.517168$$

Predicted output of the ANFIS, is (Layer 5 in Fig. 6.37; Eqn (6.99)):

$$\hat{y} = 2.278316$$

6.8 GENTIC-FUZZY SYSTEMS

Fuzzy inference systems are highly nonlinear systems with many input and output variables. The performance of an FIS depends on the knowledge base, which consists of database (membership functions of input and output variables) and rule-base (refer to Fig. 6.30). Crucial issues in the design are the tasks of selecting appropriate membership functions, and the generation of fuzzy rules. These tasks require experience and expertise. The trial-and-error approach involved in this process sometimes needs quite a lot of time, but genetic algorithms can be helpful here. We have presented the basic genetic algorithm in Appendix A.

Genetic algorithms can be introduced into fuzzy inference systems at many different levels [91]:

- tuning of membership functions, while the rule base remains unchanged;
- generating a rule-base when a set of membership functions for input/output variables remains unchanged, or
- for both of these tasks simultaneously.

We will limit our presentation to the first task, i.e., tuning of membership functions while the rule-base remains unchanged.

Tuning of Membership Functions

The fuzzy system operation can be made more efficient through suitable tuning of the fuzzy sets. The GA (genetic algorithm) alters membership functions by altering the location of characteristic points of their shapes. The information on characteristic points of the membership functions is coded in chromosomes. Once the fuzzy sets are appropriately represented in the chromosome, the GA works on the population of individuals, that is, on the population of chromosomes encompassing coded shapes of fuzzy membership functions, as per the genetic cycle consisting of the following steps:

1. Decode every individual (chromosome) of the population, recreate the set of membership functions, and construct a suitable fuzzy system. The rule-base is predefined.
2. Evaluate the performance of the fuzzy system based on the difference (error) between the system's responses and the values sought. This error defines how fit the individual (chromosome) is.
3. Select and apply genetic operators, such as crossover and mutation, and gain a new generation.

Example 6.14

Let us consider the application of GA to a fuzzy model of a manufacturing process. The process is characterized by two input variables, x_1 and x_2, and one output variable, y. The membership function distributions of the inputs and the output are shown in Fig. 6.39 and the predefined rule-base is given in Table 6.4.

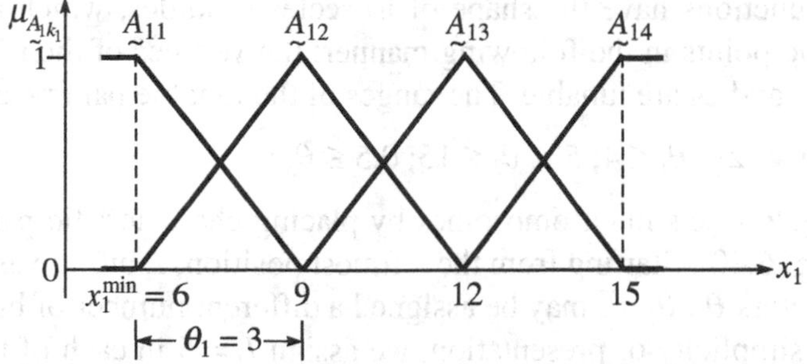

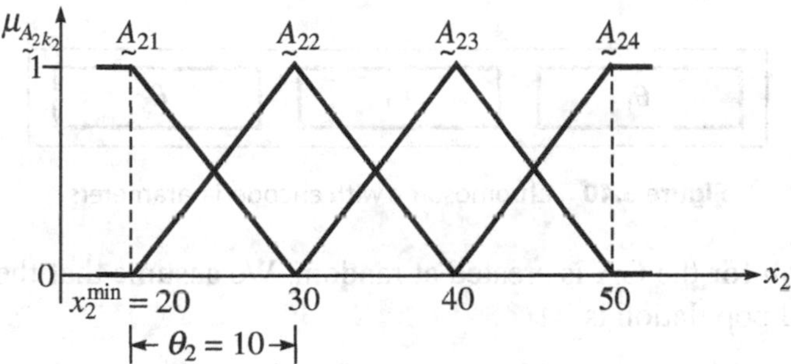

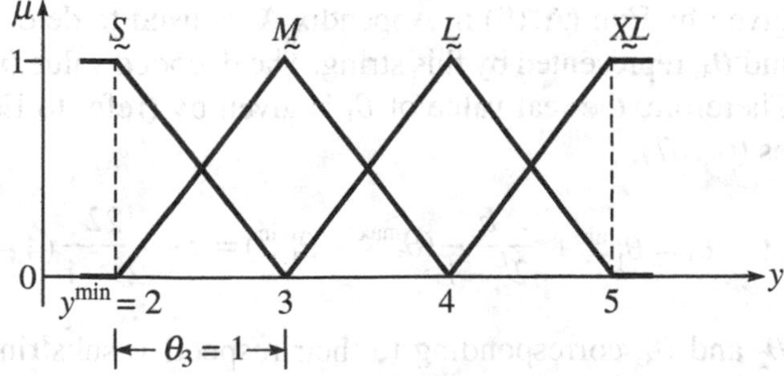

Figure 6.39 Membership distributions for Example 6.13

Table 6.4 Rule base for Example 6.13

x_1 \ x_2	$\underset{\sim}{A}_{21}$	$\underset{\sim}{A}_{22}$	$\underset{\sim}{A}_{23}$	$\underset{\sim}{A}_{24}$
$\underset{\sim}{A}_{11}$	$\underset{\sim}{S}$	$\underset{\sim}{S}$	$\underset{\sim}{M}$	$\underset{\sim}{L}$
$\underset{\sim}{A}_{12}$	$\underset{\sim}{S}$	$\underset{\sim}{M}$	$\underset{\sim}{L}$	$\underset{\sim}{L}$
$\underset{\sim}{A}_{13}$	$\underset{\sim}{M}$	$\underset{\sim}{M}$	$\underset{\sim}{L}$	$\underset{\sim}{XL}$
$\underset{\sim}{A}_{14}$	$\underset{\sim}{M}$	$\underset{\sim}{L}$	$\underset{\sim}{XL}$	$\underset{\sim}{XL}$

The membership functions have the shape of isosceles triangles, which may be described by means of characteristic points in the following manner: the vertices of the triangles are fixed, and the base-widths θ_1, θ_2, and θ_3 are tunable. The ranges of the tunable parameters are assumed to be

$$2 \le \theta_1 \le 4;\ 5 \le \theta_2 \le 15;\ 0.5 \le \theta_3 \le 1.5 \tag{6.107}$$

Let us code these fuzzy sets in chromosomes by placing characteristic parameters one by one, next to each other (Fig. 6.40). Starting from the leftmost position, L bits are assigned for parameter θ_1. Each of the parameters θ_1, θ_2, θ_3 may be assigned a different number of bits depending on their ranges. However, for simplicity of presentation, we assign $L = 5$ in each of the three cases. Thus, the GA-string is 15 bits long.

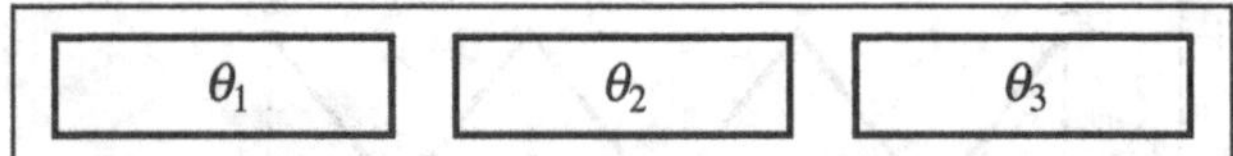

Figure 6.40 Chromosome with encoded parameters

An initial population for the GA is created at random. We assume that the first chromosome of this randomly selected population is

$$10110 \quad 01101 \quad 11011 \tag{6.108}$$

The mapping rule, given by Eqn (A.10) in Appendix A, is used to determine the real values of the parameters θ_1, θ_2, and θ_3, represented by this string. The decoded value b of the binary substring 10110 is equal to 22. Therefore, the real value of θ_1 is given by (refer to Eqn (A.10) in Appendix A, and parameter values (6.107),

$$\theta_1 = \theta_1^{\min} + \frac{b}{2^L - 1}(\theta_1^{\max} - \theta_1^{\min}) = 2 + \frac{22}{2^5 - 1}(4 - 2) = 3.419355$$

The real values of θ_2 and θ_3, corresponding to their respective substrings in Eqn (6.108), are 9.193548 and 1.370968, respectively. Figure 6.41 shows the modified membership distributions of input and output variables.

The GA optimizes the database (tunes the membership functions) with the help of a set of training examples. Assume that we are given N training examples $\{\mathbf{x}^{(i)}, y^{(i)}; i = 1, 2, \ldots, N\}$. Further, we take first training example ($i = 1$) as $\{x_1 = 10, x_2 = 28, y = 3.5\}$.

For the inputs $x_1 = 10$, $x_2 = 28$, we calculate the predicted value of the output, $\hat{y}$, of the fuzzy model when the model parameters are given by the first chromosome in the initial population. This is done using the procedure given in Section. 6.4. This will give us the absolute value of error in prediction: $e^{(1)} = |3.5 - \hat{y}|$.

From this procedure, repeated on all the training examples, we can obtain the average value of absolute errors in prediction,

$$\bar{e} = \frac{1}{N}\sum_{i=1}^{N} e^{(i)} \tag{6.109}$$

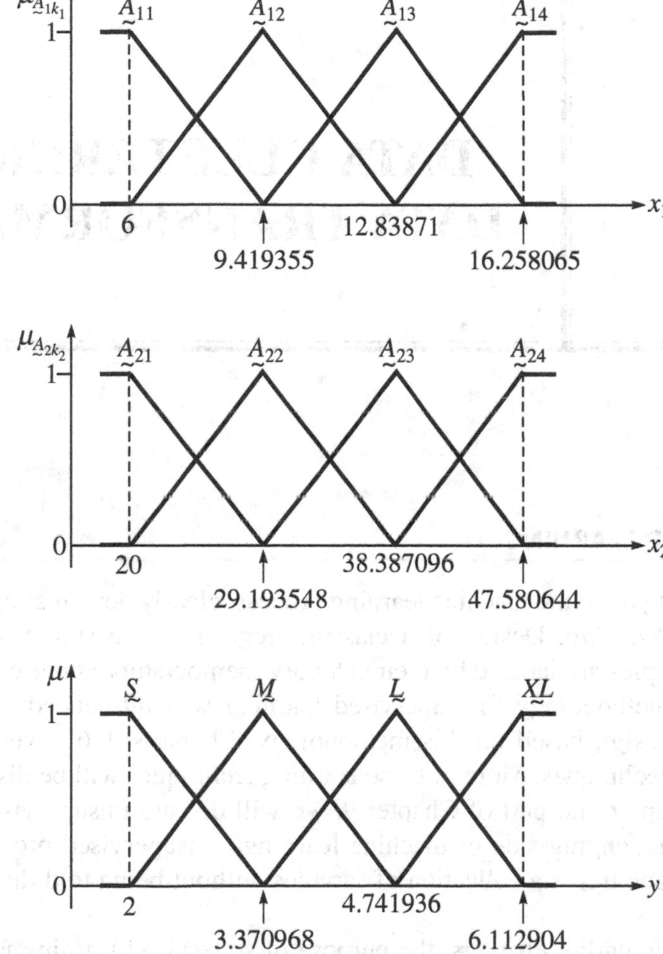

Figure 6.41 Modified membership distributions of inputs and output

Since GA is a maximization algorithm, we may choose the fitness function

$$J = \frac{1}{\bar{e} + \varepsilon} \tag{6.110}$$

where ε is a small positive number,

The population of GA-strings is then modified using different operators, such as selection, crossover and mutation, and after a few generations, the GA will be able to evolve an optimal fuzzy system (refer to Appendix A).

Chapter

7

DATA CLUSTERING AND DATA TRANSFORMATIONS

7.1 UNSUPERVISED LEARNING

Chapter 1 introduced you to the machine learning process, clearly demarcating *supervised learning* from *unsupervised learning*. Design of a classifier/regressor is a supervised learning problem wherein training samples are labeled by their category membership/numeric values.

In Chapter 2, a methodology for supervised learning was introduced; this was followed by classifier/regressor design based on this methodology. Chapters 3–6 covered a wide variety of supervised learning techniques. More of these learning techniques will be discussed in Chapter 8.

In part of this chapter and part of Chapter 9, we will discuss unsupervised learning, which is perhaps the more challenging side of machine learning. Unsupervised procedures use unlabeled samples, that is, all one has is a collection of samples without being told their categories/function values.

As we have seen in earlier chapters, the purpose of supervised learning is to answer a specific question 'phrased' as a target variable, and the various techniques to discover patterns in data to discern the value of the target. Historical data contains examples to find the best answer. Target variable is a way of putting domain expertise into the modeling process. A data miner says, 'This is what is important'. With this information, supervised techniques have enough inputs to optimize their models.

Unsupervised procedures have no such information. With these techniques, the data miner is looking for 'something interesting' in the data. Using these techniques requires more human understanding than supervised techniques.

The process of unsupervised learning is both quite different and quite similar to the process of supervised learning. The similarities are: both work with data and require exploration and understanding of the data with respect to the application domain. Both are improved by incorporating intelligent variables into the data that identify the quality of different aspects of the process in domain of interest. Because source data is usually not at the level required by the application in hand, building the data for the application requires many transformations in both the learning approaches.

Unsupervised learning differs in not having a target variable, and this poses a couple of challenges:

- Unsupervised learning is more about creative endevors—exploration, understanding and refinements that do not lend themselves to a specific set of steps, i.e., there is no specific methodology. The unsupervised learning process cannot be automated. Techniques are still not available to distinguish between the more useful and less useful. Humans play a crucial role.
- There is no right or wrong answer; no simple statistical measure that summarizes the goodness of results. Instead, descriptive statistics and visualization are key parts of the process.

When it comes to unsupervised learning methods, the requirement of dataset partitioning—into training, validation and test sets—is of little importance. Of late, researches have tried to study *semi-supervised learning*, that is, what lies between supervised and unsupervised learning. Semi-supervised learning is like a field that aims for classification but the input comprises data that is labeled as well as unlabeled. Classification is not possible in the absence of labeled data. The unlabeled data in the dataset, it seems, can actually help make classification better!

An easy way to fulfill this aim is to first use the labeled data to train a classifier. The next step is to apply the same to the unlabeled data so that it is labeled with class probabilities (using 'expectation' step of EM algorithm, as mentioned later in this chapter). The third step is to train another classifier with the help of labels for all the data (the 'maximization' step of EM algorithm). Fourth, repeat till convergence is achieved. The primary question whose answer can only be given empirically is, 'Will the repetitive parameter estimates of the EM algorithm make the classification more accurate?' This kind of combination of Naive Bayes and EM algorithms is good for document classification [18, 107].

Unlabeled data can also enhance classification performance in situations wherein there are two independent and very different viewpoints on the classification work. For example, web page documents have two perspectives—*content* and the *links* to that content from other pages. In such issues, the *co-training* concept works in the following manner.

Start by learning a different model for each viewpoint or perspective from the examples carrying labels. Then, label the unlabeled examples using each model separately. For each model, choose the example that best labels it as positive and the one that best labels as negative, and add these to the pool of labeled examples. Iterate the entire process by training both models on the augmented pool of examples till the unlabeled pool is finished [18, 108].

The next several sections of this chapter cover specific techniques associated with unsupervised learning, called *clustering* techniques. Other unsupervised techniques leading to *association rules* will be taken up in Chapter 9.

7.1.1 Clustering

Our focus in this book so far has been on *classification* and *regression* problems for the solutions using supervising learning techniques. In a training data matrix for classification problems, the outputs are known and these are *class labels*. For example, in a *credit assignment problem*, the inputs are the relevant information we have about the customer's financial history, namely—income, savings, collaterals, profession, age, past financial history, and so forth. The bank has a record of past loans containing such customer data and whether the loan was paid back or not: risk/no risk are the class labels with the data on each customer. The machine learning algorithm fits a model to the past data to be able to calculate the risk for a new application.

In a training dataset for regression problems, the outputs are known and these are numeric values $\in \Re$. Let us take an example of the regression problem: *navigation of mobile robot*. In the case of an autonomous car, the angle at which the steering should turn in order to move without bumping into any obstacle or without moving away from the path will be considered the output. Inputs, on the other hand, come from the car's sensors. In other words, inputs are given by the video camera or other things, such as the GPS. It is possible to obtain training data by closely watching and recording the car driver's movements and actions.

Ranking is an application area of machine learning that is different from classification and regression, and is sort of between the two. Consider an example. If users rate the movies they have watched as 'enjoyed/not enjoyed', this will be a binary classification problem. If they rate the movies on a scale of 1 to 10, this will be a regression problem. It will be easier for the users to say that of the movies they have watched, they like one more than the other instead of yes/no decision or a numeric value. That is, the output is *ordinal* with possible values that have a meaningful order or ranking among them, but the magnitude between successive values is not known.

Ranking has many applications. In search engines, for example, given a query, we want to retrieve the most relevant documents. If we retrieve and display the current top 10 candidates, and then the user clicks the third one skipping the first two, we understand that the third should have been ranked higher than the first and the second. Such click logs are used to train rankers. Other application areas of ranking which have been explored are information retrieval and natural language processing. Useful references for ranking applications and algorithms are [109–112].

In training datasets for *clustering* problems, outputs (class labels/numeric/ordinal) are not specified; only the feature vectors representing different objects/instances/records/situations are known. When you group a set of data objects to form several groups or *clusters*, in such a way that the components within a cluster are highly similar but differ from the components of other clusters, it is called clustering. The level of similarity and dissimilarity are evaluated on the basis of the characteristics of the variables that describe the objects or components. This assessment often involves distance measures introduced in Chapter 2.

Cluster detection is a tool for unsupervised learning, because the clustering techniques find patterns in the data without regard to any target variable. Finding clusters is rarely an end itself. After clusters have been detected, they may become objects of study.

Supervised learning, with a few exceptions, is often based on statistical techniques that have been around for many decades or centuries. These have adapted to the increased volumes of data and computer power available in the modern world, but many basic ideas remain the same. Clustering (a tool of unsupervised learning), on the other hand, is more recent and borne out of availability of lots of data and powerful computers.

Sample application examples of clustering are given as follows:

Clustering can be used for *data exploration*, to understand the structure of the data. The data may contain so much complex structure that even the best data mining techniques are unable to coax out meaningful patterns within the data. Cluster detection provides a way to learn about the structure of complex data. A natural way to make sense of complex data is to break the data into smaller clusters of data; then finding patterns within each cluster is often possible.

Image compression is an interesting application of clustering. In an image, if we decide to color code shades of the same group with a single color, say their average, then we are actually quantizing the image. If 24 bit pixels represent 16 million colors for an image, and there are shades of merely 64 main colors, we will require 6 bits for each pixel rather than 24.

Document clustering aims to classify similar documents. For instance, news reports can be further divided as those pertaining to politics, entertainment, sports, and so on.

Clustering is used in *bioinformatics* in learning sequences of amino acids that occur repeatedly in proteins; they may correspond to structural or functional elements within the sequences.

If a company has the data pertaining to past customers, along with demographic information and past transactions with the company, the clustering approach ensures that customers with similar attributes are assigned the same group. As a result, the customers of the company end up being in natural groups. This is termed, *customer segmentation.*

Clustering can be used to map data to a new space where supervised learning is easier. Clustering techniques find similarities between instances and thus group instances. It is possible to select the group mean to represent the prototype of instances in the group permitting a less complex data description. For instance, if an organization's clients fit in one of the *K* groups known as *segments*, then there will be a better understanding of the customer base letting the organization offer different strategies for different segments. This is called *customer relationship management.*

Another use of clustering methods is in *outlier detection* which is finding instances that do not lie in any of the main clusters and are exceptions. The outliers may be recording errors that should be detected and discarded in data cleansing process. However, an outlier may indicate abnormal behavior, for example, in a dataset of credit card transactions, it may indicate a fraud; in an image, it may indicate *anomalies*, for example, tumors; or it may be a *novel*, previously unseen but valid case; for example, customers who do not fall in any large group may require special attention—*churning* customers. Outlier detection is sometimes given the name *anomaly detection* or *novelty detection.*

A problem that generally occurs is that the data used to train the outlier detector is unlabeled and may contain outliers mixed with typical instances. Clustering algorithms (unsupervised learning) are useful for detecting outliers in the unlabeled data.

Spotting of that which does not normally occur is basically outlier detection. To achieve this, a kind of unsupervised learning, which is restricted, is required. This can happen by an estimation of the high-density areas; by discovering a boundary (such that it reads as a classification problem) separating high-density volumes from low-density volumes. It is then possible to use such a boundary to detect outlier. This is also known as *one-class classification.*

7.2 ENGINEERING THE DATA

In the previous chapters, we have examined an array of machine learning methods: naive Bayes classifier, *k-NN* algorithm, linear regression, logistic regression, linear discriminants, support vector machines, neural networks, fuzzy logic models. Several other learning schemes (clustering, association rules, decision trees, fuzzy decision trees) will be discussed in the current and later chapters. All are sound, robust methods, which can be applied to practical data mining problems.

Data mining applications generally have four basic types of learning—pattern recognition, numeric prediction, clustering learning and association learning. In *pattern recognition or classification* learning, classified example sets are presented and the learning scheme is expected to learn to classify examples that are not seen. In *numeric prediction or regression* learning, the result to be predicted is a numeric or continuous quantity rather than a class. *Clustering* learning seeks sets of examples that are similar or belong together. In *association* learning, any association among features is sought, not just ones that predict a particular class value. For a given data mining problem, a learning scheme that works best will have to be chosen from the range of schemes that are available for the type of learning (classification, regression, clustering, association rules) dictated by the problem. It is tempting to try out several learning schemes to make a right choice. *Ensemble learning methods* use a combination of learned models (classifiers/regressors) with the aim of creating an improved composite model. *Bagging* and *boosting* are the two such techniques for improving accuracy (refer to Section 2.5.2).

Mere selection of a learning algorithm and its running over the data collected is not enough. For data mining to be successful, much more is required. Today's real-world data are typically of huge size and their likely origin is from multiple heterogeneous sources. The low-quality data that we collect will lead to low-quality mining results. Careful *preprocessing* of the available *raw data* improves the efficiency and ease of data mining.

Integration of data from different sources into a coherent data store provides a single consistent point of data access for mining. Section 9.3 will deal with construction of *data warehouses*.

Let us look at different ways in which data can be *messaged* so that it becomes more suited to learning techniques. The essential procedures that can materially improve success on application of machine learning techniques to practical data mining issues form a type of data engineering—*engineering the data* into a form suited to the selected learning scheme. Our emphasis in this chapter is on four important processes of data preparation for successful data mining: *data cleansing*, *derived attributes*, *discretizing numeric attributes* and *attribution reduction*. All these processes result in *data transformation* to obtain high-quality mining performance.

Data transformation processes are performed using extensive *exploration* of the available data. Data *clustering* is a very useful stand-alone learning method and it is quite helpful in data exploration as well.

We examine in this chapter the *data exploration*, *data clustering*, and *data transformation* processes involved in engineering the input data to make it more amenable for learning methods. When applied before mining, these processes can substantially improve the overall quality of the patterns mined, leading to huge payoffs for decision making. We can look on them as bag of tricks that we can apply to practical data mining problems to enhance the chances of success.

The execution of these processes is not necessarily sequential. With knowledge gained in execution of a step, we may go back to the earlier steps to refine our processing before we move forward; this results in a chain of nested loops. Because so many issues are involved, we cannot set it right the first time. This is why engineering the data for mining tasks takes so long.

Although this book is not really about problems of data preparation, a feeling for the issues involved is important so that we can appreciate the complexities. With this objective in mind, our discussion is limited to more widely used techniques for applied machine learning.

7.2.1 Exploratory Data Analysis: Learning about What is in the Data

Exploratory data analysis is outside the scope of this book, but that is not because it is unimportant. In fact, books devoted exclusively to this subject have been written. Every commercial software devotes a good amount of effort on data exploration.

Exploring data is an opportunity to learn about what is in the data. Much of the learning is about the processes that generate the data—essentially data quality issues. This is to be expected. The data that data miners work with usually comes from many different source systems, which might include data warehouses, external sources of data, operational systems, marketing spreadsheets, and so on (Section 9.3 will provide the details).

What do we want to look for when exploring data? The answer to this question is difficult. In fact, to a domain expert, *data summaries*, *correlation analysis*, *cluster analysis* and *data visualization* through graphs and plots will bring out the characteristics of the data that will guide the data mining process. Exploration is thus an exercise to bring out something new, unknown to the miner, which can then be exploited to improve the success of data mining.

As said earlier, unsupervised data mining does not have a specific methodology. Unsupervised learning is more about exploration, understanding and refinements. These are creative endeavors that do not lend themselves to a specific set of steps. The big data problems that we are facing today, primarily, are the problems of unstructured and unfamiliar huge amounts of data which our conventional (first generation) data mining techniques are not able to handle. Though second-generation data mining techniques for big-data and other difficult problems are the subject of research today, data clustering and exploration techniques are providing some tools in our hand to look deeper into the problems for possible solutions.

Basic Statistical Descriptions

Basic statistical descriptions are used to get familiar with the data and its traits. These descriptions facilitate the identification of properties of the data and highlight the data values that need to be considered noise or outliers.

Three important aspects of basic statistical descriptions are as follows:

(i) The *measures of central tendency* measure the location of the center of the data distribution, particularly *mean*, *median* and *mode* described in Section 3.2. They help identify where most of the values of an attribute fall.

(ii) Second aspect of statistical description is to have an idea of *dispersion of the data*, that is, how are data spread out. The most common data dispersion measures are *variance* and *standard deviation* (Section 3.2) of the data. These measures are useful for identifying outliers.

(iii) *Correlation analysis* is another facet of statistical data descriptions. In databases containing a large number of variables, the information covered by the variables usually overlap a lot. One way of finding redundancies is to examine the *correlation matrix* (Section 3.2), which depicts the correlations between variables in pairs. Pairs with very strong correlation witness a lot of information overlap and are ideal for data reduction by eliminating one of the variables.

Using graphs and plots, basic statistical descriptions can be displayed for visual inspection of the data. Bar and pie charts, along with line graphs as well as scatter and quantile plots, are regular elements of a majority of the statistical data presentation software available.

One of the best techniques to effectively determine a relationship, trend or pattern between two numeric attributes, graphically, is the scatter plot. It offers a glimpse of bivariate data to view clusters of points and outliers or to discover if there are any possible correlation relationships. Summaries of statistical descriptive data mining techniques are available in [113, 114]. Refer to [115] for statistics-based data visualization.

Data Visualization

The objective of data visualization is to share data clearly and in an effective manner using graphs. Visualization methods can be used to explore data relationships, which cannot be clearly observed otherwise by simply taking a look at raw data. Some of the visualization methods offered by commercial software packages are pixel-oriented, geometric projection, icon-based and hierarchical visualization [17]. Refer to [116] for ground-breaking work on methods of data visualization.

For huge datasets, say, one million points, it is not uncommon to pick a random data sample and employ it to create more interpretable visualization.

Interactive visualization is preferred over static plots. Software packages for interactive visualization permit the user to interactively choose as well as alter variables on the plots, to zoom in and zoom out of different areas on the plot, and provide the user with the power to navigate through the huge volume of data.

Measuring Data Similarity and Dissimilarity

In clustering, *k-NN* classification, outlier analysis and other data mining applications, similar and dissimilar objects need to be assessed. As mentioned earlier, in clustering, data objects are so grouped/collected that while the objects in one cluster are alike, they are very unlike the objects in other clusters. Outlier analysis employs clustering-based methods to recognize potential outliers as objects with a high level of dissimilarity with others [117, 118]. Knowledge of similarities in objects can also be used in *k-NN* classification schemes.

7.2.2 Cluster Analysis: Finding Similarities in the Data

Cluster analysis groups the available data into different clusters on the basis of similarity of the objects in the dataset. Methods of clustering are applicable in the absence of class predictions tasks when the objects/instances need to be assigned to natural groups. These clusters display certain mechanism at work in the domain from which objects are drawn—a mechanism because of which certain objects resemble each other more strongly than they do with the other objects/instances. Clustering results can be expressed in various ways. The selection from among these and those not included here should be determined by the nature of the mechanisms underlining the particular clustering phenomenon. However, since these mechanisms are hardly known—their very existence is being discovered—the selection generally depends on the available clustering tools, each one of which (algorithms) would give rise to data partition with a fixed structure, in terms of location and cluster shapes.

There is a dearth of universal algorithms that can work in all situations. The issue becomes worse due to the fact that there may be thousands and millions of samples in real tasks. Clustering as a subject is still being studied and researched on, given the big-data problems that pose a challenge in this age.

While selecting the right clustering algorithm, we should make use of the knowledge of the problem the dataset describes. Data partition must usually have two features:

- *Clusters should be homogeneous within:* Data within each cluster should strongly resemble each other.
- *There should be heterogeneity between clusters:* Data should differ from one cluster to another cluster as much as possible.

The similarity of data depends on the kind of data to be clustered. As data usually describes features of objects in a numerical form, the ideal measure of similarity is by measurement of the distance between data vectors. For instance, we could use the *Euclidean norm* (refer to Section 3.2.5), the most commonly employed measurement technique to measure the likeness of objects.

The primary characteristics of the popular Euclidean distance measure are:

- It depends on scale; variables that have larger scales tend to exert a greater influence over the total distance. It is a usual practice to normalize continuous measurements before calculating the Euclidean distance. This enables conversion of all measurements to the same scale. Refer to Section 7.7.2 for details.
- It does not take into account the relationship between the measurements. In case of strong correlation of measurements, it is better to select a different distance (such as *statistical distance* or *Mahalanobis distance*). Refer to Section 3.2.5.
- Euclidean distance is sensitive to outliers (it is preferable to use more robust measures, for instance, the Mahalanobis distance).

Representing Clustering Result

When clusters are formed, the result could be expressed in the form of a diagram that shows how the instances fall into groups. For visual illustration, we use two-dimensional points, as shown in Fig. 7.1, as our data objects.

In the most basic scenario, representing a clustering result requires associating a cluster number with each object (instance), which may be portrayed through the instances: say $s^{(1)}$, $s^{(2)}$, $s^{(3)}$, $s^{(4)}$, $s^{(5)}$, $s^{(6)}$, $s^{(7)}$, $s^{(8)}$, $s^{(9)}$, $s^{(10)}$, $s^{(11)}$, in two dimensions and *partitioning* the space to indicate each cluster, as shown in Fig. 7.1(a). This *exclusive* clustering allocates each instance to a single cluster. In many cases, the point could be placed in more than one cluster. Usually, *overlapping* or *non-exclusive* clustering indicates that an instance can exist in more than one cluster at a time. This can be shown via a diagram depicting overlapping subsets representing each cluster [Fig. 7.1(b)][1]. Some algorithms associate instances with clusters in a *probabilistic* way—for every instance there exists a probability with which it belongs to each of the clusters. Figure 7.1(c) shows normal distributions;

[1] *In fuzzy clustering*, every object belongs to every cluster with a membership weight that is between 0 and 1.

probability density functions resemble mountain ranges—each cluster has a peak. If there are smaller groups or subclusters within each cluster a *hierarchy* emerges—clusters nested within each other to form a tree-like structure. Figure 7.1(d) depicts a hierarchical structure. Diagrams of this type are known as *dendrograms*, which is just another name for tree diagrams ('dendron' in Greek means 'tree').

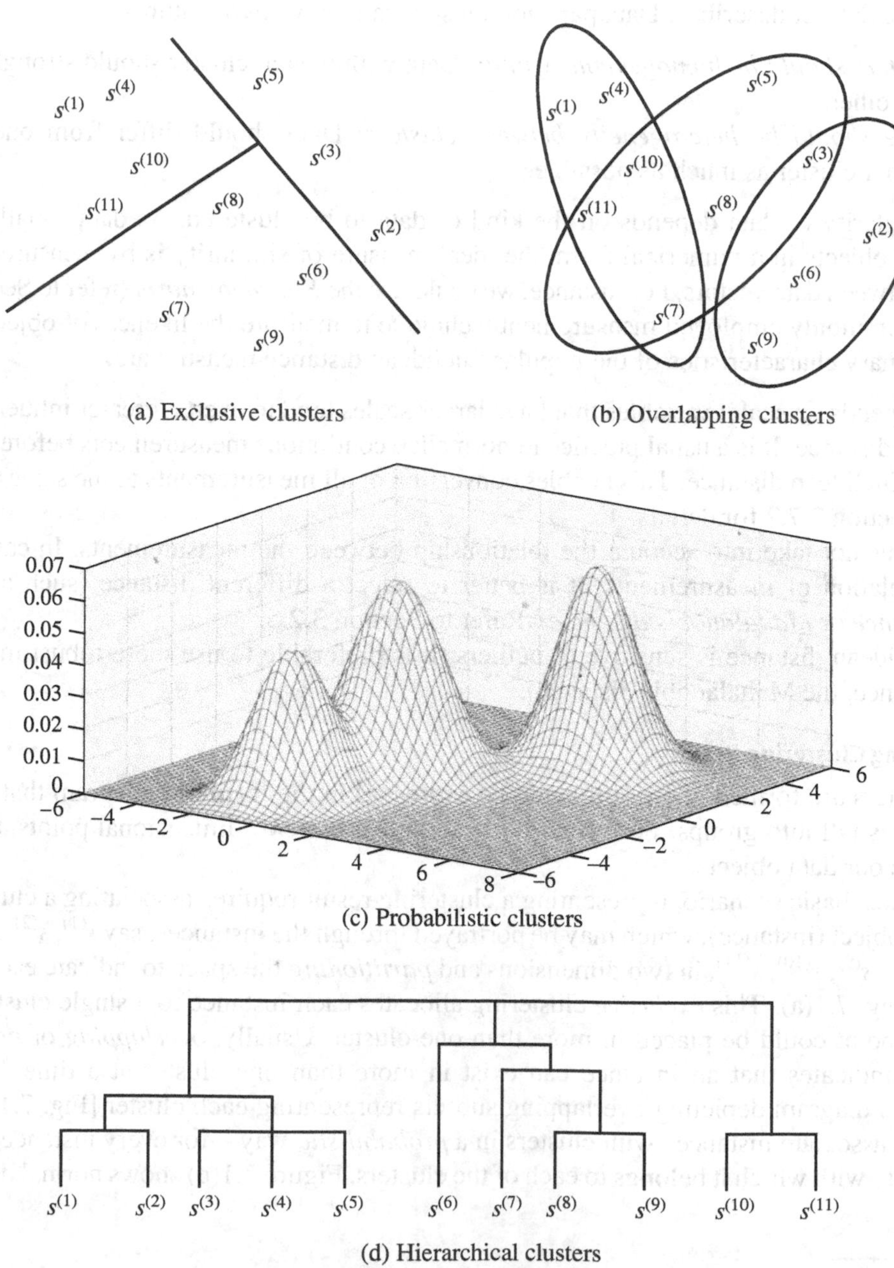

(a) Exclusive clusters
(b) Overlapping clusters
(c) Probabilistic clusters
(d) Hierarchical clusters

Figure 7.1 Different ways of representing clusters

Measuring Clustering Quality

Quantitative indices to measure clustering quality very much depend on the clustering approach. Various approaches lead to different types of clusters (Fig. 7.1). However, as we will see in the next section, the most widely used approaches, directly or indirectly, use the concepts employed in exclusive clustering (*hard clustering*) represented by Fig. 7.1(a). Our discussion that follows will be centered around this clustering. Extension of our discussion to other clustering approaches covered in this chapter will become obvious.

Cluster Validity: Following the application of a clustering technique on a dataset, the resultant clusters need to be assessed for quality. The evaluation of the result of a clustering algorithm is done using *clustering validation indices* which are used to quantitatively evaluate the result of a clustering algorithm. Many validation indices exist. Refer to [119] for a detailed survey. Two criteria that are usually considered enough to measure data partitioning quality are Compactness and Separation.

- While patterns in one cluster should have similarity, they should be dissimilar to the patterns of other clusters. *Compactness* is indicated by the variance of patterns in a cluster.
- Clusters need to be properly separated from one another. The Euclidean distance between cluster centroids indicates the cluster *separation*.

It is necessary to specify the number of clusters in advance in the majority of clustering algorithms. Conventionally, in order to determine the 'optimum' number of clusters, the algorithm needs to be run again and again, employing different values generated at random, and selecting the data partitioning resulting in the best validation measure.

It is not easy to mathematically obtain the optimum number of clusters in a dataset because this can only be possible through *a priori* knowledge and/or the absolute truth about the data, which is rarely available. In fact, many researchers have been studying the issue of arriving at an optimum number of clusters in a dataset.

In addition to mathematical measures of cluster goodness to provide some guidance on the number of clusters, we must also evaluate on subjective basis to determine their usefulness for an application. Sometimes the number of partitions (or the acceptable range for this number) is supplied by the application. For example, in the clustering problem related to customers, segmentation, how many segments a business can reasonably support is subjectively examined.

Cluster Interpretability: Let us study the characteristics of each cluster to be able to understand the clusters by:

- first getting the summary statistics from each cluster;
- then, establishing if the dataset provided has a nonrandom structure, which may result in meaningful clusters;
- and finally attempting to allocate a name or label to each cluster.

The most simple and commonly used criterion function for clustering is the *sum-of-squared-error* criterion. Let N_k be the number of samples in cluster k and μ_k be the mean of those samples:

$$\mu_k = \frac{1}{N_k}\sum_{i=1}^{N_k}\mathbf{x}^{(i)} \tag{7.1a}$$

Then the sum of squared errors is defined by,

$$J_e = \sum_{k=1}^{K}\sum_{i=1}^{N_k}\left\|\mathbf{x}^{(i)} - \mu_k\right\|^2 \tag{7.1b}$$

where K stands for number of data partitioning.

This criterion has a simple interpretation.

Cluster Stability: In case certain inputs are slightly changed, does this result in a major change in the cluster assignments?

Applications of Cluster Analysis

In cluster analysis, data objects are grouped together only on the basis of the information available in the data, which gives a description of the objects and their relationships. It is an unsupervised learning problem wherein training samples are unlabeled. That is, the problem is to see what can be done when all one has is a collection of samples without being told their categories/numeric values for regression.

It is natural to wonder why an issue as uncompromising as this should be of any interest to anyone, irrespective of whether it is possible to learn anything valuable from unlabeled samples.

Broadly speaking, cluster analysis can be used for exploring data, or it may serve as a standalone tool for machine learning/data mining applications.

Clustering for Data Exploration

- It is a useful tool for data summarization.
- The concept of outliers is linked to clusters. An outlier is an object that either comes from a remote cluster, which is small or does not belong to any cluster. Alteration of clustering algorithms can be done to ensure inclusion of *outlier detection* as a consequence of their execution.
- The fact that distinct subclasses have been found—clusters or groups of patterns, wherein there is more similarity between the members than there is with other patterns—suggests certain alternative models for designing classifier/regressor.

  For instance, several machine learning methods are complex in terms of time or space. Therefore, they are not feasible in case of huge datasets. But it is possible to apply the algorithm to a smaller dataset—comprising just prototypes of clusters—instead of the whole dataset. Based on the nature of analysis, the number of prototypes, and how accurately the data is represented by the prototypes, it is possible to compare the results with those that would have been gained in case all the data had been used.

- To get the nearest neighbors, you need to compute the pairwise distance between all points. If objects are close to their cluster prototype, then prototypes can be used to bring down the number of distance computations needed to find the nearest neighbors of an object. In case two cluster prototypes are at a distance from each other, it is not possible for the objects in the corresponding clusters to become each other's closest neighbors. As a result, finding the nearest neighbors of an object only requires computation of the distance to objects in nearby clusters.

Clustering as a Standalone Tool

Clustering is just a type of classification wherein objects are labeled with cluster or class labels. These labels, however, are drawn from the data itself, unlike supervised classification wherein the class label for each and every object is known beforehand. Therefore, cluster analysis is often called unsupervised classification. Some examples of applications are as follows:

- Clustering has been used by biologists to find groups of genes that have similar functions.
- The World Wide Web consists of billions of Web pages, and the results of a query to a search engine can return thousands of pages. Clustering can be used to group these search results into a small number of clusters, each of which captures a particular aspect of the query.
- Clustering has many applications in psychology and medicine. For example, clustering has been used to identify different types of depression. Contents of MR brain images can be clustered into subclasses for diagnostic purposes.
- In an image recognition application, subclasses or clusters can be discovered in character recognition systems, which are hand written.
- In business intelligence, a lot of customers can be organized into groups using clustering. The customers within the group will be strongly similar to each other in terms of characteristics. This helps develop business strategies meant to improve customer relationship management. For instance, customers are divided on the basis of information related to demographic and transaction history. Then, a marketing strategy is devised suited to each section. Clustering can also be used to analyze the market structure: identify groups comprising similar products as per competitive measures of similarity.
- Cluster analysis can help create balanced portfolios in finance.

Data clustering is under vigorous development. Because of the large volumes of data gathered and stored in databases, cluster analysis has of late become a very hot subject in data mining research.

Details on clustering methods are available in Sections 7.3–7.6.

7.2.3 Data Transformations: Enhancing the Information Content of the Data

Typically, data from a variety of sources must be integrated. Because the source data is usually not at the level required by the application in hand, building the data for the application requires many transformations. Preparing the data for mining by appropriate transformations depends on the nature of the data sources and the requirements of the data mining techniques. Some data preparation is almost always required and it is not unusual for data preparation to be the most time-consuming part of a data mining project. Some data preparation is required to fix problems with the source data, but much of it is designed to enhance the information content of the data. Better data means better models.

Data Cleansing

The real-world databases of modern times are extremely vulnerable to inconsistency, noise, and loss of data owing to their typically large size and also because they originate from multiple and varied sources. Data of low quality will give low-quality mining results. Preprocessing of the available raw data requires *data cleansing* to fix problems with the source data.

Data cleansing is a big challenge and consumes bulk of the effort in the data mining process. Discrepancies in data can be caused by various factors including poorly designed data collection, human-errors in data entry, deliberate errors (e.g., respondents not willing to divulge information), inconsistent data representations, errors in instrumentation that record data and system errors. Practical data mining process has to deal robustly and sensibly with errors in raw data such as missing values, inaccurate values, outliers, duplicate data, sparse data, skewed data, noisy data, anomalies because of some systematic errors, etc.

Dealing with poor-quality data cannot be captured in simple routines available off-the-shelf, though there are a number of different commercial tools that can aid in the data cleansing process. These tools employ data exploration methods to develop understanding of the data. Data cleansing process requires a great amount of trial-and-error effort using available commercial tools, and the knowledge we already have about the properties of the data (e.g., what is the domain of each attribute? What are the acceptable values for each attribute? Are there any known dependencies between attributes?, etc.)

Derived Attributes

Yet another significant aspect of the data transformation procedure pertains to the definition of new variables. These convey the information intrinsic in the data in such a way that the information becomes more useful for data mining methods. Derived variables for data preparation may involve creation of new variables through creative transformation of existing variables. When measuring variables on different scales, their standardization also becomes essential. For data mining methods that work on numerical data alone, it is necessary to numerically represent categorical data in some way.

Discretizing Numeric Attributes

Discretization of numeric attributes is absolutely essential if the task involves numeric attributes but the learning scheme chosen can only handle categorical ones. Even schemes that can handle numeric attributes often produce better results, or work faster, if the attributes are prediscretized. As has been said earlier, the converse situation in which categorical attributes must be represented numerically, also occurs, although less often.

Attribute Reduction

Datasets for analysis may contain hundreds of attributes, many of them may be irrelevant to the mining task. The role of domain expert is very important to pick out the relevant attributes (refer to Section 1.5). Leaving out relevant attributes or keeping irrelevant attributes may be detrimental, causing confusion for the mining algorithm employed. This can result in performance of the learning scheme to deteriorate. In addition, the added volume of irrelevant or redundant attributes can slow down the mining process. Feature extraction using domain knowledge is thus an important initial step in the data mining process.

Data mining success will suffer more if relevant features are left out compared to the loss in keeping irrelevant redundant attributes. This is because data transformation techniques can help eliminate the redundant attributes, but the relevant ones left out at the feature extraction stage cannot be recovered. Domain experts normally select a larger set of features, keeping all those for which they have doubts about redundancy. This larger set is then processed using transformation techniques for *reduction of the attributes*.

Transforming the data in different ways with respect to the objective, can help improve success when applying machine learning techniques.

Detailed discussion on transformation techniques will appear in Sections 7.7–7.10.

7.3 OVERVIEW OF BASIC CLUSTERING METHODS

There are many clustering algorithms in the literature. It is difficult to provide a crisp categorization of clustering methods. Nevertheless, it is useful to present a relatively organized picture of major fundamental clustering methods.

In this section, we limit our presentation to the following categories of clustering methods. The techniques based on these categories are widely used (refer to [17, 120] for broader categorization).

1. Partitional Clustering
2. Hierarchical Clustering
3. Spectral Clustering
4. Clustering using Self-Organizing Maps

7.3.1 Partitional Clustering

Partitioning is the most simple and basic version of cluster analysis, as it organizes the objects of a dataset into many groups/clusters. To ensure conciseness of the problem specification, our assumption should be that the algorithm seeks a fixed number of clusters 'K', as the user specifies. This parameter marks the beginning of partitioning techniques.

Formally, if a set of N objects is given, a partitioning technique will create K divisions of the data, where each division or partition is representative of a cluster, and $K \leq N$. In other words, it partitions the data into K groups in a way that each group must comprise a minimum of one object.

Since data frequently describes the features of objects as numbers, that is, in the numerical form, data subject to clustering will require representation by n-dimensional vectors $\mathbf{x}^{(i)}$; $i = 1, \ldots, N$, with $\mathbf{x} = \{x_j; j = 1, \ldots, n\} = [x_1\ x_2 \cdots x_n]^T$.

The set of N vectors builds a matrix $\mathbf{X}$ of dimension $N \times n$:

$$\mathbf{X} = \begin{bmatrix} x_1^{(1)} & x_2^{(1)} \cdots x_n^{(1)} \\ x_1^{(2)} & x_2^{(2)} \cdots x_n^{(2)} \\ \vdots & \vdots \quad \vdots \\ x_1^{(N)} & x_2^{(N)} \cdots x_n^{(N)} \end{bmatrix} \tag{7.2}$$

In this matrix, the rows are objects, while the columns are features or attributes.

The clusters are created by optimization of an objective partitioning criterion, such as distance-based *similarity function*, so that the objects within a cluster bear similarity with each other and bear no similarity with objects in other clusters, as far as the dataset attributes are concerned. There are many types of other criteria that help judge partition quality.

Achievement of global optimality in partitioning-based clustering can be rather prohibitive computationally. It requires analysis of all possible partitions of N objects into K clusters, which is generally a huge number. Instead, most applications resort to widely used heuristic techniques.

K-Means Algorithm

The most general of the heuristic clustering techniques is the K-means clustering. It is amongst the widely used clustering algorithms.

K-means clustering characteristically espouses *exclusive cluster separation:*

- The set of all clusters comprises all data vectors.
- Each object belongs to exactly one group.
- None of the clusters is empty and none of them contain the entire dataset $\mathbf{X}$. The clusters are not joined.

The goal of clustering algorithm is to find K points that make good *cluster centers*. These centers define the clusters. Each object is assigned to the cluster defined by the nearest cluster center. The best assignment of cluster centers could be defined as the one that minimizes the sum of distances from every point to its nearest cluster center (or the distance-squared).

Finding the optimal solution is difficult and the K-means algorithm does not attempt to do so. Instead, we follow the following iterative procedure.

K points are randomly selected as cluster centers, which gives us *cluster seeds*. All objects are allocated to their nearest cluster center as per the Euclidean distance metric. For allocation of all objects to the nearest seed, all that is required is the calculation of the distance between each object and each seed. Then comes the calculation of the centroid or mean of the objects in each cluster, which is the 'means' part of the algorithm. These centroids are considered as new cluster-center values for their respective clusters. The entire procedure is iterated with the new cluster centers. Repetition goes on until the same points are assigned to each cluster in successive rounds. At this stage, the cluster centers become stable.

The final clusters are rather sensitive to cluster seeds. Entirely different arrangements can be derived from small changes in primary random selection of the seeds, and some may be more beneficial than others. To increase the chance of discovering the solution that has the maximum benefit, the algorithm may require to be run many times with various seeds and then the best final result may be selected.

Similar situation arises in the choice of K. Often, nothing is known about the likely number of clusters, and the whole point of clustering is to find it out. A heuristic way is to try different values and choose the best.

Fuzzy *K*-Means Clustering

In case of data clustering work, it is important to define the data partition type. There is a distinction between *hard* and *soft* partitions.

The aforementioned K-means technique is an example of *hard partitioning*, which generally opts for exclusive cluster separation. In other words, each object has to be part of exactly one single group. There are many problems that do not allow such a partitioning. Algorithms that make an object belong to several clusters with varied membership degrees are helpful in case of many practical problems. This kind of partitioning can be called *soft partitioning*, and is simply an extension of hard partitioning, wherein an object may not always be grouped clearly into to one class or category.

In K-means clustering algorithm, each point $\mathbf{x}^{(i)}$ is part of only one cluster (the clusters are disconnected). In *Fuzzy K-Means Clustering* algorithms, on the other hand, the data points are allocated to many clusters with different degrees of *membership*, which is a number ranging from 0 to 1. The concept is based on the fact that data clusters generally tend to overlap to a certain level, and it is not easy to identify proper borders among them. Hence, some data vectors cannot be clearly allocated exactly to one cluster, and therefore, it becomes more sensible to assign them partially to varied clusters. There is a limit clamped on membership degrees of a specific object to different clusters—the sum total of membership degrees of this object to each of the K clusters should equal 1.

It is observed that an object existing at an almost same distance from centers of all clusters may be considered noise. It is natural for us to allocate extremely low membership degrees to noise. However, it is a must to meet the condition that the sum of all membership degrees of the said object should be equal to 1.

Probabilistic Clusters

Statistically, we can assume that clusters to be derived from cluster analysis are distributions over the data space, which can be mathematically represented using *probability density functions*. We call such clusters *probabilistic clusters*. Probabilistic clustering provides another approach to soft clustering.

If we assume that samples come from a mixture of K normal (Gaussian) distributions, we can approximate a large variety of situations. The problem then reduces to estimating the unknown mixture densities:

$$p(\mathbf{x}|\boldsymbol{\theta}) = \sum_{k=1}^{K} p(\mathbf{x}|\boldsymbol{\theta}_k) w_k \tag{7.3}$$

where $\boldsymbol{\theta} = [\boldsymbol{\theta}_1 \cdots \boldsymbol{\theta}_K]^T$ are the parameters of the *mixture density* $p(\mathbf{x}|\boldsymbol{\theta})$. The conditional densities $p(\mathbf{x}|\boldsymbol{\theta}_k)$ are called the *component densities*, and the prior probabilities w_k ($w_k = P(\text{cluster } k)$) are called the *mixing parameters*.

In many applications, *Gaussian Mixture Model* (GMM)-based clustering has been shown to be effective because it is more general than (hard) K-means clustering and (soft) Fuzzy K-means clustering methods.

Partitioning-based clustering methods are studied in depth in Sections 7.4–7.6.

7.3.2 Hierarchical Clustering

Let us now take a brief look at another clustering model which is an alternative to the partitioning-based techniques talked about earlier. Partitioning-based techniques are not based on the assumption that substructures exist in the clusters. Yet, there may be instances when data is organized in a hierarchical manner, that is, clusters have subclusters, there are subclusters within subclusters and so on. In such a scenario, it is better to have *hierarchical clustering* for effectiveness.

Given a dataset $\mathbf{X} = \{\mathbf{x}^{(1)}, \ldots, \mathbf{x}^{(N)}\} \in \Re^n$, let us consider a sequence of divisions of its elements into K clusters, where $K \in [1, N]$, an integer *not fixed a priori*. The first possible division or partition is the one that divides the data into N groups with each cluster containing only one element. The second partition divides $\mathbf{X}$ into $N - 1$ clusters by merging two closest observations into one cluster. The third partition divides $\mathbf{X}$ into $N - 2$ clusters by merging two clusters with the smallest distance (we will shortly take up the issue of measuring distance between clusters). The process goes on progressively *agglomerating* (combining) the two closest clusters until there only one cluster remains, which comprises all data samples. We consider ourselves at level l of the sequence of partitions when $K = N - l + 1$. Therefore, level one corresponds to N clusters and level N to one cluster. If, according to the property of the sequence, whenever two samples exist together in the same cluster at level l, they continue to stay together at all higher levels, then it is said to be a *hierarchical* sequence.

Usually a tree, known as *dendrogram*, represents the hierarchical clustering. Figure 7.2 represents a *dendrogram* for a dataset with ten samples. At $l = 1$, each cluster has a single pattern. At level 2, $\mathbf{x}^{(9)}$ and $\mathbf{x}^{(10)}$ are gathered in a single cluster. At last level, $l = 10$, all patterns belong to the single cluster.

The process of hierarchical clustering can be categorized into two distinct models: *agglomerative* and *divisive*. Agglomerative processes begin with N singletons and form the sequence with successive merging of clusters. Divisive processes begin with all of the samples in a single cluster and create the sequence by consecutively separating clusters. In case of agglomerative processes, simpler computation is required to move from one level to another.

Whether employing an agglomerative technique or a divisive one, the primary requirement is to measure the distance between two clusters, where each cluster is usually a set of objects. The following are the four commonly used measures for distance between clusters:

(i) The minimum distance between clusters—the distance between their two closest members. This gives the *single-linkage* clustering algorithm. Since this measure takes into consideration only the two nearest members of a pair of clusters, the procedure is sensitive to outliers. By adding merely a single new instance the whole structure can be radically changed.

(ii) The maximum distance between the clusters—the distance between the two members which are the farthest. This gives the *complete linkage* clustering algorithm. This measure is also sensitive to outliers but looks for compact clusters. Some instances, however, may end up a lot closer to other clusters than to rest of their own cluster.

(iii) The *centroid-linkage* method, which represents clusters by the *centroid* of their members (such as in *K*-means algorithm) and uses the distance between centroids.
(iv) The *average linkage* method, which calculates the average distance of all possible distances between objects in one cluster and those in the other cluster.

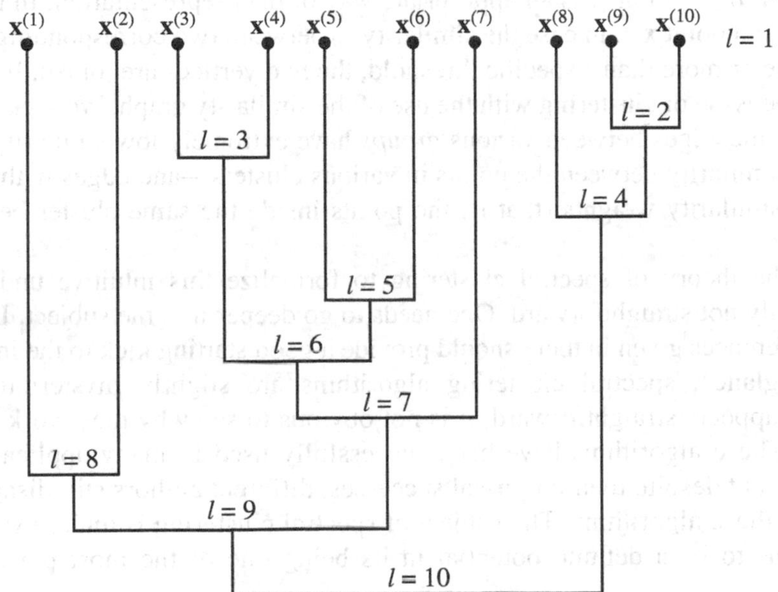

Figure 7.2 A dendrogram

Choosing from among clustering methods is done by exploitation of domain knowledge. If the probable nature of clusters is unknown, default choices are complete and average-linkage methods that yield spherical clusters.

Some observations regarding hierarchical clustering are:

- It proves to be costly and slow when it comes to large datasets, because it requires the computation and storage of $N \times N$ distance matrix (similarity matrix).
- If there is erroneous allocation of instances in the process initially, it cannot be reassigned later. In other words, the hierarchical algorithm passes through the data only once.

7.3.3 Spectral Clustering

Of late, spectral clustering has gained popularity as a modern clustering algorithm. It is not difficult to execute. It can be easily and efficiently solved using standard linear algebra software. Quite frequently, especially when the data is linked but not really compact or clustered within convex boundaries, spectral clustering works much better than other clustering algorithms, such as *K*-means and Gaussian mixtures. Refer [121] for a detailed tutorial on spectral clustering. This tutorial derives spectral clustering and offers different perspectives to working of spectral clustering.

Consider a set of data points $\mathbf{x}^{(i)}$; $i = 1, \ldots, N(\mathbf{x}^{(i)} = \{x_j^{(i)}; j = 1, \ldots, n\})$. The natural aim of clustering is to partition the data points into many classes in such a way that points in the same group bear similarity while those in varied groups bear no similarity with each other. Clustering, therefore, requires some notion of *similarity* $s_{il} \geq 0$ between all pairs of data points $\mathbf{x}^{(i)}$ and $\mathbf{x}^{(l)}$; $1 \leq i, l \leq N$.

Use of the *similarity* graph G is an appropriate way of data representation. In this graph, each *vertex* depicts a data point $\mathbf{x}^{(i)}$. In case the similarity s_{il} between two corresponding data points $\mathbf{x}^{(i)}$ and $\mathbf{x}^{(l)}$ is positive or more than a specific threshold, the two vertices are joined. It is now possible to reformulate the issue of clustering with the use of the similarity graph. We seek such a partition of the graph that the edges between various *groups* have extremely low similarity weights—this indicates the dissimilarity between the points in various clusters—and edges within a group have extremely high similarity weights (that is, the points inside the same cluster bear similarity to each other).

Diving into the theory of spectral clustering to formalize this intuitive understanding into algorithms is really not straightforward. One needs to go deeper into the subject. Luxberg tutorial [121] and the references given in there should provide a good starting kick to the interested reader.

At the first glance, spectral clustering algorithms are slightly mysterious. Though the implementation appears straightforward, it is not obvious to see why they work at all and what they really do. These algorithms have been successfully used in many applications including computer vision, but despite their empirical successes, different authors still disagree on various issues regarding these algorithms. The subject of spectral clustering is under extensive research since there seems to be a definite potential in its being one of the most powerful clustering techniques.

In this subsection, we first attempt to provide an intuitive understanding of spectral clustering and thereafter discuss one of the many proposed algorithms—the Ng-Jordan-Weiss algorithm [122].

Affinity Matrix /Adjacency Matrix

What is an Affinity? It is a metric that determines how close, or *similar*, two points are in our space $\Re^n$. Let G be a *symmetric* graph with each vertex representing data point $\mathbf{x}^{(i)}$. Each edge between two vertices i and l carries a non-negative weight $w_{il} \geq 0$. If $w_{il} = 0$, it means that the vertices corresponding to $\mathbf{x}^{(i)}$ and $\mathbf{x}^{(l)}$ are not connected by an edge. As G is symmetric, $w_{il} = w_{li}$. The *Affinity Matrix* or *Adjacency Matrix* of the graph is the matrix:

$$\mathbf{W} = \{w_{il}\} i, l = 1, \ldots, N \tag{7.4}$$

The *degree* of vertex representing $\mathbf{x}^{(i)}$ is defined as,

$$d_i = \sum_{l=1}^{N} w_{il} \tag{7.5}$$

Note that, in fact, the sum runs over all vertices adjacent to the one representing $\mathbf{x}^{(i)}$ as for all other vertices, the weight $w_{il} = 0$.

The *degree matrix* $\mathbf{D}$ is defined as the diagonal matrix with the degrees $d_1, d_2, \ldots, d_N$ on the diagonal.

Different Affinity Matrices

There are several popular constructions to transform a given set $\mathbf{x}^{(1)}, \ldots, \mathbf{x}^{(N)}$ of data points with pairwise similarities into a graph. The goal is to model the local neighborhood between the data points. In *ε-neighborhood graph*, we connect all points whose pairwise distances are smaller than ε. In *k-nearest-neighbor* graph, the vertex representing $\mathbf{x}^{(i)}$ is connected to vertex representing $\mathbf{x}^{(l)}$ if $\mathbf{x}^{(l)}$ is among the k-nearest neighbors of $\mathbf{x}^{(i)}$. Since the neighborhood relationship is not symmetric, some additional adjustments are required to construct the symmetric graph. A simple way is to ignore the directions of the edges. Another way is to connect two vertices if both $\mathbf{x}^{(i)}$ is among the k-nearest neighbors of $\mathbf{x}^{(l)}$ and $\mathbf{x}^{(l)}$ is among the k-nearest neighbors of $\mathbf{x}^{(i)}$.

We may use Gaussian similarity function

$$w(\mathbf{x}^{(i)}, \mathbf{x}^{(l)}) = w_{il} = \exp\left(-\frac{1}{\sigma^2}\left\|\mathbf{x}^{(i)} - \mathbf{x}^{(l)}\right\|^2\right) \tag{7.6}$$

to construct a graph. Note that this symmetric similarity function ($w_{il} = w_{li} > 0$) itself models local neighborhoods; the parameter σ controls the width of the neighborhoods. It controls how fast the affinity w_{il} decreases as distance between $\mathbf{x}^{(i)}$ and $\mathbf{x}^{(l)}$ increases. $w_{il} \simeq 1$ when the points $\mathbf{x}^{(i)}$ and $\mathbf{x}^{(l)}$ are close in $\Re^n$, and $w_{il} \to 0$ if the points are far apart.

Graph Laplacian Matrices

Graph Laplacian matrices are the main tools used for spectral clustering. An entire field is dedicated to studying these matrices. However, our presentation is limited to just a few widely used variants of graph Laplacians. In the following, we assume that G is an undirected weighted graph with a symmetric weight matrix $\mathbf{W}$, where $w_{il} = w_{li} \geq 0$.

The *unnormalized* graph Laplacian is defined as [123],

$$\mathbf{L} = \mathbf{D} - \mathbf{W} \tag{7.7}$$

where $\mathbf{D}$ is the diagonal degree matrix assigned to the graph vertices and $\mathbf{W}$ is the weight matrix assigned to graph edges.

A *normalized* graph Laplacian matrix is given by,

$$\mathbf{L}_N = \mathbf{D}^{-1/2}\,\mathbf{L}\mathbf{D}^{-1/2} \tag{7.8a}$$

$$= \mathbf{D}^{-1/2}\,(\mathbf{D} - \mathbf{W})\,\mathbf{D}^{-1/2}$$

$$= \mathbf{I} - \mathbf{D}^{-1/2}\,\mathbf{W}\mathbf{D}^{-1/2} \tag{7.8b}$$

The Laplacian used by Ng-Jordan-Weiss [122] is a *normalized* graph Laplacian:

$$\mathbf{L}_{NJW} = \mathbf{D}^{-1/2}\,\mathbf{W}\mathbf{D}^{-1/2}, \text{ and where } w_{ii} = 0 \tag{7.8c}$$

Left multiplying by a diagonal matrix is akin to scaling the rows, and right multiplying by a diagonal matrix is akin to scaling the columns.

The Ng-Jordan-Weiss Algorithm

Given a set of points $\mathbf{x}^{(1)}, \ldots, \mathbf{x}^{(N)}$ in $\Re^n$ that we want to cluster into K subsets:

1. Form the Affinity Matrix, $\mathbf{W} \in \Re^{N \times N}$ defined by,

$$w_{il} = \exp\left(-\frac{1}{\sigma^2}\left\|\mathbf{x}^{(i)} - \mathbf{x}^{(l)}\right\|^2\right) \text{ if } i \neq l, \quad \text{and} \quad w_{ii} = 0 \tag{7.9a}$$

2. Define $\mathbf{D}$ to be a diagonal matrix whose (i, i)-element is the sum of all the $\mathbf{W}$'s i^{th}-row elements, i.e.,

$$d_{ii} = \sum_{l=1}^{N} w_{il} \tag{7.9b}$$

3. Construct the Laplacian matrix

$$\mathbf{L}_{NJW} = \mathbf{D}^{-1/2}\,\mathbf{W}\,\mathbf{D}^{-1/2}, \text{ and where } w_{ii} = 0 \tag{7.9c}$$

4. Find $\mathbf{v}_1, \mathbf{v}_2, \ldots, \mathbf{v}_K$, the K leading eigenvectors (refer to Sections 1.9, and 7.9) and form the matrix

$$\mathbf{V} = [\mathbf{v}_1\ \mathbf{v}_2 \cdots \mathbf{v}_K] \in \Re^{N \times K} \tag{7.9d}$$

by stacking the eigenvectors in columns. The algorithm computes the K eigenvectors with the largest eigenvalues $\lambda_1, \lambda_2, \ldots, \lambda_K$ of $\mathbf{L}_{NJW}$.

5. Form the matrix $\mathbf{Z}$ from $\mathbf{V}$ by normalizing each of the $\mathbf{V}$'s rows to have unit length. That is,

$$z_{ik} = v_{ik} \Big/ \left(\sum_{k=1}^{K} v_{ik}^2\right)^{-1/2}; \; i = 1, \ldots, N; \; k = 1, \ldots, K. \tag{7.9e}$$

6. Treating each row of $\mathbf{Z}$ as a point in $\Re^K$, cluster them into K clusters via K-means or any other algorithm serving the partitioning purpose.
7. Finally, assign the original point $\mathbf{x}^{(i)}$ to the cluster k if and only if row i of the matrix $\mathbf{Z}$ was assigned to cluster k.

Points to Note

- In spectral clustering methods, the dimensionality of the new space is set to the desired number of clusters. Spectral clustering, thus, is a *dimensionality reduction method* for clustering high-dimensional data. Newly constructed space (derived from original space) should have low dimensionality [123].
- There are many dimensionality reduction methods, some of which will be presented later in this chapter. However, such methods may not be able to detect the clustering structure. The setting of clustering algorithms expects that each new dimension should be able to manifest a cluster.
- Computing eigenvectors of large matrices is costly.
- The scaling parameter σ^2 controls how rapidly the affinity w_{il} falls off with the distance between $\mathbf{x}^{(i)}$ and $\mathbf{x}^{(l)}$. Some methods have been proposed for choosing this parameter automatically. Choosing a good value of this parameter which has a profound effect on the results of spectral clustering, is usually difficult.

- Spectral clustering aims to group together the connected data, which may not always be compact within convex boundaries. A *compact* dataset comprises close points existing in the same cluster. Data points contained in different clusters are not near (Fig. 7.3(a)). However, the *connectivity* figure [Fig. 7.3(b)] indicates that the *data points within one cluster may also be far away—may be farther away than the points in various clusters.* Our aim is to transform the space in such a way that when two points are close together in the new space, they remain within the same cluster, but when they are not close to each other, they are in various separate clusters. Properties of the graph Laplacian alter the representation, improving the cluster properties in the data so that they are easily identified in the new representation. The simple K-means clustering algorithm finds it very easy to detect the clusters in a new representation.

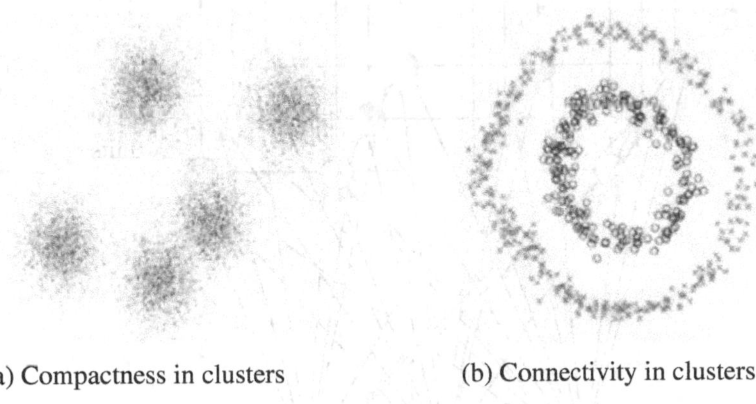

(a) Compactness in clusters (b) Connectivity in clusters

Figure 7.3

7.3.4 Clustering using Self-Organizing Maps

Self-Organizing Maps (SOMs) are a kind of neural network capable of performing unsupervised learning jobs, such as clustering. Although they were originally used for images and sounds, these networks can also recognize clusters in data. Finnish researcher, Dr Tuevo Kohonen, is the inventor of self-organizing maps, which explains why they are also known as *Kohonen networks*.

Patterns in data are identified by an unsupervised-learning network, which goes through changes in its parameters while identifying these patterns. *Self-organizing* is the process of undergoing these changes. Therefore, it can be said that self-organized learning involves regular alteration in the synaptic weights of the network, in response to input samples. The changes in weights take place according to certain learning rules. On repeatedly applying input samples to the network, there arises a significant configuration. Simply put, a sort of global order appears out of several, random and local interactions, which may finally result in some kind of similarity behaviour.

A special class of self-organizing neural networks is based on *competitive learning*. In competitive learning procedures, learning adjustments are confined to the cluster that is most similar to the pattern currently being presented. Consider, for example, the network structure shown in Fig. 7.4. The output layer (arranged in grid) consists of K units. Each of the units in the output layer is fully connected to the input terminals; the connection weights represented as $\mathbf{w}_1, \ldots, \mathbf{w}_K$, that is, $\mathbf{w}_k$ with $k = 1, \ldots, K$; $\mathbf{w}_k = \{w_{kj}; j = 1, \ldots, n\}$. The n-dimensional input pattern $\mathbf{x}^{(i)} = \{x_1^{(i)}, \ldots, x_n^{(i)}; i = 1, \ldots, N\}$ is normalized to have length $\|\mathbf{x}^{(i)}\| = 1$. Each of the cluster centers in output layer is initialized with a randomly chosen weight vector, also normalized: $\|\mathbf{w}_k\| = 1; k = 1, \ldots, K$. It is traditional, but not required, to initialize cluster centers to be K points randomly selected from data.

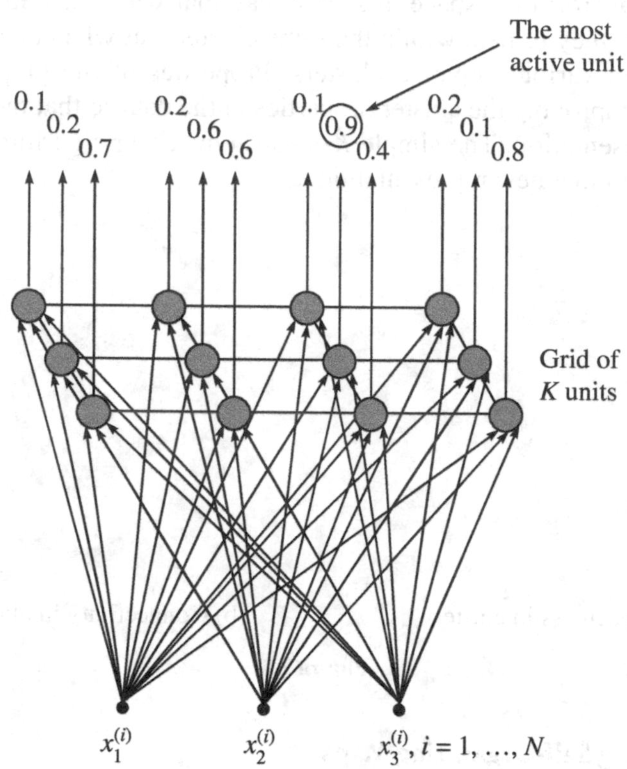

Figure 7.4 Competitive-learning architecture

When a new pattern is presented, each of the cluster units computes its *activation* $a_k = \mathbf{w}_k^T\mathbf{x}^{(i)}$. Only the unit with the largest activation (i.e., whose weight vector is closest to the new pattern) is permitted to update its weights. That is, learning is confined to the weights at the *most active unit*; the weight vector at this unit is updated to be more like the input pattern:

$$\mathbf{w}_k(t + 1) = \mathbf{w}_k(t) + \eta(\mathbf{x}^{(i)} - \mathbf{w}_k(t)) \tag{7.10}$$

where η is a learning rate ('t' is the iteration index).

Note that whereas the clustering algorithms we have presented so far, typically have all the data present before clustering begins (i.e., are *off-line*), the competitive learning is an *on-line* algorithm where clustering is performed on-line as the data streams in.

The fact that only most active unit is modified for a given input $\mathbf{x}^{(i)}$, results in *hard competitive learning*. The *Self-Organizing Map* (SOM), developed by Kohonen, is *soft competitive learning network*, wherein not only the weights for the most active units are adjusted, but the weights for units in its immediate *neighborhood* are also adjusted to strengthen their response to inputs [124].

The SOM is an artificial system emulating specific brain mappings. The cerebral cortex, which comprises an extremely complex structure of billions of neurons and even more synapses, dominates the brain. The cortex bears no uniformity or homogeneity. It consists of areas identifiable by how thick their layers are, and by the neurons they contain. These neurons are related to various sensory inputs as they are responsible for auditory, visual, sensory, motor and other human activities. It will not be wrong to say that each sensory input is mapped to a corresponding area of the cerebral cortex. Therefore, the cortex is a self-organizing computational map within our brain.

Kohonen came up with the *principle of topographic map formation*, according to which the location of an output neuron (in terms of space) in the topographic map matches a specific trait of the input pattern.

The fundamental idea of SOM is that inputs (belonging to an event space) are received by a simple network of adaptive elements; the signal representations are automatically mapped onto a set of outputs, in a way that the responses achieve the same topological order as that of the events.

Figure 7.5 shows a conventional feature-mapping architecture. The inputs to the network can be written in the vector form as,

$$\mathbf{x} = [x_1 \, x_2 \cdots x_n]^T$$

and the synaptic weight vector of the neuron k in the two-dimensional array is given by,

$$\mathbf{w}_k = [w_{k1} \, w_{k2} \cdots w_{kn}]^T; \; k = 1, 2, \ldots, K \tag{7.11a}$$

where K is the total number of output neurons in the array. The *best match* of the input vector $\mathbf{x}$ with the synaptic weight vector $\mathbf{w}_k$ is determined from the Euclidean distance between the input and the weight vector. The output unit with the *smallest* Euclidean distance, i.e.,

$$\min_{\forall k} \|\mathbf{x} - \mathbf{w}_k\| \tag{7.11b}$$

is the *most active neuron q*.

The next step in the *Kohonen's algorithm* is to update the synaptic weight vector associated with the *most active* neuron and the neurons within a defined neighborhood N_q of this neuron. The learning rule is given by,

$$\mathbf{w}_k(t+1) = \begin{cases} \mathbf{w}_k(t) + \eta(\mathbf{x}(t) - \mathbf{w}_k(t)) & \text{if } k \in N_q \\ \mathbf{w}_k(t) & \text{if } k \notin N_q \end{cases} \tag{7.11c}$$

where η is the learning rate parameter, and t is the iteration index.

The neighborhood of the most active neuron may include neurons within one, two or even three positions on either side. Generally, training a SOM begins with the neighborhood of a fairly large size. Then, as training proceeds, the neighborhood size gradually decreases.

Initialization of the network weights can be carried out by either randomly initializing them, or selecting a set of weights that reflect some *a priori* knowledge about the input data, that is, information regarding the possible distribution of the output clusters. Stopping conditions can be, for example, based on total number of specified iterations or based on monitoring the weight changes.

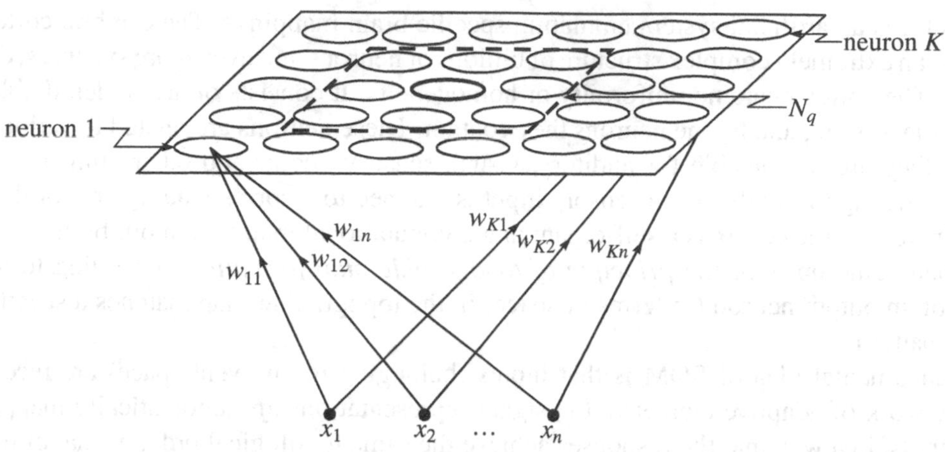

Figure 7.5 Conventional feature-mapping architecture of SOM

Given is a set of N patterns $\{\mathbf{x}^{(i)}; i = 1, 2, \ldots, N\}$ to be divided into K clusters. On each iteration, a pattern $\mathbf{x}^{(i)}$ is presented to the network. Repeated presentations (that is, large number of iterations) of the input patterns to the network are normally required before the map begins to 'unfold'.

A competitive learning network performs an on-line clustering process on the input patterns. A cluster center's position is specified by the weight vector connected to the corresponding output unit. The update formula in Eqn (7.11c) implements a sequential scheme for finding the cluster centers of a dataset. When an input $\mathbf{x}^{(i)}$ is presented to the network, the weight vector closest to $\mathbf{x}^{(i)}$ rotates towards it. Consequently, weight vectors move toward those areas where most inputs appear, and, eventually, the weight vectors become the cluster centers for the dataset. When the process is complete, the input data are divided into disjoint clusters such that similarities between individuals in the cluster are larger than dissimilarities in different clusters. Here the dissimilarity measure of the Euclidean distance is used as a metric of similarity. Obviously, other metrics can be used, and different selections lead to different clustering results.

7.4 *K*-MEANS CLUSTERING

The K-means clustering algorithm unambiguously partitions the given data into K clusters. The data describes the features of the objects in numerical form; data subject to clustering will be represented by n-dimensional vectors $\mathbf{x}^{(i)}$; $i = 1, \ldots, N$, with $\mathbf{x} = \{x_j; j = 1, \ldots, n\} = [x_1\ x_2 \cdots x_n]^T$. The set of N vectors creates data matrix $\mathbf{X}$ of dimension $N \times n$:

$$\mathbf{X} = \begin{bmatrix} x_1^{(1)} & x_2^{(1)} & \cdots & x_n^{(1)} \\ \vdots & \vdots & & \vdots \\ x_1^{(N)} & x_2^{(N)} & \cdots & x_n^{(N)} \end{bmatrix} \tag{7.12}$$

Matrix rows are the objects and the columns are features (attributes). For example, in the case of medical diagnostics, objects may be identified with the patients, while features will be identified with symptoms of a disease or with results of laboratory analysis of those patients.

In the K-means clustering, which may be referred to as *hard* clustering, the object entirely belongs or does not belong to a given cluster. The objective of data clustering is data partitioning into K clusters C_k so that,

$$\bigcup_{k=1}^{K} C_k = \mathbf{X} \tag{7.13a}$$

$$C_k \cap C_l = \varnothing;\ 1 \le k \ne l \le K \tag{7.13b}$$

$$\varnothing \subset C_k \subset \mathbf{X};\ 1 \le k \le K \tag{7.13c}$$

Condition (7.13a) means that the set of all clusters contains all data vectors and each object belongs to exactly one cluster. Condition (7.13b) means that clusters are disjoint, and condition (7.13c) means that none of the clusters is empty nor contains the whole dataset.

When executing K-means algorithm, we compute the *distance* between each vector $\mathbf{x}^{(i)}$; $i = 1, \ldots, N$, and the *cluster center* μ_k; $k = 1, \ldots, K$. If N_k represents the number of samples in cluster k, then,

$$\mu_k = \frac{1}{N_k} \sum_{i=1}^{N_k} x^{(i)} \tag{7.14}$$

That is, μ_k is the mean of all samples in cluster k.

Distance measure to represent similarity of data, may be defined in different ways (the most commonly used distance measure is the Euclidean norm). By using various distance measures, we can obtain different shapes of clusters. The partitioning based on *Euclidean norm* distance measure results in *spherical-shaped clusters* (hyperspheres). The clusters generated by *Mahalanobis norm* are *hyperellipses*.

In this section, we present K-means algorithm using Euclidean norm, defined as follows. The distance d_{ik} between data point $\mathbf{x}^{(i)}$ and cluster μ_k is given by,

$$d_{ik} = \sqrt{\sum_{j=1}^{n} (x_j^{(i)} - \mu_{kj})^2} \tag{7.15a}$$

In the vector notation,

$$d_{ik} = \left[(\mu_k - \mathbf{x}^{(i)})^T (\mu_k - \mathbf{x}^{(i)})\right]^{1/2} = \left\|\mathbf{x}^{(i)} - \mu_k\right\| \tag{7.15b}$$

The number of clusters K is selected on the basis of external considerations (e.g., prior knowledge, practical limitations, and so on) or we can attempt some different values for K and calculate the subsequent clusters. With repeated trials, the quality of clustering should become better. For this it is important for intracluster similarity to be high (distances low) and intercluster similarity to be low (distances high).

The goal of K-means clustering algorithm is twofold:

- To find the centers μ_k of the clusters C_k; $k = 1, \ldots, K$.
- To determine the clusters (data points N_k belonging to cluster k) of each center in the dataset.

The second goal can easily be achieved once we accomplish the first goal. Given cluster centers μ_k, a data point $\mathbf{x}^{(i)}$ in the dataset belongs to the cluster whose center is the closest, i.e.,

$$\mathbf{x}^{(i)} \in C_k \text{ if } \|\mathbf{x}^{(i)} - \mu_k\| < \|\mathbf{x}^{(i)} - \mu_l\|;\ k = 1, \ldots, K;\ l \neq k \tag{7.16}$$

In order to achieve the first goal, i.e., finding the cluster centers, we have to establish a criterion that can be used to search for these cluster centers. The simplest and most widely used criterion function is the *sum-of-squared-error* criterion:

$$J = \sum_{k=1}^{K} \sum_{i=1}^{N_k} \left\|\mathbf{x}^{(i)} - \mu_k\right\|^2 \tag{7.17}$$

where N_k is the number of samples in the cluster k, and μ_k is the vector of cluster center to be identified. This criterion is useful because a set of *true* cluster centers will give a *minimal* J for a given dataset $\mathcal{D}$.

To improve the similarity of data samples in each cluster, we can minimize J with respect to μ_k by setting $\frac{\partial J}{\partial \mu_k} = \mathbf{0}$. Thus the optimal location of μ_k is the mean of the samples in the cluster.

$$\mu_k = \frac{1}{N_k} \sum_{i=1}^{N_k} \mathbf{x}^{(i)} \tag{7.18}$$

The error J must decrease monotonically with K because the squared error can be reduced each time K is increased. If the N data samples are really grouped into K^* well separated clusters satisfying conditions (7.13a–7.13c), we would expect to see $J(K)$ decrease rapidly until $K = K^*$, decreasing much more slowly thereafter until it reaches zero at $K = N$.

$$J(1) = \sum_{\mathbf{x} \in \mathcal{D}} \|\mathbf{x} - \mu\|^2$$

where μ is the sample mean of the full data $\mathcal{D}$. We now partition the data into two, $\mathcal{D}_1$ and $\mathcal{D}_2$, so as to minimize,

$$J(2) = \sum_{k=1}^{2} \sum_{\mathbf{x} \in \mathcal{D}_k} \|\mathbf{x} - \mu_k\|^2$$

where μ_k is the mean of the samples in $\mathcal{D}_k$.

$$J(2) < J(1)$$
$$J(2)/J(1) < 1$$

$J(K)/J(1) \simeq 1$ indicates poor performance, and $J(K)/J(1) \simeq 0$ indicates good performance. K-means clustering algorithm should, therefore, yield $J(K)/J(1)$ a small value (a threshold).

The following steps show the execution of the K-means clustering algorithm:

1. Initialize algorithm.
2. Determine the membership of objects based on their distance from the cluster centers.
3. Establish new cluster centers by calculating the average of the distances of the objects belonging to various clusters.
4. Check the stopping criterion.

The algorithm initialization consists in the choice of the number of clusters K and the initial location of their centers. The initial locations of the K cluster centers $\mu_1^{(0)}, \mu_2^{(0)}, \ldots, \mu_K^{(0)}$ may be chosen at random. Alternatively, these initial parameters $\mu_k^{(t)}$ $(t = 0)$; $k = 1, \ldots, K$, may be identical with K vectors $\mathbf{x}^{(i)}$ chosen at random (t is the iteration index).

The algorithm stopping criterion is most frequently an appropriately small change in the number of elements of the clusters, that is, $\sum_{k=1}^{K} \left| N_k^{(t+1)} - N_k^{(t)} \right| < \varepsilon$, where ε is a fixed constant. Alternatively, we may check the change in the cluster centers location, i.e., $\sum_{k=1}^{K} \left\| \mu_k^{(t+1)} - \mu_k^{(t)} \right\| < \varepsilon$. The K-means algorithm may give various results depending on the initial location of the cluster centers.

The algorithm may be extended to include trial-and-error search for the value of K. A summary of the algorithm for a given choice of K is shown in Table 7.1.

Table 7.1 Summary of K-means clustering algorithm

Step 1: Initialize $\mathbf{X}$, K, $\mu_1^{(0)}, \ldots, \mu_K^{(0)}$. Set $t = 1$.
Step 2: Classify N samples according to nearest μ_k: $\mathbf{x}^{(i)} \in \mu_k$ cluster if $$\left\| \mathbf{x}^{(i)} - \mu_k^{(t-1)} \right\| < \left\| \mathbf{x}^{(i)} - \mu_l^{(t-1)} \right\| \text{ for each } l \neq k$$ Identify $N_k^{(t-1)}$; $k = 1, \ldots, K$.
Step 3: Recompute μ_k: $$\mu_k^{(t)} = \frac{1}{N_k^{(t-1)}} \sum_{i=1}^{N_k^{(t-1)}} \mathbf{x}^{(i)}, k = 1, \ldots, K$$
Step 4: If stopping criterion is satisfied, stop; otherwise go to step 2. Set $t = t + 1$.
Step 5: Return $\mu_1, \mu_2, \ldots, \mu_K$

7.5 FUZZY *K*-MEANS CLUSTERING

The fuzzy *K*-means clustering algorithm generalizes the *K*-means clustering algorithm to allow a data point to partially belong to multiple clusters. Therefore, it produces a *soft* partition for a given dataset. To do this, the objective function of (hard) *K*-means has been extended in two ways:

1. The fuzzy membership degrees in clusters have been incorporated into the objective function. The parameter μ_{ik} denotes the membership of $\mathbf{x}^{(i)}$ into cluster k and $\mu_{ik} \in [0, 1]$; $i = 1, \ldots, N$; $k = 1, \ldots, K$.

   For better incorporation, the condition

   $$\sum_{k=1}^{K} \mu_{ik} = 1 \tag{7.19}$$

   must be valid for $i = 1, 2, \ldots, N$.

   Note that in *K*-means algorithm, we have used μ to denote the cluster center. However, earlier in Chapter 6, we have used μ to denote the fuzzy membership degrees. Therefore, only for this section, we deviate a little from our earlier nomenclature. We continue to use μ for fuzzy membership degrees and use γ to denote the cluster centers.

2. An additional parameter m has been introduced as a weight component in the fuzzy membership; the weighted fuzzy membership degrees used in the objective function take the values μ_{ik}^m.

   The parameter $m \in [1, \infty)$. It is chosen to 'turn out' the noise in the data. The higher the value of m is used, less number of data points whose membership values are uniformly low, will contribute to the objective function. Consequently, such points tend to be ignored in determining the cluster centers and membership degrees.

   The extended (refer to Eqn (7.17)) objective function is,

   $$J = \sum_{k=1}^{K} \left(\sum_{i=1}^{N} \mu_{ik}^m \left\| \mathbf{x}^{(i)} - \boldsymbol{\gamma}_k \right\|^2 \right) \tag{7.20}$$

In order to minimize this objective function, $\boldsymbol{\gamma}_k$ and μ_{ik} are chosen in such a way that higher membership grades are assigned to the data closer to the cluster center and lesser grades are assigned to the data far away from the cluster center.

Like *K*-means, fuzzy *K*-means also tries to find a good partition by searching for prototypes $\boldsymbol{\gamma}_k$ that minimize the objective function J. Unlike *K*-means, however, the fuzzy *K*-means algorithm also needs to search for membership functions μ_{ik} that minimize J. To accomplish these two objectives, a necessary condition for local/global minimum of J is derived below. (Note that with a high value of m, the probability of getting stuck at local minima diminishes. A typical value for m is 1.5 or 2.)

We have to minimize the objective function,

$$J = \sum_{k=1}^{K} \left(\sum_{i=1}^{N} \mu_{ik}^m \left\| \mathbf{x}^{(i)} - \boldsymbol{\gamma}_k \right\|^2 \right)$$

Let,

$$\mu_{ik} = (w_{ik})^2 \text{ and } \left\| \mathbf{x}^{(i)} - \boldsymbol{\gamma}_k \right\|^2 = d_{ik} \tag{7.21a}$$

This gives,

$$J = \sum_{k=1}^{K}\left(\sum_{i=1}^{N}(w_{ik})^{2m}\right)d_{ik} \tag{7.21b}$$

Let $\lambda_1, \lambda_2, ..., \lambda_N$ be the Lagrange parameters (refer to Chapter 4) and $L(\mathbf{w}, \boldsymbol{\lambda})$ be the Lagrangian, given as,

$$L(\mathbf{w}, \boldsymbol{\lambda}) = \sum_{k=1}^{K}\sum_{i=1}^{N}(w_{ik})^{2m}d_{ik} + \lambda_i\left(\sum_{k=1}^{K}(w_{ik})^2 - 1\right) \tag{7.22}$$

(Note that $\sum_{k=1}^{K}\mu_{ik} = 1$ is the constraint given in Eqn (7.19))

To obtain the optimal values of the parameters $\mathbf{w}$ and $\boldsymbol{\lambda}$ that minimize $L(\cdot)$, the gradient in both the sets of variables must vanish.

$$\frac{\partial L(\mathbf{w},\boldsymbol{\lambda})}{\partial \lambda_i} = \sum_{k=1}^{K}(w_{ik})^2 - 1 = 0 \tag{7.23a}$$

$$\frac{\partial L(\mathbf{w},\boldsymbol{\lambda})}{\partial w_{ik}} = 2m(w_{ik})^{2m-1}d_{ik} + 2\,w_{ik}\lambda_i = 0 \tag{7.23b}$$

From Eqn (7.23b), we get,

$$(w_{ik}^*)^{2(m-1)} = \frac{-\lambda_i^*}{md_{ik}} \text{ or } (w_{ik}^*)^2 = \left(\frac{-\lambda_i^*}{md_{ik}}\right)^{\frac{1}{m-1}} \tag{7.23c}$$

where '*' stands for optimal values.

Summing the equations over l where l varies from 1 to K:

$$\sum_{l=1}^{K}(w_{il}^*)^2 = \left(\frac{-\lambda_i^*}{\sum_{l=1}^{K}md_{il}}\right)^{\frac{1}{m-1}} \tag{7.24}$$

Hence, we get from Eqns (7.23a) and (7.24),

$$(-\lambda_i^*)^{\frac{1}{m-1}} = \left(\sum_{l=1}^{K}md_{il}\right)^{\frac{1}{m-1}} \tag{7.25a}$$

From Eqn (7.23c), we have,

$$(-\lambda_i^*)^{\frac{1}{m-1}} = (w_{ik}^*)^2 (md_{ik})^{\frac{1}{m-1}} \tag{7.25b}$$

From Eqns (7.25a) and (7.25b), we get,

$$(w_{ik}^*)^2 (md_{ik})^{\frac{1}{m-1}} = \left(\sum_{l=1}^{K} md_{il}\right)^{\frac{1}{m-1}}$$

or

$$(w_{ik}^*)^2 = \frac{\left(\sum_{l=1}^{K} md_{il}\right)^{\frac{1}{m-1}}}{(md_{ik})^{\frac{1}{m-1}}} = \frac{\sum_{l=1}^{K} (d_{il})^{\frac{1}{m-1}}}{(d_{ik})^{\frac{1}{m-1}}}$$

$$= \frac{(d_{ik})^{-\frac{1}{m-1}}}{\sum_{l=1}^{K} (d_{il})^{-\frac{1}{m-1}}} = \frac{\left(\frac{1}{d_{ik}}\right)^{\frac{1}{m-1}}}{\sum_{l=1}^{K} \left(\frac{1}{d_{il}}\right)^{\frac{1}{m-1}}} \tag{7.26a}$$

$$= \frac{\left(\frac{1}{\|\mathbf{x}^{(i)} - \boldsymbol{\gamma}_k\|^2}\right)^{\frac{1}{m-1}}}{\sum_{l=1}^{K} \left(\frac{1}{\|\mathbf{x}^{(i)} - \boldsymbol{\gamma}_l\|^2}\right)^{\frac{1}{m-1}}} \tag{7.26b}$$

The optimal value of $\boldsymbol{\gamma}_k^*$ that minimizes J is the weighted mean of all vectors in cluster k:

$$\boldsymbol{\gamma}_k^* = \frac{\sum_{i=1}^{N} \mathbf{x}^{(i)} (\mu_{ik})^m}{\sum_{i=1}^{N} (\mu_{ik})^m} \tag{7.26c}$$

Because Eqns (7.26b) and (7.26c) rarely have an analytic solution, the cluster means and member functions are estimated iteratively. In the following, we describe fuzzy K-means algorithm for iterative solutions.

Fuzzy *K*-Means Algorithm

The following two equations serve as the foundation of the fuzzy *K*-means algorithm.

$$\mu_{ik} = \frac{\left(\dfrac{1}{\left\|\mathbf{x}^{(i)} - \gamma_k\right\|^2}\right)^{\frac{1}{m-1}}}{\sum_{l=1}^{K}\left(\dfrac{1}{\left\|\mathbf{x}^{(i)} - \gamma_l\right\|^2}\right)^{\frac{1}{m-1}}}$$

$$\gamma_k = \frac{\sum_{i=1}^{N}(\mu_{ik})^m \mathbf{x}^{(i)}}{\sum_{i=1}^{N}(\mu_{ik})^m}$$

The two equations are dependent on each other.

A summary of the algorithm is shown in Table 7.2.

Table 7.2 Summary of fuzzy K-means clustering algorithm

Step 1: Initialize $\mathbf{X}$, K, m, $\gamma_k^{(0)}$; $k = 1, \ldots, K$. Set $t = 1$

Step 2: Compute $\mu_{ik}^{(t-1)}$:

$$\mu_{ik}^{(t-1)} = \frac{\left(\dfrac{1}{\left\|\mathbf{x}^{(i)} - \gamma_k^{(t-1)}\right\|^2}\right)^{\frac{1}{m-1}}}{\sum_{l=1}^{K}\left(\dfrac{1}{\left\|\mathbf{x}^{(i)} - \gamma_l^{(t-1)}\right\|^2}\right)^{\frac{1}{m-1}}}; \; i = 1, \ldots, N; \; k = 1, \ldots, K$$

Step 3: Compute $\gamma_k^{(t)}$:

$$\gamma_k^{(t)} = \frac{\sum_{i=1}^{N}\left[\mu_{ik}^{(t-1)}\right]^m \mathbf{x}^{(i)}}{\sum_{i=1}^{N}\left[\mu_{ik}^{(t-1)}\right]^m}; \; k = 1, \ldots, K.$$

Step 4: If stopping criterion is satisfied, stop; otherwise go to step 2. Set $t = t + 1$.

Step 5: Return $\gamma_1, \gamma_2, \ldots, \gamma_K$

The algorithm stopping criterion is the same as in the case of *K*-means algorithm.

The fuzzy *K*-means algorithm, like *K*-means, may give various results depending on the initialization. The shape of clusters depends on the adopted distance measure. Refer to [125] for a thorough discussion on fuzzy clustering.

Example 7.1

Consider the dataset of Table 7.3.

Table 7.3 Dataset for Example 7.1

Objects *i*	x_1	x_2
1	2	12
2	4	9
3	7	13
4	11	5
5	12	7
6	14	4

Assume $K = 2$. Suppose we set the parameter m in fuzzy *K*-means algorithm at 2 and the initial prototypes to $\boldsymbol{\gamma}_1^{(0)} = [5 \quad 5]^T, \boldsymbol{\gamma}_2^{(0)} = [10 \quad 10]^T$.

The initial membership functions μ_{ik} of the two clusters are calculated as follows:

$$\mu_{11}^{(0)} = \frac{\left(\dfrac{1}{\left\|\mathbf{x}^{(1)} - \boldsymbol{\gamma}_1\right\|^2}\right)^{\frac{1}{2-1}}}{\sum\limits_{l=1}^{2}\left(\dfrac{1}{\left\|\mathbf{x}^{(1)} - \boldsymbol{\gamma}_l\right\|^2}\right)^{\frac{1}{2-1}}} = \mu_{ik}^{(0)};\ i = 1, k = 1$$

$$= \frac{\dfrac{1}{3^2 + 7^2}}{\dfrac{1}{3^2 + 7^2} + \dfrac{1}{8^2 + 2^2}} = \frac{1}{\dfrac{58}{58} + \dfrac{58}{68}} = 0.5397$$

$$\mu_{12}^{(0)} = \frac{1}{\dfrac{68}{58} + \dfrac{68}{68}} = 0.4603 = \mu_{ik}^{(0)};\ i = 1, k = 2$$

$$\mu_{21}^{(0)} = \frac{1}{\dfrac{17}{17} + \dfrac{17}{37}} = 0.6852 = \mu_{ik}^{(0)};\ i = 2, k = 1$$

$$\mu_{22}^{(0)} = \frac{1}{\frac{37}{17}+\frac{37}{37}} = 0.3148 = \mu_{ik}^{(0)};\ i=2,\ k=2$$

$$\mu_{31}^{(0)} = \frac{1}{\frac{68}{68}+\frac{68}{18}} = 0.2093;\ \mu_{32}^{(0)} = \frac{1}{\frac{18}{68}+\frac{18}{18}} = 0.7907$$

$$\mu_{41}^{(0)} = \frac{1}{\frac{36}{36}+\frac{36}{26}} = 0.4149;\ \mu_{42}^{(0)} = \frac{1}{\frac{26}{36}+\frac{26}{26}} = 0.5806$$

$$\mu_{51}^{(0)} = \frac{1}{\frac{53}{53}+\frac{53}{13}} = 0.197;\ \mu_{52}^{(0)} = \frac{1}{\frac{13}{53}+\frac{13}{13}} = 0.803$$

$$\mu_{61}^{(0)} = \frac{1}{\frac{82}{82}+\frac{82}{52}} = 0.3881;\ \mu_{62}^{(0)} = \frac{1}{\frac{52}{82}+\frac{52}{52}} = 0.6119$$

Membership functions of the initial prototypes of the two clusters indicate that $\mathbf{x}^{(1)}$ and $\mathbf{x}^{(2)}$ are more in the first cluster while the remaining points are more in the second cluster.

The fuzzy K-means algorithm then updates the prototypes:

$$\gamma_1^{(1)} = \frac{\sum_{i=1}^{6}(\mu_{i1}^{(0)})^2\mathbf{x}^{(i)}}{\sum_{i=1}^{6}(\mu_{i1}^{(0)})^2}$$

$$= \frac{(0.5397)^2\begin{bmatrix}2\\12\end{bmatrix}+(0.6852)^2\begin{bmatrix}4\\9\end{bmatrix}+(0.2093)^2\begin{bmatrix}7\\13\end{bmatrix}+(0.4194)^2\begin{bmatrix}11\\5\end{bmatrix}+(0.197)^2\begin{bmatrix}12\\7\end{bmatrix}+(0.3881)^2\begin{bmatrix}14\\4\end{bmatrix}}{(0.5397)^2+(0.6852)^2+(0.2093)^2+(0.4194)^2+(0.197)^2+(0.3881)^2}$$

$$= \begin{bmatrix}6.6273\\9.1484\end{bmatrix}$$

$$\gamma_2 = \frac{(0.4603)^2\begin{bmatrix} 2 \\ 12 \end{bmatrix} + (0.3148)^2\begin{bmatrix} 4 \\ 9 \end{bmatrix} + (0.7909)^2\begin{bmatrix} 7 \\ 13 \end{bmatrix} + (0.5806)^2\begin{bmatrix} 11 \\ 5 \end{bmatrix} + (0.803)^2\begin{bmatrix} 12 \\ 7 \end{bmatrix} + (0.6119)^2\begin{bmatrix} 14 \\ 4 \end{bmatrix}}{(0.4603)^2 + (0.3148)^2 + (0.7909)^2 + (0.5806)^2 + (0.803)^2 + (0.6119)^2}$$

$$= \begin{bmatrix} 9.7374 \\ 8.4887 \end{bmatrix}$$

The updated prototype γ_1 is moved closer to the center of the cluster formed by $\mathbf{x}^{(1)}$, $\mathbf{x}^{(2)}$, and $\mathbf{x}^{(3)}$; while the updated prototype γ_2 is moved closer to the cluster formed by $\mathbf{x}^{(4)}$, $\mathbf{x}^{(5)}$, and $\mathbf{x}^{(6)}$. This is illustrated in Fig. 7.6.

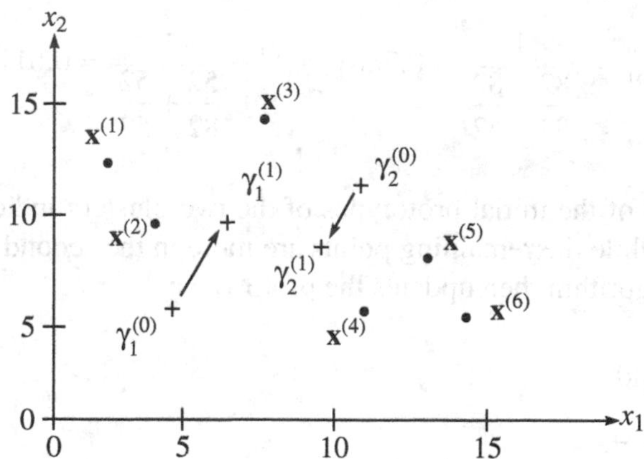

Figure 7.6 Illustration of an iteration in fuzzy *K*-means algorithm.

7.6 EXPECTATION-MAXIMIZATION (EM) ALGORITHM AND GAUSSIAN MIXTURES CLUSTERING

We first describe the abstract form of EM algorithm as it is often given in the literature. We then develop the EM parameter estimation procedure for finding the parameters of a mixture of Gaussian densities. We try to emphasize intuition rather than mathematical rigor.

7.6.1 EM Algorithm

In order to help readers develop an intuitive understanding of what the EM algorithm is, what it does, and what the goal is, we first give the background of the algorithm.

Given a set of observed data (an independent random sample), alongside a proposed model, how can we estimate the model parameters?

The *maximum likelihood* approach [4] is often used to answer this question. That is, for all possible parameters that our distribution might have, which of those are *most likely*?

We call the 'most likely' estimate that we choose from the maximum likelihood approach, *the maximum likelihood estimate*. It is computed as follows:

- For each of our observed data points, we compute a *likelihood* as a function of the parameters we seek to estimate. This likelihood is just the value of the corresponding probability density function, evaluated at that particular data point.
- We then compute the *likelihood function*, which is the combination of the likelihoods for all the data points we observed.
- We then seek to estimate the set of parameters which will maximize the likelihood function.

Let $\mathbf{X} = \{\mathbf{x}^{(1)}, \ldots, \mathbf{x}^{(N)}\}$ denote the observed data in a set of N independently drawn instances. $\mathbf{x} = [x_1\ x_2 \cdots x_n]^T$ is the n-dimensional feature vector. $\mathbf{X}$ is thus a random variable. The probability of receiving some measurements $\mathbf{x}^{(i)}$ is given by the probability density function

$$p(\mathbf{x}^{(i)}|\Theta)$$

where Θ denotes the set of parameters governing the distribution.

The probability of receiving the whole series of measurements is then

$$p(\mathbf{x}|\Theta) = \prod_{i=1}^{N} p(\mathbf{x}^{(i)}|\Theta)$$

Since the measurements are independent, the *likelihood function* is defined as a function of Θ:

$$\mathcal{L}(\Theta|\mathbf{X}) = \prod_{i=1}^{N} p(\mathbf{x}^{(i)}|\Theta) \tag{7.27}$$

Note that the likelihood function can be viewed as a function with the parameters as a function of the data, rather than the data as a function of the parameters. We assume that the data is fixed but parameters can vary.

In the maximum likelihood problem, our goal is to find Θ that maximizes $\mathcal{L}$. That is, we wish to find Θ^* where

$$\Theta^* = \underset{\Theta}{argmax}\ \mathcal{L}(\Theta|\mathbf{X}) \tag{7.28}$$

Often we maximize $\log(\mathcal{L}(\Theta|\mathbf{X}))$ instead because it is analytically easier. Because the logarithm is monotonically increasing, the Θ^* that maximizes the *log-likelihood*, also maximizes *likelihood:*

Log-likelihood function

$$\log(\mathcal{L}(\Theta|\mathbf{X})) = \log\left(\prod_{i=1}^{N} p(\mathbf{x}^{(i)}|\Theta)\right)$$

$$= \sum_{i=1}^{N} \log p(\mathbf{x}^{(i)}|\Theta) \tag{7.29}$$

If $p(\mathbf{x}^{(i)}|\Theta)$ is a well-behaved differentiable function of Θ, Θ^* can be found by standard methods of differential calculus; for example, normal (Gaussian) distribution. For many problems, however, it is not possible to find such analytical expressions, and we must resort to more elaborate techniques. There are plenty of problems that make finding an analytical solution difficult or impossible. A common example is incomplete or missing data; other examples include 'hidden' or 'latent' variables. This is where the EM algorithm comes in.

The EM algorithm is an iterative method for approximating the maximum of the likelihood function.

In particular, the EM algorithm comes in when we have unobserved data/variables. The unobserved data problem occurs when the data indeed has missing values due to problems with or limitations of the observation process. The other problem occurs when optimizing the likelihood function is analytically intractable but when the likelihood function can be simplified by assuming the existence of and values for additional but missing (or *hidden*) variables. The latter problem is more common in the computational pattern recognition community.

In the general setting of the algorithm, let $\mathbf{X} = \{\mathbf{x}^{(1)}, \ldots, \mathbf{x}^{(N)}\}$ denote the *observed* data in a set of N independently drawn instances, and $\mathbf{Z} = \{\mathbf{z}^{(1)}, \ldots, \mathbf{z}^{(N)}\}$ denote the *unobserved* data in the same instances. We assume that a complete dataset $\mathbf{S} = (\mathbf{X}, \mathbf{Z})$ exists, and also assume a joint density function

$$p(\mathbf{S}|\Theta) = p(\mathbf{X}, \mathbf{Z}|\Theta)$$

With this new density function, we can define a new likelihood function

$$\mathcal{L}(\Theta|\mathbf{S}) = \mathcal{L}(\Theta|\mathbf{X}, \mathbf{Z}) = p(\mathbf{X}, \mathbf{Z}|\Theta)$$

Note that this function is, in fact, a random variable since the missing information $\mathbf{Z}$ is unknown, random, and presumably governed by an underlying distribution.

The EM algorithm first finds the *expected value* of the complete-data log-likelihood function log $p(\mathbf{X}, \mathbf{Z}|\Theta)$ with respect to the unknown data $\mathbf{Z}$, given the observed data $\mathbf{X}$ and the current parameter estimates. That is, we define

$$Q(\Theta|\Theta^{(t)}) = \mathbb{E}[\log p(\mathbf{X}, \mathbf{Z}|\Theta)|\mathbf{X}; \Theta^{(t)}] \tag{7.30}$$

where the left-hand side denotes that $Q(\Theta|\Theta^{(t)})$ is a function of Θ with $\Theta^{(t)}$ assumed fixed; and the right-hand side denotes that expected value is over the missing features assuming $\Theta^{(t)}$ are the true parameters describing the (full) distribution. The simplest way to interpret this equation is the following:

The parameter vector, $\Theta^{(t)}$, is the current (best) estimate for the full distribution; Θ is a candidate vector for an improved estimate. Given such a candidate Θ, the right-hand side of Eqn (7.30) calculates the log-likelihood of the data, including the unknown features $\mathbf{Z}$, marginalized with respect to the current best distribution, which is described by $\Theta^{(t)}$. Different candidates will, of course, lead to different such likelihoods. EM algorithm will select the best such candidate and call it $\Theta^{(t+1)}$—the one corresponding to the largest $Q(\Theta|\Theta^{(t)})$.

The EM algorithm bounces back and forth between the two processes.

1. Given the current parameters $\Theta^{(t)}$ and the observed data, estimate the missing/hidden features, i.e., obtain $Q(\Theta|\Theta^{(t)})$ given by Eqn (7.30).

   The evaluation of this expectation is called the *Expectation-Step* (*E-Step*) of the algorithm. This process is initiated by assuming that the true parameter vector is given by $\Theta^{(t)}$. We index Θ by 't' because ultimately the EM algorithm results in an iteration procedure; so we use 't' as an iteration index.
2. The second step in EM algorithm is the *Maximization-Step* (*M-step*) through which we maximize the expectation we computed in the first step:

$$\Theta^{(t+1)} = \underset{\Theta}{argmax}\ Q(\Theta|\Theta^{(t)}) \tag{7.31}$$

These two steps are repeated as necessary. Each iteration is guaranteed to increase the log-likelihood and the algorithm is guaranteed to converge to the local maximum of the likelihood function. If we let threshold T representing the convergence criterion, the iteration process will continue until

$$Q(\Theta^{(t+1)}|\Theta^{(t)}) - Q(\Theta^{(t)}|\Theta^{(t-1)}) \le T \tag{7.32}$$

As presented above, it is not clear how exactly to 'code up' the algorithm. The details of the steps required to compute the given quantities are very much dependent on the particular application. In fact, some reviewers have commented that the term 'algorithm' is not appropriate for this Expectation-Maximization iteration scheme, since its formulation is too general. One has to develop an 'algorithm' for the specific application in hand based on the general Expectation-Maximization iteration scheme.

7.6.2 Gaussian Mixture Models

We now turn to the specific application: *finding maximum likelihood mixture densities parameters via EM*.

The mixture-density parameter estimation problem is probably one of the most widely used application of the EM algorithm in the computational pattern recognition community.

The K-means clustering algorithm, as we have seen in an earlier section, positions instances into disjoint clusters deterministically. Sometimes, a brittleness is associated with such schemes that make 'hard' judgments because of no enough evidence to make a completely firm decision on clustering. *Probabilistic clustering methods* assign instances to clusters probabilistically, i.e., each instance has a certain amount of probability of belonging to each cluster [1,6].

The foundation for probabilistic clustering is a statistical model called *mixture model*. A mixture model is a set of K probability distributions, representing K clusters, that govern the attribute values for members of that cluster. In other words, each distribution gives the probability that a particular instance would have a certain set of attribute values if it were *known* to be a member of that cluster. Any particular instance belongs to one and only one of the clusters, but it is not known which one.

Our focus here is on the most widely used *Gaussian Mixture Model* (GMM), a set of Gaussian (or normal) distributions with means and covariances. The clustering problem is to take a set of instances and a prespecified number of clusters (Gaussian distributions) and work out each cluster's mean and covariance, and the population distribution between the clusters.

Let $\mathbf{x} = \{x_j; j = 1, \ldots, n\}$ be an n-dimensional vector to be modeled using a Gaussian mixture distribution. Let us assume that the model has K subclasses (clusters). Then the following parameters are required to completely specify the kth subclass; $k = 1, \ldots, K$.

ω_k – the probability that a data sample $\mathbf{x}^{(i)}$; $i = 1, \ldots, N$, belongs to subclass k. $\omega_k \geq 0$ with $\sum_{k=1}^{K} \omega_k = 1$. This parameter gives the population distribution between clusters; $\omega_k = N_k/N$; where N_k = number of samples belonging to k^{th} cluster.

$\boldsymbol{\mu}_k$ – the n-dimensional *mean vector* for subclass k.

$\boldsymbol{\Sigma}_k$ – the $n \times n$ *covariance matrix* for subclass k.

We use $\boldsymbol{\omega}$, $\boldsymbol{\mu}$, and $\boldsymbol{\Sigma}$ to denote the parameter sets $\{\omega_k\}_{k=1}^{K}$, $\{\boldsymbol{\mu}_k\}_{k=1}^{K}$ and $\{\boldsymbol{\Sigma}_k\}_{k=1}^{K}$, respectively. The complete set of parameters for a GMM is then given by,

$$\{K,(\boldsymbol{\omega},\boldsymbol{\mu},\boldsymbol{\Sigma})\} = \{K,\boldsymbol{\omega},\boldsymbol{\theta}\};\ \boldsymbol{\omega} = \{\omega_k\}_{k=1}^{K};\ \boldsymbol{\theta} = \{\boldsymbol{\theta}_k\}_{k=1}^{K},\ \boldsymbol{\theta}_k = \{\boldsymbol{\mu}_k, \boldsymbol{\Sigma}_k\}$$

The joint Gaussian (normal) density has the form [refer to Eqns (3.23)-(3.24)]:

$$p(\mathbf{x}|\boldsymbol{\theta}) = \frac{1}{(2\pi)^{n/2}|\boldsymbol{\Sigma}|^{1/2}} \exp\left[-\tfrac{1}{2}(\mathbf{x}-\boldsymbol{\mu})^T \boldsymbol{\Sigma}^{-1}(\mathbf{x}-\boldsymbol{\mu})\right] \tag{7.33a}$$

This is the general form of ***multivariate normal density function.***

A special *naive* case is where the components x_j of $\mathbf{x}$ are statistically independent; and *cov* $(x_j, x_l) = 0$ for $j \neq l$, *var* $(x_j) = \sigma_j^2\ \forall\ j$. Then the covariance matrix is diagonal:

$$\boldsymbol{\Sigma} = \begin{bmatrix} \sigma_1^2 & 0 & \cdots & 0 \\ 0 & \sigma_2^2 & \cdots & 0 \\ \vdots & \vdots & & \vdots \\ 0 & 0 & \cdots & \sigma_n^2 \end{bmatrix};\ \boldsymbol{\Sigma}^{-1} = \begin{bmatrix} \frac{1}{\sigma_1^2} & 0 & \cdots & 0 \\ 0 & \frac{1}{\sigma_2^2} & \cdots & 0 \\ \vdots & \vdots & & \vdots \\ 0 & 0 & \cdots & \frac{1}{\sigma_n^2} \end{bmatrix}$$

Further simplification occurs when we assume that $\sigma_j^2 = \sigma^2\ \forall\ j = 1, \ldots, n$. Then the covariance matrix becomes

$$\boldsymbol{\Sigma} = \sigma^2\mathbf{I};\ \boldsymbol{\Sigma}^{-1} = \frac{1}{\sigma^2}\mathbf{I};\ \mathbf{I} \text{ is } n \times n \textit{ unity matrix}$$

We will present the results for the general case of covariance matrix $\mathbf{\Sigma}$. For the subclass k (k^{th} cluster in Gaussian mixtures), $k = 1, \ldots, K$, the normal density function will be expressed as,

$$p(\mathbf{x}|\boldsymbol{\theta}_k) = \frac{1}{(2\pi)^{n/2}|\mathbf{\Sigma}_k|^{1/2}} \exp\left[-\tfrac{1}{2}(\mathbf{x}-\boldsymbol{\mu}_k)^T \mathbf{\Sigma}_k^{-1}(\mathbf{x}-\boldsymbol{\mu}_k)\right] \tag{7.33b}$$

Our problem here involves a mixture of K different Gaussian distributions, and we cannot observe which instances are generated by which distribution. Thus, we have a prototypical example of a problem with 'hidden' or 'latent' variables. Here, the EM algorithm comes in because of the *unobserved latent variables*.

To use the EM algorithm for this kind of problems, we think of a full description of each instance as $\{\mathbf{x}^{(i)}, \mathbf{z}^{(k,i)}\}$; $i = 1, \ldots, N$; $k = 1, \ldots, K$, where $\mathbf{x}^{(i)} = \{x_j^{(i)}\}$; $j = 1, \ldots, n$, are the observed values of the ith instance and where $\mathbf{z}^{(k,i)} = \{z_{ki}\}$; $i = 1, \ldots, N$; $k = 1, \ldots, K$, indicate which of the K normal distributions was used to generate the value of $\mathbf{x}^{(i)}$. In particular, z_{ki} has a value of 1 if $\mathbf{x}^{(i)}$ was created by the kth normal distribution, and 0 otherwise. Here, $\mathbf{x}^{(i)}$ are the observed variables in the description of the instance and $\mathbf{z}^{(k,i)}$ are the hidden variables.

By this choice, $\mathbf{z}^{(k,i)}$ belongs to a matrix $\mathbf{Z}$ whose entry z_{ki} is equal to one if and only if GMM component k produced measurement i—otherwise it is zero. Note that each column of matrix $\mathbf{Z}$ contains exactly one entry equal to one. By this choice, we introduce additional information into the process: $\mathbf{z}^{(k,i)}$ tells us the probability density function that underlines certain measurement. We cannot really measure this hidden variable, and it is not important. At this stage, we just assume that we have it, do some calculations and see what happens.

We assume the following mixture model:

$$p(\mathbf{x}|\boldsymbol{\omega}, \boldsymbol{\theta}) = \sum_{k=1}^{K} \omega_k p(\mathbf{x}|\boldsymbol{\theta}_k) \tag{7.34}$$

where the parameters are $\{\omega_1, \ldots, \omega_K; \boldsymbol{\theta}_1, \ldots, \boldsymbol{\theta}_K\} = \{\omega_k ; \boldsymbol{\theta}_k\}_{k=1}^{K}$ such that $\sum_{k=1}^{K} \omega_k = 1$, and each $p(\mathbf{x}|\boldsymbol{\theta}_k)$ is a density function parameterized by $\boldsymbol{\theta}_k$. In other words, we assume we have K component densities mixed together with K mixing coefficients ω_k.

The incomplete-data log-likelihood expression for this mixture density from the data $\mathbf{X}$ is given by,

$$\begin{aligned}\log(\mathcal{L}(\boldsymbol{\omega}, \boldsymbol{\theta}|\mathbf{X})) &= \log \prod_{i=1}^{N} p(\mathbf{x}^{(i)}|\boldsymbol{\omega}, \boldsymbol{\theta}) \\ &= \sum_{i=1}^{N} \log\left(\sum_{k=1}^{K} \omega_k p(\mathbf{x}^{(i)}|\boldsymbol{\theta}_k)\right)\end{aligned}$$

which is difficult to optimize because it contains log of the sum. If we consider $\mathbf{X}$ as incomplete, and assume the existence of unobserved data items $\mathbf{Z}$ whose values inform us which component density 'generated' each data item, the likelihood expression is significantly simplified. That is, we assume that $z_{ki} = 1$ if the ith sample was generated by the kth mixture component, otherwise $z_{ki} = 0$.

If we know the values of **Z**, the likelihood becomes:

$$\log(\mathcal{L}(\boldsymbol{\omega}, \boldsymbol{\theta}|\mathbf{X}, \mathbf{Z}) = \sum_{i=1}^{N} \log\left(\sum_{k=1}^{K} z_{ki}\omega_k p(\mathbf{x}^{(i)}|\boldsymbol{\theta}_k)\right) \tag{7.35}$$

Remember z_{ki} is zero for all but one term in inner sum. However, we do not know the value of k for each sample $\mathbf{x}^{(i)}$. If we assume z_{ki} is a random variable, we can proceed. We must first derive an expression for the distribution of the unobserved data. Let us first guess the parameters for the mixture density, i.e., we guess that $\{\omega_1^{(t)}, \ldots, \omega_K^{(t)}; \boldsymbol{\theta}_1^{(t)}, \ldots, \boldsymbol{\theta}_K^{(t)}\}$ ($t = 0$; t is the iteration index) are the appropriate parameters for the likelihood $\mathcal{L}(\boldsymbol{\omega}^{(t)}; \boldsymbol{\theta}^{(t)} |\mathbf{X}, \mathbf{Z})$. Given$\{\boldsymbol{\omega}^{(t)}; \boldsymbol{\theta}^{(t)}\}$, we can easily compute $p(\mathbf{x}^{(i)}|\boldsymbol{\theta}_k^{(t)})$ for each i and k. In addition, the mixing parameters, ω_k, can be thought of as prior probabilities of each mixture component, i.e., $\omega_k = P(\text{component } k)$. Therefore, using Baye's rule, we can compute:

$$\begin{aligned} P(\text{component } k|\mathbf{x}^{(i)}; \boldsymbol{\omega}^{(t)}; \boldsymbol{\theta}^{(t)}) &= \frac{p(\mathbf{x}^{(i)} | \boldsymbol{\theta}_k^{(t)})P(\text{component } k)}{p(\mathbf{x}^{(i)} | \boldsymbol{\theta}^{(t)})} \\ &= \frac{p(\mathbf{x}^{(i)} | \boldsymbol{\theta}_k^{(t)})P(\text{component } k)}{\sum_{l=1}^{K} p(\mathbf{x}^{(i)} | \boldsymbol{\theta}_l^{(t)})P(\text{component } l)} \\ &= \frac{\omega_k^{(t)} p(\mathbf{x}^{(i)} | \boldsymbol{\theta}_k^{(t)})}{\sum_{l=1}^{K} \omega_l^{(t)} p(\mathbf{x}^{(i)} | \boldsymbol{\theta}_l^{(t)})} \end{aligned} \tag{7.36}$$

Inserting Gaussian distribution (7.33b) in Eqn (7.36), we get,

$$P(\text{component } k|\mathbf{x}^{(i)}; \boldsymbol{\omega}^{(t)}; \boldsymbol{\mu}^{(t)}; \boldsymbol{\Sigma}^{(t)})$$

$$= \frac{\omega_k^{(t)}[\Sigma_k^{(t)}]^{-1/2} \exp[(-1/2)(\mathbf{x}^{(i)} - \boldsymbol{\mu}_k^{(t)})^T [\Sigma_k^{(t)}]^{-1}(\mathbf{x}^{(i)} - \boldsymbol{\mu}_k^{(t)})]}{\sum_{l=1}^{K} \omega_l^{(t)}[\Sigma_l]^{-1/2} \exp[(-1/2)(\mathbf{x}^{(i)} - \boldsymbol{\mu}_l^{(t)})^T [\Sigma_l^{(t)}]^{-1}(\mathbf{x}^{(i)} - \boldsymbol{\mu}_l^{(t)})]} \tag{7.37}$$

Note that *expected value* of z_{ki}, $\mathbb{E}[z_{ki}]$, is just the probability that instance $\mathbf{x}^{(i)}$ was generated by k^{th} normal distribution, given the parameters of the distributions, and the mixing parameters, i.e., $P(\text{component } k|\mathbf{x}^{(i)}; \boldsymbol{\omega}^{(t)}; \boldsymbol{\theta}^{(t)})$. Let us denote this probability by the variable h_{ki}, which is equal to $\mathbb{E}[z_{ki}]$. We see that z_{ki} is a 0/1 random variable, while the expected value h_{ki} of this hidden variable is the posterior probability that $\mathbf{x}^{(i)}$ is generated by component k of the Gaussian mixture. Because this is the probability, it is between 0 and 1 and is a 'soft' label as opposed to the 0/1 'hard' label of K-means. In K-means clustering, $\mathbf{x}^{(i)}$ either belongs to cluster k or it does not, while in Gaussian mixtures clustering, $\mathbf{x}^{(i)}$ may belong to more than one cluster with different probabilities. Gaussian mixtures clustering is, therefore, a 'soft clustering' technique (as is Fuzzy K-means) while K-means is a 'hard clustering' technique. From Eqn (7.37) we have,

$$h_{ki} = \frac{\omega_k[\Sigma_k]^{-1/2} \exp[(-1/2)(\mathbf{x}^{(i)} - \boldsymbol{\mu}_k)^T[\Sigma_k^{-1}](\mathbf{x}^{(i)} - \boldsymbol{\mu}_k)]}{\sum_{l=1}^{K} \omega_l[\Sigma_l]^{-1/2} \exp[(-1/2)(\mathbf{x}^{(i)} - \boldsymbol{\mu}_l)^T[\Sigma_l^{-1}](\mathbf{x}^{(i)} - \mu_l)]} \tag{7.38}$$

We have thus obtained the *marginal density* of the hidden variables by assuming their existence and making an initial guess on the parameters.

In this case, Eqn (7.30) takes the form (refer to Eqn (7.35)),

$$Q(\boldsymbol{\omega}; \boldsymbol{\mu}, \boldsymbol{\Sigma}|\boldsymbol{\omega}^{(t)}; \boldsymbol{\mu}^{(t)}, \boldsymbol{\Sigma}^{(t)}) = Q(\boldsymbol{\omega}; \boldsymbol{\theta}|\boldsymbol{\omega}^{(t)}; \boldsymbol{\theta}^{(t)})$$

$$= \mathbb{E}\left[\sum_{i=1}^{N} \log\left(\sum_{k=1}^{K} z_{ki}\omega_k p(\mathbf{x}^{(i)}|\boldsymbol{\theta}_k)|\mathbf{X}; \boldsymbol{\omega}^{(t)}; \boldsymbol{\theta}^{(t)}\right)\right] \tag{7.39a}$$

This can now be rewritten as,

$$Q(\boldsymbol{\omega}; \boldsymbol{\theta}|\boldsymbol{\omega}^{(t)}; \boldsymbol{\theta}^{(t)}) = \mathbb{E}\left[\sum_{i=1}^{N}\sum_{k=1}^{K} z_{ki} \log(\omega_k p(\mathbf{x}^{(i)}|\boldsymbol{\theta}_k)|\mathbf{X}; \boldsymbol{\omega}^{(t)}; \boldsymbol{\theta}^{(t)})\right] \tag{7.39b}$$

since z_{ki} is zero for all but one term in the inner sum for which $z_{ki} = 1$.

Note that the above expression for $Q(\cdot)$ is a linear function of z_{ki}. In general, for any function $f(z)$ that is a linear function of z, the following equality holds:

$$\mathbb{E}[f(z)] = f(\mathbb{E}[z])$$

This general fact about linear functions allows us to write

$$Q(\boldsymbol{\omega}; \boldsymbol{\theta}|\boldsymbol{\omega}^{(t)}; \boldsymbol{\theta}^{(t)}) = \sum_{i=1}^{N}\sum_{k=1}^{K} \mathbb{E}[z_{ki}|\mathbf{X}; \boldsymbol{\omega}^{(t)}; \boldsymbol{\theta}^{(t)}]\log(\omega_k p(\mathbf{x}^{(i)}|\boldsymbol{\theta}_k))$$

$$= \sum_{i=1}^{N}\sum_{k=1}^{K} h_{ki}^{(t)} \log\omega_k + \sum_{i=1}^{N}\sum_{k=1}^{K} h_{ki}^{(t)} \log p(\mathbf{x}^{(i)}|\boldsymbol{\theta}_k) \tag{7.40}$$

To maximize this expression, we can maximize the term containing ω_k and the term containing $\boldsymbol{\theta}_k$ independently since they are not related. Note that $h_{ki}^{(t)}$ is constant given the parameters $\{\boldsymbol{\omega}^{(t)}; \boldsymbol{\theta}^{(t)}\}$.

Optimizing with respect to ω_k

This is a constrained optimization problem; the constraint is $\sum_{k=1}^{K} \omega_k = 1$. We solve this problem by the method of Lagrange multipliers (refer to Section 4.4).

Introducing the Lagrange multiplier λ, we solve the following equation:

$$\frac{\partial}{\partial \omega_k}\left[\sum_{i=1}^{N}\sum_{k=1}^{K} h_{ki}\log \omega_k + \lambda\left(\sum_{k=1}^{K}\omega_k - 1\right)\right] = 0$$

This gives,

$$\sum_{i=1}^{N}\frac{1}{\omega_k}h_{ki} + \lambda = 0$$

Summing both sides over k, we get,

$$\sum_{i=1}^{N}\sum_{k=1}^{K} h_{ki} + \lambda\sum_{k=1}^{K}\omega_k = 0 \tag{7.41}$$

For each value of i, the inner sum equals one; therefore,

$$N + \lambda = 0 \text{ or } \lambda = -N$$

resulting in (from Eqn (7.41)),

$$\omega_k = \frac{1}{N}\sum_{i=1}^{N} h_{ki} \tag{7.42}$$

Optimizing with respect to $\boldsymbol{\theta}_k$

For the Gaussian mixture model, $\boldsymbol{\theta}_k$ consists of mean $\boldsymbol{\mu}_k$ and covariance $\boldsymbol{\Sigma}_k$ of the kth component. Taking the derivative of the second term on the right-hand side of Eqn (7.40) with respect to $\boldsymbol{\mu}_k$ and setting it equal to zero, we get,

$$\frac{\partial}{\partial \boldsymbol{\mu}_k}\left[\sum_{i=1}^{N}\sum_{k=1}^{K} h_{ki}\log p(\mathbf{x}^{(i)}|\boldsymbol{\mu}_k, \boldsymbol{\Sigma}_k)\right] = \mathbf{0}$$

This gives,

$$\sum_{i=1}^{N} h_{ki}\frac{\partial}{\partial \boldsymbol{\mu}_k}\log p(\mathbf{x}^{(i)}|\boldsymbol{\mu}_k, \boldsymbol{\Sigma}_k) = \mathbf{0}$$

By inserting the Gaussian distribution (7.33b), we arrive at,

$$\sum_{i=1}^{N} h_{ki}\frac{\partial}{\partial \boldsymbol{\mu}_k}\log\left[\frac{1}{(2\pi)^{n/2}|\boldsymbol{\Sigma}_k|^{1/2}}\exp\left[-\tfrac{1}{2}(\mathbf{x}-\boldsymbol{\mu}_k)^T\boldsymbol{\Sigma}_k^{-1}(\mathbf{x}-\boldsymbol{\mu}_k)\right]\right] = \mathbf{0}$$

Ignoring constant terms (since they disappear after taking derivatives), we get,

$$\sum_{i=1}^{N} h_{ki}\frac{\partial}{\partial \boldsymbol{\mu}_k}\left[-\tfrac{1}{2}(\mathbf{x}-\boldsymbol{\mu}_k)^T\boldsymbol{\Sigma}_k^{-1}(\mathbf{x}-\boldsymbol{\mu}_k)\right] = \mathbf{0}$$

This gives,

$$\boldsymbol{\mu}_k = \frac{\sum_{i=1}^{N} h_{ki}\mathbf{x}^{(i)}}{\sum_{i=1}^{N} h_{ki}} \tag{7.43}$$

Similar procedure yields [6],

$$\boldsymbol{\Sigma}_k = \frac{\sum_{i=1}^{N} h_{ki}(\mathbf{x}^{(i)} - \boldsymbol{\mu}_k)(\mathbf{x}^{(i)} - \boldsymbol{\mu}_k)^T}{\sum_{i=1}^{N} h_{ki}} \tag{7.44}$$

Summarizing, the estimates of new parameters in terms of the old parameters are as follows (given initial parameters: $\omega_k^{(t)}$; $\boldsymbol{\mu}_k^{(t)}$, $\boldsymbol{\Sigma}_k^{(t)}$; $t = 0$):

$$h_{ki}^{(t)} = \frac{\omega_k^{(t)}[\Sigma_k^{(t)}]^{-1/2} \exp[(-1/2)(\mathbf{x}^{(i)} - \boldsymbol{\mu}_k^{(t)})^T[\Sigma_k^{(t)}]^{-1}(\mathbf{x}^{(i)} - \boldsymbol{\mu}_k^{(t)})]}{\sum_{l=1}^{K} \omega_l^{(t)}[\Sigma_l^{(t)}]^{-1/2} \exp[(-1/2)(\mathbf{x}^{(i)} - \boldsymbol{\mu}_l^{(t)})^T[\Sigma_l^{(t)}]^{-1}(\mathbf{x}^{(i)} - \boldsymbol{\mu}_l^{(t)})]} \tag{7.45a}$$

$$\omega_k^{(t+1)} = \frac{1}{N}\sum_{i=1}^{N} h_{ki}^{(t)} \tag{7.45b}$$

$$\boldsymbol{\mu}_k^{(t+1)} = \frac{\sum_{i=1}^{N} h_{ki}^{(t)}\mathbf{x}^{(i)}}{\sum_{i=1}^{N} h_{ki}^{(t)}} \tag{7.45c}$$

$$\boldsymbol{\Sigma}_k^{(t+1)} = \frac{\sum_{i=1}^{N} h_{ki}^{(t)}(\mathbf{x}^{(i)} - \boldsymbol{\mu}_k^{(t+1)})(\mathbf{x}^{(i)} - \boldsymbol{\mu}_k^{(t+1)})^T}{\sum_{i=1}^{N} h_{ki}^{(t)}} \tag{7.45d}$$

EM is initialized by *K*-means. After a few iterations of *K*-means, we get the estimates for the centers $\boldsymbol{\mu}_k$, and using the instances covered by each center, we estimate the $\boldsymbol{\Sigma}_k$; N_k/N gives us the ω_k. We run EM using Eqns (7.45) from that point on. The algorithm proceeds by using the newly derived parameters as the guess for the next iteration.

The Gaussian Mixture Model (GMM) consists of cluster centers along with the formulas that define the Gaussian probability distributions for each of the clusters. These formulas are defined in terms of parameters that have been optimized in the course of training the model. In the end, the model is just a collection of Gaussian distributions.

For new instances, the algorithm applies these formulas to the data featuring new instances, in the following steps [19]:

- Calculate the likelihood that the instance belongs to each cluster by applying the appropriate formula for each cluster.
- Normalize the likelihoods to obtain the probabilities of membership in each cluster (that is, divide each one by the sum of all of them, so the total sums up to one).
- Optionally, assign cluster membership based on the highest probability.

Thus, the data can be assigned to one of the K clusters by applying GMM.

Introduction to mixture-models and EM algorithm can be found in recent books [6, 8, 126, 127].

7.7 SOME USEFUL DATA TRANSFORMATIONS

Data often calls for general transformations of a set of attributes selected for the data mining problem. It might be useful to define new attributes by applying specified mathematical functions to the existing ones. New variables derived from existing ones express the information inherent in the data in ways that make the information more useful; thereby improving model performance.

Too many attributes are both a blessing and a curse. More variables generally mean more descriptive information is available that can be used to build better models. On the other hand, they are a curse because of the increased risk of high correlation among variables (redundancy of information, the increased risk of overfitting the data) and sparseness of the dataset. Transformations that lead to attribute reduction are very useful for deriving new attributes.

In the following, we give an overview of some standard transformations routinely carried out in data mining. In fact, many of these transformations have been discussed in earlier chapters.

7.7.1 Data Cleansing

Real-world data tend to be incomplete, noisy, and inconsistent. Data cleansing routines attempt to fill in missing values, smooth out noise while identifying outliers, and correct inconsistencies in the data. This section provides an overview of basic methods for data cleansing. Data cleansing is discussed in a number of books, including [128].

Missing Values: Most datasets encountered in practice contain missing values. They may occur for several reasons, such as malfunctioning measurement equipment, changes in experimental design during data collection, human-errors in data entry, and deliberate errors (e.g., respondents not willing to divulge information).

If the number of instances with missing values is small, those instances might be omitted. However, if we have a large number of attributes, even a small proportion of missing values can affect a lot of instances. With only 30 variables (attributes), if only 5% of values are missing (spread randomly and independently among patterns and variables), almost 80% of the instances would have to be omitted from the analysis (the chance that a given instance would escape having a missing value is $(0.95)^{30} = (0.215)$).

An alternative to omitting instances with missing values is to replace the missing value with an imputed value, based on the other values for that variable across the instances. For example, we might substitute the mean of the variable across all samples for the missing value of that variable, or attribute mean for all samples belonging to the same class as the tuple with missing value may be substituted for missing value. Note that using such techniques will understate the variability in a dataset, introducing bias. However, we can assess the variability and performance of our data mining technique using the validation set.

An alternative to dropping samples with missing values or imputing the missing values is to examine the importance of the attribute with a large number of missing values. If it is not very crucial, it can be dropped. When such an attribute is deemed crucial, the best solution is to invest in obtaining the missing data.

It is important to note that in some cases, a missing value may not imply an error in the data. If missing values mean that an operator has decided not to make a particular measurement, then that may convey a great deal more than the mere fact that the value is unknown. Human judgment may be required for individual cases or to determine a special rule to deal with the situation.

Although we can try our best to clean the data after it is seized, good design of databases should help minimize the missing values or errors in the first place.

Noisy Data: Noise is a random error or variance in a measured variable. Random errors (noise) is a commonly observed problem in datasets that requires 'smoothing' out the data. Various smoothing techniques are employed by commercial tools available for cleansing the data.

Outliers: The more data we are dealing with, the greater the chance of encountering *inaccurate values* resulting from measurement error, data-entry error, or the like. Inaccurate values often deviate significantly from the pattern that is apparent in the remaining values. Sometimes, however, inaccurate vales are hard to find, particularly without specialist domain knowledge.

Values that lie far away from the bulk of the data are called *outliers*. The term *far away* is deliberately left vague because what is or is not called an outlier is basically an arbitrary decision. Analysts use rules of thumb such as 'anything over three standard deviations away from the mean is an outlier', but no statistical test can tell us whether such an outlier is the result of an error. The purpose of identifying outliers is usually to call attention to values that need further review. Judgments are made by someone with domain knowledge.

In any case, if the number of samples with outliers is very small, they might be treated as noise/missing values. Clustering techniques could be used to identify outliers. Sometimes data visualization gives a clear indication of presence of outliers.

Sparse Data: Sometimes data matrix is *sparse*, most attributes have a value of 0 for most of the samples. For example, in text mining, samples are documents; data matrix columns and rows represent words of documents, and the numbers indicate how many times a particular word appears in a document. Most documents have a rather small vocabulary, so most entries are 0.

Sparse data presents special challenges. Many learning techniques are unable to make use of very sparse data. One possible way out is the use of encoding schemes to represent information from a large number of sparse variables by a small number of dense ones.

Other Types of Errors: It is, in fact, difficult to list all the possible types of errors in data. *Anomalies* because of some *systematic errors* appear sometimes. Only a deep semantic knowledge of what is going on will be able to explain systematic data errors. Short of consulting a human expert, there is really no way of telling whether a particular sample is an error or whether it just does not fit.

Duplicate data presents another source of error. Most machine learning algorithms will produce different result if some of the samples in data files are duplicated because repetition gives them more influence on the result.

When information on the same topic is collected from multiple sources, the various sources often represent the same data different ways. The *inconsistency* in data can occur due to inconsistent definitions of fields, units of measurement, time periods, and so on.

Finally, data goes stale. We need to consider whether the data we are using is still current.

We have, in fact, neither listed all possible errors that can appear in data, nor the solution for correcting discrepancies generated by the listed errors. We have simply given an overview of the complexities involved.

Data cleansing is a time-consuming and labor-intensive procedure but one that is absolutely necessary for successful data mining. Data experts need to be consulted to explain anomalies. Data exploration normally helps even non-experts in getting to know the data. Various simple visualizations often help with the task of understanding the meaning of different attributes, the conventions used in coding them, the significance of missing values, measurement noise, outliers, and the presence of systematic errors.

7.7.2 Derived Attributes

Creating derived variables is about defining new variables that express the information inherent in the data in ways that make the information more useful to machine learning techniques. It is thus an art of *making the data mean more*.

Derived variables allow learning models to incorporate human insights into the modeling process, and allow learning model to take advantage of important characteristics already known about the data being processed. Derived variables definitely improve model performance.

Standardizing Numeric Variables: Some algorithms require that the data be normalized before the algorithm can be implemented effectively.

Attributes are often normalized to lie in a fixed range, say, from zero to one, by dividing all values by the maximum value encountered or by subtracting the minimum value and dividing by the range between the maximum and the minimum values. Another normalization technique is to calculate mean and standard deviation of the attribute values, subtract the mean from each value, and divide the result by the standard deviation. This process is called *standardizing* a numeric variable and results in a set of values whose mean is zero and standard deviation is one.

To see why this might be necessary, consider the case of clustering. Clustering typically involves calculating a distance measure that reflects how far each instance is from a cluster center or from other instances. With multiple variables, different units will be used. If one variable has units in the thousands and everything else is in tens, the variable with value in thousands will dominate the distance measure.

Replacing Categorical Variables with Numeric Ones: We are encountered with the situations in which categorical attributes must be represented numerically. Some learning algorithms—notably the *k-NN* method, regression methods, neural networks, clustering—handle only attributes that are numeric. How can they be extended to categorical attributes?

A common mistake of novice data miners is to replace categorical attributes with arbitrary numbers. The problem is that this enumeration creates spurious information that data mining algorithms have no way of ignoring.

Another popular approach is to create a separate binary variable for each category. A '1' indicates that the category is present and a '0' indicates that it is not. This works well when we have only a few categories in the data.

Often, the most satisfactory approach to categorical variables is to replace each one with a handful of numeric variables that capture important attributes of the categories. Domain knowledge plays a very important role in this conversion process.

Fourier/Wavelet Transforms: Mathematical transformations are applied to signals to obtain further information from that signal that is not readily available in raw form. Signal processing techniques result in creative data transformations in machine learning.

There are a number of transformations that can be applied, among which the *Fourier transformations* are probably by far the most popular. *Discrete Fourier Transform* (DFT) is a frequently used technique in machine learning for image processing, computer vision, time-series data, and data cleansing applications.

Another linear signal processing technique, the *wavelet transform*, is closely related to DFT.

Transformations based on the Nature of Datasets/Learning Scheme: Given raw data may be a set of images, documents, audio clips, video clips, graphs, numerical measurements/observations, etc. These are a set of objects/instances/patterns. In an input data file, we represent them as N instances characterized by their numerical/categorical values on a fixed set of n features or attributes. The transformation of raw data to numeric form calls for techniques that depend on the type of input. Textual input, audio input and image input call for their own specialized conversions, described later in Section 9.5. Conversion of time-series data into standard data matrix has already been presented in Section 1.4.

7.7.3 Discretizing Numeric Attributes

Some classification and clustering methods deal with categorical attributes only, and cannot handle ones measured on a numeric scale. To use them on general datasets, numeric attributes must first be 'discretized' into a smaller number of distinct ranges.

There are two basic approaches to the problem of discretization. One is to quantize each attribute in the absence of any knowledge of the classes of the instances in the training set—so-called *unsupervised discretization*. The other is to take the classes into account when discretizing—*supervised discretization*.

Unsupervised Discretization: The obvious way of discretizing a numeric attribute is to divide its range into a predetermined number of intervals (*bins*). For example, attribute values can be discretized by applying equal-width or equal-frequency *binning* and then replacing each bin value by the bin mean or median. In both of these binning methods, the total range is divided into user-specified k intervals simultaneously. In equal-width binning, the continuous range of a feature is evenly divided into intervals that have an equal width, while in equal-frequency binning, an equal number of continuous values are placed in each bin.

The term 'arity' in the discretization context means the number of intervals/partitions, k. There is a trade-off between arity and its effect on the accuracy of classification and other tasks.

When arity is set to k, the maximum number of *cut-points* is $k - 1$. The term 'cut-point' refers to a real value within the range of continuous values that divides the range into two intervals, one interval is less than or equal to the cut-point and the other interval is greater than the cut-point.

Supervised Discretization: Binning methods mentioned above may not give good results in cases where the distribution of the continuous variables is not uniform. Furthermore, the approach is vulnerable to outliers as they affect the ranges significantly. Supervised discretization methods overcome this shortcoming, wherein class information is used to find the proper intervals caused by cut-points. Different methods have been devised to use class information for finding meaningful intervals.

Supervised discretization methods commonly apply 'entropy' measure to find a potential cut-point to split a range of continuous values into two intervals. These methods recursively binarize ranges or subranges until a *stopping criterion* is met. Many of these methods use ad hoc stopping conditions. Many findings reported in the literature point toward MDLP (Minimum Descriptive Length Principle (Section 3.9)) being identified as the first choice for discretization as it provides a more principled way of determining when the recursive splitting should stop.

In the next section, we will discuss the MDLP approach for discretization. The method we will present is a *top-down* approach. Top-down methods start with an empty set of cut-points, and keep on adding new ones to the list by 'splitting' intervals as the discretization progresses. An alternative group of discretization methods use *bottom-up* approach, that starts with the complete list of all the continuous values of the feature as cut-points, and removes some of them by 'merging' intervals as the discretization progresses. The majority of the methods are found in the splitting category.

Choosing a suitable discretization method is generally a complex matter, and largely depends on user's need and other considerations. Entropy (MDLP) is generally the first choice when no specific consideration dictates a particular choice. Refer to [129] for a comprehensive survey of discretization methods.

Some transformations depend intuitively on the semantic of the particular machine learning scheme. We have encountered such transformations in earlier chapters, for example, Fisher linear discriminants (Section 3.8) and kernel methods for support vector machines (Chapter 4).

Sometimes, it is useful to add noise to data, perhaps to test the robustness of a learning algorithm.

7.7.4 Attribute Reduction Techniques

Many techniques for transformation of data to reduce the number of attributes are available. These techniques can be categorized based on two factors:

- The first factor is whether or not the technique uses the target variable to select the input variables.
- The second factor is whether or not the technique uses a subset of the original variables or derives new variables from them that maximize the amount of information.

The advantage in keeping the original variables is understandability. Presumably, the original variables in the data are easier to understand than those generated automatically by some variable reduction technique.

Although there are many techniques for attribute reduction, we focus on only two of them in this chapter: principal components and rough set theory.

Principal Components Analysis (PCA), discussed in Section 7.9, is a useful technique for reducing the number of attributes in the model by analyzing the input variables. It is especially valuable when we have subset of measurements that are highly correlated. In that case, it provides a few new variables that are weighted linear combinations of the original variables that retain the explanatory power of the full original set. PCA-based attribute reduction procedure does not use the target variable (unsupervised learning).

Rough Sets-Based Attribute Reduction, discussed in Section 7.10, uses the target variable (supervised learning) and retains a subset of original variables.

A survey of feature selection methods is given in [130]. The technique based on Laplacian eigenmaps, presented earlier in Section 7.3.3, is a powerful method of attribute reduction, which uses the idea of feature embedding such that given pairwise similarities are preserved [123].

7.8 ENTROPY-BASED METHOD FOR ATTRIBUTE DISCRETIZATION

In general, a discretization is simply a *logical* condition in terms of one or more continuous-valued attributes that serves to partition the data into at least two subsets. In many soft-computing algorithms, a continuous-valued attribute is typically handled by partitioning its range into subranges, i.e., a test based on some logical condition is devised that quantizes the continuous range. Supervised discretization provides useful classification information with respect to the classes to which the examples in the attribute's range belong.

In this section, we present a *binarization* procedure for discretization of continuous-valued features. Patterns are described by a fixed set of attributes x_j; $j = 1, \ldots, n$. The output variable y is a Boolean-valued function (binary classification problems) defined over the set S of patterns $\{s^{(i)}\} \equiv \{\mathbf{x}^{(i)}\}$; $i = 1, \ldots, N$. That is, y takes on values y_q; $q = 1, 2$. If we assume $y_1 \equiv 0$ and $y_2 \equiv 1$, then $y : S \rightarrow [0, 1]$. The training data is described by the dataset $\mathcal{D}$ of N patterns with corresponding observed outputs:

$$\mathcal{D} = \{s^{(i)}, y^{(i)}\} = \{\mathbf{x}^{(i)}, y^{(i)}\}; i = 1, 2, \ldots, N \tag{7.46}$$

For each continuous-valued attribute x_j, we select the 'best' cut-point T_{x_j} from its range of values by evaluating every candidate cut-point in the range of values. The examples are first sorted by increasing value of the attribute x_j, and midpoint between each successive pair of values in the sorted sequence is evaluated as potential cut-point. Thus, for each continuous-valued attribute, $N-1$ evaluations will take place (assuming that examples do not have identical attribute values).

A cut-point T_{x_j} for x_j will partition the patterns in $\mathcal{D}$ into two subsets satisfying the conditions V_{x_j} (value of x_j) $\leq T_{x_j}$, and $V_{x_j} > T_{x_j}$, respectively, thereby creating a binary discretization. Many different criteria may be used to evaluate the potential cut-points. We consider here the one which uses a clever idea for evaluating a potential cut-point, borrowed from the world of information theory (revisiting Section 3.9 will be helpful). If a partition $\mathcal{D}_1$ of dataset $\mathcal{D}$ has all patterns of one class, say y_1, it is entirely *pure* because no additional information is needed to classify the patterns of this partition. On the other hand, the partition is *impure* if additional information is needed to classify the patterns. Information theory has a measure for this impurity, called *entropy*, which measures how disorganized a partition is with respect to the class information.

Imagine selecting a pattern at random from the subset $\mathcal{D}_1$ and announcing that it belongs to class y_q. The message has the probability,

$$P_q = \frac{freq(y_q, \mathcal{D}_1)}{|\mathcal{D}_1|} \tag{7.47}$$

where $freq(y_q, \mathcal{D}_1)$ stands for the number of patterns in $\mathcal{D}_1$ that belong to the class y_q, and $|\mathcal{D}_1|$ denotes the total number of patterns in $\mathcal{D}_1$.

The expected information needed to classify a pattern in $\mathcal{D}_1$ is given by,

$$Info(\mathcal{D}_1) = -\sum_{q=1}^{2} P_q \log_2(P_q) \tag{7.48}$$

A log function to the base 2 is used because information is encoded in *bits* (refer to Section 3.9). $Info(\mathcal{D}_1)$ is just the average amount of information needed to identify the class label of a pattern in $\mathcal{D}_1$. Note that, at this point, the information we have is based solely on the proportions of patterns in each class in the data subset $\mathcal{D}_1$; in all calculations involving Eqn (7.48), we define $0\log_2 0 = 0$.

$Info(\mathcal{D}_1)$ is the *entropy* of $\mathcal{D}_1$—a measure of how disorganized (impure) $\mathcal{D}_1$ is with respect to class information. Notice that the entropy is 0 if all members of $\mathcal{D}_1$ belong to the same class. For example, if all members are of Class 1 ($P_1 = 1$), then $P_2 = 0$ and $Entropy(\mathcal{D}_1) = -1 \log_2 1 - 0 \log_2 0 = 0$. The entropy is 1 when the collection contains an equal number of Class 1 and 2 examples ($P_1 = P_2 = \frac{1}{2}$). If the collection contains unequal number of Class 1 and 2 examples, the entropy is between 0 and 1.

How much more information would we still need for a perfect classification after partitioning on x_j? This amount is called the *expected information requirement* for classifying a pattern in $\mathcal{D}$ based on partitioning by T_{x_j}. It is given by,

$$Info(\mathcal{D}, T_{x_j}) = \frac{|\mathcal{D}_1|}{|\mathcal{D}|} \times Entropy(\mathcal{D}_1) + \frac{|\mathcal{D}_2|}{|\mathcal{D}|} \times Entropy(\mathcal{D}_2) \tag{7.49a}$$

This is the entropy of the partitions induced by T_{x_j}, denoted as $Entropy(\mathcal{D}, T_{x_j})$.

$$Entropy(\mathcal{D}, T_{x_j}) = \frac{|\mathcal{D}_1|}{|\mathcal{D}|} \times Entropy(\mathcal{D}_1) + \frac{|\mathcal{D}_2|}{|\mathcal{D}|} \times Entropy(\mathcal{D}_2) \tag{7.49b}$$

The cut-point T_{x_j} for which $Entropy(\mathcal{D}, T_{x_j})$ is minimal amongst all the candidate cut-points, is taken as the best cut-point. This determines binary discretization of $\mathcal{D}$ by attribute x_j. The process of determining a cut-point is recursively applied to each partition obtained, until some stopping criterion is met, such as when the entropy on all candidate cut-points is less than a small threshold, or when the number of partitions is greater than a threshold.

The entropy and information requirement measures described here are also used for decision tree induction [141]. These measures are revisited in greater detail in Chapter 8.

Example 7.2

The weather problem is an entirely fictitious tiny dataset (Table 7.4); it supposedly concerns the conditions that are suitable for playing tennis. The patterns in the dataset are characterized by four features: *outlook*, *temperature*, *humidity*, and *wind*. The outcome is whether to play tennis; the output variable (class label) is *PlayTennis*. The attributes Outlook and Wind have categorical values and the attributes Temperature and Humidity have continuous numeric values.

Table 7.4 The weather data

Patterns	Outlook x_1	Temperature (°F) x_2	Humidity (%) x_3	Wind x_4	PlayTennis y
$s^{(1)}$	sunny	85	85	weak	no
$s^{(2)}$	sunny	80	90	strong	no
$s^{(3)}$	overcast	83	86	weak	yes
$s^{(4)}$	rain	70	96	weak	yes
$s^{(5)}$	rain	68	80	weak	yes
$s^{(6)}$	rain	65	70	strong	no
$s^{(7)}$	overcast	64	65	strong	yes
$s^{(8)}$	sunny	72	95	weak	no
$s^{(9)}$	sunny	69	70	weak	yes
$s^{(10)}$	rain	75	80	weak	yes
$s^{(11)}$	sunny	75	70	strong	yes
$s^{(12)}$	overcast	72	90	strong	yes
$s^{(13)}$	overcast	81	75	weak	yes
$s^{(14)}$	rain	71	91	strong	no

The dataset $\mathcal{D}$ in Table 7.4 has two categorical/nominal attributes: Outlook and Wind, and two continuous/numeric attributes: Temperature and Humidity. To see entropy-based discretization procedure working in practice, we consider here discretization of numeric attribute Temperature. This attribute has the following numeric values:

85 80 83 70 68 65 64 72 69 75 75 72 81 71

Ascending-order sorting gives (repeated values have been collapsed together):

64	65	68	69	70	71	72	75	80	81	83	85
yes	no	yes	yes	yes	no	no	yes	no	yes	yes	no
						yes	yes				

It is common to place candidate cut-points halfway between the values, although something might be gained by adopting a more sophisticated policy (one such policy will be described shortly in this section). There are 11 candidate cut-points. The entropy of the partitions induced by each of the 11 candidate cut-points is calculated using Eqn (7.49b). For example,

$$Entropy(\mathcal{D}, T_{x_2} = 71.5) = \frac{|\mathcal{D}_1|}{|\mathcal{D}|} \times \text{Entropy}(\mathcal{D}_1) + \frac{|\mathcal{D}_2|}{|\mathcal{D}|} \times \text{Entropy}(\mathcal{D}_2)$$

$$|\mathcal{D}| = 14, |\mathcal{D}_1| = 6, |\mathcal{D}_2| = 8$$

$$Entropy(\mathcal{D}_1) = -P_1 \log_2 P_1 - P_2 \log_2 P_2;\ P_1 = \frac{4}{6},\ P_2 = \frac{2}{6}$$

$$Entropy(\mathcal{D}_2) = -P_1 \log_2 P_1 - P_2 \log_2 P_2;\ P_1 = \frac{5}{8},\ P_2 = \frac{3}{8}$$

$$Entropy(\mathcal{D}, T_{x_2} = 71.5) = 0.939 \text{ bits}$$

This represents the expected information required to specify the individual values of *yes* and *no* given the cut-point. We seek a discretization that makes the partitions as pure as possible, hence, we choose a cut-point where $Entropy(\mathcal{D}, T_{x_2})$ value is smallest (this is the same as splitting where expected information requirement for classifying a pattern in $\mathcal{D}$ is the smallest).

It can easily be verified that $Entropy\ (\mathcal{D}, T_{x_2} = 84)$ is the smallest value (0.827 bits), which separates off just the final value, a *no* pattern, from the preceding list.

Invoking the algorithm again on the lower range of temperature, from 64 to 83, yields the minimum at 80.5 (0.800 bits), which splits off the next two values, both *yes* patterns. Again invoking the algorithm on the lower range, now from 64 to 80, produces a minimum at 77.5 (0.801 bits), splitting off another *no* pattern. Continuing the process, the next minimum for the temperature range 64 to 75 at 73.5 (0.764 bits) splits off two *yes* patterns; next to this one at 70.5 (0.796 bits) for the temperature range 64 to 72 splits off two *no*'s and a *yes*, and finally for the range 64 to 70, the minimum is at 66.5 (0.4 bits).

The fact that the recursion only ever occurs in the first interval of each split is an artifact of this example; in general, both the upper and the lower intervals will have to be split further.

Cut-Point Optimization

It has been demonstrated that the cut-point selection criterion described above constitutes a powerful heuristic when used to guide the search for a good decision tree (Chapter 8) to classify a set of training examples. It does not guarantee optimal results but it is a good method for determining which attributes are relevant to the classification task at hand.

One of the main problems with this selection criterion is that it is relatively expensive. It must be evaluated $(N - 1)$ times for each attribute (assuming that the N samples have distinct values). Machine learning programs are designed to work with large sets of training data, so N is typically very large.

Fayyad and Irani [131] have proved that regardless of how many classes there are (binary or multiclass problems) and how they are distributed, the *cut-point will always occur on the boundary between two classes*. This leads to a useful optimization: a cut-point that minimizes the entropy will never occur between two patterns of the same class; it is only necessary to consider potential divisions that separate patterns of different classes.

This cut-point optimization result helps us improve the efficiency of the algorithm without changing its function at all. Since the cut-point must occur on a boundary, we only need to evaluate boundary points between classes instead of evaluating possibly all $N - 1$ candidate cut-points. Of course, savings in computation depend on the class changes from one example to the next in the sorted sequence of examples.

Example 7.3

We revisit the previous example (Example 7.2) for discretizing the temperature attribute of the weather data whose values are

↓	↓			↓	↓	↓	↓	↓		↓	
64	65	68	69	70	71	72	75	80	81	83	85
yes	no	yes	yes	yes	no	no	yes	no	yes	yes	no
						yes	yes				

There are 11 possible positions for the breakpoint; we have considered all of these candidate cut-points in the previous example.

As per the cut-point optimization result described above, breakpoints are not allowed to separate items of the same class. This reduces the number of candidate cut-points to 8. Also, these candidate cut-points are the boundary values, not the middle values of the intervals. It can easily be ascertained that the candidate cut-points are: 64, 65, 70, 71, 72, 75, 80, 83.

$$Entropy(\mathcal{D}, T_{x_2} = 64) = \frac{|\mathcal{D}_1|}{|\mathcal{D}|} \times Entropy(\mathcal{D}_1) + \frac{|\mathcal{D}_2|}{|\mathcal{D}|} \times Entropy(\mathcal{D}_2)$$

$$|\mathcal{D}| = 14, |\mathcal{D}_1| = 13, |\mathcal{D}_2| = 1$$

$$Entropy(\mathcal{D}_1) = -P_1 \log_2 P_1 - P_2 \log_2 P_2; \; P_1 = \frac{8}{13}, \; P_2 = \frac{5}{13}$$

$$= 0.4310 + 0.5302 = 0.9612$$

$$Entropy(\mathcal{D}_2) = -P_1 \log_2 P_1 - P_2 \log_2 P_2;\ P_1 = \frac{1}{1},\ P_2 = \frac{0}{1} = 0$$

$$\text{Entropy}(\mathcal{D}, T_{x_2} = 64) = \frac{13}{14} \times 0.9602 + \frac{1}{14} \times 0 = 0.8925 \text{ (bits)}$$

It can easily be verified that $Entropy(\mathcal{D}, T_{x_2} = 83)$ is the smallest value:

$$Entropy(\mathcal{D}, T_{x_2} = 83) = \frac{13}{14}\left(\frac{-9}{13}\log_2\frac{9}{13} - \frac{4}{13}\log_2\frac{4}{13}\right) + \frac{1}{14}(0)$$

$$= \frac{13}{14}(0.3672 + 0.5232) = 0.8268 \text{ (bits)}$$

7.9 PRINCIPAL COMPONENTS ANALYSIS (PCA) FOR ATTRIBUTE REDUCTION

Principal Components Analysis (PCA) is a valuable result from applied linear algebra and is abundantly used for reducing the number of variables (features/attributes) in the dataset that are correlated. The reduced set of variables are weighted linear combinations of the original variables that (approximately) retain the information content of the original dataset. The linear combinations of the original variables lose physical meaning (if any) attached to the original variables.

The goal of PCA is to transform data in the best possible way. The most important question is: What does '*best transformation*' of the data mean? Let us build an intuitive answer to this question.

The potential problems with the data are *noise* and *redundancy*. A simple way to quantify the redundancy between individual attributes is through the covariance matrix **Σ**. Covariance matrix describes all relationships between pairs of attributes in our dataset. If our goal is to reduce redundancy, we would like each variable to co-vary as little as possible with other variables. More precisely, to remove redundancy, we would like the covariance between different variables to be zero. Evidently, in an 'optimized' matrix, all off-diagonal terms in **Σ** are zero. Therefore, removing redundancy diagonalizes **Σ**.

There are many methods for diagonalizing **Σ**; PCA selects the easiest method (based on linear algebra), perhaps accounting for its widespread application.

Let us now look at the other potential problem in a given dataset, the noise. There exists no absolute scale for noise but rather all noise is measured relative to the signal. A common measure is *signal-to-noise ratio* (SNR), which is the ratio of variances:

$$\text{SNR} = \frac{\sigma^2_{\text{signal}}}{\sigma^2_{\text{noise}}} \tag{7.50}$$

A high SNR (>>1) indicates high precision data, while a low SNR indicates noise contaminated data. The data with high SNR will have attributes with larger associated variances, representing interesting dynamics. The data with low SNR will have attributes with lower variances, representing noise.

We have discussed all aspects of the goal PCA; what remains is the linear algebra solution. The solution is based on transformation of the covariance matrix to a diagonal form with larger associated variances.

We can visualize the data with n attributes $x_1, x_2, \ldots, x_n$, as a cloud of N points $\mathbf{x}^{(i)}$; $i = 1, \ldots, N$, $\mathbf{x} = [x_1\ x_2 \cdots x_n]^T$ is an n-dimensional vector space.

The $N \times n$ data matrix $\mathbf{X}$ is given as,

$$\mathbf{X} = \begin{bmatrix} (\mathbf{x}^{(1)})^T \\ \vdots \\ (\mathbf{x}^{(N)})^T \end{bmatrix} \tag{7.51}$$

where $(\mathbf{x}^{(i)})^T$ are row vectors representing data samples. Each row of $\mathbf{X}$ corresponds to values of all the n attributes of a particular data sample, and each column of $\mathbf{X}$ corresponds to all the N values of a particular attribute. The data matrix $\mathbf{X}$ is in a nice table-like format.

PCA is an *unsupervised* method as it does not use the output information. We are interested in finding a mapping from the inputs in the original n-dimensional space to a new $(k < n)$-dimensional space with the minimum loss of information.

Let us first review the basic statistics covered in Section (3.2); the review will be helpful in understanding the process of Principle Components Analysis. We assume here that the reader has working knowledge of linear algebra (refer to Section 1.9).

The *mean* (*average*) of the attribute x_j is given by,

$$\mu_j = \frac{\sum_{i=1}^{N} x_j^{(i)}}{N}\ ;\ j = 1, \ldots, n \tag{7.52a}$$

Variance (a measure of the spread of the data) is given by,

$$var\,(x_j) = \sigma_j^2 = \frac{\sum_{i=1}^{N} (x_j^{(i)} - \mu_j)(x_j^{(i)} - \mu_j)}{N} \tag{7.52b}$$

$$= \frac{\sum_{i=1}^{N} (x_j^{(i)} - \mu_j)^2}{N}\ ;\ j = 1, \ldots, n \tag{7.52c}$$

Covariance (a measure to find out how attributes vary *with respect to each other*) is given by,

$$cov\,(x_j, x_l) = \sigma_{jl} = \frac{\sum_{i=1}^{N} (x_j^{(i)} - \mu_j)(x_l^{(i)} - \mu_l)}{N}\ ;\ j, l = 1, \ldots, n \tag{7.52d}$$

For the attribute vector $\mathbf{x} = [x_1 \; x_2 \cdots x_n]^T$ with mean $\boldsymbol{\mu} = [\mu_1 \; \mu_2 \cdots \mu_n]^T$, the *covariance matrix*

$$\boldsymbol{\Sigma} = \frac{\sum_{i=1}^{N} (\mathbf{x}^{(i)} - \boldsymbol{\mu})(\mathbf{x}^{(i)} - \boldsymbol{\mu})^T}{N}$$

$$= \begin{bmatrix} \sigma_1^2 & \sigma_{12} & \cdots & \sigma_{1n} \\ \sigma_{21} & \sigma_2^2 & \cdots & \sigma_{2n} \\ \vdots & \vdots & & \vdots \\ \sigma_{n1} & \sigma_{n2} & \cdots & \sigma_n^2 \end{bmatrix} \tag{7.52e}$$

Note that $\boldsymbol{\Sigma}$ is an $n \times n$ *square* matrix.

Since $\sigma_{jl} = \sigma_{lj}$ as seen by the formula (7.52d), the covariance matrix $\boldsymbol{\Sigma}$ is a *symmetric* matrix ($\boldsymbol{\Sigma} = \boldsymbol{\Sigma}^T$).

When we subtract the mean from each of the data dimensions, we get the transformed data in *mean-deviation form*. For PCA, working with the data whose mean is zero, is more convenient. From now onwards, we assume that the given dataset is a zero-mean dataset. We continue to use the same notation $\mathbf{x}^{(i)}$ for the zero-mean dataset as well. For this dataset,

$$\boldsymbol{\Sigma}_{\mathbf{X}} = \frac{\sum_{i=1}^{N} \mathbf{x}^{(i)} (\mathbf{x}^{(i)})^T}{N} = \frac{1}{N} \mathbf{X}^T \mathbf{X} \tag{7.53}$$

where $\mathbf{X}$ is $N \times n$ data matrix given by (7.51).

As we will shortly see, for linear algebra solution, working with an $n \times N$ data matrix is more convenient. This can, of course, be achieved by taking $\mathbf{X}^T$ as the data matrix. However, we will prefer to define new data matrix $\hat{\mathbf{X}}$ derived from the given data matrix $\mathbf{X}$:

$$\hat{\mathbf{X}} = \mathbf{X}^T \tag{7.54a}$$

In terms of $\hat{\mathbf{X}}$, the covariance is given by,

$$\boldsymbol{\Sigma}_{\hat{\mathbf{X}}} = \frac{1}{N} \hat{\mathbf{X}} \hat{\mathbf{X}}^T \tag{7.54b}$$

The potential problems with the data are redundancy and noise. Let us first look at the linear algebra solution for reducing redundancy. To reduce redundancy, we would like each variable (feature) to co-vary as little as possible with other variables. More precisely, we would like the covariance between features to be zero. That is, we look for a mapping,

$$\underset{(n \times N)}{\hat{\mathbf{Z}}} = \underset{(n \times n)}{\mathbf{W}} \; \underset{(n \times N)}{\hat{\mathbf{X}}} \tag{7.55a}$$

such that,

$$\begin{aligned}\Sigma_{\hat{\mathbf{Z}}} &= \frac{1}{N}\hat{\mathbf{Z}}\,\hat{\mathbf{Z}}^T \\ &= \frac{1}{N}(\mathbf{W}\,\hat{\mathbf{X}})\,(\mathbf{W}\,\hat{\mathbf{X}})^T \\ &= \frac{1}{N}\mathbf{W}\,\hat{\mathbf{X}}\hat{\mathbf{X}}^T\,\mathbf{W}^T \\ &= \mathbf{W}\,\Sigma_{\hat{\mathbf{X}}}\,\mathbf{W}^T \end{aligned} \tag{7.55b}$$

is diagonal (in the transformed space, the covariance matrix has zero off-diagonal terms). Thus, diagonalizing $\Sigma_{\hat{\mathbf{X}}}$ using transformation matrix $\mathbf{W}$ removes redundancy. Selecting transformation matrix $\mathbf{W}$ to be a matrix whose each row is an *eigenvector* of $\Sigma_{\hat{\mathbf{X}}}$ results in diagonal matrix $\Sigma_{\hat{\mathbf{Z}}}$, as follows (refer to Section 1.9):

For the $n \times n$ square matrix $\Sigma_{\hat{\mathbf{X}}}$ (refer to Eqn (1.37c)),

$$\mathbf{E}^{-1}\Sigma_{\hat{\mathbf{X}}}\,\mathbf{E} = \Lambda$$

$$= \begin{bmatrix} \lambda_1 & 0 & \cdots & 0 \\ 0 & \lambda_2 & \cdots & 0 \\ \vdots & \vdots & & \vdots \\ 0 & 0 & \cdots & \lambda_n \end{bmatrix} \tag{7.56a}$$

where,

$$\mathbf{E} = [\mathbf{e}_1\ \mathbf{e}_2 \cdots \mathbf{e}_n] \tag{7.56b}$$

The columns $\mathbf{e}_1, \mathbf{e}_2, \ldots, \mathbf{e}_n$ of $\mathbf{E}$ are eigenvectors of $\Sigma_{\hat{\mathbf{X}}}$ associated with its eigenvalues $\lambda_1, \lambda_2, \ldots, \lambda_n$, respectively.

Since the $n \times n$ square matrix $\Sigma_{\hat{\mathbf{X}}}$ is symmetric, its eigenvectors $\mathbf{e}_1, \mathbf{e}_2, \ldots, \mathbf{e}_n$ are orthonormal. This means that $\mathbf{E}$ is an orthonormal matrix, and by the result (Eqn (1.30)),

$$\mathbf{E}^T = \mathbf{E}^{-1}$$

we get,

$$\mathbf{E}^T\,\Sigma_{\hat{\mathbf{X}}}\mathbf{E} = \Lambda, \text{ a diagonal matrix} \tag{7.57}$$

Thus, the symmetric covariance matrix $\Sigma_{\hat{\mathbf{X}}}$ can be diagonalized by selecting the transformation matrix $\mathbf{W}$ in Eqn (7.55b) to be a matrix whose each row is an eigenvector of $\Sigma_{\hat{\mathbf{X}}}$. By this selection,

$$\mathbf{W} \equiv \mathbf{E}^T \tag{7.58}$$

This was the first goal for PCA. Let us now pay attention to the other goal: reducing noise. We see from Eqns (7.55b), (7.57), and (7.58),

$$\Sigma_{\hat{\mathbf{Z}}} = \mathbf{E}^T\,\Sigma_{\hat{\mathbf{X}}}\mathbf{E} = \Lambda \tag{7.59}$$

The variances in the transformed domain are, therefore, given by,

$$\begin{bmatrix} \sigma_{1\hat{z}}^2 & 0 & \cdots & 0 \\ 0 & \sigma_{2\hat{z}}^2 & \cdots & 0 \\ \vdots & \vdots & & \vdots \\ 0 & 0 & \cdots & \sigma_{n\hat{z}}^2 \end{bmatrix} = \begin{bmatrix} \lambda_1 & 0 & \cdots & 0 \\ 0 & \lambda_2 & \cdots & 0 \\ \vdots & \vdots & & \vdots \\ 0 & 0 & \cdots & \lambda_n \end{bmatrix} \tag{7.60a}$$

or

$$\sigma_{j\hat{z}}^2 = \lambda_j;\ j = 1, \ldots, n \tag{7.60b}$$

The largest eigenvalue, thus, corresponds to maximum variance. Therefore, the *first principal component* is the eigenvector of the covariance matrix $\Sigma_{\hat{\mathbf{x}}}$ associated with the largest eigenvalue. The *second principal component* should also maximize the variance, which implies that the eigenvector associated with the second largest eigenvalue is the second principal component. It follows that other principal components are given by eigenvectors associated with eigenvalues with decreasing magnitude.

Envision how PCA might work. Once eigenvectors and eigenvalues are found from the covariance matrix, the next step is to order the eigenvalue-eigenvector pairs by the magnitude of the eigenvalues: from the largest to the smallest (a real symmetric matrix has all real and nonnegative eigenvalues). This gives us eigenvalue-eigenvector pairs in order of significance. Remembering that $\sum_{j=1}^{n} \lambda_j = \sum_{j=1}^{n} \sigma_{j\hat{z}}^2$ (Eqn (7.60a)), some eigenvalues have little contribution to the variance and may be discarded. We take the leading k components that explain more than, for example, 90% of variance. When λ_j are sorted in descending order, the *proportion of variance* explained by the k principal components is,

$$\frac{\lambda_1 + \lambda_2 + \cdots + \lambda_k}{\lambda_1 + \lambda_2 + \cdots + \lambda_n} \tag{7.61}$$

If the features are highly correlated, there will be a small number of eigenvectors with large eigenvalues and k will be much smaller than n, and a large reduction of dimensionality may be attained. If the features are not correlated, k will be as large as n, and there is no gain through PCA.

The final step in PCA is to derive the new dataset—transformation of the original dataset (refer to Eqn (7.55a)):

$$\underset{(k \times N)}{\hat{\mathbf{Z}}} = \underset{(k \times n)}{\mathbf{W}} \underset{(n \times N)}{\hat{\mathbf{X}}} \tag{7.62}$$

We originally have n-dimensional data, and so we calculate n eigenvalues and n eigenvectors. We then choose only the first k ($k \le n$) eigenvectors from the list of eigenvalue-eigenvector pairs arranged in order of descending magnitude of eigenvalues. The k eigenvectors are called *principle components*, arranged in order of significance. We save them as row vectors of the transformation matrix $\mathbf{W}$. Equation (7.62) then gives us the transformed and reduced dataset in a table-like format. Refer to [132] for an online procedure for doing PCA.

We have used N for normalization of variance (Eqn (7.52c)) and covariance (Eqn (7.52d)). This, however, gives a biased estimation of variances (refer to Section 3.8), particularly when the sample size N is small. The normalization by N would have given unbiased estimation if we were calculating the variances of the *entire population*. For a subset of entire population (sample of N instances), the correction is to use $N - 1$ instead of N in the formula for sample variance [4]. This correction reduces the bias error due to finite sample count. When the sample size is large, this correction may not be applied.

Example 7.4

Table 7.5 gives a set of normalized data (mean subtracted), and Fig. 7.7 shows a plot of this data for the toy example. The zero-mean dataset has 10 samples ($N = 10$), and it has two attributes/features ($n = 2$).

The dataset may be expressed as a 2 × 10 data matrix

$$\hat{\mathbf{X}} = \begin{bmatrix} 0.69 & -1.31 & 0.39 & 0.09 & 1.29 & 0.49 & 0.19 & -0.81 & -0.31 & -0.71 \\ 0.49 & -1.21 & 0.99 & 0.29 & 1.09 & 0.79 & -0.31 & -0.81 & -0.31 & -1.01 \end{bmatrix}$$

Table 7.5 A zero-mean dataset

$s^{(i)}$	x_1	x_2
$s^{(1)}$	0.69	0.49
$s^{(2)}$	–1.31	–1.21
$s^{(3)}$	0.39	0.99
$s^{(4)}$	0.09	0.29
$s^{(5)}$	1.29	1.09
$s^{(6)}$	0.49	0.79
$s^{(7)}$	0.19	–0.31
$s^{(8)}$	–0.81	–0.81
$s^{(9)}$	–0.31	–0.31
$s^{(10)}$	–0.71	–1.01

The covariance matrix

$$\Sigma_{\hat{\mathbf{X}}} = \frac{1}{N-1} \hat{\mathbf{X}} \hat{\mathbf{X}}^T$$

$$= \begin{bmatrix} 0.617 & 0.615 \\ 0.615 & 0.717 \end{bmatrix}$$

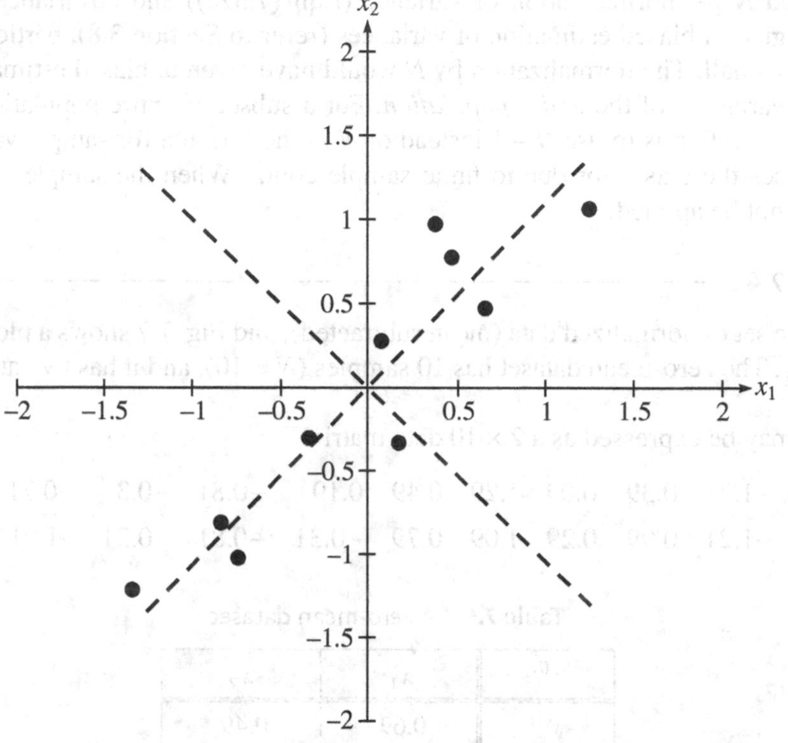

Figure 7.7 A plot of the dataset of Table 7.5 with the eigenvectors of the covariance matrix

For the square symmetric covariance matrix $\Sigma_{\hat{\mathbf{x}}}$, the eigenvalues and eigenvectors are:

$$\lambda_1 = 0.049; \lambda_2 = 1.284$$

$$\mathbf{e}_1 = \begin{bmatrix} -0.735 \\ 0.678 \end{bmatrix}; \mathbf{e}_2 = \begin{bmatrix} -0.678 \\ -0.735 \end{bmatrix}$$

Note that these are orthonormal eigenvectors; shown as dotted lines in Fig. 7.7. The eigenvectors provide us information about the patterns in the data.

Ordering eigenvalues from highest to lowest gives us the principle components in order of significance. In our example, principle components are:

$$\mathbf{e}_2^T = [-0.678 \quad -0.735]$$

$$\mathbf{e}_1^T = [-0.735 \quad -0.678]$$

Given our example set of data, and the fact that we have two components, we have two choices.

We can either form a transformation matrix with both of the components:

$$\mathbf{W} = \begin{bmatrix} -0.678 & -0.735 \\ -0.735 & 0.678 \end{bmatrix}$$

or we can choose to leave out one corresponding to smaller eigenvalue—the less significant component:

$$\mathbf{W} = [-0.678 \quad -0.735]$$

What needs to be done now is to generate the transformed data matrix:

$$\hat{\mathbf{Z}} = \mathbf{W}\hat{\mathbf{X}}$$

For the choice of $k = n = 2$, we use the transformation matrix given by Eqn (7.55a). The transformed dataset

$$\hat{\mathbf{Z}} = \begin{bmatrix} -0.828 & 1.778 & -0.992 & -0.274 & -1.676 & -0.913 & 0.099 & 1.145 & 0.438 & 1.224 \\ -0.175 & 0.143 & 0.384 & 0.130 & -0.209 & 0.175 & -0.350 & 0.046 & 0.018 & -0.163 \end{bmatrix}$$

The transformed data $\hat{\mathbf{Z}}$ has been plotted in Fig. 7.8. The original data $\hat{\mathbf{X}}$ was in terms of axes x_1 and x_2. Through transformation matrix $\mathbf{W}$, we have changed the *basis vectors* of our data; from the axes x_1 and x_2 to the directions of principle components.

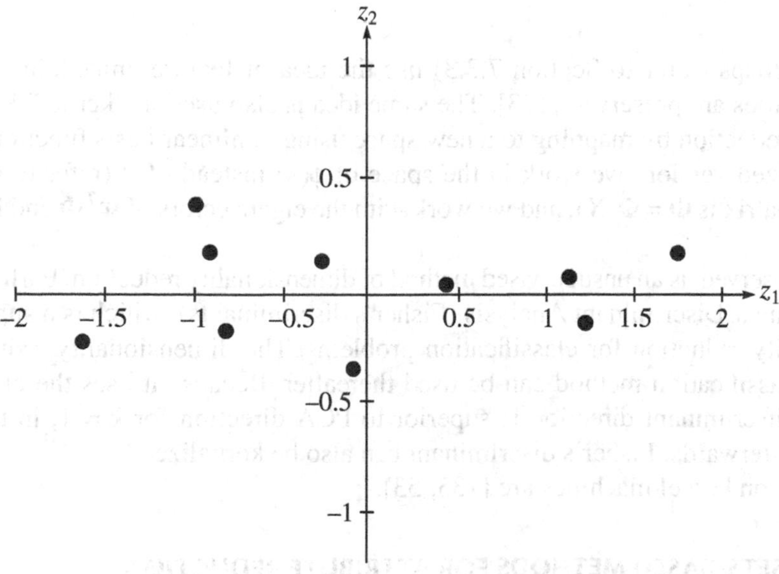

Figure 7.8 A plot of data obtained by applying PCA using both eigenvectors.

In the case when the transformed data matrix has reduced dimensionality ($k = 1 < n$), we use the transformation matrix given by Eqn (7.62).

$$\hat{\mathbf{Z}} = [\ -0.828 \quad 1.778 \quad -0.992 \quad -0.274 \quad -1.676 \quad -0.913 \quad 0.099 \quad 1.145 \quad 0.438 \quad 1.224\ ]$$

As expected, it has only single dimension (Fig. 7.9).

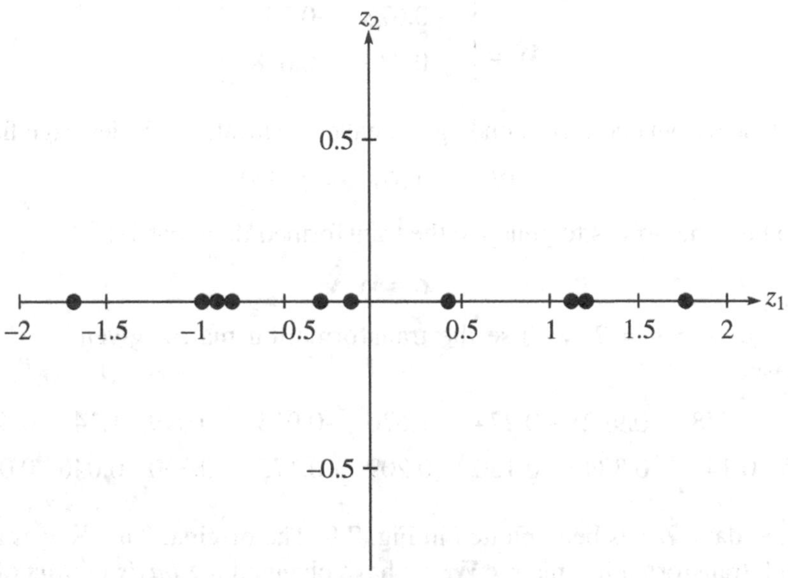

Figure 7.9 A plot of data obtained by applying PCA using one eigenvector.

Kernel PCA

Laplacian eigenmaps (refer to Section 7.3.3) use the idea of feature embedding such that given pairwise similarities are preserved [123]. The same idea is also used in "kernel" PCA: a nonlinear dimensionality reduction by mapping to a new space using nonlinear basis functions.

In the kernalized version, we work in the space of $\phi(\mathbf{x})$ instead of $\mathbf{x}$ (refer to Chapter 4). The projected data matrix is $\mathbf{\Phi} = \mathbf{\Phi}(\mathbf{X})$, and we work with the eigenvectors of $\mathbf{\Phi}^T\mathbf{\Phi}$ and hence the kernel matrix $K(\cdot)$.

PCA, as we observed, is an unsupervised method of dimensionality reduction. Earlier in Section 3.8, we presented Linear Discriminant Analysis (Fisher's discriminants), which is a supervised method for dimensionality reduction for classification problems. The dimensionality reduction is from n to 1, and any classification method can be used thereafter. Because it uses the class information, Fisher's linear discriminant direction is superior to PCA direction for $k = 1$, in terms of ease of discrimination afterwards. Fisher's discriminant can also be kernalized.

Classic books on kernel machines are [133, 53].

7.10 ROUGH SETS-BASED METHODS FOR ATTRIBUTE REDUCTION

The *theory of rough sets*, developed by Z. Powlak and his co-workers in the early 1980s [134], has become a recognized and widely researched mathematical approach to imperfect knowledge. The conceptual foundation of the theory is the consideration that all perception is subject to *granularity*. Classification on abstract levels seems to be a key issue in reasoning, learning and decision making.

Rough set theory deals with the classificatory analysis of data tables. Consider, for example, a group of patients suffering from a certain disease. With every patient, a data file is associated containing information like body temperature, blood pressure, name, age, and others. All patients revealing the same symptoms are *indiscernible* (similar) in view of the available information and may be thought of as representing a *granule* (disease unit) of medical knowledge. These granules are called *elementary sets* (*concepts*) within the given training data. All the data tuples forming an elementary set are indiscernible.

Elementary sets can be considered as elementary building blocks of knowledge. Elementary concepts can be combined into compound concepts, i.e., concepts that can be uniquely determined in terms of elementary concepts. Any union of elementary sets is a *crisp set*, and any other sets are referred to as *rough* (*imprecise*). Due to granularity of knowledge, rough sets cannot be characterized by available knowledge. Therefore, with every rough set we associate two crisp sets, called its *lower* and *upper approximation*. Intuitively, the *lower approximation* of a rough set consists of all elements that *surely* belong to the set, whereas the *upper approximation* of the rough set consists of all elements that *possibly* belong to the set. The difference between the upper and the lower approximations is a *boundary region*. It consists of all elements that cannot be classified uniquely to the set or its complement by employing available knowledge. All the data tuples forming the boundary region are *discernible* (*not similar*).

The concept of lower and upper approximations of a rough set is illustrated in Fig. 7.10 in a two-dimensional universe of discourse. Each rectangular region represents an elementary set (granule of knowledge). The lower and upper approximations, and the boundary region for a set (shown in grey shade) in two-dimensional universe of discourse are graphically displayed.

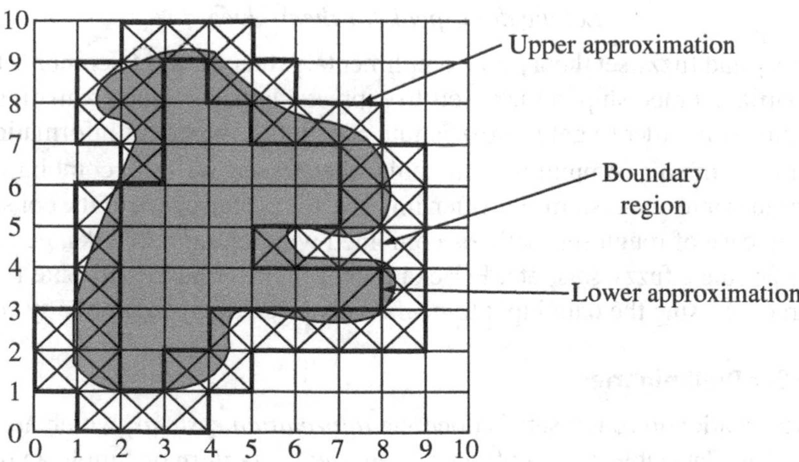

Figure 7.10 Set approximations in two-dimensional universe of discourse

Given the real-world data, it is common that some *decision classes* (*target concepts*) cannot be distinguished in terms of the available attributes. Rough sets can be used to approximately or 'roughly' define such classes. *Decisions rules* can be generated for each class. Typically, a *decision table* is used to represent the rules.

Rough sets can also be used for attribute subset selection (or *feature reduction*, where attributes that do not contribute to the classification of the given training data can be identified and removed) and *relevance analysis* (where the contribution or *significance* of each attribute is assessed with respect to the classification task). Rough set theory applies to discrete-valued attributes. Continuous-valued attributes must therefore be discretized before its use.

Basic problems of data analysis which can be tackled employing the rough set approach are the following:

- Characterization of target concepts in terms of available knowledge (attribute values).
- Analysis of relevance of attributes.
- Reduction of attributes.
- Generation of decision rules.

Rough set theory has simple algorithms to answer these questions and enables a straight-forward interpretation of obtained results. In this section, we use rough set theory for relevance analysis of attributes, and attribute reduction.

Rough set theory is generally regarded as part of the 'soft computing' paradigm. However, while other soft computing methods require additional model assumptions such as prior probabilities, fuzzy functions, etc., rough set theory is unique in the sense that it is 'non-invasive', i.e., it uses only the information given by the data, and does not rely on other model assumptions. The numerical value of *imprecision* is not pre-assumed—as it is done in probability theory or fuzzy sets—but is calculated on the basis of approximations, which are the fundamental concepts used to express imprecision of knowledge. The main motto of rough set theory is

Let the data speak for themselves

Rough set theory and fuzzy set theory are complimentary. It is natural to combine the two models of uncertainty (partial membership in fuzzy sets to express vagueness, approximations in rough sets to express coarseness) in order to get more accurate account of imperfect information.

It is the experience of soft-computing community that *hybrid systems*, combining different soft computing techniques into one system, can often improve the performance of the constructed system. This is also true in case of rough set methods combined with neural networks, genetic algorithms, statistical inference rules, fuzzy sets, etc. For example, rough-set based attribute reduction can be very useful in preprocessing the data input to neural networks, fuzzy inference systems, etc.

7.10.1 Rough Set Preliminaries

Knowledge representation in rough sets is done via *information systems*, which are a form of data table. Each row of the data table represents a *case*, an *event*, a *pattern* or simply an *object*, and each column is represented by an *attribute*. More precisely, the pair (U,X) constitutes an information system I, where U is a non-empty finite set of objects called the *universe*, and X is a non-empty finite set of attributes. As done earlier in the book, we will represent the objects by the index $i = 1, 2, \ldots, N$; and attributes can be interpreted as variables x_j; $j = 1, 2, \ldots, n$. Therefore, $U = \{1, 2, \ldots, N\}$ is the universe of objects, and $X = \{x_1, x_2, \ldots, x_n\}$ is the set of attributes. With every attribute $x_j \in X$, we associate a set V_{x_j} of its *values*, called the *domain* of x_j. Attribute x_j may take a finite value from a finite set of d_j discrete values $v_{1x_j}, v_{2x_j}, \ldots, v_{d_jx_j}$ In such a case, $V_{x_j} = \{v_{1x_j}, v_{2x_j}, \ldots, v_{d_jx_j}\} = \{v_{kx_j}; k = 1, 2, \ldots, d_j\}$, and $v_{kx_j} \in V_{x_j}$. By $(x_1^{(i)}, \ldots, x_n^{(i)})$ we denote a data pattern defined by the object with index i and attributes from X. A data pattern $i \in U$ of I is the value set $\{x_1^{(i)}, \ldots, x_n^{(i)}\}$ where $x_j^{(i)} \in V_{x_j}$ for $j = 1, 2, \ldots, n$. The elements x_j of X are called the *condition* attributes.

For supervised classification problems, an outcome (i.e., class of each pattern) is known *a priori*. This knowledge is expressed by one distinguished attribute, called the *decision* attribute. Information systems of this kind are called *decision systems*. More precisely, the tuple (U, X, y) constitutes a decision system $\mathcal{D}$, where U is a finite set of objects, X is a finite set of condition attributes, and y is the decision attribute. Any such decision system can be represented by a data table (*decision table*) whereas rows correspond to observations/patterns belonging to U and columns correspond to attributes belonging to $X \cup y$. The decision attribute may take any of the values given by the set $V_y = \{1, 2, \ldots, M\}$ for a multiclass problem. The output $y^{(i)}$ for a pattern i will take a value $y_q^{(i)}$; $q = 1, 2, \ldots, M$, from the set V_y.

Table 7.6 shows an example of a decision system: U consists of six objects (*patients*); X consists of three condition attributes (symptoms); x_1: *Headache*, x_2: *Muscle-pain*, x_3: *Temperature*; and there is one decision attribute y: *Flu*. The sets V_{x_j} of attribute values consist of observations on the patients. Decision attribute classifies each object to either Flu: yes; $y_1 = 1$ or Flue: no: $y_2 = 0$. Note that for binary classification, values $\{0,1\}$ have been used.

Table 7.6 Tutorial dataset

i = Patient index	x_1 : Headache	x_2 : Muscle-pain	x_3 : Temperature	y : Flu
1	no	yes	high	yes
2	yes	no	high	yes
3	yes	yes	very high	yes
4	no	yes	normal	no
5	yes	no	high	no
6	no	yes	very high	yes

Basic Granules of Knowledge about the Universe

Now we shall present two definitions that are very important in the rough set theory.

Indiscernibility relation: Suppose we are given two finite non-empty sets U and X, where U is the universe, and X is set of attributes. Each attribute $x_j \in X$ takes a value from finite set of values $v_{kx_j} \in V_{x_j}$; $k = 1, 2, \ldots, d_j$. V_{x_j} is the domain of x_j.

Any subset Z of X determines a binary relation on U which will be called an *indiscernibility relation*, and is defined as follows:

$$\text{IND}(Z) = \{(i, l) \in U \times U \mid \forall\, x_j \in Z, V_{x_j}(i) = V_{x_j}(l)\} \tag{7.63}$$

where $V_{x_j}(i)$ denotes the value of attribute $x_j \in Z$ for element $i \in U$.

Objects $i \in U$ and $l \in U$ are *indiscernible* (similar) from each other by attributes from Z. Obviously, IND(Z) is an *equivalence relation*, called the *Z-indiscernibility relation*.

Equivalence Classes: The equivalence relation divides a set in which it is defined, into a family of disjoint sets called *equivalence classes of the relation*. The *family* of all equivalence classes of IND(Z), i.e., partitions determined by Z, will be denoted by *U*/IND(Z). Equivalence classes form basic granules of knowledge about the universe.

A group of objects having the property that each attribute $x_j \in Z$ takes the same value v_{kx_j}; $k = 1, \ldots, d_j$, forms an equivalence class. For $Z = x_j$ (one attribute),

$$U/\text{IND}(x_j) = \{i \in U \mid x_j^{(i)} = v_{kx_j}\} \tag{7.64}$$

Equivalence classes are also generated by considering multiple attributes $Z \subseteq X$ at a time:

$$U/\text{IND}(Z) = \{i \in U \mid \forall x_j \in Z, x_j^{(i)} \text{ have similar values}\} \tag{7.65}$$

Equivalence classes of the relation IND(Z) (or blocks of the partition *U*/IND(Z)) are referred to as *Z-elementary sets.* In the rough set approach, the elementary sets are the basic building blocks (concepts) of our knowledge about reality.

The indiscernibility relation will be used next to define set approximations—the basic concepts of rough set theory. Before that, a tutorial example will be helpful to appreciate the definitions given before.

Example 7.5

Consider tutorial dataset given in Table 7.6. From this table, we observe that patients 2, 3, and 5 are indiscernible with respect to the attribute x_1. Also, patients 1, 4, and 6 are indiscernible with respect to this attribute. Therefore for $Z = x_1$, we have,

$$U/\text{IND}(Z) = U/\text{IND}(x_1) = \{\{1, 4, 6\}, \{2, 3, 5\}\}$$

With respect to attributes $Z = \{x_1, x_2\}$, patients 1, 4, 6 are indiscernible. Also patients 2, 5 are indiscernible. Therefore,

$$U/\text{IND}\,(x_1, x_2) = \{\{1, 4, 6\}, \{2, 5\}, \{3\}\}$$

Consider now $Z = \{x_1, x_2, x_3\} = X$. Patients 2 and 5 are indiscernible with respect to the whole set X:

$$U/\text{IND}(X) = U/\text{IND}\,(x_1, x_2, x_3) = \{\{1\}, \{2, 5\}, \{3\}, \{4\}, \{6\}\}$$

One can thus compute elementary sets generated by X or any subset of X.

Set Approximations

An important issue in data analysis is discovering the dependency of the decision attribute y on the set of condition variables $Z \subseteq X$. Attribute y depends totally on the set of attributes Z, denoted $Z \Rightarrow y$, if all values of attribute y are uniquely determined by values of attributes from Z (i.e., there exists a functional dependency between values of y and Z). Rough set interpretation of total dependency of y on Z is that *all* elements of the universe U can be uniquely classified to blocks of the partition of U generated by decision attribute y, employing condition attributes Z. The family of all partitions generated by classification labels y_q; $q = 1, 2, \ldots, M$, will be denoted as $U\backslash y$.

$$U\backslash y = \{B^{(1)}, \ldots, B^{(M)}\};\ B^{(q)} \subset U;\ B^{(q)} \cap B^{(m)} = \varnothing;\ \cup B^{(q)} = U;\ q, m = 1, 2, \ldots, M; \qquad (7.66)$$
$$B^{(q)} = \{i \in U | y^{(i)} = y_q\}$$

denotes the set of elements of the universe U through q^{th} classification label. This is referred to as *equivalence class* of U generated by classification label y_q.

Equation (7.66) gives a set of equivalence classes generated by decision attribute y employing condition attributes Z. Generation of those equivalence classes where all elements of U can be uniquely identified to these classes is possible only if there are no ambiguities in the dataset. This is, however, not true in real-world situations. The concept given by Powlak is to approximate each class by a pair of exact sets, called the 'lower' and 'upper' approximations. *Lower approximation* is the set of objects which *certainly* are identified to a given class, while *upper approximation* is the set of objects which can be *possibly* classified to a given class.

Equation (7.66) gives a set of equivalence classes $B^{(1)}, \ldots, B^{(M)}$ generated by decision attributes y_q; $q = 1, \ldots, M$, employing condition attributes Z. The lower approximation of an equivalence class B in $Z(Z \subseteq X)$, denoted as $\underset{\sim}{B}(Z)$, is defined as the union of all elementary sets in space Z which are contained in B. We will refer to this as *Z-lower approximation*. More formally, Z-lower approximation

$$\underset{\sim}{B}(Z) = \cup \{C \in U/\text{IND}(Z) \mid C \subseteq B\} \qquad (7.67)$$

The upper approximation of an equivalence class B in $Z(Z \subseteq X)$, denoted as $\tilde{B}(Z)$, is defined as the union of all elementary sets in space Z which have at least one element in common with B. We will refer to this as *Z-upper approximations*:

$$\tilde{B}(Z) = \cup \{C \in U/\text{IND}(Z) | C \cap B \neq \varnothing\} \qquad (7.68)$$

The set difference $\tilde{B}(Z) - \underset{\sim}{B}(Z)$ will be referred to as the *Z-boundary region*.

The *accuracy* of approximation

$$\mu_Z(B) = \frac{\text{card}\ \underset{\sim}{B}(Z)}{\text{card}\ \tilde{B}(Z)} \qquad (7.69)$$

(The *cardinality* of a set is the number of samples contained in the set)

Obviously, $0 \le \mu_Z(B) \le 1$. If $\mu_Z(B) = 1$, the set Z is crisp with respect to B (attributes of Z are precise with respect to equivalence class B for a label y_q); otherwise, if $\mu_Z(B) < 1$, Z is rough with respect to B.

Example 7.6

Objects 2 and 5 in Table 7.6 have the same values of condition attributes x_1, x_2, and x_3, but different values of decision attribute y. Objects 2 and 5 are, thus, indiscernible with respect to condition attributes and are discernible with respect to decision attribute. These objects contribute to our lack of knowledge about elements of U. The imprecision in rough set approach is expressed by *boundary region* of the dataset. *Flu* (decision attribute y = yes) cannot be characterized with respect to the attributes (symptoms) *Headache* (x_1), *Muscle-pain* (x_2), and *Temperature* (x_3). Therefore, patients 2 and 5 are the boundary-line cases which cannot be properly classified in view of the available knowledge.

The patients 1, 3 and 6 display symptoms which enable us to classify them with certainty as *Flu*, and patient 4 for sure does not have flu in view of the displayed symptoms. Thus, the *lower approximation* of the set of patients having flu with respect to the three symptoms (attributes x_1, x_2, x_3) is the set of patients {1, 3, 6}.

The *upper approximation* of the set of patients is {1, 2, 3, 5, 6} (Note that *boundary region* of the set is given by patients {2, 5}).

Now consider the concept 'no-flu' (decision attribute y = no). Patient 4 does not have flu and patients 2 and 5 cannot be properly classified in view of the available knowledge. Thus the *lower approximation* of the concept 'no-flu' is the set of patients {4}, whereas the *upper approximation* is the set {2, 4, 5}; and the *boundary region* of this concept is the set {2, 5}—the same as in the previous case.

Let us now express these observations in terms of rough-set terminology.

$$Z = \{x_1, x_2, x_3\} = X$$

From Example 7.5, we have,

$$U/\text{IND}(X) = \{\{1\}, \{2, 5\}, \{3\}, \{4\}, \{6\}\}$$

The equivalence classes of U determined by decision attribute y_q, $q = 1, 2$:

$$U\backslash y = \{B^{(1)}, B^{(2)}\}$$

$$B^{(1)} = \{1, 2, 3, 6\}; B^{(2)} = \{4, 5\}$$

Elementary sets {1}, {3}, {6} from family of $U/\text{IND}(X)$, are contained in class $B^{(1)}$, and elementary set {4} is contained in class $B^{(2)}$; elementary set {2, 5} is contained in none of the classes. Since $\{1\} \cup \{3\} \cup \{6\} \cup \{4\} \neq U$, all the elements of the universe cannot be uniquely classified by blocks of partition $U\backslash y$, i.e., equivalence classes generated by decision attribute employing $X = \{x_1, x_2, x_3\}$. Therefore y depends partially on X.

The lower approximation

$$\underset{\sim}{B}^{(1)}(X) = \{1\} \cup \{3\} \cup \{6\} = \{1, 3, 6\}$$

$$\underset{\sim}{B}^{(2)}(X) = \{4\}$$

The upper approximation

$$\tilde{B}^{(1)}(X) = \{1, 2, 3, 5, 6\}$$

$$\tilde{B}^{(2)}(X) = \{2, 4, 5\}$$

The accuracy of approximation

$$\mu_X(B^{(1)}) = \frac{\text{card } \underset{\sim}{B}^{(1)}(X)}{\text{card } \tilde{B}^{(1)}(X)} = \frac{3}{5}$$

$$\mu_X(B^{(2)}) = \frac{\text{card } \underset{\sim}{B}^{(2)}(X)}{\text{card } \tilde{B}^{(2)}(X)} = \frac{1}{3}$$

Since $0 < \mu_X(B) < 1$ ($\neq 1$), the attribute set X is rough with respect to $B^{(1)}$ and $B^{(2)}$. It means that the decision attributes $y_q = \{0, 1\}$ can be characterized partially and not uniquely.

7.10.2 Analysis of Relevance of Attributes

A decision system expresses the knowledge about the underlying problem domain in the form of raw measurements. The decision system may be unnecessarily large, in part, because it has redundancy in at least two ways. The similar (indiscernible) objects may be represented several times, and/or some of the condition attributes may be superfluous or dependent. Superfluous attributes have no effect on the classification performance of the information system, and dependent attributes lead to redundancy because of a lack of discriminatory power. Reduction of data matrix is thus an important data-preprocessing step.

Rough sets can be used for *relevance analysis*, where the contribution or *significance* of each attribute is assessed with respect to the classification task. The analysis of relevance of each attribute is then used for attribute subset selection or *attribute reduction*, where attributes that do not contribute to the classification task can be identified and removed.

Some additional definitions will help us develop a systematic method for relevance analysis. When all the elements of the universe cannot be uniquely classified to blocks of the partition $U\backslash y$, i.e., to equivalence classes $B^{(1)}, \ldots, B^{(M)}$, employing Z, y depends *partially* on Z. We say y depends on Z to a *degree* $\gamma (0 \leq \gamma < 1)$, denoted $Z \overset{\gamma}{\Rightarrow} y$, where γ expresses the ratio of the elements of the universe which can be classified to blocks of the partition $U\backslash y$ employing variables Z to the total elements in the universe. The concepts, related to ambiguity in the datasets, can be captured in rough set terminology as follows.

Z-Positive Region

The set $\text{POS}_Z(B)$, called the Z-positive region of B, is the set of those samples which can, with certainty, be classified in the set B.

The positive region of a set is equal to its lower approximation:

$$\text{POS}_Z(B) = \underset{\sim}{B} \tag{7.70}$$

If $B^{(1)}, \ldots, B^{(M)}$ are equivalence classes of the universe determined by decision y, then the set $\underset{\sim}{B}^{(1)}(Z) \cup \cdots \cup \underset{\sim}{B}^{(M)}(Z)$ is called the Z-positive region of y and is denoted by $\text{POS}_Z(y)$

Degree of Dependence of *y* on Z

We say that y depends upon Z in a degree $\gamma (0 \leq \gamma < 1)$ if,

$$\gamma(Z, y) = \frac{\text{card POS}_Z(y)}{\text{card } U} \tag{7.71}$$

Significance of Attribute $x_j \in X$

The theory of rough sets introduces the notion of *dependency* between features (attributes) of the decision system: condition attributes and decision attributes. Thanks to that, we can check whether it is necessary to know the values of all features in order to unambiguously describe the object belonging to the set U.

Dependency degree of decision attribute y on the set of condition attributes X is defined as

$$k = \gamma(X, y) = \frac{\text{card POS}_X(y)}{\text{card } U} \tag{7.72}$$

where X is the complete set of condition attributes. It is now possible to define the *significance* of an attribute $x_j \in X$. This is done by calculating the change in dependency after removing the attribute x_j from the set of considered condition attributes.

- We say that x_j is *dispensable* in X if,

$$\gamma(X, y) = \gamma(X - x_j, y) \tag{7.73}$$

  Otherwise, x_j is *indispensable* in X.
- Set X is *independent* if all its attributes are indispensable.
  The *normalized coefficient of significance*

$$\sigma_{X,y}(x_j) = \frac{\gamma(X, y) - \gamma(X - x_j, y)}{\gamma(X, y)} \tag{7.74}$$

- The coefficient of significance plays an important role in relevance analysis of attributes. The zero value obtained for a given attribute x_j indicates that this attribute may be deleted from the set X of condition attributes. The higher the value of $\sigma_{X,y}(x_j)$, which corresponds to higher change $\gamma(X, y) - \gamma(X - x_j, y)$ in dependency, the more significant x_j is.

Example 7.7

Let us reconsider the dataset in Table 7.6. From the previous example, we have,

$$\underset{\sim}{B}^{(1)}(X) = \{1, 3, 6\};\ \underset{\sim}{B}^{(2)}(X) = \{4\}$$

Therefore,

$$\text{POS}_X(y) = \underset{\sim}{B}^{(1)}(X) \cup \underset{\sim}{B}^{(2)}(X) = \{1, 3, 4, 6\}$$

The decision attribute y depends on condition attributes X in a degree

$$\gamma(X, y) = \frac{\text{card POS}_X(y)}{\text{card } U} = \frac{4}{6}$$

$$= \frac{\text{Elements of the universe which can be classified to blocks of partition } U\backslash y}{\text{Total number of elements in the universe}}$$

Let us now determine significance of each attribute $x_j; j = 1, 2, 3$. This is given by the coefficient of significance:

$$\sigma_{X,y}(x_j) = \frac{\gamma(X, y) - \gamma(X - x_j, y)}{\gamma(X, y)}$$

The coefficient of significance for the attribute x_3,

$$\sigma_{X,y}(x_3) = 1 - \frac{\gamma(X - x_3, y)}{\gamma(X, y)} = 1 - \frac{1/6}{4/6} = 0.75$$

We get the following coefficients for x_1 and x_2:

$$\sigma_{X,y}(x_1) = 0; \quad \sigma_{X,y}(x_2) = 0$$

This shows that attribute x_3 is not dispensable having a significance coefficient of 0.75; while attributes x_1 and x_2 can be dispensed with as they do not provide any significant information for the classification.

The variable x_3, thus, has strong significance in the information system presented in Table 7.6.

7.10.3 Reduction of Attributes

Rough set theory defines strong and weak relevance for variables. For a given dataset, a set of strongly relevant variables forms a *core*. A set of variables satisfactory to describe concepts of a given dataset, including a core and possibly some weakly relevant variables, forms *reduct*.

Subset Z of X is a *reduct* of X if Z is independent (i.e., all its attributes are indispensable) and IND(Z) = IND(X).

Thus, a reduct is a set of attributes that preserves partition. It means that a reduct is the subset of attributes that enables the same classification of the elements of the universe as the whole set of attributes. In other words, attributes that do not belong to a reduct are superfluous with regard to classification of elements of the universe.

A reduct is defined as a subset X_R of the condition attributes X, such that

$$\gamma(X_R, y) = \gamma(X, y) \tag{7.75}$$

A given dataset may have many attribute reduct sets; the set of all reducts is given by,

$$R = \{X_R \mid X_R \subseteq X; \gamma(X_R, y) = \gamma(X, y)\} \tag{7.76}$$

The intersection of all the sets in R is the *core*, the elements of which are those attributes which cannot be eliminated without introducing more contradictions to the dataset.

In rough set-based attribute reduction, a reduct with minimum cardinality is searched for; in other words, an attempt is made to search for the element $R_{\min}$ (the reduct with minimal cardinality) from R—the set of all reducts.

$$R_{\min} = \{X_R \mid X_R \in R; \forall\, Y \in R, \text{card } X_R \le \text{card } Y\} \tag{7.77}$$

A basic way of achieving this is to calculate the dependencies of all possible subsets of X. Any subset Z with $\gamma(Z, y) = 1$ is a reduct; the smallest subset (i.e., with minimal cardinality) with this property is a minimal reduct $R_{\min}$.

However, for large datasets this method is impractical and an alternative strategy is required. QUICKREDUCT algorithm attempts to calculate a minimal reduct without exhaustively generating all possible subsets. It starts off with an empty set and adds in turn those attributes that result in the greatest increase in γ until this produces its maximum possible value for the dataset (usually 1). However, it has been proved that this method does not always generate a *minimal* reduct as γ is not perfect heuristic. It does result in a close-to-minimal reduct, which is still useful in greatly reducing dataset dimensionality.

In addition to γ being a non-optimal heuristic, the algorithm also suffers due to the information lost in the discretization procedure. A method using fuzzy-rough sets should therefore be more informed.

Example 7.8

Example dataset given in Table 7.7 has 8 objects ($i = 1, \ldots, 8 \in U$), four features $\{x_1, x_2, x_3, x_4\} \in X$ which are the condition attributes, and one decision attribute y_q; $q = 1, 2, 3$.

Table 7.7 Example dataset

$i \in U$	x_1	x_2	x_3	x_4	y
1	1	0	2	2	0
2	0	1	1	1	2
3	2	0	0	1	1
4	1	1	0	2	2
5	1	0	2	0	1
6	2	2	0	1	1
7	2	1	1	1	2
8	0	1	1	0	1

For the dataset, we can easily make the following computations.

$U/\text{IND}(x_1) = \{\{2, 8\}, \{1, 4, 5\}, \{3, 6, 7\}\}$

$U/\text{IND}(x_2) = \{\{1, 3, 5\}, \{2, 4, 7, 8\}, \{6\}\}$

$U/\text{IND}(x_3) = \{\{3, 4, 6\}, \{2, 7, 8\}, \{1, 5\}\}$

$U/\text{IND}(x_4) = \{\{5, 8\}, \{2, 3, 6, 7\}, \{1, 4\}\}$

$B^{(1)} = \{1\}$; $B^{(2)} = \{3, 5, 6, 8\}$; $B^{(3)} = \{2, 4, 7\}$

None of the elementary sets from family of $U/\text{IND}(x_1)$ is contained in $B^{(1)}$, $B^{(2)}$ or $B^{(3)}$; therefore,

$$\underset{\sim}{B}^{(1)}(x_1) = \underset{\sim}{B}^{(2)}(x_1) = \underset{\sim}{B}^{(3)}(x_1) = 0$$

This gives,

$$\text{POS}_{x_1}(y) = 0; \; \gamma(x_1, y) = 0$$

Only the elementary set {6} from the family of U/IND(x_2) is contained in $B^{(2)}$; therefore $\underset{\sim}{B}^{(2)}(x_2) = \{6\}$. It can be verified that $\underset{\sim}{B}^{(1)}(x_2) = \underset{\sim}{B}^{(3)}(x_2) = 0$.

This gives,

$$\text{POS}_{x_2}(y) = \{6\};\ \gamma(x_2, y) = 1/8$$

For the attribute x_3, we get,

$$\underset{\sim}{B}^{(1)}(x_3) = \underset{\sim}{B}^{(2)}(x_3) = \underset{\sim}{B}^{(3)}(x_3) = 0;\ \text{POS}_{x_3}(y) = 0;\ \gamma(x_3, y) = 0$$

For the attribute x_4,

$$\underset{\sim}{B}^{(1)}(x_4) = 0;\ \underset{\sim}{B}^{(2)}(x_4) = \{5, 8\};\ \underset{\sim}{B}^{(3)}(x_4) = 0;\ \text{POS}_{x_4}(y) = \{5, 8\}$$

$$\gamma(x_4, y) = 2/8.$$

On similar lines, we obtain the following:

$$U/\text{IND}(x_1, x_2) = \{\{1, 5\},\{2, 8\},\{3\},\{4\},\{6\},\{7\}\}$$

$$U/\text{IND}(x_1, x_3) = \{\{2, 8\},\{3, 6\},\{1, 5\},\{4\},\{7\}\}$$

$$U/\text{IND}(x_1, x_4) = \{\{1, 4\},\{3, 6, 7\},\{2\},\{5\},\{8\}\}$$

$$U/\text{IND}(x_2, x_3) = \{\{1, 5\},\{2, 7, 8\},\{3\},\{4\},\{6\}\}$$

$$U/\text{IND}(x_2, x_4) = \{\{1\},\{2, 7\},\{3\},\{4\},\{5\},\{6\},\{8\}\}$$

$$U/\text{IND}(x_3, x_4) = \{\{3, 6\},\{2, 7\},\{1\},\{4\},\{5\},\{8\}\}$$

$$U/\text{IND}(x_1, x_2, x_3) = \{\{2, 8\},\{1, 5\},\{3\},\{4\},\{6\},\{7\}\}$$

$$U/\text{IND}(x_1, x_2, x_4) = \{\{1\},\{2\},\{3\},\{4\},\{5\},\{6\},\{7\},\{8\}\}$$

$$U/\text{IND}(x_2, x_3, x_4) = \{\{1\},\{2, 7\},\{3\},\{4\},\{5\},\{6\},\{8\}\}$$

$$U/\text{IND}(x_1, x_2, x_3, x_4) = \{\{1\},\{2\},\{3\},\{4\},\{5\},\{6\},\{7\},\{8\}\}$$

$$\text{POS}_{x_1, x_2}(y) = \{3, 6, 7, 4\}$$

$$\text{POS}_{x_1, x_3}(y) = \{3, 6, 4, 7\}$$

$$\text{POS}_{x_1, x_4}(y) = \{5, 8, 2\}$$

$$\text{POS}_{x_2, x_3}(y) = \{3, 6, 4\}$$

$$\text{POS}_{x_2, x_4}(y) = \{1, 2, 3, 4, 5, 6, 7, 8\}$$

$$\text{POS}_{x_3, x_4}(y) = \{1, 3, 5, 6, 8, 2, 4, 7\}$$

$$\text{POS}_{x_1, x_2, x_3}(y) = \{3, 6, 7, 4\}$$

$$\text{POS}_{x_1, x_2, x_4}(y) = \{1, 2, 3, 4, 5, 6, 7, 8\}$$

$$\text{POS}_{x_2, x_3, x_4}(y) = \{1, 2, 3, 4, 5, 6, 7, 8\}$$

$$\text{POS}_{x_1, x_2, x_3, x_4}(y) = \{1, 2, 3, 4, 5, 6, 7, 8\}$$

The degree of dependency of decision attribute y on a set of conditional attributes:

$$\gamma(X, y) = \frac{\text{card POS}_X(y)}{\text{card } U}$$

$$\gamma(x_1, x_2; y) = 4/8 = 1/2$$

$$\gamma(x_1, x_3; y) = 4/8 = 1/2$$

$$\gamma(x_1, x_4; y) = 3/8$$

$$\gamma(x_2, x_3; y) = 3/8$$

$$\gamma(x_2, x_4; y) = 8/8 = 1$$

$$\gamma(x_3, x_4; y) = 8/8 = 1$$

$$\gamma(x_1, x_2, x_3; y) = 4/8 = 1/2$$

$$\gamma(x_1, x_2, x_4; y) = 8/8 = 1$$

$$\gamma(x_1, x_3, x_4; y) = 1$$

$$\gamma(x_2, x_3, x_4; y) = 1$$

$$\gamma(x_1, x_2, x_3, x_4; y) = 1$$

Note that $\gamma < 1$ shows that all the objects i cannot be classified into the decision attribute, given condition attributes X. For example, $\gamma(x_2, x_3; y) = 3/8$ shows that out of eight objects, only three can be classified into decision attribute y, given condition attributes x_2 and x_3. The other five objects represent contradictory information.

The most basic solution to the problem of finding a *reduct* is to simply generate *all* possible subsets and retrieve those with high degree of dependency. In our example, the maximum dependency degree is 1, and the set of all the reducts are:

$$R = \{\{x_2, x_4\}, \{x_3, x_4\}, \{x_1, x_2, x_4\}, \{x_1, x_3, x_4\}, \{x_2, x_3, x_4\}, \{x_1, x_2, x_3, x_4\}\}$$

Reducts of minimal cardinality of the condition attribute are $\{x_2, x_4\}$ and $\{x_3, x_4\}$. Therefore,

$$R_{\min} = \{x_2, x_4\}, \{x_3, x_4\}$$

If $\{x_2, x_4\}$ is chosen, then the dataset can be reduced as in Table 7.8. Clearly, each object can be uniquely classified according to remaining attribute values.

$$\begin{aligned} \text{Core} &= \{x_2, x_4\} \cap \{x_3, x_4\} \\ &= \{x_4\} \end{aligned}$$

Table 7.8 Reduced dataset

$i \in U$	x_2	x_4	y
1	0	2	0
2	1	1	2
3	0	1	1
4	1	2	2
5	0	0	1
6	2	1	1
7	1	1	2
8	1	0	1

More on Rough Sets

Concise summaries of rough set theory in data mining are given in [135] and [136]. For use of rough sets in feature selection and expert system design, refer to [135], [137] and [138]. Algorithms to reduce the computational intensity in finding reducts have been proposed in [139].

Chapter

8

DECISION-TREE LEARNING

8.1 INTRODUCTION

A decision tree can be said to be a map of a reasoning process. It uses a structure resembling that of a tree to describe a dataset and solutions can be visualized by following different pathways through the tree.

It is a hierarchical set of rules explaining the way in which a large set of data can be divided into smaller data *partitions*. Each time a split takes place, the components of the resulting partitions become increasingly similar to one another with regard to the target.

If we had to select a classification method capable of performing well across a wide range of situations without the analyst needing to put in effort, and easy for the customer to understand, the tree methodology would be the preferred choice. Several types of decision-tree learning techniques are available with varying needs and abilities. Decision-tree learning is usually best suited to problems with the following features:

- Patterns are described by a fixed set of attributes $x_j; j = 1, 2, \ldots, n$, and each attribute x_j takes on a small number of disjoint possible values (categorical or numeric) v_{lxj}; $l = 1, 2, \ldots, d_j$.
- The output variable y is a Boolean-valued function (binary classification problems) defined over the set S of patterns $\{s^{(i)}\} \equiv \{\mathbf{x}^{(i)}\}$; $i = 1, 2, \ldots, N$. That is, y takes on values y_q ; $q = 1, 2$. If we assume $y_1 \equiv 0$ and $y_2 \equiv 1$, then $y : S \rightarrow [0, 1]$.
- The training data is described by the dataset $\mathcal{D}$ of N patterns with corresponding observed outputs:

$$\mathcal{D} = \{s^{(i)}, y^{(i)}\} = \{\mathbf{x}^{(i)}\ y^{(i)}\}; i = 1, 2, \ldots, N$$

Several practical problems have been observed with these features. Therefore, decision-tree learning can be applied to problems, such as learning to categorize medical patients according to their ailments/conditions, equipment defects according to their faults, loan applications according to their payment failures, and so on.

Extensions of the basic decision-tree learning algorithm permit the handling of continuous-valued attributes, and learning functions possessing more than two likely output values (multiclass classification problems). Extensions also allow learning target functions with continuous-valued outputs (regression problems), though the application of decision trees in this setting is not very common.

J. Ross Quinlan, a machine learning researcher (during the late 1970s and early 1980s), proposed a decision-tree algorithm called ID3 [140]. It is so called because it was the third procedure in a series of 'interactive dichotomizer' processes. Later, Quinlan's C4.5 (ID3's successor), became the benchmark with which newer supervised learning algorithms are often compared [141]. In 1984, L. Breiman, J. Friedman, R. Olshen, and C. Stone, all statisticians, together authored the book *Classification and Regression Trees* (CART) [142]. ID3 and CART were inventions independent of one another, but at almost the same time, they followed a similar model of learning decision trees from training examples.

Thereafter, there was an active spell of work on decision tree induction. Quinlan's C4.5 was succeeded by his C5.0. Its decision-tree induction seems to be essentially the same as that used by C4.5, and tests show some differences but negligible improvements. However, its speed of generating rules is high and it employs a different method.

As already mentioned, a decision tree is a hierarchical set of rules. Rule-based classification systems have a drawback—they include sharp cutoffs for attributes. For instance, consider the following rule for approval of customer credit application:

IF (years_employed ≥ 2) AND (income ≥ 50K) THEN credit = approved

According to the rule, applications of customers who have been employed for a minimum of two years will be eligible for credit if their income is $50,000. They will not be eligible if their income is say, $49,500. Such hard thresholds may appear to be unfair (refer to Chapter 6). To overcome this drawback, various researchers have come up with the *fuzzy decision tree induction algorithms*, bringing in softness in these conditions or thresholds. The evaluation of classification capabilities of fuzzified attributes is done by all fuzzy decision tree induction methods with the help of certain appropriate measure of uncertainty consistent with the human information processing, such as vagueness and ambiguity. It is possible to build fuzzy decision trees by incorporating this measure in *crisp decision tree induction algorithms*.

Virtually all tree-based methods incorporate the fundamental techniques of decision-tree learning. There are strengths and weaknesses of individual methods, of course. Our focus in this chapter is *not* on providing details of many different decision tree algorithms reported in the literature/commercially available [143], but on the basic building blocks.

We will incorporate the following basic characteristics of decision-tree building:

- A decision tree is a *hierarchical model* for supervised learning. An algorithm starts with a learning set of instances (patterns) and their associated class labels. The training set is partitioned into smaller subsets in a sequence of recursive splits as the tree is being built. The tree-building follows a *top-down* hierarchical approach.
- The tree-learning algorithms are said to be *greedy*, because at every stage, beginning at the root with the complete dataset, we search for the *best* split (nonbacktracking).

- For a given training set, there exist many trees which code it with no error, and for simplicity, we are interested in finding the smallest tree where tree size is measured as the number of nodes in the tree. This is achieved through local search procedures based on heuristics that give reasonable trees (accuracy-complexity trade-off) in reasonable time.
- *Divide-and-conquer*, a frequently used heuristic, is the tree-building strategy.

This chapter opens up with an example of a decision tree for a binary classification task. This simple example with categorical attribute values highlights the way decision trees can provide insight into a decision problem, and how easy decision trees are to understand. The chapter continues with more technical detail on how to create decision trees. It ends up with a discussion on fuzzy decision trees.

8.2 EXAMPLE OF A CLASSIFICATION DECISION TREE

The decision tree shown in Fig. 8.1 was created from entirely fictitious tiny weather dataset which supposedly concerns the conditions that are suitable for playing tennis. The sample is shown in Table 8.1. The input variables are: x_1 = *Outlook*, x_2 = *Temperature*, x_3 = *Humidity*, and x_4 = *Wind*; and the target variable y = *PlayTennis*. The task is to predict the value of *PlayTennis* for an arbitrary Saturday morning, based on the values of its attributes.

The target function to be learnt is

$$\hat{y} : S \rightarrow [0, 1]$$

$\hat{y}^{(i)} \equiv \hat{y}(s^{(i)}) = 1$, if *PlayTennis* = *Yes*; $\hat{y}^{(i)} = 0$, if *PlayTennis* = *No*

The set S of patterns has 14 data samples:

$$S = \{s^{(i)}\}; i = 1, \ldots, 14$$

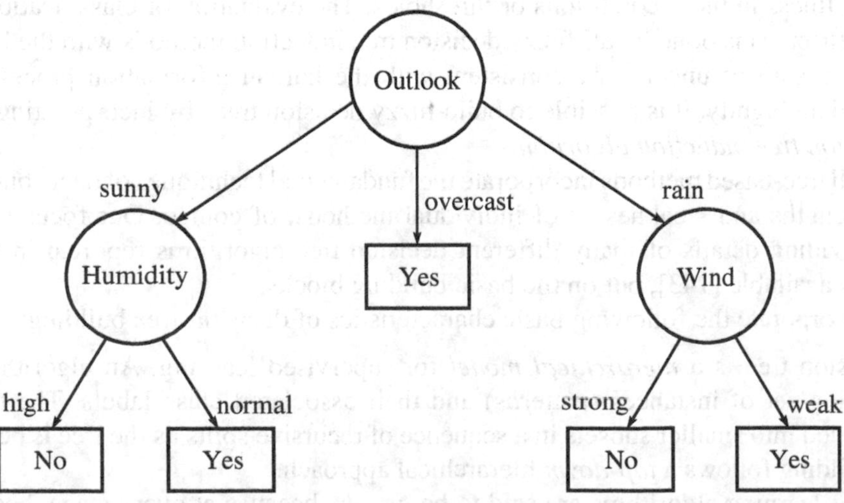

Figure 8.1 Decision tree for weather data

Table 8.1 The weather data

Instance	Outlook x_1	Temperature x_2	Humidity x_3	Wind x_4	PlayTennis y
$s^{(1)}$	sunny	hot	high	weak	No
$s^{(2)}$	sunny	hot	high	strong	No
$s^{(3)}$	overcast	hot	high	weak	Yes
$s^{(4)}$	rain	mild	high	weak	Yes
$s^{(5)}$	rain	cool	normal	weak	Yes
$s^{(6)}$	rain	cool	normal	strong	No
$s^{(7)}$	overcast	cool	normal	strong	Yes
$s^{(8)}$	sunny	mild	high	weak	No
$s^{(9)}$	sunny	cool	normal	weak	Yes
$s^{(10)}$	rain	mild	normal	weak	Yes
$s^{(11)}$	sunny	mild	normal	strong	Yes
$s^{(12)}$	overcast	mild	high	strong	Yes
$s^{(13)}$	overcast	hot	normal	weak	Yes
$s^{(14)}$	rain	mild	high	strong	No

The circle at the top of the diagram in Fig. 8.1 is the *root node*, which contains all the training data used to grow the tree. The tree begins from the root node and grows downwards by dividing the data at each level into new *daughter nodes*. The root or *parent node* comprises the whole data and the *internal nodes* or daughter nodes carry the respective data subsets. All nodes are connected by *branches* shown by the line segments in the figure. The nodes that are at the end of the branches are called *terminal nodes* or *leaf nodes*, shown by boxes in the figure. The leaf nodes in this figure are *class labels.*

The classification of instances by decision trees is done by sorting them down the tree from the root to some leaf node, which offers the classification of the instance. Each node in the tree stipulates a test of some *attribute* of the instance, and each branch descending from that node tallies with one of the possible values of this attribute. Classification of an instance begins at the root node of the tree, where the specified attribute is tested and then it moves down the tree branch corresponding to the value of the attribute in the example stated. This procedure is iterated for the subtree rooted at the new node.

Consider the instance (from the training set): (*Outlook* = *sunny*, *Temperature* = *hot*, *Humidity* = *high*, *Wind* = *strong*). This instance would be sorted down the left-most portion of the decision tree; the tree predicts that

$$PlayTennis = No$$

The tests on attributes *Outlook* and *Humidity* have determined the class.

Consider another example (*not* from the training set): (*Outlook* = *overcast*, *Temperature* = *cool*, *Humidity* = *normal*, *Wind* = *weak*). The tree predicts that

$$PlayTennis = Yes$$

Here, only the test on *Outlook* attribute has determined the class.

If there are sufficient attributes, a decision tree can be constructed that will accurately classify each object in the training set, and generally, there are several such correct decision trees for a learning task. *Induction* is all about moving beyond the training set, that is, constructing a decision tree that accurately categorizes not just the patterns from the training set but from other unseen patterns also. To make this happen, the decision tree has to capture certain meaningful relationships between the class of the object and the values of its attributes. Suppose, there is a set of examples of *y*—the task of induction is to return a hypothesis (decision tree) that approximates *y*. Learning is tough, conceptually speaking, as it is difficult to decide whether a specific hypothesis is a 'good' approximation of *y*. A hypothesis is good if it is able to predict unseen examples accurately, that is, is able to generalize well.

If one is asked to select between two hypotheses (decision trees), both of which are correct over the training set, it makes sense to opt for the simpler one because the likelihood of its capturing structure inherent in the problem is more. Therefore, it is expected that the simpler tree will accurately classify more objects outside the training set. Seeking the most simple hypothesis (smallest decision tree) that generalizes well is the goal, but it is an interactable problem. We can do a good job with some heuristic effort.

Figure 8.1 shows a decision tree with each branch growing just deep enough to correctly classify the training examples. This is quite a reasonable strategy, but may not work always. At times, it may result in difficulties if there is noise in the data or if the number of training examples is too small to give rise to a representative sample of the tree target function. Whatever the case may be, this strategy can generate trees that *overfit* the training examples. A decision tree (hypothesis) is said to overfit the training examples if there is another decision tree, which does not fit the training examples that well, but actually ends up performing better over the entire distribution of instances (i.e., including instances beyond the training set).

The practical difficulty of 'overfitting' is of significance in decision tree learning (similar to other learning techniques). We will talk about the various approaches for avoiding overfitting in decision-tree learning later.

The fundamental idea of the decision-tree learning algorithm is to test the most significant attribute first. The most important attribute is the one that impacts the classification of an example the most. This will ensure that we obtain the accurate classification with a small number of tests—wherein all paths in the tree are short and the tree, in general, is a small one.

The most important attribute is placed at the root node. Then, we create a branch for each likely value of this attribute. This *divides* the dataset into smaller subsets—one subset for each value of the attribute. The procedure can now be iterated *recursively* for each branch or daughter node, with the help of only those examples that actually reach the branch. At any point of time, if all instances at a node end up with the same classification, it indicates that the leaf node has been reached and it is time to stop developing that portion of the branch. Therefore, tree construction is a *divide-and-conquer* procedure.

The only thing left is how to determine which attribute to split on, given a set of examples with different classes. Referring to Table 8.1, there are four candidates for the most important attribute (root node): *Outlook*, *Temperature*, *Humidity*, and *Wind*. These four attributes produce the trees at the top level as shown in Fig. 8.2. Which is the best choice?

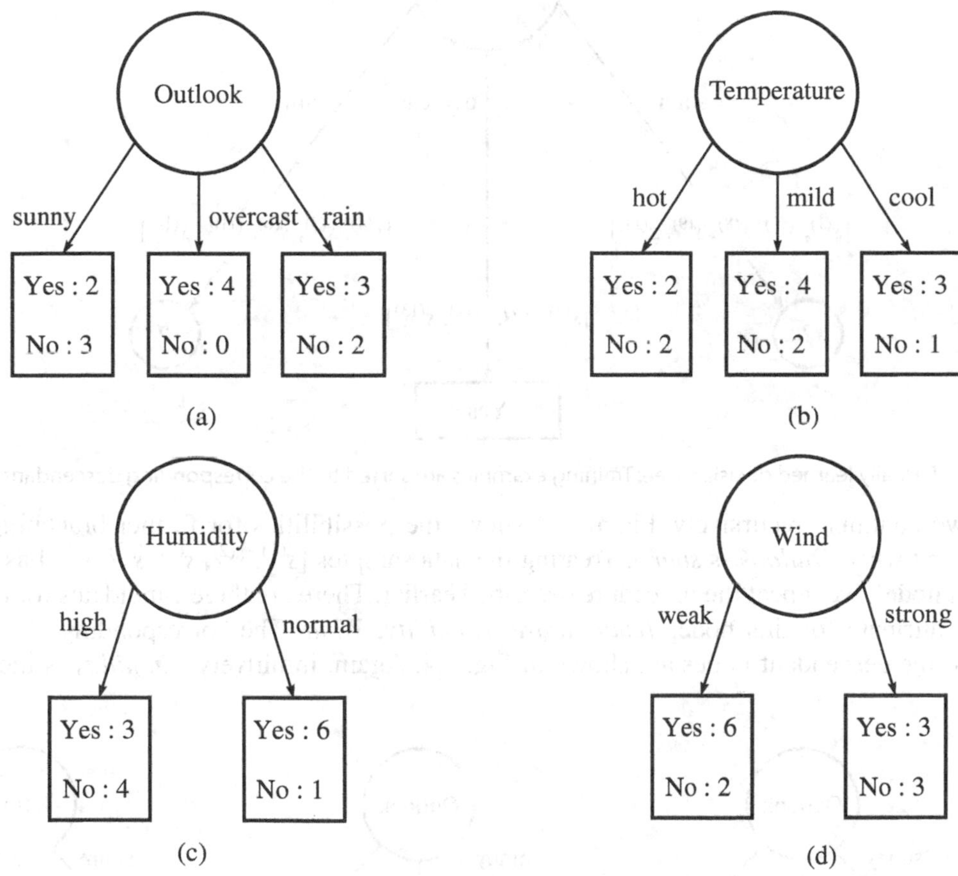

Figure 8.2 Tree stumps for the weather data

Because we seek small trees, we would like the leaf nodes (all instances at the node having the same classification, *Yes* or *No*) to come up as soon as possible. Therefore, we would choose the attribute whose daughter nodes achieve this objective as close as possible (so that further splits are reduced). When evaluating the attribute *Outlook*, we see that the number of *Yes* and *No* classes in daughter nodes are [2, 3], [4, 0], and [3, 2], respectively. For *Temperature*, the numbers are [2, 2], [4, 2], and [3, 1]. For *Humidity*, we have [3, 4] and [6, 1]; and for *Wind*, [6, 2] and [3, 3]. Therefore, we select *Outlook* as the splitting attribute at the root of the tree. Intuitively, this is the best one to use because it is the only choice for which one daughter node becomes a leaf node. This gives it a considerable advantage over other choices. The next best choice is *Humidity* because it produces a daughter node that has large number of instances of only one classification. Figure 8.3 gives the data distribution at the root node and daughter nodes when *Outlook* is selected as the root node.

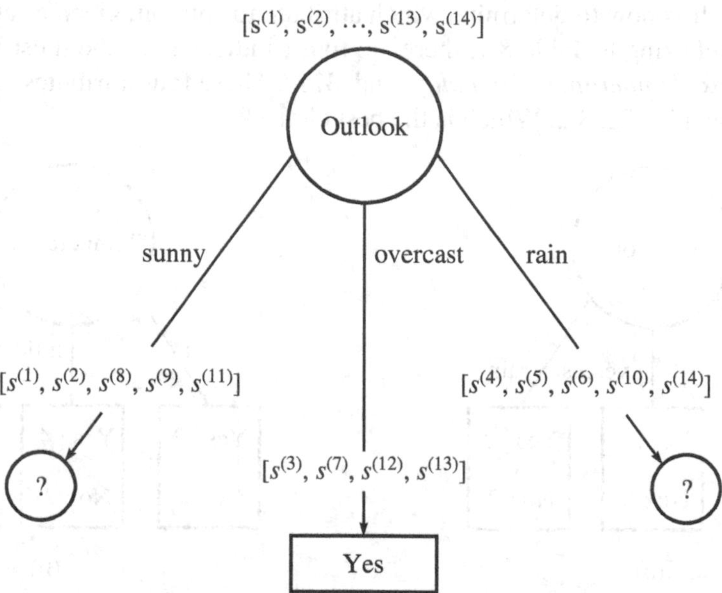

Figure 8.3 Partially learned decision tree: Training examples are sorted to the corresponding descendant nodes

Then we continue, recursively. Figure 8.4 shows the possibilities for further branching at the node reached when *Outlook* is *sunny*. Treating the data samples $[s^{(1)}, s^{(2)}, s^{(8)}, s^{(9)}, s^{(11)}]$ as if these were 'root node', we repeat the procedure described earlier. There are three candidates for the most important attribute for this node: *Temperature*, *Humidity*, *Wind*. The corresponding *Yes* and *No* classes for the descendant nodes are shown in Fig. 8.4. Again, intuitively, *Humidity* is the best to select.

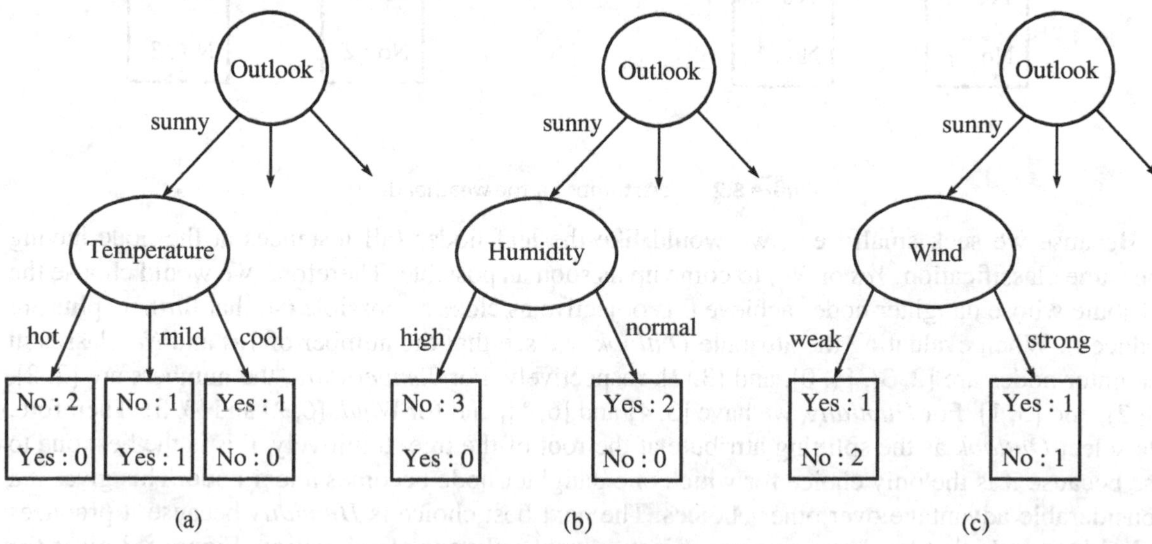

Figure 8.4 Expanded tree stumps for the weather data

Continued application of the same idea leads to the decision tree of Fig. 8.1 for the weather data.

The scheme with the help of which attributes are selected in decision tree learning is designed to reduce the depth of the final tree. The objective is to choose the attribute that provides as far as possible the precise classification of the examples. A perfect attribute splits the examples into sets that are either all *Yes* or all *No*. An attribute will be of no use if the example sets are left with somewhat the same proportion of *Yes* and *No* examples as the original set.

Then, all that is required is a formal measure of 'fairly good' and 'really useless.' A measure with its maximum value when the attribute is perfect, and its minimum value when the attribute is useless, is a candidate for our choice. One suitable measure is the expected *impurity reduction* provided by the attribute.

The process of growing decision trees repeatedly *splits* the data into smaller and smaller groups in such a way that each new set of daughter nodes has greater *purity* than its ancestors with respect to the target variable. A pure node has maximum homogeneity (e.g., leaves with all *Yes* or *No*). An impure node has heterogeneity, which is maximum when *Yes* and *No* are in equal numbers. The root node in Fig. 8.1 is an impure node having nine data samples with *Yes* as output value and five data samples with *No* as output.

The measure to evaluate a potential split is *impurity* of the target variable in the daughter nodes. High impurity means that the distribution of the target in the daughter nodes is similar to that of parent node, whereas low impurity means that members of a single class predominate. The best split is the one that decreases impurity in the daughter nodes by the greatest amount.

The next section is devoted to measures of impurity for evaluating splits in decision tree.

8.3 MEASURES OF IMPURITY FOR EVALUATING SPLITS IN DECISION TREES

An impurity measure is a heuristic for selection of the splitting criterion that best separates a given dataset $\mathcal{D}$ of class-labeled training tuples into individual classes. If we divide $\mathcal{D}$ into smaller partitions as per the outcome of the splitting criterion, each partition should ideally be *pure*, with all the tuples falling into each partition belonging to the same class. We choose the criterion that is closest in terms of outcome to this ideal scenario.

Evaluation of potential splits may be done with the help of various criteria. Alternate splitting criteria may result in trees appearing very different from one another, but having similar performance. Diverse impurity measures choose different splits, but since all the measures attempt to seize the same idea, the resulting models end up behaving in the similar manner. In this section, we explain three popular impurity measures—*information gain, gain ratio*, and *Gini index*.

8.3.1 Information Gain/Entropy reduction

Information gain uses a clever idea for defining impurity, borrowed from the world of information theory (revisiting Section 3.9 will be helpful). If a leaf is entirely pure, then the classes in the leaf can be described very easily—there is only one. On the other hand, if a leaf is highly impure, then describing the classes in the leaf is much more complicated. Information theory has a measure for this, called *entropy*, which measures how disorganized a system is.

The root node holds the entire dataset $\mathcal{D}$ which describes the patterns $s^{(1)}, s^{(2)}, \ldots, s^{(N)}$ with their corresponding classes y_1 or y_2 (for a binary classification task). Imagine selecting a pattern at random from the dataset $\mathcal{D}$ and announcing that it belongs to class y_q. This message has the probability

$$P_q = \frac{freq\,(y_q, \mathcal{D})}{|\mathcal{D}|} \tag{8.1}$$

where $freq\,(y_q, \mathcal{D})$ stands for the number of patterns in $\mathcal{D}$ that belong to class y_q and $|\mathcal{D}|$ denotes the total number of patterns in $\mathcal{D}$ ($|\mathcal{D}| = N$).

The expected *information* needed to classify a pattern in $\mathcal{D}$ is given by

$$Info\,(\mathcal{D}) = -\sum_{q=1}^{2} P_q \log_2(P_q) \tag{8.2}$$

A log function to the base 2 is used because information is encoded in *bits* (refer to Section 3.9). *Info* ($\mathcal{D}$) is just the average amount of information needed to identify the class label of a pattern in $\mathcal{D}$. Note that at this point, the information we have is solely based on the proportions of patterns in each class. *Info* ($\mathcal{D}$) can also be expressed as *entropy* of $\mathcal{D}$, denoted as *Entropy* ($\mathcal{D}$).

$$Entropy\,(\mathcal{D}) = -\sum_{q=1}^{2} P_q \log_2 P_q \tag{8.3}$$

Associated with root node of the decision tree, *Info*($\mathcal{D}$) represents the expected amount of information that would be needed to specify whether a new instance should be classified as y_1 or y_2, given that the example reached the node. *Info*($\mathcal{D}$) is 0 if all patterns in $\mathcal{D}$ belong to the same class ($P_1 = 0, P_2 = 1$): $-P_1 \log_2 P_1 - P_2 \log_2 P_2 = 0$ (note that $0\log_2 0 = 0$). *Info* ($\mathcal{D}$) is 1 when the collection $\mathcal{D}$ contains an equal number of Class 1 and Class 2 patterns $\left(P_1 = \frac{1}{2}, P_2 = \frac{1}{2}\right)$ representing maximum heterogeneity (randomness) in the dataset: $-P_1 \log_2 P_1 - P_2 \log_2 P_2 = 1$. If the collection $\mathcal{D}$ contains unequal number of Class 1 and Class 2 patterns, *Info*($\mathcal{D}$) is between 0 and 1. It is, thus, a measure of *impurity* of the collection of examples. More the impurity (more the heterogeneity in the dataset), more the entropy, more the expected amount of information that would be needed to classify a new pattern; more the *purity* (more the homogeneity in the dataset), less the entropy, less the expected amount of information that would be needed to classify a new pattern.

To illustrate, we consider training set of Table 8.1 (Weather Data). It has nine examples of class *Yes*, and five examples of class *No*. Therefore,

$$Info(\mathcal{D}) = Entropy(\mathcal{D}) = -\tfrac{9}{14}\log_2 \tfrac{9}{14} - \tfrac{5}{14}\log_2 \tfrac{5}{14}$$

$$= 0.94 \text{ bits}$$

Root node with dataset $\mathcal{D}$ will therefore be a highly impure node.

The training set $\mathcal{D}$ contains instances that belong to a mixture of classes (high entropy). In this situation, the idea of 'divide-and-conquer' strategy is to divide $\mathcal{D}$ into subsets of instances that are, or seem to be, heading towards single-class collection of instances.

Suppose we select attribute x_j for the root node. x_j has distinct values v_{lx_j}; $l = 1, \ldots, d_j$, as observed from the training data $\mathcal{D}$. Attribute x_j can be used to split data $\mathcal{D}$ into l; $l = 1, \ldots, d_j$, partitions or subsets $\{\mathcal{D}_1, \mathcal{D}_2, \ldots, \mathcal{D}_{d_j}\}$, where $\mathcal{D}_l$ contains those patterns in $\mathcal{D}$ that have values v_{lx_j} of x_j. These partitions would correspond to branches grown from the node. Ideally, we would like this partitioning to produce an exact classification, i.e., we would like each partition to be pure. However, it is quite likely that partitions will be impure (i.e., a partition may contain a collection of patterns from different classes rather than a single class). How much more information we still need (after the partitioning) in order to arrive at an exact classification? This amount is measured by

$$Info(\mathcal{D}, x_j) = \sum_{l=1}^{d_j} \frac{|\mathcal{D}_l|}{|\mathcal{D}|} \times Info(\mathcal{D}_l) \tag{8.4}$$

The term $|\mathcal{D}_l|/|\mathcal{D}|$ acts as the *weight* of lth partition.

$Info(\mathcal{D}_l)$ is given by

$$Info\ (\mathcal{D}_l) = -\sum_{q=1}^{2} P_{ql} \log_2 P_{ql} \tag{8.5}$$

where P_{ql} is the probability that the arbitrary sample in subset $\mathcal{D}_l$ belongs to class y_q, and is estimated as

$$P_{ql} = \frac{freq\ (y_q, \mathcal{D}_l)}{|\mathcal{D}_l|} \tag{8.6}$$

$Info(\mathcal{D}, x_j)$ is expected information required to classify a pattern from $\mathcal{D}$ based on the partitioning by x_j. The smaller the expected information (still) required, the greater the purity of the patterns.

The basic idea is to pick the attribute x_j; $j = 1, \ldots, n$, that goes as far as possible toward providing exact classification of the patterns. A fairly good attribute divides the data into subsets with each subset $\mathcal{D}_l$ containing large number of examples belonging to the same class y_q (low entropy). A really useless attribute leaves the subsets of data with roughly the same proportion of class examples as in the original dataset (high entropy).

Reconsider the dataset of Table 8.1: x_1 = *Outlook*, x_2 = *Temperature*, x_3 = *Humidity*, x_4 = *Wind*; v_{1x_1}= sunny, v_{2x_1}= *overcast*, v_{3x_1}= *rain*; v_{1x_2}= *hot*, v_{2x_2}= *mild*, v_{3x_2}= *cool*; v_{1x_3}= *high*, v_{2x_3}= *normal*; v_{1x_4}= *weak*, v_{2x_4}= *strong*. For these four choices of attribute at the root node, the tree stumps for the weather data are shown in Fig. 8.2.

Consider the attribute x_1 = *Outlook* (Tree stump of Fig. 8.2a). Refer to Figs (8.3) and (8.2a). Five patterns belong to the value *sunny*; out of which two belong to *Yes* class and three belong to *No*. Four patterns belong to the value *overcast*; all of them belong to *Yes* class. Five patterns belong to the value *rain*; out of which three belong to *Yes* class and two belong to *No*.

$$Info(\mathcal{D}, x_1) = \sum_{l=1}^{d_1} \frac{|\mathcal{D}_l|}{|\mathcal{D}|} \times Info(\mathcal{D}_l);\ d_1 = 3$$

$$Info(\mathcal{D}_l) = -\sum_{q=1}^{2} P_{ql} \log_2 P_{ql};\ P_{ql} = \frac{freq(y_q, \mathcal{D}_l)}{|\mathcal{D}_l|}$$

Therefore,

$$Info(\mathcal{D}_1) = -\tfrac{2}{5}\log_2 \tfrac{2}{5} - \tfrac{3}{5}\log_2 \tfrac{3}{5} = 0.97$$

$$Info(\mathcal{D}_2) = 0$$

$$Info(\mathcal{D}_3) = -\tfrac{3}{5}\log_2 \tfrac{3}{5} - \tfrac{2}{5}\log_2 \tfrac{2}{5} = 0.97$$

$$Info(\mathcal{D}, x_1) = \tfrac{5}{14} \times 0.97 - \tfrac{5}{14} \times 0.97 = 0.693$$

Similarly, for other tree stumps of Fig. 8.2, we obtain

$$Info(\mathcal{D}, x_2) = 0.911\ (\textit{Temperature})$$

$$Info(\mathcal{D}, x_3) = 0.788\ (\textit{Humidity})$$

$$Info(\mathcal{D}, x_4) = 0.892\ (\textit{Wind})$$

We see that expected required information for classification is the least if we select attribute *Outlook* for the root node. *Humidity* is the next best choice.

Information gain is defined as the difference between the original information requirement (i.e., based on the partition of classes in the entire dataset $\mathcal{D}$) and the new requirement (i.e., obtained after partitioning on x_j). That is,

$$Gain(\mathcal{D}, x_j) = Info(\mathcal{D}) - Info(\mathcal{D}, x_j) \tag{8.7}$$

In other words, $Gain(\mathcal{D}, x_j)$ tells us how much would be gained by branching on x_j. It is the expected reduction in information requirement (expected reduction in entropy) by partitioning on x_j. The attribute x_j with the highest information gain, $Gain(\mathcal{D}, x_j)$, is chosen as the splitting attribute at the root node. This is equivalent to saying that we want to partition on the attribute x_j that would do the best classification so that the amount of information still required (i.e., $Info(\mathcal{D}, x_j)$) to finish classification task is minimal.

We have selected x_1 = *Outlook* as the splitting attribute at the root node, for which

$$Gain(\mathcal{D}, x_1) = 0.94 - 0.693 = 0.247$$

Gains for other attributes are:

$$Gain(\mathcal{D}, x_2) = 0.029$$

$$Gain(\mathcal{D}, x_3) = 0.152$$

$$Gain(\mathcal{D}, x_4) = 0.048$$

Obviously, *Outlook* provides the maximum gain.

The same strategy is applied recursively to each subset of training instances. Figure 8.4 shows the possibilities for a further branching at the node reached when *Outlook* is *sunny*.

The information gain for the three attributes at the daughter nodes are:

$$Gain\ (Temperature) = 0.571$$
$$Gain\ (Humidity) = 0.971$$
$$Gain\ (Wind) = 0.020$$

Therefore, we select *Humidity* as the splitting attribute at this point. There is no need to split these nodes any further; so this branch has reached the leaf nodes. Continued application of the same idea leads to the decision tree of Fig. 8.1 for the weather data.

Note that information gain, *Gain* $(\mathcal{D}, x_j)$, measures the *expected reduction in entropy*, caused by partitioning the patterns in dataset $\mathcal{D}$ according to the attribute x_j.

$$Gain\ (\mathcal{D}, x_j) = Entropy(\mathcal{D}) - Entropy\ (\mathcal{D}, x_j) \tag{8.8a}$$

$$= Entropy(\mathcal{D}) - \sum_{l=1}^{d_j} \frac{|\mathcal{D}_l|}{|\mathcal{D}|} \times Entropy(\mathcal{D}_l) \tag{8.8b}$$

where

$$Entropy\ (\mathcal{D}) = -\sum_{q=1}^{2} P_q \log_2 P_q ; P_q = \frac{freq\ (y_q, \mathcal{D})}{|\mathcal{D}|} \tag{8.8c}$$

and

$$Entropy\ (\mathcal{D}_l) = -\sum_{q=1}^{2} P_{ql} \log_2 P_{ql} ; P_{ql} = \frac{freq\ (y_q, \mathcal{D}_l)}{|\mathcal{D}_l|} \tag{8.8d}$$

When the output in dataset $\mathcal{D}$ belongs to M distinct classes, i.e., y takes on values y_q; $q = 1, 2, \ldots, M$, the defining equations for *Gain/EntropyReduction* become:

$$Gain(\mathcal{D}, x_j) = Info(\mathcal{D}) - \sum_{l=1}^{d_j} \frac{|\mathcal{D}_l|}{|\mathcal{D}|} \times Info(\mathcal{D}_l) \tag{8.9a}$$

where

$$Info(\mathcal{D}) = -\sum_{q=1}^{M} P_q \log_2 P_q ; P_q = \frac{freq(y_q, \mathcal{D})}{|\mathcal{D}|} \tag{8.9b}$$

and

$$Info(\mathcal{D}_l) = -\sum_{q=1}^{M} P_{ql} \log_2 P_{ql} ; P_{ql} = \frac{freq(y_q, \mathcal{D}_l)}{|\mathcal{D}_l|} \tag{8.9c}$$

The first term in Eqn (8.8a) is just the entropy of the original dataset $\mathcal{D}$, which represents the level of randomness in the dataset with respect to target variable. The second term in Eqn (8.8a) is the expected value of entropy after $\mathcal{D}$ is partitioned using attribute x_j. The expected entropy described by this term is simply the sum of the entropies of each subset $\mathcal{D}_l$, weighted by fraction of patterns $|\mathcal{D}_l|/|\mathcal{D}|$, that belong to $\mathcal{D}_l$ (Eqn (8.8b)). *Gain* $(\mathcal{D}, x_j)$ is, therefore, the expected reduction in entropy caused by the partitioning.

$$Entropy\ Reduction\ (\mathcal{D}, x_j) = Entropy(\mathcal{D}) - \sum_{l=1}^{d_j} \frac{|\mathcal{D}_l|}{|\mathcal{D}|} \times Entropy(\mathcal{D}_l) \tag{8.10a}$$

where

$$Entropy(\mathcal{D}) = -\sum_{q=1}^{M} P_q \log_2 P_q; P_q = \frac{freq\,(y_q, \mathcal{D})}{|\mathcal{D}|} \tag{8.10b}$$

and

$$Entropy(\mathcal{D}_l) = -\sum_{q=1}^{M} P_{ql} \log_2 P_{ql}; P_{ql} = \frac{freq(y_q, \mathcal{D}_l)}{|\mathcal{D}_l|} \tag{8.10c}$$

8.3.2 Gain Ratio

For ID3, a decision-tree tool developed by Quinlan, the selection of partitioning was made on the basis of the *information gain/entropy reduction*. It gave quite good results, and became part of several commercial data mining software packages. It, however, ran into trouble for applications having some attributes x_j with large number of possible values v_{lx_j}; $l = 1, ..., d_j$; giving rise to multiway splitting with many daughter nodes. Just by breaking the larger dataset into large number of small subsets, the number of classes represented in each node tends to go down; so each daughter node increases in purity. The information gain criterion has, thus, a serious deficiency—it has a strong bias in favor of attributes with large number of values. The attribute with large number of values will get selected at root itself and may lead to all leaf nodes, resulting in a too simple hypothesis model unable to capture the structure of the data.

C4.5, a successor of ID3, uses an extension of *information gain*, known as *gain ratio*, which attempts to overcome this bias. It applies a kind of normalization to information gain using a 'split information' value defined analogously with $Info(\mathcal{D}, x_j)$ as

$$SplitInfo(\mathcal{D}, x_j) = -\sum_{l=1}^{d_j} \frac{|\mathcal{D}_l|}{|\mathcal{D}|} \log_2 \frac{|\mathcal{D}_l|}{|\mathcal{D}|} \tag{8.11}$$

This value is representative of the potential information derived from the division of the dataset, $\mathcal{D}$, into d_j partitions matching with the d_j values of the attribute x_j. For each value of x_j, the number of tuples possessing that value is considered with respect to the total number of tuples in $\mathcal{D}$. This is different from information gain, which measures the information with respect to classification obtained on the basis of same partitioning.

The *gain ratio* is defined as

$$GainRatio(\mathcal{D}, x_j) = \frac{Gain\,(\mathcal{D}, x_j)}{SplitInfo(\mathcal{D}, x_j)} \tag{8.12}$$

The attribute with the maximum gain ratio is selected as the splitting attribute. Note that *SplitInfo* term discourages the selection of attributes with many uniformly distributed values d_j. *SplitInfo* is high when d_j is large.

One practical issue that arises in using *GainRatio* in place of *Gain* to select attributes is that the denominator in Eqn (8.12) can be zero or very small when $|\mathcal{D}_l| \simeq |\mathcal{D}|$ for one of the $\mathcal{D}_l$. This either makes the *GainRatio* undefined or very large for such attributes. A standard *fix* is to choose the attribute that maximizes the gain ratio, provided that the information gain for that attribute is at least as great as the average information gain for all the attributes examined. That is, a constraint is added whereby the information gain of the attribute selected must be large.

Returning to the tree stumps of the weather data in Fig. 8.2, $x_1 = Outlook$ splits the dataset into three subsets of size 5, 4, and 5 (refer to Fig. 8.3); and thus the *SplitInfo* is given by

$$\begin{aligned} SplitInfo(\mathcal{D}, x_1) &= -\sum_{l=1}^{3} \frac{|\mathcal{D}_l|}{|\mathcal{D}|} \log_2 \frac{|\mathcal{D}_l|}{|\mathcal{D}|} \\ &= -\tfrac{5}{14}\log_2 \tfrac{5}{14} - \tfrac{4}{14}\log_2 \tfrac{4}{14} - \tfrac{5}{14}\log_2 \tfrac{5}{14} \\ &= 1.577 \end{aligned}$$

without paying any attention to the classes involved in the subsets. We can normalize the *information gain* by dividing by the *split info* value to get the *gain ratio*:

$$\begin{aligned} GainRatio(\mathcal{D}, x_1) &= \frac{Gain(\mathcal{D}, x_1)}{SplitInfo(\mathcal{D}, x_1)} \\ &= \frac{0.247}{1.577} = 0.156 \end{aligned}$$

The results of these calculations for the tree stumps of Fig. 8.2 are summarized as follows:

Outlook : $Gain(\mathcal{D}, x_1) = 0.247$, $SplitInfo(\mathcal{D}, x_1) = 1.577$, $GainRatio(\mathcal{D}, x_1) = 0.156$

Temperature : $Gain(\mathcal{D}, x_2) = 0.029$, $SplitInfo(\mathcal{D}, x_2) = 1.362$, $GainRatio(\mathcal{D}, x_2) = 0.019$

Humidity : $Gain(\mathcal{D}, x_3) = 0.152$, $SplitInfo(\mathcal{D}, x_3) = 1.000$, $GainRatio(\mathcal{D}, x_3) = 0.152$

Wind : $Gain(\mathcal{D}, x_4) = 0.048$, $SplitInfo(\mathcal{D}, x_4) = 0.985$, $GainRatio(\mathcal{D}, x_4) = 0.049$

Outlook still comes out on top, but *Humidity* now is a much closer contender because it splits the data into two subsets instead of three.

8.3.3 Gini Index

Another popular splitting criterion is named Gini, after the 20th century Italian statistician and economist Corrado Gini. Gini index is used in CART.

$$Gini(\mathcal{D}) = 1 - \sum_{q=1}^{M} P_q^2 \tag{8.13}$$

where P_q is the probability that a tuple in $\mathcal{D}$ belongs to class y_q, and is estimated by

$$P_q = \frac{freq(y_q, \mathcal{D})}{|\mathcal{D}|}$$

Gini index considers a binary split for each attribute. Let us first consider the case where x_j is continuous-valued attribute having d_j distinct values v_{lx_j}; $l = 1, 2, \ldots, d_j$. It is common to take mid-point between each pair of (sorted) adjacent values as a possible split-point (It is a simple policy, although something might be gained by adopting a more sophisticated policy. One such policy will be discussed in the next section). The point giving the minimum Gini index for the attribute x_j is taken as its split-point.

For a possible split-point of x_j, $\mathcal{D}_1$ is the number of tuples in $\mathcal{D}$ satisfying $x_j \le$ *split-point*, and $\mathcal{D}_2$ is the set of tuples satisfying $x_j >$ *split-point*. The reduction in impurity that would be incurred by a binary split on x_j is

$$\Delta Gini(x_j) = Gini(\mathcal{D}) - Gini(\mathcal{D}, x_j) \tag{8.14a}$$

$$Gini(\mathcal{D}, x_j) = \frac{|\mathcal{D}_1|}{|\mathcal{D}|} Gini(\mathcal{D}_1) + \frac{|\mathcal{D}_2|}{|\mathcal{D}|} Gini(\mathcal{D}_2) \tag{8.14b}$$

The attribute that maximizes the reduction in impurity (or equivalently, has the minimum Gini index) is selected as the splitting attribute. Then one of these two parts ($\mathcal{D}_1$, $\mathcal{D}_2$) is divided in a similar manner by choosing a variable again and a split value for the variable. This process is continued till we get *pure* leaf nodes (refer to Example 8.3).

Let us now consider the case where x_j is a categorical attribute, for example, categorical variable *Outlook* with categories {*sunny, overcast, rain*} occurring in $\mathcal{D}$ of Table 8.1. To determine the best binary split on *Outlook*, we examine all possible subsets that can be formed using categories of *Outlook*: {*sunny, overcast, rain*}, {*sunny, overcast*},{*sunny, rain*}, {*overcast, rain*}, {*sunny*}, {*overcast*}, {*rain*}, and {}. We exclude the powerset {*sunny, overcast, rain*}, and the empty set {} from consideration since, conceptually, they do not represent a split. Therefore, there are $2^{\gamma} - 2$ possible ways to form two partitions of dataset $\mathcal{D}$, based on the binary splits on x_j having γ categorical values. Each of the possible binary splits is considered; the subset that gives the minimum Gini index for attribute x_j is selected as its splitting subset.

8.4 ID3, C4.5, AND CART DECISION TREES

Virtually all tree-based classification techniques incorporate the fundamental techniques described in earlier sections. Our discussion so far was based on the core ideas of implementation of ID3 decision tree. ID3 is intended for use with nominal (categorical) inputs only. If the problem involves real-valued variables, they are first binned into intervals, each interval being treated as an unordered nominal attribute. Every split has a branching value v_{lx_j}, $l = 1, 2, \ldots$, where v_{lx_j} are discrete attribute bins of the variable x_j chosen for splitting. In practice, these are seldom binary and thus a *GainRatio* impurity should be used.

The C4.5 algorithm, a successor and refinement of ID3, is the most popular in tree-based classification methods. In C4.5, multiway splits for categorical variable are treated the same way as in ID3. Continuous-valued attributes have been incorporated by dynamically defining new discrete-valued attributes that partition the continuous-attribute values into a binary set of intervals. In particular, a new attribute z_j is created that is *true* if $z_j < v_{z_j}^{\text{th}}$ and *false* otherwise. The only question is how to select the best value for the threshold $v_{z_j}^{\text{th}}$.

It has been found that an entropy impurity works acceptably in most cases and is a natural default. However, CART, which uses Gini index as impurity measure, provides a general framework that can be instantiated in various ways to produce different classification (and regression) trees.

The CART approach restricts the splits to binary values for both categorical and continuous-valued attributes. Thus, CART is a *binary tree*.

In the following, we consider an example of implementation for each of the three decision trees.

Example 8.1

The ID3 Decision Tree

In the collection of data $\mathcal{D}$ of Table 8.2, there are 12 instances. The attributes x_1, x_2, x_3, and x_4 have two/three distinct unordered numeric values; these attributes will be treated as nominal attributes. The variable y gives the class information.

To select an attribute as the root of the decision tree, we need to compute the information gain of each of the four attributes x_j; $j = 1, 2, 3, 4$.

$$Info(\mathcal{D}) = Entropy(\mathcal{D}) = -\sum_{q=1}^{2} P_q \log_2 P_q;\ P_q = \frac{freq(y_q, \mathcal{D})}{|\mathcal{D}|}$$

The dataset $\mathcal{D}$ of Table 8.2 has nine examples of class y_1 and three examples of class y_2. Therefore,

$$Info(\mathcal{D}) = Entropy(\mathcal{D}) = -\tfrac{9}{12}\log_2 \tfrac{9}{12} - \tfrac{3}{12}\log_2 \tfrac{3}{12}$$

Attribute x_1 has two nominal values: $v_{1x_1} = 1$, and $v_{2x_1} = 2$. x_1 can be used to split data $\mathcal{D}$ into two partitions or subsets $\{\mathcal{D}_1, \mathcal{D}_2\}$ where $\mathcal{D}_l$; $l = 1, 2$, contains those examples in $\mathcal{D}$ that have values v_{lx_1} of x_1. These partitions would correspond to branches grown from the root node.

Table 8.2 Toy data with numerical values of attributes

$s^{(i)}$	x_1	x_2	x_3	x_4	y
1	1	2	2	1	1
2	1	2	3	2	1
3	1	2	2	3	1
4	2	2	2	1	1
5	2	3	2	2	2
6	1	3	2	1	1
7	1	2	3	1	2
8	2	3	1	2	1
9	1	2	2	2	1
10	1	1	3	2	1
11	2	1	2	2	2
12	1	1	2	3	1

$$Info(\mathcal{D}, x_1) = \sum_{l=1}^{2} \frac{|\mathcal{D}_l|}{|\mathcal{D}|} \times Info(\mathcal{D}_l)$$

$$Info(\mathcal{D}_l) = -\sum_{q=1}^{2} P_{ql} \log_2 P_{ql}; P_{ql} = \frac{freq(y_q, \mathcal{D}_l)}{|\mathcal{D}_l|}$$

From data $\mathcal{D}$ of Table 8.2, we find that

$$freq(y_1, \mathcal{D}_1) = 7, freq(y_2, \mathcal{D}_1) = 1$$

$$freq(y_1, \mathcal{D}_2) = 2, freq(y_2, \mathcal{D}_2) = 2$$

Therefore,

$$Info(\mathcal{D}_1) = -\tfrac{7}{8} \log_2 \tfrac{7}{8} - \tfrac{1}{8} \log_2 \tfrac{1}{8}$$

$$Info(\mathcal{D}_2) = -\tfrac{2}{4} \log_2 \tfrac{2}{4} - \tfrac{2}{4} \log_2 \tfrac{2}{4}$$

$$Info(\mathcal{D}, x_1) = \tfrac{8}{12}\left(-\tfrac{7}{8} \log_2 \tfrac{7}{8} - \tfrac{1}{8} \log_2 \tfrac{1}{8}\right) + \tfrac{4}{12}\left(-\tfrac{2}{4} \log_2 \tfrac{2}{4} - \tfrac{2}{4} \log_2 \tfrac{2}{4}\right)$$

$$= 0.696 \text{ bits}$$

$$Gain(\mathcal{D}, x_1) = Info(\mathcal{D}) - 0.696$$

The information gains of the other three attributes are calculated on similar lines:

$$Gain(\mathcal{D}, x_2) = Info(\mathcal{D}) - \left[\tfrac{3}{12}\left(-\tfrac{2}{3} \log_2 \tfrac{2}{3} - \tfrac{1}{3} \log_2 \tfrac{1}{3}\right)\right.$$

$$\left. + \tfrac{6}{12}\left(-\tfrac{5}{6} \log_2 \tfrac{5}{6} - \tfrac{1}{6} \log_2 \tfrac{1}{6}\right) + \tfrac{3}{12}\left(-\tfrac{2}{3} \log_2 \tfrac{2}{3} - \tfrac{1}{3} \log_2 \tfrac{1}{3}\right)\right]$$

$$= Info(\mathcal{D}) - 0.784$$

$$\text{Gain } (\mathcal{D}, x_3) = Info(\mathcal{D}) - \left[\tfrac{8}{12}\left(-\tfrac{6}{8} \log_2 \tfrac{6}{8} - \tfrac{2}{8} \log_2 \tfrac{2}{8}\right) + \tfrac{3}{12}\left(-\tfrac{2}{3} \log_2 \tfrac{2}{3} - \tfrac{1}{3} \log_2 \tfrac{1}{3}\right)\right]$$

$$= Info(\mathcal{D}) - 0.771$$

$$Gain(\mathcal{D}, x_4) = Info(\mathcal{D}) - \left[\tfrac{4}{12}\left(-\tfrac{3}{4} \log_2 \tfrac{3}{4} - \tfrac{1}{4} \log_2 \tfrac{1}{4}\right) + \tfrac{6}{12}\left(-\tfrac{4}{6} \log_2 \tfrac{4}{6} - \tfrac{2}{6} \log_2 \tfrac{2}{6}\right)\right]$$

$$= Info(\mathcal{D}) - 0.729$$

Hence, attribute x_1 is chosen as the root node of the decision tree (Fig. 8.5), after which, two branches of root node are tested respectively in the same manner on the other three attributes, and a whole decision tree can be built as depicted in Fig. 8.6.

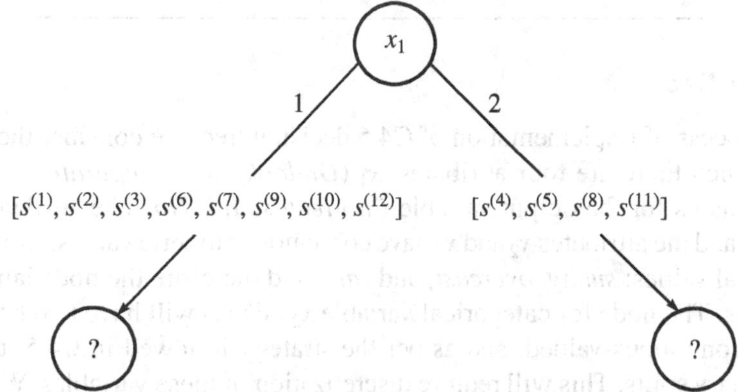

Figure 8.5 Partially learned decision tree

It is clear from the decision tree that there are seven leaves; so seven classification rules in total can be obtained. It should be mentioned here (the reader is encouraged to verify) that during the test of branch $x_1 = 1$, the information gains of attributes x_3 and x_4 are equal; but the number of classification rules would be eight if attribute x_4 were selected (In Fig. 8.6, x_3 has been selected). Also note that branch $x_3 = 1$ does not appear on the path $x_1 = 1$ because in the training data, when $x_1 = 1$, x_3 never takes the value 1. Similar argument applies for $x_4 = 3$ branch missing on the path $\{x_1 = 1, x_3 = 3\}$; and the branch $x_3 = 3$ missing on the path $\{x_1 = 2, x_2 = 3\}$.

If the test example has $x_1 = 1, x_3 = 1$ (missing branch with respect to training data), then what happens on test data? $\mathcal{D}$ contains no cases; the decision tree is a leaf and the class associated is the most frequent class of data $\mathcal{D}$ with $x_1 = 1$.

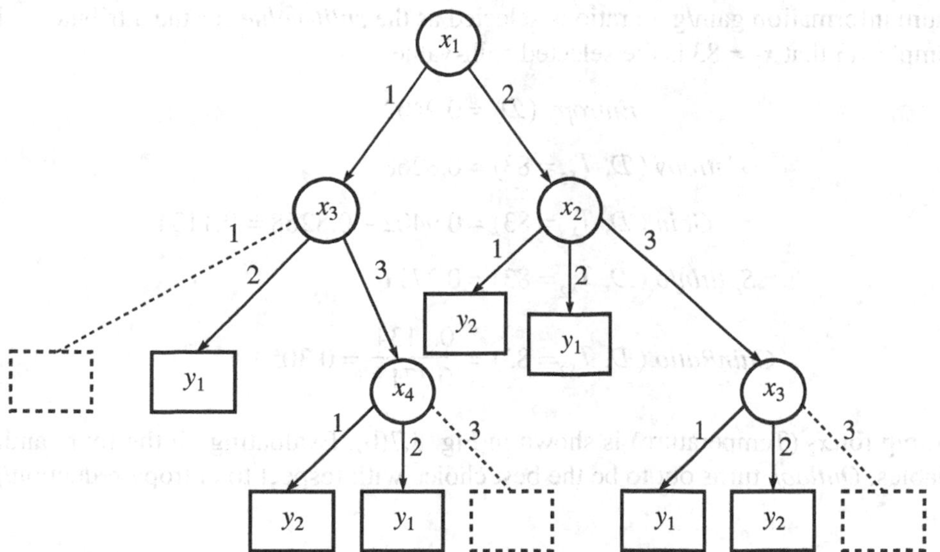

Figure 8.6 Decision tree for data of Table 8.2

Example 8.2

The C4.5 Decision Tree

To illustrate the process of implementation of C4.5 decision tree, we consider the (toy) training set of Table 7.3, in which there are four attributes: x_1 (*Outlook*), x_2 (*Temperature*), x_3 (*Humidity*), x_4 (*Wind*); and two classes for the output variable *PlayTennis*: *Yes*, *No*. The attributes x_1 and x_4 have categorical values, and the attributes x_2 and x_3 have continuous numeric values. Attribute x_1 (*Outlook*) has three categorical values: *sunny*, *overcast*, and *rain*, and therefore the node labeled *Outlook* will have three branches. The node for categorical variable x_2 (*Wind*) will have two branches. The other two variables are continuous-valued, and as per the strategy followed in C4.5, the corresponding nodes will have binary splits. This will require discretization of these variables. We will use here the entropy-based method for attribute discretization discussed earlier in Section 7.8.

There are four choices of attribute at the root node: *Outlook*, *Temperature*, *Humidity*, and *Wind*. Tree stumps for the attribute *Outlook* (x_1) are shown in Fig. 8.7(a); a three-way split corresponding to the three categorical values of x_1.

Temperature (x_2) has continuous numeric values. For this attribute, we will select the best cut-point T_{x_2} from its range of values by evaluating *every candidate cut point*. Examples are first sorted by increasing value of the attribute, and interval between each successive pair of values in the sorted sequence gives a potential cut-point. However, as stated in Section 7.8, the cut-point that minimizes the entropy will never occur between two patterns of the same class. Therefore, it is necessary to consider potential divisions that separate patterns of different classes. For the weather data of Table 7.3, this gives us eight potential cut-points: {64, 65, 70, 71, 72, 75, 80, 83}. Note that boundary points of the intervals between classes have been taken as the potential cut-points (Example 7.3). Entropy for each of these cut-points is evaluated and the one that results in maximum information gain/gain ratio is selected as the *split-value* for the attribute x_2. It follows from Example 7.3 that $x_2 = 83$ is the selected split-value.

$$Entropy\ (\mathcal{D}) = 0.9402$$

$$Entropy\ (\mathcal{D},\ T_{x_2} = 83) = 0.8268$$

$$Gain\ (\mathcal{D},\ T_{x_2} = 83) = 0.9402 - 0.8268 = 0.1134$$

$$SplitInfo\ (\mathcal{D},\ T_{x_2} = 83) = 0.3711$$

$$GainRatio\ (\mathcal{D},\ T_{x_2} = 83) = \frac{0.1134}{0.3711} = 0.3053$$

Tree stump for x_2 (Temperature) is shown in Fig. 8.7(b). Evaluating all the four candidate root node variables, *Outlook* turns out to be the best choice with respect to entropy reduction/gain ratio measure.

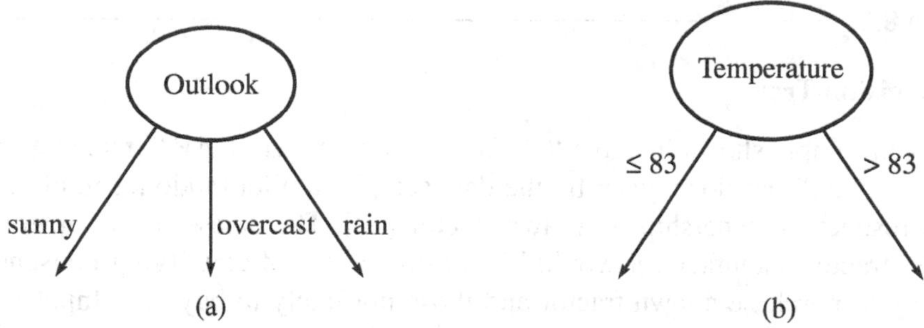

Figure 8.7 Tree stumps for weather data of Table 7.3

Table 8.3 shows the dataset for the branch *sunny*, obtained from the data $\mathcal{D}$ of Table 7.3. Repeating the process described above on this dataset, we select *Humidity* as the daughter node with split-value = 70.

Table 8.3 Dataset corresponding to the branch *sunny*.

$s^{(i)}$	Temperature	Humidity	Wind	Class
1	75	70	strong	Yes
2	80	90	strong	No
3	85	85	weak	No
4	72	95	weak	No
5	69	70	weak	Yes

The fully-grown C4.5 decision tree from the data of Table 7.3 is shown in Fig. 8.8.

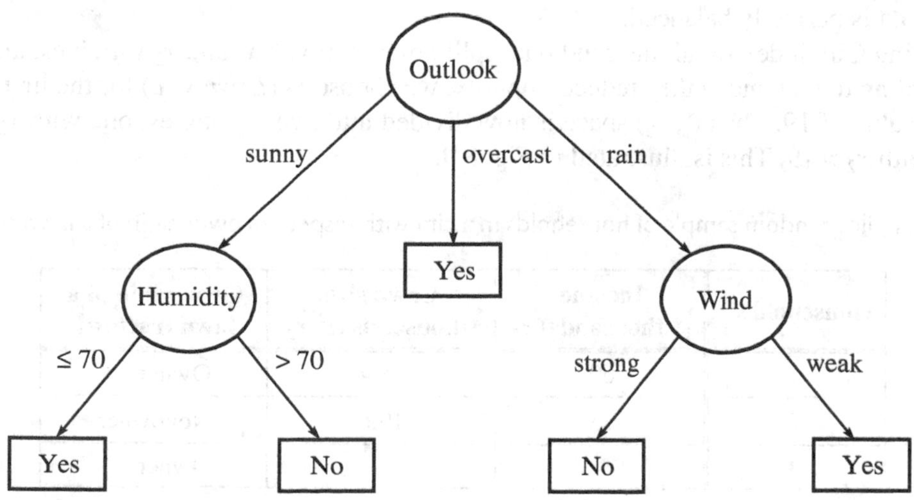

Figure 8.8 Decision tree for the weather data of Table 7.3

Example 8.3

The CART Decision Tree

Consider the data sample shown in Table 8.4, for training a binary CART tree. Note that there are $N = 24$ points in two dimensions given by the data sample; a pilot random sample of households in a city with respect to ownership of a lawn tractor [20]. The dataset is set up for a predictive model—a lawn-tractor manufacturer would like to find a way of classifying households in a city into those likely to purchase a lawn tractor and those not likely to buy one. Input variables, x_1 = *Income* and x_2 = *Lawn size*, are recorded for 24 households, and the target variable y = *Ownership of a lawn tractor*, is assigned to each household. The dataset is balanced, containing equal numbers of *Owner/Nonowner* households.

When searching for a binary split on a continuous-valued input variable, midpoints between the consecutive values may be treated as candidate values for the split. The candidate split points for the variable x_1 (*Income*) are {38.1, 45.3, 50.1, ..., 109.5}, and those for x_2 (*Lawn size*) are {14.4, 15.4, 16.2, ..., 23}. We need to rank the candidate split points according to how much they reduce impurity (heterogeneity) in the resulting subsets after the split. Note that the total dataset $\mathcal{D}$ given in Table 8.4 has the highest impurity. With respect to Gini index as impurity measure,

$$Gini(\mathcal{D}) = 1 - \sum_{q=1}^{2} P_q^2;\ P_q = \frac{freq(y_q, \mathcal{D})}{|\mathcal{D}|}$$

$$= 1 - (0.5)^2 - (0.5)^2$$

$$= 0.5$$

It can easily be verified that the Gini index impurity measure is at its peak when $P_q = 0.5$, i.e., when the data is perfectly balanced.

Calculating Gini index for all the candidate split points for both x_1 and x_2 variables, and ranking them according to how much they reduce impurity, we choose x_2 (*Lawn size*) for the first split with a splitting value of 19. The (x_1, x_2) space is now divided into two rectangles, one with $x_2 \le 19$ and the other with $x_2 > 19$. This is illustrated in Fig. 8.9.

Table 8.4 A pilot random sample of households in a city with respect to ownership of a lawn tractor

Household $s^{(i)}$	Income ($ thousands) x_1	Lawn size (thousands ft^2)x_2	Ownership of a lawn tractor y
1	60	18.4	Owner
2	75	19.6	Nonowner
3	85.5	16.8	Owner

Contd.

4	52.8	20.8	Nonowner
5	64.8	21.6	Owner
6	64.8	17.2	Nonowner
7	61.5	20.8	Owner
8	43.2	20.4	Nonowner
9	87	23.6	Owner
10	84	17.6	Nonowner
11	110.1	19.2	Owner
12	49.2	17.6	Nonowner
13	108	17.6	Owner
14	59.4	16	Nonowner
15	82.8	22.4	Owner
16	66	18.4	Nonowner
17	69	20	Owner
18	47.4	16.4	Nonowner
19	93	20.8	Owner
20	33	18.8	Nonowner
21	51	22	Owner
22	51	14	Nonowner
23	81	20	Owner
24	63	14.8	Nonowner

$$
\begin{aligned}
Gini(\mathcal{D}, x_2) &= \frac{|\mathcal{D}_1|}{|\mathcal{D}|} \times Gini(\mathcal{D}_1) + \frac{|\mathcal{D}_2|}{|\mathcal{D}|} \times Gini(\mathcal{D}_2) \\
&= \tfrac{12}{24} \times Gini(\mathcal{D}_1) + \tfrac{12}{24} \times Gini(\mathcal{D}_2) \\
&= \tfrac{12}{24}\left(1 - \left(\tfrac{3}{12}\right)^2 - \left(\tfrac{9}{12}\right)^2\right) + \tfrac{12}{24}\left(1 - \left(\tfrac{9}{12}\right)^2 - \left(\tfrac{3}{12}\right)^2\right) \\
&= \tfrac{1}{2} \times 0.375 + \tfrac{1}{2} \times 0.375 = 0.375
\end{aligned}
$$

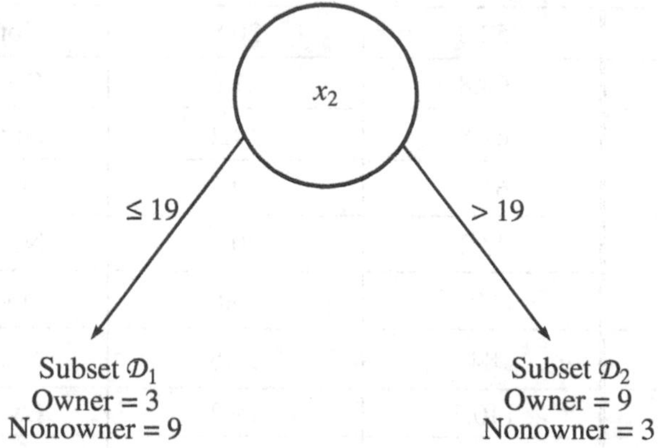

(a) Tree stumps after the first split

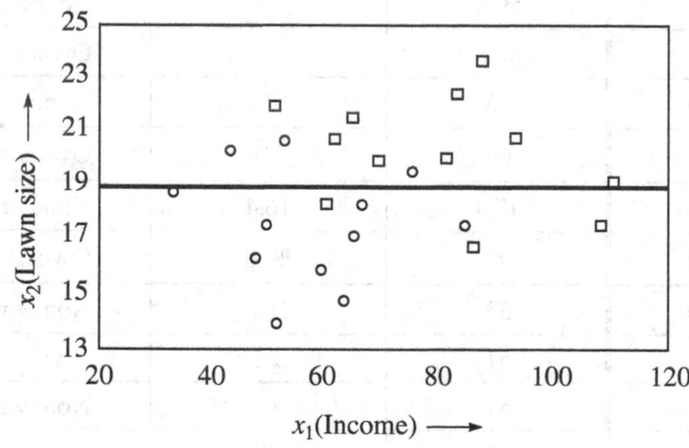

(b) Scatter plot after the first split

Figure 8.9 First level of recursive partitioning for data of Table 8.4

Thus, the Gini impurity index decreased from 0.5 before the split to 0.375 after the split. Each of the rectangles created by the split is more homogeneous than the rectangle before the split. The upper rectangle contains points that are mostly *Owners* and the lower rectangle contains mostly *Nonowners*.

By comparing the reduction in impurity across all possible splits in all possible attributes, the next split is chosen. The next split is found to be on the x_1 (*Income*) variable at the value 84.75. Figure 8.10 shows that once again the tree procedure has actually chosen to split a rectangle to increase the purity of the resulting rectangles. The left lower rectangle ($x_1 \leq 84.75$, $x_2 \leq 19$) has all points that are *Nonowners* with one exception, whereas the right lower rectangle ($x_1 > 84.75$, $x_2 \leq 19$) consists exclusively of *Owners*.

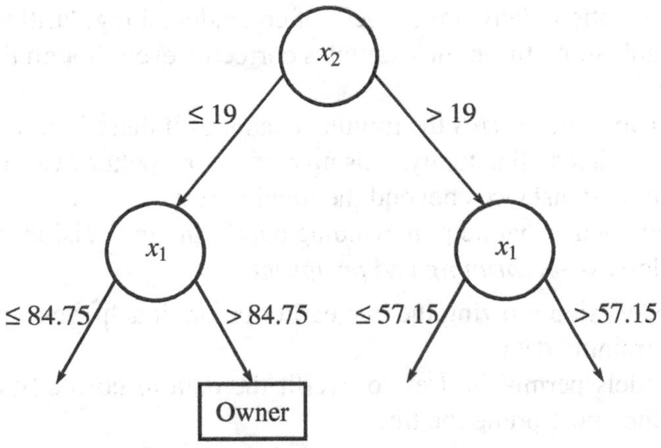

(a) Tree stumps after first three splits

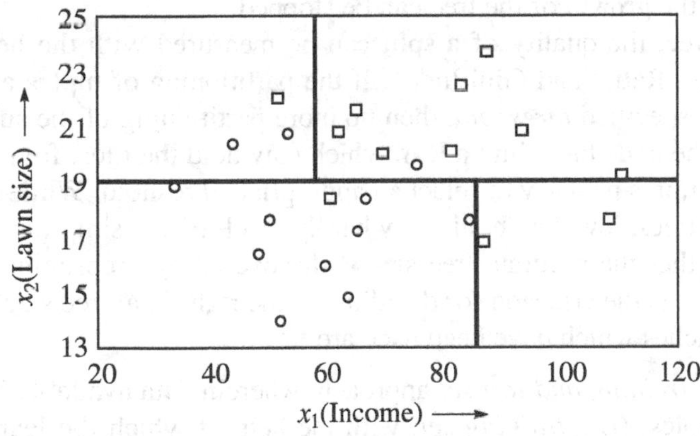

(b) Scatter plot after first three splits

Figure 8.10 First two levels of recursive partitioning for data of Table 8.4

If we continue partitioning till all the branches hit leaf nodes, each rectangle will have data points from just one of the two classes.

8.5 PRUNING THE TREE

The fundamental algorithm for decision trees continues to grow the tree by splitting nodes as long as new divisions generate daughter nodes that increase purity. Such a tree has undergone optimization for the training set. Therefore, elimination of any leaf nodes will simply lead to an increase in the error rate of the tree on the training set. But this certainly does not mean that the entire tree with pure leaf nodes also performs the best on new data!

The algorithm explained earlier grows each branch of the tree just to a depth sufficient for perfect classification of the training examples. This strategy can result in problems when there is random noise in the data (Noise, in fact, exists in all real-world datasets). Problems can result even when the training data is bereft of noise, particularly when small numbers of training data are associated with leaf nodes. In such a situation, there are chances of coincidental regularities taking place, wherein certain attributes are able to partition the examples correctly, even though they are unrelated to the actual target function.

It is said that a decision tree *overfits* the training examples if there is some other tree that doesn't fit the training examples that well actually ends up performing better over the entire distribution of instances (that is, including instances beyond the training set).

There are many approaches that help in avoiding overfitting in decision tree learning. These can be divided into two classes—*prepruning* and *postpruning*.

- *Prepruning* models stop growing the tree earlier before it achieves a point where it perfectly categorizes the training data.
- *Postpruning* models permit the tree to overfit the data to correctly categorize the training examples, and then post-prune the tree.

Even though the prepruning approach may appear to be more direct, the postpruning approach has seen more success when practiced. This is because the first approach finds it difficult to estimate accurately as to when the growth of the tree can be stopped.

While building a tree, the quality of a split can be measured with the help of measures like Information Gain, Gain Ratio, and Gini Index. If the partitioning of tuples at a node results in a division that meets prespecified *threshold*, then no more partitioning of the subset takes place. On stopping the growth, the node turns into a leaf, which may hold the most frequent class among the subset tuples. However, it is not easy to select an appropriate threshold. While high thresholds may lead to oversimplified trees, low thresholds may hardly result in any simplification at all.

Irrespective of whether the accurate tree size is discovered by prepruning or postpruning, the main question is, 'What is the criterion for deciding on the right final tree size?'

Some of the approaches which have been used are:

- Make use of the *training and test set* approach, wherein data available is separated into two groups of examples: (i) a *training set*, with the help of which the learned decision tree is formed, and (ii) a separate *test set*, with the help of which the evaluation of accuracy of this tree over subsequent data is done, especially, evaluation of the effect of pruning this tree (refer to Section 2.6). The approach is effective when huge volumes of data are available.
- Instead of pruning trees on the basis of estimated error rates, we can prune trees according to *adjusted error rate*, which is equal to the misclassification error of a tree plus a penalty factor on the tree size.

The CART algorithm develops *binary trees* and goes on splitting as long as new divisions can be discovered that enhance purity [142]. Within a complicated tree, there are several simpler subtrees, and each of them is representative of a different trade-off between how complex and how accurate a model is. The CART algorithm recognizes a group of such subtrees as candidate models. On applying these candidate subtrees to the test set, the tree showing the lowest test set misclassification error is chosen as the final model. The chosen model may have leaf nodes, which are impure, whereas the original tree had pure leaf nodes.

The identification of candidate subtrees by the CART algorithm is done via a procedure of repeated pruning. The aim is to first prune those branches that offer the least additional predictive power per leaf. These least useful branches are identified by CART with the help of a concept known as *cost complexity*, which is a measure that increases each node's misclassification error on training set by the imposition of a complexity penalty on the basis of the number of leaves on the tree. Cost complexity measure is used to identify weak branches and mark them for pruning.

For a tree T that has $L(T)$ leaf nodes, the cost complexity can be written as

$$CC(T) = E_{rr}(T) + \alpha L(T)$$

where $E_{rr}(T)$ is the fraction of training data observations that are misclassified by tree T and α is a penalty factor for tree size. When $\alpha = 0$, the fully grown unpruned tree is the best tree. If α is too large, the best tree may merely be the tree with the root node. Therefore, the idea is to begin with a full-grown tree and then raise the penalty factor slowly till a point is reached where the cost complexity of the full-grown tree is more than that of the subtree obtained by replacing an internal node at the next higher level with a leaf node. The same process is then repeated on the subtree. The process continues and as a result, a succession of trees is generated with a decreasing number of nodes. From the sequence of trees, it appears natural to select the one that gives the least misclassification error on the test dataset.

The trees grown by C4.5 resemble those grown by CART (although, unlike CART, C4.5 makes multiway divisions on categorical variables).

C4.5 makes use of a technique known as *pessimistic pruning*, which is like the cost complexity technique—it also makes decisions related to pruning subtree with the help of error rate estimates [141]. However, the pruning strategy is rather different because C4.5 makes no use of a test set to select from among candidate subtrees. The data employed to grow the tree is also used to decide the manner in which the tree should be pruned. Remember, an estimation of accuracy or error, on the basis of the training set is excessively optimistic and, hence, strongly biased. Therefore, the pessimistic pruning technique regulates the error rates received from the training set through the addition of a penalty, so that the bias incurred can be countered.

The postpruning technique in C4.5 is based on unstable statistical assumptions, and often does not prune enough. However, it is very fast and therefore, popular. But in several applications, it is useful to spend more computational effort on achieving compactness in a decision tree. In such situations, it may be more appropriate to use the conventional cost-complexity pruning technique from the CART learning system [143].

8.6 STRENGTHS AND WEAKNESSES OF DECISION-TREE APPROACH

In this section, we present important strengths and weaknesses of tree methodology. Some extensions to the basic building blocks described so far in this chapter (variants of basic decision-tree technique) will also be introduced.

Feature Extraction using Trees

A decision tree does its own feature extraction. It only uses the necessary variables, and after the tree is built, certain features may not be used at all. Features closer to the root are more important globally. For example, the decision tree given in Fig. 8.1 uses x_1, x_3, and x_4 but not x_2.

When faced with dozens or hundreds of unfamiliar variables, we can use a decision tree to direct our attention to a useful subset; with the most important variables usually showing up at the top of the tree. In fact, decision trees are often used as a tool for selecting variables for use with another modeling technique: we build a tree and then take only those features used by the tree as inputs to another learning method.

In general, decision trees do a reasonable job of extracting a small number of fairly independent features, but because each splitting decision is made independently, it is possible for different nodes to choose correlated variables.

There is no need for transformation of variables for decision free learning; any monotone transformation will result in the same tree.

Rule Extraction from Trees

Interpretability is yet another benefit of decision trees. The conditions carried by decision nodes are easy to comprehend. Each trail from the root to the leaf corresponds to *conjunction* of tests, as it is essential to satisfy all these conditions to reach the leaf. Together, these paths can be laid down as a set (*rule-base*) of IF-THEN rules. For instance, the decision tree of Fig. 8.8 can be written down as the following set of rules:

$R1$: IF (Outlook is sunny) AND (Humidity $\leq$ 70) THEN Class = Yes

$R2$: IF (Outlook is sunny) AND (Humidity > 70) THEN Class = No

$R3$: IF (Outlook is overcast) THEN Class = Yes

$R4$: IF (Outlook is rain) AND (Wind is strong) THEN Class = No

$R5$: IF (Outlook is rain) AND (Wind is weak) THEN Class = Yes

There is a rule associated with each leaf—conjunction of the individual decisions from the root node through the tree to the specific leaf. Therefore, a set of rules can describe the entire tree—one for each leaf. This complete set of rules can be easily comprehended by users. They definitely have a more transparent logic than the use of weights in neural networks.

To find a high level of accuracy in a decision tree, a method known as *rule post-pruning*, has been used with great success in practice. A variant of this pruning technique is used by C4.5. Conversion into rules permits distinguishing among the various contexts in which a decision node is employed. Since each separate path through the decision-tree node generates a decision rule, the pruning decision pertaining to that attribute test can be created differently for each path. On the contrary, if the tree itself was pruned, there could be two selections—total removal of the decision node or its retention in the original form.

The steps involved in rule post-pruning are:

1. Induce the decision tree from the training set, and grow the tree allowing overfitting until the training data is fit as well as possible.
2. Change the learned tree into an equivalent set of rules through the creation of one rule for each path from the root node to the leaf node.
3. Given a specific rule, each of its condition is considered for removal by temporarily deleting it, figuring out which of the training examples are now covered by this rule, calculating from

this a pessimistic estimate of the error rate of the new rule, and comparing this with the pessimistic estimate for the original rule. In case the new rule is found to be better, remove that condition and proceed, looking for other conditions to remove. After all the rules have been pruned in this manner, check for duplicates, if any, and delete them from the rule set.

4. Sort the pruned rules on the basis of their estimated accuracy, and take them into account in this sequence during classification of subsequent instances.

Note: Once the rules are pruned, it may be impossible to write them back as a tree.

Regression Trees

A *regression tree* is built in a manner almost similar to that of the classification tree. The only difference is that the impurity measure required for classification is substituted by a measure suitable for regression. In regression, the goodness of split is measured by the mean square error from the estimated value [6].

A regression tree can only generate as many distinct output values as there are leaf nodes in the tree (a discrete approximation of continuous function). Using methods like neural networks for regression problems seems more natural. Decision trees are more often used for classification than for regression.

Multivariate Trees

So far we have considered *univariate trees* wherein at each internal node, the test makes use of only one of the input dimensions. In case the used input dimension x_j is discrete, taking one of d_j possible values, the decision node checks the value of x_j and takes the corresponding branch, implementing a d_j-way split. In a *multivariate tree*, at a decision node, all input dimensions can be used, and thus it is more general [6].

Multivariate tree induction techniques gained popularity in recent times. You will find a review and comparison on many datasets in reference [144]. Several studies have shown that the univariate trees are rather precise and capable of interpretation, and the additional complexity brought by multivariate nodes is hardly justified. Refer to [145] for a recent survey.

Decision Tree is a Nonlinear Classifier

A decision tree uses a hierarchical approach for supervised learning, wherein we split the input space into *local regions* via a sequence of recursive splits. It comprises internal decision nodes and terminal nodes. A test function is implemented by each decision node, with discrete outcomes labeling the branches. Considering an input at each node, a test is applied and one of the branches is taken on the basis of the outcome. The procedure begins at the root and is recursively repeated until a leaf node is reached.

In a univariate tree, at each internal node, the test uses only one of the input dimensions x_j. In case of a binary division, a decision node splits the input space into two areas. Decision nodes occurring successively on the path from the root to the leaf, further split these areas into two with the help of other attributes and create splits orthogonal to each other. The leaf nodes define hyperrectangles in the input space (see Fig. 8.10).

Decision trees are a way of carving the space into regions, each of which can be labeled with a class. Any new pattern that falls into one of the regions is classified accordingly. This horizontal and vertical splitting of the attribute space results in a nonlinear model for the classification task.

However, there is an element of weakness as well in this nonlinear model created by a univariate decision tree. Since the splits are done on single attributes rather than in combinations of attributes, the tree is likely to miss relationships between attributes, in particular, linear structures. A classification tree is, therefore, expected to have lower performance than methods such as SVM, in some applications.

Decision Tree is a Non-parametric Model

In *parametric* estimation, we define a model over the whole input space and learn its parameters from all of the training data. Then, we use the same model and the same parameter set for any test input. The decision tree is a *non-parametric* technique, similar to *k-NN* discussed in Chapter 3, but many differences exist as follows:

- Each leaf node corresponds to a 'bin'; however, bins may not necessarily comprise an equal number of training instances, such as in *k-NN*.
- The bins are not divided on the basis of similarity in input space (some distance measure), but by supervised output information through entropy. Also, it is possible to find bin (leaf) much faster with a smaller number of comparisons.
- Once the decision tree is constructed, it does not store all the training datasets. It only stores the structure of the tree; this implies that the space complexity is much less, as opposed to *k-NN* method that stores all training examples.

Handling Missing Values and Outliers

The best part about decision trees is that they are capable of handling missing values in attributes by using *null* as an allowable value [146]. This approach is preferred to throwing out the data with missing values or attempting to impute missing values. Throwing out data may produce a biased set as the data with missing values is probably not a random population sample. Replacement of missing values with imputed values runs the risk that important information provided by the fact that the value is missing, will be ignored in the model.

Trees are intrinsically robust to outliers since we choose a path (a split) based on *ordering* of the values of a pattern and not on the absolute magnitudes. But their oversensitivity to changes in data can cause extremely different splits.

When are Decision Trees Appropriate?

In addition to the performance issues raised earlier in this section, computational complexity issues need to be considered. From a computational perspective, trees may be relatively costly to grow, owing to the multiple sorting in computing all possible splits on each variable. Pruning of the tree with the help of the test set adds further computation time. Since decision trees need a huge dataset for construction of a good and accurate classifier, the computational issues are rather significant.

An essential practical benefit of trees is the transparent rules generated by them. This kind of transparency is often beneficial in managerial applications. Decision tree classifies are better suited when such a requirement is important.

This book primarily focuses on pattern recognition, on the basis of feature vectors of real-valued and discrete-valued numbers. In these cases, there has been a natural measure of distance between vectors. For example, in the nearest-neighbors classifier, the notion forms the core of the method—while for neural networks, the notion of similarity emerges when two input vectors suitably 'close' result in similar outputs. Most practical pattern recognition techniques address problems of this nature, where feature vectors are real-valued and there exists some notion of *metric*.

Tree-based classifiers handle *nonmetric data*, such as discrete descriptions that do not have any natural notion of similarity or ordering. Tree-based classifiers are, therefore, an important tool in pattern recognition research [147].

The general recommendation is that a decision tree should be tested so that its accuracy can be considered to be the benchmark for other algorithms, before they are used.

8.7 FUZZY DECISION TREES

There exist many techniques for decision analysis; some of the frequently applied ones have been presented in earlier chapters. We have observed that each technique has its own strengths and weaknesses. In data mining, decision tree analysis is one of the most popular techniques for making classification decisions in pattern recognition applications [148].

The decision tree analysis approach has been effectively applied to various fields—financial management, business rule management, banking and insurance environmental science and even medical science.

A very important practical strength of trees is the transparent rules they generate. Such transparency is often useful in managerial applications. When faced with dozens or hundreds of unfamiliar variables, we can use a decision tree to direct our attention to a useful subset, with the most important variables usually showing up at the top of the tree. There is no need for transformation of variables in the decision tree learning. Decision trees are able to handle missing values in attributes and they are intrinsically robust to outliers.

Of course, there are weaknesses as well in decision tree analysis. Since the splits in widely-used univariate trees are done on a single attribute rather than in combinations of attributes, and also only horizontal and vertical splitting of the attribute space is achieved, a decision tree is likely to miss linear/nonlinear relationships between attributes having a potential of giving better performance. Computational issues also need to be considered. A strong weakness of decision tree is in its instability; it is recognized as highly unstable classifier with respect to minor perturbations in training set. Because of the gracefulness of the fuzzy sets and approximate reasoning methods used, *fuzzy decision trees* can deal with noise and uncertainties in the data collected [149–151].

Fuzzy decision tree induction requires fuzzification of all the attributes in the training set. The fuzzification process can be performed manually by experts or can be carried out automatically using some sort of clustering algorithm. The *fuzzy K-means clustering* algorithm, described in Section 7.5, gives a simple procedure to generate a set of membership functions on numerical data.

As in the case of crisp decision trees, we will focus on basic building blocks for inducing fuzzy decision trees. We will use *minimal fuzzy entropy* criterion to select expanded attributes.

The fuzzy decision tree learning procedure comprises the following steps:

- Fuzzification of attributes in the training set
- Induction of a fuzzy decision tree
- Simplification of the decision tree
- Classification of new instances

Let a set of learning patterns $\{s^{(i)}; i = 1, ..., N\}$ be described by a collection of attributes $\{x_j, j = 1, ..., n\}$. Decision attribute, y, classifies each pattern to a single class from M classes of concern. That is, y takes on values y_q; $q = 1, \ldots, M$. The training data is described by the dataset $\mathcal{D}$ of N patterns with corresponding observed outputs:

$$\mathcal{D} = \{s^{(i)}, y^{(i)}\}; i = 1, \ldots, N$$

$$s^{(i)} = \{x_j^{(i)}\}; j = 1, \ldots, n$$

For a crisp dataset $\mathcal{D}$ (refer to Eqn (8.10b)),

$$Entropy(\mathcal{D}) = -\sum_{q=1}^{M} P_q \log_2 P_q;\ P_q = \frac{freq(y_q, \mathcal{D})}{|\mathcal{D}|}$$

Crisp dataset $\mathcal{D}$ can be viewed as a specific case of fuzzy dataset with unity membership grade for each input example, i.e., $\mu_{\mathcal{D}}(s^{(i)}) = 1; i = 1, \ldots, N$. Sometimes, *a priori* knowledge about the data provides membership information on all input examples $s^{(i)}$. This *a priori* knowledge can be effectively utilized in the fuzzy decision tree learning process.

Consider a fuzzy dataset $\mathcal{D}$ with membership grades $\mu_{\mathcal{D}}(s^{(i)})$, $i = 1, \ldots, N$ $(0 < \mu_{\mathcal{D}} \le 1)$ given for all input examples $s^{(i)}$. Each class $q = 1, \ldots, M$, can be modeled as a fuzzy set with membership degree of $y^{(i)}$, $i = 1, \ldots, N$, defined as follows:

$$\mu_q(y^{(i)}) = \begin{cases} 1, & \text{if } y^{(i)} \text{ belongs to } q^{\text{th}} \text{ class} \\ 0, & \text{otherwise} \end{cases} \tag{8.15}$$

The *membership* concerning the qth class is defined as

$$\beta_q = \frac{\sum_{i=1}^{N} min\{\mu_{\mathcal{D}}(s^{(i)}), \mu_q(y^{(i)})\}}{\sum_{i=1}^{N} \mu_{\mathcal{D}}(s^{(i)})};\ 0 \le \beta_q \le 1 \tag{8.16}$$

Fuzzy entropy for the fuzzy dataset $\mathcal{D}$, is defined as

$$Entropy\ (\mathcal{D}) = -\sum_{q=1}^{M} \beta_q \log_2 \beta_q \tag{8.17}$$

The root node holds the entire dataset $\mathcal{D}$ which describes the patterns $s^{(1)}, s^{(2)}, \ldots, s^{(N)}$, with their corresponding membership grades $\mu_{\mathcal{D}}(s^{(i)})$, and the class $q = 1, 2, \ldots$, or M. Entropy ($\mathcal{D}$) is a measure of the impurity of the root node.

Now we have to divide the dataset $\mathcal{D}$ into fuzzy subsets $\mathcal{D}_1, \ldots, \mathcal{D}_k, \ldots, \mathcal{D}_{K_j}$, on some attribute $x_j; j = 1, \ldots, n$. Each attribute x_j is represented by a set of fuzzy linguistic terms $\{ F_{kx_j}, k = 1, \ldots, K_j\}$; F_{kx_j} refers to the kth fuzzy set of x_j and K_j is equal to the number of fuzzy sets on x_j. $\mu_k(x_j^{(i)})$ is the membership degree of the ith value of attribute x_j on the fuzzy set F_{kx_j}. The fuzzy sets F_{kx_j} would correspond to the branches grown from the root node.

The entropy after partitioning $\mathcal{D}$ on x_j is given by (the definitions run parallel to decision tree induction, review of earlier sections will be helpful)

$$Entropy\,(\mathcal{D}, x_j) = \sum_{k=1}^{K_j} w_k \times \text{Entropy}\,(\mathcal{D}_k) \tag{8.18a}$$

$$Entropy\,(\mathcal{D}_k) = -\sum_{q=1}^{M} \beta_{qk} \log_2 \beta_{qk} \tag{8.18b}$$

$$\beta_{qk} = \frac{\sum_{i=1}^{N} min\,\{\mu_k(x_j^{(i)}), \mu_q(y^{(i)})\}}{\sum_{i=1}^{N} \mu_k(x_j^{(i)})} \tag{8.18c}$$

$$w_k = \frac{\sum_{i=1}^{N} \mu_k(x_j^{(i)})}{\sum_{k=1}^{K_j}\left(\sum_{i=1}^{N} \mu_k(x_j^{(i)})\right)} \tag{8.18d}$$

$\mu_k(x_j^{(i)})$ represents the membership grade of ith value of x_j in the fuzzy set F_{kx_j}. w_k represents the weight of the kth fuzzy set of jth attribute; β_{qk} is the membership concerning the qth class induced by fuzzy set F_{kx_j}; and $\mathcal{D}_k$ is the data subset resulting due to partitioning induced by fuzzy set F_{kx_j}. For each data sample $s^{(i)}$ in $\mathcal{D}_k$, the membership value $\mu_{\mathcal{D}_k}(s^{(i)})$ is the product of its membership value $\mu_{\mathcal{D}}(s^{(i)})$ in $\mathcal{D}$ and membership value $\mu_k(x_j^{(i)})$ of ith value of x_j in the fuzzy set F_{kx_j}.

Information gain, Gain ($\mathcal{D}, x_j$), measures the *expected reduction in entropy* caused by partitioning the patterns in dataset $\mathcal{D}$ according to the attribute x_j:

$$Gain(\mathcal{D}, x_j) = Entropy(\mathcal{D}) - Entropy(\mathcal{D}, x_j) \tag{8.19}$$

We calculate information gain, $Gain(\mathcal{D}, x_j)$, for all attributes $x_j; j = 1 \ldots, n$, and select the one that yields maximum gain. The dataset $\mathcal{D}$ is then divided into fuzzy subsets $\mathcal{D}_k$, $k = 1, \ldots, K_j$ and K_j new nodes for these subsets are generated. We label the fuzzy sets F_{kx_j}; $k = 1, \ldots, K_j$, to edges (branches) that connect the new internal nodes to the root node.

Replacing $\mathcal{D}$ by $\mathcal{D}_k$; $k = 1, \ldots, K_j$, further tree expansion is carried out. If the recursive partitioning continues until all the sample data in each leaf node belong to one class, the resulting tree will be poor in accuracy. In order to improve the accuracy, the learning is stopped early (in crisp decision trees, pruning is more common). As a result, two thresholds are defined.

Fuzziness control threshold θ_{th1}: If the proportion in a node data of a class q is greater than or equal to a threshold θ_{th1}, stop expanding the tree from that node. For example, if in a node data, the proportion of Class 1 examples is 90%, Class 2 examples is 10% and θ_{th1} is 85%, then stop expanding that node.

Leaf decision threshold θ_{th2}: If the number of data samples in a node is less than a threshold θ_{th2}, stop expanding. For example, a dataset has 600 examples where θ_{th2} is 2%. If the number of samples in a node is less than 12 (2% of 600), then stop expanding that node.

Example 8.4

With the help of (toy) dataset given is Table 8.5, we describe the complete procedure of fuzzy decision tree building.

The data sample $\mathcal{D}$ in Table 8.5 corresponds to a sample of $s^{(i)}$; $i = 1, \ldots, 8$, beachgoers. The classification problem is to predict *sunburns* at the beach. The attributes in the data sample are *Height* (x_1), *Weight*(x_2), and *Hair Color* (x_3); the class labels y are *sunburned* (y_1) and *not_sunburned* (y_2). Based on *a priori* knowledge, a membership grade $\mu_{\mathcal{D}}(s^{(i)})$ is given for all input examples $s^{(i)}$; $i = 1, \ldots, 8$.

Table 8.5 Sunburn at the beach

Person $s^{(i)}$	$\mu_{\mathcal{D}}(s^{(i)})$	Height x_1	Weight x_2	Hair Color x_3	Sunburned (y)
$s^{(1)}$	1	160	60	Blond	yes
$s^{(2)}$	0.8	180	80	Black	no
$s^{(3)}$	0.2	170	75	Black	no
$s^{(4)}$	0.7	175	60	Red	yes
$s^{(5)}$	1	160	75	Black	no
$s^{(6)}$	0.3	175	60	Red	no
$s^{(7)}$	1	165	60	Blond	no
$s^{(8)}$	0.5	180	70	Blond	yes

Note that the input examples $s^{(4)}$ and $s^{(6)}$ are *inconsistent*; the decision is different for the same values of the attributes. Ideally, the crisp decision tree growing procedure should terminate once all leaf nodes are pure, i.e., once they hold all examples possessing the same classification. However, such an ideal situation cannot be reached for the given dataset. (Data cleansing step in preprocessing of data often gets rid of inconsistencies. In the given dataset, an additional feature: x_4 = *lotion used or not*, may result in a more complete description of the classification problem.)

Fuzzification to create the decision tree: F_{kx_j} represents fuzzy sets for the attribute x_j; $k = 1, 2, \ldots, K_j$. For the given data sample, $x_1 = Height$, $x_2 = Weight$, $x_3 = Hair\ Color$. Through this simple toy dataset, we will illustrate the manual process for fuzzification.

We assume $K_1 = 3$ (*Low, Middle, High*), $K_2 = 3$ (*Light, Middle, Heavy*), and $K_3 = 2$ (*Light, Dark*). The membership functions for the fuzzy sets $F_{1x_1}(Low)$, $F_{2x_1}(Middle)$, and $F_{3x_1}(High)$ of x_1 (*Height*) are assumed to be triangular and are graphically represented in Fig. 8.11. Table 8.6 gives the membership values $\mu_1(x_1^{(i)})$, $\mu_2(x_1^{(i)})$, and $\mu_3(x_1^{(i)})$ for x_1 values in the training data.

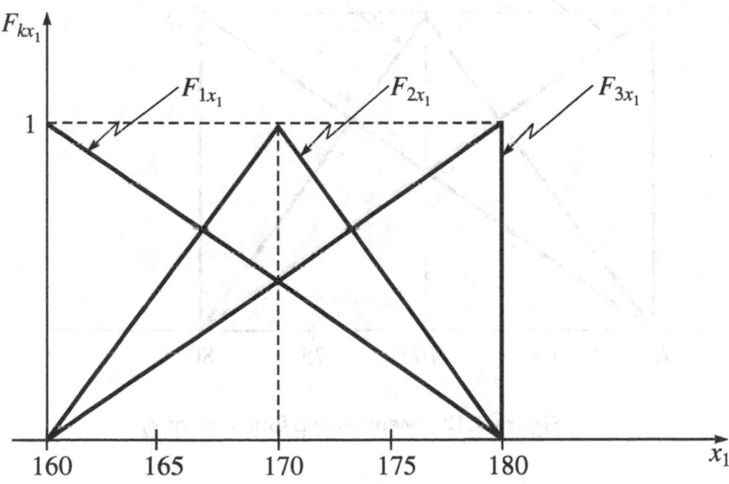

Figure 8.11 Membership functions of x_1

Table 8.6 Membership values for x_1

$s^{(i)}$	x_1(Height)	$\mu_1(x_1^{(i)})$ (Low)	$\mu_2(x_1^{(i)})$ (Middle)	$\mu_3(x_1^{(i)})$ (High)
$s^{(1)}$	160	1	0	0
$s^{(2)}$	180	0	0	1
$s^{(3)}$	170	0.5	1	0.5
$s^{(4)}$	175	0.2	0.5	0.8
$s^{(5)}$	160	1	0	0
$s^{(6)}$	175	0.2	0.5	0.8
$s^{(7)}$	165	0.8	0.5	0.2
$s^{(8)}$	180	0	0	1

Now consider the attribute $x_2(Weight)$. The membership functions for the fuzzy sets $F_{1x_2}(Light)$, $F_{2x_2}(Middle)$, and $F_{3x_2}(Heavy)$, of x_2 are again assumed to be triangular; Fig. 8.12 shows a graphical representation. Table 8.7 gives the membership values for $\mu_1(x_2^{(i)})$, $\mu_2(x_2^{(i)})$, and $\mu_3(x_2^{(i)})$ for x_2 values in the training data.

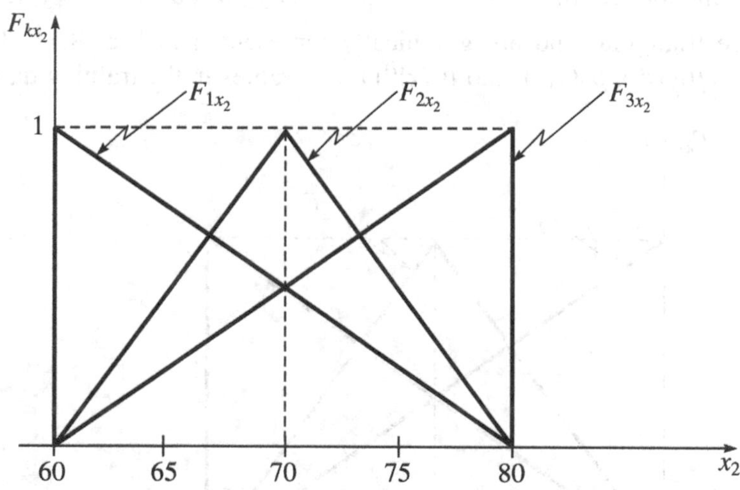

Figure 8.12 Membership functions of x_2

Table 8.7 Membership values for x_2

$s^{(i)}$	x_2 (Weight)	$\mu_1(x_2^{(i)})$ (Low)	$\mu_2(x_2^{(i)})$ (Middle)	$\mu_3(x_2^{(i)})$ (Heavy)
$s^{(1)}$	60	1	0	0
$s^{(2)}$	80	0	0	1
$s^{(3)}$	75	0.2	0.5	0.8
$s^{(4)}$	60	1	0	0
$s^{(5)}$	75	0.2	0.5	0.8
$s^{(6)}$	60	1	0	0
$s^{(7)}$	60	1	0	0
$s^{(8)}$	70	0.5	1	0.5

$x_3(Hair\ Color)$ has categorical values, unlike x_1 and x_2 which have numerical values. Triangular membership functions, $F_{1x_3}(Light)$, and $F_{2x_3}(Dark)$ of x_3 gives us the values of $\mu_1(x_3^{(i)})$ and $\mu_2(x_3^{(i)})$

for x_3. Table 8.8 gives the membership values; x_3 = Red being between x_3 = Blond (*Light*) and x_3 = Black (*Dark*), suitable values of $\mu_1(x_3^{(i)})$ and $\mu_2(x_3^{(i)})$ have been assumed for x_3 = Red.

Table 8.8 Membership values for x_3

$s^{(i)}$	x_3(Hair color)	$\mu_1(x_3^{(i)})$ (Light)	$\mu_2(x_3^{(i)})$ (Dark)
$s^{(1)}$	Blond	1	0
$s^{(2)}$	Black	0	1
$s^{(3)}$	Black	0	1
$s^{(4)}$	Red	0.3	0.6
$s^{(5)}$	Black	0	1
$s^{(6)}$	Red	0.3	0.6
$s^{(7)}$	Blond	1	0
$s^{(8)}$	Blond	1	0

Table 8.9 shows membership values for all the three attributes $x_1^{(i)}$, $x_2^{(i)}$, and $x_3^{(i)}$ given in Tables 8.6–8.8, after multiplication with $\mu_{\mathcal{D}}(s^{(i)})$.

Table 8.9 Membership values for expansion of root node

$s^{(i)}$	x_1	$\mu_1(x_1^{(i)})$	$\mu_2(x_1^{(i)})$	$\mu_3(x_1^{(i)})$	x_2	$\mu_1(x_2^{(i)})$	$\mu_2(x_2^{(i)})$	$\mu_3(x_2^{(i)})$	x_3	$\mu_1(x_3^{(i)})$	$\mu_2(x_3^{(i)})$
$s^{(1)}$	160	1	0	0	60	1	0	0	Blond	1	0
$s^{(2)}$	180	0	0	0.8	80	0	0	0.8	Black	0	0.8
$s^{(3)}$	170	0.1	0.2	0.1	75	0.04	0.1	0.16	Black	0	0.2
$s^{(4)}$	175	0.14	0.35	0.56	60	0.7	0	0	Red	0.21	0.42
$s^{(5)}$	160	1	0	0	75	0.2	0.5	0.8	Black	0	1
$s^{(6)}$	175	0.06	0.15	0.24	60	0.3	0	0	Red	0.09	0.18
$s^{(7)}$	165	0.8	0.5	0.2	60	1	0	0	Blond	1	0
$s^{(8)}$	180	0	0	0.5	70	0.25	0.5	0.25	Blond	0.5	0

Fuzzy Entropy and Information Gain: Next we have to calculate the fuzzy entropy and information gain of the fuzzy data in Table 8.9, to expand the tree.

$$\beta_1 = \frac{\sum_{i=1}^{8} min\,\{\mu_{\mathcal{D}}(s^{(i)}), \mu_1(y^{(i)})\}}{\sum_{i=1}^{8} \mu_{\mathcal{D}}(s^{(i)})} = \frac{1+0+0+0.7+0+0+0+0.5}{1+0.8+0.2+0.7+1+0.3+1+0.5} = \frac{2.2}{5.5}$$

$$\beta_2 = \frac{0+0.8+0.2+0+1+0.3+1+0}{5.5} = \frac{3.3}{5.5}$$

$$\begin{aligned} Entropy(\mathcal{D}) &= -\beta_1 \log_2 \beta_1 - \beta_2 \log_2 \beta_2 \\ &= -\frac{2.2}{5.5}\log_2\frac{2.2}{5.5} - \frac{3.3}{5.5}\log_2\frac{3.3}{5.5} \\ &= 0.971 \end{aligned}$$

Entropy after partitioning $\mathcal{D}$ on x_1

$$\beta_{11} = \frac{\sum_{i=1}^{8} min\,\{\mu_1(x_1^{(i)}), \mu_1(y^{(i)})\}}{\sum_{i=1}^{8} \mu_1(x_1^{(i)})} = \frac{1+0+0+0.14+0+0+0+0}{1+0+0.1+0.14+1+1+0.06+0.8+0} = \frac{1.14}{3.1}$$

$$\beta_{21} = \frac{1.96}{3.1}$$

$$Entropy(\mathcal{D}_1) = -\beta_{11}\log_2 \beta_{11} - \beta_{21}\log_2 \beta_{21} = 0.949$$

$$\beta_{12} = \frac{0.35}{1.2}, \beta_{22} = \frac{0.85}{1.2}$$

$$Entropy(\mathcal{D}_2) = 0.8709$$

$$\beta_{13} = \frac{1.06}{2.4}, \beta_{23} = \frac{1.34}{2.4}$$

$$Entropy(\mathcal{D}_3) = 0.990$$

$$w_1 = \frac{\sum_{i=1}^{8} \mu_1(x_1^{(i)})}{\sum_{k=1}^{3}\left(\sum_{i=1}^{8} \mu_k(x_1^{(i)})\right)} = \frac{3.1}{3.1+1.2+2.4} = \frac{3.1}{6.7}$$

$$w_2 = \frac{1.2}{6.7}$$

$$w_3 = \frac{2.4}{6.7}$$

$$\begin{aligned} Entropy(\mathcal{D}, x_1) &= w_1 \times Entropy(\mathcal{D}_1) + w_2 \times Entropy(\mathcal{D}_2) + w_3 \times Entropy(\mathcal{D}_3) \\ &= \frac{3.1}{6.7} \times 0.949 + \frac{1.2}{6.7} \times 0.8709 + \frac{2.4}{6.7} \times 0.990 \\ &= 0.950 \end{aligned}$$

$$\begin{aligned} Gain(\mathcal{D}, x_1) &= Entropy(\mathcal{D}) - Entropy(\mathcal{D}, x_1) \\ &= 0.971 - 0.950 \\ &= 0.021 \end{aligned}$$

Entropy after partitioning $\mathcal{D}$ on x_2

Following the procedure given above for x_1, we obtain

$$Entropy(\mathcal{D}, x_2) = 0.8542$$
$$Gain(\mathcal{D}, x_2) = 0.1168$$

Entropy after portioning $\mathcal{D}$ on x_3

$$Entropy(\mathcal{D}, x_3) = 0.8072$$
$$Gain(\mathcal{D}, x_3) = 0.1638$$

The Information Gain of the attribute x_3 (*Hair Color*) has the highest value. We use it to expand the tree.

Figure 8.13 shows the partitioning of data at the root node. The root node is x_3 (*Hair Color*) that has a fuzzy set of all $\mathcal{D}$ with membership values $\mu_{\mathcal{D}}(s^{(i)})$. We generate two sub-nodes with the input examples where the membership values $\mu_{\mathcal{D}_k}$ at these sub-nodes are the product of the original membership values $\mu_{\mathcal{D}}(s^{(i)})$ and the membership values of F_{1x_3} and F_{2x_3} of the attribute x_3. The example is omitted if its membership value $\mu_{\mathcal{D}_k}$ is null. The node datasets in Fig. 8.13 follow from Table 8.9.

Thresholds: If the proportion of a node dataset of a class q is greater than or equal to threshold θ_{th1}, we stop expanding the tree from that node. Also, if the number of data samples in a node is less than threshold θ_{th2}, we stop expanding that node.

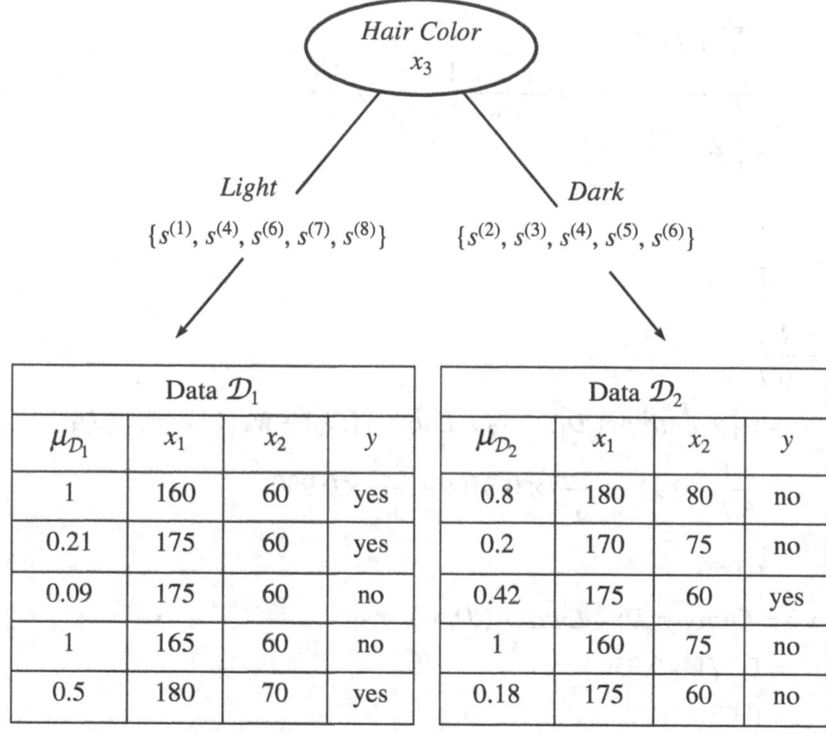

Data $\mathcal{D}_1$			
$\mu_{\mathcal{D}_1}$	x_1	x_2	y
1	160	60	yes
0.21	175	60	yes
0.09	175	60	no
1	165	60	no
0.5	180	70	yes

Data $\mathcal{D}_2$			
$\mu_{\mathcal{D}_2}$	x_1	x_2	y
0.8	180	80	no
0.2	170	75	no
0.42	175	60	yes
1	160	75	no
0.18	175	60	no

Figure 8.13 Expansion of tree at root node

The level of these thresholds has great influence on the result of the tree. We define them at different levels in our experiment to find optimal values.

Furthermore, if there are no more attributes for classification, the algorithm does not create a new node.

Building fuzzy decision tree: The procedure followed for root node needs to be followed for the two sub-nodes with $\mathcal{D}$ replaced with $\mathcal{D}_1 = \mathcal{D}_{\text{sub1}}$ and $\mathcal{D}_2 = \mathcal{D}_{\text{sub2}}$, respectively. We leave the calculations as an exercise for the reader; only the results are given.

Dataset $\mathcal{D}_{\text{sub1}}$ has three examples of Class 1, and two examples of Class 2. Membership values β_q are 0.6 and 0.4, respectively, for $q = 1, 2$. Dataset $\mathcal{D}_{\text{sub2}}$ has one example of Class 1, and four examples of Class 2. Membership values β_q are 0.1615 and 0.8385, respectively, for $q = 1, 2$.

We see that proportion of dataset for Class 2 in $\mathcal{D}_{\text{sub2}}$ is high; with appropriate choice of θ_{th1}, we may declare $\mathcal{D}_{\text{sub2}}$ as a leaf node. Membership values of β_q confirm this decision.

Sub-node $\mathcal{D}_{\text{sub1}}$ needs further expansion. We now have two attributes for expansion: x_1 and x_2. We calculate entropy and information gain corresponding to these attributes. Referring to $\mathcal{D}_{\text{sub1}}$ as $\mathcal{D}$, we repeat the procedure illustrated earlier.

$$Entropy(\mathcal{D}) = -\frac{1.71}{2.8}\log_2\frac{1.71}{2.8} - \frac{1.09}{2.8}\log_2\frac{1.09}{2.8} = 0.9643$$

$$Entropy(\mathcal{D}, x_1) = 0.8901$$
$$Gain(\mathcal{D}, x_1) = 0.0742$$
$$Entropy(\mathcal{D}, x_2) = 0.7610$$
$$Gain(\mathcal{D}, x_2) = 0.2033$$

Weight (x_2) is selected as the sub-node.

Continuing with the same procedure, we get a fuzzy decision tree for the given dataset. Figure 8.14 shows the tree with membership values at the leaf nodes.

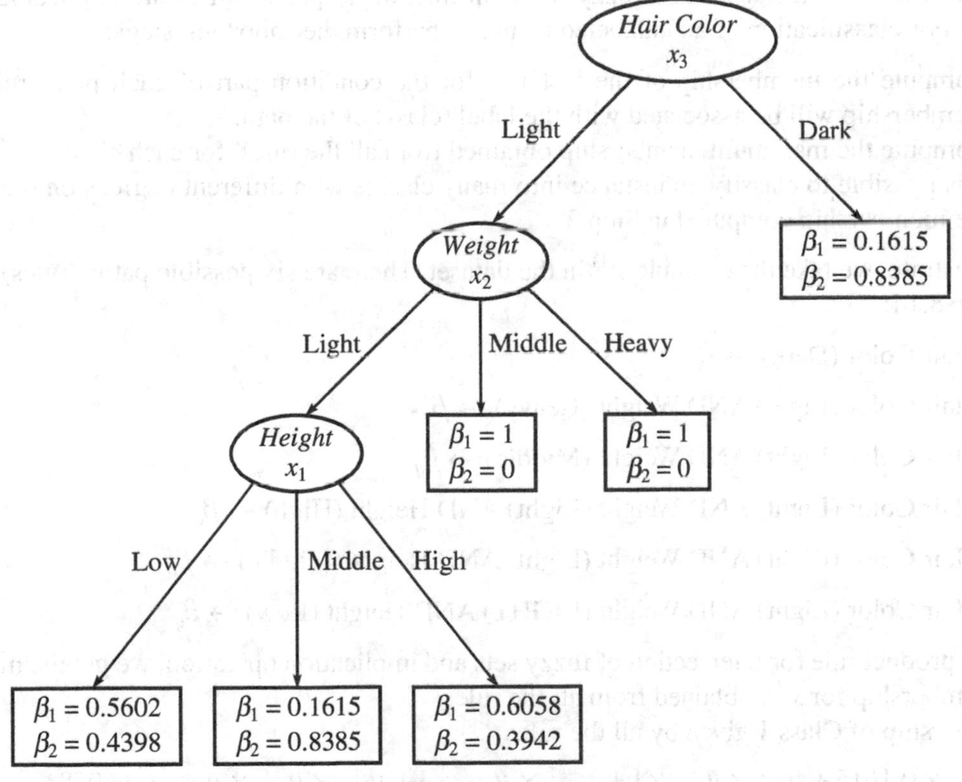

Figure 8.14 A fuzzy decision tree for dataset of Table 8.5

Fuzzy classification rules: A fuzzy classification rule can be written in the form (refer to Fig. 8.14)

IF (x_3 is *Light*) AND (x_2 is *Middle*) THEN (Class is 1)

IF (x_3 is *Light*) AND (x_2 is *Heavy*) THEN (Class is 1)

IF (x_3 is *Light*) AND (x_2 is *Light*) AND (x_1 is *Low*) THEN (Class is 1 with membership 0.5602)

⋮

A fuzzy classification rule with composite fuzzy set F_E can be written as

IF (**x** is F_E) THEN (Class is q with membership β_q)

The composite fuzzy set F_E is called the *fuzzy evidence*. For the third example rule given above,

$$F_E = F_1(x_3) \cap F_1(x_2) \cap F_1(x_1)$$

"**x** is F_E" is the premise or antecedent of the rule, and "Class is q with membership β_q" is the consequent of the rule.

Classification of new instances: A crisp decision tree consists of a single path (rule) which can be applied for every instance. In a fuzzy decision tree, many paths (rules) are applicable for one instance. For classification of an unlabeled instance, perform the following steps:

1. Compute the membership of the instance for the condition part of each path (rule). This membership will be associated with the label (class) of the path.
2. Compute the maximum membership obtained from all the rules, for each class.
3. It is possible to classify an instance into many classes with different degrees on the basis of the membership computed in Step 2.

To illustrate, we take the example $s^{(1)}$ in the dataset. There are six possible paths (rules), as seen from Fig. 8.14:

$R1$: Hair Color (Dark) $\rightarrow \beta_q$

$R2$: Hair Color (Light) AND Weight (Heavy) $\rightarrow \beta_q$

$R3$: Hair Color (Light) AND Weight (Middle) $\rightarrow \beta_q$

$R4$: Hair Color (Light) AND Weight (Light) AND Height (High) $\rightarrow \beta_q$

$R5$: Hair Color (Light) AND Weight (Light) AND Height (Middle) $\rightarrow \beta_q$

$R6$: Hair Color (Light) AND Weight (LIGHT) AND Height (Low) $\rightarrow \beta_q$

Using product rule for intersection of fuzzy sets and implication operation, we get the maximum class membership for $s^{(1)}$, obtained from all the rules.

Membership of Class 1 given by all the rules

$$= \mu_{F_{2x3}} \times 0.1615 + \mu_{F_{1x3}} \times \mu_{F_{3x2}} \times 1 + \mu_{F_{1x3}} \times \mu_{F_{2x2}} \times 1 + \mu_{F_{1x3}} \times \mu_{F_{1x2}} \times \mu_{F_{3x1}} \times 0.6058$$

$$+ \mu_{F_{1x3}} \times \mu_{F_{1x2}} \times \mu_{F_{2x1}} \times 0.1615 + \mu_{F_{1x3}} \times \mu_{F_{1x2}} \times \mu_{F_{1x1}} \times 0.5602$$

$$= 0 \times 0.1615 + 1 \times 0 \times 1 + 1 \times 0 \times 1 + 1 \times 1 \times 0 \times 0.6058 + 1 \times 1 \times 0 \times 0.1615 + 1 \times 1 \times 1 \times 0.5602$$

$$= 0.5602$$

Membership of Class 2 given by all the rules

$$= 0.4698$$

References [149–153] are useful for detailed study of the subject.

Chapter

9 BUSINESS INTELLIGENCE AND DATA MINING: TECHNIQUES AND APPLICATIONS

9.1 AN INTRODUCTION TO ANALYTICS

In today's public and private organizations, decision making is an ongoing process. These decisions are usually quite significant when it comes to the impact on the organization, whether in the short-term or long-term. The decision-making process involves people as well as rules at different levels in the organization's hierarchy. A person's capability to make the right decisions at the right time plays a significant role in an organization's performance and competitive strength.

In many organizations, decision making uses easy and spontaneous techniques based on the experience, awareness of the application domain, and the existing information, and other such elements. This model is not ideal for unstable conditions owing to repeated and fast changes that take place in the economic environment. In fact, decision-making processes in today's organizations are rather complicated and dynamic. Therefore, they cannot be handled properly using an intuitive model. Rather, they require a scientific approach.

A lot has been invested in business infrastructure in the last fifteen years, due to which the data collection capability has become better across the enterprise. Each and every facet of business is now geared up/equipped to collect data, be it marketing, operations, supply-chain management, manufacturing or even customer management. Also, information related to industry happenings, movement of competition, market trends, and other external events is now easily available. Due to ease of accessing data, many meaningful scenarios and opportunities have arisen. The question is, whether it is possible to transform such diverse data into meaningful information and knowledge, which can be employed in making the right decisions at the right time, in order to gain an edge over competition [154].

'Analytics' is the term widely used in business for data-driven analyses employed in the decision-making process. The term refers to the application of different methods of analysis of data to solve problems. It cannot be called a technology in itself, rather, a set of 'data science' processes and techniques are used together to obtain information, analyze it, forecast diverse possible actions towards a solution, and quantify the impact of these likely decisions of the future on optimizing business [155, 156].

A world that is becoming more and more complex with each passing day, rapidly increasing data, collection, and the urgency to overtake competition, have all made the use of *data analytics* (analytics-driven data analyses) very significant within organizations. With data analytics, enterprises are able to get a '360 degrees' view of their operations and customers. The insight thus gained is useful in directing and optimizing, and automating their decision making to fulfill their organizational goals. *Business analytics* (data-driven analyses of business data) helps in strategic, operational and tactical decision making through the *industry verticals* including telecom, retail, and financial services.

There are many subgroups in the field of analytics. Figure 9.1 depicts the field along with its subfields.

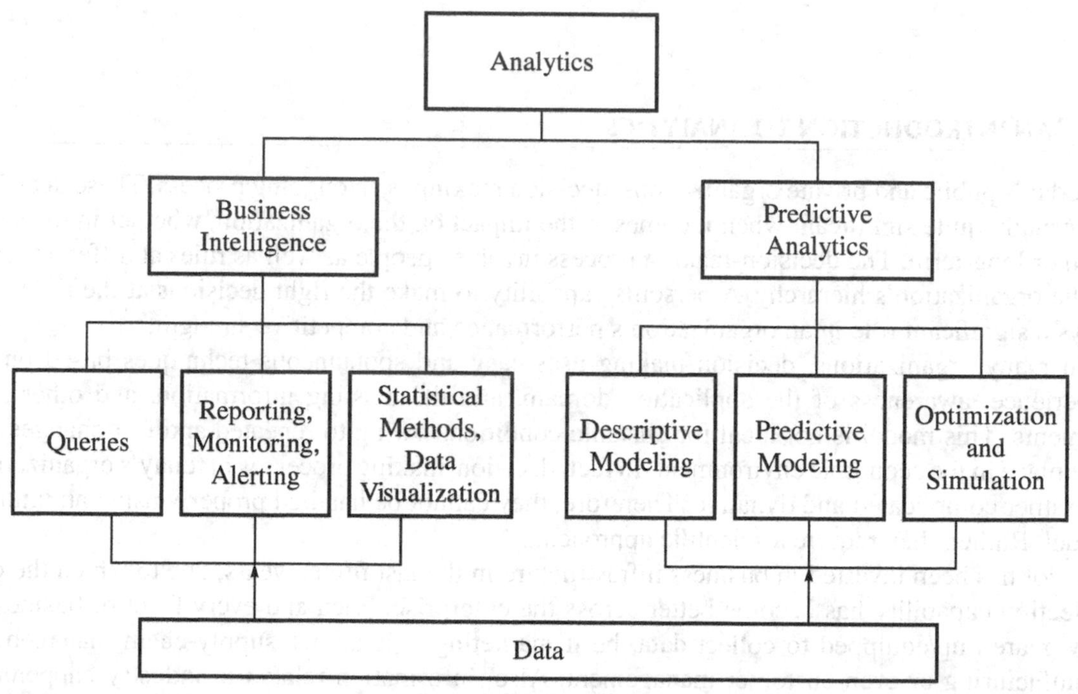

Figure 9.1 Analytics and relevant sub-fields

Analytics covers the skills, technologies, applications, and practices that are required for repeatedly exploring and investigating data to gain insight and drive business planning. Its two primary areas are *business intelligence* and *predictive analytics*.

Business Intelligence (*Insight into the past*)

A group of architectures and technologies capable of transforming raw data into information that can be used for business purposes by making it meaningful, is called *Business Intelligence* (BI). Using basic BI tools, businesses can easily interpret huge volumes of data, and measure past performance. They cannot only summarize and describe various facets of business but also help business managers comprehend all that is happening within the organization and with the business.

Computation of *summary statistics* is the basic building block of data analysis. We should select summary statistics paying attention to the business problem, and the way the data being summarized is distributed. It depicts things, such as total stock in inventory, average dollars spent per customer, year over year alterations in sales, and so on. *Reports* reveal historical information pertaining to the organization's production, financial status, operations, sales, inventory, and customers.

The term 'business intelligence' is often inclusive of the creation and maintenance of *data warehouses*. It is essential to collect and integrate the varied/diverse data stored in various primary and secondary sources, most of which are part of operational systems. A significant amount of effort is needed for the purification and integration of the data sources.

Data warehouses gather and merge together data from various transaction-processing systems, from across an enterprise, with its own database. It is possible for analytical systems to access data from data warehouses. Data warehousing is a facilitating technology for exploring data. BI tools that help explore data comprise query and reporting systems, as well as statistical and data visualization techniques. These are called *passive* methodologies as the decision makers are required to create prior hypotheses and then employ the tools of analysis to arrive at the answers and confirm what they had originally perceived.

Query is a particular request for a subset of data or a request for statistics from data, framed in the form of a technical language and posed to a database system. Tools capable of answering these queries are generally *front ends* to database systems, based on *Structured Query Language* (SQL) or with a *Graphical User Interface* (GUI) to facilitate the framing of queries. Database queries that use SQL/GUI are ideal for an analyst who is already aware of what could be an interesting population of data and wishes to explore this population or confirm a hypothesis pertaining to it. This activity is basically different from data mining as it does not involve the prediction of any patterns, or models.

Query tools are usually capable of executing sophisticated logic, which includes computing summary statistics over subpopulations, sorting, joining together multiple tables and related data, and more. *On-Line Analytical Processing* (OLAP) offers a user-friendly GUI for the purpose of querying voluminous data collections in order to explore data. 'On-line' processing takes place in real-time; therefore, analysts and those who are responsible for making decisions can derive answers to their queries fast and with great efficiency.

Besides the querying and visualization of data, traditional BI environments make it possible to implement *rule-based alerts* to inform decision-makers about important events or changes. However, the actions in BI are always defined through human interaction and are performed by humans.

Predictive Analytics (*Understanding the Future*)

Predictive Analytics goes beyond BI by using sophisticated quantitative techniques (for example, descriptive modeling, predictive modeling, simulation, and optimization) to produce insights that traditional approaches to BI (such as query, reporting, summary statistics, data visualization) are unlikely to discover. It deals with automatic discovery of meaningful patterns in structured as well as unstructured data.

Predictive analytics is capable of predicting what can happen. These analytics pertain to the understanding of the future, and offer data-based insights to companies. They also estimate the possible future outcome. Predictive analytics can be employed across the organization—to predict client behavior and buying patterns, and also to identify trends in sales-related activities. They also facilitate the prediction of demands for inputs from the supply chain, operations, and inventory. Financial services use the credit score produced by predictive analytics to determine the likelihood of customers making timely credit payments in future.

The main sub-fields of predictive analytics are *descriptive modeling*, *predictive modeling*, and *optimization and simulation*. Descriptive modeling refers to the use of *active* methodologies to extract information and knowledge from data. Unlike the tools described for Business Intelligence, models of an active kind do not require decision makers to formulate prior hypothesis to be later verified. Their purpose is, instead, to expand the existing knowledge and use it for predictive modeling. Exploratory data analysis includes basic statistical descriptions (measures of central tendency of data: mean, median and mode; measures of dispersion of data: variance, standard deviation), correlation analysis, cluster analysis, and data visualization through graphs and plots. Active data exploration is an exercise to bring out something new, unknown to the analyst, which can then be exploited to improve the success of predictive modeling.

Predictive modeling is the practice of analyzing data to make statistically accurate predictions about the future. In business environments, predictive models automatically find and exploit patterns found in historical and transactional data in order to extrapolate to future events, and by that means, predict the most likely future. Models describing these patterns capture relationships among many factors that human beings are unable to handle. This allows, for example, the identification of previously unknown risks and opportunities.

Predictive modeling gives out a number of possible solutions for the business problem. Optimization and simulation sub-field attempts to quantify the effect of future decisions based on possible solutions to the problem, in order to determine possible outcomes before the decisions are actually made. Big business organizations have been efficiently employing these methods to optimize production, scheduling, and inventory in supply chain to deliver the right products on time and optimize the customer experience.

Even on successful adoption of these analytical techniques, the decision is taken by decision-makers, on the basis of the informal and unstructured information available to alter and adopt the recommendations and conclusions using analytics.

9.1.1 Machine Learning, Data Mining, and Predictive Analytics

Methodologies and technologies from both statistics and computer science have played an important role in the development of predictive analytics. The main contributions to the discipline of predictive analytics come from *Machine Learning* and *Data Mining*.

The set of techniques for the extraction of predictive models from data is now known as *machine learning*. *Data mining* is actually derived from machine learning as a research field, which focuses on concerns raised in research on real-world applications. As such, research concentrated on commercial applications and business-related problems of data analysis tend to drift in the direction of data mining. Both the fields are related to each other, and also share methods and algorithms. They pertain to the analysis of the data to seek informative patterns.

Looking forward, the main challenge ahead is applications. Applications-related issues will not come from machine learning experts but from people who work with the data and the problems from which it arises. Machine learning research will respond to the challenges thrown by new applications and will create new opportunities in decision making.

Even though machine learning forms the crux of the data mining process, it also includes other significant steps — data formatting, data cleansing, data exploration (visualization, summarization), and use of human expert knowledge to frame the inputs for the machine learning algorithm.

The ever-increasing complexities of the world, the explosion of data, and the urgency to overtake competition have pushed the enterprises to seek a 360 degrees view of their operations and customers—insights into the past and understanding of the future. *Predictive analytics* tools predict different possible actions towards a solution and quantify the effects of these possible future decisions on business optimization. Machine learning/data mining technologies are central to the toolkit of predictive analytics.

The three fields: machine learning, data mining, and predictive analytics, do not have precise boundaries for the spectrum of their functions; their functions overlap. Any quantitative description of the three terms: machine learning, data mining, predictive analytics, is really not meaningful. The qualitative description of these terms given here should suffice—understanding the essence of these tools for data-driven decision making is all that is important [157–160].

9.1.2 Basic Analytics Techniques

The following are the commonly used basic technologies in analytics.

- Data warehousing
- Naive Bayes classifier, *k*-Nearest Neighbor algorithm, linear and logistic regression
- Support vector machines
- Neural networks
- Fuzzy inference systems
- Decision trees
- Data clustering
- Data transformations
- Learning associations
- Optimization algorithms

All of these techniques have earlier been covered in this book except the following two:

1. Data warehousing
2. Learning associations

To make our toolkit complete and ready for business environments, we give a brief presentation of these two techniques in this chapter; Section 9.3 on data warehousing and Section. 9.4 on mining frequent patterns and association rules.

In addition to industry verticals like Financial Services analytics, Sales analytics, Customer Relationship Management (CRM) analytics, Telecom analytics, Manufacturing analytics, and so forth, recent advances have led to the newest and the hottest trends in business analytics—*text analytics* and *multimedia analytics*

- **Text analytics** allows users to derive insights from unstructured text data collections.

  Unstructured text data is transformed into a structured format that can be used as the base for predictive and descriptive analytics.
- **Multimedia analytics** uses various technologies to derive insights from unstructured data that comes together in multiple modalities including images, audio, video, and text.

In the last ten years, the Internet has opened up many new avenues and presented great opportunities for businesses and organizations to reach out to their customers, promote their products, close deals and carry on transactions. In a well-established business approach, companies fully run and regulate their image and reputation on the Web via the content that appears on their websites.

The focus of discussions pertaining to various facets of a company's product portfolio nowadays is shifting from individual company websites to blogs, forums, and other collaborative sites—in other words, the *social media*. This new media allows everyone to post comments and opinions about companies and their products, which may impact the perceptions and buying behavior of innumerable (likely) buyers. Marketing companies are concerned about this because it is not easy to check the spread of negative information, nor is it easy to find out or identify the same in large spaces, such as forums, blogs, and social networking sites.

While conventional marketing techniques have to face a major challenge in the form of expanding user-generated content in the blogosphere, it also offers opportunities to marketing companies to design their strategy taking advantage of social media. However, to do this, there is a need for new thought processes, modern automated analytics-based capabilities that define the *social media analytics*, a discipline that is fast emerging [161].

Technologies that can process unstructured data provide better analytics models. We give an overview of text, image and audio analytics in Section 9.5. Section 9.6 describes various business applications of analytics, and lastly an overview of big data issues and emerging analytical methods (*big-data analytics*) is given in Section 9.6.

Data science is yet another commonly used term today. Meaning *data-driven science*, the term refers to an interdisciplinary field which pertains to scientific processes, techniques, and systems used for the extraction of knowledge from structured or unstructured data. It is a collective term for all the methods used in an effort to extract insights and information from data—the methods to manipulate data, analyze data with statistics and machine learning, communicate data with information visualization, and work with Big Data.

Data analytics is the science that examines raw data in order to draw conclusions about that information. The focus of data analytics is on inference.

We conclude this section with the comment that various terms introduced here have no precise boundaries for the spectrum of their functions; their functions overlap. Any quantitative description of these terms is really not meaningful. The qualitative description given here should suffice.

9.2 THE CRISP-DM (CROSS-INDUSTRY STANDARD PROCESS FOR DATA MINING) MODEL

Data mining requires the application of science and technology in significant amounts, and productive application involves art as well. It involves a process that identifies a structure in the problem and allows consistency, repeatability, and objectiveness.

The CRoss-Industry Standard Process for Data Mining (CRISP-DM) codifies the data mining process. Details of CRISP-DM can be downloaded free of cost from www.crisp-dm.org, the CRISP-DM Consortium's website.

Figure 9.2 depicts the process. From the diagram, it is quite clear that iteration is the rule and not the exception. Going through the process once without obtaining a solution to the problem is not actually a failure. The entire process explores the data, and once the first iteration is over, the analysts are more aware, and therefore, the next iteration will be a lot more informed.

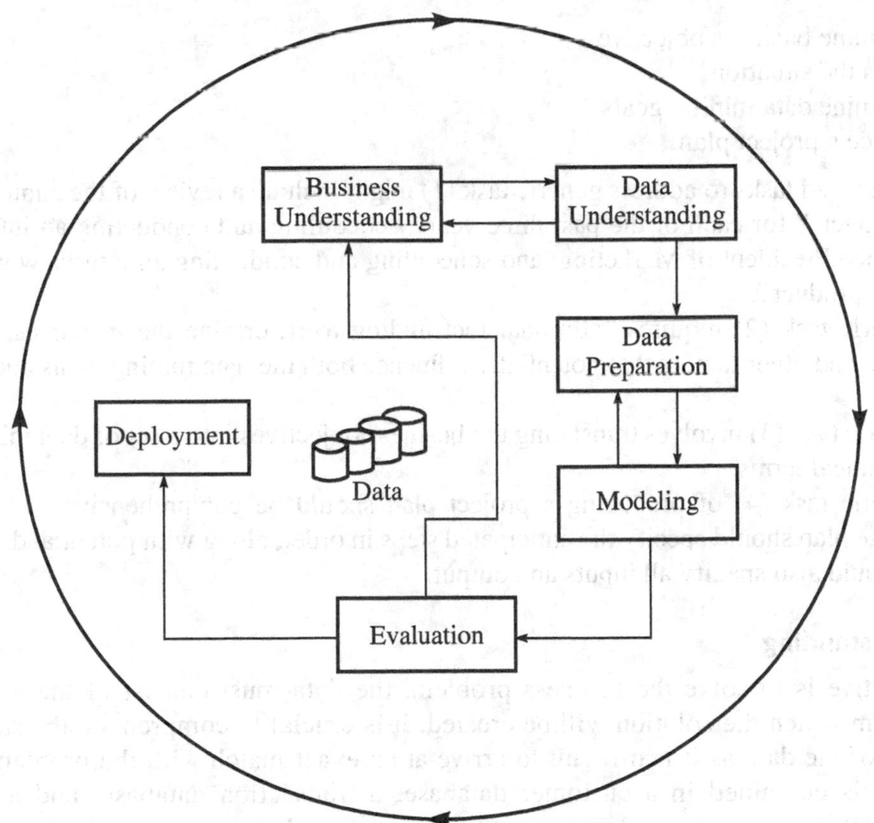

Figure 9.2 The CRISP-DM Process Model

It would be ideal to consider the data mining process as a set of nested loops rather than a feedforward process. The steps do have a natural order, but completely finishing with one before moving on to the next is neither needed nor desired. Things learned in later steps will result in the revisiting of the earlier ones.

There are six phases of the CRISP-DM process model (Fig. 9.2). Underneath each phase will be several *generic tasks*—high-level descriptions of activities that need to be carried out for nearly all projects. Each of these generic tasks is then made project-specific through the specification of one or more *specialized tasks*.

Let us now discuss the phases in detail.

Business Understanding

In the beginning, it is important to understand the business problem that needs to be solved. While this may appear quite obvious, business projects rarely show up as pre-packaged and clear data mining problems. Many a time reorganizing the problem and designing a solution is a repetitive process of discovery—cycles within a cycle, as depicted in Figure 9.2.

The generic tasks of this phase are:

1. Determine business objectives
2. Assess the situation
3. Determine data mining goals
4. Produce a project plan

The specialized tasks to address generic task (1) might include a review of the annual marketing plan for product *X* for each of the past three years; scheduling and conducting an interview with Executive Vice President of Marketing; and scheduling and conducting an interview with Product Manager for product *X*.

The generic task (2) requires additional fact-finding to determine the resources, constraints, assumptions, and other factors that potentially influence both the data mining goals and the project plan.

The generic task (3) involves translating the business objectives into a set of data mining project goals in technical terms.

The generic task (4) of producing a project plan should be comprehensive and detailed. In particular, the plan should specify the anticipated steps in order, along with potential dependencies. The plan should also specify all inputs and output.

Data Understanding

If the objective is to solve the business problem, the data must consist of the available raw material from which the solution will be created. It is crucial to comprehend the strengths and boundaries of the data as it is difficult to arrive at an exact match with the problem. Different information is contained in a customer database, a transaction database, and a marketing-response database. They may also pertain to varying overlapping populations and may have different levels of reliability.

The costs of data also commonly differ. While certain data will be almost free, others can be obtained with some effort. Certain data can be bought, while other data is non-existent and can only be collected using entire ancillary projects. A crucial component of the data-understanding stage is the estimation of the costs and benefits of each data source before determining whether it is feasible to invest any further. Once all datasets are obtained, some extra effort is needed to collate them.

The four generic tasks related to this phase are:

1. Collection of initial data
2. Description of data
3. Exploration of data
4. Verification of data quality

The first generic task pertains to the location, assessment, and obtaining of the required data from internal as well as external sources.

The second generic task involves the examination of 'surface' properties of the data collected—the format of the data (e.g., relational database table versus a text file); the amount of data (in terms of the number of variables and records); data types (e.g., categorical or numeric), coding schemes, and definitions of the variables of the data. A key objective of this task is to determine whether the collected data will answer the identified data mining question(s).

The generic task (3) involves the use of frequency distributions, cross-tabulations, means, correlations, and other simple descriptive statistics of the variables of interest, along with a number of other variables of interest initially thought to influence the variables of interest. In addition, graphical tools such as histograms, scatter plots, and other simple plots of the data are useful at this stage. These analyses may help refine the data description, lead to a better understanding of the potential data quality issues, and help gain a basic understanding of the nature of relationships between different variables that will be of use in modeling phase.

The generic task (4) addresses the important questions of whether the data is complete (i.e., all the relevant cases we hope to examine are covered or not), whether all the variables are correct (e.g., variables that should be all numbers may contain some character entries), whether there are missing values in the data, etc.

Data Preparation

While the analytic methods are powerful, they inflict particular requirements on the data they use. They often require data to be in a different form from the one naturally provided and require some data conversion. The stage of preparing data often progresses with data understanding, involving manipulation and conversion of data into forms capable of giving better results.

The five generic tasks of this phase are:

1. Select data
2. Clean data
3. Construct data
4. Integrate data
5. Format data

The generic task (1) relates to both what variables to have in the dataset to be used in actual data mining, as well as the nature of data records to be used in analysis (e.g., skewed data in classification problems).

The generic task (2) involves how to deal with data with missing values (removing or inferring missing values). Variables with values that are likely to be incorrect must also be dealt with.

The generic task (3) typically involves creating new variables through transforming original variables. Attribute reduction techniques, if employed, result in new variables in the reduced dataset. Also, transformation may involve creation of new variables through conversion of data —categorical to numerical and vice versa, discretizing continuous numeric values of variables, normalizing or scaling so that numerical values are compatible, are some conversions that may be of interest in specific problems. The other thing this task may involve is creating completely new records.

The generic task (4) involves merging different data tables together (say the transaction history of a customer, that is contained in the transactions database table, with the customer's personal information, contained in the customer information table) in order to create a single dataset that can be used by the data mining tools.

The generic phase (5) primarily refers to potential minor changes in the structure of variables or the order of variables in the database so that they match with what is expected by a particular data mining method.

Modeling

In the modeling phase, the actual models are constructed and assessed. The generic tasks associated with this phase are:

1. Select modeling technique(s)
2. Generate a test design
3. Build model
4. Assess model

Modeling has been the subject of earlier chapters and we have observed that selection of an appropriate modeling method(s) is dependent on the nature of the application. However, for most applications, there is more than one appropriate method. In generic task (1), a decision is made as to which of the possible methods that can be used should be used. The decision could be made to use all possible tools and then select the model that is best among the set of possible models as part of the generic task (4), which, in turn, relies on the testing procedures developed in generic task (2).

The generic task (2) needs to be done prior to building any models (generic task (3)). The main purpose of the test environment is to assess the quality and validity of different models. This is typically accomplished by taking the dataset created in the data preparation phase and dividing it into three different samples. The first of these samples, the training sample, is used to actually build the model(s). The second sample, the validation sample, is used to examine the accuracy and validity of a particular model, and to provide a basis for comparing the accuracy of different models. Based on the validation sample, the best model can be selected. However, to get an unbiased estimate of the likely impact of the use of this best model, a third sample, a holdout or test sample, is needed to make this assessment (which actually occurs in the evaluation phase of the process). The generic task (3) is where the data mining methods discussed in earlier chapters are applied to the dataset. In the generic task (4), the focus is on assessing a model on its technical merits as opposed to its business merits. The main concern at this point is the accuracy and generality of the model. An assessment with respect to the business problem being addressed is done in the evaluation phase.

The assessment of a model can result in the conclusion that the model can be improved upon, and also suggest ways of making an improvement, resulting in a new model building task (nested loop in Fig. 9.2).

Evaluation

The evaluation stage aims to thoroughly assess the data mining results and to make sure of their validity and reliability before proceeding further. This can be partially achieved for predictive models, by making use of the holdout for developing an estimation of the returns from use of the model. This evaluation phase also ensures that the approach fulfills the original business goals.

Remember, the main aim is to support the decision-making process for the business, and the process began with focus on the business problems that require resolution. A data mining solution is usually just a part of the larger solution, and requires evaluation. In addition, even if the approach passes rigid technical evaluation tests, certain external factors may render it unfeasible.

Different stakeholders have different interests in business decision-making that can be accomplished or supported by resultant approaches/models. In several cases, these stakeholders require to 'sign off' on the use of the approaches/models, and therefore, require to be satisfied by the quality of the model's decisions. Very often, stakeholders try to find out whether the approach is really useful or whether it is doing harm, and whether it is good enough to avoid huge mistakes.

A complete evaluation framework is essential as obtaining detailed information on the model's performance may be tough or quite impossible.

The generic tasks of this phase are:

1. Evaluate results
2. Review process
3. Determine next steps

The generic task (1) activities have been discussed in the prior paragraphs.

The generic task (2) is really a quality assurance assessment, which addresses concerns such as: Was the model correctly built? Were the variables defined in a way that is consistent with the variables available in the actual database? Will the variables used in this analysis be available for future analyses?

In generic task (3), the project team needs to decide whether to finish the project and move on to deployment (if appropriate) or to revisit certain phases of the project in an attempt to improve upon them.

Deployment

To position a model in a production system, it has to be re-coded for the production environment, so that it has better speed and is compatible with the current system. This could be a costly process and may require significant expenditure.

Irrespective of the success or failure of deployment, the procedure often goes back to the Business Understanding phase. The data mining process results in a lot of insight into the business problem and the challenges of its solution. If repeated a second time, it can give a better solution.

Failure in deployment is not necessary for the cycle to begin again. The evaluation stage may indicate that results are not satisfactory for deployment, and the problem definition requires to be adjusted or there is a need to obtain different data (see Fig. 9.2).

In order to successfully deploy a data-mining based solution, four generic tasks may (depending on the type of project) need to be undertaken:

1. Plan deployment
2. Plan monitoring and maintenance
3. Produce a final report
4. Review project

The generic task (1) involves winning confidence of all stakeholders. This requires assessment of possible deployment pitfalls.

The generic task (2) involves assessment of what changes could occur in future which would trigger an examination of the deployed model (the entrance of a major new competitor, or a sharp rise in interest rates, may trigger an assessment of model's current predictive accuracy). In addition, a maintenance schedule to periodically test whether a model is still accurate, along with criteria to determine the point at which a model needs to be 'refreshed' (i.e., rebuilt using more recent data), needs to be developed.

The generic task (3) requires producing a final written report at the end of the project.

The generic task (4) is an assessment of what went both right and wrong with the project.

Data mining is, in fact, a continuous process for exploring large amounts of data to discover meaningful patterns. The process does not have a beginning and an end; it is ongoing. Data mining starts with data, then through analysis, inspires action, which in turn, creates data that begets more data mining.

We have used the term 'data mining' in describing the model of Fig. 9.2. It may be noted that this figure represents a business analytics process model.

9.3 DATA WAREHOUSING AND ONLINE ANALYTICAL PROCESSING

In Section 9.1, we highlighted the importance of analytics in today's competitive business world. The technologies which support winning analytics were also listed. It was pointed out that data warehousing is crucial for today's requirements of data-driven decision making.

Since the main focus of this book has been on applied machine learning algorithms, the target here is to provide only an overview of data warehousing technology. This will help the reader to appreciate the relevance of the subject. Lot of literature giving a detailed account of the subject is available. The reader may refer to [17, 19] as an initial seed.

9.3.1 Basic Concepts

Since the time computers were introduced to the data processing centers a few decades ago, there are hardly any *operational systems* that have not been computerized. Almost all of them are capable of generating huge volumes of data along the way. Record of each product bought by a customer, each and every bank transaction, each visit to a web page, every credit card purchase, every package delivered, each telephone call exists in a minimum of one operational system. Computer-based automation has led to new markets and revolutionized current ones: online retailing, social networking, automated tellers, just-in-time inventory control, frequent flier/buyer clubs, to name a few.

Data-Base Management Systems (DBMS), comprising a *database* along with a group of software programs, offer mechanisms to store interrelated data and manage data access. Online operational database systems perform online transactions and query processing. These systems are known as *On-Line Transaction Processing* (OLTP) systems, and include a majority of daily/routine operations of an organization pertaining to purchase, inventory, manufacture, banking, payroll, and accounts. A typical organization has many different operational databases.

OLTP systems offer answers to frequent queries, such as 'What is the address of customer Jones?'; 'What is the total price of all the products in the box?'; or 'What is the balance in account no. 980?'

The most widely used query language is SQL (**S**tructured **Q**uery **L**anguage), which facilitates data manipulation and also helps retrieve stored data from the database. It calculates the aggregate functions such as averages, sum, min, max, and count. For data analysis, the queries that can be asked are, 'List all the items sold in the previous quarter', 'Give the total branchwise sales last month', 'Show the number of sales transactions in the month of November?', 'List the total sales achieved by each sales person?', 'Name the salesperson with the maximum sales', and so on.

Huge volumes of data can be collected from transactional systems. A fast-food restaurant is able to sell hundreds of thousands of meals annually, while a chain of supermarkets has hundreds of thousands of transactions daily. Big banks process millions of checks, credit cards, and debit cards purchases every day. Large websites experience millions of hits every day. A web-based ad server can keep track of billions of views daily. An OLTP system is able to manage current data, which is extremely detailed and can be easily employed to make decisions. With rising competition in every industry, business executives require data science techniques for systematic organization, understanding and use of data for strategic decision making. *Data warehousing* offers tools and architectures for storing and managing decision-support summary data.

A *data warehouse* is a database system created by integrating various sources for the purpose of decision support. It is maintained in isolation from the operational databases of organizations.

For a retailer, a data warehouse may cover information from the database of the market, or the database of a supplier, or that of the customers. In other words, a data warehouse may include information related to not just sales, but customer billing, customer service, and so on. The data in the payroll database may not exist in the data warehouse if it is not found to be essential for decision support.

Merely dumping data from diverse databases onto one disk cannot lead to the creation of a data warehouse. Diverse operational systems employ diverse styles of keeping records, various conventions, varying time periods, diverse degrees of aggregation, and will have various types of error. There is a need for the assembly, integration and cleaning of data. To build data warehouses is an expensive exercise—manual intervention is required as well as a thorough understanding of the operational databases.

A data warehouse, therefore, offers a single steady point of access to an organization's data for decision support. Instead of focusing on day-to-day operations and transaction processing, it concentrates only on data analysis and data modeling for making decisions. It offers a modest and compact view of specific subject issues as it excludes useless data in the decision support procedure. Data warehouse systems are called *On-Line Analytical Processing* (OLAP) systems.

Decision support tasks need various kinds of queries. For instance, 'What is the monthly region-wise sales of products?' or 'What is the difference in sales this year and last year?' are typical queries for decision support. 'Online analytical processing' is the term used for the employment of databases to obtain data summaries, with the main mechanism being aggregation.

An OLTP system is used to manage current data, which is typically extremely detailed for use in decision making. An OLAP system is used for management of huge volumes of historic data, for facilitation of summarization and aggregation, and for storage and management of information at varying levels of granularity. With these features, it is convenient to use the data to make informed decisions.

The requirements of OLTP and OLAP from the database management system are different. The former requires the latest and updated data, and lets the queries make changes to the database, permits simultaneous execution of various transactions without disturbing each other, and expects fast responses, and so on. The queries and updates, themselves, in case of OLTP are rather simple.

On the other hand, the OLAP queries tend to be rather complex, and are executed one at a time. OLAP queries do not make changes to the data, and while they may seek information about previous sales, it is not essential to have the latest sale details. Given the differences in requirements, it is sensible to use diverse types of database systems to handle the two applications.

The design of data structures by OLAP systems takes into account the users and their reporting requirements. It begins with the definition of a set of *dimensions*—say month, geography, and product. It stores important measures, say total sales and total discounts pertaining to each combination of dimensions. Additionally, OLAP systems offer handy analysis functions, which are difficult/impossible to express in SQL.

OLAP tools, therefore, support *multidimensional data analysis* and decision making with functionalities, such as summarization, consolidation and aggregation. To set up the multidimensional structure, data analysis and an analysis of user requirements is a must. This is usually achieved by specialists who are aware of the data and the tool, through a procedure known as dimensional modeling. Although some initial investment is required to design and load an OLAP system, the result offers informative and rapid access to users. Response times are usually measured in seconds facilitating exploration of data by the users.

In the past, several organizations have spent millions of dollars to build enterprise-wide data warehouses. Many are of the opinion that with increasing competition in every industry, data warehousing is a weapon that one must possess.

9.3.2 Databases

We live in the information age, where everyone believes that information ensures success. Therefore, we gather a range of data: from business transactions and scientific data to satellite images, text reports, as well as military intelligence. Creating structured *databases* and effective *Data-Base Management Systems* (DBMS) have proven to be essential assets when it comes to managing huge volumes of data, particularly to retrieve specific information efficiently from a large collection whenever required.

Today, we are playing with information that is too much for us to manage or deal with. Mere *information retrieval* will not suffice for decision-making any more. Faced with large volumes of data, we have created the need for automatic data summarization. We have made it essential to extract the 'essence' of stored information and find patterns in raw data.

While the spread of DBMS has led to more information gathering, the introduction of data mining has encouraged information harvesting. Due to the varying nature of complex *data types*, different types of data are stored in different types of repositories. It is certainly not realistic to hope for an adaptable data mining technique, which will give satisfactory mining results on all types of repositories in an efficient and effective manner. Different types of data repositories demand diverse algorithms and techniques. Some examples of the schemes are as follows:

- **Flat Files:** These are the most commonly found data source for data mining algorithms, particularly at the research level. A flat file is a simple *data table*—a collection of *N* data objects, each comprising a fixed set (*n*) of data fields (attributes). The values of attributes could be numerical or categorical.
- **Relational Databases:** A set of tables, with each of them having been assigned a unique name which represents an entity, is called a relational database. Each table comprises a group of attributes (table columns) and generally stores a huge set of tuples or rows of the table—the data corresponding to the entity. Each of the tuples in a relational table is representative of an object which is recognized by a unique *key* and is described by a set of attribute values.

A semantic data model, such as an *Entity-Relationship* (ER) data model, is often built for relational databases. It denotes the database as a set of entities and their relationships.

For instance, a store is typically described by the following relational tables—*customer, item, employee*, and *branch*. The entity, *customer*, consists of a collection of attributes (cust-ID, name, address, age, occupation, annual_ income, credit_ information, category, and so on). In the same way, each of the entities *item*, *employee*, and *branch* comprises a set of attributes that describe their properties. Tables can even be employed to represent relationships that exist among several entities. In this particular instance, these include *purchases* (customer purchase items, creating a sales transaction managed by an employee), *items_ sold* (in a given transaction), and *works_ at* (the branch the salesperson belongs to).

- **Transactional Databases:** Each product bought by a customer, each bank deal, each visit to a web page, each credit card purchase, each flight sector, each package delivered, and each phone call is recorded in at least one *operational* system. Such data pertaining to transaction levels forms the raw material for comprehension of customer behavior.

It is possible to store transactions in a table—one record per transaction. The transactional database for a store, for instance, may contain just two columns—one for *trans_ID*, and one for *list_ of items_ IDs*. The transactional database may have additional tables associated with it, which contain other information regarding the sale, such as the date of transaction, the customer ID number, the ID number of the salesperson, and of the branch at which the sale occurred, and so on.

- **Data warehouses:** These provide architectures and business tools for executives to enable them to organize, comprehend and make use of data in a systematic way for strategic decision-making. Organizations utilize data warehouses to support business activities, such as to increase customer focus through an analysis of customer buying behavior; by repositioning products and managing product portfolios through a comparison of the sales performance quarterly, yearly, or region-wise; by an analysis of operations and looking for sources of profit, and so on. Data warehousing is significant when it comes to integrating heterogeneous databases. Data from operational databases and external data feeds are stored in warehouses once data has been extracted, cleaned, integrated and transformed. A lot of effort has been invested in the database industry and research community to attain this objective. Data warehouses are widely employed in banking and financial services, customer goods and retail distribution sectors, and controlled manufacturing such as demand-driven production.

For years now, the business world has been familiar with generation of automated reports, which fulfill business requirements. Production of such reports is the main task of IT departments. Even small alterations in these reports require modification of the code. As a result, there is considerable delay between a user making a request for changes, and the user viewing the new information, measured in weeks and months. Organizations are drifting away from such old technology.

Query generation packages can be bought off the shelves and are popularly used to access data. These are capable of generating queries in SQL and talking to not just local but remote data sources as well using a standard protocol. Business analysts are generally able to produce the reports they need, and the response time is measurable in minutes or hours. Compared to the report-generation packages of the past, these respond faster but they continue to make it tough to exploit the data.

OLAP systems (data warehouses) are a considerably improved version of other systems as these systems design the data structure keeping in mind the users and their reporting requirements.

We will soon see how data is organized in a warehouse in a multidimensional structure, around primary subjects—customer, supplier, product, sales, and so on. OLAP operations (roll-up, and drill-(down, across, through), slice-and-dice, and pivot (rotate)) permit users to navigate through summarized and detailed data. This reduces response times to business queries to seconds, making the decision-making process very efficient.

The data warehouse evolves with continuous usage. In the beginning, the data warehouse is primarily utilized to generate reports, answer queries, perform fundamental statistical analyses, analyze summarized and detailed data, and reporting, using crosstabs, tables, charts or graphs. This is referred to as usage of data warehouse for *information processing*. Later, the data warehouse is utilized for *analytical processing* with the use of OLAP operations. Finally, the data warehouse may be used to make strategic decisions employing *data mining* tools.

Data mining can gain from SQL in terms of selecting, transforming and consolidating data. It goes beyond what SQL could provide, such as prediction, comparison, detection, deviation, and so on.

The functionalities of OLAP and data mining are also disconnected. OLAP tools aim to simplify and support interactive data analysis, whereas the objective of data mining tools is to automate as much of the process as possible, but at the same time permit the users to direct the process. Data mining involves more automated and deeper analysis than OLAP, as it has broader applications.

Researchers of database systems have established well-known principles in data models, query languages, query processing and optimization techniques, data storage, indexing and accessing methodologies, and ensuring consistency and security of the information stored even if systems develop snags, crash or attempts are made to access them by unauthorized means.

- **Multimedia Databases:** Video data, image data, and even audio and text data are all part of multimedia databases. Since multimedia objects may need gigabytes of storage, there will be a need for specialized storage as well as search methodologies. These methods require integration with standard mining techniques. Promising models include creating multimedia data cubes, extracting multiple features from multimedia data, and matching patterns based on similarity.

  Data mining from multimedia repositories may require computer vision, computer graphics, image interpretation, and natural language processing methodologies.

- **Spatial Databases:** Over and above the usual data, spatial databases store geographical information, such as maps, and global and regional positioning. For data mining algorithms, such spatial databases are full of challenges.

- **Time-Series Databases:** Time-series databases comprise sequential data (time-related attributes), and generally have new data flowing in continuously, for instance, stock market data. This, at times, leads to the need for versatile real-time analysis. In such databases, data mining usually involves the study of trends and correlations between the evolution of different variables, as well as the predictions of trends and movements of the variables in time.

- **World Wide Web:** The World Wide Web is the most diverse and dynamic repository available. Conceptually, the World Wide Web consists of three primary parts: the content of the Web, encompassing available documents available; the structure of the Web, covering the hyperlinks and the relationships between documents; and the utilization of the Web, describing how the resources are accessed and when. It is possible to add a fourth dimension pertaining to the dynamic nature or evolution of the documents. Data mining in the World Wide Web, or *Web Mining*, attempts to address all these problems.

9.3.3 Data Warehousing: A General Architecture, and OLAP Operations

Data warehouses often adopt a three-tier architecture [17], as described in Fig. 9.3.

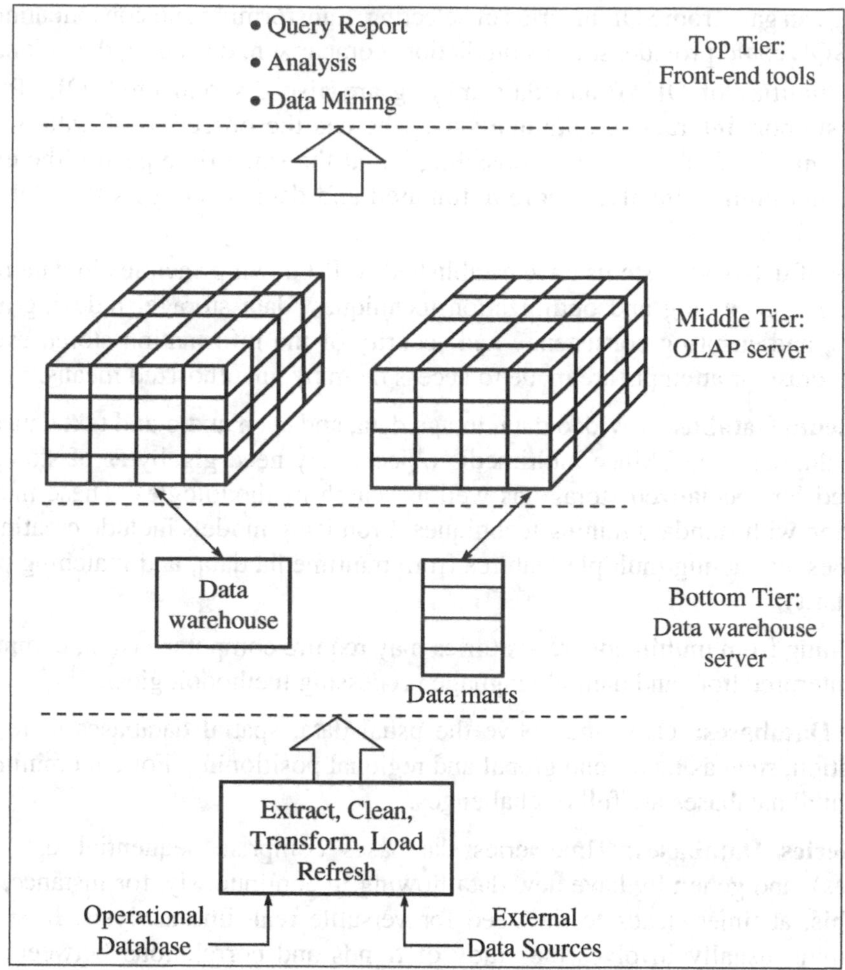

Figure 9.3 A three-tier data warehousing architecture [17]

The Bottom Tier

The bottom tier is a *data warehouse server* that is usually a relational database accessed through some variant of SQL. This *central repository* is the heart of the data warehouse.

The origin of data lies in the *source systems*, which are operational systems, and external data feeds. Their design ensures efficiency in operation, not decision support. Data warehouse systems extract data, clean data and also perform data integration and transformation using back-end devices/utilities. Back-end devices and utilities carry out load and refresh functions also in order to shift data from source systems to the data warehouse or the bottom level.

The Middle Tier

The middle tier is an OLAP server implemented with the use of a *multidimensional data model.* This approach looks at data in the form of a *data cube*. Cubes are conveniently stored in relational databases with the help of a data structure known as the *star schema.*

A data cube permits multidimensional views of data and is defined by *dimensions* and *facts*. Generally speaking, dimensions are entities or viewpoints according to which an organization wishes to maintain records. For instance, a retail store may construct a sales data warehouse for the purpose of maintaining sales records of the store in terms of dimensions: *item*, *time*, *location*, *branch*, *supplier*, etc. These dimensions facilitate the tracking of periodic sales of products, branch wise or location wise.

It is possible for each dimension to have diverse attributes. For instance, *item* as a dimension may possess the attributes *item_name*, *brand*, and *type*. Such attributes are stored in reference tables, known as *dimension tables*.

Figure 9.4 depicts a 3D data cube as per *item*, *time*, and *location*. Each grouping of entity (dimension) values (one value for every entity) defines a *cell* of the multidimensional array. Each cell contains what is representative of a value of the *fact*—the value of the target variable, which corresponds to the central theme we wish to analyze. The values of facts are numeric because a primary goal of multidimensional data analysis is to find aggregate quantities, such as totals or averages. Examples of facts for a sales data warehouse include *dollars_sold* (sale amount in dollars), *units_sold* (number of units sold), and *amount_budgeted*. The data cube in Fig. 9.4 depicts *dollars_sold* as the fact. Each cell comprises a value for *dollars_sold* for that *item*, during that quarter (*time*), in the store in that city (*location*).

Although, we think of cubes as 3D geometric structures, in data warehousing the data cube is *n*-dimensional. A simple 2D data cube is, in fact, a table or spreadsheet. Suppose we want to view the data according to two dimensions: item and time, for a particular location. The data will be represented by a table with *dollars_sold* (the selected fact) values displayed.

Now suppose we want to view the data with three dimensions: *item*, *time*, and *location*. The 3D data is representative of a series of 2D tables. The conceptual representation of the data can be done in the form of a data cube, as shown in Fig. 9.4.

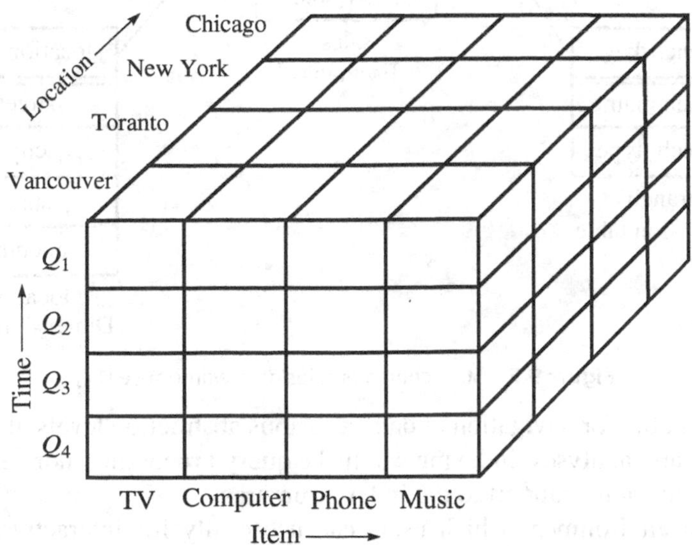

Figure 9.4 A 3D cube representation [17] (dimensions: *item, time,* and *location*; fact value in the cell: *dollars_sold*)

By adding the fourth dimension, for instance, *branch*, we can consider 4D cubes as being a series of 3D cubes: one cube for branch 1, another for branch 2, and so on. By continuing this way, we can represent any n-dimensional data as a series of $(n - 1)$-dimensional cubes. The data cube is a metaphor for multidimensional storage of data. The actual physical storage of such data may be different from the way it is logically represented.

Cubes can be conveniently stored in relational databases with the use of a data structure known as the *star schema*. One benefit of the star schema is its utilization of the standard database technology to attain the power of OLAP.

In a star schema (Fig. 9.5), the data warehouse comprises a *fact table*, which is simply a huge central table and a group of attendant *dimensional tables*, one table for each dimension. The schema graph looks much like a starburst, wherein the *dimension tables* are represented in a radial pattern around the central *fact table*, which lists the names of the facts, as well as the keys to each of the related *dimensional tables*.

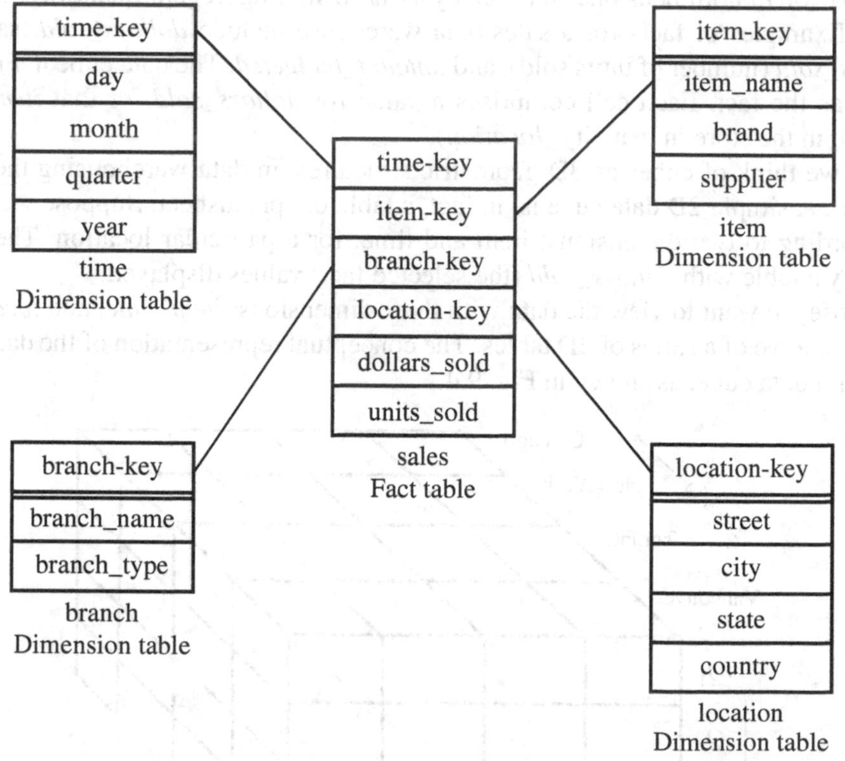

Figure 9.5 Star schema of sales data warehouse [17]

Data cubes are suitable for navigation of data at various abstraction levels, thereby offering rapid interactive querying and analyses. Indexing can make query processing more efficient. The bulk of data warehouse systems lend support to the index structures.

OLAP creates an environment which users can use easily for interactive data analysis. The following are certain typical OLAP operations for multidimensional data. At the center of Fig. 9.6 is a data cube for a retail store (refer to Fig. 9.4).

The **roll-up** operation achieves aggregation on a data cube by *climbing up a concept hierarchy* (from low-level concept of *city* to higher-level concept of *country*), instead of grouping the data by city; the resulting cube groups the data by country. Figure 9.6 shows roll-up climbing up a concept hierarchy for *location* from the level of *city* to the level of *country*.

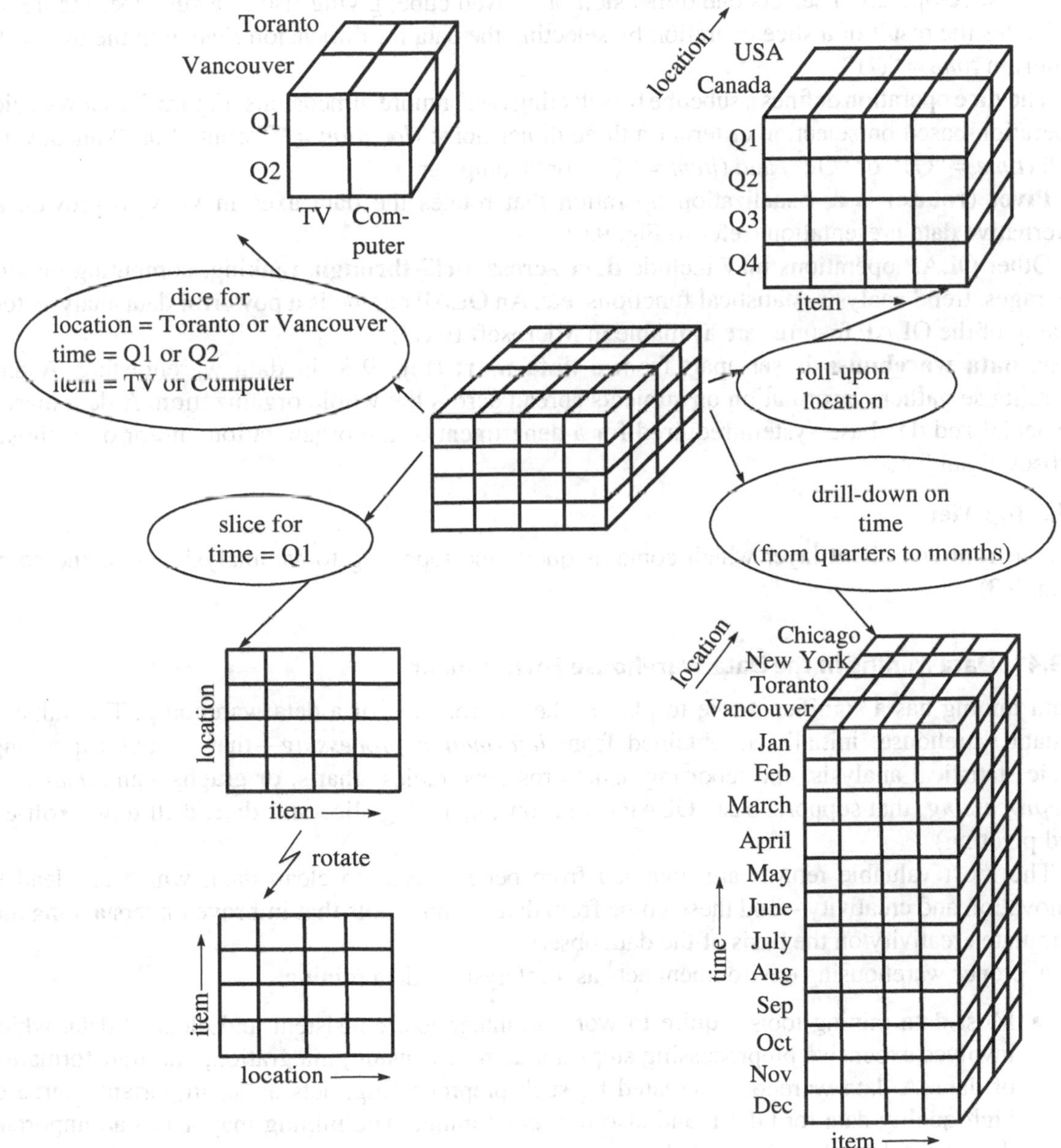

Figure 9.6 Examples of typical OLAP operations [17]

The navigation of **drill-down** takes place from less detailed to more detailed data, which is the opposite of roll-up. This can be achieved by *stepping down a concept hierarchy* for a dimension. Figure 9.6 depicts the result of a drill-down operation by stepping down a concept hierarchy for *time* from the level of *quarter* to the more detailed level of *month*.

The **slice** operation selects one dimension of a given cube, giving rise to a subcube. Figure 9.6 indicates the result of a slice operation by selecting the data for dimension *time* with the use of the criterion *time* = "*Q*1".

The **dice** operation defines a subcube by selecting two or more dimensions. Figure 9.6 shows a dice operation based on selection criteria on three dimensions: (*location* = "Toronto" or "Vancouver") and (*time* = "Q1" or "Q2") and (*item* = "TV" or "Computer").

Pivot (rotate) is a visualization operation that rotates the data axes in view to provide an alternative data presentation (refer to Fig. 9.6).

Other OLAP operations may include **drill-across, drill-through**, ranking, computing moving averages, trend analysis, statistical functions, etc. An OLAP engine is a powerful data analysis tool (many of the OLAP features are available in Microsoft Excel).

A **data warehouse** is set apart from a **data mart** (Fig. 9.3) in data warehousing. A data warehouse gathers information on subjects spread across the whole organization. A data mart is a specialized database system required for a department of the organization and/or on a chosen subset of subjects.

The Top Tier

It is a front-end client layer which contains query and reporting tools, analysis tools, and so on (Fig. 9.3).

9.3.4 Data Mining in the Data Warehouse Environment

Data mining has a significant role to play in the environment of a data warehouse. The value of a data warehouse, initially, is obtained from *information processing*—that supports querying, basic statistical analysis, and reporting using crosstabs, tables, charts, or graphs—and *analytical preprocessing* (that supports basic OLAP operations, including slice-and-dice, drill-down, roll-up, and pivoting).

The most valuable returns are obtained from better access to clean data, which can lead to innovation and creativity—and these come from data mining tools that improve understanding and stimulate creativity on the basis of the data observed.

A proper warehousing environment acts as a catalyst for data mining.

- Most data mining tools require to work on integrated, consistent and cleaned data, which involves expensive preprocessing steps including cleansing, integration, and transformation of data. A data warehouse, created by such preprocessing, acts as an important source of high-quality data for OLAP and also for data mining. Data mining may act as an important device for data cleansing and also for data integration.
- Exploratory analysis is required for effective data mining. OLAP and data visualization processes offered by the data warehouse environment serve to improve the power and flexibility of data mining.

- Data mining algorithms are capable of taking advantage of big volumes of data in a data warehouse. While seeking patterns that identify rare events, possessing huge amounts of data guarantees that there is an appropriate amount of data for analysis.

We can force the data into main memory, and access efficiently by the data mining algorithms or use special-purpose algorithms for disk access.

Some applications may demand typically much smaller dataset for mining. We can create a pseudo dataset that is statistically relevant; it can be accessed (e.g., in main memory) by the data mining algorithms.

Figure 9.7 shows how data mining and database technology interact. Data warehousing deals with the issue of managing data, which is characteristically not addressed clearly in most descriptions of data mining algorithms. It is the usual practice to run the data mining algorithms repeatedly several times, as part of the overall data mining process, using various approaches, different variables, and so on, in an exploratory way before settling on a final result. The iterative process comprises:

1. *Data cleansing and data integration*—This encompasses removal of noisy and unrelated data and a combination of several data sources, often diverse (construction of data warehouse).
2. *Data selection*—This encompasses identification of useful data related to the analysis and its retrieval from data warehouse.
3. *Data transformation*—This involves transformation of the chosen data into appropriate forms suited to the process of pattern recognition.
4. *Pattern recognition and evaluation*—This is an important step wherein clever methods are used for the extraction of patterns which are potentially useful.
5. *Knowledge representation*—The knowledge discovered is presented in visuals, for better interpretation by the user.

Data warehousing is different from data mining, but the former complements the latter and vice versa. Data mining is not limited to analysis of data stored in data warehouses. It may also analyze transactional, spatial, textual and multimedia data which are not easy to model using existing multidimensional database technology.

9.4 MINING FREQUENT PATTERNS AND ASSOCIATION RULES

Pattern classification and *Regression* are both concerned with model building using *supervised learning* methods. Another data mining application is concerned with *cluster analysis*. Unlike classification and regression, clustering analyzes data objects using *unsupervised learning* methods. All of these data mining applications have been discussed in the earlier chapters of the book.

Here we are concerned with yet another application of data mining: *frequent patterns detection* and generating *association rules*, using *unsupervised learning*. Mining frequent patterns originated with the study of customer transactions databases to determine association between purchases of different items/service offerings. This popular area of application is called *market basket analysis*, which studies customers' buying habits for products that are purchased together. This application is commonly encountered in online *recommender systems* where customers examining product(s) for possible purchase are shown other products that are frequently purchased in conjunction with the desired product(s); display from Amazon.com, for example.

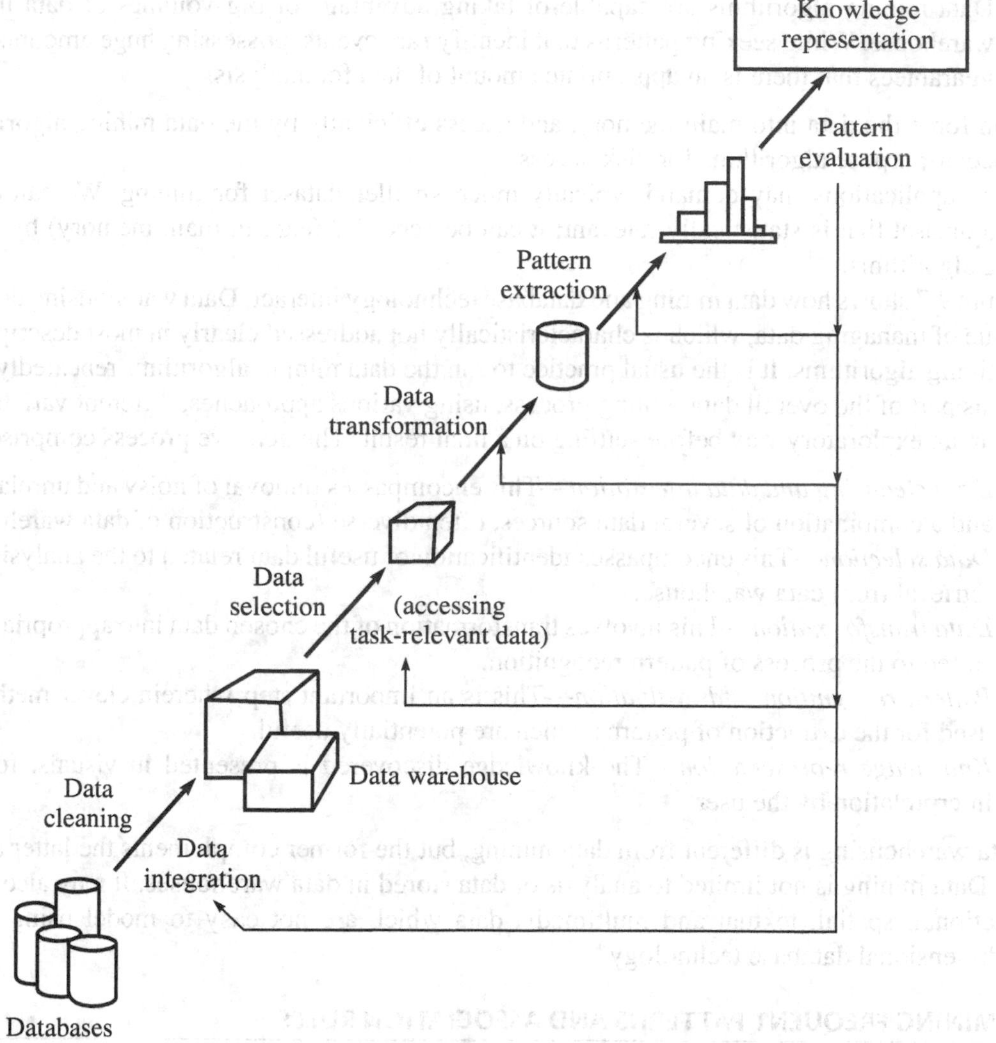

Figure 9.7 Interaction of data mining and database technology

Other than market basket data, association analysis can also be applied to web mining, medical diagnosis, text mining, scientific data analysis, and other application domains. A medical researcher wishes to find out about the symptoms that match the confirmed diagnosis. While analyzing Earth science data, the association patterns often disclose interesting links among the ocean, land and atmospheric pressures. The association approach, in terms of text documents, can be used to discover word co-occurrence relationships (used to find linguistic patterns). Association analysis can be used to discover web usage patterns.

Even though association analysis is generally applicable to a wider variety of datasets, for illustrative purposes, our discussion will focus mainly on market basket data. A manager of a retail outlet would want to know details about his customers' purchase habits—to be specific, 'which are the products sought after by the customers on a visit to the outlet?'. To arrive at the answer, market basket analysis may be done on the retail data of customer dealings at the outlet. Marketing can be planned, and advertising strategies formulated on the basis of the results. The results can also be utilized to design new catalogs or to modify store layouts. With the help of market basket analysis, retailers can plan which products need to be offered on sale at discounted rates.

There are two stages in the market basket analysis. The first stage proceeds by *mining frequent patterns* in customers' transactions (sets of items/service offerings that are frequently purchased together). The second stage takes frequent patterns and generates *association rules* from them. In this section, we first introduce the basic concepts of the two-stage procedure, and then present the basic methods for detection of frequent patterns and generation of association rules.

9.4.1 Basic Concepts

Market basket data is transaction data that typically describes the *transactions* (purchase of one or more items or service offerings), and some additional information such as product/service description, information about the salesperson and branch, customer identity, date and time of purchase, cash or credit card payment, and so on. For the purpose of analysis, the products or service offerings are referred to as *items*, and the set of items sold together in transactions as *itemsets*. *Frequent itemsets* are sets of items that are frequently sold together. Detection of such frequent patterns from transactional data is our task.

Let $X = \{x_j\}; j = 1, ..., n$ be the set of items available at a store. Let $\mathcal{D}$, the *task-relevant data*, be a set of database transactions $\{s^{(i)}\}$; $i = 1, 2, \ldots, N$; where each transaction $s^{(i)}$ is a non-empty itemset $X^{(i)}$ such that $X^{(i)} \subseteq X$. Table 9.1 illustrates an artificial example of basket data with $N = 9$, and $n = 5$.

Table 9.1 An artificial example of basket data

Transactions $s^{(i)}$	Items purchased
$s^{(1)}$	x_1, x_2, x_5
$s^{(2)}$	x_2, x_4
$s^{(3)}$	x_2, x_3
$s^{(4)}$	x_1, x_2, x_4
$s^{(5)}$	x_1, x_3
$s^{(6)}$	x_2, x_3
$s^{(7)}$	x_1, x_3
$s^{(8)}$	x_1, x_2, x_3, x_5
$s^{(9)}$	x_1, x_2, x_3

Each of these transactions gives information about which products are purchased with which other products. The mining task is to detect frequent patterns—itemsets that are frequently purchased together.

The toy example of basket data, provided in Table 9.1, can be depicted in binary format as in Table 9.2. While the rows are representative of the transactions of individual customers, the columns depict the items in the store. A '1' in the location (i, j) implies that customer i bought item j; and a '0' is indicative of the fact that the item was not bought. Each item x_j is a Boolean variable, representative of the existence or absence of the items inside a basket. Each basket can then be depicted using a Boolean vector of values assigned to these variables.

The binary-format general representation of data is offered by a data matrix with N rows (corresponding to transactions/baskets), and n columns (corresponding to the items held in the store). Matrices such as this may be large with N in millions and n in the tens of thousands; and is generally very sparse because a typical basket holds very few items.

The existence of an item in a transaction is taken to be rather more significant than its non-existence. Therefore, an item is *asymmetric* binary variable. Each transaction comprises a subset of items selected from X; the number of items existing in an itemset may range from 1 to n. If the itemset has k items, it is called *k-itemset*.

Table 9.2 Binary format representation of basket data given in Table 9.1

Transactions $s^{(i)}$	x_1	x_2	x_3	x_4	x_5
$s^{(1)}$	1	1	0	0	1
$s^{(2)}$	0	1	0	1	0
$s^{(3)}$	0	1	1	0	0
$s^{(4)}$	1	1	0	1	0
$s^{(5)}$	1	0	1	0	0
$s^{(6)}$	0	1	1	0	0
$s^{(7)}$	1	0	1	0	0
$s^{(8)}$	1	1	1	0	1
$s^{(9)}$	1	1	1	0	0

A major challenge in mining frequent itemsets from a large dataset is computational complexity. Any given set of transactional data is going to have many possible itemsets. To find frequent patterns, the brute-force method requires executing a linear scan of the data once for every possible itemset. For a few hundred products, the number of possible itemsets quickly climbs to millions. The challenge is to detect frequent itemsets in a large data matrix in a relatively efficient computational manner.

Let us represent detected frequent pattern sets by X_{fl}; $l = 1, \ldots, m$. Let Z and Y be two sets of items. A frequent itemset X_{fl} is said to contain Z and Y if $Z \subset X_{fl}$, $Y \subset X_{fl}$, $Z \neq \varnothing$, $Y \neq \varnothing$, and $Z \cap Y = \varnothing$. An *association rule* is an implication of the form 'If a customer purchases Z set of items, then the customer also purchases Y set of items'. This implication can be captured in the form of rules

$$\text{IF } Z \text{ THEN } Y$$

or (9.1)

$$Z \Rightarrow Y$$

(The presence of items in antecedent Z implies the presence of items in consequent Y)

The *rule structure* is rather simple and easy to interpret, which explains why the association rules are generally found to be appealing. However, such rules consist of a relatively weak form of knowledge. They are merely co-occurrence patterns in the observed data instead of strong statements that characterize the population as a whole. Indeed, in a sense that the term 'rule' usually indicates a causal interpretation (from the left hand to the right-hand side), the term 'association rule' is, strictly speaking, a misnomer since these patterns are inherently correlational but need not be causal.

9.4.2 Measures of Strength of Frequent Patterns and Association Rules

Association rule mining consists of finding *frequent itemsets first*, from which *strong* association rules are generated. Frequent itemsets are often interesting themselves, and there is no need to generate association rules from them. In other cases, the goal is to turn the frequent itemsets into association rules.

The *search* for detecting frequent patterns is the major goal of association rule mining algorithms. Data mining can discover a very large number of patterns/rules; in some cases, the number can reach millions. Algorithms for mining are designed specifically to operate on very large sparse datasets in a relatively efficient manner in terms of storage and computation requirements. These algorithms explicitly try to minimize the number of linear scans through the database. Restricting to 'strong' patterns/rules is not only helpful in reducing computational complexity but also is more *useful* from users' perspective.

Let us defer discussion of how to mine frequent itemsets/association rules, and instead ask another question: Which patterns/rules are useful or interesting? Whether the knowledge discovered is useful or interesting, is very subjective and depends upon the application and the user. In the following, we present traditionally used *pattern evaluation measures* that reflect the *strength* of patterns/rules in terms of usefulness.

The transactional datasets have usually *skewed distributions*; most of the k-itemsets have relatively low to moderate frequencies, but a small number of them have very high frequencies. Restricting to high-frequency itemsets that apply to relatively large number of baskets (called patterns with high *coverage* or *support*), results in strong patterns and therefore strong rules may be derived. If Z and Y are two nonempty subsets of frequent itemset X_{fl} such that $Z \cap Y = \varnothing$ and $Z \cup Y = X_{fl}$, then the coverage or support of X_{fl} is also coverage or support of the rule $Z \Rightarrow Y$.

In addition to coverage or support, another traditionally used measure for a strong association rule is its *accuracy* (also called *confidence*) which measures the proportion of baskets to which the rule applies.

It is common for association rule mining algorithms to use a support-confidence framework. Even though this framework helps to remove unwanted exploration of a certain number of rules that are not interesting to the users, several of the generated rules are yet not interesting. It is possible to supplement the support-confidence framework with additional interesting measures on the basis of correlation analysis. One such common measure is *Lift*.

Support: The support of a rule measures the number of transactions including both the antecedent and consequent itemsets. It is referred to as ' support' as it is used to measure the degree to which the data 'supports' the validity of the rule.

Given an association rule:

$$\text{IF } Z \text{ THEN } Y$$

or

$$Z \Rightarrow Y$$

the antecedent itemset is Z and consequent itemset is Y. *Frequency* of itemset Z, $f_r(Z)$, is the number of transactions (baskets) that contain Z. *Support_count* is $f_r(Z \cup Y)$, i.e., number of baskets that contain both Z and Y. We define the *support* of association rule (9.1) as

$$\text{Support}(Z \Rightarrow Y) = \frac{f_r(Z \cup Y)}{N} \tag{9.2}$$

Confidence: Besides support, there is another measure to convey the level of uncertainty about the IF-THEN rule. This is called the *confidence* of the rule—a measure, which makes a comparison between the co-occurrence of the antecedent and consequent itemsets in the database, and the occurrence of the antecedent itemsets. Confidence is the ratio of the number of transactions including all the antecedent and consequent itemsets (the support_count) to the number of transactions including all the antecedent itemsets.

$$\text{Confidence}(Z \Rightarrow Y) = \frac{f_r(Z \cup Y)}{f_r(Z)} \tag{9.3}$$

To view the relationship between support and confidence from a different perspective, let us express these measures in terms of probability functions. We can express support as the (estimated) probability that a transaction selected randomly from the database will contain all items in the antecedent and the consequent, i.e., $f_r(Z \cup Y)/N$ can be viewed as empirical estimate of $P(Z \cup Y)$. On similar lines, confidence may be expressed as the (estimated) conditional probability that a transaction selected randomly will include the item in the consequent *given* that the transaction includes all the items in the antecedent, i.e., we can view

$$f_r(Z \cup Y)/f_r(Z) = \frac{f_r(Z \cup Y)/N}{f_r(Z)/N} \tag{9.4a}$$

as an empirical estimate of

$$P(Y \mid Z) = \frac{P(Z \cup Y)}{P(Z)} \tag{9.4b}$$

Note that instead of a simple frequency-based estimate, we could use more robust estimate of probability functions $P(Z \cup Y)$ and $P(Y|Z)$ for small sample sizes. However, since association rules are typically used in applications with very large datasets, the simple frequency-based estimates will be quite sufficient in such cases.

Also note that the inference made by an association rule does not necessarily imply causality. Causality requires knowledge about cause and effect attributes in the data; a model is derived from this data (like a decision tree) that predicts what an unknown output will be for a given input. Association rules do not predict uniquely what the consequent item will be for a given antecedent itemset. Various rules are in force (not a single rule) for an antecedent itemset that give various values of the consequent.

Therefore, there is no consistent and global description of the dataset offered by the set of association rules. However, the collection of association rules or frequent itemsets can be considered as providers of an alternative representation of the original dataset. In the condensed form, this representation will be of use.

The collection of association rules can be very easily condensed by discarding the rules with support and confidence measures lesser than certain thresholds.

The model structure for association rules is the set of all possible conjunctive probabilistic rules that satisfy the constraint that the support and the confidence are greater than or equal to the user-specified minimum support (denoted by *minsup*) and minimum confidence (denoted by *minconf*), respectively. These thresholds constitute the 'score function' used in association rule searching (minsup = 0.1 means we want only those rules that cover at least 10% of the data. Minconf = 0.9 means that we want rules that are at least 90% accurate). A pattern gets a score of 1 if it satisfies both the threshold conditions, and a score of 0 otherwise. The goal is to find all rules (patterns) with a score of 1.

Lift: Yet another way to gauge the power of an association rule is by comparing the confidence of the rule with a benchmark value, wherein it is assumed that the antecedent and consequent itemsets are independent. When independence is assumed, the support would be $P(Z \cup Y) = P(Z) \times P(Y)$; and the benchmark confidence would be $P(Z) \times P(Y)/P(Z)$.

Lift (also called *improvement*) is the ratio of the confidence of the rule and the benchmark confidence.

$$\text{Lift}(Z \Rightarrow Y) = \frac{\text{Confidence}}{\text{Benchmark confidence}} \tag{9.5}$$

A lift value greater than 1.0 suggests that there is some usefulness to the rule. The larger the lift, the greater the strength of association.

9.4.3 Frequent Item Set Mining Methods

Several efficient algorithms have been developed for frequent itemset mining, which may result in association rules. The *Apriori algorithm* [162] is an important algorithm used to mine frequent itemsets. The algorithm is named on the basis of fact that it makes use of *prior knowledge* of frequent itemset characteristics/features. It explores the *Apriori property* that *all nonempty subsets of a frequent itemset must also be frequent*. Variations that include hashing and reducing transactions (a transaction which does not hold any frequent k-itemsets cannot contain any frequent $(k + 1)$-itemsets, and can be removed from further data scans) can be employed to ensure the efficiency of the process. Partitioning (mining on each partition and then combining the results) and sampling (mining on a data subset) of the data are some of the other variations capable of reducing the number of data scans.

As we will shortly see, the Apriori algorithm is a generate-and-test method; it generates *candidate itemsets* using Apriori property, and then tests the candidates with respect to *minsup*, the minimum support threshold. Though there is significant performance gain in comparison to determining the support of all possible combinations of items, candidate generation process can be quite costly.

Is it possible to design a technique that mines the complete set of frequent itemsets without an expensive candidate generation procedure? One technique in this endeavor is known as *frequent patterns growth*, or merely *FP growth*; it mines the entire set of itemsets without candidate generation. It encodes the dataset with the help of a compact data structure referred to as an *FP-tree*, and directly takes out the frequent itemsets from this structure.

In the following section, we limit our discussion to basic Apriori algorithm employing support-confidence framework for pattern evaluation [162].

Apriori Algorithm

Apriori makes use of a repetitive approach called *level-wise* search, wherein $(k + 1)$-itemsets are explored using k-itemsets. First, the set of frequent 1-itemsets is discovered by scanning the database to gather the count for each item, and gathering the items that satisfy minimum support. The set that results is depicted by S_1. S_1 is then used to find S_2—the set of frequent 2-itemsets that satisfy minimum support, with the help of which S_3 is found, and so on, until we frequent itemsets of all sizes has been generated. To look for each frequent k-itemset, one complete scan of the database is necessary.

To make the level-wise generation of frequent itemsets more efficient, the search space is reduced using Apriori property. According to Apriori property, the nonempty subsets of a frequent itemset should also be frequent. This property is formed on the basis of following observation. In case an itemset Z does not fulfill the minimum support threshold, *minsup*, then Z is not frequent. On adding an item x_j to the itemset Z, the occurrence of the resultant itemset $Z \cup x_j$ cannot be more frequent than Z. Therefore, $Z \cup x_j$ is not frequent either. To comprehend this, examine how S_k for $k \geq 2$ is found with the help of S_{k-1}. It follows a two-step process comprising *join* and *prune*.

1. **Join:** To find S_k, a set of candidate k-itemsets is generated by joining S_{k-1} with itself, represented as $S_{k-1} \bowtie S_{k-1}$. The *candidate itemset*, obtained after the join step, is denoted by C_k, i.e.,

$$C_k = S_{k-1} \bowtie S_{k-1}$$

To see how the 'join' step works, let X_{f1} and X_{f2} be two itemsets in the set of frequent itemsets S_{k-1}. The jth item, x_j, in the itemset X_{fl} is denoted as $X_{fl}(j)$, e.g., $X_{fl}(k-2)$ refers to the second last item in X_{fl}.

- Frequent itemsets X_{f1} and X_{f2} of S_{k-1} are joinable if their first $(k-2)$ items are in common, i.e., $X_{f1}(1) = X_{f2}(1)$, $X_{f1}(2) = X_{f2}(2)$, …, $X_{f1}(k-2) = X_{f2}(k-2)$.

- Another condition imposed on the join step is

$$X_{f1}(k-1) < X_{f2}(k-1)$$

That is, if $(k-1)$th item in X_{f1} is x_p; $p = 1, \ldots, n$, and $(k-1)^{\text{th}}$ item in X_{f2} is x_q; $q = 1, \ldots, n$, then $p < q$. This condition ensures that no duplicates are generated.

The resulting itemset formed by joining X_{f1} and X_{f2} is $\{X_{f1}(1), X_{f1}(2), \ldots, X_{f1}(k-1), X_{f2}(k-1)\}$.

2. **Prune:** C_k is a superset of S_k; its members may or may not be frequent, but all of the frequent k-itemsets are included in C_k. A database scan to determine the frequency of each candidate in C_k would result in the determination of S_k (i.e., all candidates having support no less than *minsup* are frequent by definition, and therefore belong to S_k).

But, C_k can be large, which could involve weighty computation. With the use of Apriori property as shown below, the size of C_k can be reduced.

No $(k-1)$-itemset, which is not frequent, can form a subset of a frequent k-itemset. Therefore, if any $(k-1)$-item subset of a candidate k-itemset is not in S_{k-1}, then the candidate cannot be frequent either and hence, can be deleted from C_k. This testing of subset does not require time, and can be performed by maintaining a hash tree of all frequent itemsets.

Let us now demonstrate the concepts using the basket data of Table 9.1 [17].

1. In the first iteration of the algorithm, each item x_j; $j = 1, \ldots, n$, is a member of candidate 1-itemsets C_1. The algorithm simply scans all the transactions in order to count the number of occurrences, $f_r(x_j)$, of each item (Fig. 9.8).

Scan $\mathcal{D}$ for frequency →

C_1

Itemset	Frequency
x_1	6
x_2	7
x_3	6
x_4	2
x_5	2

Compare candidate support with *minsup* →

S_1

Itemset	Support
x_1	6/9
x_2	7/9
x_3	6/9
x_4	2/9
x_5	2/9

Figure 9.8

2. We assume that *minsup* = 2/9 = 22%. The set of frequent 1-itemsets S_1 consists of the candidate itemsets satisfying *minsup* (Fig. 9.8).
3. To determine the set of frequent 2-itemsets S_2, the algorithm uses the join step $C_2 = S_1 \bowtie S_1$; C_2 is the set of candidate 2-itemsets.

C_2 consists of $\binom{|S_1|}{2}$ 2-itemsets, i.e., $\binom{5}{2} = \dfrac{5!}{2!\,3!} = \dfrac{5 \times 4}{2} = 10$ 2-itemsets.

During the prune step, no candidates are removed from C_2 as each subset of the candidates is frequent. In the next step, scanning of the transactions in $\mathcal{D}$ takes place followed by the accumulation of the frequency of each candidate itemset in C_2 (Fig. 9.9).

4. The set of frequent 2-itemsets S_2 consists of candidate 2-itemsets with support ≥ 2/9 (Fig. 9.9).

Generate C_2 candidates from S_1 and scan $\mathcal{D}$ for frequency →

C_2

Itemset	Frequency
$\{x_1, x_2\}$	4
$\{x_1, x_3\}$	4
$\{x_1, x_4\}$	1
$\{x_1, x_5\}$	2
$\{x_2, x_3\}$	4
$\{x_2, x_4\}$	2
$\{x_2, x_5\}$	2
$\{x_3, x_4\}$	0
$\{x_3, x_5\}$	1
$\{x_4, x_5\}$	0

Compare candidate support with *minsup* →

S_2

Itemset	Support
$\{x_1, x_2\}$	4/9
$\{x_1, x_3\}$	4/9
$\{x_1, x_5\}$	2/9
$\{x_2, x_3\}$	4/9
$\{x_2, x_4\}$	2/9
$\{x_2, x_5\}$	2/9

Figure 9.9

5. The generation of the set of candidate 3-itemsets C_3 is done as follows:

$$\text{Join} : C_3 = S_2 \bowtie S_2$$

$$C_3 = [\{x_1, x_2\}, \{x_1, x_3\}, \{x_1, x_5\}, \{x_2, x_3\}, \{x_2, x_4\}, \{x_2, x_5\}]$$
$$\bowtie [\{x_1, x_2\}, \{x_1, x_3\}, \{x_1, x_5\}, \{x_2, x_3\}, \{x_2, x_4\}, \{x_2, x_5\}]$$
$$= [X_{f1}, X_{f2}, X_{f3}, X_{f4}, X_{f5}, X_{f6}] \bowtie [X_{f1}, X_{f2}, X_{f3}, X_{f4}, X_{f5}, X_{f6}]$$

- X_{f1} is not joinable with X_{f1}, since their last item-numbers are equal; we require that last item-number of the first set to be less than that of the second set.
- $X_{f1} \bowtie X_{f2} = \{x_1, x_2\} \bowtie \{x_1, x_3\} = \{x_1, x_2, x_3\}$
- $X_{f1} \bowtie X_{f3} = \{x_1, x_2\} \bowtie \{x_1, x_5\} = \{x_1, x_2, x_5\}$
- $X_{f1} \bowtie X_{f4} = \{x_1, x_2\} \bowtie \{x_2, x_3\}$; not joinable because their first item-numbers are not equal
- $X_{f1} \bowtie X_{f5} = \{x_1, x_2\} \bowtie \{x_2, x_4\}$; not joinable because their first item-numbers are not equal
- $X_{f1} \bowtie X_{f6} = \{x_1, x_2\} \bowtie \{x_2, x_5\}$; not joinable because their first item-numbers are not equal

- $X_{f2} \bowtie X_{f1} = \{x_1, x_3\} \bowtie \{x_1, x_2\}$; not joinable because the last item-number of the first set is not less than that of the second set
- $X_{f2} \bowtie X_{f2} = \{x_1, x_3\} \bowtie \{x_1, x_3\}$; not joinable because the last item-number of the first set is not less than that of the second set
- $X_{f2} \bowtie X_{f3} = \{x_1, x_3\} \bowtie \{x_1, x_5\} = \{x_1, x_3, x_5\}$

$$\vdots$$

$$C_3 = [\{x_1, x_2, x_3\}, \{x_1, x_2, x_5\}, \{x_1, x_3, x_5\}, \{x_2, x_3, x_4\}, \{x_2, x_3, x_5\}, \{x_2, x_4, x_5\}]$$

Prune: All nonempty subsets of frequent itemsets must also be frequent. Do any of the candidates have a subset that is not frequent?

- The 2-item subsets of $\{x_1, x_2, x_3\}$ are $\{x_1, x_2\}$, $\{x_1, x_3\}$, and $\{x_2, x_3\}$. All of these subsets are members of S_2. Therefore, we keep $\{x_1, x_2, x_3\}$ in C_3.
- The 2-item subsets of $\{x_1, x_2, x_5\}$ are members of S_2; so we keep $\{x_1, x_2, x_5\}$ in C_3.
- The 2-item subsets of $\{x_1, x_3, x_5\}$ are $\{x_1, x_3\}$, $\{x_1, x_5\}$, and $\{x_3, x_5\}$. $\{x_3, x_5\}$ is not a member of S_2, and so it is not frequent. Therefore, we remove $\{x_1, x_3, x_5\}$ from C_3.
- Similarly, we remove $\{x_2, x_3, x_4\}$, $\{x_2, x_3, x_5\}$ and $\{x_2, x_4, x_5\}$ from C_3.

After pruning, C_3 becomes

$$C_3 = [\{x_1, x_2, x_3\}, \{x_1, x_2, x_5\}]$$

Next, transactions in $\mathcal{D}$ are scanned and the frequency of each candidate itemset in C_3 is accumulated. The scan results are: frequency of $\{x_1, x_2, x_3\} = 2$; frequency of $\{x_1, x_2, x_5\} = 2$.

6. The set of frequent 3-itemsets S_3 consists of candidate 3-itemsets with support $\geq 2/9$. This gives

$$S_3 = \{\{x_1, x_2, x_3\}, \{x_1, x_2, x_5\}\} = \{X_{f1}, X_{f2}\}$$

7. The algorithm uses $S_3 \bowtie S_3$ to generate candidate set of 4-itemsets C_4. The join step results in $\{x_1, x_2, x_3, x_5\}$. The prune step results in C_4 = empty set, since its subset $\{x_2, x_3, x_5\}$ is not frequent.

The algorithm terminates, having found all the frequent itemsets.

9.4.4 Generating Association Rules from Frequent Itemsets

As said earlier, association rule mining works in two stages.

Stage 1: *Generation of all frequent itemsets:* A frequent itemset is an itemset with transaction support above *minsup*.

Stage 2: *Generation of all confident association rules from frequent itemsets:* A confidence association rule is a rule with confidence above *minconf.*

The first stage usually presents the computational challenge for most association analysis data, as was discussed earlier in Section 9.4.3 while describing the Apriori algorithm. In the second stage, the computation of the confidence is quite simple.

For the generation of rules for each frequent itemset X_{fl}, the nonempty subsets in X_{fl} are used. For each subset Y, we output a rule of the form

$$Z \Rightarrow Y; Z = X_{fl} - Y \tag{9.6a}$$

$$\text{Confidence}\,(Z \Rightarrow Y) = \frac{f_r(Z \cup Y)}{f_r(Z)} = \frac{f_r(X_{fl})}{f_r(Z)} \tag{9.6b}$$

All the frequencies or counts required to compute confidence are available. This is because if X_{fl} is frequent, then its nonempty subsets are also frequent, and its support_count is already recorded in the mining process (it is possible to store frequent itemsets in hash tables along with their counts so that they can be rapidly accessed). Therefore, no data scan is required to generate rules.

But this strategy of generating rules is not efficient. Designing an efficient algorithm leads us to observe that support_count of X_{fl} in the confidence computation aforementioned, does not alter with the change of Y. For a rule $X_{fl} - Y \Rightarrow Y$ to sustain, all rules of the form $X_{fl} - Y_{\text{sub}} \Rightarrow Y_{\text{sub}}$ will also hold, where Y_{sub} is the nonempty subset of Y, as the support_count of $(X_{fl} - Y_{\text{sub}})$ will have to be less than or equal to the support_count of $(X_{fl} - Y)$.

Therefore, considering a given frequent itemset X_{fl}, if a rule with consequent Y holds, then so will the rules with consequents that are subsets of Y. Hence, from the frequent itemset X_{fl}, we first generate all rules with one item in consequent. Then with the help of the consequents of these rules, we generate all likely consequents with two items, which can appear in a rule, and so on. This way we can build up from single-consequent rules to candidate double-consequent rules, from double-consequent rules to candidate triple-consequent rules, and so on. Each candidate rule should be checked against the hash table to ensure it has more than the minimum confidence specified. However, this usually involves checking far fewer rules than the brute-force technique. Interestingly, this technique of building up candidate (k+1)-consequent rules from actual k-consequent rules is identical to building up candidate (k+1)-itemsets from actual k-itemsets, as explained earlier.

Let us look at an example based on the transactional data represented earlier in Table 9.1. The data contains frequent itemset $X_{f2} = \{x_1, x_2, x_5\}$ (refer to Section 9.4.2). From this frequent itemset, we get the following candidate association rules (nonempty subsets of X_{f2} are: $\{x_1, x_2\}$, $\{x_1, x_5\}$, $\{x_2, x_5\}$, $\{x_1\}$, $\{x_2\}$, $\{x_5\}$).

Rule 1: $\{x_1, x_2\} \Rightarrow x_5$, with confidence $= \dfrac{f_r(X_{f2})}{f_r(\{x_1, x_2\})} = \dfrac{2}{4}$

Rule 2: $\{x_1, x_5\} \Rightarrow x_2$, with confidence $= \dfrac{f_r(X_{f2})}{f_r(\{x_1, x_5\})} = \dfrac{2}{2}$

Rule 3: $\{x_2, x_5\} \Rightarrow x_1$, with confidence $= \dfrac{f_r(X_{f2})}{f_r(\{x_2, x_5\})} = \dfrac{2}{2}$

Rule 4: $x_1 \Rightarrow \{x_2, x_5\}$, with confidence $= \dfrac{f_r(X_{f2})}{f_r(x_1)} = \dfrac{2}{6}$

$$\text{Rule 5: } x_2 \Rightarrow \{x_1, x_5\}, \text{ with confidence} = \frac{f_r(X_{f2})}{f_r(x_2)} = \frac{2}{7}$$

$$\text{Rule 6: } x_5 \Rightarrow \{x_1, x_2\}, \text{ with confidence} = \frac{f_r(X_{f2})}{f_r(x_5)} = \frac{2}{2}$$

If the minimum confidence sought is 70%, only the second, third, and last rules will be reported.

In practice, the amount of computation needed to generate association rules critically depends upon specified *minsup*. The *minconf* has less influence as it does not impact the number of data scans required to be made. We would, at times, wish to get a certain number of rules—say 50—with the maximum possible support at a prespecified minimum confidence level. One method of doing this is to start by laying down the support to be quite high and then successively reducing it, followed by re-execution of the whole two-stage process for each of the support values and repetition of the same till the desired number of rules have been generated.

For more details, refer to [17, 163].

9.5 INTELLIGENT INFORMATION RETRIEVAL SYSTEMS

Retrieval of text-based information from a large database has traditionally been termed as *information retrieval*. The proliferation of multimedia data over the years has created a huge demand for image-based, audio-based, and video-based information from large databases, in addition to text-based information. We will, therefore, use the term 'information retrieval systems' for retrieval of information from text, image, audio, and video databases.

Owing to plenty of text information, there are several text-based information retrieval systems available, such as on-line library catalog systems, on-line document management systems, and so on. Retrieving of information has of late become a subject of discussion, more so with the arrival of text search engines on the Internet. Retrieval techniques are used to look for documents on the Web through queries (The widely popular example is that of Google system).

Information retrieval, as a subject, has developed along with database systems over several years. Database systems have concentrated on querying and transaction processing of structured data. On-line transaction and query processing are the primary jobs of on-line operational database systems. On-Line Transaction Processing (OLTP) systems cover almost all the daily/routine operations of an organization, which include accounting, payroll, purchasing, banking, manufacturing and inventory. On-Line Analytical Processing (OLAP) systems are capable of organizing and presenting data in different formats so that the needs of different users, in terms of data analysis and decision-making can be accommodated. A query is merely a particular request made in a technical language, for a particular data subset or for statistics from data. A query is posed to a database system for the desired processing. These queries can be formulated using Structured Query Language (SQL) or a Graphical User Interface (GUI) (refer to Section 9.3).

In contrast, data mining is used to come up with queries as a pattern or regularity in data in information retrieval systems. An information retrieval problem may typically be the location of relevant documents in a document pool, on the basis of a user's query. This is quite often related to certain keywords which describe the need for an information, although it could be an example relevant document. Such *content-based* retrieval systems are referred to as *intelligent information retrieval systems* [23].

The common approach used for intelligent information retrieval systems is the representation of objects in the database and querying, both as vectors in a high-dimensional space, and use of a proper measure for computation of the similarity between the query vector and the object representation vector. The objects can then be ranked based on the similarity values.

The content-based retrieval systems, which operate with domain specific constraints, are likely to make a real impact in terms of commercial activities. One practical example of such retrieval systems is the content-based image retrieval from Biometric databases. Here, the user has a pattern of interest (fingerprint, for example) and wishes to find similar patterns in the dataset. The notion of *similarity* is critical in content-based retrieval systems; however, an exact match is not the target. Problems of this nature take the following form:

"Find the k objects in the database that are most similar to the specific query pattern".

In the context of a database, the conventional idea of a query is an operation that returns a set of records (or entities) that *precisely* match a set of specifications demanded. Database management systems, traditionally, have been designed to answer specific queries in an efficient manner.

There are several applications wherein we are interested in queries that tend to be *more general* and *less precise*. For instance, in the medical context, we may have the demographic details of a patient—age, gender, nationality, and so on; X-ray reports, blood test details, and information pertaining to other routine physical tests, along with biomedical time-series. It will help the physician diagnose better if he has information about the treatment approach taken for similar patients from the database of the hospital, and the treatment and results, therein. The challenge is to determine the *similarity* among patients on the basis of various data types (in this case, multivariate time-series and image data). However, the concept of a precise match is irrelevant here, as the probability of any other patient matching the measurements of the patient in question, exactly, will be very low.

Mixed-media information retrieval systems is a research subject of current interest. Substantial progress has, however, been made in image-retrieval and text-retrieval systems.

The subject of Information Retrieval is presently at a stage where we are dealing with first-generation machine learning algorithms, which are of significant value in a range of real-world information-retrieval applications that involve text data. To some extent, success has been achieved with regard to image/audio data. However, these first-generation algorithms are not without important limitations. Typically, they work with the assumption that it is possible to transform the text/image/audio data into an $N \times n$ structured data matrix with numeric values alone. It is assumed that the data has been gathered with care into a single database keeping in mind a particular data mining task for a particular retrieval need.

The next ten years are expected to produce advances of high magnitude. Such advancement could be driven by the development of new algorithms that accommodate dramatically more varied sources and kinds of unstructured text/image/audio/mixed media data for information retrieval applications (refer to Section 9.7). This section aims to familiarize the reader with the various types of text, image, audio, video, and audiovisual data, emphasizing features of these data types for some limited retrieval applications.

Measuring Accuracy of Information Retrieval (IR) Systems

There are different measures available for evaluating the performance of retrieval systems. The measures are based on how good the system is in retrieving the relevant information from the given set of objects and a query. The main objective of the evaluation is to measure the accuracy or ability to take the right classification decision. We have earlier described in Section 2.8, the well-known Receiver-Operating Characteristics (ROC) used to characterize the performance of binary classifiers. The IR community has traditionally been using the *Precision-Recall Performance curves*, which are essentially equivalent (except for relabeling of the axes) to the ROC curves.

Precision and Recall: Suppose we have to evaluate the performance of a particular retrieval algorithm in response to a specific query Q on an independent dataset. The objects in the test data have already been classified as 'relevant' or 'irrelevant' to the query Q. The assumption is that the test dataset has not been used to tune the performance of the retrieval algorithm. We can consider the retrieval algorithm as mere classification of the objects in the dataset (in terms of relevance to Q), wherein the true class labels are concealed from the algorithm, but are known for test purposes.

Relevant to Q: +ve class
Not relevant to Q: –ve class

Suppose that the test dataset has N objects.

$N = TP + FP + FN + TN$ = total number of labeled objects

TP = number of correct classifications of +ve test examples ('true positive')

FN = number of incorrect classifications of +ve test examples ('false negative')

FP = number of incorrect classifications of –ve test examples ('false positive')

TN = number of correct classifications of –ve test examples ('true negative')

Table 9.3 summarizes the performance of a retrieval system. This form of reporting classification results is sometimes referred to as *confusion matrix*. Through confusion matrix, it becomes convenient to introduce the terms *precision* and *recall*, which are more suitable measures (than *accuracy*) in such applications, because they measure how precise and how complete the classification is on the +ve class (the class we are interested in).

Table 9.3 Performance of a retrieval system

		Retrieved Objects	
		Classified +ve	Classified –ve
Actual Objects	Actual +ve	*TP*	*FN*
	Actual –ve	*FP*	*TN*

Precision refers to the fraction of retrieved objects which are relevant, whereas *recall* is the fraction of relevant objects which are retrieved. Precision is the number of relevant objects retrieved divided by the total number of objects retrieved by that search (number of correctly classified +ve examples divided by the total number of examples classified as +ve). Recall is the number of relevant objects retrieved divided by the total number of relevant objects in the dataset (number of correctly classified +ve examples divided by the total number of +ve examples in the dataset).

$$\text{Precision} = \frac{TP}{TP + FP} \tag{9.7a}$$

$$\text{Recall} = \frac{TP}{TP + FN} \tag{9.7b}$$

If the algorithm makes use of a distance measure for *ranking* the set of objects, then it is typically parameterized by a *threshold*. Therefore, K_T objects will be returned by the algorithm as +ve class. K_T is a trade-off parameter. With a rise in the number of retrieved objects $K_T = TP + FP$ (i.e., as we increase the threshold and allow the algorithm to declare more objects to be relevant), Recall is expected to increase (in the limit, we can return all objects, in which case Recall = 1), while Precision is expected to fall (as K_T is increased, it will be tougher to return only relevant objects). If we run the retrieval system for different values of the threshold parameter, we will obtain a set of pairs of Precision-Recall points.

Practically speaking, rather than performance evaluation relative to a single query Q, we estimate the average recall-precision performance over a set of queries.

Typical plot of recall-precision of various retrieval algorithms relative to the same dataset and set of queries is shown in Fig. 9.10. Algorithm A has the maximum level of precision for low recall values and low precision for high recall values, while algorithm B has lower precision for low recall values, and higher precision for high recall values, compared to algorithm A. A choice cannot clearly be made between A and B unless we are operating at a particular recall value.

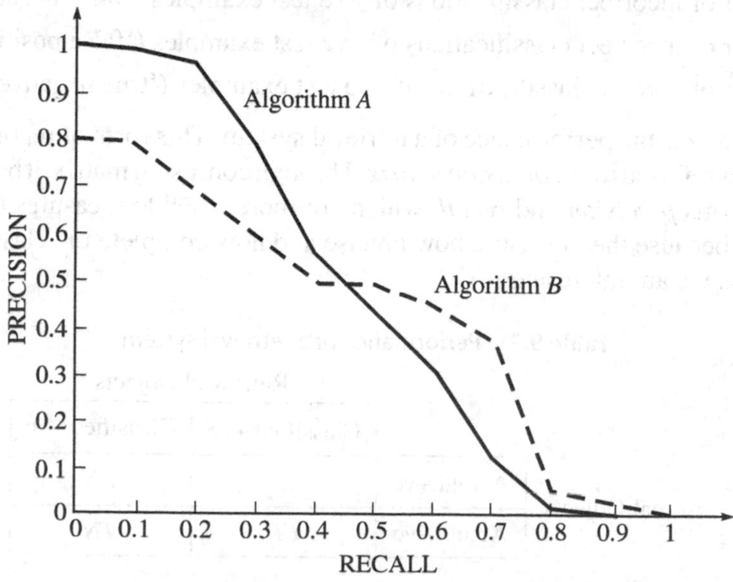

Figure 9.10 Precision-Recall of different algorithms

High recall-low precision returns many results but most of the predicted labels are incorrect. High precision-low recall returns very few results but most of its predicted labels are correct. High scores for both precision and recall indicate that the classifier is returning accurate results as well as returning a majority of all positive results.

There are a number of schemes we can use to summarize precision-recall performance by a single parameter. *F-score* is the commonly used scheme. *F*-score is the harmonic mean of precision and recall:

$$F = \frac{2}{\dfrac{1}{\text{Precision}} + \dfrac{1}{\text{Recall}}} = \frac{2(\text{Precision})\,(\text{Recall})}{\text{Precision} + \text{Recall}} \tag{9.8}$$

The harmonic mean of two numbers tends to be closer to the smaller of the two. Thus for *F*-score to be high, both Precision and Recall should be high.

9.5.1 Text Retrieval

Text documents are information sources and machine learning techniques are used to retrieve useful text from large pools of documents. A lot of research on text retrieval is concentrated on searching for general *representations* for documents that support both the following:

- The capability of retention of as much of the semantic content of the data as possible
- The efficient computation of similarity measures between queries and documents

The text documents should be represented to match the input format of the learning algorithm. As a majority of the learning algorithms employ the attribute-value representation, it becomes necessary to transform a document into a vector space model [164].

In vector space models widely used in a variety of text applications, documents and queries are represented as vectors of *terms* within documents in a collection. A *term* can be a word, word-pair or phrase within a document. A list of terms (features/attributes) needs to be identified and their numeric values (*term weights*) determined to generate a vector space model.

The process of identifying terms (extraction of features) can by itself be quite nontrivial, including issues such as how to define terms. In-depth domain knowledge, assisted by developed software tools for the purpose, is necessary to do a meaningful extraction of features. The following presentation has the objective of providing only a basic understanding of the steps involved in preprocessing of documents for feature extraction.

Steps of preprocessing include:

(a) **Tokenization**

A token is a basic element of a text document. Set of characters between white spaces are treated as tokens. A token often corresponds to a word, but not always, as illustrated: Examples of token: "computer", "mining", "6457", "PDP-11/45", "AT and T", "up/down". Since complex words PDP-11/45, AT and T, up/down, have related subparts, we consider them as single token.

(b) **Lexical Analysis**

- Numbers and terms comprising digits are deleted in conventional information retrieval systems, except certain specific types, such as dates, times, and other prescribed types with regular expressions.
- Hyphens are deleted as a general rule with certain exceptions. In certain words, hyphens form an integral part of the word, e.g., PDP-11/45, and are therefore retained.
- Punctuations are handled just the way as hyphens are.
- All the letters are usually transformed into either upper case or lower case.

(c) **Stop-words Removal**

Stop-words are those words which have no information and have a non-linguistic view. Their removal means removal of non-information words from the documents, such as 'a', 'an', 'the', this, 'that', 'I', 'you', 'she', 'he', 'again', 'almost', 'before', 'after'. Stop-words are very common and their removal does not affect the meaning of the sentences.

(d) **Stemming**

Different forms of the same word are often used in documents for grammatical purposes, such as *categorize, categorizes* and *categorizing*. Additionally, there are families of derivatively related words which have same meaning—*democracy, democratic* and *democratization*. There are often situations where it would be useful for a search for one of these words to return documents containing another word in the set.

The goal of stemming is to reduce inflectional forms and sometimes derivatively related forms of a word to a common root called a *stem*. For instance, the words *computer, compute, computing, computation*, may be mapped to the stem 'compute'; *car, cars, car's, cars*', may be mapped to 'car'.

The most common algorithm of stemming English is PORTER STEMMER [165].

(e) **Phrases**

In addition to the words and word-pairs, phrases can also be used as features.

(f) **Term-Weighting**

Varying (largely ad hoc) suggestions have been made about how the term 'weights' (values of attributes/features) should be set to make retrieval better. These weights should ideally be selected in a way that more relevant documents are placed on a higher rank than the irrelevant or less relevant ones.

A weighting scheme, called the *TF-IDF weighting*, has proven to be very useful in practice [23]. The *TF* (*term frequency*) is the number of occurrences of the term in a document. The information bagged by term frequency pertains to how prominent a word is within a given document. The higher the term frequency (the more often the word occurs), the more is the probability of that word describing the content of the document.

However, in case of a term featuring regularly in several documents in the document pool, the use of *TF* weights to retrieve may not have much discriminative power. The *IDF* (*Inverse-Document-Frequency*) weight facilitates improvement of discrimination. It is defined as the log of the inverse of the fraction of documents in the whole collection containing the term j. If N is the total number of documents, and N_j is the number of documents containing term j, then

$$IDF \text{ weight} = \log (N/N_j) \tag{9.9}$$

The use of the logarithm is motivated by the desire to make the weight relatively insensitive to number of documents N.

The *TF-IDF* weight is simply the product of *TF* and *IDF* weights for a particular term in a particular document.

Each individual document $s^{(i)}$; $i = 1, 2, \ldots, N$, is represented by a *term vector*:

$$\mathbf{x}^{(i)} = [x_1^{(i)} \ x_2^{(i)} \cdots x_n^{(i)}]$$

comprising of n terms $x_j^{(i)}$; $j = 1, 2, \ldots, n$. Each vector $\mathbf{x}^{(i)}$ can be considered as a *surrogate* document for the original document. The whole set of vectors can be represented as an $N \times n$ matrix.

When a document is represented as an n-dimensional term vector, the word order of the original document gets lost, along with the syntactic information, for instance, the structure of the sentence. In spite of such information loss, term vectors can be quite effectively used in retrieval applications.

A query is also represented as an $(n \times 1)$ vector and compared with the document vectors. The most relevant documents for a query are expected to be those characterized by vectors closest to the query, i.e., the documents which bear similarity to the query. A commonly used measure of similarity in this context is the *cosine distance*, which is defined as

$$d_c(s^{(i)}, s^{(k)}) = \frac{\sum_{j=1}^{n} x_j^{(i)} x_j^{(k)}}{\sqrt{\sum_{j=1}^{n} (x_j^{(i)})^2 \sum_{j=1}^{n} (x_j^{(k)})^2}} \tag{9.10}$$

This is the cosine of the angle between two vectors. A cosine value of 0 implies that the two vectors are at right angles to each other (orthogonal) and, therefore, have no match. The closer the cosine value to 1, the smaller the angle and the greater the match between two vectors.

The conventional distance measures we have explored earlier may not be appropriate for sparse numeric data. Term-weight vectors are typically very long (thousands of attributes) and sparse (have many 0 values). Many 0 values in common does not make term weight vectors similar. Cosine similarity measure ignores zero-matches.

Classifiers that are both accurate and efficient for high-dimensional vectors (n in thousands), representing large collections of documents (N may be 50,000 or more), are usually the methods of choice. SVM and Naive Bayes tend to work well for text applications.

The 'bag-of-words' approach of representing documents, in practice, is the most naive approach where even the basic grammatical constructs are ignored. In this approach, documents are considered as merely collection of their words. At the other end of the continuum of approaches for representing documents is 'Natural Language Processing', where, essentially, the computer understands the words, the relationships among them, and the concepts they represent in the real world [166].

Language is obviously a critical component of how people communicate and how information is stored in the business world and beyond. Some estimates say that up to 80 percent of important information for business is stored in the form of text. Text is data, so data mining process is useful in finding and exploiting useful information in text (the application referred to as *text analytics*). There are useful text analytics applications that are not exactly data mining; however, these methods are often incorporated into text analytics solutions. Some examples are spell checking and grammar checking, translation from one human language to another, search engines, summarizing documents, etc. Important information in text data from Social Media is opening up a new challenge for text analytics (refer to Section 9.7).

Text analytics has been studied extensively in information retrieval with many textbooks and survey articles [167–171]

9.5.2 Image Retrieval

Many decades of research in pattern recognition and computer vision has undoubtedly shown that the performance of humans in visual understanding and recognition is extremely difficult to replicate with computer algorithms. Particular problems, such as biometric recognition, can be handled with success, but research is still to be undertaken on general purpose image-representation systems. The capacity to extract semantic content from raw image data remains relatively unique to the brain. Therefore, unsurprisingly, most existing techniques for image retrieval are dependent on relatively low-level visual clues.

For similarity searching in image data, we take into account two primary families of retrieval systems:

- *Description-based retrieval systems*, which retrieve objects on the basis of image descriptions, such as keywords, captions, size, and time of creation.
- *Content-based retrieval systems*, which support retrieval of objects on the basis of the image content, such as color histogram, texture, pattern, image topology, and shape of the objects and their layouts and locations within the image.

In a content-based retrieval system, there are often two kinds of queries: *image sample-based queries*—to find all of the images that are similar to the given image sample; and *image feature specification queries*—to return images that are close to the specified features (this set of features is unable to capture the description of the whole image data but provides salient clues).

The results of description-based retrieval are typically of poor quality. Recent progress in web-based image clustering and classification techniques has made the quality of description-based web image retrieval much better. Image-surrounded text information along with web linkage information can help in the extraction of proper description, and group images which describe a similar theme together.

Content-based retrieval has wide applications—medical diagnosis, weather forecasts, TV production, web search engines for images, and *e*-commerce.

Some systems, such as QBIC (Query By Image Content)—developed by researchers at IBM—support both sample-based and image feature specification queries. There are also systems which support retrieval based on both content and description.

To a certain extent, the image retrieval issue is like solving the general image representation problem—the problem of extraction of semantic content from image data.

Color-Histogram Based Representation: A digital image is usually the result of a discretization process (sampling) of a continuous image function, and it is stored in the computer as a two-dimensional array $\{x, y\}$, with $x = 0, 1, \ldots, N_x - 1$; and $y = 0, 1, \ldots, N_y - 1$, i.e., it is stored as $N_x \times N_y$ array (only monochrome images will be considered). Every (x, y) element of the array corresponds to a *pixel* (picture element or image element) of the image, whose brightness or intensity is equal to $I(x, y)$.

Furthermore, $I(x, y)$ can take one of the integer values 0, 1, …, $2^m - 1$, when quantized using m-bits. If m is 8, this gives brightness levels ranging between 0 and 255, which are usually displayed as black and white, respectively, with shades of gray in between. Small values of m give fewer available levels, reducing the available contrast in an image; *8-bit pixels* is a common choice in many applications.

Selection of an appropriate value for the image size, $N_x \times N_y$, is more difficult and complex. We seek a size large enough to resolve the needed level of spatial detail in the image. If the size is very small, the image will be coarsely quantized, and some details will get lost. Larger values of size give more detail but require more storage space and longer processing time.

In this approach of image representation, information about shape, image topology, or texture is not included. Therefore, two images with similar color combination but containing diverse shapes or textures may be recognized as similar, even if they are totally unrelated semantically.

Multi-Feature Composed Representation: In this model, the image representation is inclusive of a composition of various features: color histogram, shape, image topology, and texture. This is a highly popular model in practice.

It is essential to consider that (with existing methods at least), we can realistically work only with a limited idea of semantic content, based on relatively simple 'low-level' measurements—color, texture, and simple geometric properties of objects, such as edges and corners. 'High-level' feature extraction deals with searching for shapes in computer images. Extracting a shape implies locating their position, orientation as well as size [172–174].

Video Data

While capturing a scene at various times, 3D elements are mapped into corresponding pixels in the images. Therefore, if image features are not blocked, they can be connected to each other and motion can be featured as a group of displacements in the image plane. The displacement tallies with the movement of the object in the scene and is referred to as the *optical flow*. Suppose you consider an image and its optical flow, you should be capable of constructing the *next* frame in the image sequence.

A *video shot* is a set of frames or images wherein the video content from one frame to the next does not alter all of a sudden. The key frame is the one that is most represented in a video shot. It is possible to analyze each key frame with the help of the image feature extraction and analysis techniques. The sequence of key frames will then be used to define the sequence of the events taking place in the video clip. Therefore, detecting key shots and extracting key frames from video clips becomes the most important task in video mining and retrieval.

It is still the initial stages of video data processing. There are many research issues to be solved before it becomes general practice [175].

9.5.3 Audio Retrieval

Audio information retrieval systems are capable of performing various human-computer interactions, including voice mining, dialog, and question answering. The primary point to focus on when it comes to speech is that the sounds generated by a human are filtered by the shape of the vocal tract including tongue, teeth, and so on. The sound emitted is determined by this shape. The shape of the vocal tract manifests itself in the envelope of the short-time power spectrum. There is a vibration in an organ in the ear, at varying spots as per the frequency of the incoming sounds. Based on the spot that vibrates (which wobbles small hair), various nerves send information to the brain that certain frequencies are present. Usually, the organ is unable to differentiate between two closely-spaced frequencies. This effect becomes more prominent as the frequencies increase. One discriminatory factor is the energy in different frequency areas. The bases of automatic (machine) speech-recognition systems are the attributes or traits of the audio signals related to human auditory system's response.

Let us look at the popular approach to the process of feature extraction:

The digitized speech is read into an array $x[n]$; $n = 0, \ldots, N-1$ (N samples). The values correspond to audio intensity (amplitudes).

Speech signals are non-stationary and exhibit quasi-stationary (statistically stationary) behavior at short durations. For this reason, the audio signal is split into distinct 'frames' of 20–40 msec; 25 msec is standard. This means the frame length for a 16 kHz signal is $0.025 \times 16000 = 400$ samples. Frame step is usually something like 10 msec (160 samples), which allows some overlap to the frames. The first 400 sample frame starts at sample 0, the next 400 sample frame starts at sample 160, etc., until the end of the speech file is reached.

The conversion of each frame of N samples from the time domain to the frequency domain requires use of Discrete Fourier Transform (DFT), which is capable of converting the convolution of the glottal pulse and the vocal tract impulse response in the time domain into frequency domain. From DFT, power spectrum for each frame is calculated. We generally keep 257 power spectrum coefficients.

The Discrete Fourier Transform

$$X[k] = \sum_{n=0}^{N-1} x[n]\, e^{-j2\pi kn/N} \quad \text{for } 1 \le k \le K,\ K \text{ specifies the } K\text{-point DFT.}$$

$$= \sum_{n=0}^{N-1} x[n] \cos\left(\frac{2\pi kn}{N}\right) - j \sum_{n=0}^{N-1} x[n] \sin\left(\frac{2\pi kn}{N}\right) \tag{9.11}$$

Power Spectrum

$$P[k] = Re(X[k])^2 + Im(X[k])^2 \tag{9.12}$$

Next, we determine the amount of energy existing in different frequency regions, using *mel filter bank*. This is a group of 20–40 (26 is standard) triangular filters applied to the power spectrum estimate. The magnitude response of each triangular filter equals unity at the center frequency and reduces linearly to zero at the center frequency of two adjacent filters. The first is an extremely narrow filter, which indicates the amount of energy present near 0 Hz. With the frequencies going up, the filters go on becoming wider, and we become less bothered about vibrations. We focus only on how much energy exists at each spot. The *Mel-Frequency Scale* shows us precisely how our filters are spaced and how wide they can be made. The term 'mel' is derived from 'melody' and indicates that pitch comparisons form the basis of scale. A popular formula for frequency to 'mel' scale conversion is:

$$f_{\text{mel}} = 2595 \log_{10}(1 + f/700) \tag{9.13}$$

To calculate filter bank energies, we multiply each filter with the power spectrum and then add up the coefficients. Once this is performed, we are left with 26 numbers that give us an indication of how much energy was in each filter.

We take the log of each of the 26 energies (this is also motivated by human hearing). This leaves us with 26 *log filter bank energies*. Taking the *Discrete Cosine Transform* (DCT) of the 26 log filter bank energies $y[k]$; $k = 0, \ldots, K - 1$ $(K = 26)$ gives us 26 *cepstral coefficients*.

$$C[r] = \sum_{k=0}^{K-1} 2y[k] \cos\left(\frac{\pi}{2K} r(2k+1)\right) \tag{9.14}$$

Only lower 12 of the 26 coefficients are kept. The resulting features (12 numbers for each frame) are called *Mel Frequency Cepstral Coefficients* (MFCCs).

Mel Frequency Cepstral Coefficients are the features widely used in automatic speech and speaker recognition. They were introduced by Davis and Mermelstein in the 1980s and have been the state-of-the-art ever since.

Cepstrum is the Fourier transformation of the logarithm of a spectrum. It is therefore the 'spectrum of a spectrum'. In the original paper, the authors coined the word 'cepstrum' by reversing the first four letters of 'spectrum'.

Refer to [176, 177] for speech recognition.

Audiovisual Data

Other than still images, an incommensurable amount of audiovisual information is now available in digital form in digital archives, on the World Wide Web, in broadcast data streams and in personal and professional databases. The demands for effective content-based data mining and retrieval techniques for audio and video data are high. Typical examples include searching for and multimedia editing of particular video clips in a TV studio, identifying suspicious individuals or scenes in surveillance videos, exploring patterns and outliers in weather radar readings, looking for specific events in a personal multimedia source, and searching for a certain melody or tune in audio albums.

To help record, search and analyse audio and video information, there are certain standards for describing and compressing multimedia information. Examples of typical video compression schemes are: MPEG (developed by **M**oving **P**ictures **E**xperts **G**roup) and JPEG (**J**oint **P**hotographic **E**xperts **G**roup).

Multimedia analytics has been studied extensively. Some useful references are [178–181].

It is essential for the readers to recall that this presentation on information retrieval was not aimed at training the readers to deal with projects on the subject, but at creating excitement for further exploring the subject. This section focuses on the fact that both the domain knowledge and knowledge of machine learning algorithms are critical for the solution of real-life problems. This fact becomes all the more essential for text, image, audio, and video data; rather, domain knowledge, to change raw data into structured data matrix with numeric values, becomes the primary part of the project.

9.6 APPLICATIONS AND TRENDS

Machine learning is a growing new technology used for knowledge mining from data. It is a technology that many people are starting to get serious about. Now we have a first generation of machine learning algorithms—for learning decision trees, fuzzy rules, neural networks, Bayesian classifications, and logistic regressions—which have shown their significant value in various real-world mining applications. Different companies across the globe now offer commercial applications of these algorithms (www.kdnuggets.com) in addition to efficient interfaces to commercial databases and well-designed user interfaces.

9.6.1 Data Mining Applications

Several data mining systems have been created for *domain-specific* applications, including finance, the retail industry, telecommunications, healthcare, and so on. These systems are aimed at integrating domain-specific knowledge with data analysis methods and offer mission-specific solutions.

Financial Services

Banks as well as financial institutions usually provide different banking services (e.g., current and savings accounts for business or individual customers), credit (e.g., business, mortgage and automobile loans), and investment services (e.g., mutual funds). Some provide insurance services and stock investment services as well. Financial data gathered by banks and the financial industry are very often relatively thorough in terms of reliability and quality. This helps systematic analysis of data and also mining of data. A few typical examples are [17]:

- Designing and constructing data warehouses for multidimensional analysis of data to examine the general properties of financial data, and data mining.
- Predicting loan payment and analyzing customer credit policy.

- Classifying and clustering customers for targeted marketing.
- Detecting crimes such as money laundering.

The Retail Industry

The retail industry is one of the main areas where data mining is applied, because it gathers large volumes of data pertaining to sales, customer shopping history, transportation of goods, consumption, and service. The amount of data gathered keeps growing fast, particularly because of the increasing availability, convenience, and popularity of businesses conducted on the web, or *e-commerce*. In these modern times, many businesses offer the convenience of buying goods online to their customers, through their websites.

Retail data mining is capable of identifying the purchase behaviors, discovering customer shopping patterns and styles, improving customer service quality, achieving better customer retention, designing better policies for transporting and distributing goods, and reducing business expenditure.

A few typical examples are [17]:

- Designing and constructing data warehouses for retail data encompassing a wide spectrum of dimensions—sales, customer relations, human resources, goods transportation, consumption, and services—to help in efficient data mining.
- Multidimensional analysis of sales, customers, products, time, and region.
- Analysis of the effectiveness of sales campaigns using ads, coupons and different types of discounts and bonuses.
- Analysis of customer loyalty.
- Mining of associations from sales records reveals that a customer who purchases a particular item is likely to purchase another set of products. This associative information can be used to promote sales, on the basis of the principle that if users shared similar interests in the past, it is highly probable that they will exhibit similar behavior in the future too. *Recommender systems* employ data mining methods to make personalized product recommendations during live customer transactions, on the basis of the opinions of other customers.
- Fraudulent analysis and the identification of unusual patterns.

The Telecommunication Industry

The rapid evolution of the telecommunication industry has led to the availability from local and long-distance telephone services to several other complete communication services—fax, pager, cellular phone, internet messenger, images, e-mail, computer and web data transmission, and similar data traffic. The expansion of the telecommunication market due to the integration of telecommunication, computer network, internet, and numerous other means of communication and computing, has happened quite rapidly. It is also becoming more competitive, which is creating a huge demand for data mining so as to help comprehend the business involved, recognize telecommunication patterns, detect fraudulent activities, improve use of resources, and service quality. A few typical examples in which data mining can improve telecommunication services are:

- Analysis of multiple dimensions of telecommunication data—calling-time, duration, location of caller, location of caller, and type of call—with the help of OLAP and visualization devices on large data in telecommunication data warehouses.
- Detection and analysis of fraudulent patterns and identification of rare patterns—recognize potentially fraudulent users and their usage patterns, detect attempts to obtain fake entry into customer accounts, and learn uncommon patterns that may require special attention, such as busy-hour frustrated call attempts, switch and route congestion patterns, and periodic calls from automatic dial-out equipment (like fax machines) that has not been programmed properly.
- Analysis of sequential pattern and multidimensional association for the promotion of telecommunication services.
- With more and more integration of mobile telecommunication, web and information services, and mobile computing, there is a likelihood of data mining playing a significant role in the improvement of services to beat competition.

The Insurance Industry

The insurance databases possess a wealth of data comprising a potential goldmine of important business information in the ever-altering insurance environment. Data mining is capable of facilitating insurance firms in making crucial business decisions and turning the newly discovered knowledge into business practices which can be acted upon—development of products, marketing, analysis of claim distribution, management of asset liability and solvency analysis. To be more specific, data mining can perform the following tasks:

- Identify risk factors that are important for predicting profits, claims, and losses
- Customer level analysis
- Developing new product lines
- The selection of policies for reinsurance
- Estimating outstanding claims provision
- Predicting fraudulent behavior

Healthcare Services

A few typical examples in which data mining can improve healthcare services are: evidence based medicine, comparative effectiveness research, clinical analytics, fraud/waste/abuse management, etc.

Supply Chain Management

Conventional data mining methods have been employed in the banking, insurance, retail, and telecommunication sectors. They are necessary to understand the needs, preferences, and behaviors of customers. They are also important in pricing, promotion and product development.

There are several opportunities and applications of data mining in addition to the obvious ones. One such area is 'supply chain management'. Disturbances in the market emerge from factors such as rapid introduction and customization of products, tough design specifications, and customer shifts. This makes it important to maintain continuous contact with customers and suppliers through integration of supply chain. Given the competition in the global business environment, the agility of an organization's supply chain will directly affect its ability to produce and deliver innovative products to their customers not only in a timely manner but also in a cost-effective fashion. Regular monitoring of the quality delivered and customer expectations can help bring down costs through just-in-time purchasing, scheduling and distribution.

e-Governance

Data mining has also been popularly used for analysis of government and institutional administrative data.

Customer Relationship Management (CRM)

Establishing a business around *customer relationship* is a revolutionary change for most organizations. Banks have usually been known to concentrate on the maintenance of the spread between the rate paid by them to bring in money and the rate charged by them to loan out money. Telephone organizations have always focused on the connection of calls through the network. Insurance organizations have focused on claims processing, investment management, and maintenance of loss ratio. Progressive and product-focused organizations are now adopting *customer-centric* models: trying to understand each individual customer and making use of that understanding to make it more convenient (and more lucrative) for the customer to conduct business with them instead of with the competitors. This change in focus from broad market segments to individual customers requires changes in all sectors: financial services, retail industry, telecommunication industry, insurance industry, and the related businesses. The emergence of a new wave of big data (Section 9.7) has made customer-centric approach almost essential, with text-mining supporting traditional data mining. Customer-centric applications include [19]:

- *Memory-Based Reasoning (MBR) and Collaborative Filtering*

  Analogous situations in the past form the basis of MBR results. More information can be added through collaborative filtering, with the use of similarities among customers, as well as their preferences. One product of collaborative filtering is *recommendation systems*.

- *Using Survival Analysis to understand Customers*

  Survival analysis is crucial to the understanding of customers and their life cycles. Despite the origin and terminology being from medical research and failure analysis in manufacturing, the ideas are customized for marketing—measuring the likelihood of customers leaving, customer value calculations, predicting customer levels, and so on.

- *Using Link Analysis for understanding relationships and connections*

  Who has made friends with whom on Facebook? Who makes a call to whom on the telephone? Which medicines are prescribed by which doctor to which patients? Who visits whose blogs and on what subjects? … These relationships form a treasure trove of information and are all visible in data. This need is addressed by the method known as link analysis.

- *Graph theory*

  It is that branch of mathematics which forms the basis of link analysis. It represents relationships between various objects in the form of edges in a graph. This has worked well in the identification of authoritative sources of information on the web by an analysis of the links between its pages; analysis of telephone call patterns to seek influential customers capable of recruiting new subscribers from competing networks; comprehension of doctors' referral patterns to detect insurance fraud; and so on.
- *Building Customer Signatures*

  The process of preparation of data for mining requires building of customer signatures, that is, collecting the traces left by customers on making purchases, visiting websites, contacting call centers, paying bills, responding to offers, and interacting with the organization in other ways. These dispersed bits of data are made meaningful in the form of customer signatures, which are useful for training data mining models. Building customer signature is, thus, the process of finding customers in all kinds of scattered data, and bringing it to one place for data mining.
- *Analytics on text data*

  Language is obviously a critical component of how people communicate and how information is stored in the business world and beyond. A huge volume of valuable business information is stored as text—e-mails from customers, notes by customer service representatives, transcripts by physicians, dealings (voice-to-text translation) of customer-service calls, website comments, articles and stories in newspapers and magazines, professional reports, and so on.

  Perhaps the most common technique for text mining is to convert unstructured data into a structured one, i.e., extracting structured features. Extracting derived variables is usually a matter of looking for specific patterns in the text. We have earlier described 'bag-of-words' approach (Section 9.5) for representing documents in a structured format.

  Deriving variables from text is one application area, but not the only one. Often we do not know the exact features we are looking for. For example, in a study of analyzing incoming e-mails, the goal may be to determine whether the e-mail comment is a complaint or a compliment. Other examples appear in the media: news articles blogs, and the like.

  There are many useful text analysis applications that are not exactly data mining. Thus text mining encompasses various applications of analytics on text, some of which are more properly considered data mining, and some of which are just useful applications of text analytics. Some examples of text analytics applications, that are not exactly data mining, are: spell checking and grammar checking, translations from one human language to another, searching the documents, summarizing documents, and *sentiment analysis*, where the goal is to understand the attitude of the writer toward the subject.

  The toughest part of text mining is the management of text itself. At the extreme are two different methodologies—the bag-of-words model, which looks at documents as unordered lists of words, and at the other extreme, the *Natural Language Processing* (NLP) approach, which considers the meanings and grammatical features of the language.

Some estimates say that up to 80 per cent of important information for business is stored in the form of text. Text data is mostly unstructured and is the major source for 'big data analytics' discussed in Section 9.7.

9.6.2 Data Mining Trends

Looking ahead, the major challenge of the future is applications. If data exists, things can be learned from it. If there is excess data, the mechanics of learning will have to be automatic. Applications will not come from machine learning experts, but from those who handle data and issues related to the same. New applications will pose challenges and machine learning research will respond to these creating new opportunities in business decision making.

The diverse nature of data, and data-mining tasks and models present several challenges in data mining. Some of the significant tasks for data mining researchers and application developers include the development of efficient and effective data mining techniques, systems and services; interactive and integrated data mining environments, and utilization of data mining methods for the resolution of sophisticated application problems. Let us look at some of the data-mining trends that reflect the overcoming of these challenges [182].

The first generation of machine learning algorithms—for learning decision trees, Bayesian methods, rules, neural networks, and logistic regressions—that have proved to be important in various data mining applications, work on the assumption that the data comprises only numeric and symbolic features, and no text, image, or audio features. The assumption is that data has been gathered with care, into a single database, which has specific data mining purpose. The first-generation algorithms are of extreme significance when the data is structured [34].

Today, data mining is used in a vast array of areas. Numerous commercial data mining systems and services are available. Many challenges, however, still remain. Probably, the most difficult challenge is mining large volumes of data we are collecting today that is semi-structured or unstructured. Such mining includes mining multimedia data, graphs and networks, spatiotemporal data, cyber-physical system data, web data, and data streams.

In Section 9.5, the use of first-generation algorithms for text, image, and audio data was discussed. However, that discussion was limited to specific tasks of data mining where the data can be preprocessed in a structured format. Mining multi-structured or unstructured multimedia data is a frontier for research community.

Some of the trends in data mining applications are as follows that reflect the persuit of these challenges [17, 18].

Mining Sequence Data

A *sequence* is referred to as an ordered list of events. *Time-series data sequences* comprise numeric data, which is recorded at equal time intervals; for instance, data generated by stock markets, and scientific and medical observations. *Symbolic sequence data* comprises event sequences or nominal data sequences, which are not usually observed at equal time intervals. Examples are, customer shopping sequences, web click streams and *biological sequences* (e.g., DNA and protein sequences). Since biological sequences hold extremely complex semantic meaning and present several challenging research issues, maximum investigations are conducted in the field of *bioinformatics*.

Mining Graphs and Networks

Graphs are representative of a more general class of structures than sets, sequences and trees. There exists a broad range of graph applications on the web and in social networks, information networks, biological networks, bioinformatics, chemical informatics, computer vision, and multimedia and text retrieval. Therefore, graph and network mining have become very important and a subject of heavy research.

Mining Spatiotemporal Data

Spatial data mining finds patterns and knowledge in data pertaining to geospace, stored in geospatial data sources. *Spatiotemporal data* pertain to space as well as time. Examples of spatiotemporal data mining are uncovering of weather patterns, forecasting earthquakes and hurricanes, and determination of global warming trends. Spatiotemporal data mining has grown significantly because mobile phones, GPS devices, internet-based map services, and so on, have become very popular.

Mining Cyber-Physical System Data

A *cyber-physical system* comprises a huge volume of interacting physical and information elements; for instance, a patient care system that connects a patient monitoring system to a network of patient/medical information, and an emergency handling system; a transportation system connected to a transportation monitoring network (containing several sensors and video cameras) with a traffic information and control system.

Mining Multimedia Data

Multimedia data mining refers to the detection of interesting patterns in multimedia databases, that is, image data, graphics, video data, audio data, text data and *hypertext* data (comprising text, text markups, and linkages). This interdisciplinary field assimilates image processing, computer vision, data mining, and pattern recognition.

Mining Web Data

The World Wide Web is a huge, widely distributed, global information center for news, ads, consumer information, financial management, education, government and e-commerce. Any information one seeks is available on www. It is a rich and dynamic source of information on web page contents with hypertext structures and multimedia, hyperlink information, and access usage information; thus offering plenty of sources for mining data.

Mining Data Streams

Stream data implies data which flows into a system in large quantities, changes dynamically, and comprises multidimensional characteristics. It is not possible to store such data in traditional databases. This presents difficulties for the efficient mining of stream data.

Data stream mining is significant as it forms the basis of several applications—sensor network processing, network traffic, web searches, and so on. Its objective is the determination of patterns of structures of continuous data, which may be later employed to deduce likely occurrence of events.

Recommender Systems

The customers of today have millions of products and services to choose from, while shopping online. *Recommender systems* make the work easier by making product recommendations to consumers—which consumers are likely to be interested in. The recommender systems are personalized according to diverse user preferences. This is a significant factor when it comes to effective recommendation.

Recommender systems may make use of either:

(i) *content-based model*
(ii) *collaborative model*
(iii) hybrid model (a mix of content-based and collaborative approaches)

The content-based model recommends items that are similar to the ones preferred by the user or queried in the past. It relies on product traits and textual product descriptions. The *collaborative filtering* approach may take into account a user's social environment. Its recommendation of items is based on the opinions of other users whose tastes or preferences are similar to that of the user. The historical data that is reflective of preferences may contain product ratings, web click logs, reviews or tags.

Adversarial Situations

A prime example of adversarial situations is the *junk email. Spam filtering* is thus an important application of machine learning.

Sadly, several other adversarial situations exist in our world today. Closely related to junk email is the *search engine spam*, wherein sites try to trick Internet search engines into displaying them prominently in the lists of search results. There are also computer virus wars, wherein designers of viruses and virus protection software respond to one another's innovations.

Computer network security is a constantly escalating war. Protectors toughen networks, operating systems, and applications, and attackers seek vulnerable areas. This has encouraged *intrusion detection and prevention*, making it a crucial element of networked systems. An intrusion is the set of actions threatening the integrity, confidentiality or availability of network resources. Privacy-preserving data mining is a significant issue as there is a growing need for storage of personal data for users.

Data Mining for Big Data

Scaling to complex, extremely large datasets—the *big data analytics*—is probably the most debated current research issue. We are now in the arena of 'big data'—the *more* is just not *more*, but *more* is *different*. The next section presents some highlights of this debate.

Just like other areas of technology, data mining stands on an ever changing landscape. The old section of the landscape is undergoing a redefinition as the new sections continue to emerge.

9.7 TECHNOLOGIES FOR BIG DATA

In this section, an overview of *big-data analytics* is presented. Many books are now available on this emerging field. The overview here is based on the references [183–185].

Big data is often referred to as the data that exceeds the capability of widely-used hardware environments and software tools that capture, manage, and process it within an acceptable time period for its users. As per this definition, the qualification for big data will alter with technological advancement. What is big data presently, will not be so later. This definition also indicates that the composition of big data will vary from industry to industry and organization to organization, according to the varying tools and technologies and their varying capability.

Big data definitely involves huge volumes of data, but that isn't all there is to it. Big data also possesses high *velocity* (i.e., the rate of transmission and receipt of data), and a *variety* of data sources compared to conventional data sources. While the focus is mostly on its size, the challenging part of big data actually is the absence of structure in it as compared to conventional data, which is structured and possesses a fixed file format (refer to Section 1.6). Big data sources rarely have any control on format. Its data is *unstructured*, with majority of the data being at-least *semi-structured* following a logical flow and format, but the format is not easy to use.

Let us look at certain specific examples of big data. *Web data* can be said to be the most commonly used and recognized source of big data. There are several other sources of big data, with their own important users. While some are well-known others are relatively unclear.

Let us look at a representative cross-section of big data sources, to help readers understand the breadth and types of available big data, as well as the breadth of analysis that the data enables.

Web Data

Online businesses are already high data oriented. Potentially every industry but particularly those with a lot of customer data—such as retail, travel and transportation, telecommunications, media and entertainment, and financial services, are collecting and analyzing the detailed transaction histories of their customers. Several companies make the assumption that transactional history offers the closest possible view to a '360-degree' view of their customers. Web integration with the inclusion of online transactions, facilitate such a view.

Traditional web analytics vendors offer operational reporting on click-through rates, traffic sources, and summary statistics on the basis of web data. The goal has to move to the combining of *customer-level web behavior data* with other cross-channel customer data. This means, going well beyond click-through reports and summaries of page views.

Customer-level web behavior data is basically a new source of information. Any action taken by customers in the purchasing procedures must be captured if possible; this means, a detailed event history from any customer touch point. Websites, mobile apps, and social media are some of the popular touch points of the modern era. This data can provide information related to customer preferences, future intentions and motivations. Knowledge of this kind permits a new level of interactions with customers, which drives further business.

Just like the integration of web transactions with traditional transactions caused a revolution in the power and depth of analysis, these new sources of customer-level web behavior data have taken analytics to an all new level. With the ability of storing and processing data that we witness today, it is quite possible to succeed, and several progressive organizations have already demonstrated this by application of the data.

Web data can be applied by organizations to improve current analytics, enable new analytics, and make business better. Here are a few examples:

- Prediction of the next best offer to each customer.
- Flagging of those customers who are at the risk of cancelling their accounts so that proactive action may be taken for prevention. *Churn* is a significant issue facing the industry and *Churn models* help solve this issue.
- The web behavior permits identification of customers who are presently interested in buying. It also makes it possible to score and rank customers by the probability of taking action. Then appropriate *customer segments* are generated on the basis of those ranks so that the customers can be reached out to.
- Improved assessment, paid search and online advertising results in another high-impact analysis enabled with customer-level web behavior data.

Text Data

Language is obviously a critical component of how people communicate and how information is stored in the business world. There exist e-mails, text messages, tweets, social media postings, real-time chats, and audio recordings which have been translated into text. Text data is one of the least structured and the largest sources of big data available today.

A widely common use of text analysis today is *sentiment analysis*, which examines the general direction of opinion across a vast number of people to offer information on the market—what it says, what it thinks, and what it feels about a company. Text analysis is useful in sorting through complaints, repair notes, and other remarks from customers and, therefore, an organization will be able to identify and rectify issues faster. It can also be applied to detect fraud.

Social Network Data

The entire set of cell phone calls or text message records that a cellular carrier captures, is a huge dataset usually employed for various purposes. But *social network analysis* looks into many degrees of association and not merely one. It examines who the calls were between, and goes even deeper. A more complete picture of a social network can be obtained by examining as many layers as the analysis systems can possibly tackle. The urge for navigating from customer to customer and call to call only adds to the process and nature of analysis. The same idea is applicable to social networking sites, such as LinkedIn or Facebook.

Social networking analysis has changed the emphasis from profitability of accounts to profitabillity of network. It is capable of offering insights into a customer's total influence and value, which can totally alter the way a customer is viewed by a company.

Time and Location Data

With the arrival of the Global Positioning Systems (GPS) and cellular phones, time and location information has come to be a growing source of data. It is a type of big data, which has a maximum privacy-sensitivity.

Time and location information on assets can be obtained with the help of Radio Frequency IDentification (RFID) tag, which is a tiny tag fixed on objects, such as shipping pallets or product packages. When an RFID reader emits a signal, the RFID tag sends a response in the form of information. A primary use of RFID data is in asset tracking.

Time and location data is going to continue to grow in adoption, application, and impact.

Sensor Data

There are several complex machines and engines across the globe—aircrafts, military vehicles, trains, construction tools and devices, drilling equipment, and so on. To maintain the smooth functioning of such equipment is a must since they are expensive. Recently, embedded sensors have begun to be used for the monitoring of secondwise status of the equipment.

Smart grids form the next generation of electrical power infrastructure. They have extremely refined monitoring, communication, and generation systems. Various sensors and monitors keep track of diverse facets of the power grid itself and the electrical flow through it.

An application in auto insurance industry involves putting a sensor, or black box, into a car to capture information about what is happening with the car. This black box can measure any number of things depending on how it is configured. It can monitor speed, mileage driven, or heavy braking. This data helps insurance companies better understand customer risk levels and set insurance rates.

How is Big Data Different?

Clearly, big data is a totally new source of data and not merely an extended collection of traditional data.

Big data is at times generated automatically using a machine in an automated manner. A sensor embedded in an engine, for instance, throws out data related to its surroundings even if no-one demands it. We seize all this data, and only bother about it during the analysis stage.

The design of big data sources is far from user friendly. For instance, in case of text streams from social media sites, users cannot be asked to follow specific standards of grammar, or sentence ordering or vocabulary. We will receive what people do when they make a posting. It is not easy to work with such unstructured data.

Huge volumes of big data streams may not have much value. We capture all that is possible to make sure that nothing is missed but also make the process of big data analysis more painful.

Will big data remain a wild west of crazy formats, unrestricted streams, and without definition? Probably not. Over a period of time, standards will be established.

9.7.1 Emerging Analytic Methods

As with any new topic attracting much attention, there are all kinds of claims regarding how big data will alter everything about the way analysis is performed and the way it is used. It is actually not the case; the hype goes much beyond reality.

Just because big data has much volume, arriving rapidly, and from a range of sources in complex formats, it does not gain in importance compared to other data. Perhaps the most interesting bit about big data is how it will affect a business when combined with other data from an organization.

It is crucial that organizations do not develop a distinct big data strategy from their conventional data strategy. Organizations require to focus on developing a cohesive strategy wherein big data is merely another aspect of an enterprise data strategy.

New evolving *big data sources* pertaining to customers, from a variety of newly evolving touch points, such as web browsers, mobile applications, kiosks, social media sites (Facebook, Twitter, YouTube, LinkedIn, and others), are the *game-changing* sources. Amazing new frontiers are shooting up, which are capable of revolutionizing organizations' customer insights and the effect those insights will have on their business. The challenge is to analyze the behavioral data at the customer's level.

This data is mostly unstructured or semi-structured text data. Unstructured data has no format to it, while semi-structured data has a logical flow and format to it, but the format is not user-friendly. The traditional data comes in a fixed file format.

Analytic professionals have used a range of tools and techniques over the years that have made it possible for them to prepare data for analysis, execution of analytic algorithms, and assessment of the results. Unsurprisingly, the depth and functionality of these tools and techniques have gone up with time. The need for new tools and methods is being felt by analytic professionals to handle big data problems. When combining new tools and techniques with the evolved scalability, organizations will be positioned perfectly well to tackle big data challenges.

At times, analytic professionals will just do a lot more of the old and tried tools and methodologies as a result of today's new scalability. However, new techniques to address new business issues are continuously emerging. While tools and techniques will go on evolving, let us look at some that are worth consideration today.

The biggest problem with big data may not really pertain to analytics, but to data preparation for analysis. Within a big data feed, there will be certain information possessing long-term strategic value, some information that will be employed for immediate and tactical use, and some data that will be useless. A major part of the big data challenge is the determination of the category into which the pieces fall. Analytic processes may need filters for the removal of portions of big data stream that are hardly required. During the data processing, there will be other filters, which will filter data to the user actions that require examination for the business issues that need to be addressed.

Traditional structured data does not require as much effort in this area since it is specified, understood, and structured in advance. Therefore, the *extract*, *transform*, and *load* processes for data preparation require bigger effort for big data problems. Complex filtering algorithms will be developed to siphon off the meaningful pieces from a raw stream of big data.

Data exploration is an important element of traditional analytic process. It is going to get more important for big data problems. Typically unstructured data itself is not analyzed. Rather, unstructured data is processed in a way that applies some sort of structure to it. Then those structured results are what is analyzed. Data exploration becomes an important step before we proceed to the analysis phase.

Data exploration also helps in delivering small, quick results for business decision making. Small wins demonstrated by data exploration will confirm to the organization that progress is being made; that will build support for further efforts.

Modern visualization tools allow the analytic professionals to interactively explore data in a visual paradigm, and to share the complicated results of the analysis with the non-technical business people. In the world of analytics, visualization implies charts, graphs and tables displaying data. Over a period of time, graphical interfaces have been created to let users perform many actions via point-and-click environments. Today's tools let several tabs of graphs and charts be connected to the underlying data. More importantly, the tabs, graphs and charts can be connected to each other. Tools now include more sturdy graphics, workflow diagrams, and other applications that require manipulation of data. In addition, the ability to link to big data, interweave the visuals, and search and drill down at will, gives rise to something with a lot of power.

Very often, professional analysts will merely perform a lot more of the same long-standing analytic techniques to handle new scalability issues. One trend is the use of *commodity models*— one that has been produced without bothering about making use of every bit of the predictive power. The objective of a commodity model is not to obtain the best model, but to obtain a *good-enough model* that will quickly produce a better result than in the absence of a model. The most important consideration is to ensure that we create a process capable of generating good-enough models.

A plain and straight forward technique of applying long-standing algorithms to a large dataset, is to divide the data into portions of smaller size and learn separate models for each, combining the results by averaging. In a multi-structured dataset, data can be divided into structured subsets. Either a parallel *bagging-like scheme* or a sequential *boosting-like scheme* can be used to build a model. There is more to *Ensemble methods* than selecting the best performer from a set of models. It is really about combining the results of various models to obtain the single final answer, by averaging or a much more complex formula. Due to the evolution of analytic tools, the use of ensemble approaches has increased. Without an ideal way for management of work flow and tying of the results of various models together, ensemble modeling will prove to be a tedious procedure.

One of the fastest growing techniques used by businesses today is the application of analytics on text. The growth of text analytics can be attributed to the wealth of new sources of text data. Of late, everything from e-mails to social media commentary from sites such as Facebook and Twitter, to online inquiries, to text messages, to call center conversations is captured in bulk. It is not possible to ignore this kind of unstructured data. It is a widely used type of big data, and text analytics tools and techniques have actually evolved.

An important theme here is that typically, *unstructured data itself is not analyzed; rather, unstructured data is processed in a way that applies some sort of structure to it.* Then those structured results are what is analyzed.

Thus applications of analytics on text include:

- Data mining on structured text data, and
- Representing documents in the computer to facilitate mining.

Representing documents in the computer is the first challenge in text mining. There is a continuum of approaches for helping the computer understand the documents. At one end is the 'bag-of-words' approach, where documents are considered merely a collection of their words (refer to Section 9.5). At the other end is the 'understanding' approach, where an attempt is made to actually understand the document and what each word specifically means. Technically, this is called *semantics*.

The bag-of-words approach treats the documents as unordered lists of their words. This approach is the most naive approach, where even basic grammatical constructs are ignored. The other is the *natural language processing* approach, which takes into account the meanings and grammatical features of the language. There are many variations of both methods. A handful of extensions to the bag-of-words approach make it more useful

The next step is mining the resulting text data. Singular Value Decomposition (SVD), naive Bayesian methods, nearest-neighbor approaches and clustering are some of the techniques that are common in the world of text analysis.

Big data will not really change the goals and purpose of analysis, or procedure of the analysis itself. The issues addressed will definitely evolve with big data. However, at the end of the day, analysts will merely explore new and extraordinarily huge datasets to reveal important trends and patterns. Big data will definitely drive new and innovative analytics, and force analytic professionals to go on expressing their creativity within their scalability limitations.

Deep Learning Big Data allows extraction of high-level, complex abstractions as data representations through a hierarchical learning process. A key benefit of deep learning in big data analysis is that it can learn from massive amounts of unsupervised data. This makes it a valuable tool for Big Data Analytics where huge amounts of raw data are uncategorized.

The present state-of-the-art in deep learning algorithms:

- needs large amount of data to understand it perfectly
- heavily depends on high-end machines; accelerated computing using Graphics Processing Unit (GPU) together with a CPU (Deep learning algorithms inherently do a large amount of matrix multiplication operations. These operations can be efficiently optimized using a GPU because GPU is built for this purpose)
- takes a long time to train; about two weeks to train completely from scratch
- lacks interpretability (this factor is the main reason deep learning is still thought as an option for several times before its use in industry)

Although deep learning is used by Google in its voice-recognition and image-recognition algorithms, by Amazon to make decisions regarding what and what not to purchase next, and by researchers to make forecasts for the future, the knowledge obtained from (and offered by) deep learning algorithms has remained mostly untapped in the context of Big Data Analytics. Deep learning algorithms are capable of facilitating research, and have become the primary focus of data science.

9.7.2 Emerging Technologies for Higher Levels of Scalability

The traditional methods will not work for handling big data; updating technologies to provide a higher level of scalability is necessary. There are multiple technologies available that address different aspects of making use of big data in analytic processes. Some of these advances are quite new. Analytic and data management environments are converging. In-database processing is replacing much of the traditional off-line analytic processing.

In traditional architectures, databases are built for each specific purpose or team, and single-purpose databases (often called 'data marts') are spread all over an organization. Analytic processing occurs in the *analytic environment* (analyst's *desktop*, for example); relevant data is pulled from data marts into analytic environment to create variables required for an analysis. Leading organizations now see value in moving the analysis to the data, and not data to the analysis; analytic processing occurs in Enterprise Data Warehouse where the entire data of the organization has been consolidated. That is, analysis is done where all the data is together; need not be pulled together just for analysis. The user's machine just submits the request to *in-database analytic environment*.

Enterprise data warehouses (as well as data marts) are scalable using *Massively Parallel Processing* (MPP) technology. MPP systems behave like separate computers connected by a very high-speed network (MPP is also referred to as *grid computing*). Data is spread out into these independent processing units; each has its own memory and its own disk storage, breaking the analysis job into pieces and allowing the different sets of processing units to run the process concurrently. It gets a little more complicated in cases where data must be moved from one processing unit to another as part of the requirements of the query, but MPP systems are built to handle that in a very, very fast way.

SQL, the inherent language of an MPP system, is known for its efficiency for a range of requirements. SQL has undergone evolution. Several core data preparation tasks can be translated into SQL. It can be easily generated for several common analytic algorithms as well.

User-defined functions, coded in languages such as C++ or Java, compile code into new database functions capable of being called from an SQL query. For a user, the analytic function will act as the original analytic tool and will run parallelly on the database with efficiency.

The MPP systems help build and deploy advanced analytic procedures. For analytic professionals to use an enterprise data warehouse or data mart more effectively, it is a must for analysts to have a workspace with a group of resources that allow them to experiment with and reshape data in whatever fashion they require. Various terms are employed for this dedicated workspace—*analytics sandbox* (the word 'sandbox' is derived from the sandboxes where children play reshaping the sand as they wish to create anything they want to), *agile analytics cloud* and *data lab*, and so on; we will use the term 'sandbox' for the concept.

A sandbox is used by a very small set of users. Analytic dataset will be created within the sandbox, in the format needed for specific analysis in hand. For generating the same, data needs to be transformed, aggregated and combined. It will replicate a flat file structure with a single record for each customer, location, product, or whatever entity type is being analyzed. The analytic dataset bridges the gap between efficient storage and convenience of use. Storage structures efficiently store and retrieve data, but make advanced analytics efforts rather complicated. Analytic tools generally need data in a simple flat file format.

Sandbox users will be permitted to load their own data for short periods as part of the project, even if that data does not form part of the official enterprise data model. The shelf life of data in a sandbox will be constrained. The objective is not to establish a lot of permanent data. The data required during a project is built as per need and deleted once the project is over.

The sandbox's involvement should stop the moment things develop into ongoing user-managed processes or production processes. Sandbox is separated from the production database.

There are many kinds of sandbox environments—internal, external and hybrid. A section of an enterprise data warehouse is reserved to act as an internal sandbox. In case of an external sandbox, a physically separate analytic sandbox is created to test and develop analytic procedures. A hybrid sandbox combines an internal and an external sandbox, with positives and negatives of each. The decision is dependent on several factors: taking advantage of available hardware resources and existing infrastructure, cost-effectiveness, workload management, simplicity, and so on.

As time goes by, new technologies will emerge. A trend that is becoming popular is putting more advanced analytic functions directly into the database, making it possible to use a data mart or data warehouse as an analytic sandbox. Enhancing databases is becoming possible partly due to improvements in hardware that are speeding up databases, and partly due to more functionality in the SQL language such as analytic functions, grouping sets, better scalar factions, and user-defined functions.

New demands on ad hoc analyses or ad hoc queries for applications flooded with big data are throwing new challenges, and the technology is moving forward to face these challenges.

Cloud computing as a concept is much talked about today. *Cloud* is a non-dedicated set of computers linked to the Internet, which can be used, whenever required, to handle small to very huge processing jobs with the help of the Internet to distribute data and computing tasks to several computers across the world, but with a centralized infrastructure. An actual cloud, whether public or private, requires to comply with the three following criteria:

- Enterprises do not incur any costs, infrastructure or capital. They only incur operational costs, on a pay-per-use basis without any contractual obligations.
- Cloud hardware resources are *elastic*, that is, they may easily increase or decrease any time. Therefore, it is possible to scale up or down the capacity dynamically and promptly.
- The underlying hardware can exist anywhere geographically, hidden from the user.

When it comes to working with big data, it will be wise to augment the sandbox with **Map Reduce** environment. Map Reduce is a parallel programming framework. It is complementary to existing technologies.

Several decades earlier, a programming language called APL, introduced the ideas of *scalar extension* and *reduce operator*. Later these ideas were picked up by other programming languages including Lisp. In Lisp, scalar extension is known as *map*. The idea of map operator (scalar extension) was that any function defined on a scalar (word used for single value) could be applied to the elements of an array of any shape and number of dimensions. The idea of reduce operator was to use a function to reduce an array's number of dimensions by one.

Several decades later, researchers at Google revisited these Lisp primitives and realized that map and reduce could be implemented on massively parallel processing systems to process very large volumes of data. Their efforts in building the infrastructure to support grid computing, and code to implement map and reduce, led to the invention of MapReduce.

MapReduce is a programming framework. The most common implementation of MapReduce goes by the name of **Hadoop**, named by the project's founder Dave Cutter who named it after his son's favorite toy, a stuffed elephant. This open-source platform is available on Amazon's elastic cloud computing infrastructure.

Increasingly, Hadoop platform is expanding to include higher-level SQL-like languages such as Hive, for manipulating data. Hadoop supports MapReduce but it also supports other technologies as well. On the other hand, MapReduce can be implemented using platforms other than Hadoop. For example, Teradata's Aster platform has patented SQL MapReduce implementation that allows MapReduce processes to be executed as part of an SQL query. Massively parallel processing (MPP) systems, sandboxes, clouds, MapReduce, Hadoop, SQL and SQL-like higher-level languages, all have a role to play in an evolving analytic environment for big data.

MapReduce is a tool that is helping organizations handle the unstructured and semi-structured sources of data that are not easy to analyze with traditional tools. In a relational database, the data is already in tables of rows and columns, and has well-defined relationships. This is not always true with raw data streams of multiple types of data, such as text, images, web logs, sensor data, etc. This is where MapReduce can really be powerful. The visual of the MapReduce framework is shown in Fig. 9.11.

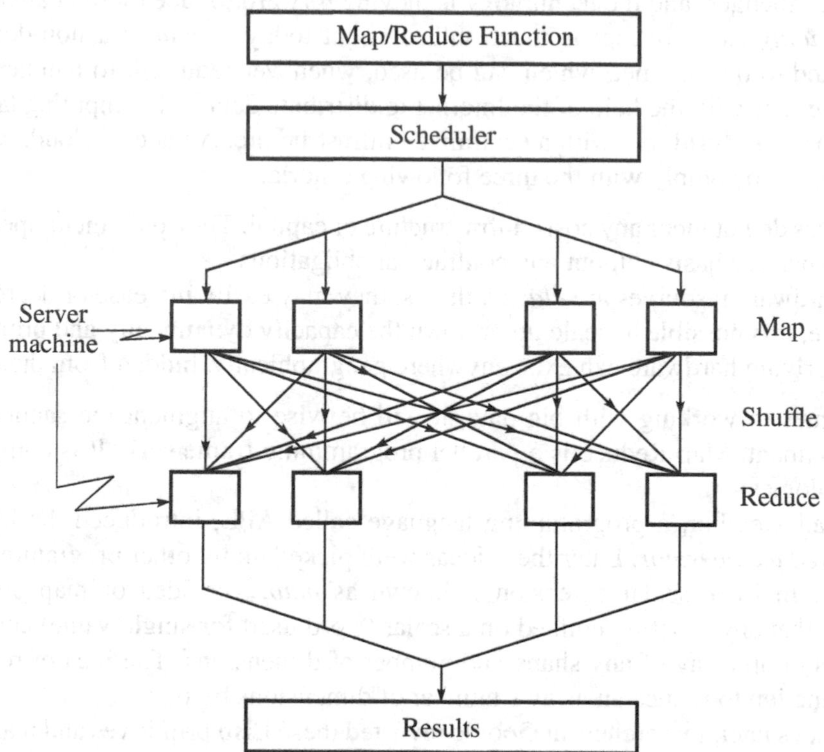

Figure 9.11 MapReduce programming framework

The data is first distributed by the *scheduler* to each of the *server nodes* with the help of a simple file copy technique. The data is, therefore, inside a file of a certain format decided by the user before the MapReduce procedure begins.

Then, the programmer submits two programs to the scheduler—**map** and **reduce**. The former program finds the data on the disk and executes the logic contained in it. This happens independently on each of the server nodes, which do not interact nor possess knowledge of each other. After the map step, the next stage is shuffling, during which the results from map are distributed by hashing so that the same concept map results are sent to the same reduce node. The reduce program then summarizes and aggregates the final results.

There can be thousands of map and reduce tasks running across thousands of server machines. That is how MapReduce becomes so strong. When huge streams of data are available, data has to be broken into pieces, which are then distributed among server machines. Fully parallel processing takes place because each machine works independently.

MapReduce will grow in terms of impact and become more popular with big data becoming a bigger part of what organizations require to handle. By dividing the work into small bits, it gets done faster and more economical than via other options. Once the immediate task is accomplished, the most significant pieces of data can be loaded onto a database to be strategically analyzed.

Massively parallel relational databases, clouds, and MapReduce all have to play significant roles in an analytic ecosystem. A combination of the three can help maximize results. When these technologies interact and work together, each can improve the others if used properly. Grid computing configurations can also sit alongside any of the resulting scenarios when we combine these technologies.

Appendix

A GENETIC ALGORITHM (GA) FOR SEARCH OPTIMIZATION

Because of complexity of biological processes, there has hardly been any progress made to comprehend these processes, and whatever understanding has been achieved is limited and crude. Yet, even this limited understanding has managed to give the much required push to the emulation of some human learning behaviors with the help of systems science and mathematics.

Neural networks and *fuzzy* logic are the two disciplines which have given us the mathematical tools for *cognitive information*, which is a field of much importance. With the continual development of these disciplines and enhanced understanding of biological processes, we are hopefully getting closer to the 'true' emulation of human learning behaviors.

Just as the neural networks and fuzzy logic, *evolutionary computation*—the area which encompasses *genetic algorithms*, *evolution strategies* and *genetic programming*—are methods of solving problems, that have biological processes as their inspiration. The natural evolution process has, over a period of millions of years, given rise to adaptable, specialized species that are extremely fit for their environments. In 1858, Darwin first suggested the theories of *evolution* and *natural selection*, which explained what he observed of the plants and animals in the world of nature.

It was Darwin's observation that with the introduction of variations into a population, in each new generation, those individuals who are not adequately fit have the tendency to die-off in trying to compete for food. And this fight for survival, which is won only by the fittest, results in enhancements in the species (This is known as *survival of the fittest* principle). The concept of natural selection helped to provide explanation about the manner in which the species have been able to adopt themselves to changing environments, and also how, as a result, species that are alike in terms of the ability to adapt may have developed or evolved.

Evolutionary computation simulates evolution on a computer. Such a simulation gives rise to a set of optimization algorithms, which are generally based on a basic set of rules. Here, we will see the way in which the insights gained from studying evolution can be applied to optimization problems.

The evolutionary approach has its basis in computational models of natural selection and genetics. They are referred to as *evolutionary computation*, an umbrella term, which is a combination of *genetic algorithms*, *evolution strategies*, and *genetic programming*. All these methods present a simulation of evolution with the help of the processes of *selection*, *mutation*, and *reproduction*. Our focus in this appendix is on Genetic Algorithms (GAs), which are the simplest and have been successfully applied to many real-life optimization problems. Most of the other evolutionary algorithms can be seen as variants of GAs.

Detailed presentation of only the *basic genetic algorithm* will be made. We have already examined its application to the learning of neural networks (Section 5.10), and to the structural and parametric adaptations of fuzzy systems (Section 6.8).

Optimization may not appear as a task of mining data, but it is seen that several data mining problems can be reformulated as optimization problems. All that is needed is a method of generating candidate solutions and a technique of measuring how fit a solution is or the 'fitness' of a potential solution. For example, *predicting* the level of inventory required for an item is a characteristic data mining problem. It is possible to rephrase the same problem as that of *optimization:* reducing the discrepancy that exists between demand and level of inventory to the minimum [19]. A field where genetic algorithms have been established to be reasonably effective is in the case of problems that involve scheduling. These problems are outside the range of traditional data mining problems; however, they are not only interesting but also demonstrate the supremacy of genetic algorithms (refer to [5] for maintenance scheduling problem). Genetic algorithms have also been used to facilitate the classification of comments sent to an airline [19]. Feedback received from customers directly is a potent information source for businesses, such as mobile phone companies, international airline processes, and many others. Classification of comments is an example of *the problem of message routing*.

There are three features of *optimization problems:*

- A set of parameters capable of adjustments.
- An objective function that pools the parameters into a one single value.
- A set of constraints on the parameters.

The objective is to search for the parameters capable of maximizing or minimizing the objective function subject to the constraints. It is a difficult task even for the fastest of computers to conduct a thorough search of all the combinations of parameters that fulfill the constraints; even for a very miniscule number of parameters, the number of combinations is way too large to explore.

For the problems that can be represented as well-behaved continuous functions and have extrema (maxima or minima), the extrema can be found analytically using calculus. Techniques based on calculus have been studied at length. Indirect techniques for extremization look for local extrema by solving the generally nonlinear set of equations that result from setting the gradient of the objective function equal to zero (the derivative of the objective function describes the rate at which it is increasing or decreasing; at the extremum value, the function is neither increasing nor decreasing, so the derivative is zero). On the contrary, direct techniques are *search* methods that look for local optima by riding on the function and progressing in a direction pertaining to the local gradient. This is basically the concept of *hill climbing:* to seek out the local best, climb the function in the steepest direction permissible. Hill-climbing algorithms, such as gradient descent, Newton's technique, and conjugate gradient, begin with an estimated solution and iteratively enhance it with the help of local search.

The scope of techniques based on calculus is local. The optima they look for are in the region of the present point. It is clear that beginning the search process in the locality of the lower peak will lead us to miss the primary event (the higher peak). After the lower peak is attained, more improvement has to be searched for via random restart or other tricks. Another issue with calculus-based techniques is that they rely on the fact that derivatives exist (very uncommon in the business world). Even if we permit numerical estimation of derivatives, it is a severe limitation. The real world of search is flooded with irregularities and vast multimodel (i.e., comprising several "hills") noisy search spaces; techniques that rely on restrictive needs of continuity and derivative existence, are adequate only for a limited domain of problems. To attack more complex problems, other techniques will be required. We will study the way genetic algorithms help tackle complex problems.

Many successful applications have been described in the GA literature. However, there are several cases wherein GAs do not perform well. When faced with a potential application, how is it possible to understand whether a GA is an appropriate technique to employ? There is no clear answer, although researchers do share the intuitions that if the space for exploration is large enough, is far from smooth and unimodel, or is not understood appropriately; or if the fitness function is noisy; and if the task does not need a global optimum to be found—i.e., if arriving at a fairly "good" solution quickly is sufficient—a GA will have an opportunity to surpass other techniques.

These intuitions, definitely do not predict exactly when a GA will serve as an efficient search process competent enough to match other processes. Research on this facet of genetic algorithms has still to produce definite answers.

A.1 A SIMPLE OVERVIEW OF GENETICS

Knowledge of biological terminology, though not necessary, may help better appreciation of genetic algorithms.

Life is dependent on proteins, comprising 20 basic units known as amino acids. The *chromosomes* in a cell's nucleus are strands of DNA (Deoxyribo Nucleic Acid) carrying the blueprints of the proteins the cell requires. Conceptually, we may divide a chromosome into *genes*—functional blocks of DNA, each encoding a specific protein. In a very rough manner, one can consider a gene as encoding a *trait*, say, eye color. The various likely 'settings' for a trait (e.g., green, brown or blue eyes) are known as *alleles*. Each gene is found at a specific position or *locus* on the chromosome. Genes are capable of expressing themselves in destructive ways, which result in the death of the resulting organism. Healthy organisms are able to survive to produce offspring and pass on their DNA to the next generation.

The total collection of genetic material (all chromosomes within a cell collectively) is called the *genome* of the organism. *Genotype* is the term referring to the specific set of genes encompassed within a genome. The genotype is responsible—under fetus and later development—for giving rise to the *phenotype* of the organism—its physical and mental traits, for instance, eye color, height, brain size, and intelligence.

Through sexual reproduction, surviving organisms pass on their DNA to the next generation. During sexual reproduction, the DNA from one survivor is really mixed with the DNA from another via the process known as *recombination* (or *crossover*). Crossover creates two new genomes from two present ones by joining together pieces of each one.

Offsprings are a lot like their parents. They may represent a novel combination of parental traits, but they are subject to *mutation*, in which single nucleotides (elementary bits of DNA) are changed from the parents to offspring; mutation may cause the chromosomes of offsprings to be different from those of their biological parents. Mutation is quite rare in nature. The resulting change in the gene occasionally represents a significant improvement in fitness, although more often than not, the results are harmful.

All these processes combined with that of *natural selection*, over several generations, can give rise to organisms that are extremely adaptable to their environment: the *evolution* process. The process of natural selection permits only the individuals who are most fit in the population to live to pass on their genetic material on to the next generation.

Therefore, evolution can be considered a process that results in the maintenance or increase of population's survival capability and the ability to reproduce in a particular environment. This ability is known as *evolutionary fitness*. Even though it is not possible for fitness to be directly measured, it is possible to estimate it based on the ecology and functional morphology of the organism in its environment. Evolutionary fitness can also be considered a measure of the organism's ability to anticipate environmental changes. Therefore, the fitness can be seen as the quality that natural life is optimizing. Natural evolution is an endless, uninterrupted process.

A.2 GENETICS ON COMPUTERS

In 1858, with the presentation of his theory of evolution, Charles Darwin [186], marked the launch of a revolution in biology. Darwin's classical theory of evolution combined with the theory of natural selection and the theory of genetics, are now representative of the *new-Darwinian* paradigm, which is based on the processes of reproduction, mutation, competition and selection.

Is it possible for a computer to simulate the natural evolution process? There exist many different techniques of evolutionary computation today, and all of them simulate natural evolution. This is usually done through the creation of a population of individuals, assessing their fitness, generating a new population through genetic operations, and repeating this procedure several times. However, there are various means of doing evolutionary computing. Here, the concepts are presented using the *basic genetic algorithm*; most other evolutionary algorithms, for instance, genetic programming [187] and evolution strategies [188], can be simply considered as variants.

In the 1970s, John Holland, founder of evolutionary computation, introduced the genetic algorithms concept [189], and this became popular through the efforts of David Goldberg [190]. Holland aimed to get computers to perform all that nature did. It is possible to represent his GA through a series of procedural steps that move from one population of 'artificial chromosomes' to another population. It employs natural selection, and methods inspired by genetics, that is, crossover and mutation.

Encoding and *evaluation* are the two instruments connecting a GA to the issue that it is offering a solution for. In Holland's work, encoding is done through the representation of chromosomes as strings of binary digits (1s and 0s). *Chromosome* is a term used to refer to a candidate solution to an optimization problem. The 'genes' are short blocks of neighboring bits encoding a specific element of the candidate solution. An 'allele' in a bit string is either 0 or 1. We show one such string in Fig. A.1. It is a *binary string* of length 12 bits for representing a chromosome. In the context of parametric optimization, the parameter θ of the problem is coded as a finite length string (chromosome). In case of multiparameter function optimization, the parameter vector $\boldsymbol{\theta} = [\theta_1 \; \theta_2 \; \cdots \; \theta_n]^T$ is coded as a chromosome; the bits encoding a particular parameter θ_j, might be considered to be a gene.

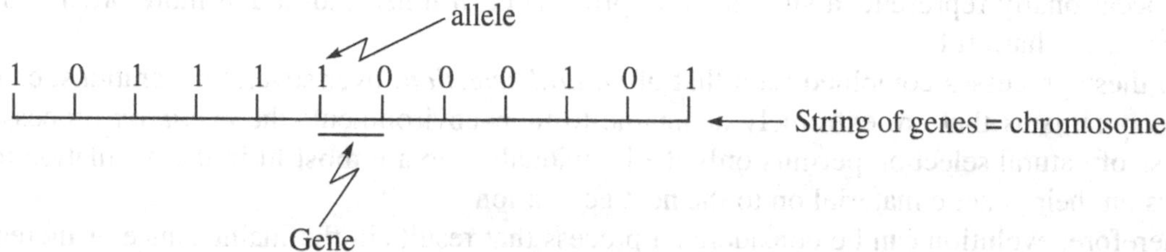

Figure A.1 Binary string for representing a chromosome

An evaluation function is employed for measuring how fit the chromosome is in terms of its performance to solve the problem. The *fitness function* helps in measuring how fit a chromosome is to be able to survive in a population of chromosomes. The GA will look for maximization of the fitness function $J(\boldsymbol{\theta})$ through the evolution of successive generations of chromosomes (binary strings) that $\boldsymbol{\theta}$ represents.

The basic GA is depicted in Fig. A.2. It consists of a population of strings or chromosomes, and three evolutionary operators—*selection*, *crossover*, and *mutation*. Each chromosome encodes a candidate solution to the problem in question, and its associated fitness is dependent on the

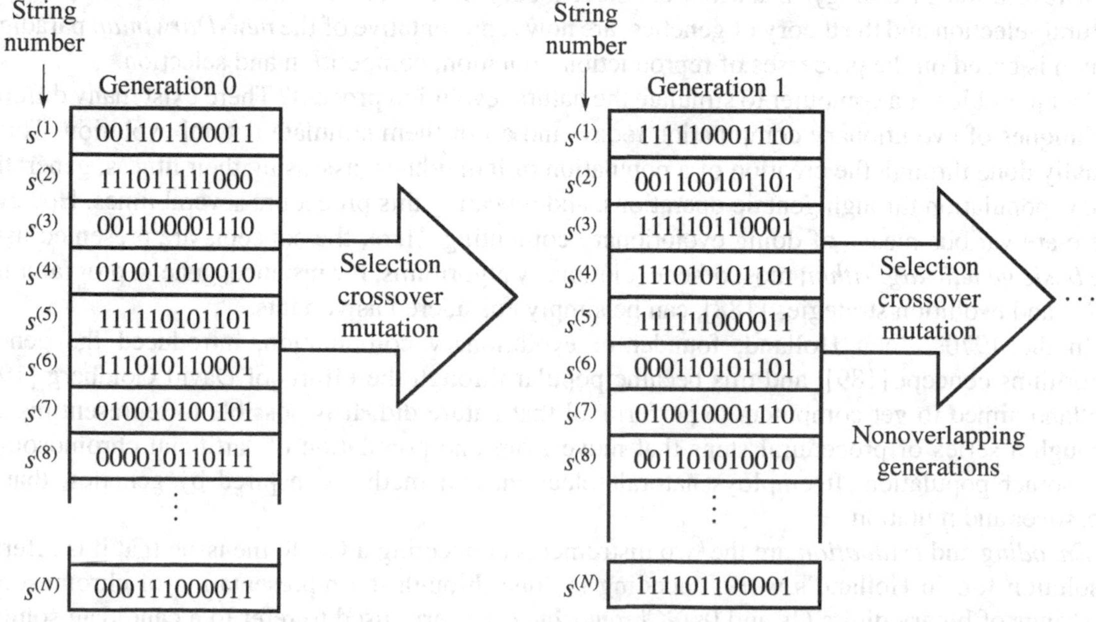

Figure A.2 The basic GA

application. The original population is randomly generated. A population with a high level of fitness evolves via many generations. Iterating selection, crossover and mutation processes gives rise to distinct generations. Once a new generation is evolved, the old can be discarded. New generations are evolved until some stopping criterion is met (this is unlike natural populations). Also, the size of the population stays constant for generations, which is not the case in natural populations. Therefore, there is no chance of the extinction of the GA population (which would undoubtedly not be the optimum solution!).

Genetic Operators

Let us now explain the genetic operators and their importance. They will be described in the form of a traditional GA free of problem-specific alterations.

Selection: Essentially, as per Darwin, the individuals who are most qualified survive to mate. The term 'most qualified' can be quantified using the fitness value of a chromosome. In the *fitness proportionate selection*, the fitness function assigns a fitness to candidate solutions or chromosomes. This fitness level is used to associate a probability of selection with each individual chromosome. The fitness of a chromosome is divided by the total fitness of all chromosomes in a generation, to obtain its probability of selection.

The selection operation could be imagined to be like a roulette wheel in a casino. Imagine a roulette wheel wherein each candidate solution is representative of a pocket on the wheel; the size of the pockets is proportional to the selection probability. Selection of N chromosomes from the population is carried out by playing N games on the roulette wheel, wherein every candidate is drawn independently. A random selection is made just the way the roulette wheel is rotated. Chromosomes that are fitter will end up with more copies in the 'mating pool', that is, the collection of chromosomes chosen for mating; hence, chromosomes with larger-than-average fitness will embody a greater portion of the next generation. At the same time, owing to the probabilistic nature of the selection process, it is possible that certain chromosomes that are relatively less fit, may find their way into the mating pool.

Crossover: In biological terms, we consider crossover as mating, which at a basic biological level involves the process of *combining chromosomes*. We randomly pair off the chromosomes in the mating pool (i.e., form pairs to mate). The moment two chromosomes are chosen to mate, two offspring are created using the crossover operator.

To crossover chromosomes of parents, we choose a "cross site" randomly, chosen between one and $(L-1)$ bits where L is the length of the chromosome. After this, all the bits to the right of the cross site of a string are exchanged with those of the other. This procedure is illustrated in Fig. A.3. In this example, the cross site is position five on the string, and therefore, the last seven digits are swapped between the two strings. A repetition of this procedure is done for each pair of strings in the mating pool. This is the most simple one-point crossover system, wherein cross site is chosen randomly.

Mutation: The mutation operation is usually performed on the chromosomes in the mating pool after the crossover operation.

In a binary-coded GA, it is possible to mutate by flipping a bit. Mutation gives rise to random incremental changes in the offspring generated through crossover, as shown in Fig. A.3.

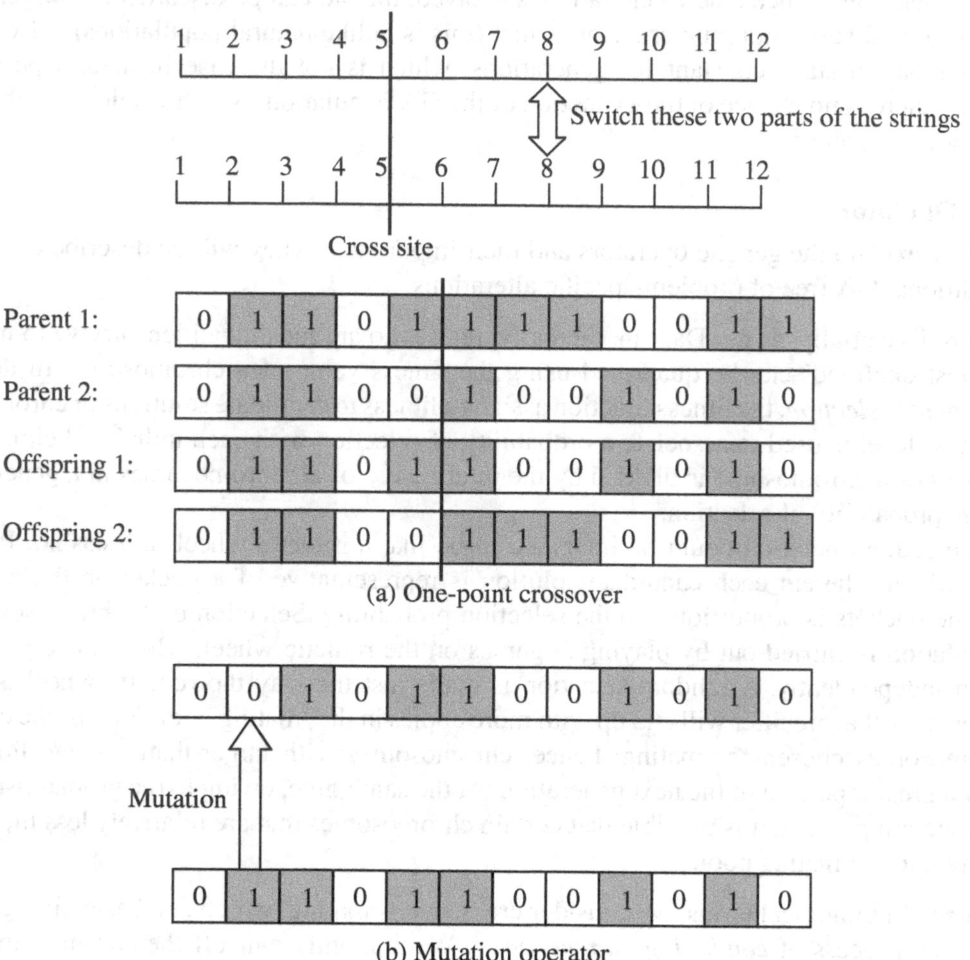

Figure A.3 Crossover and mutation operations in reproduction phase

When mutation is carried out by itself, in the absence of any crossover, it is equivalent to a random search, comprising incremental alteration of the present solution, and acceptance in case improvement is seen. But when employed in GA, its behavior alters drastically. In a GA, the important role of mutation is that it replaces the gene values, which the population loses during the process of selection, so that mating can be tried out in a new context, or it provides the gene values that were absent in the original population. Therefore, mutation makes it possible to reach the whole search space in spite of a limited population size.

The mutation probability is defined as the likelihood of mutation of each gene. It keeps control of the rate at which new gene values are injected in the population. If this rate is very less, then several gene values that would have been otherwise useful are never tried out. If it is too much, then excess of random disturbance will take place, and the offspring will end up losing their resemblance to the parents. The algorithm's capability to learn from the history of the search will, thus, vanish.

During reproduction, the crossover operator exchanges components of the two single chromosomes, whereas the mutation operator alters the gene value in certain arbitrarily selected position of the chromosome. Due to this, after several successive reproductions, the chromosomes that are not very fit vanish altogether, becoming extinct. The fit ones that manage to survive slowly go on to control the population. The approach or model is rather basic. Still, even crude reproduction processes possess the ability to solve certain difficult problems

Genetic algorithms are a category of stochastic search algorithms formed on the basis of biological evolution. For an unambiguously defined problem to be solved, a basic GA applies certain simple steps for arriving at a solution to the search optimization problem. We will now examine these simple steps of genetic algorithm.

A.3 THE BASIC GENETIC ALGORITHM

We consider here a simple example to illustrate how the basic genetic algorithm works. Let us find the maximum value of the function

$$J(\theta) = 15\theta - \theta^2 \tag{A.1}$$

where parameter θ varies between 0 and 15. For simplicity, we assume that θ takes only integer values.

We can solve this trivial optimization problem by calculus. The derivative of $J(\theta)$ is $(15 - 2\theta)$. At the maximum, the derivative is 0; therefore $\theta = 7.5$ results in maxima of $J(\theta)$. Maximum value of $J(\theta)$ is $(15 \times 7.5 - (7.5)^2) = 56.25$. How can the genetic algorithm be used to solve this simple problem? In GA, evolution is simulated with the following steps.

Encoding

The first decision is how to represent candidate solutions. Basic GA requires the natural parameter set of the problem to be *coded* as a finite-length string of binary bits 0 and 1. A bit-string is a combination of 0s and 1s, which represents the value of a number in binary form. An *m*-bit string can accommodate all the integers up to the value $2^m - 1$ (conversion from decimal (base 10) to binary (base 2)). The problem under consideration requires 4-bit strings to represent candidate solutions; the bit-string is the computerized genetic material called a *chromosome*. *Encoding* results in the following chromosomes for the population:

{0001, 0010, 0011, 0100, 0101, 0110, 0111, 1000, 1001, 1010, 1011, 1100, 1101, 1110, 1111}

Fitness Function

A fitness function takes a chromosome (binary string) as an input and returns a number that is a measure of chromosome's performance on the problem to be solved. GA is a maximization routine; the fitness function must be a non-negative figure of merit.

If the optimization problem is to maximize objective function $J(\theta) \geq 0$, then this function itself may be used as the objective function. It is often necessary to transform the underlying natural objective function to a fitness function through a mapping. If the optimization problem is to *minimize* the cost function $\bar{J}(\theta)$, then the following cost-to-fitness transformation may be used:

$$J(q) = \frac{1}{\bar{J}(\theta) + \varepsilon} \tag{A.2}$$

where ε is a small positive number. Minimization of $\bar{J}$ can be achieved through maximization of J.

Another way to define the fitness function is to let

$$J(\theta) = -\bar{J}(\theta) + \max_{\theta} |\bar{J}(\theta)| \tag{A.3}$$

The minus sign in front of $\bar{J}(\theta)$ term turns the minimization problem to a maximization problem, and $\max_{\theta} |\bar{J}(\theta)|$ term is needed to shift the function up so that $\bar{J}(\theta)$ is always positive.

A Fitness function can be any nonlinear, non-differentiable, discontinuous positive function because the algorithm only needs a fitness value assigned to each string.

The GAs work by evolving successive generations of chromosomes that get progressively more and more *fit*; that is, the successive generations have chromosomes that are better and better solutions to the original problem. As with many optimization techniques on complex problems, GAs are not guaranteed to produce the best value. However, by creating successive generations of solutions that get better and better, they have proven successful in practice.

The role of the fitness function in GA is similar to that of the environment in the evolution of nature. Interaction of the individual with the environment provides a measure of fitness to reproduce. In the same way, the manner in which a chromosome (binary string) interacts with a fitness function provides a measure of the fitness of the GA while reproducing.

The fitness function for the problem in hand may be defined as

$$J(\theta) = 15\theta - \theta^2 \tag{A.4}$$

Initialization of Population

The basic element processed by GA is the string—a binary coding of a parameter (set) of the search space. We start with a randomly selected initial population of such chromosomes (strings); each chromosome in the population represents a point in the search space, and hence, a possible solution to the problem. Each string is then decoded to obtain its fitness value, which determines the probability of the chromosome being acted upon by the genetic operators. The population then evolves, and a new population is created through the application of genetic operators. The new generation is expected to perform better than the previous generation (better fitness values). The new set of strings is again decoded and evaluated, and another generation is created using the basic genetic operators. This process is continued until we reach the *terminal conditions* (to be described shortly).

The initial population of the chromosomes for our problem may appear similar to what is depicted in Table A.1; comprising randomly generated 1s and 0s filled in six 4-bit strings. An actual practical problem would essentially have a population of several thousand chromosomes.

Survival of only the most fit species is possible in natural selection. Only the fittest can breed and pass on their genes to the generation that follows. An approach similar to this is used by GAs.

However, there is no change in the size of the chromosome population in successive generations which is not the case in nature. This helps in efficient and economical computation.

Table A.1 Initial randomly generated population of chromosomes

String number	String	θ	J	$J/\Sigma J$	
$s^{(1)}$	1100	12	36	0.165	
$s^{(2)}$	0100	4	44	0.202	
$s^{(3)}$	0001	1	14	0.064	$Av\ J = 36$
$s^{(4)}$	1110	14	14	0.064	$Max\ J = 56$
$s^{(5)}$	0111	7	56	0.257	
$s^{(6)}$	1001	9	54	0.248	

For the problem in hand, Table A.1 gives an initial population of six strings, the corresponding *decoded* value of the parameter θ for each string and the fitness value for each string. One example of decoding (conversion) from binary (base 2) to decimal (base 10)) follows:

$$(1100)_2 = 1 \times 2^3 + 1 \times 2^2 + 0 \times 2^1 + 0 \times 2^0 = (12)_{10} \tag{A.5}$$

The average fitness of initial population is 36; pretty good, because the actual maximum is 56.25. Evolution can improve it further.

Let us represent the decision variable by $\theta(k)$ where the iteration index k stands for the generation number. The parameter θ for various strings in Table A.1 has been assigned initial values corresponding to randomly selected strings; these values thus stand for generation 0, and are denoted as $\theta(0)$.

String in a population may be represented as $s^{(i)}(k)$; $i = 1, \ldots, N$. For the initial population in Table A.1, $N = 6$, and $k = 0$.

The fitness function for our problem is

$$J(\theta(k)) = 15\theta(k) - (\theta(k))^2 \tag{A.6}$$

The entries in Table A.1 correspond to $k = 0$.

Selection

In the process of natural selection, the most qualified (fittest) creatures survive to mate. Fitness is determined by a creature's ability to survive predators, pestilence, and other obstacles to adulthood and subsequent reproduction. In our unabashedly artificial setting, we quantify "most qualified" via a chromosome's fitness $J(\theta(k))$. The fitness function is the final arbiter of the string-creature's life or death. Selecting strings according to their fitness values means that the strings with a higher value have a higher probability of contributing one or more offspring in the next generation.

Selection is a process in which good-fit strings in the population are selected to form a mating pool. A chromosome is selected for mating pool according to the probability proportional to its fitness value. The probability for selecting the ith string $s^{(i)}$ is

$$P^{(i)} = \frac{J(\theta^{(i)}(k))}{\sum_{i=1}^{N} J(\theta^{(i)}(k))} \tag{A.7}$$

For the initial population of six strings in Table A.1, the probability of selection is given in the column $J/\Sigma J$ of Table A.1:

$$P^{(1)} = 0.165, \quad P^{(2)} = 0.202, \quad P^{(3)} = 0.064, \quad P^{(4)} = 0.064, \quad P^{(5)} = 0.257, \quad P^{(6)} = 0.248$$

Survival is based on randomly choosing chromosomes for the next generation in such a way that the chance of a chromosome being selected is proportional to its fitness.

Chromosome selection process may be carried out by first creating a number line between 0 and 1. Each chromosome in generation 0 is allotted a portion of the number line proportional to its probability of selection. For the example in hand, $s^{(1)}$ gets 16.5% of the line, $s^{(2)}$ gets 20.2%, $s^{(3)}$ gets 6.4%, $s^{(4)}$ gets 6.4%, $s^{(5)}$ gets 25.7% and $s^{(6)}$ gets 24.8% (Fig. A.4). Chromosomes $s^{(5)}$ and $s^{(6)}$ (the most fit chromosomes) occupy the largest segments, whereas the chromosomes $s^{(3)}$ and $s^{(4)}$ (the least fit) have much smaller segments. To select a chromosome for mating, a random number is generated in the interval [0, 1], and the chromosome whose segment spans this number, is selected. In our example, we have an initial population of six chromosomes. Thus to establish the same size population in the next generation, the random number is generated six times. Notice that, in general, this procedure produces more copies of the fitter chromosomes and fewer of the less fit.

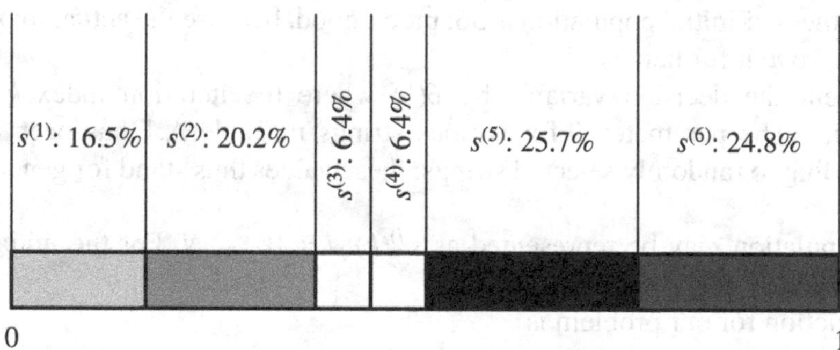

Figure A.4 The selection strategy (similar to a roulette wheel in casino)

The first two random number generations for selection might select chromosomes $s^{(6)}$ and $s^{(2)}$ to become parents, the second pair of random number generations might choose chromosomes $s^{(1)}$ and $s^{(5)}$, and the last two might select $s^{(2)}$ and $s^{(5)}$, resulting in the mating pool shown in Fig. A.5.

$$\begin{aligned}
s^{(6)} &: 1 \quad 0 \quad 0 \quad 1 \\
s^{(2)} &: 0 \quad 1 \quad 0 \quad 0 \\
s^{(1)} &: 1 \quad 1 \quad 0 \quad 0 \\
s^{(5)} &: 0 \quad 1 \quad 1 \quad 1 \\
s^{(2)} &: 0 \quad 1 \quad 0 \quad 0 \\
s^{(5)} &: 0 \quad 1 \quad 1 \quad 1
\end{aligned}$$

Figure A.5 Mating pool

Notice that due to the probabilistic nature of the selection process, it is possible that some relatively unfit strings may end up in the mating pool. For the problem in hand, string $s^{(3)}$ or $s^{(4)}$ also might have been selected in the process of random number generations for selection; it is just the luck of the draw (we do not worry too much about this issue; relatively unfit strings will not survive too long, i.e., too many generations, because of the selection operator). If the population size were a large and random number generated many times, the average results would be closer to the expected values.

Reproduction Phase; Crossover

The next operator applied to the surviving chromosomes is crossover. Crossover, which is analogous to sexual reproduction in nature, creates two new chromosomes from two existing ones by gluing together pieces of each one. There are many popular approaches to crossover; we will consider here the simplest approach: the *single-point crossover.* Single-point crossover starts with two parent chromosomes and a random position, the crossover point, where the two chromosomes 'break'. The crossover operator exchanges all bits to the right of the cross site of one string with those of the other.

This process is pictured in Fig. A.6. In this example, the cross site is position two on the string, and hence, we *swap* the last two bits between the two strings (clearly, the cross site is a random number between one, and the number of bits in the string minus one). As a result of the crossover, two offspring are created, each having a piece of their genetic code inherited from each of their parents.

Figure A.6 Crossover operation example

Just as in nature, one has no reason to expect offspring to be any more or less fit than their parents; just different. The chromosomes of offspring include traits inherited from both parents. When a particular combination turns out to have a high fitness value, it is likely to be replicated in future generations. When the solution space is very large, as it would be in typical applications of genetic algorithms, there is high probability that some of the offspring will happen to be very fit, and favored by selection. This is how crossover improves the average fitness of the population over time. If good chromosomes are not created by crossover, they will not survive too long because of the selection operator.

If a pair of chromosomes does not crossover, then chromosome *cloning* takes place, and the offsprings are created as exact copies of each parent.

Applying crossover to a population proceeds by selecting pairs of chromosomes and randomly deciding whether they split and swap. The crossover probability P_c is usually chosen to be high enough to cause many new candidate solutions to be tried, but not so high that promising traits that may be represented by substrings of the chromosome are broken up before they can demonstrate their value. A value near 0.7 generally produces good results. For pairs selected for split and swap, a random position for cross site is chosen and the offspring of the original chromosomes replace them in the next generation.

For the example under consideration, the randomly generated pairs for mating are: $s^{(6)}$ and $s^{(2)}$; $s^{(1)}$ and $s^{(5)}$; and $s^{(2)}$ and $s^{(5)}$. The chromosomes $s^{(6)}$ and $s^{(2)}$ could be crossed over after the second bit, chromosomes $s^{(1)}$ and $s^{(5)}$ after the first bit, and chromosomes $s^{(2)}$ and $s^{(5)}$ may not cross over. The chromosomes after this crossover process, are shown in Fig. A.7.

1 0 0 0

0 1 0 1

1 1 1 1

0 1 0 0

0 1 0 0

0 1 1 1

Figure A.7 Chromosomes after a crossover process

Why does the GA need to perform crossover besides the fact that the crossover facilitates the modeling of the mating part of the evolution process? Essentially, the crossover operation perturbs the parameters close to good values to attempt to explore better solutions to the optimization problem. It has the tendency to assist in performing a localized search around the much fitter strings (because on an average, the strings in the generation k mating pool are much fitter than those found in the generation k population).

In a GA, the process for crossover first demands the specification of the *crossover probability* P_c, followed by the execution of certain steps, as follows:

(i) Random pairing off of the strings in the mating pool. In case of odd number of strings in the mating pool, the string that occurs last is simply paired off with another string, which has already been paired off.

(ii) Consider a chromosome pair formed in step (i). Generate a random number $r \in [0, 1]$.

  (a) If $r < P_c$, then crossover takes place.

  (b) If $r > P_c$, then the crossover will not take place; hence, we do not modify the strings.

(iii) Repeat step (ii) for each pair of strings in the mating pool.

Reproduction Phase; Mutation

In nature, mutation rarely takes place, as it is the outcome of miscoded genetic material being passed on to an offspring from its parent. The change that results in the gene is occasionally representative of an important improvement in fitness over the current population, though the results cause damage quite often.

Why does the GA perform mutation besides the fact that this operator facilitates the modeling of the mutation in a biological system? Essentially, its role is to provide a guarantee that the search algorithm is not trapped on a local optimum. The selection and crossover operations may end up being stagnated at any homogenous set of solutions. All chromosomes appear same under such conditions, and therefore, no improvement is possible in the average level of fitness of the population. But the solution may seem to become optimal, merely because the search algorithm is unable to move ahead. Mutation offers random excursions into new areas of the search space. We may get lucky and undergo mutation to a good solution. It is a process that attempts to ensure that we are not trapped at local optima, and are able to explore other areas of the search space to seek a global optimum. But, it is to be noted that when the initial population offers a decent coverage of the search space, successive generations progress rapidly towards the optimal solution with the help of selection and crossover; there is a likelihood of modifications brought in by mutation being harmful. Generally, the *mutation probability* P_m is selected to be very small, with a typical range being 0.001 to 0.01, as this guarantees that not all the strings in the mating pool undergo mutation, so that there is no loss of any progress made in search. Owing to the extremely low probability of mutation, the harmful effects, if any, fail to last for more than one or two generations.

In artificial genetic systems (GA), mutation is realized by inverting a randomly chosen bit in a string. For example, the third string (chromosome) in Fig. A.7 might be mutated in its second bit, and fifth string in its third bit, as shown in Fig. A.8.

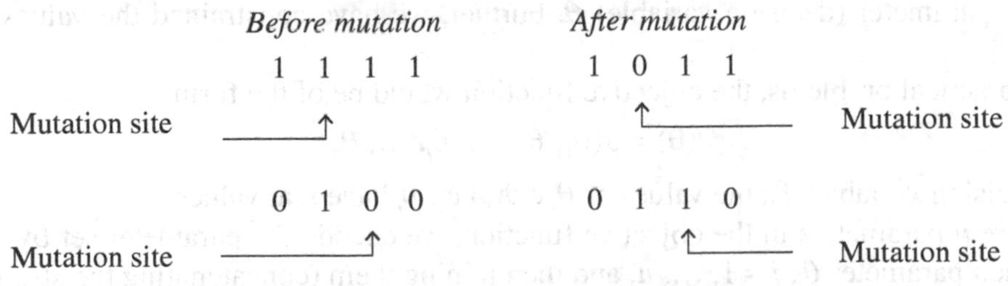

Figure A.8 Mutation operation examples

Terminal Conditions

We get the generation $k + 1$ population as a result of the selection, crossover and mutation operations on generation k. The iteration of these steps results in successive generations being produced, and this way we model evolution (an extremely basic model).

While the process of biological evolution keeps going on, maybe even forever (competition and selection processes usually occur in the natural world, wherein the limitation of space and resources restrict the expansion of populations of various species), we prefer to terminate our artificial one and explore the following:

(i) To determine the population string—say $\theta^*(k)$—that maximizes the fitness function, we require knowledge of the generation number k where the fittest string was present (not essentially in the last generation). A computer code, which implements the GA, keeps track of the highest J value, and the generation number and string that achieved this value of J.

(ii) The value of the fitness function is $J(\theta^*(k))$. How can then the genetic algorithm be terminated? There are several ways for termination of a GA; many of these resemble the termination conditions employed for conventional optimization algorithms. To introduce a few of these, let $\varepsilon > 0$ be a small number and $K_1 > 0$ and $K_2 > 0$ be integers. Termination of a GA can be done in any of the following ways:

(i) By stopping the algorithm after K_1 generations.

(ii) By stopping the algorithm post the occurrence of at least K_2 generations, and at least for K_1 iterations, the maximum (or average) value of J for all population members has risen by ε, and not more.

(iii) By stopping the algorithm the moment J takes on a value higher than a fixed value.

These possibilities are not difficult to implement on a computer. However, at times, you may wish to watch the evolution of parameters and make a decision pertaining to when the algorithm should be stopped. After many generations (typically several hundred), the population evolves to a *near-optimal* solution.

Multi-Parameter Fitness Functions

For the simple illustrative problem considered so far, the objective function

$$J(\theta) = 15\theta - \theta^2$$

has a single parameter (decision variable) θ. Further, we have constrained the values of θ to be integers.

In more practical problems, the objective function would be of the form

$$J(\boldsymbol{\theta}) = J(\theta_1, \theta_2, \ldots, \theta_j, \ldots, \theta_n) \tag{A.8}$$

having n decision variables θ_j; the values of $\theta_j \in \Re$, i.e., θ_j have real values.

If there are n parameters in the objective function, we encode the parameter set by creating bit string for each parameter θ_j; $j = 1, \ldots, n$, and then joining them (concatenating the strings). If each parameter θ_j is encoded as an m-digit binary number, then a chromosome is a string of $m \times n$ binary digits. We start with a randomly selected population of such chromosomes; each chromosome in the population represents a point in the search space, and hence a possible solution to the problem. Each sub-string is then decoded to obtain the fitness value.

Consider, for example, an objective function $J(\theta_1, \theta_2)$. The problem specification may impose different values of minimum and maximum for θ_1 and θ_2; we assume here that the minimum value to which we would expect θ_j; $j = 1, 2$, to go would be –2, and the maximum would be 5.

Therefore,

$$\theta_{j\min} = -2, \quad \text{and} \quad \theta_{j\max} = 5; \quad j = 1, 2$$

The first step is to represent the problem variables as a chromosome. In other words, we represent parameters θ_1 and θ_2 as a concatenated binary string. We take the string length to be 12. The first six encode the parameter θ_1, and the next six encode the parameter θ_2. The strings (000000, 000000) and (111111, 111111) represent the points $(\theta_{1\min}, \theta_{2\min})$, and $(\theta_{1\max}, \theta_{2\max})$, respectively, in the parameter space for the parameter set (θ_1, θ_2). Decoding of (000000) and (111111) to decimal form gives 0 and 63 ($2^L - 1$; $L = 6$), respectively. Since ($\theta_{j\min}$, $\theta_{j\max}$) = (–2, 5), the best resolution would be obtained if decoding of (000000) and (111111) to decimal form gives –2 and 5, respectively.

How is decoding done?

Consider a string (a concatenation of two sub-strings):

$$000111 : 010100 \text{ (decimal values: } 7 : 20) \tag{A.9}$$

representing a point in the parameter space for the set (θ_1, θ_2). The decimal value of the substring (000111) is 7 and that of (010100) is 20. A sub-string may be mapped to the value of the parameter θ_j with fixed range {$\theta_{j\min}$, $\theta_{j\max}$}, by the mapping

$$\theta_j = \theta_{j\min} + \frac{b}{2^L - 1}(\theta_{j\max} - \theta_{j\min}) \tag{A.10}$$

where b is the number in decimal form that is being represented in binary form, L is the length of the bit sub-string (i.e., the number of bits in the sub-string), and $\theta_{j\min}$ and $\theta_{j\max}$ are user-specified constants, which depend on the problem in hand.

For the string given in (A.9), the mapping gives the values:

$$\theta_1 = \theta_{1\min} + \frac{b}{2^L - 1}(\theta_{1\max} - \theta_{1\min}) = -2 + \frac{7}{2^6 - 1}(5 - (-2)) = -1.22$$

$$\theta_2 = \theta_{2\min} + \frac{b}{2^L - 1}(\theta_{2\max} - \theta_{2\min}) = -2 + \frac{20}{2^6 - 1}(5 - (-2)) = 0.22$$

The next step is to calculate the fitness of each chromosome. Using decoded values of θ_1 and θ_2 as inputs in the fitness function $J(\theta_1, \theta_2)$, the GA calculates the fitness of each chromosome (a concatenation of sub-string representing θ_1 and θ_2).

Note that the larger is the string length, better is the resolution of encoding, resulting in better solution accuracy. The length of the bit strings GA uses is based on the handling capacity of the computer being used for realization of GA, i.e., how long a string the computer can manipulate at an optimum level.

As an exercise example, consider the problem of minimizing the function

$$\bar{J}(\theta_1, \theta_2) = (\theta_1^2 + \theta_2 - 11)^2 + (\theta_2^2 + \theta_1 - 7)^2 \tag{A.11}$$

in the interval $0 \le \theta_1, \theta_2 \le 6$. The true solution [3, 2] to the problem is having a function value equal to zero.

Take up this problem to explain the steps involved in GA: maximizing the function

$$J(\theta_1, \theta_2) = \frac{1.0}{1.0 + \bar{J}(\theta_1, \theta_2)}; 0 \le \theta_1, \theta_2 \le 6 \tag{A.12}$$

Step 1: Take 10 bits to code each variable.

Step 2: Take population size equal to total string length, i.e., 20. Create a random population of strings.

Step 3: Consider the first string of the initial random population. Decode the two sub-strings and determine the corresponding parameter values. What is the fitness function value corresponding to this string? Similarly, for other strings, calculate the fitness values.

Step 4: Select good strings in the population to form the mating pool.

Step 5: Perform crossover on random pairs of strings (the crossover probability P_c is 0.8).

Step 6: Perform bitwise mutation with probability 0.05 for every bit.

The resulting population is the new population. This completes one iteration of GA, and the generation count is incremented by 1.

Summary of the Basic Genetic Algorithm

On-line view of the basic genetic algorithm is summarized in Table A.2 [5].

Table A.2 Summary of the basic genetic algorithm

Generate a population of chromosomes of size N: strings $s^{(1)}, s^{(2)}, \ldots, s^{(N)}$; string $s^{(i)}$ representing the parameter set $\boldsymbol{\theta}^{(i)} = \{\theta_1^{(i)}, \theta_2^{(i)}, \ldots, \theta_j^{(i)}, \ldots, \theta_n^{(i)}\}$

Step 1: Calculate the fitness of each chromosome:

$$J(\boldsymbol{\theta}^{(1)}), J(\boldsymbol{\theta}^{(2)}), \ldots, J(\boldsymbol{\theta}^{(N)})$$

Step 2: If the termination criterion is satisfied, go to step 9; otherwise go to step 3.

Step 3: Select a pair of chromosomes for mating.

Step 4: With the crossover probability P_c, exchange parts of the two selected chromosomes and create two offspring.

Step 5: With the mutation probability P_m, randomly change the bit values in the two offspring chromosomes.

Step 6: Place the resulting chromosomes in the new population.

Step 7: If the size of the new population is N, then go to step 8; otherwise, go to step 3.

Step 8: Replace the current chromosome population with the new population, and go to step 1.

Step 9: Terminate optimization.

A.4 BEYOND THE BASIC GENETIC ALGORITHM

Researchers have been exploring improvements in genetic algorithms in all directions possible. Certain improvements have occurred in the basic algorithm; others alter the algorithm to present a better model of genetic activity in the natural world.

The simple genetic algorithm previously described has room for improvement in several areas. An overview of the research trends is given below.

Schema Theorem

Although convenient for many problems, bit strings are not the only representation of chromosomes. In reference [5], bit strings have been used to represent genomes in a *maintenance scheduling problem*, but for *the traveling salesman problem*, a vector of integers has been used to represent genomes. Vector of integers has also been used in a message routing problem in reference [19].

One of the reasons that bit strings are often preferred has to do with the way sub-strings that match a particular pattern or *schema* represent whole families of potential solutions.

John Holland in 1975 [189] introduced the notation of *schema*, which came from the Greek word meaning 'form'. A schema is a set of bit strings of 1s and 0s and asterisks, where each asterisk (∗) can assume either value 1 or 0. The 1s and 0s represent the fixed positions of a schema, while asterisks represent 'wild cards'. For example, the schema

1 ∗ ∗ 0

stands for a set of 4-bit strings. Each string in this set begins with 1 and ends with 0. The strings are called *instances* of the schema. A chromosome matches a schema when the fixed positions in the schema match the corresponding positions in the chromosome. For example, the schema

1 ∗ ∗ 0

matches the following set of 4-bit chromosomes.

1 1 1 0
1 1 0 0
1 0 1 0
1 0 0 0

The number of defined bits (non-asterisks) in a schema is called the *order* of the schema. The distance between the outermost bits of a schema is called *defining length*.

When they run, GAs manipulate schemata (plural of the word schema). As per the *Schema Theorem*, short and low-order schemata possessing fitness beyond average, increase in population from one generation to the other. Simply put, low-order schemata are the *building blocks* that lay the foundation for the GAs to operate. The building blocks that are most fit survive from generation to generation, mingling with each other to create genomes that are better, superior and more fit.

According to the Schema Theorem, genetic algorithms are really searching through the possible schemata to identify the building blocks fit enough to survive from generation to generation. By processing as many schema as possible, from generation to generation, it is possible to find the fittest building blocks.

The Schema theorem lends us insight into the reason for genomes to work better when there are just two symbols (0s and 1s) in the representation. Consider the schema [∗ 0 0]. Only two chromosomes, [0 0 0] and [1 0 0], process this schema when two symbols are used for representation of chromosomes. If there are four symbols, 0, 1, 2, and 3, in the representation, then four chromosomes: [0 0 0], [1 0 0], [2 0 0], and [3 0 0], process this schema. Because genetic algorithms search for the best schemata using a given population size, the additional chromosomes do not help this search.

Rigorous presentation of the details of Schema Theorem is beyond the scope of this Appendix.

Diploid Chromosomes

The genetics employed till now is based on the most basic chromosomes existing in nature: *haploid* chromosomes comprising just a single-strand of genes. These occur in single-cell organisms that are not complex. In the more complex organisms, the chromosomes are two-stranded (occur in arrays of pairs), or *diploid*, as in case of our own DNA. In the case of diploid sexual reproduction, genes are exchanged between pairs of chromosomes of the two parents to give rise to a *gamete* (a single chromosome) in each parent. These gametes from the two parents pair up to give rise to a complete set of diploid chromosomes. On the other hand, the exchange of genes takes place between the single-strand chromosomes of the two parents, in case of haploid sexual reproduction.

The algorithmic features of diploid chromosomes are similar to that of haploid chromosomes, as it is possible to treat diploid chromosomes as two chromosomes bound together. Selection, crossover, and mutation progress in the same manner. The only way they differ is in the fact that there are now two *alleles* for each gene and not just one. If they match, it is fine, but if they don't, the fitness function makes use of the *dominant* allele. For example, if an allele for blue eyes pairs up with an allele for brown eyes, the latter will win. In the language of genetics, the alleles for brown eyes are *expressed* instead of the blue eyes. A solution has been found by researchers—they include information related to dominance in the alleles themselves.

New Models

Certain improvements in genetic algorithms change the algorithm so that they offer a better model of genetic activity in the natural world. An overview of the popular new models follows.

- A model for the simulation of natural evolution was suggested in the early 1960s, in Germany [188]. This model is referred to as an *evolution strategy*. The genetic algorithm differs from evolution strategy because it uses both crossover and mutation, unlike the evolution strategy which employs mutation alone. Additionally, evolution strategies need no representation of the problem in a coded form.
- *Genetic Programming* is an extension of the traditional genetic algorithm, but the objective of genetic programming is not merely the evolution of a bit-string representation of a certain problem but the evolution of the computer code that offers a solution to the problem. This means, genetic programming can lead to the creation of computer programs as the solution, whereas genetic algorithms give rise to strings of binary numbers representing the solution. Genetic programming applies the same evolutionary model as genetic algorithm [187].

Other Learning Approaches Inspired by Nature

In addition to evolutionary computation, many other learning approaches have been developed, which are inspired by nature. Out of these approaches, *swarming* and *immune systems* are widely researched and used methods (refer to Section 1.7). The thread that ties together learning based on evolution process, swarm intelligence and immune systems is that all have been applied successfully to a variety of optimization problems. Interested readers may find references [26–29] useful.

Appendix

B REINFORCEMENT LEARNING (RL)

B.1 INTRODUCTION

Reinforcement learning (RL) is a learning paradigm pertaining to *learning to control a system to maximize a numerical performance measure that expresses a long-term objective*. It is of immense interest owing to the huge volume of applications it can help address. The application of reinforcement learning in learning to play games has proved to be very impressive. Job-shop scheduling (seeking a schedule of jobs fulfilling temporal and resource limitations) is one of the several operation-related problems that RL can effectively address. Other problems include inventory control, maintenance problems, targeted marketing, vehicle routing, fleet management, elevator control, etc. RL is capable of addressing a lot of information theory problems, for instance, optimal coding, packet routing, and optimization of channel allocation or sensor networks. Another significant category of problems emanates from finance. These include optimal portfolio management and option pricing. RL is employed to handle problems in control engineering: optimal control of chemical or mechanical systems in process control and manufacturing applications. The problem of controlling robots is part of the latter [31].

Reinforcement learning finds basis in the logical concept that if an action is followed by a satisfactory state of affairs, or by an improvement in the state of affairs (as established by a clearly defined measure), then the inclination to produce that action becomes stronger, that is, *reinforced*. Allowing actions to depend on state information, brings in the feedback aspect. Therefore, an RL learning system is one which results in an improvement in performance by interacting with its environment. Through interaction, the learning system receives feedback in the form of a scalar reward (or penalty)—a *reinforcement signal*, consistent with the response. No supervision is given to the learning system regarding what action is to be taken. Rather, trials lead to the discovery of the action which will give the most reward. The actions influence not just the *immediate reward* but also the situation that follows, and hence, all successive rewards. These two features—learning by trial-and-error and *cumulative reward*—are the two essential distinguishing traits of reinforcement learning. Even though the initial performance may be weak, through sufficient environmental interaction it will ultimately be able to learn an effective strategy that maximizes the cumulative reward.

Suppose, we wish to build a machine that learns to play chess [6]. It is not possible for us to use a supervised learner as there is no best move; the appropriateness of a move is dependent on the moves that follow. A single move does not count but a *sequence* of moves is the best if the game is won after playing them (cumulative reward). There is a decision maker called the *agent* (the game-player, playing against an opponent) that is placed in an *environment* (the chess board). At any time, the environment is in a certain *state* (the state of the chess board), and the agent has a set of possible *actions* (legal moves of pieces on the chess board). Once an action is selected and performed, the state is altered. A *sequence of actions* is required for the solution to the task; the learning agent learns the best sequence of actions to solve a problem, where 'best' is quantified as a sequence possessing the maximum cumulative reward (winning the game). Feedback in the form of immediate reward following each action (move on the chess board) is used to select actions in order to ensure maximum cumulative reward.

Another example we take here is a robot (agent) placed in a maze (environment) [6]. The robot can move in one of the four compass directions without hitting the walls (set of possible actions). At any time, the robot is in a certain position in the maze, that is, one of the possible positions (environment is in a certain state that is one of the possible states). The robot performs a sequence of moves and when it reaches the exit, only then does it get a reward (cumulative reward). In this case, there is no opponent, but we can have preference for shorter trajectories, implying that in this case we play against time.

Reinforcement learning framework formulation to solve *sequential decision problems*, generally considers, the reinforcement learning problem as a straightforward framing of the problem of learning from interaction to attain a goal. The learner and the decision-maker is called *an agent*. Everything beyond that is known as the *environment*. There is continuous interaction amongst them, with the agent choosing the actions and the environment providing responses to them and presenting new situations (*states* of the environment) to the agent. Figure B.1 is a diagram of a generic agent, which perceives its environment via *sensors* and *acts* upon it using *effectors*. RL means learning to map states to actions so as to maximize a numerical *reward*. The agent is not

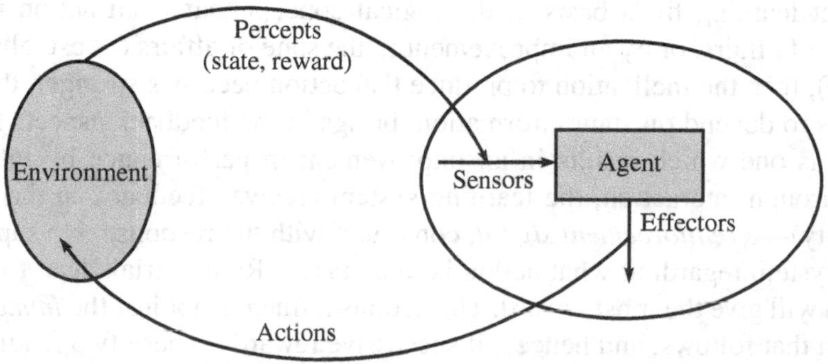

Figure B.1 A generic agent

informed regarding the actions to be taken; rather, it should discover the actions that give the maximum reward, by trying them out. To achieve huge rewards, an RL agent has to necessarily opt for actions that have been tried out earlier and found to be efficient in reward production. In order to find such actions, it has to test actions that have not been chosen before. The agent has to *exploit* the knowledge it already possesses by being *greedy* to maximize reward, but it has to also *explore* so as to choose better actions in the future. The problem is that exclusively exploring or exploiting will only lead to failure at the task. The agent has to attempt various actions and progressively opt for the ones that seem to be most efficient. Even if the agent's performance is not good enough initially, with adequate environmental interaction, it will ultimately learn an effective *policy* for reward maximization.

Reinforcement learning is not the same as supervised learning. It could be considered as "learning with a critic", which is different from "learning with a teacher". A *critic* is not the same as a teacher as it does not show us what is to be done. It only tells us how well we have done in the past; the critic gives no advance information. After several actions are taken and rewards received, it is desired to assess the individual actions performed earlier, and identify the moves that resulted in the winning of the reward, so that they can be recorded and recalled in the future. An RL program actually learns to generate a *value* for immediate states or actions; that is, how well they lead us to the goal. The moment an immediate reward mechanism of this kind is learned, the agent can perform the local actions to maximize it.

Until now we have assumed in the book that the instances that constitute a sample are *iid* (independently and identically drawn). This assumption is, however, not valid for applications where successive instances are dependent. Processes where there are a sequence of observations can't be modeled as simple probability distributions.

The problems which require a sequence of actions are ideally described in the *Markovian Decision Processes* (MDPs) framework of [6], wherein the sequence is characterized as being generated by a parametric random process. The states can be observed in an MDP model. At any time t we are aware of the state, and as the system progresses from one state to another, we obtain observation sequence, that is, a sequence of states and actions. In a *Hidden Markov Model* (HMM), the states cannot be observed.

In real-life applications, we usually come across two situations: observable states, and partially observable states. In certain applications, the agent is not aware of the exact state. It has the sensors that return an observation, which the agent then employs for an estimation of the state. Suppose there is a robot navigating in a room. It may be unaware of its precise location in the room, or what else exists in the room. The robot could have a camera for the purpose of recording sensory observations. This does not really convey to the robot its precise state, but provides some indication regarding its probable state. This setting resembles an MDP, except that once the action is taken, the new state is not known, but there is a sensor observation that is a stochastic function of earlier state and action. The solution to this *partially observable* MDP is somewhat similar to MDP: from sensor observation, the state (or rather the probability distribution for the states) can be inferred and then acted upon.

In our brief presentation in this appendix, we will assume observable states, and use MDP model to model the behavior of the agent.

The standard approach for solving MDPs is with the help of *dynamic programming*, which converts the problem of identifying a good agent into the problem of seeking a good *value function*. But, other than the simplest case when MDP has very few states and actions, dynamic programming is not feasible. The RL algorithms discussed here can be considered as a way of converting the infeasible dynamic programming techniques into feasible algorithms to make them applicable to large-scale problems. There are just two primary ideas that permit RL algorithms to attain this goal [31]:

- One primary idea is to employ *samples* that can represent the dynamics of the control system in a compact manner.
- Another key idea behind RL algorithms is to make use of powerful *function approximation* techniques (neural networks, for example) to efficiently represent value functions. Using neural networks helps control RL in terms of the realistic problems possessing large state- and action spaces. A neural network possesses the *generalization* property; experience with a restricted subset of state space is typically generalized to give rise to a good approximation over a much bigger subset.

The two ideas fit together properly. Samples may focus on a small subset of space they belong to, from which function approximation methods generalize to bigger spaces. *It is the understanding of the interplay between dynamic programming, samples, and function approximation, which is at the heart of the design, analysis, and application of RL algorithms.*

Recent advances relating reinforcement learning to dynamic programming are providing solid mathematical foundation; mathematical results that guarantee optimality in the limit for an important class of reinforcement learning systems are now available (the property that we lack in case of supervised learning, which is an empirical science—the asymptotic effectiveness of the learning systems has been validated only empirically).

In this appendix, we will concentrate on those RL algorithms that are built on the foundation of the powerful theory of dynamic programming. RL discussed here is also referred to as *neuro-dynamic programming* or *approximate dynamic programming*. The term neuro-dynamic programming is derived from the fact that, in several cases, RL algorithms are used with artificial neural networks.

There are several software packages which support the development and testing of RL algorithms. The most notable could be the RL-GLUE (http://glue.rl-community.org) and RL-LIBRARY (http://library.rl-community.org) packages.

The coverage of reinforcement learning in this appendix is to be regarded as an introduction to the subject; a springboard to advanced studies [30–32]. The inclusion of the topic has been motivated by the observation that reinforcement learning has the potential of solving many nonlinear control problems.

B.2 ELEMENTS OF REINFORCEMENT LEARNING

The learning decision maker is known as the *agent*. There is an interaction of the agent with the *environment*, which comprises everything beyond the agent. The agent is equipped with sensors that decide on its *state* in the environment and takes an *action* that modifies its state. When the agent takes an action, the environment offers a *reward* (Fig. B.2(a)). Time is discrete: $t = 0, 1, 2, \ldots$. When the agent is in state s_t, takes an action a_t, the clock ticks, reward $r(s_t) \in \Re$ is received, and the agent moves to the next state s_{t+i}.

Beyond the agent and the environment, it is possible to find four main sub-elements of a reinforcement learning system—a *policy*, a *reward function*, a *value function* and *horizon* of decisions. A *policy* defines the learning agent's behavior at a given time. A *policy* is, roughly, a mapping of the environment from perceived states to actions that need to be implemented in those states. A *reward function* defines *immediate reward* for an action that is accountable for the present state of the environment. It maps environmental states to a scalar, a *reward*, which indicates the inherent desirability of the state. While a reward function implies whatever is good in the immediate sense, a *value function* lays down what is good in the long term. The *value* of a state is the *cumulative reward* that an agent can expect to receive in the future as an outcome of the sequence of its actions, beginning from that state. While rewards establish the immediate, intrinsic desirability of environment states, values imply the desirability of states in the long run, after considering the states that are expected to follow, and the rewards obtainable in those states. An agent's only aim is the maximization of the cumulative reward (value) attained in the long term.

The value function is dependent on the existence of a *finite* or an *infinite horizon* for making decisions. A finite horizon indicates a *fixed* time after which nothing really matters, as the game is kind of finished. In case of a finite horizon, the optimal action for a given state may be different at different times, that is optimal policy for a finite horizon is *nonstationary*.

In the absence of a fixed time limit, on the contrary, the reason for any different behavior in the same state at different times does not arise. Therefore, the optimal action is dependent solely on the present state, and the policy is *stationary*. Policies for finite-horizon case are complex, whereas policies for the infinite-horizon are much simpler.

'Infinite horizon' does not imply that all state sequences are infinite; it merely means that no deadline is fixed. There will not be any infinite sequences if the environment comprises *terminal states* and if it is known with certainty that the agent will eventually reach one.

Our focus in this chapter is on reinforcement learning solutions to control problems (Fig. B.2(b)). The controller (agent) has a set of sensors to observe the state of the controlled process (environment); the learning task is to learn a control strategy (policy) for choosing control signals (actions) that achieve minimization of a performance measure (maximization of cumulative reward).

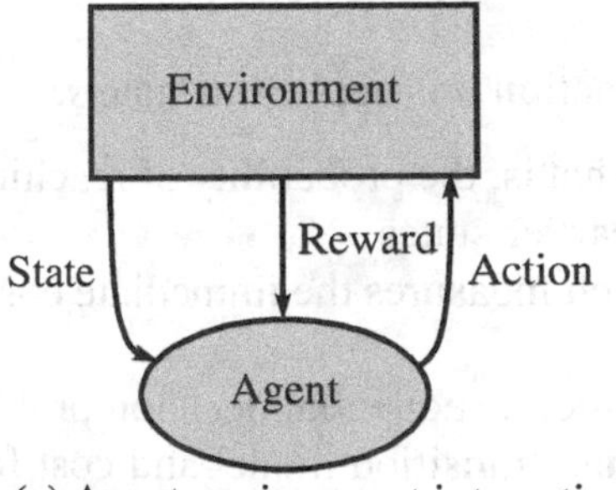

(a) Agent-environment interaction

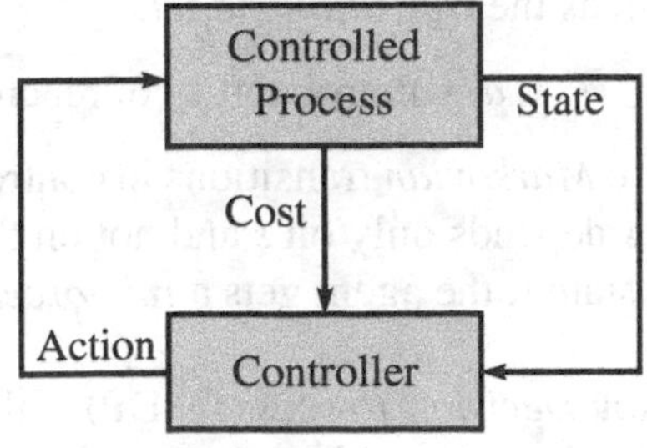

(b) The basic reinforcement learning scenario for control problems

Figure B.2

In control problems, we minimize a performance measure; frequently referred to as *cost function*. The reinforcement learning control solution seeks to minimize the long-term accumulated cost the controller incurs over the task time. The general reinforcement learning solution seeks to maximize the long-term accumulated reward the agent receives over the task time. Since in control problems, reference of optimality is a cost function, we assign *cost* to the reward structure of the reinforcement learning process; *the reinforcement learning solution then seeks to minimize the long-term accumulated cost the agent incurs over the task time*. The value function of the reinforcement learning process is accordingly defined with respect to cost structure.

The stabilizing control problems are all infinite-horizon problems. Here also, we will limit our discussion to this class of control problems.

Some reinforcement learning systems have one more element—a *model* of the environment. This replicates the behavior of the environment. For instance, considering a state and action, the model may predict the subsequent next state and next cost.

RL systems were initially clearly model-free, trial-and-error learners. Over time, it became quite clear that RL techniques are similar to dynamic programming techniques, which make use of models. *Adaptive dynamic programming* has emerged as a solution method for reinforcement learning problems wherein the agent learns the models through trial-and-error interaction with the environment, and then uses these models in dynamic programming methods.

In control engineering, vector **x** is used to represent the *state* of a physical system: $\mathbf{x} = [x_1\ x_2 \cdots x_n]^T$, where x_j: $j = 1, \ldots, n$, are *state variables* of the system. State **x**, a vector of real numbers, is a point in the state space. In reinforcement learning (RL) framework, we will represent the state by '*s*'; thus *s* is a point in the *n*-dimensional state space. Similarly, the vector **u** is used for control. We will represent this by the action '*a*' in our RL framework.

If the environment is *deterministic*, then an agent's action *a* will transit the state of the environment from *s* to *s*′ deterministically; there is no probability involved. If the environment is *stochastic*, then transition of *s* to *s*′ under action *a* will be different each time action *a* is applied in state *s*. This is captured by a probabilistic model. If the environment is deterministic, but uncertain, then also transition of *s* to *s*′ under action *a* will not be unique each time action *a* is applied is state *s*. Since uncertainty in environments is the major issue leading to complexity of the control problem, we will be concerned with probabilistic models.

(1) A specification of the outcome probabilities for each admissible action in each possible state is known as the *transition model*.

$$P(s, a, s'): \text{probability of reaching state } s' \text{ if action } a \text{ is applied in state } s.$$

(2) There are *Markovian* transitions in control problems. That is, the probability of reaching state *s*′ from *s* depends only on *s* and not on the history of earlier states.

(3) In each state *s*, the agent gets a *reinforcement* $r(s)$, which measures the immediate cost of the action.

(4) A *Markov Decision Process* (MDP) is the specification of a sequential decision problem for a completely observable environment, with a Markovian transition model and cost for each state

(5) The RL framework is based on the Markov decision processes.

Off-line Learning and On-line Learning

Dynamic programming is a general-purpose technique used to identify optimal control strategies for nonlinear and stochastic dynamic systems. It addresses the problem of designing closed-loop policies off-line, assuming that a precise model of the stochastic dynamic system is available. The off-line process of design generally results in a computationally efficient technique which determines each action as a function of the observed system state.

There are two practical issues related to the use of dynamic programming:

(1) For many real-world problems, the number of possible states and admissible actions in each state are so large that the computational requirements of dynamic programming are overwhelming ('curse of dimensionality').

(2) Dynamic programming algorithms require accurate model of the dynamic system; this prior knowledge is not always available ('curse of modeling').

Over the past three decades, the focus of researchers has been to develop methods capable of finding high-quality approximate solutions to problems where exact solutions via classic dynamic programming are not attainable in practice due to high computational complexity and lack of accurate knowledge of system dynamics. In fact, *reinforcement learning* is a field that represents this stream of activities. All of the reinforcement learning can be viewed as attempts to achieve the same effect as dynamic programming, only with less computation and without assuming a perfect model of the dynamic system. By focusing on computational effort along behavioral patterns of interactions with the environment, and by using function approximation (neural network) for generalization of experience to states not reached through interactions, reinforcement learning can be used no-line for problems with large state spaces and with lack of accurate knowledge of system dynamics.

RL and use of dynamic programming to arrive at solutions for sequential decision problems are closely related as follows:

(i) the environment in both is characterized by a set of states, a set of admissible actions, and a cost function;

(ii) both aim to identify a decision policy that reduces the cumulative cost over time to a minimum.

However, there is a significant difference. While solving a sequential decision problem with the help of dynamic programming, the agent (apparently the designer of the control system) has a complete (albeit stochastic) model of the environmental behavior. With this information, the agent can calculate the optimal control policy pertaining to the model. In RL, the set of states, and the set of admissible actions are known *a priori*. However, the effects of actions on the environment and on the cost are unknown. Therefore, the agent cannot compute an optimal policy *a priori* (off-line). Rather, the agent should learn an optimal policy by experimenting in the environment. Therefore, RL system is an on-line system.

In an on-line learning system, the learner moves around in a real environment observing the outcomes. In such a situation, the main concern is generally the number of real-world actions that the agent should perform to converge to a computational agreeable policy (instead of the number of cycles, as in off-line learning). The reason is that for many practical problems, the costs in time and in dollars of performing actions dominate the computational costs.

An *adaptive dynamic programming* agent learns the transition model of the environment by interacting with the environment. It then plugs the transition model and the observed costs in the dynamic programming algorithms. Adaptive dynamic programming is, therefore, an on-line learning system.

The process of learning the model itself is not difficult when it is possible to fully observe the environment. In the most simple case, we can represent the transition model as a table of probabilities. We track the frequency of each action outcome, and make an estimation of the transition probability $P(s, a, s')$ from the frequency with which state s' is attained during the execution of the action a in state s.

Temporal Difference Learning: An idea that is central and novel to RL, is undoubtedly *Temporal Difference* (TD) learning. Temporal difference learning can be considered as a version of dynamic programming, with the only difference that TD techniques can learn on-line in real-time, from raw experience without a model of the environment's dynamics. TD techniques do not make an assumption of complete knowledge of the environment; they need only *experience*—sample sequences of states, actions and costs from actual environmental interaction. Learning from *actual* experience is outstanding as it needs no prior knowledge of the environment's dynamics, but still can attain optimal behavior.

The main benefit of dynamic programming is that, if a problem can be specified in terms of Markov decision process, then its analysis can be done and an optimal policy achieved *a priori*. The two primary drawbacks of dynamic programming are as follows:

(i) for various tasks, it is not easy to specify the dynamic model; and
(ii) as dynamic programming establishes a fixed control policy *a priori*, it fails to offer a mechanism for adapting the policy to compensate for disturbances and/or modeling errors (nonstationary dynamics).

Reinforcement learning has complimentary benefits as follows:

(i) it needs no prior dynamical model of any type, but learns by experience gathered directly from the environment; and
(ii) to a certain level, it can keep track of the dynamics of nonstationary systems.

The primary drawback of RL is that, typically, several trials (repeated experiences) are needed to learn an optimal control strategy, particularly if the system begins with a weak initial policy.

It appears that the respective drawbacks of these two approaches can be taken care of through their integration. That is, if a complete, possibly incorrect, model to the task is available *a priori*, model-based techniques (including dynamic programming) may be employed for the development of initial policy for an RL system. A reasonable initial policy can substantially improve the system's initial performance and reduce the time required to reach an acceptable level of performance. Alternatively, if an adaptive RL element is added to an otherwise model-based fixed controller, an inaccurate model can be compensated for.

In this appendix, we limit our discussion to *naive* reinforcement learning systems. Our focus is on temporal difference learning. We begin with an introduction to dynamic programming, and then using this platform, develop temporal difference methods of learning.

B.3 BASICS OF DYNAMIC PROGRAMMING

We first define a general formulation of the problem of learning sequential control strategies. To do so, we consider building a learning controller for stabilization of an inverted pendulum. Figure B.3 shows an inverted pendulum with its pivot mounted on a cart. The cart is driven by an electric motor. The motor drives a pair of wheels of the cart; the whole cart and the pendulum become the 'load' on the motor. The motor at time t exerts a torque $T(t)$ on the wheels. The linear force applied to the cart is $u(t)$; $T(t) = Ru(t)$, where R is the radius of the wheels.

The pendulum is obviously unstable. It can, however, be kept upright by applying a proper control force $u(t)$. This somewhat artificial system example represents a dynamic model of a space booster on take off—the booster is balanced on top of the rocket engine thrust vector.

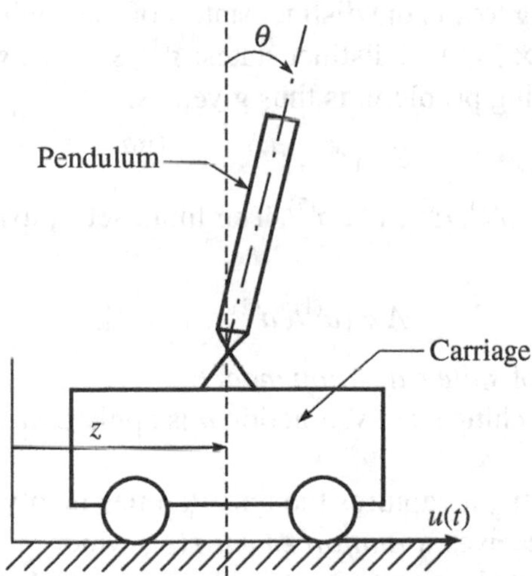

Figure B.3 Inverted pendulum system

In the *reinforcement learning control* setting, the controller, or *agent*, has a set of sensors to observe the *state* of its *environment* (the dynamic system: inverted pendulum mounted on a cart). For example, a controller may have sensors to measure angular position θ and velocity $\dot{\theta}$ of the pendulum, and horizontal position z and velocity $\dot{z}$ of the cart; and actions implemented by applying a force of u newtons to the cart. Its task is to learn control strategy, or *policy*, for choosing actions that achieve its goals.

A common way of obtaining approximate solutions for continuous state and action tasks is to quantize the state and action spaces, and apply finite-state dynamic programming (DP) methods. The methods we explore later in this appendix, make learning tractable on the realistic control problems with continuous state spaces (infinitely large set of quantized states).

Suppose that our stabilization problem demands that the pendulum must be kept within ± 12° from vertical, and the cart must be kept within ± 2.4m from the center of the track.

We define the following finite sets of possible states S and available actions A.

States →	1	2	3	4	5	6
Pend. angle(deg); θ	< -6	-6 to -1	-1 to 0	0 to 1	1 to 6	> 6
Pend. velocity; $\dot{\theta}$	< -50	-50 to 50	> 50			
Cart position(m); z	< -0.8	-0.8 to 0.8	> 0.8			
Cart velocity; $\dot{z}$	< -0.5	-0.5 to 0.5	> 0.5			

Actions →	1	2	3	4	5	6	7
Apply force of u newtons	-10	-6	-2	0	2	6	10

Define: $x_1, = \theta$, $x_2, = \dot{\theta}$, $x_3 = z$, $x_4 = \dot{z}$. Vector $\mathbf{x} = [x_1\ x_2\ x_3\ x_4]^T$ defines a point in the state space; the distinct point corresponding to $\mathbf{x}$ is the distinct state s of the environment (pendulum on a cart). Therefore, there are $6 \times 3 \times 3 \times 3 = 162$ distinct states: $s^{(1)}, s^{(2)}, \ldots, s^{(162)}$, of our environment. The finite set of states, in our learning problem, is thus given as,

$$S : \{s^{(1)}, s^{(2)}, \ldots, s^{(162)}\}$$

The action set size is seven: $a^{(1)}, a^{(2)}, \ldots, a^{(7)}$. The finite set of available actions in our learning problem, is thus given as,

$$A : \{a^{(1)}, a^{(2)}, \ldots, a^{(7)}\}$$

We assume the knowledge of *state transition model*:

$P(s, a, s')$: probability of reaching state s' if action a is applied in state s; for all $s \in S$, and for all $a \in A$

Note that our model is stochastic; it captures the uncertainties involved in the environment.

In each state s, the agent receives a *reinforcement* $r(s)$, which measures the immediate *cost* of action. For the particular inverted pendulum example, a cost of '–1' may be assigned to *failure states* ($\theta > 12°$; $\theta > -12°$), and a cost of '0' may be assigned to every other state. Note that cost structure for a learning problem is an important design parameter. It controls the convergence speed of a learning algorithm. The functions $P(\cdot)$ and $r(\cdot)$ are part of the environment.

The specification of a sequential design problem for a fully observable (the agent knows where it is) environment with a Markovian decision model and cost for each state, is a *Markov Decision Process* (MDP). An MDP is defined by the tuple (S, A, P, r) where S is the set of possible states the environment can occupy; A is the set of admissible actions the agent may execute to change the state of the environment, P is the state transition probability, and r is the cost function. We assume that

$$S : \{s^{(1)}, s^{(2)}, \ldots, s^{(N)}\};\ A : \{a^{(1)}, a^{(2)}, \ldots, a^{(M)}\}$$

where N represents the total number of distinct states of the environment, and M represents the total number of admissible actions in each state.

Let us now consider the structure of solution to the problem. Any fixed action sequence (open-loop structure) will not solve the problem because due to uncertainties in the behavior of the environment, the agent might end up in a failure state; i.e., the scheme lacks the robustness properties. Therefore, a solution must specify what the agent should do far *any* state that the environment might reach. The resulting feedback loop is a source of a measure of internal/external disturbances. A solution of this kind is called a *policy*. We usually denote a policy by π.

A stationary policy π for an MDP is a mapping $\pi: S \rightarrow \Omega(A)$, where $\Omega(A)$ is the set of all probability distributions over A. $\pi(a, s)$ stands for the probability that policy π chooses action a in state s. Since each action $a^{(1)}, a^{(2)}, \ldots, a^{(M)}$ is a candidate for state s, policy $\pi(a, s)$ for s is a set of action-selection probabilities associated with $a^{(1)}, \ldots, a^{(M)}$; their sum equals one.

A stationary deterministic policy π is a policy that commits to a single action choice per state, that is, a mapping $\pi: S \rightarrow A$ from states to actions. In this case, $\pi(s)$ *indicates the action that the agent takes in state s*. For every MDP, there exists an *optimal deterministic policy*, which minimizes the *expected, total discounted cost* (to be defined shortly) from any initial state. It is therefore, sufficient to restrict the search for the optimal policy only within the space of deterministic policies.

The next question we must decide is how to calculate the *value of a state*. Recall that the value of a state is the *cumulative cost* an agent can expect to incur over the future as a result of sequence of its actions, starting from that state. A sequence of actions for a given task will force the environment through a sequence of states. Let us call it *environment trajectory* of a given task. In an infinite-horizon problem, the number of actions for a task is not fixed; therefore, number of distinct states in an environment trajectory is not fixed. A typical state sequence in a trajectory may be expressed as $\{s_0, s_1, s_2, \ldots\}$ where, each s_t; $t = 0, 1, 2, 3, \ldots$, could be any of the possible environment states $s^{(1)}, \ldots, s^{(N)}$.

Given the initial state s_t and the agent's policy π, the agent selects an action $\pi(s_t)$, and the result of this action is next state s_{t+1}. The state transition model, $P(s, a, s')$, gives a probability that the next state s_{t+1} will be $s' \in S$, given that the current state $s_t = s$ and the action $a_t = a$. Since each state $s^{(1)}, s^{(2)}, \ldots, s^{(N)}$ is a candidate to be the next state s', the environment simulator gives a set of probabilities: $P(s_t, a_t, s^{(1)}), \ldots, P(s_t, a_t, s^N)$; their sum equals one. Thus, a given policy π generates not one state sequence (environment trajectory), but a whole range of possible state sequences, each with a specific probability determined by the transition model of the environment.

The quality of a policy is, therefore, measured by the *expected value* (cumulative cost) of a state, where the expectation is taken over all possible state sequences that could occur. For MDPs, we can define the 'value of a state under policy π' formally as,

$$V^{\pi}(s) = E_{\pi}\left\{\sum_{t=0}^{\infty} \gamma^t r(s_t)\right\} \tag{B.1}$$

where $E_{\pi}\{\cdot\}$ denotes the expected value given that the agent follows policy π. This is a *discounted cost* value function; the *discount factor* γ is a number between 0 and 1 $(0 \le \gamma < 1)$.

Note that

$$\sum_{t=0}^{\infty} \gamma^t r(s_t) \le \sum_{t=0}^{\infty} \gamma^t r_{\max} = r_{\max}/(1-\gamma)$$

Thus, the infinite sequence converges to a finite limit when costs are bounded and $\gamma < 1$.

The discount factor γ determines the relative value of delayed versus immediate costs. In particular, costs incurred t steps into the future are discounted exponentially by a factor of γ^t. Note that if we set $\gamma = 0$, only the immediate cost is considered. If we set γ closer to 1, future costs are given greater emphasis relative to the immediate cost. The meaning of γ substantially less than 1 is that future costs matter to us less than the costs paid at this present time. The discount factor is an important design parameter in reinforcement learning scheme.

The final step is to show how to choose between policies. An *optimal policy* is a policy that yields the lowest expected value. We use π^* to denote an optimal policy.

$$\pi^* = \arg\min_{\pi} E_{\pi}\left[\sum_{t=0}^{\infty} \gamma^t r(s_t)\right] \tag{B.2}$$

The 'arg min' notation denotes the values of π at which $E_{\pi}[\cdot]$ is minimized. $\pi^*(s)$ is, thus, a solution (obtained off-line) to the sequential decision problem. Given π^*, the agent decides what to do in real time by observing the current state s and executing the action $\pi^*(s)$. This is the simplest kind of agent, selecting fixed actions on the basis of the current state. A reinforcement learning agent, as we shall see shortly, is *adaptive*; it improves its policy on the basis of on-line, real-time interactions with the environment.

In the following, we describe algorithms for finding optimal policies of the dynamic programming agent.

B.3.1 Finding Optimal Policies

The dynamic programming technique rests on a very simple idea known as the *principle of optimality* [32].

An optimal policy has the property that whatever the initial state and initial decisions are, the remaining decisions must constitute an optimal policy with regard to the state resulting from the previous decisions.

Consider a state sequence (environment trajectory) resulting from the execution of optimal policy $\pi^*: \{s_0, s_1, s_2, \ldots\}$ where each s_t: $t = 0, 1, 2, \ldots$, could be any of the possible environment states $s^{(1)}, s^{(2)}, \ldots, s^{(N)}$. The index t represents *stages of decisions* in the sequential decision problem.

The dynamic programming algorithm expresses a generalization of the principle of optimality. It states that *the optimal value of a state is the immediate cost for that state plus the expected discounted optimal value of the next state, assuming that the agent chooses the optimal action*. That is, the optimal value of a state is given by

$$V^*(s) = r(s) + \gamma \min_{a} \sum_{s'} P(s, a, s')V^*(s') \tag{B.3}$$

This is one form of the *Bellman optimality equation* for V^*. For finite MDPs, this equation has a unique solution.

The Bellman optimality equation is actually a system of N simultaneous *nonlinear* equations in N unknowns, where N is the number of possible environment states. If the dynamics of the environment ($P(s,a,s')$) and the immediate costs underlying the decision process ($r(s)$) are known, then, in principle, one can solve this system of equations for V^* using any one of the variety of methods for solving systems of nonlinear equations. Once one has V^*, it is relatively easy to determine an optimal policy:

$$\pi^*(s) = \arg\min_a \sum_{s'} P(s, a, s')V^*(s') \tag{B.4}$$

Note that $V^*(s) = V^{\pi^*}(s)$:

$$V^*(s) = \min_\pi V^\pi(s) \text{ for all } s \in S \tag{B.5}$$

The solution of Bellman optimality equation (B.3) directly gives the values V^* of states with respect to optimal policy π^*. From this solution, one can obtain optimal policy using Eqn (B.4).

Equation (B.5) suggests an alternative route to finding optimal policy π^*. It uses *Bellman equation for* V^π, given below.

$$V^\pi(s) = r(s) + \gamma \sum_{s'} P(s, \pi(s), s')\, V^\pi(s') \tag{B.6}$$

Note that this equation is a system of N simultaneous *linear* equations in N unknowns, where N is the number of possible environment states (Eqns (B.6) are same as Eqns (B.3) with 'min' operator removed). We can solve these equations for $V^\pi(s)$ by standard linear algebra methods.

Given an initial policy π_0, one can solve (B.6) for $V^{\pi_0}(s)$. Once we have V^{π_0}, we can obtain improved policy π_1, using the strategy given by Eqn (B.4):

$$\pi_1(s) = \arg\min_a \sum_{s'} P(s, a, s')V^{\pi_0}(s') \tag{B.7}$$

The process is continued:

$$\pi_0 \rightarrow V^{\pi_0} \rightarrow \pi_1 \rightarrow V^{\pi_1} \rightarrow \pi_2 \rightarrow \cdots \rightarrow \pi^* \rightarrow V^*$$

It is certain that each policy will be a strict improvement over the previous one (unless it is already optimal). As a finite MDP has only a limited number of policies, this procedure will have to end up as an optimal policy π^* and optimal value function V^* in a limited number of iterations.

Thus, given a complete and accurate model of MDP in the form of knowledge of the state transition probabilities $P(s, a, s')$ and immediate costs $r(s)$ for all states $s \in S$ and all actions $a \in A$, it is possible—at least in principle—to solve the decision problem off-line. There is one problem: the Bellman Eqn (B.3) is nonlinear because of the 'min' operator; solution of nonlinear equations is problematic. The Bellman Eqn (B.6) is linear and therefore, can be solved relatively quickly. For large state spaces, time might be prohibitive even in this relatively simpler case.

In the following, we describe basic forms of two dynamic programming algorithms: *value iteration* and *policy iteration*—a step towards answering the computational complexity problems of solving Bellman equations.

B.3.2 Value Iteration

As employed for solving Markov decision problems, *value iteration* is a successive approximation process that solves the Bellman optimality Eqn (B.3), where the fundamental operation is 'backing up' estimates of optimal state values. We can solve Eqn (B.3) with the help of a simple iterative algorithm:

$$V_{(l+1)}(s) \leftarrow r(s) + \gamma \min_{a} \sum_{s'} P(s, a, s') V_l(s') \tag{B.8}$$

The algorithm begins with arbitrary guess $V_0(s)$ for each $s \in S$. The sequence of $V_1(s)$, $V_2(s)$, …, is then obtained. The algorithm converges to the optimal values $V^*(s)$ as the number of iterations l approaches infinity (We use the index l for the stages of iteration algorithm, whereas we have used earlier the index t to denote the stages of decisions in the sequential decision problem). In practice, we stop once the value function changes by a small amount. Then a *greedy policy* (choosing the action with the lowest estimated cost) with respect to the optimal set of values is obtained as an optimal policy.

The computation (B.8) is done off-line, i.e., before the real system starts operating. An optimal policy, that is, an optimal choice of $a \in A$ for each $s \in S$, is computed either simultaneously with V^*, or in real time, using Eqn (B.4).

A sequential implementation of iteration algorithm (B.8) requires temporary storage locations so that all the iteration-$(l + 1)$ values are computed based on the iteration-l values. The optimal values V^* are then stored in a *lookup table*. In addition to a problem of the memory needed for large tables, there is another problem of time needed to accurately fill them. Suppose there are N states, and M is the largest number of admissible actions for any state, then each iteration comprising backing up the value of each state precisely once, needs about $M \times N^2$ operations. For the huge state sets, typically seen in control problems, it is tough to attempt to complete even one iteration, leave alone iterate the procedure till it ends up in V^* (curse of dimensionality).

The iteration of *synchronous* DP algorithm defined in (B.8) *backs up* the value of every state once to produce the new approximate value function. We call this kind of operation as *full backup*; it is based on all possible next states rather than on a sample next state. We think of the backups as being done in a *sweep* through the state space.

Asynchronous DP algorithms are not organized in terms of systematic sweep of the entire set of states in each iteration. These algorithms back up the values of the states in any random order, with the help of whatever values of other states are available. The values of certain states may be backed up many times before the values of others are backed up once. To converge accurately, however, an asynchronous algorithm will have to continue to back up the values of all the states.

Of course, evading sweeps need not imply that we can get away with less computation. It simply means that our algorithm does not require to be locked into any hopelessly long sweep before it can progress. We can attempt to leverage this flexibility by choosing the states to which backups are applied to improve the algorithm's progress rate. We can attempt to order the backups to let value information propagate efficiently from one state to another. Some states may not require the backing up of their values as often as other states. Some state orderings give rise to faster convergence as compared to others, depending on the problem.

B.3.3 Policy Iteration

A *policy iteration algorithm* operates by alternating between two steps (the algorithm begins with arbitrary initial policy π_0):

(i) *Policy evaluation step*

Given the current policy π_k, we perform policy evaluation step that computes $V^{\pi_k}(s)$ for all $s \in S$, as the solution of the linear system of equations (Bellman equation)

$$V^{\pi_k}(s) = r(s) + \gamma \sum_{s'} P(s, \pi_k(s), s')V^{\pi_k}(s') \tag{B.9}$$

in the N unknowns $V^{\pi_k}(s)$.

To solve these equations, an iteration procedure similar to the one used in value iteration algorithm (given by (B.8)) may be used.

$$V_{l+1}^{\pi_k}(s) \leftarrow r(s) + \gamma \sum_{s'} P(s, \pi_k(s), s')V_l^{\pi_k}(s') \tag{B.10}$$

(ii) *Policy improvement step*

Once we have V^{π_k}, we can obtain improved policy π_{k+1} (refer to Eqn (B.7)) as follows:

$$\pi_{k+1}(s) = \arg\min_{a} \sum_{s'} P(s, a, s')V^{\pi_k}(s') \tag{B.11}$$

The two-step procedure is repeated with policy π_{k+1} used in place of π_k, unless we have $V^{\pi_{k+1}}(s) \approx V^{\pi_k}(s)$ for all s; in which case, the algorithm is terminated with optimal policy $\pi^* = \pi_k$.

Policy iteration algorithm can be viewed as an *actor-critic system* (Fig. B.4). In this interpretation, the policy evaluation step is viewed as the work of a *critic*, who evaluates the performance of the current policy π_k, i.e., generates an estimate of the value function V^{π_k} from states and reinforcement supplied by the environment as inputs. The policy improvement step is viewed as the work of an *actor*, who takes into account the latest evaluation of the critic, i.e., the estimate of the value function, and acts out the improved policy π_{k+1}.

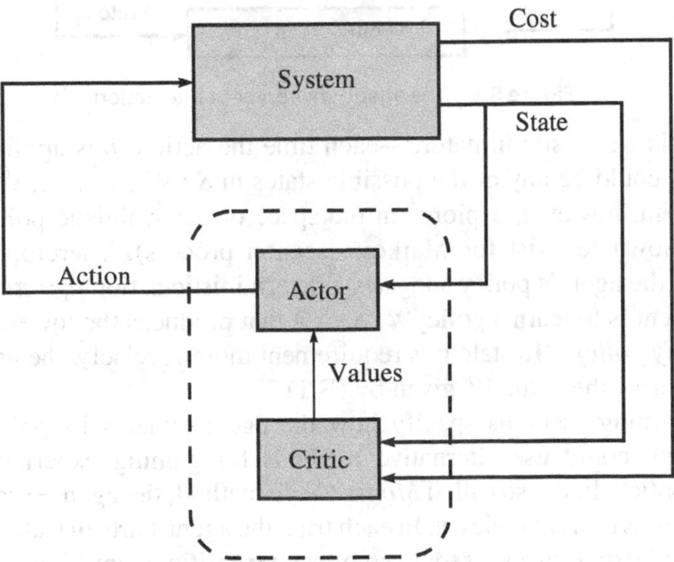

Figure B.4 The actor-critic architecture

The algorithm described till now needs updation of the values/policy for all states at once. But this is not strictly required. In fact, it is possible for us to pick any subset of states, on each iteration, and carry out updating to that subset. This algorithm is known as *asynchronous policy iteration*. Considering specific conditions on the initial policy and value function, asynchronous policy iteration is certain to converge to an optimal policy. The liberty to select any states to work on implies that it is possible to design highly effective heuristic algorithms—for instance, algorithms focussing on updation of the values of states which have a likelihood of being reached by a good policy.

B.4 TEMPORAL DIFFERENCE LEARNING

The novel aspect of learning that we address now is that it assumes the agent that does *not* have knowledge of $r(s)$ and $P(s, a, s')$, and therefore it cannot learn solely by simulating actions with environment model (off-line learning not possible). It has no choice but to interact with the environment and learn by observing consequences.

Figure B.5 gives a general setting of the agent-environment interaction process. Time advances by discrete unit length quanta; $t = 0, 1, 2, \ldots$. At each time step t, the agent senses the current state $s_t \in S$ of the environment, chooses an action $a_t \in A$, and performs it. The environment responds by giving the agent a cost $r_t = r(s_t)$, and by producing the succeeding state $s_{t+1} \in S$.

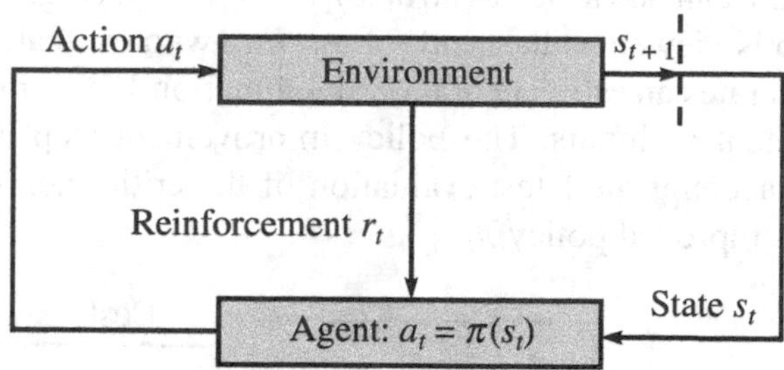

Figure B.5 The agent-environment interaction

The environment is stochastic in nature—each time the action a_t is applied in the state s_t, the succeeding state s_{t+1} could be any of the possible states in $S : s^{(1)}, s^{(2)}, \ldots, s^{(N)}$. For the stochastic environment, the agent, however, explores in the space of deterministic policies (a deterministic optimal policy is known to exist for Markov decision process). Therefore, for each observed environment state s_t, the agent's policy suggests a deterministic action $a_t = \pi(s_t)$.

The task of the agent is to learn a policy $\pi : S \rightarrow A$ that produces the lowest possible cumulative cost over time (*greedy policy*). To state this requirement more precisely, the agent's task is to learn a policy π that minimizes the value V^{π} given by (B.1).

Reinforcement learning methods specify how the agent updates its policy as a result of its experience. The agent could use alternative methods for gaining experience and using it for improvement of its policy. In the so called *Monte Carlo* method, the agent executes a set of *trials* in the environment using its current policy π. In each trial, the agent starts in state $s^{(i)}$ (any point $s^{(1)}, \ldots, s^{(N)}$ of state space) and experiences a sequence of state transitions until it reaches a terminal state.

In infinite-horizon discounted cost problems under consideration, terminal state corresponds to the *equilibrium state*. A learning episode (trial) is infinitely long, because the learning is continual. For the purpose of viewing the *infinite-horizon* problem in terms of *episodic learning*, we may define a stability region around the equilibrium point and say that the environment has terminated at a *success state* if the state continues to be in stability region for a prespecified time period. [In a real-time control, any uncertainty (internal or external) will pull the system out of stability region and a new learning episode begins.] Failure states (situations corresponding to 'the game is over and it is lost') if any, are also terminal states of the learning process.

In a learning episode, agent's precepts supply both the current state and the cost incurred in that state. Typical state sequences (environment trajectories) resulting from trials might look like this:

(1) $(s^{(1)})_{r(1)} \to (s^{(5)})_{r(5)} \to (s^{(9)})_{r(9)} \to (s^{(5)})_{r(5)} \to (s^{(9)})_{r(9)} \to (s^{(10)})_{r(10)} \to$

$\to (s^{(11)})_{r(11)} \to (s^{(\text{SUCCESS})})_{r(\text{SUCCESS})}$

(2) $(s^{(1)})_{r(1)} \to (s^{(5)})_{r(5)} \to (s^{(9)})_{r(9)} \to (s^{(10)})_{r(10)} \to (s^{(11)})_{r(11)} \to$

$\to (s^{(7)})_{r(7)} \to (s^{(11)})_{r(11)} \to (s^{(\text{SUCCESS})})_{r(\text{SUCCESS})}$

(3) $(s^{(1)})_{r(1)} \to (s^{(2)})_{r(2)} \to (s^{(3)})_{r(3)} \to (s^{(7)})_{r(7)} \to (s^{(\text{FAILURE})})_{r(\text{FAILURE})}$

Note that each state percept is subscripted with the cost incurred. The objective is to use the information about costs to learn the expected value $V^{\pi}(s)$ associated with each state. The value is defined to be the expected sum of (discounted) costs incurred if policy π is followed (refer to Eqn (B.2)).

When a nonterminal state is visited, its value is estimated based on what happens after that visit. Thus, the value of a state is the expected total cost from that state onward, and each trial (episode) provides *samples* of the value for each state visited. For example, the first trial in the set of three given above, provides one sample of value for state $s^{(1)}$:

(i) $r(1) + \gamma r(5) + \gamma^2 r(9) + \gamma^3 r(5) + \gamma^4 r(9) + \gamma^5 r(10) + \gamma^6 r(11) + \gamma^7 r(\text{SUCCESS})$;

two samples of values for state $s^{(5)}$:

(i) $r(5) + \gamma r(9) + \gamma^2 r(5) + \gamma^3 r(9) + \gamma^4 r(10) + \gamma^5 r(11) + \gamma^6 r(\text{SUCCESS})$;

(ii) $r(5) + \gamma r(9) + \gamma^2 r(10) + \gamma^3 r(11) + \gamma^4 r(\text{SUCCESS})$;

two samples of values for state $s^{(9)}$:

(i) $r(9) + \gamma r(5) + \gamma^2 r(9) + \gamma^3 r(10) + \gamma^4 r(11) + \gamma^5 r(\text{SUCCESS})$;

(ii) $r(9) + \gamma r(10) + \gamma^2 r(11) + \gamma^3 r(\text{SUCCESS})$;

and so on.

Therefore, at the end of each episode, the algorithm computes the observed total cost for each state visited, and the estimated value is accordingly updated for that state simply by maintaining a running average for each state in a table. In the limit of infinitely many trials, the sample average will converge to the true expectation of Eqn (B.2).

The Monte Carlo method differs from dynamic programming in the following two ways:

(i) First, it operates on *sample experience*, and thus can be used for direct learning without a model.
(ii) Second, it does not build its value estimates for a state on the basis of estimates of the possible successor states (refer to Eqn (B.6)); it must wait until the end of the trial to determine the update in value estimates of states. In dynamic programming methods, the value of each state equals its own cost plus the discounted expected value of its successor states.

The *Temporal Difference* (TD) learning techniques are a combination of sampling of Monte Carlo, and the value estimation scheme of dynamic programming. TD techniques update value estimates on the basis of the cost of one-step real-time transition and learned estimate of successor state, without waiting for the final outcome. Typically, when a transition occurs from state s to state s', we apply the flowing update to $V^{\pi}(s)$:

$$V^{\pi}(s) \leftarrow V^{\pi}(s) + \eta \left(r(s) + \gamma V^{\pi}(s') - V^{\pi}(s)\right) \tag{B.12}$$

where η is the learning parameter.

Because the update uses the difference in values between successive states, it is called the temporal-difference or TD equation. TD methods have an advantage over dynamic programming methods in that they do not require a model of the environment. Advantage of TD methods over Monte Carlo is that they are naturally implemented in an on-line fully incremental fashion. With Monte Carlo methods, one must wait until the end of a sequence, because only then is the value known, whereas with TD methods, one need only wait one time step.

Note that the update (B.12) is based on *one* state transition that just happens with a certain probability, whereas in (B.6), the value function is updated for all states simultaneously using all possible next states, weighted by their probabilities. This difference disappears when the effects of TD adjustments are averaged over a large number of transitions. The interaction with the environment can be repeated several times by restarting the experiment after success/failure state is reached. For one particular state, the next state and received reinforcement can be different each time the state is visited. As the frequency of each successor in the set of transitions is approximately proportional to its probability, TD can be considered as a crude but effective approximation to dynamic programming.

The TD Eqn (B.12) is, in fact, approximation of policy-evaluation step of policy iteration algorithm of dynamic programming (refer to previous section for a recall), where the agent's policy is fixed and the task is to learn the values of states. This, as we have seen, can be done without a model of the system. However, improving the policy using (B.11) still requires the model.

One of the most significant breakthroughs in RL was the development of model-free TD control algorithm, called *Q*-learning.

B.4.1 Q-learning

In addition to recognizing the inherent relationship between RL and dynamic programming, Watkins [30, 32] made a significant contribution to RL by suggesting a new algorithm named *Q-learning*. The importance of *Q*-learning is that when applied to a Markov decision process, it can be shown to converge to the optimal policy, under appropriate conditions. Q-learning is the first RL algorithm seen to be convergent to the optimal policy for decision problems which involve cumulative cost.

The Q-learning technique learns an *action-value* representation rather than learning value function. We will make use of the notation $Q(s, a)$ *to denote the value of doing action a in state s.* Q-function is directly related to value function as follows:

$$V(s) = \min_a Q(s, a) \tag{B.13}$$

Q-functions may seem like just another way of storing value information, but they have a very imported property: *a TD agent that learns a Q-function does not need a model for either learning or action selection.* For this reason, Q-learning is called a *model-free* method.

The connections between Q-learning and dynamic programming are strong: Q-learning is motivated directly by value-iteration, and its convergence proof is based on a generalization of the convergence proof for value-iteration.

We can use the value-iteration algorithm (B.8) directly as an update equation for an iteration process that calculates exact Q-values, given an estimated model:

$$Q_{l+1}(s, a) \leftarrow r(s) + \gamma \sum_{s'} P(s, a, s')\left[\min_{a'} Q_l(s', a')\right] \tag{B.14}$$

It converges to the optimal Q-values, $Q^*(s, a)$.

Once one has $Q^*(s,a)$ for all $s \in S$ and all $a \in A$, it is relatively easy to determine an optimal policy:

$$\pi^*(s) = \arg\min_a Q^*(s, a) \tag{B.15}$$

This does, however, require that a model is given [or is learned (adaptive dynamic programming)], because Eqn (B.14) uses $P(s, a, s')$.

The temporal-difference approach, on the other hand, requires no model. The update equation for TD Q-learning is (refer to Eqn (B.12))

$$Q(s, a) \leftarrow Q(s, a) + \eta\left(r(s) + \gamma\left[\min_{a'} Q(s', a')\right] - Q(s, a)\right) \tag{B.16}$$

which is calculated whenever action a is executed in state s leading to s'.

The Q-learning algorithm (B.16) takes a back-up of the Q-value for just a single state-action pair at each time step of control, where the state-action pair comprises the observed present state and the action actually executed. Particularly, make an assumption that at time step t in real-time control, the agent observes state s_t and has the estimated Q-values created by all the preceding stages of real-time Q-learning (the estimates stored in a *lookup table* with one entry for each state-action pair). We depict these estimates by $Q_t(s, a)$ for all admissible state-action pairs. The agent selects an action $a_t \in A$ using this information available in lookup table:

$$a_t = \arg\min_a Q_t(s_t, a)$$

After executing a_t, the agent receives the immediate cost $r_t = r(s_t)$ while the environment state changes to s_{t+1}. The Q-values in the lookup table are then updated as follows:
For the state-action pair (s_t, a_t):

$$Q_{t+1}(s_t, a_t) = Q_t(s_t, a_t) + \eta_t\left[r_t + \gamma\left(\min_a Q_t(s_{t+1}, a)\right) - Q_t(s_t, a_t)\right] \tag{B.17a}$$

For other admissible state-action pairs, the Q-values remain unchanged:

$$Q_{t+1}(s, a) = Q_t(s, a) \; \forall (s, a) \neq (s_t, a_t) \tag{B.17b}$$

Watkins [30, 32] has shown the Q-learning system that

(1) decreases its learning parameter at an appropriate rate (e.g., $\eta_t = 1/t^{\beta}$, where $0.5 < \beta < 1$); and
(2) visits each state-action pair infinitely often, is guaranteed to converge to an optimal policy.

Convergence, thus requires that *the agent selects actions in such a fashion that it visits every possible state-action pair infinitely enough*. This means, if action a is an admissible action from state s, then over a period of time the agent will have to execute action a from state s again and again with nonzero frequency as the length of its action sequence approaches infinity.

In the Q-learning algorithm shown in Eqns (B.17), the strategy for the agent in state s_t at time step t is to choose the action a that minimizes $Q_t(s_t, a)$, and thus, exploits the present approximation of Q^* by following a *greedy policy*. But, with this strategy, the agent stands the risk that it will overcommit to actions discovered with low Q-values during early stages, while failing to *explore* other actions that can have even lower values. In fact, the convergence condition needs each state-action transition to take place infinitely often. This will not take place if the agent follows the greedy policy all the time.

The Q-learning agent should necessarily follow the policy of *exploring* and *exploiting*: the former, exploration, makes sure that all admissible state-action pairs are visited enough to satisfy the Q-learning convergence condition, and the latter, exploitation, tries to minimize the cost by adopting a greedy policy.

Many exploration schemes have been employed in the RL literature. While the most basic one is to behave greedily most of the time, but every once a while, with small probability ε, instead choose a random action, uniformly, and independent of the action-value estimates. We call techniques that make use of this near-greedy action-selection rule *ε-greedy techniques*.

B.4.2 Generalization

Till now the assumption has been that the Q-values the agent learns are represented in a tabular form with a single entry for each state-action pair. This is a clear and instructive case, but it is certainly restricted to tasks possessing small numbers of states and actions. The issue is not merely the memory required for large tables, but the computational time required to experience all the state-action pairs to generate data to fill the tables in an accurate manner.

Very few decision and control problems in the real world fit into *lookup table representation* strategy for solution; the number of possible states and actions in the real world is often much too large to accommodate the computational and storage requirements. The problem is more severe *when state/action spaces include continuous variables*—to use a table, they should be discretized to finite size, which may cause errors. The only way of learning anything about these tasks is to *generalize* from past experienced states, to states which have never been seen. Simply put, experience with a finite subset of state space can be meaningfully generalized to give rise to a good approximation over a bigger subset.

Luckily, generalization from examples has been studied at length, and we do not require to devise altogether novel techniques to use in *Q*-learning. To a great extent, we require to simply create a combination of *Q*-learning and off-the-shelf architectures for the purpose of inductive generalization—frequently referred to as *function approximation*, as it takes examples from desired *Q*-function and tries to generalize from them for the construction of an approximation of the function. Function approximation is done with the help of *supervised learning*. In principle, any of the techniques examined in this field can be employed in *Q*-learning.

In parametric methods, the tabular (exact) representation of the real-valued functions $Q(s,a)$ is replaced by a generic parametric function approximation $\hat{Q}(s,a;\mathbf{w})$ where $\mathbf{w}$ are the adjustable parameters of the approximator. Learning $Q(s,a)$ for all $s \in S$ and $a \in A$ amounts to learning parameters $\mathbf{w}$ of $\hat{Q}(s,a;\mathbf{w})$. The new version of *Q*-learning Eqn (B.16) is

$$w_j \leftarrow w_j + \eta\left(r(s) + \gamma\left[\min_{a'} Q(s',a';\mathbf{w})\right] - Q(s,a;\mathbf{w})\right)\frac{\partial Q(s,a,\mathbf{w})}{\partial w_j} \tag{B.18}$$

This update rule can be shown to converge to the closest possible approximation to the true function when the function approximator is *linear* in the parameters.

Unfortunately, all bets are off when *nonlinear* function approximators—such as neural networks—are used. For many tasks, *Q*-learning fails to converge once a nonlinear function approximator is introduced. Fortunately, however, the algorithm does converge for large number of applications. The theory of *Q*-learning with nonlinear function approximator still contains many open questions; at present, it remains an empirical science.

For *Q*-learning, it makes more sense to use an incremental learning algorithm that updates the parameters of function approximator after each trial. Alternatively, examples may be collected to form a training set and leaned in batch mode, but it slows down learning as no learning happens while a sufficiently large sample is being collected.

We give an example of *neural Q-learning*. Let $\hat{Q}_t(s,a;\mathbf{w})$ denote the approximation to $Q_t(s,a)$ for all admissible state-action pairs, computed by means of a neural network at time step t. The state s is input to the neural network with parameter vector $\mathbf{w}$ producing the output $\hat{Q}_t(s,a;\mathbf{w})$ $\forall\, a \in A$. We assume that the agent uses the training rule of (B.17) after initialization of $\hat{Q}(s,a;\mathbf{w})$ with arbitrary finite values of $\mathbf{w}$.

Treating the expression inside the square bracket in (B.17a) as the error signal involved in updating the current value of parameter vector $\mathbf{w}$, we may identify the target (desired) value of $\hat{Q}_t$ at time step t as,

$$Q_t^{\text{target}}(s_t, a_t;\mathbf{w}) = r_t + \gamma\left(\min_a Q_t(s_{t+1}, a;\mathbf{w})\right) \tag{B.19}$$

At each iteration of the algorithm, the weight vector $\mathbf{w}$ of the neural network is changed slightly in a way that brings the output $\hat{Q}_t(s_t, a_t;\mathbf{w})$ closer to the target $Q_t^{\text{target}}(s_t, a_t;\mathbf{w})$ for the current (s_t, a_t) pair. For other state-action pairs, *Q*-values remain unchanged (Eqn (B.17b)).

B.4.3 Sarsa-learning

The Q-learning algorithm is an *off-policy* TD method: the learned action-value function Q directly approximates Q^*, the optimal action-value function, independent of the policy being followed; optimal action for state s is then obtained from Q^*. The Q-learning is motivated by value iteration algorithm in dynamic programming.

The alternative approach, motivated by policy iteration algorithm in dynamic programming, is an *on-policy* TD method. The distinguishing feature of this method is that it attempts to evaluate and improve the same policy that it uses to make decisions.

Earlier is this section on TD learning, we considered transitions from state to state and learned the value of states (Eqn (B.12)) when following a policy π. The relationship between states and state-action pairs is symmetrical. Now we consider transition from state-action pair to state-action pair and learn the value of state-action pairs, following a policy π. In particular, for on-policy TD method, we must estimate $Q^\pi(s, a)$ for the current policy π and for all states $s \in S$ and actions $a \in A$. We can learn Q^π using essentially the same TD method used in Eqn (B.12) for learning V^π:

$$Q^\pi(s, a) \leftarrow Q^\pi(s, a) + \eta(r(s) + \gamma Q^\pi(s', a') - Q^\pi(s, a)) \tag{B.20}$$

where a' is the action executed in state s'.

This rule uses every element of the quintuple of events, (s, a, r, s', a'), that make up a transition from one state-action pair to the next. This quintuple (State-Action-Reinforcement-State-Action) gives rise to the name SARSA for this algorithm. Unlike Q-learning, here the agent's policy does matter. Once we have $Q^{\pi_k}(s, a)$, improved policy can be obtained as follows:

$$\pi_{k+1}(s) = \arg\min_a Q^{\pi_k}(s, a) \tag{B.21}$$

Since tabular (exact) representation is impractical for large state and action spaces, function approximation methods are used. Approximations in the policy-iteration framework can be introduced at the following two places:

(i) *The representation of the Q-function*: The tabular representation of the real-valued function $Q^\pi(s, a)$ is replaced by a generic parametric function approximation $\hat{Q}^\pi(s, a; \mathbf{w})$ when $\mathbf{w}$ are the adjustable parameters of the approximator.

(ii) *The representation of the policy*: The tabular representation of the policy $\pi(s)$ is replaced by a parametric representation $\hat{\pi}(s; \boldsymbol{\theta})$ where $\boldsymbol{\theta}$ are the adjustable parameters of the representation.

The difficulty involved in use of these approximate methods within policy iteration is that the off-the-shelf architectures and parameter adjustment methods cannot be applied blindly; they have to be fully integrated into the policy-iteration framework.

The introduction to the subject of reinforcement learning given in this appendix had the objective of motivating the readers for advanced studies [30–32]. Reinforcement learning has been successfully applied to many real-life decision problems. Vast amount of literature is available on the subject. Going through the available case studies, and experimenting with some, will provide a transition to the next improved state of learning.

DATASETS FROM REAL-LIFE APPLICATIONS FOR MACHINE LEARNING EXPERIMENTS

In various chapters of the book, it has been repeatedly emphasized that there is no generally applicable systematic methodology for the design of intelligent systems *guaranteed* to result in high performance. However, despite the lack of a general systematic design procedure, machine learning design methodology does provide a way to design intelligent systems for a wide variety of applications. Once the methodology is understood, it provides 'a way to at least get a solution, and often a way to quickly get an acceptable solution', for many types of machine learning problems. One must keep in mind that machine learning has significant functional capabilities (recall the *universal approximation property*) and therefore, with *enough work*, the designer should be able to achieve just about anything *that is possible* in terms of performance (up to the computational limits of the computer on which the *machine learning experiment* is carried out). The problem is that just because the performance can be achieved does *not* mean that it is easy to obtain, or that the current framework in which we are designing will work. Ultimately, the design of machine learning experiments is nothing more than a heuristic technique.

Learning by Doing

The best way to learn the basics of how to design intelligent systems is to do machine learning experiments ourselves—and for a variety of applications. Even if we focus on *one* application, a (somewhat) systematic design methodology seems to emerge. While the procedure is typically linked to the application-specific concepts and parameters, and is, therefore, generally not applicable to other applications, it does often provide a very nice framework in which the designer can think about how to attempt a solution for a problem from a new application domain. However, to be an 'effective' designer, we should gain experience on wide variety of machine learning experiments from real-life application domains.

The approach into the field of machine learning taken in this book has been to focus on machine learning theory, and provide a platform for the hands-on experience through 'self-study machine learning projects.'

Self-Study Machine Learning Projects

The general idea is that 'small projects' are designed and executed after learning machine-learning theory from this book. The projects should target specific questions that generally still bother your (the readers of this book) minds. The projects should be 'small' in terms of number of data samples and number of features representing these data, to ensure that they soon get completed with learning benefits extracted, to move on to the next project. The projects should require small resources: a desktop or laptop connected to the internet; not requiring exotic software, web infrastructure, or third-party data or service.

We recommend that you need not write the code (the goal is to learn something and not to create a unique source). This is not to undermine the significance of writing your own codes (in fact, in this book, we have presented the techniques in a way that provides a comfortable environment for writing your own codes), but only to accelerate the understanding of machine learning. Open-source software tools may be used for these projects (refer to Section 1.10). As a 'machine-learning professional' you will be surely curious enough to develop your own codes, to achieve better performance than that achievable by using available software tools. You will embed in your codes, your expertise on machine-learning theory.

There are excellent libraries of data sources available (refer to Section 1.10). You may browse and choose a dataset for your project that matches your immediate interest and curiosity.

The following steps for a machine learning experiment are recommended:

- Browse the repositories containing a large number of datasets frequently used by machine learning researchers for bench-marking purposes. Choose an appropriate dataset for your project for the application area of your interest. By appropriate we mean, it should be helpful in doing a small self-study course project. For example, datasets with image, text or audio data require conversion to numerical form for carrying out machine learning project. This requires deeper domain knowledge. Take your knowledge on application domain into consideration while selecting a dataset. Probably, selecting one which is already in numerical form may be a choice worth considering.
- Browse the Internet and download relevant articles/papers by machine learning researchers using the dataset you have selected. Acquire the required knowledge from these sources to prepare a platform for launching machine learning experiment.
- Select a suitable software tool matching your background.
- Perform the experiments, using the knowledge acquired from this book, to assess and compare the performances of various learning algorithms in practice. Analyze the results of these experiments.

Some proposals for self-study projects are given below for helping you to get initiated on these steps.

Datasets for Some Realistic Problems

In the following, we present data structure for machine learning applications, with description of datasets for some realistic problems. This will help the reader appreciate better the real-world environment for machine learning. The sources of the datasets and relevant papers describing work done on these datasets are also given. Students may consider using these datasets for their self-study projects. However, exploration of the dataset websites will give them many alternative choices for the projects.

Experiment 1: Breast Cancer Diagnosis

In *medical diagnosis*, the inputs are the relevant information we have about the patient and the classes are the illnesses. The inputs contain the patient's age, gender, past medical history and current symptoms. A wrong decision may lead to a wrong or no treatment, and in cases of doubt, it is preferable that the classifier reject and defer decision to a human expert.

We consider here the *breast cancer diagnosis problem*. Cancer is a group of diseases in which cells in the body grow, change, and multiply out of control. Breast cancer refers to the erratic growth of cells that originate in the breast tissue. A group of rapidly dividing cells may form a lump or mass or extra tissue. These masses are called *tumors*. Tumors can either be cancerous (*malignant*) or non-cancerous (*benign*). Malignant tumors penetrate and destroy healthy body tissues.

Breast cancer is the second (exceeded only by lung cancer) largest cause of death among women. At the same time, it is also among the most curable cancer types if it is diagnosed early. Mortality rate from breast cancer has decreased in recent years with an increased emphasis on diagnostic techniques and more effective treatments. There is no doubt that evaluation of data taken from patients and decision of experts are the most important factors in diagnosis. However, machine learning techniques for classification also help experts to a great deal.

We present here structure of the Wisconsin Breast Cancer (WBC) dataset from the UCI Machine Learning Repository (https://archive.ics.uci.edu/ml/datasets/Breast+Cancer+Wisconsin+(Original)), made available in public domain by Dr. William H. Wolberg of the Department of Surgery of the University of Wisconsin Medical School. The dataset contains 699 samples from needle aspirates from human breast cancer tissue. [A breast *Fine Needle Aspiration* (FNA) is a quick and simple procedure to perform, which removes some fluid or cells from a breast lesion or cyst with a fine needle. The sample of fluid or cells is smeared on a glass slide which is examined by a specialist doctor under a microscope.]

Image-processed features for breast cancer diagnosis are objective and precise compared to subjective evaluation. A small region of each breast FNA is digitized; an image-analysis program using curve-fitting is used to determine boundaries of the nuclei.

The details of the features listed in the dataset (each of which is represented as an integer between 1 and 10) are given below.

x_1 : Clump Thickness

x_2 : Uniformity of Cell Size

x_3 : Uniformity of Cell Shape

x_4 : Marginal Adhesion

x_5 : Single Epithelial Cell Size

x_6 : Bare Nuclei

x_7 : Bland Chromatin

x_8 : Normal Nucleoli

x_9 : Mitoses

y(class) : 0 (benign), 1 (malignant)

- The features of uniformity of cell in the clump, capture the variation of cancer cells in size and shape.
- In the case of marginal adhesion, the normal cells tend to stick together, where cancer cells tend to lose this ability. So loss of adhesion is a sign of malignancy.
- In the single epithelial cell size, the size is related to the uniformity mentioned above. Epithelial cell that is significantly enlarged may be a malignant cell.
- The bare nuclei is a term used for nuclei that are not surrounded by cytoplasm (the rest of the cell). These are typically seen in benign tumors.
- The bland chromatin describes a uniform 'texture' of the nucleus seen in benign cells. In cancer cells, the chromatin tends to be coarser.
- The normal nucleoli are small structures seen in the nucleolus. In normal cells, the nucleolus is usually very small if visible. In cancer cells, the nucleoli become more prominent, and sometimes there are more of them.
- Mitoses is nuclear division plus cytokines and produce two identical daughter cells during prophase. It is the process in which the cell divides and replicates. Pathologists can determine the grade of cancer by counting the number of mitoses.

A few samples from Breast Cancer dataset, structured in the form of data matrix of Table 1.2, are shown in Table E1.1.

References helpful for experiment on this dataset are [191, 192].

Table E1.1 Samples from Breast Cancer Dataset

Features and Class / Patients $s^{(i)}$	x_1	x_2	x_3	x_4	x_5	x_6	x_7	x_8	x_9	y
$s^{(1)}$	5	1	1	1	2	1	3	1	1	0
$s^{(2)}$	5	4	4	5	7	10	3	2	1	0
$s^{(3)}$	8	10	10	8	7	10	9	7	1	1
$s^{(4)}$	8	7	5	10	7	9	5	5	4	1
.	.	.	.	.	.	.	.	.	.	.
.	.	.	.	.	.	.	.	.	.	.
.	.	.	.	.	.	.	.	.	.	.
.	.	.	.	.	.	.	.	.	.	.

The reader might not have understood the features to his/her satisfaction. This has been intentionally done to highlight the point that *domain knowledge* for the application is extremely important for any real-life problem. We have talked about domain knowledge in various chapters of the book.

Another point to keep in mind is that the dataset for machine learning is a set of FNA images. The dataset in numerical form has been obtained by processing these images to extract appropriate features. Feature extraction from image database requires specific image-processing techniques for each application; features must have discriminating power from one image to another. We have given an overview of some of these techniques in Section 9.5.

Experiment 2: Optical Recognition of Handwritten Digits

Recognition of both printed and handwritten characters is a typical domain where machine learning has been successfully applied. In fact, *Optical Character Recognition* (OCR) systems are today among the commercial applications of learning systems.

OCR is the ability of a computer to translate character images into a text. This is an application example where there are multiple classes, as many as there are characters we would like to recognize. Especially interesting is the case when the characters are handwritten. People have different handwriting styles; characters may be written small or large, slanted, with a pen or pencil, and there are many possible images corresponding to the same character. We do not have a formal description of 'A' that covers all 'A's and none of the non-'A's. Not having it, we take samples from writers and learn the definition of A-ness from these examples. Though we do not know what it is that makes an image of an 'A', we are certain that all those distinct 'A's have something in common, which is what a learning system extracts from the examples [6].

To capture the character images, we can use a scanner. It either passes light-sensitive sensors over the illuminated surface of a page or moves a page through the sensors. The scanner processes the image by dividing it into hundreds of pixel-sized boxes per inch and representing each box by either 1 (if the box is filled) or 0 (if the box is empty). The resulting matrix of dots is called a *bit map*. Bit maps can be stored, displayed and printed by a computer. The patterns of dots have to be recognized as characters by the computer. This is the job for machine learning.

For simplicity, we limit our discussion to the recognition of digits from 0 to 9. We start with printed digits recognition, and then carry on to handwritten digits recognition.

The application areas of digits recognition systems are recognizing zip codes on mail for postal mail sorting, processing bank cheque amounts, numeric entries in forms (for example, tax forms) and so on.

Recognition of Printed Digits [5]

In commercial applications, where a high resolution is required, each digit is represented by 16×16 bit map. Here, for demonstration, we consider 5×9 bit map as shown in Fig. E2.1. The data matrix, showing input vectors **x** representing the bit maps, and output vector (classes y_q; $q = 1, \ldots, 10$), is given in Table E2.1. Input vector **x** for a digit comes from 9 rows of the pixel matrix of the digit (Fig. E2.1), each row having 5 binary values. Therefore, **x** is 45×1 vector.

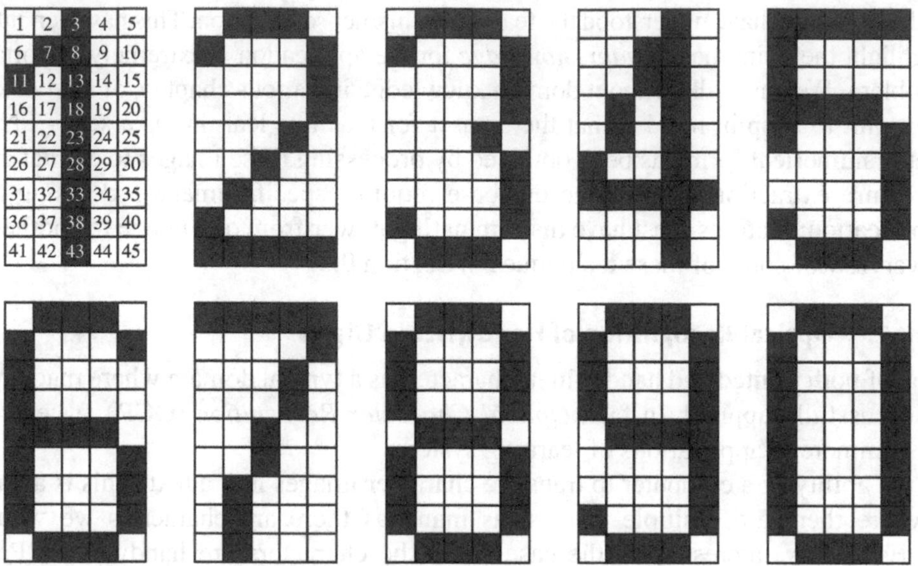

Figure E2.1 Bit maps for printed digits

Table E2.1 Data matrix for printed digits

Digits $s^{(i)}$	Input Patterns $x_1\ x_2$				. . .				$x_{44}\ x_{45}$	Output y_q
$s^{(1)}$	00100	01100	10100	00100	00100	00100	00100	00100	00100	1
$s^{(2)}$	01110	10001	00001	00001	00010	00100	01000	10000	11111	2
$s^{(3)}$	01110	10001	00001	00001	00010	00001	00001	10001	01110	3
$s^{(4)}$	00010	00110	00110	01010	01010	10010	11111	00010	00010	4
$s^{(5)}$	11111	10000	10000	11110	10001	00001	00001	10001	01110	5
$s^{(6)}$	01110	10001	10000	10000	11110	10001	10001	10001	01110	6
$s^{(7)}$	11111	00001	00010	00010	00100	00100	01000	01000	01000	7
$s^{(8)}$	01110	10001	10001	10001	01110	10001	10001	10001	01110	8
$s^{(9)}$	01110	10001	10001	10001	01111	00001	00001	10001	01110	9
$s^{(10)}$	01110	10001	10001	10001	10001	10001	10001	10001	01110	10

Recognition of Handwritten Digits

Optdigits dataset is a collection of handwritten samples from the UCI Machine Learning Repository (https://archive.ics.uci.edu/ml/datasets/Optical+Recognition+of+Handwritten+Digits). There are 5620 samples with 64 attributes in this dataset. All input attributes are integers in the range 0–16 while class codes are in the range 0–9. The samples have been collected from 43 people.

The size of a digit varies but is typically around 40 × 60 pixels. To normalize the size of the characters, the characters were made to fit in 32 × 32 bit maps after removing the extraneous marks in the image. To reduce the dimensionality of the attributes, the 32 × 32 bit maps were divided into non-overlapping blocks of 4 × 4, and the number of 1's (representing 'white') counted in each block. This generates an input matrix of 8 × 8 where each element is an integer in the range 0-16 (Fig. E2.2). each block is considered a feature, resulting in 64 features: $x_1, \ldots, x_{64}$. Data for 10 samples, each sample corresponding to a specific digit from 0 to 9, is given in Table E2.2. References helpful for experiment on this dataset are [193, 194].

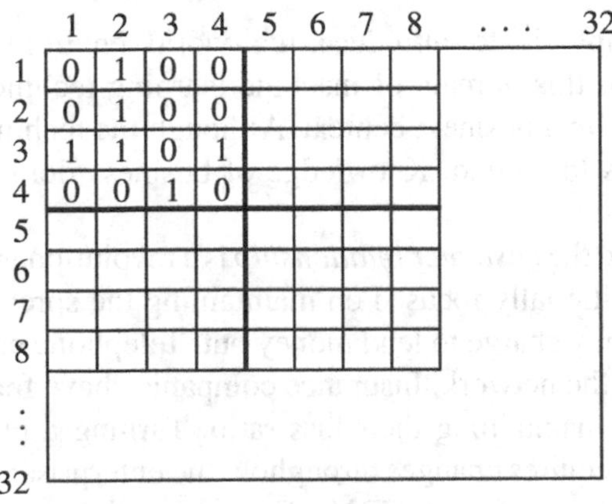

Figure E2.2 Bit map for a handwritten digit

Table E2.2: Samples from optdigits dataset

Data / Features	Digit 1	Digit 2	Digit 3	Digit 4	Digit 5	Digit 6	Digit 7	Digit 8	Digit 9	Digit 0
x_1	0	0	0	0	0	0	0	0	0	0
x_2	0	0	0	0	0	0	0	0	0	1
x_3	0	11	7	0	1	5	8	14	0	6
x_4	3	16	11	3	11	14	15	16	4	15
x_5	16	10	11	11	13	4	16	8	13	12
⋮	⋮	⋮	⋮	⋮	⋮	⋮	⋮	⋮	⋮	⋮
x_{64}	0	6	0	0	0	0	0	0	0	0

In today's world, *biometric recognition* (authentification of people using their physiological and/or behavioral characteristics) using machine learning, is being extensively used. Examples of physiological characteristics are: images of face, fingerprint, iris, and palm. Examples of behavioral characteristics are: dynamics of signature, and voice. Forgeries are possible with usual identification methods—photo, printed signature, or password. When there are many different (uncorrelated) inputs, forgeries would be more difficult. Machine learning is used both in separate recognizers for these different inputs and in combination of their decisions to get an overall accept/reject decision.

Domain knowledge for each of these characteristics is very important for application of machine learning techniques. An overview of *image* and *audio* processing to generate numeric/categorical attributes/features for machine learning algorithms was given in Section 9.5.

We have assumed a simple input based on collection of random dots for digit recognition. However, a character image is not just a collection of random dots; it is a collection of strokes that we can capture by a learning program.

Experiment 3: Prediction of Success of Bank Telemarketing

The purpose of this example is to introduce real-world environment for machine learning applications in business. For this domain of machine learning techniques, the techniques do not exist in a vacuum; they exist in a business context. Although the techniques are interesting in their own right, they are a means to an end. Knowledge of business domain is crucial for success of machine learning techniques.

Building a business around the *customer relationship* is a revolutionary change in today's business world [19]. Banks have traditionally focused on maintaining the spread between the rate they pay to bring money in and rate they charge to lend money out. Telephone companies have concentrated on connecting calls through the network. Insurance companies have focused on processing claims, managing investments, and maintaining their loss ratio. Turning a product-focused organization into a customer-centric one requires changes throughout the enterprise; more so in *marketing*, *sales*, and *Customer Relationship Management* (CRM). Forward-looking companies are moving toward the goal of understanding each customer individually and using that understanding to make it easier (and more profitable) for the customer to do business with them rather than with competitors. These same firms are learning to look at the value of each customer so that they know which ones are worth investing money and effort to hold on to and which ones should be allowed to depart. With the inflow of large amounts of customer data, machine learning has become an indispensable tool for present day large business organizations.

This example is about the role of machine learning to improve *direct marketing campaigns*. Advertising can be used to reach prospects about whom nothing is known as individuals. Direct marketing requires at least a tiny bit of additional information such as name and address or a phone number or e-mail address.

Companies use direct marketing when targeting segments of customers to make a sale of merchandise or service. Centralized customer remote interactions in a *Contact Center* eases operational management of campaigns. Such centers allow communicating with customers through various channels; telephone (fixed-line or mobile) being one of the most widely used. Marketing operationalized through contact center is called *telemarketing* due to the remoteness characteristic. Contacts can be divided into inbound or outbound, depending on which side triggered the contact (client or contact center), with each case posing different challenges (e.g., outbound calls are often considered more intrusive).

Machine learning offers intelligent decision-support systems that can automatically predict the result of a phone call to sell a merchandise or service. Such decision-support system is valuable to assist managers in prioritizing and selecting the next customers to be contacted during marketing campaigns. Machine learning analyzes the probability of success and leaves to managers only the decision on how many customers to contact. As a consequence, the time and costs of such campaigns would be reduced. Also, by performing fewer and more effective phone calls, client stress and intrusiveness would be diminished.

The goal of machine learning is to model the previous campaign's customer behavior to analyze what combination of factors make a customer more likely to accept the merchandise/service. This will serve as the basis for the design of a new campaign (In what is closely related to the telemarketing phone calls approach, a mass media (e.g., radio and television) marketing campaign could affect the selling of a merchandise or service).

In the following, we present the structure of a publically available dataset. The data is related with direct marketing campaign of a Portuguese banking institution (https://archive.ics.uci.edu/ml/datasets/Bank+Marketing). The marketing campaigns were based on phone calls. Often more than one contact to the same client was required in order to assess if the product (bank term-deposit) would be or would not be subscribed. The classification goal is to predict if the client will subscribe a term deposit or not (variable y). The dataset contains 45211 observations capturing 16 attributes/features.

Attribute Information

x_1 : age (numeric)

x_2 : type of job(categorical)—"admin", "unknown", "unemployed", "management", "housemaid", "entrepreneur", "student", "blue-collar", "self-employed", "retired", "technician", "services"

x_3 : marital status (categorical)—"married", "divorced", "single"; "divorced" means divorced or widowed

x_4 : education (categorical)—"unknown", "secondary", "primary", "tertiary"

x_5 : credit in default? (categorical)—"yes", "no"

x_6 : average yearly balance, in euros (numeric)

x_7 : has housing loan? (categorical)—"yes", "no"

x_8 : has personal loan? (categorical)—"yes", "no"

x_9 : contact communication type (categorical)—"unknown", "telephone", "cellular"

x_{10} : last contact day of the month (numeric)

x_{11} : last contact month of the year (categorical)—"jan", "feb", ..., "nov", "dec"

x_{12} : last contact duration in seconds (numeric)

x_{13} : number of contacts performed during this campaign and for this client (numeric); includes last contact

x_{14} : number of days that passed by after the client was last contacted from a previous campaign (numeric); –1 means that the client was not previously contacted

x_{15} : number of contacts performed before this campaign and for this client (numeric)

x_{16} : outcome of the previous marketing campaign (categorical)—"unknown", "other", "failure", success"

Output Variable (Desired Target)

y = has the client subscribed to term deposit? (categorical)—"yes", "no"

There are three kinds of attributes:

- Numerical : x_1 (age), x_6 (average yearly balance), x_{10} (last contact day of the month), x_{12} (last contact duration), x_{13} (number of contacts), x_{14} (number of days that passed by after the last contact), x_{15} (number of contacts performed before the campaign). Each attribute has a certain numeric *range*.
- Categorical : x_2 (type of job), x_3 (marital status), x_4 (education), x_9 (contact type), x_{10} (last contact month), x_{16} (outcome of previous campaign). Each attribute belongs to a *set* of categorical values.
- Binary categories : x_5 (credit in default), x_7 (housing loan), x_8 (personal loan), y (subscription to term deposit). Each attribute belongs to 'yes', 'no' binary categories.

Some classification methods, as we have seen, accept categorical attributes; for others, we turn categorical attributes with more than two categories into *dummy variables* first.

Table E3.1 gives 15 samples of the dataset for illustration. The features of this real-life application are shown in the table. Table E3.2 shows dummy variables for two categorical features: x_8 and x_{11}. On similar lines, dummy variables are created for all categorical features; and the categorical features are replaced with numeric features given by the corresponding dummy variables (MATLAB tool box function *dummyvar* creates dummy variables).

Table E3.1 Bank marketing data

$s^{(i)}$	x_j and y																
	x_1	x_2	x_3	x_4	x_5	x_6	x_7	x_8	x_9	x_{10}	x_{11}	x_{12}	x_{13}	x_{14}	x_{15}	x_{16}	y
$s^{(1)}$	43	technician	single	secondary	no	593	yes	no	unknown	5	may	55	1	−1	0	unknown	no
$s^{(2)}$	34	services	single	secondary	no	16	yes	no	cellular	20	nov	340	1	−1	0	unknown	no
$s^{(3)}$	36	unemployed	married	secondary	no	−872	yes	yes	cellular	20	nov	153	1	183	1	failure	no
$s^{(4)}$	35	blue-collar	married	secondary	no	667	no	no	cellular	20	nov	178	1	−1	0	unknown	no
$s^{(5)}$	59	admin	married	secondary	no	2343	yes	no	unknown	5	may	1042	1	−1	0	unknown	yes
$s^{(6)}$	32	blue-collar	married	secondary	no	305	yes	no	cellular	20	nov	73	1	−1	0	unknown	no
$s^{(7)}$	50	blue-collar	married	primary	no	2590	yes	no	telephone	20	nov	281	2	195	6	failure	no
$s^{(8)}$	30	self-employed	single	tertiary	no	−174	no	no	cellular	20	nov	80	1	−1	0	unknown	no
$s^{(9)}$	30	technician	single	secondary	no	925	no	no	cellular	20	nov	240	1	−1	0	unknown	no
$s^{(10)}$	55	management	married	tertiary	no	10065	no	no	cellular	20	nov	197	5	177	3	failure	no
$s^{(11)}$	34	management	married	tertiary	no	273	yes	no	cellular	20	nov	308	1	188	1	success	no
$s^{(12)}$	42	entrepreneur	divorced	tertiary	yes	2	yes	no	unknown	5	may	380	1	−1	0	unknown	no
$s^{(13)}$	58	retired	married	primary	no	121	yes	no	unknown	5	may	50	1	−1	0	unknown	no
$s^{(14)}$	35	services	single	primary	no	167	no	yes	cellular	11	jul	614	2	−1	0	unknown	yes
$s^{(15)}$	48	housemaid	married	secondary	no	4	no	no	cellular	18	aug	68	4	−1	0	unknown	no

Table E3.2 Sample dummy variables

x_8		x_{11}											
x_8d_1	x_8d_2	$x_{11}d_1$	$x_{11}d_2$	$x_{11}d_3$	$x_{11}d_4$	$x_{11}d_5$	$x_{11}d_6$	$x_{11}d_7$	$x_{11}d_8$	$x_{11}d_9$	$x_{11}d_{10}$	$x_{11}d_{11}$	$x_{11}d_{12}$
0	1	0	0	0	0	1	0	0	0	0	0	0	0
0	1	0	0	0	0	0	0	0	0	0	1	1	0
1	0	0	0	0	0	0	0	0	0	0	0	1	0
0	1	0	0	0	0	0	0	0	0	0	0	1	0
0	1	0	0	0	0	1	0	0	0	0	0	0	0
0	1	0	0	0	0	0	0	0	0	0	0	1	0
0	1	0	0	0	0	0	0	0	0	0	0	1	0
0	1	0	0	0	0	0	0	0	0	0	0	1	0
0	1	0	0	0	0	0	0	0	0	0	0	1	0
0	1	0	0	0	0	0	0	0	0	0	0	1	0
0	1	0	0	0	0	0	0	0	0	0	0	1	0
0	1	0	0	0	0	1	0	0	0	0	0	0	0
0	1	0	0	0	0	1	0	0	0	0	1	0	0
1	0	0	0	0	0	0	0	1	0	0	0	0	1
0	1	0	0	0	0	0	0	0	1	0	0	0	0

References helpful for experiment on this dataset are [195, 196].

Experiment 4: Forecasting Stock Market Index Changes

In business, forecasts are needed for marketing, production, purchasing, manpower, and financial planning. Furthermore, top management needs forecasts for planning and implementing long-term strategic objectives and for planning capital expenditure. More specifically, marketing managers use sales forecasts to (1) determine optimal sales force allocations, (2) set sales goals, and (3) plan promotions and advertising. Production managers need forecasts in order to schedule production activities, order materials, establish inventory levels, and plan shipments. Financial managers must estimate the future cash inflow and outflow. The personnel department requires forecasts in planning for human resources in business.

Some powerful methods to develop models and schemes for business forecasting have emerged. One direction is based on discovering certain laws governing the market under consideration from first principles and then building nonlinear mathematical models. Lacking first principles, as encountered in many practical situations, time-series analysis of historical observations (data) seems to be a natural alternative.

In this example, we consider a financial forecasting problem. Time series forecasting has applications in weather, biomedical, engineering, and other areas as well.

Recently, a lot of interesting work has been done in the area of applying machine learning algorithms for analyzing price patterns and predicting *stock* prices and *index* changes. Most stock traders now a days depend on Intelligent Trading Systems which help them in predicting prices based on various situations and conditions, thereby helping them in making instantaneous investment decisions.

Stock prices are considered to be very dynamic and susceptible to quick changes because of underlying nature of financial domain. Though it is very hard to replace the expertise that an experienced trader has gained, an accurate prediction can directly result into profits for investment firms (buying a stock before the price rises, or selling it before its value declines). Accuracy of prediction of at least *trend* (rise or fall) has a direct relationship with the profits made.

The EMH (*Efficient Market Hypothesis*) hypothesizes that the future stock price is completely unpredictable given the past trading history of the stock. With the advent of more powerful computing infrastructure, trading companies now build very efficient algorithmic trading systems that can exploit the underlying price patterns when a huge amount of data points are available on hand. Therefore machine learning techniques have a potential of seriously challenging the EMH.

The seemingly random character of the share market time series is due to many factors that influence share prices. Financial market modeling is a difficult task because of the ever-changing dynamics of the fundamental driving factors, partly uncontrollable. However, there is evidence to suggest that financial markets are partly *random* and partly *ordered*. Financial time series forecasting is concerned with exploiting the ordered part of the share market.

Factors affecting the market are many and modeling all these factors at once is well out of the reach today. Hence, there is a need to select the most relevant factors for a given time series. This is (possibly) the most important preprocessing part and relies heavily on expert knowledge.

Indicators (z_l; $l = 1, \ldots, m$, time series) that are frequently used in the technical analysis of stock prices are:

- *Moving Average* (MA): The average of the past P values till today.
- *Exponential Moving Average* (EMA): Gives more weightage to the most recent values while not discarding the old observations entirely.
- *Rate of Change* (ROC): The ratio of the current price and P quotes earlier (P is generally 5 to 10 days).
- *Relative Strength Index* (RSI): Measures the relative size of recent upward trends against the size of the downward trends within the specified time interval (usually 9–14 days).

Several studies examine the cross-sectional relationship between stock returns and macroeconomic variables, and find that these variables have some power to predict stock returns. Table E4.1 outlines an array of macroeconomic variables for forecasting S&P500 stock index. The data on these variables is available in public domain (www.ibiblio.org/pub/archives/misc.invest/historical-data/index/stocks/sp500). The weekly data covers the horizon from January 1980 to December 1992. There are 679 samples.

Table E4.1 List of potential input and output variables

Input variables	
1. S&P High	2. S&P Low
3. NYSE Advancing Issues	4. NYSE Declining Issues
5. OTC Advancing Issues	6. OTC Declining Issues
7. NYSE New Highs	8. NYSE New Lows
9. OTC New Highs	10. OTC New Lows
11. NYSE Total Volume	12. NYSE Advancing Volume
13. NYSE Declining Volume	14. OTC Total Volume
15. OTC Advancing Volume	16. OTC Declining Volume
17. S&P Earnings	
18. Short-term Interest Rates in the Three-Month Treasury Bill Yield	
19. Long-term Interest Rates in the 30-Year Treasury Bond Yield	
20. Gold	21. S&P Close
Output variable **K-step (weeks) ahead prediction of S&P Close**	
• The Standard & Poor' 500, abbreviated as the S&P500, is an American stock market index based on the market capitalizations of 500 large companies • NYSE: The New York Stock Exchange • OTC: Over-The-Counter; refers to stocks that trade via a dealer network as opposed to on a centralized exchange.	

In some studies, S&P500 forecasting uses the following macroeconomic variables:

1. S&P Close (z_1).
2. NYSE Advancing Issues (z_2).
3. NYSE Declining Issues (z_3).
4. NYSE New Highs (z_4).
5. NYSE New Lows (z_5).

6. Short-term Interest Rates in the Three-Month Treasury Bill Yield (z_6).
7. Long-term Interest Rates in the 30-Year Treasury Bond Yield (z_7).

The chosen inputs for machine learning are:

- x_1 : Weekly S&P closing price.
- x_2 : The ratio of the number of advancing issues and declining issues in the week for the stocks in the New York Stock Exchange (NYSE); breadth indicator for the stock market.
- x_3 : The ratio of the number of new highs and new lows achieved in the week for NYSE market; indicating strength of an upward or downward trend.
- x_4 : Short-term interest rates in the Three-Months Treasury Bill Yield.
- x_5 : Long-term interest rates in the 30-Year Treasury Bond Yield.
- $x_6 - x_{10}$: are defined below.

For each of the five inputs x_1–x_5, following function is used to highlight Rate-Of-Change (ROC) features:

For the variable x,

$$\text{ROC}n(t) = (x(t) - \text{BA}(t-n))/(x(t) + \text{BA}(t-n))$$

where BA stands for Block-Average; BA(t – n) is a five-unit block average of adjacent values centered around the value n periods ago.

To make a prediction three weeks into the future, we will take data at least as far back as three weeks:

$$\text{BA}(t) = \frac{x(t-2) + x(t-1) + x(t) + x(t+1) + x(t+2)}{5}$$

$$\text{ROC3}(t) = \frac{x(t) - \text{BA}(t-3)}{x(t) + \text{BA}(t-3)}$$

Table E4.2 gives a sample of the data only for those macroeconomic variables that have been used for features required for a machine learning algorithm. Table E4.3 gives a sample of the features.

References helpful for experiment on this dataset are [197, 198].

Table E4.2 Data for a subset of macroeconomic variables

Date	z_1	z_2	z_3	z_4	z_5	z_6	z_7
1/4/1980	106.52	1246	296	47	17	12.11	9.64
1/11/1980	109.92	965	585	149	7	11.94	9.73
1/18/1980	111.07	713	809	80	19	11.9	9.8
1/25/1980	113.61	660	832	119	33	12.19	9.93

(Contd.)

2/1/1980	115.12	778	669	71	34	12.04	10.2
2/8/1980	117.95	878	656	185	63	12.09	10.48
2/15/1980	115.41	382	1130	14	104	12.31	10.96
2/22/1980	115.04	374	1155	24	220	13.16	11.25
2/29/1980	113.66	929	554	19	106	13.7	12.14
3/7/1980	106.9	349	1235	0	360	15.14	12.1
3/14/1980	105.43	604	875	2	172	15.38	12.01
3/21/1980	102.31	478	983	5	179	15.05	11.73
3/28/1980	100.68	1401	267	2	172	16.53	11.67
4/3/1980	102.15	718	730	6	51	15.04	12.06
4/11/1980	103.79	963	615	9	29	14.42	11.81
4/18/1980	100.55	796	715	6	41	13.82	11.23
4/25/1980	105.16	662	779	9	19	12.73	10.59
5/2/1980	105.58	747	651	13	7	10.79	10.42
5/9/1980	104.72	511	996	9	19	9.73	10.15
5/16/1980	107.35	835	622	27	4	8.6	9.7
5/23/1980	110.62	1207	388	52	2	8.95	9.87
5/30/1980	111.24	837	610	21	6	7.68	9.86
6/6/1980	113.2	913	550	65	3	8.04	9.77

Table E4.3 Features obtained from given time-series data

$s^{(i)}$	x_1	x_2	x_3	x_4	x_5	x_6	x_7	x_8	x_9	x_{10}	y
$s^{(1)}$	117.950	1.338	2.937	12.090	10.480	0.029	– 0.130	– 0.396	0.002	0.030	-3.637
$s^{(2)}$	115.410	0.338	0.135	12.310	10.960	0.008	–0.550	– 0.961	0.011	0.044	-7.374
$s^{(3)}$	115.040	0.324	0.109	13.160	11.250	0.002	– 0.472	– 0.919	0.042	0.045	-8.354

(Contd.)

$s^{(i)}$	x_1	x_2	x_3	x_4	x_5	x_6	x_7	x_8	x_9	x_{10}	y
$s^{(4)}$	113.660	1.677	0.179	13.700	12.140	–0.008	0.359	– 0.817	0.052	0.069	– 9.986
$s^{(5)}$	106.900	0.283	0.000	15.140	12.100	–0.038	–0.548	– 1.000	0.089	0.047	–5.819
$s^{(6)}$	105.430	0.690	0.012	15.380	12.010	–0.038	–0.069	– 0.966	0.073	0.027	– 3.111
$s^{(7)}$	102.310	0.486	0.028	15.050	11.730	–0.042	–0.153	– 0.514	0.038	0.002	1.447
$s^{(8)}$	100.680	5.247	0.012	16.530	11.670	–0.038	0.767	– 0.699	0.066	– 0.007	– 0.129
$s^{(9)}$	102.150	0.984	0.118	15.040	12.060	– 0.018	– 0.261	0.437	– 0.004	0.005	2.947
$s^{(10)}$	103.790	1.566	0.310	14.420	11.810	0.001	0.009	0.804	– 0.034	– 0.004	1.725
$s^{(11)}$	100.550	1.113	0.146	13.820	11.230	– 0.011	– 0.234	0.209	– 0.050	– 0.027	4.147
$s^{(12)}$	105.160	0.850	0.474	12.730	10.590	0.016	– 0.377	0.588	– 0.081	–0.050	2.083
$s^{(13)}$	105.580	1.147	1.857	10.790	10.420	0.015	– 0.260	0.795	– 0.147	– 0.048	4.774
$s^{(14)}$	104.720	0.513	0.474	9.730	10.150	0.006	– 0.376	– 0.102	– 0.157	– 0.050	6.226
$s^{(15)}$	107.350	1.342	6.750	8.600	9.700	0.016	0.128	0.824	– 0.177	– 0.056	5.449

Experiment 5: System Identification using Gas Furnace Data

In sub-section 1.5.1, NARMA (Nonlinear Auto-Regressive Moving Average) model (refer to Eqn (1.6a))

$$y(t + 1) = f(y(t), y(t - 1), \ldots, y(t - n))$$

was used. For dynamic systems with output $y(t)$ and external input $u(t)$, a NARMAX (Nonlinear Auto-Regressive Moving Average with Exogeneous variable) model is used:

$$y(t + 1) = f(y(t), \ldots, y(t - n), u(t), \ldots, u(t - m)); n \geq m$$

Here, 't' counts the multiple sampling periods so that $y(t)$ specifies the present output, $y(t - 1)$ signifies the output observed at the previous sampling instant, etc. Note that the output of the dynamic system has been described as a function of number of past inputs and outputs. (Time-invariant nonlinear dynamic systems with scalar input and scalar output are considered here; extension to the case of vector input and vector output is straightforward.) Machine learning has been employed for inferring the values of output of dynamic systems from past observations on output and input.

The dataset used for approximating $f(\cdot)$ has information on n past inputs and m past outputs of the dynamic system. An experiment on the dynamic system is conducted to produce a set of examples of how the dynamic system to be identified responds to various inputs. The experiment is particularly important in relation to nonlinear modeling; one must be extremely careful to collect a set of data that describes how the system behaves over its entire range of operation. The issues like sampling frequency, input signals, preprocessing of data (filtering, etc.) must be considered in relation to acquisition of data.

Another important factor is the *leg space*, i.e., number of delayed signals used as regressors. A wrong choice of lag space may have a disastrous impact on some *system identification* applications. Too small obviously implies that essential dynamics will not be modeled, but too large can often be a problem (issues discussed in various chapters). Applying physical insight can often guide the proper lag space. If one has no idea regarding the lag space, it is sometimes possible to determine it empirically.

The input structure for nonlinear system identification consists of number of past inputs and outputs (refer to Fig. E5.1) where $\hat{y}$ is the predicted value of the output y at sampling instant t (note that t is a number; the physical sampling instant $t \times T$, where T is the sampling interval in secs.).

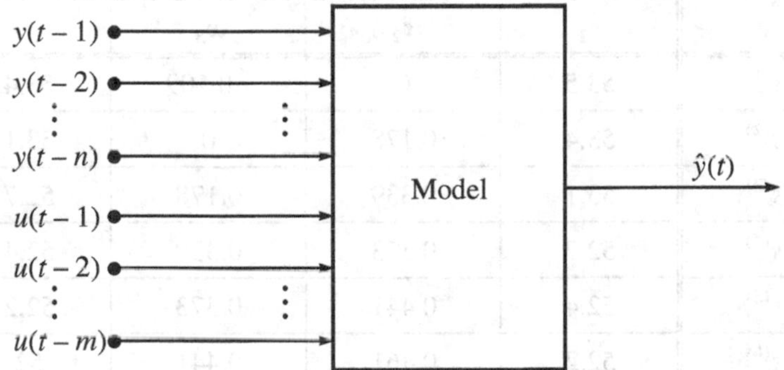

Figure E5.1 Input structure for identification of single-input single-output nonlinear dynamic system

In the following, we present a well-known case study—nonlinear system identification using the gas furnace data. The dataset was recorded from a combustion process of a methane-air mixture. The input measurement $u(t)$ is gas flow rate into the furnace and the output measurement $y(t)$ is CO_2 concentration in outlet gas. The sampling interval is 9 sec. The task of machine learning is to provide prediction of the CO_2 concentration in $\hat{y}(t)$.

The dataset (http://openmv.net/info/gas-furnace) consists of 296 rows and two columns, corresponding to 296 input-output measurements. 'InputGasRate' column represents the flow rate of the methane gas in a gas furnace, and 'CO_2' column represents the concentration of carbon dioxide in the gas mixture flowing out of the furnace under a steady air supply. A small portion of the dataset (20 samples) is shown in Table E5.1. This dataset when transformed into the structure of Table 1.2, looks like the one given in Table E5.2. This corresponds to the model structure suggested for good prediction results:

$$\begin{aligned}\hat{y}(t) &= f(y(t-1), u(t-3), u(t-4)) \\ &= f(x_1, x_2, x_3)\end{aligned}$$

That is to predict CO_2 concentration in $y(t)$, we consider the inputs three and four steps before, resulting in features: $x_1 = y(t-1)$, $x_2 = u(t-3)$, $x_3 = u(t-4)$.

References helpful for experiment on this dataset are [199, 200].

Table E5.1 Gas furnace data

InputGasRate	– 0.109	0	0.178	0.339	0.373	0.441	0.461	0.348	0.127	– 0.18
CO_2	53.8	53.6	53.5	53.5	53.4	53.1	52.7	52.4	52.2	52
InputGasRate	– 0.588	– 1.055	– 1.421	– 1.52	– 1.302	– 0.814	– 0.475	– 0.193	0.088	0.435
CO_2	52	52.4	53	54	54.9	56	56.8	56.8	56.4	55.7

Table E5.2 Features obtained from gas furnace time-series data

$s^{(i)}$	x_j and y			
	x_1	x_2	x_3	y
$s^{(1)}$	53.5	0	–0.109	53.4
$s^{(2)}$	53.4	0.178	0	53.1
$s^{(3)}$	53.1	0.339	0.178	52.7
$s^{(4)}$	52.7	0.373	0.339	52.4
$s^{(5)}$	52.4	0.441	0.373	52.2
$s^{(6)}$	52.2	0.461	0.441	52
$s^{(7)}$	52	0.348	0.461	52
$s^{(8)}$	52	0.127	0.348	52.4
$s^{(9)}$	52.4	– 0.18	0.127	53
$s^{(10)}$	53	– 0.588	–0.18	54
$s^{(11)}$	54	–1.055	–0.588	54.9
$s^{(12)}$	54.9	–1.421	–1.055	56
$s^{(13)}$	56	–1.52	–1.421	56.8
$s^{(14)}$	56.8	–1.302	–1.52	56.8
$s^{(15)}$	56.8	–0.814	–1.302	56.4
$s^{(16)}$	56.4	–0.475	– 0.814	55.7

PROBLEMS

P1.1: (a) It is said that guiding principle of machine learning is, "exploit tolerance for imprecision, uncertainty and partial truth". Explain the principle with the help of examples.

(b) Give one example for which machine learning approach is appropriate and one for which it seems inappropriate. Give justification for each.

P1.2: (a) A machine is said to *learn from experience* with respect to some *class of tasks* if its performance, as measured by a *performance measure*, improves with experience.

Pick up one machine learning problem. Describe it by stating as precisely as possible the learning task, the measure of performance and the task experience. Propose a target function to be learned.

(b) Give two examples of machine learning problems, one with task experience in the form of experimental data, and the other with task experience in the form of structured human knowledge.

P1.3: What are the different forms of learning? Give one example of each form that brings out the basic characteristics of that form of learning and its role to solve real-life problems.

P1.4: Explain the following forms of learning:

(i) Supervised learning
(ii) Unsupervised learning
(ii) Reinforcement learning
(iv) Evolutionary leaning

P1.5: Describe the following learning tasks with one real-life application of each:

(i) Classification
(ii) Regression
(iii) Learning associations
(iv) Clustering

P1.6: (a) Take a sample univariate time series data. Employing NARMA model for the time series, transform this data into a standard data matrix required for machine learning algorithms.

(b) Repeat for multivariate time series data.

P2.1: Through simple regression and classification examples, show that minimizing training error given by empirical risk function is not a successful solution to our learning task. What is the successful solution? What is it that we need for empirical risk minimization over the training dataset to lead to a successful solution to the learning task?

P2.2: Give mathematical equations for true risk function and empirical risk function. Why for real-world problems, finding optimal machine by true risk minimization is mostly not possible? What is the way out?

P2.3: (a) What is inductive learning? How is it different from deductive learning?

(b) What is generalization performance? Why is it a fundamental problem in inductive learning?

P2.4: Several results explore ways to improve generalization performance of a learning algorithm with respect to the task it addresses. Explain how this issue has been studied in terms of

(i) Bias and variance

(ii) VC model

(iii) Occum's razor principle

(iv) Minimum description length principle

P2.5:
- Higher the complexity of a hypothesis function (more flexible function with large number of free parameters), the lower is the approximation error.
- Simpler (inflexible; less number of free parameters) models lead to higher approximation error.

With respect to these observations, explain how improving generalization error requires a trade-off between hypothesis complexity and approximation error.

P2.6: To achieve good generalization performance, we require bias and variance to be low at the same time. Illustrate through an example bias-variance trade-off for regression.

P2.7: Describe capacity (of a function) concept of VC theory in terms of VC dimension. How does it lead to an induction principle called structural risk minimization?

P2.8: Occum's Razor Principle and Overfitting Avoidance are extensively exploited for heuristic search in inductive learning. Explain how? How is overfitting avoidance related to generalization performance?

P2.9: Training accuracy is 100% for a design done by you. Will you be proud of your design? Justify your answer?

P2.10: Before being used, a machine learning system should be evaluated in many aspects. Describe the aspects of accuracy, robustness, computational complexity and speed, interpretability and scalability.

P2.11: Describe regularization approach for finding a hypothesis function of complexity consistent with the data.

P2.12: Machine learning involves searching through space of possible hypotheses. List the practical hypothesis classes you are aware of.

P2.13: Explain the concept of ensemble learning, and the techniques used for this purpose.

P2.14: Describe *K*-fold cross-validation procedure for training and evaluation of a learning machine.

What is the difference in roles assumed by the validation partition and the test partition of a dataset?

P2.15: In many classification applications, wrong decisions—namely, false-positives and false-negatives—have cost, and the costs for the two wrong decisions may be different. How do we evaluate a classifier taking relative costs into consideration? Give examples of such applications.

P2.16: (a) Why do we need "confusion matrix" for evaluation of a machine trained for a pattern recognition task?

(b) What is ROC curve? How does it help in comparing the performance of classifiers?

P2.17: The design of a supervised learning system for classification and regression problems usually entails the repetition of a number of different activities: data collection, features choice, model choice, training, and evaluation. Present an overview of the design cycle and explain some of the issues that frequently arise.

P2.18: Table P2.18 shows a confusion matrix for medical data where the data values are *yes* or *no* for a class label attribute, *cancer*.

Table P2.18

Predicted Class / Actual Class	Cancer = "yes"	Cancer = "no"
yes	90	210
no	140	9560

The numbers within the four cells indicate number of patients. Based on the given information, answer the following:

(a) Determine the misclassification rate. Can we use misclassification rate as a performance measure for the given data? Justify your answer.

(b) For the given confusion matrix, determine sensitivity and specificity.

(c) Which is more important class (positive class) out of the two, and why? Which measure has better ability to detect the important class members correctly: misclassification rate; sensitivity; specificity.

P2.19: Give one example each of confusion matrix of class-imbalanced data and balanced data. Show that for balanced data, misclassification rate is an acceptable measure of performance, while it is not when the data is imbalanced. Describe the measures that can be used for judging performance of imbalanced data.

P2.20: The dataset in P3.8 is a toy dataset. In a real-life dataset, the observations given in Table P 2.20 were made.

Table P2.20

Actual Class \ Predicted Class	buys_computer = "yes"	buys_computer = "no"
buys_computer = "yes"	6954	46
buys_computer = "no"	412	2588

The numbers within the four cells indicate number of customers. Based on the given information, answer the following.

(a) Determine the misclassification error for the data given. Can we use misclassification error as a performance measure? Justify your answer.

(b) For the given confusion matrix, determine sensitivity, and specificity. How is accuracy related to these measures? Comment on whether training data is balanced or unbalanced. Justify your answer.

P2.21: Multiple models learned from the data may be combined to improve classification accuracy. Outline the basic features of the following popular ensemble methods:

(a) Bagging

(b) Boosting and AdaBoost

(c) Random Forests

How can we improve classification accuracy of class-imbalanced data?

P3.1: A hospital tested the body-fat data for 18 randomly selected adults; the results are given in Table P3.1.

(a) Calculate the mean, variance and standard deviation of each attribute.

(b) Calculate covariance between two attributes, and the correlation coefficient.

Table P3.1

x_1: age	23	23	27	27	39	41	47	49	50
x_2: % fat	9.5	26.5	7.8	17.8	31.4	25.9	27.4	27.2	31.2
x_1	52	54	54	56	57	58	58	60	61
x_2	34.6	42.5	28.8	33.4	30.2	34.1	32.9	41.2	35.7

P3.2: Given the following two-dimensional data points:

$$\mathbf{x} = \begin{bmatrix} x_1 \\ x_2 \end{bmatrix} = \begin{bmatrix} 1.5 \\ 1.7 \end{bmatrix}, \begin{bmatrix} 2 \\ 1.9 \end{bmatrix}, \begin{bmatrix} 1.6 \\ 1.8 \end{bmatrix}, \begin{bmatrix} 1.2 \\ 1.5 \end{bmatrix}, \begin{bmatrix} 1.5 \\ 1.0 \end{bmatrix}$$

(a) Given a new data point $x = \begin{bmatrix} 1.4 \\ 1.6 \end{bmatrix}$ as a query, rank the data points based on similarity with the query using Euclidean distance, and Manhattan distance.

(b) Normalize the dataset to make the norm of each data point equal to 1. Use Euclidean distance on the transformed data to rank the data points. Comment on the result.

P3.3: Consider the dataset given in Table P3.3

Table P3.3

Records	Age	Income ($)
1	56	156,000
2	65	99,000
3	32	192.000
4	49	57,000

Answer the following questions.

(a) Using Euclidean distance as a measure of similarity, show which two records are farthest from each other?

(b) Normalize the data. Does normalization change the result obtained in part (a)?

P3.4: (a) Describe Bayes Rule. Explain approximations that lead to naive Bayes classifier.

(b) Why is naive Bayes classifier called "naive"?

(c) Naive Bayes classification could depend on *Maximum-a-Posteriori* or *Maximum-Likelihood* criteria. What is the difference between the two?

P3.5: Table P3.5 provides a set of 14 training examples of the target concept: PlayTennis, where each day is described by the attributes: Outlook, Temperature, Humidity, and Wind. Use the naive Bayes classifier and the training data from this table to classify the following instance: (Outlook = sunny, Temperature = cool, Humidity = high, Wind = strong)

Table P3.5

Day	Outlook	Temperature	Humidity	Wind	PlayTennis
D1	sunny	hot	high	weak	no
D2	sunny	hot	high	strong	no
D3	overcast	hot	high	weak	yes
D4	rain	mild	high	weak	yes

(Contd.)

Day	Outlook	Temperature	Humidity	Wind	PlayTennis
D5	rain	cool	normal	weak	yes
D6	rain	cool	normal	strong	no
D7	overcast	cool	normal	strong	yes
D8	sunny	mild	high	weak	no
D9	sunny	cool	normal	weak	yes
D10	rain	mild	normal	weak	yes
D11	sunny	mild	normal	strong	yes
D12	overcast	mild	high	strong	yes
D13	overcast	hot	normal	weak	yes
D14	rain	mild	high	strong	no

P3.6: An auditing firm has data on 1000 companies that it investigated in the past. For each company, they have information on whether a company is *fraudulent* or *truthful*, and whether *legal charges* have been filed against it or not. The counts from the data are shown in Table P3.6.

A company has just been charged with fraudulent financial reporting. Using naive Bayes classification technique, compute the probabilities to each of the two classes.

Table P3.6

	legal charges	no legal charges
fraudulent	50	50
truthful	180	720

P3.7: Consider the 10 companies listed in Table P3.7. For each company, we have information on whether or not charges were filed against it, whether it is a small or a large company, and whether (after investigation) it turned out to be fraudulent or truthful in financial reporting.

A "small" company has just been "charged" with fraudulent financial reporting. Using naive Bayes classification technique, compute the probability that the company is "fraudulent".

Through the dataset in Table P3.7, show how exact Bayes classifications differ from naive Bayes.

Table P3.7

Company $s^{(i)}$	x_1: charges filed	x_2: company size	y: status
1	yes	small	truthful
2	no	small	truthful
3	no	large	truthful

(Contd.)

4	no	large	truthful
5	no	small	truthful
6	no	small	truthful
7	yes	small	fraudulent
8	yes	large	fraudulent
9	no	large	fraudulent
10	yes	large	fraudulent

P3.8: Table P3.8 presents a training set, randomly selected from the customer database of a store. The data samples are described by attributes: age (x_1), income (x_2), student (x_3), and credit-rating (x_4). The continuous-valued attributes have been transformed to categorical values. Class label attribute: buys_ computer (y), has two distinct values.

The sample we wish to classify is **x**: (x_1 = "≤ 30", x_2 = "medium", x_3 = " yes", x_4 = "fair"), Using naive Bayes classifier, predict the class label for this sample. Steps (algorithm) used in prediction must be clearly given.

Table P3.8

Data samples $s^{(i)}$	**x_1 : age**	**x_2 : income**	**x_3 : student**	**x_4 : credit_rating**	**y : buys_computer**
$s^{(1)}$	≤ 30	high	no	fair	no
$s^{(2)}$	≤ 30	high	no	excellent	no
$s^{(3)}$	31...40	high	no	fair	yes
$s^{(4)}$	> 40	medium	no	fair	yes
$s^{(5)}$	> 40	low	yes	fair	yes
$s^{(6)}$	> 40	low	yes	excellent	no
$s^{(7)}$	31...40	low	yes	excellent	yes
$s^{(8)}$	≤ 30	medium	no	fair	no
$s^{(9)}$	≤ 30	low	yes	fair	yes
$s^{(10)}$	> 40	medium	yes	fair	yes
$s^{(11)}$	≤ 30	medium	yes	excellent	yes
$s^{(12)}$	31...40	medium	no	excellent	yes
$s^{(13)}$	31...40	high	yes	fair	yes
$s^{(14)}$	> 40	medium	no	excellent	no

(**Hint:** Note that attribute values for "age" are discrete bins. You may view them as categorical values: young, middle-aged, senior).

P3.9: *Graphical models* represent the interaction between variables visually and have the advantage that inference over a large number of variables can be decomposed into a set of local calculations involving a small number of variables making use of conditional independencies. Graphical models, also called *belief networks*, are composed of nodes and arcs between the nodes.

Draw the Bayesian belief network that represents the conditional independence assumptions of the naive Bayes classifier for the *PlayTennis* problem of P3.5.

P3.10: Consider the data given in Table 8.4 of Chapter 8. It is desired to categorize a new household with \$ 60,000 income and lawn size 20,000 square feet.

(i) Among the households in the training set, find the one closest (using Euclidean distance similarity measure) to the new household.

(ii) Choosing $k = 3$, find the three nearest households, and categorize the new household using majority voting.

P3.11: In the case of normally distributed classes, discriminant functions are linear (straight lines, planes, and hyperplanes for two-, three-, and n-dimensional feature vectors, respectively) when the covariances matrices of corresponding classes are equal. Confirm this by deriving discriminant functions for a binary classification problem.

Given:

$$P(\mathbf{x}|y_q) = \frac{1}{(2\pi)^{n/2}|\Sigma|^{1/2}} \exp\left(-\tfrac{1}{2}(\mathbf{x}-\boldsymbol{\mu}_q)^T \Sigma^{-1}(\mathbf{x}-\boldsymbol{\mu}_q)\right); q = 1, 2$$

prove that linear discriminant functions

$$g_q(\mathbf{x}) = \boldsymbol{\mu}_q^T \Sigma^{-1} \mathbf{x} - \tfrac{1}{2}\boldsymbol{\mu}_q^T \Sigma^{-1} \boldsymbol{\mu}_q + \ln P(y_q); q = 1, 2$$

and decision boundary $g(\mathbf{x}) = g_1(\mathbf{x}) - g_2(\mathbf{x}) = 0$ is given by

$$g(\mathbf{x}) = \mathbf{w}^T \mathbf{x} + w_0 = 0$$

$$\mathbf{w}^T \mathbf{x} + w_0 = (\boldsymbol{\mu}_1^T - \boldsymbol{\mu}_2^T)\Sigma^{-1}\mathbf{x} - \tfrac{1}{2}(\boldsymbol{\mu}_1^T \Sigma^{-1}\boldsymbol{\mu}_1 - \boldsymbol{\mu}_2^T \Sigma^{-1}\boldsymbol{\mu}_2) + \ln \frac{P(y_1)}{P(y_2)}$$

(**Hint:** Use Eqns (3.61)–(3.62))

P3.12: Patterns $\mathbf{x} \in \Re^2$ drawn from two equiprobable normally distributed classes are given in Table P3.12. Find the decision boundary between the two classes.

Table P3.12

$s^{(i)}$	x_1	x_2	y_1
$s^{(1)}$	1	2	0
$s^{(2)}$	2	2	0
$s^{(3)}$	2	3	0
$s^{(4)}$	3	1	0
$s^{(5)}$	3	2	0

$s^{(i)}$	x_1	x_2	y_2
$s^{(1)}$	6	8	1
$s^{(2)}$	7	8	1
$s^{(3)}$	8	7	1
$s^{(4)}$	8	8	1
$s^{(5)}$	7	9	1

P3.13: Perform two iterations of the gradient algorithm to find the minima of $E(\mathbf{w}) = 2w_1^2 + 2w_1w_2 + 5w_2^2$. The starting point is $\mathbf{w} = [2 \ -2]^T$. Draw the contours and show your learning path graphically.

P3.14: In a linear regression problem, the data has been rearranged in the following vector form:

$$\underset{(1 \times N)}{\mathbf{y}^T} = \underset{(1 \times (n+1))}{(\bar{\mathbf{w}}^T)} \times \underset{((n+1) \times N)}{(\mathbf{X})}$$

Give an expression for $\bar{\mathbf{w}}$ in terms of $\mathbf{X}$ and $\mathbf{y}$.

P3.15: Show that logistic regression is a nonlinear regression problem. Is it possible to treat logistic discrimination in terms of equivalent liner regression problem? Justify your answer.

P3.16: Parameter k in *k-NN* algorithm could be a very large value or a very small value. Give the drawbacks, if any, of each choice.

P3.17: Consider the dataset given in Table P 3.17. It is desired to transform the data to one-dimensional data using Fisher linear discrimant analysis. Find the one-dimensional transformed data. What is the significance of Fisher linear discriminants for classification problems?

Table P3.17

$s^{(i)}$	x_1	x_2	Class
$s^{(1)}$	1	2	1
$s^{(2)}$	2	3	1
$s^{(3)}$	3	3	1
$s^{(4)}$	4	5	1
$s^{(5)}$	5	5	1
$s^{(6)}$	1	0	2
$s^{(7)}$	2	1	2
$s^{(8)}$	3	1	2
$s^{(9)}$	3	2	2
$s^{(10)}$	5	3	2
$s^{(11)}$	6	5	2

P4.1: Let $\{\mathbf{x}^{(1)}, ..., \mathbf{x}^{(N)}\}$ be a finite number of linearly separable samples in n dimensions. Derive and explain the perceptron algorithm that will find a separating hyperplane $\mathbf{w}^T\mathbf{x} + w_0 = 0$ in a finite number of steps, with zero misclassification error.

P4.2: By minimizing $\|\mathbf{x} - \mathbf{x}^{(i)}\|^2$ subject to the constraint $g(\mathbf{x}) = 0$, show that the distance from the hyperplane $g(\mathbf{x}) = \mathbf{w}^T\mathbf{x} + w_0 = 0$ to the point $\mathbf{x}^{(i)}$ is $|g(\mathbf{x}^{(i)})|/\|\mathbf{w}\|$.

Also prove that the projection of $\mathbf{x}^{(i)}$ on to the hyperplane is given by

$$\mathbf{x}_P = \mathbf{x}^{(i)} - \frac{g(\mathbf{x}^{(i)})}{\|\mathbf{w}\|^2}\,\mathbf{w}$$

P4.3: Consider the case of a hyperplane for linearly separable patterns, which is defined by the equation

$$\mathbf{w}^T\mathbf{x} + w_0 = 0$$

where $\mathbf{w}$ denotes the weight vector, w_0 denotes the bias; and $\mathbf{x}$ denotes the input vector. The hyperplane is said to correspond to a canonical pair $(\mathbf{w}, w_0)$ if, for the set of input patterns $\mathbf{x}^{(i)}$; $i = 1, ..., N$, the additional requirement

$$\min_{i=1,2,\ldots,N} |\mathbf{w}^T\mathbf{x}^{(i)} + w_0| = 1$$

is satisfied. Show that this requirement leads to a margin of separation between the two classes equal to $2/\|\mathbf{w}\|$.

P4.4: Find the distance from the point

$$\mathbf{x} = [1 \quad 1 \quad 1 \quad 1 \quad 1]^T$$

to the hyperplane

$$x_1 - x_2 + x_3 - x_4 + x_5 + 1 = 0$$

P4.5: (a) You have a choice of handling a binary classification task using number of misclassifications as the performance measure, and maximizing the margin between the two classes as the performance measure. On what factors does your decision depend?

(b) You have a choice of handling a binary classification task using (i) linear SVM and (ii) perceptron algorithm. On what factors does your decision depend?

P4.6: Using the technique of Lagrange multipliers, find the maximum of the function

$$f(\mathbf{x}) = x_1^2 + 4x_2^2$$

subject to the constraint

$$x_1 + 2x_2 = 6$$

P4.7: Using KKT conditions, find the minimum of the function

$$f(\mathbf{x}) = (x_1 - 1)^2 + (x_2 - 2)^2$$

subject to the following constraints:

$$x_2 - x_1 = 1$$
$$x_1 + x_2 \leq 2$$
$$x_1 \geq 0, x_2 \geq 0$$

Check your result graphically.

P4.8: Consider the following data:

Class 1: $[1\ \ 1]^T$, $[2\ \ 2]^T$, $[2\ \ 0]^T$

Class 2: $[0\ \ 0]^T$, $[1\ \ 0]^T$, $[0\ \ 1]^T$

Plot the six training points and construct by inspection the weight vector for the optimal hyperplane, and the optimal margin. What are the support vectors?

Construct now the solution in the dual space, and compare the two results.

P4.9: Consider the two-class classification task that consists of the following points:

Class 1: $[1,\ 1]^T$, $[1,\ -1]^T$

Class 2: $[-1,\ 1]^T$, $[-1,\ -1]^T$

(a) Using the simple geometry of the problem, compute the SVM linear classifier and the margin hyperplanes (lines). Determine and margin and the possible support vectors.
(b) Consider another linear classifier $x_1 + x_2 = 0$. Determine the margin hyperplanes (lines) and the margin.
(c) Set up mathematical formulation of the problem: define the Lagrangian function in the primal space and give the KKT conditions

P4.10: For the optimization of the separating hyperplane for nonseparable patterns, formulate the primal and the dual problems.

P4.11: It is said that support vectors constitute a small percentage of data, and only these vectors are used to compute weight of the decision function. What are support vectors? Identify these vectors in hard-margin and soft-margin SVM formulations.

P4.12: A three-dimensional input vector $\mathbf{x} = [x_1\ x_2\ x_3]^T$ is mapped into the feature vector:

$$\mathbf{z}(\mathbf{x}) = [\phi_1(\mathbf{x})\ \phi_2(\mathbf{x})\ \phi_3(\mathbf{x})\ \phi_4(\mathbf{x})\ \phi_5(\mathbf{x})\ \phi_6(\mathbf{x})\ \phi_7(\mathbf{x})\ \phi_8(\mathbf{x})\ \phi_9(\mathbf{x})]^T$$
$$= [x_1\ x_2\ x_3\ (x_1)^2\ (x_2)^2\ (x_3)^2\ x_1x_2\ x_1x_3\ x_2x_3]^T$$

Show that a linear decision hyperplane in feature space $\Re^m (m = 9)$ corresponds to a nonlinear hypersurface in the original input space $\Re^n (n = 3)$.

P4.13: The points $\mathbf{x} = [1\ \ 1]^T$ and $[-1\ \ -1]^T$ are in category 1 (class y_1) and points $\mathbf{x} = [1\ \ -1]^T$ and $[-1\ \ 1]^T$ are in category 2 (class y_2). Confirm that this classification problem cannot be solved using a linear discriminant operating directly on the features.

Following the approach of SVMs, we transform the features to map them to higher dimension space where they can be linearly separated. One such simple mapping is to a six-dimensional space by $1, \sqrt{2}x_1, \sqrt{2}x_2, \sqrt{2}x_1x_2, x_1^2$, and x_2^2. Using analytic techniques, show that

(i) the optimal hyperplane is $g(x_1, x_2) = x_1 x_2 = 0$, and

(ii) the margin is $\sqrt{2}$.

What are the support vectors?

P4.14: Repeat problem P4.13 using the same mapping to six-dimensional space but with the following four points:

Class 1: $[1\ \ 5]^T$, $[-2\ \ -4]^T$

Class 2: $[2\ \ 3]^T$, $[-1\ \ 5]^T$

P4.15: Prove that a Gaussian kernel maps the points $\mathbf{x} \in \Re^n$ to infinite dimensional space.

(**Hint:** $K(\mathbf{x}^{(i)}, \mathbf{x}^{(j)}) = \exp(-\sigma \|\mathbf{x}^{(i)} - \mathbf{x}^{(j)}\|^2)$

$= \exp(-\sigma \|\mathbf{x}^{(i)}\|^2) \exp(-\sigma \|\mathbf{x}^{(j)}\|^2) \exp(2\sigma (\mathbf{x}^{(i)})^T \mathbf{x}^{(j)})$

Expand each term by Taylor series expansion).

P4.16: A support vector machine (SVM) has the inner-product kernel given by

$$K(\mathbf{x}, \mathbf{x}^{(i)}) = \exp\left(-\frac{1}{2\sigma^2} \|\mathbf{x} - \mathbf{x}^{(i)}\|^2\right)$$

The width σ^2 is specified *a priori* by the user. Outline the SVM classification algorithm.

P4.17: Describe One-Against-All (OAA), and One-Against-One (OAO) classification schemes.

We usually transform a multi-class SVM classification problem into a set of equivalent OAA/OAO binary classification problems. Why? What are the constraints of today's multiclass-based SVM variants?

P4.18: Explain what do we mean by a kernel classifier. Give an example to show how kernels are useful. Also give three examples of kernel functions.

P4.19: *Software exercise*

Train an SVM classifier with the data given in Table P4.17. What is the equation of the separating hyperplane, the margin, and the support vectors?

Table P4.17

$s^{(i)}$	x_1	x_2	y
$s^{(1)}$	–3.0	–2.9	1
$s^{(2)}$	0.5	8.7	1
$s^{(3)}$	2.9	2.1	1
$s^{(4)}$	–0.1	5.2	1
$s^{(5)}$	–4.0	2.2	1
$s^{(6)}$	–1.3	3.7	1

(*Contd.*)

$s^{(7)}$	–3.4	6.2	1
$s^{(8)}$	–4.1	3.4	1
$s^{(9)}$	–5.1	1.6	1
$s^{(10)}$	1.9	5.1	1
$s^{(11)}$	–2.0	–8.4	–1
$s^{(12)}$	–8.9	0.2	–1
$s^{(13)}$	–4.2	–7.7	–1
$s^{(14)}$	–8.5	–3.2	–1
$s^{(15)}$	–6.7	–4.0	–1
$s^{(16)}$	–0.5	–9.2	–1
$s^{(17)}$	–5.3	–6.7	–1
$s^{(18)}$	–8.7	–6.4	–1
$s^{(19)}$	–7.1	–9.7	–1
$s^{(20)}$	–8.0	–6.3	–1

P4.20: In SVM regressor algorithms, two important design parameters are parameter C and ε-insensitivity zone. Describe the trade-off these parameters deal with.

In a regression problem, we normally use sum-of-error-squares criterion for design. Describe the criterion used by SVM-regressor algorithm.

P4.21: *Software exercise*

Sinc function is one of the commonly used dataset for testing nonlinear regression algorithms. This function is given by the following equation:

$$y = \frac{\sin \pi x}{\pi x}$$

(a) Generate 50 data points from this function in the range [–3, 3]. Also add Gaussian noise to the data.

(b) Train an SVM regressor with the data generated as per (a). Use suitable parameters required for training the regressor. Comment on the results.

P4.22: *Software exercise*

Model (reconstruct) the simple relation $y = \sin x$ (known to you but not to the learning machine), using a support vector machine, having a set of ten data pairs from measurements corrupted by 25% of normally distributed noise with zero mean.

Analyze the influence of an insensitivity zone on regression quality.

P4.23: Parameter $C \geq 0$ trades off the complexity as measured by the norm of weight vector and data misfit as measured by number of nonseperable points. Show this through the formulation of basic SVM.

Parameter C in basic SVM is replaced by parameter $\nu \in [0, 1]$ in one of its variants. Discuss the advantages and limitations of this replacement.

P4.24: One of the limitations of SVM algorithm is computational complexity, arising due to quadratic optimization problem. What kind of variants of basic SVM are addressing this issue?

P5.1: Construct a multilayer perception network with six input terminals, a hidden-layer of four neurons, and an output layer of two neurons. What is the hidden layer for, and what does it hide?

Write a functional relationship between output output vector **y** and input vector **x**, showing explicitly the parameters involved.

P5.2: What are the main problems with the backpropagation learning algorithm? How can learning be accelerated in multilayer neural networks?

P5.3: A neural network typically starts out with random initial weights; hence, it produces essentially random predictions when presented with its first case. What is the key ingredient by which the network evolves to produce a more accurate prediction?

P5.4: In logistic discrimination, we model the ratio of posterior probabilities of the two classes y_1 and y_2 (binary classification problems). Assuming log of this ratio to be linear:

$$\log \frac{P(y_1|\mathbf{x})}{P(y_2|\mathbf{x})} = \log \frac{P(y_1|\mathbf{x})}{1 - P(y_1|\mathbf{x}))} = \text{logit}\ (P(y_1|\mathbf{x})) = \mathbf{w}^T\mathbf{x} + w_0$$

Show how do we obtain the estimate of $P(y_1/\mathbf{x})$ using

(a) Maximum likelihood method;

(b) Training a single neuron with logistic (log-sigmoid) function.

P5.5: Consider the following sigmoidal activation functions:

(i) $\sigma(a) = \dfrac{1}{1 + e^{-\lambda a}}$ (ii) $\sigma(a) = \dfrac{1 - e^{-\lambda a}}{1 + e^{-\lambda a}}$

where λ is a positive constant.

Show that in both cases, the derivative $\sigma'(a)$ can be written simply in terms of $\sigma(a)$.

P5.6: Define log-sigmoid and tan-sigmoid activation functions used in neural networks. Show that log-sigmoid is related to tan-sigmoid by a simple transformation.

P5.7: Explain the difference between batch training and incremental training protocols for training neural networks.

How is incremental training used for training feedforward neural networks related to online training?

P5.8: Describe the following practical techniques for improving backpropagation-learning performance in MLP networks:

(i) choice of activation functions; (ii) scaling the input; (iii) avoiding overfitting; (iv) adding momentum term; (v) weight decay; (vi) number of hidden layers; (vii) number of hidden units; (ix) initializing weights; (x) learning rates; (xi) incremental or batch training?; and (xii) stopping criterion.

P5.9: Consider the error function

$$E(\overline{\mathbf{w}}) = \frac{1}{2}\sum_{i=1}^{N}\sum_{q=1}^{M}(y_q^{(i)} - \hat{y}_q^{(i)})^2 + \gamma\sum_{i,j} w_{ji}^2$$

Derive the gradient descent update rule for minimization of E. What is the purpose of adding an additional term which is a function of weights, to the error-minimization criterion? (**Hint:** One method of simplifying the network and avoiding overfitting is to apply the heuristic that the weights should be small)

P5.10: One common strategy in machine learning is to learn nonlinear functions with a linear machine: we first transform the data into a new feature space by a fixed nonlinear mapping, and then use a linear machine in the new feature space.

One frequently seeks to identify the smallest set of m functions $\phi_l(\mathbf{x})$ that still conveys the essential information contained in the original attributes, and allows the use of linear learning in ϕ-space.

(i) In neural networks, commonly used ϕ_l-functions are sigmoidal functions; and in RBF networks, these are Gaussian functions; both of these have universal approximation property. It is commonly held that feedforward neural networks with sigmoidal activation functions are representatives of global approximation schemes, while RBF networks are representatives of local approximation schemes. Describe this fundamental difference between the two.

(ii) In RBF networks, we need a judicial selection of minimum number of m functions so that patterns which are not linearly separable in x-space, become linearly separable in ϕ-space. We can easily achieve this feature by letting $m \to \infty$ but it increases computational complexity. In support vector machines (SVMs), we let $m \to \infty$. Explain how computational complexity is taken care of in case of SVMs.

P5.11: Show through a schematic diagram, a typical structure of RBF networks, clearly indicating the involved learning parameters. Explain a method of training such a network.

P5.12: A Gaussian function can be used as a kernel function in SVM regression algorithm. A Gaussian function is used as an activation function in RBF networks. Are the roles of the Gaussian function in two applications same? If not, explain the different roles.

P5.13: The learning environment for an ADALINE comprises a training set of N data samples: $\{\mathbf{x}^{(i)}, y^{(i)}; i = 1, ..., N\}$, consisting of input vector $\mathbf{x} = [x_1\ x_2 \cdots x_n]^T$, and output y. Develop a gradient descent algorithm for batch training. The network weights are $w_1, w_2, ..., w_n$; and the bias parameter is w_0.

P5.14: Class y_1 consists of the two-dimensional vectors $[0.2\ \ 0.7]^T$, $[0.3\ \ 0.3]^T$, $[0.4\ \ 0.5]^T$, $[0.6\ \ 0.5]^T$, $[0.1\ \ 0.4]^T$; and class y_2 of $[0.4\ \ 0.6]^T$, $[0.6\ \ 0.2]^T$, $[0.7\ \ 0.4]^T$, $[0.8\ \ 0.6]^T$, $[0.7\ \ 0.5]^T$.

Design the sum of error squares linear classifier $w_1x_1 + w_2x_2 + w_0 = 0$ using batch processing Least Squares Estimation approach.

P5.15: Consider a two-layer feedforward neural network with two inputs x_1 and x_2, one hidden unit (output z) and one output unit (output y). This network has five weights (w_1, w_2, w_0, v_1, v_0) where (w_0, v_0) represent the bias terms for the two units with sigmoidal activation function for hidden layer and linear function for the output.

Initialize these weights to the values (0.1, 0.1, 0.1, 0.1, 0.1); then determine their values after each of the first two training iterations of the backpropagation algorithm. Assume learning rate $\eta = 0.3$, incremental weight updates, and the following training examples:

x_1	x_2	y
1	0	1
0	1	0

P5.16: Consider the network of Figs 5.14–5.15 in Chapter 5, with the M output sigmoidal units replaced with m output linear units. An input signal $\mathbf{x} = [x_1\ x_2 \cdots x_n]^T$ comprising features, augmented by a constant input component (bias) is applied to this network. The network output comprises $\hat{\mathbf{y}} = [\hat{y}_1\ \ \hat{y}_2 \cdots \hat{y}_m]^T$. The learning environment comprises a set of N data points $\{\mathbf{x}^{(i)}, \mathbf{y}^{(i)}; i = 1, ..., N\}$; $\mathbf{x} \in \Re^n$ and $\mathbf{y} \in \Re^m$. The outputs y_l; $l = 1, ..., m$, correspond to the unknown functions $y_l = f_l(\mathbf{x})$; and $\hat{y}_l$ represents the approximation of the function.

Explain a method of training the network for the multi-output regression task. Write the equations for gradient-descent incremental training.

P5.17: Consider a four-input single-node perceptron with a bipolar sigmoidal function (tan-sigmoid)

$$\sigma(a) = \frac{2}{1+e^{-a}} - 1$$

where 'a' is the activation value for the node.

(a) Derive the weight update rule for $\{w_i\}$ for all i. The learning rate $\eta = 0.1$. Input variables: x_i; $i = 1, 2, 3, 4$. Desired output is y.

(b) Use the rule in part (a) to update the perceptron weights incrementally for one epoch. The set of input and desired output patterns is as follows:

$$\mathbf{x}^{(1)} = [1\ \ -2\ \ 0\ \ -1]^T, \qquad y^{(1)} = -1$$

$$\mathbf{x}^{(2)} = [0\ \ 1.5\ \ -0.5\ \ -1]^T, \quad y^{(2)} = -1$$

$$\mathbf{x}^{(3)} = [-1\ \ 1\ \ 0.5\ \ -1]^T, \quad y^{(3)} = 1$$

The initial weight vector is chosen as

$$\mathbf{w}_0^T = [1\ \ -1\ \ 0\ \ 0.5]$$

The perceptron does not possess a bias term.

(c) Use the training data and initial weights given in part (b) and update the perceptron weights for one epoch in batch mode.

P5.18: We are given the two-layer backpropagation network shown in Fig. P5.18.

(a) Derive the weight update rules for $\{v_l\}$ and $\{w_{li}\}$ for all i and l. Assume that activation function for all the nodes is a unipolar sigmoid function

$$\sigma(a) = \frac{1}{1 + e^{-a}}$$

where 'a' represents the activation value for the node. The learning constant $\eta = 0.1$. The desired output is y.

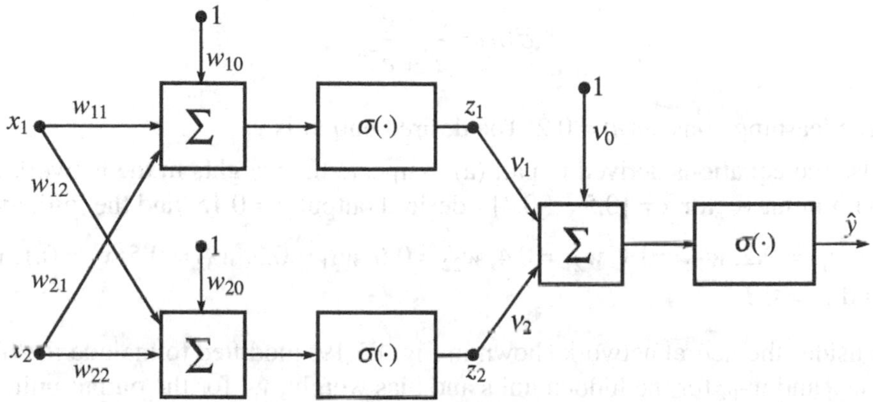

Figure P5.18

(b) Use the equations derived in part (a) to update the weights in the network for one step with input vector $\mathbf{x} = [1\ 0]^T$, desired output $y = 1$, and the initial weights:

$$w_{10} = 1,\ w_{11} = 3,\ w_{21} = 6,\ w_{20} = -6,\ w_{12} = 4,\ w_{22} = 5$$

$$v_0 = -3.92,\ v_1 = 2, \text{ and } v_2 = 4$$

(c) As a check, compute the error with the same input for initial weights and updated weights and verify that the error has decreased.

P5.19: We are given two-layer backpropagation network shown in Fig. P5.19.

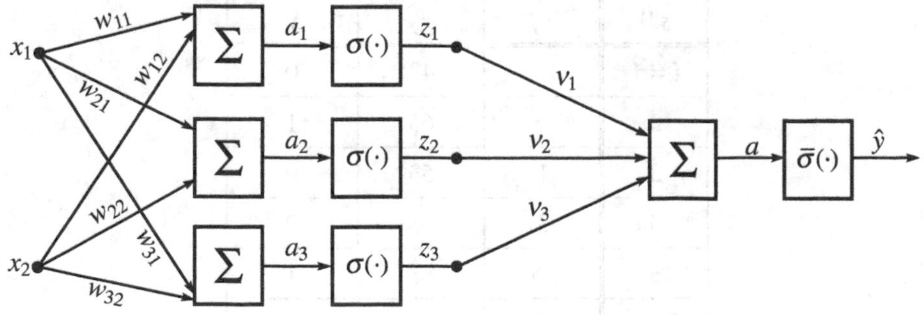

Figure P5.19

(a) Derive the weight update rules in incremental mode for $\{v_l\}$ and $\{w_{li}\}$ for all i and l; the iteration index is k. Assume that the activation function for all nodes in the hidden layer is

$$\sigma(a_l) = \frac{1}{1+e^{-a_l}}$$

and the activation function for the node in the output layer is

$$\bar{\sigma}(a) = \frac{e^a - e^{-a}}{e^a + e^{-a}}$$

The learning constant $\eta = 0.2$. The desired output is y.

(b) Use the equations derived in part (a) to update the weights in the network for one step with input vector $\mathrm{x} = [0.5 \quad -0.4]^T$, desired output $y = 0.15$, and the initial weights:

$$w_{11} = 0.2,\ w_{12} = 0.1,\ w_{21} = 0.4,\ w_{22} = 0.6,\ w_{31} = 0.3,\ w_{32} = 0.5;\ v_1 = 0.1,\ v_2 = 0.2$$

and $v_3 = 0.1$.

P5.20: Reconsider the neural network shown in Fig. P5.19, modified to include the bias weights: w_{10}, w_{20} and w_{30}, for the hidden units and bias weight, v_0, for the output unit. All the bias weights vary in the range 0.0 to 1.0.

A binary-coded GA is used to update connection weights including biases. Extend the procedure given in Example 5.3 in Chapter 5, to this modified network.

P5.21: Table P5.21 shows hypothetical bank data on consumers' use of credit card facilities of the bank; the variables are x_1 (number of years that the customer has been with the bank), x_2 (customer's salary in thousands of dollars), and y (equal to 0 if balance was paid off at the end of each month: and 1 if customer left an unpaid credit-card balance at the end of at least one month in the prior year).

Table P5.21

$s^{(i)}$	x_1	x_2	y
$i = 1$	4	43	0
2	18	65	1
3	1	53	0
4	3	95	0
5	15	88	1
6	6	112	1

Figure P5.21 shows an example of a typical neural net that could be used for predicting the consumers' use of credit card facilities, with logistic activation for all neurons. Illustrate one pass through the network for the first data in the table. Assume the following initial weights:

$w_{11} = 0.05$, $w_{12} = -0.01$, $w_{13} = 0.02$, $w_{21} = 0.01$, $w_{22} = 0.03$, $w_{23} = -0.01$
$w_{10} = -0.3$, $w_{20} = 0.2$, $w_{30} = 0.05$
$v_1 = 0.01$, $v_2 = 0.05$, $v_3 = 0.015$, $v_0 = -0.015$

How will you use this regression neural network for the classification task?

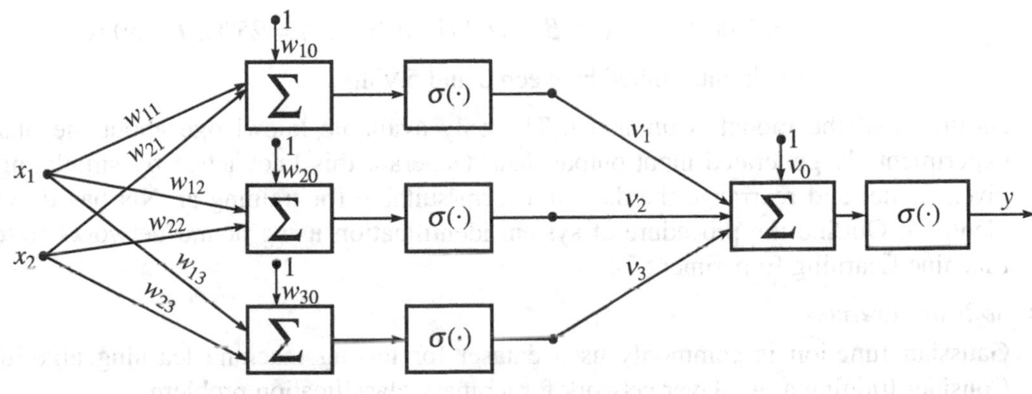

Figure P5.21

P5.22: *Simulation exercise*

Linear single input-single output dynamic systems can be described by the following generic discrete equation:

$$y(k) = a_1\, y(k-1) + a_2\, y(k-2) + \cdots + a_n\, y(k-n) + b_1 u(k-1) + b_2 u(k-2) + \ldots + b_n\, u(k-n)$$

where $y(k-i)$ and $u(k-i)$ are past outputs and inputs.

The system identification problem of determining a's and b's can be viewed as a regression problem (refer to Machine Learning Experiment 5).

Consider the identification of the following second-order system:

$$y(k) = 1.615\, y(k-1) - 0.7788\, y(k-2) + 0.08508\, u(k-1) + 0.07824\, u(k-2)$$

(a) Generate 50 data pairs from this function.

(b) Structure of a linear neuron for identification of a second-order system is shown in Fig. P5.22. Use this structure to estimate parameters a_1, a_2, b_1, and b_2.

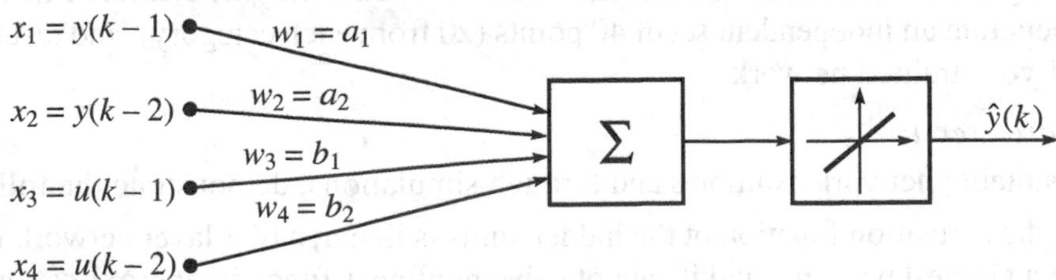

Figure P5.22

P5.23: *Simulation exercise*

Plant model of a temperature control system is

$$y(k+1) = F\,y(k) + \frac{g}{1+\exp[0.5y(k)-40]}u(k) + (1-F)Y_0$$

$$F = e^{-\alpha T};\quad g = \frac{\beta}{\alpha}[1 - e^{-\alpha T}]$$

$$\alpha = 1.00\,151 \times 10^{-4},\ \beta = 8.67973 \times 10^{-3},\ Y_0 = 25°\text{C},\ T = 30 \text{ sec}$$

u = input, limited between 0 and 5 volts.

Assume that the model is unknown. The only available knowledge about the plant is experimentally generated input-output data. Generate this knowledge by simulating the given model and rearrange the data in a form suitable for training an NN-based system identifier. Outline the procedure of system identification using neural networks (refer to Machine Learning Experiment 5).

P5.24: *Software exercise*

Gaussian function is commonly used dataset for testing machine learning algorithms. Consider training a two-layer network for a binary classification problem.

(a) Generate a training set of 100 points, 50 from each category, from the two-dimensional Gaussian distributions:

(i) Class y_1; $\boldsymbol{\mu} = \begin{bmatrix} 0 \\ 0 \end{bmatrix}, \Sigma = \begin{bmatrix} 1 & 0 \\ 0 & 1 \end{bmatrix}$

(ii) Class y_2; $\boldsymbol{\mu} = \begin{bmatrix} 1 \\ 0.5 \end{bmatrix}, \Sigma = \begin{bmatrix} 3 & 1 \\ 1 & 2 \end{bmatrix}$

(b) Train a network with different number of hidden units, $2 \le m \le 10$, and one output unit (all neurons with log-sigmoid activation functions). If m^* are the number of units selected by you, justify your choice.

(c) Re-initialize the network with m^* hidden units, and comment on the effects of different initial weights.

(d) Initial hidden layer weights must be different from each other; verify this by initializing the weights to have identical values and observing the effects of this choice.

(e) Generate an independent set of 40 points (20 from each category), and test the accuracy of your trained network.

P5.25: *Software exercise*

Take suitable network examples and through simulations, demonstrate the following:

(i) If the activation function of the hidden units is linear, a two-layer network is equivalent to a single-layer one, and it cannot solve nonlinear function approximation problems.

(ii) Preprocessing data (for example, by subtracting off the mean and scaling standard deviation in each dimension) can lead to significant reduction in time of learning.

P5.26: *Software exercise*

Consider the training data given in Table P5.26.

(a) Construct a neural network with two sigmoidal hidden units and one sigmoidal output unit.

(b) Initialize all weights randomly in the range {–1, 1} and train a regressor $\hat{y} = \hat{f}(\mathbf{x})$. Assume $\eta = 0.1$. Use incremental training protocol. Plot a learning curve—the training error as a function of epoch. Explain and justify the stopping criterion you use.

(c) Can you use the trained network for a classification problem wherein y in the table represents two categories of output? If yes, explain the technique you use, and the weakness of the technique, if any.

(d) How can we build a better classifier by changing the structure of the network to one with more than one output units?

Table P5.26

$s^{(i)}$	x_1	x_2	x_3	y
$s^{(1)}$	1.58	2.32	–5.8	1
$s^{(2)}$	0.67	1.58	–4.78	1
$s^{(3)}$	1.04	1.01	–3.63	1
$s^{(4)}$	–1.49	2.18	–3.39	1
$s^{(5)}$	–0.41	1.21	–4.73	1
$s^{(6)}$	1.39	3.16	2.87	1
$s^{(7)}$	1.20	1.40	–1.89	1
$s^{(8)}$	–0.92	1.44	–3.22	1
$s^{(9)}$	0.45	1.33	–4.38	1
$s^{(10)}$	–0.76	0.84	–1.96	1
$s^{(11)}$	0.21	0.03	–2.21	0
$s^{(12)}$	0.37	0.28	–1.8	0
$s^{(13)}$	0.18	1.22	0.16	0
$s^{(14)}$	–0.24	0.93	–1.01	0
$s^{(15)}$	–1.18	0.39	–0.39	0
$s^{(16)}$	0.74	0.96	–1.16	0

(*Contd.*)

$s^{(i)}$	x_1	x_2	x_3	y
$s^{(17)}$	–0.38	1.94	–0.48	0
$s^{(18)}$	0.02	0.72	–0.17	0
$s^{(19)}$	0.44	1.31	–0.14	0
$s^{(20)}$	0.46	1.49	0.68	0

P5.27: The emerging subject 'deep learning' was out of the scope of this book. With the background of machine learning developed through this book/other sources, it will be a logical step for the reader to self-learn this subject though available literature and software tools.

P6.1: (a) In the following, we suggest a membership function for fuzzy description of the set 'real numbers *close* to 2':

$$\underset{\sim}{A} = \{x, \mu_{\underset{\sim}{A}}(x)\}$$

where

$$\mu_{\underset{\sim}{A}}(x) = \begin{cases} 0 & ; \quad x < 1 \\ -x^2 + 4x - 3 & ; \quad 1 \le x \le 3 \\ 0 & ; \quad x > 3 \end{cases}$$

Sketch the membership function (arc of a parabola) and determine its supporting interval, and α-cut interval for $\alpha = 0.5$.

(b) Sketch the piecewise quadratic membership function

$$\mu_{\underset{\sim}{B}}(x) = \begin{cases} 2(x-1)^2 & ; \quad 1 \le x < 3/2 \\ 1 - 2(x-2)^2 & ; \quad 3/2 \le x < 5/2 \\ 2(x-3)^2 & ; \quad 5/2 \le x \le 3 \\ 0 & ; \quad \text{otherwise} \end{cases}$$

and show that it also represents 'real number *close* to 2'. Determine its support, and α-cut for $\alpha = 0.5$.

P6.2: (a) The well-known *Gaussian distribution* in probability is defined by

$$f(x) = \frac{1}{\sigma\sqrt{2\pi}} e^{-\frac{1}{2}\left(\frac{x-\mu}{\sigma}\right)^2} \quad ; -\infty < x < \infty$$

where μ is the mean and σ is the standard deviation of the distribution. Construct a normal, convex membership function from this distribution (select parameters μ and σ) that represents 'real numbers *close* to 2'. Find its support, and α-cut for $\alpha = 0.5$. Show that the membership function

$$\mu_{\underset{\sim}{A}}(x) = \frac{1}{1 + (x-2)^2}$$

also represents 'real numbers *close* to 2'. Find its support and α-cut for $\alpha = 0.5$.

P6.3: Consider the piecewise quadratic function

$$f(x) = \begin{cases} 0 & ; \quad 1 \le x < 3/2 \\ 2\left(\dfrac{x-a}{b-a}\right)^2 & ; \quad a \le x < \dfrac{a+b}{2} \\ 1 - 2\left(\dfrac{x-b}{b-a}\right)^2 & ; \quad \dfrac{a+b}{2} \le x < b \\ 1 & ; \quad b \le x < c \end{cases}$$

Construct a normal, convex membership function from $f(x)$ (select parameters a, b and c) that represents the set 'tall men' on the universe $\{3, 9\}$. Determine the crosspoints and support of the membership function.

P6.4: (a) Writer an analytical expression for the membership function $\mu_{\underset{\sim}{A}}(x)$ with supporting interval $[-1, 9]$ and α-cut interval for $\alpha = 1$ given as $[4, 5]$.

(b) Define what we mean by a normal membership function and a convex membership function. Is the function described in (a) above (i) normal, (ii) convex?

P6.5: (a) Let the fuzzy set A be the linguistic 'warm' with membership function

$$\mu_{\underset{\sim}{A}}(x) = \begin{cases} 0 & ; \quad x < a_1 \\ \dfrac{x-a_1}{b_1-a_1} & ; \quad a_1 \le x \le b_1 \\ 1 & ; \quad b_1 \le x \le b_2 \\ \dfrac{x-a_2}{b_2-a_2} & ; \quad b_2 \le x \le a_2 \\ 0 & ; \quad x \ge a_2 \end{cases}$$

$$a_1 = 64°\text{F},\ b_1 = 70°\text{F},\ b_2 = 74°\text{F},\ a_2 = 78°\text{F}$$

(i) Is $\underset{\sim}{A}$ a normal fuzzy set?

(ii) Is $\underset{\sim}{A}$ a convex fuzzy set?

(iii) Is $\underset{\sim}{A}$ a singleton fuzzy set?

If answer to one or more to these is 'no', then given an example of such a set.

(b) For fuzzy set $\underset{\sim}{A}$ described in part (a), assume that $b_1 = b_2 = 72°\text{F}$.

Sketch the resulting membership function and determine its support, crosspoints and α-cuts for $\alpha = 0.2$ and 0.4.

P6.6: Consider two fuzzy sets $\underset{\sim}{A}$ and $\underset{\sim}{B}$; membership functions $\mu_{\underset{\sim}{A}}(x)$ and $\mu_{\underset{\sim}{B}}(x)$ are shown in Fig. P6.6.

The fuzzy variable x is temperature.

Sketch the graph of $\mu_{\bar{\underset{\sim}{A}}}(x)$, $\mu_{\underset{\sim}{A} \cap \underset{\sim}{B}}(x)$ and $\mu_{\underset{\sim}{A} \cup \underset{\sim}{B}}(x)$.

Which t-norm and t-conorm have you used?

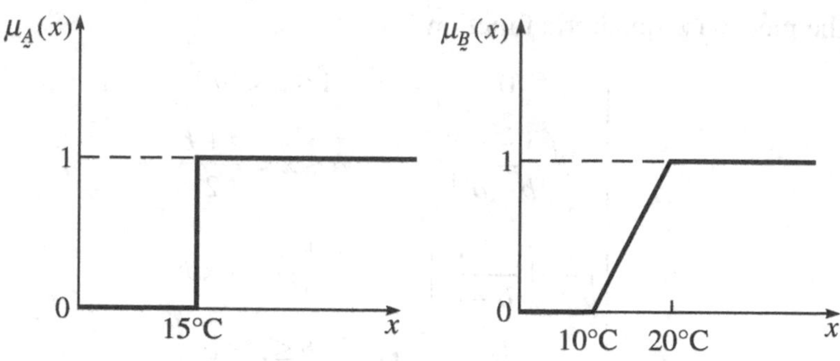

Figure P6.6

P6.7: Assume the membership function of the fuzzy set $\underset{\sim}{A}$, *big pressure*, is

$$\mu_{\underset{\sim}{A}}(x) = \begin{cases} 1 & ; \quad x \geq 5 \\ 1 - \dfrac{5-x}{4} & ; \quad 1 \leq x \leq 5 \\ 0 & ; \quad \text{otherwise} \end{cases}$$

Assume the membership function of the fuzzy set $\underset{\sim}{B}$, *small volume*, is

$$\mu_{\underset{\sim}{B}}(y) = \begin{cases} 1 & ; \quad y \leq 1 \\ 1 - \dfrac{y-1}{4} & ; \quad 1 \leq y \leq 5 \\ 0 & ; \quad \text{otherwise} \end{cases}$$

Find the truth values of the following propositions:

(i) 4 is big pressure.
(ii) 3 is small volume.
(iii) 4 is big pressure AND 3 is small volume.
(iv) 4 is big pressure → 3 is small volume.

Explain the conjunction and implication operations you have used for this purpose.

P6.8: Consider a fuzzy relation $\underset{\sim}{R}$ described by the relationship '*x* is approximately equal to *y*' by means of the following membership grade:

$$\mu_{\underset{\sim}{R}}(x, y) = max\,(1 - 0.5|x - y|, 0)$$

(i) Determine $\mu_{\underset{\sim}{A}}(x)$ for the fuzzy set $\underset{\sim}{A}$ = "approximately 5". Show the membership plot for $\underset{\sim}{A}$.
(ii) Show that the membership value

$$\mu_{\underset{\sim}{R}}(x, y) = e^{-(x-y)^2}$$

also describes the relation '*x* is approximately equal to *y*'.

P6.9: Two fuzzy sets 'young man' and 'tall man' are given in Fig. P6.9, defined on different universes of discourse 'age' and 'height'. This figure also shows a discretization of the fuzzy sets. Using MIN operator for intersection, determine a relation $\underset{\sim}{R}$ (find $\mu_{\underset{\sim}{R}}$ (age, height)) for the concept 'young and tall' man.

Graphically obtain a surface which represents the membership function $\mu_{\underset{\sim}{R}}$ (age, height) of the relation given by the relation matrix $\underset{\sim}{R}$ (you may use MATLAB or any other software tool).

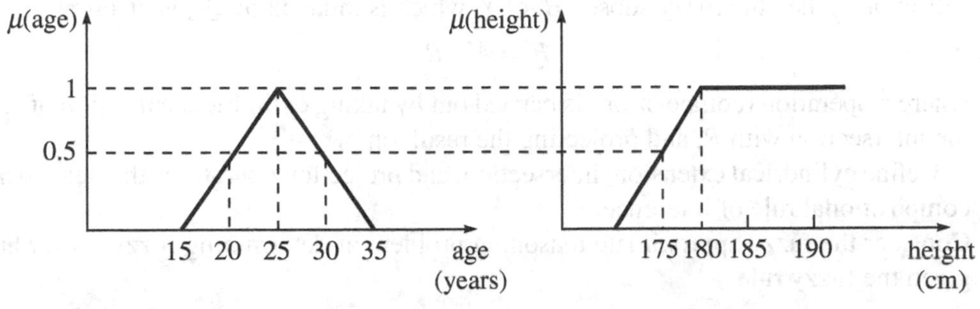

Figure P6.9

P6.10: Describe the three operations: cylindrical extension, intersection, and projection, involved in composition of fuzzy relations.

P6.11: The following function was used to calculate membership values for the set 'healthy'; x stands for BMI (Body Mass Index) values. A membership value of 1 is *healthy*, a membership value of 0 is *not-healthy*, and a membership value between 0 and 1 is the degree of membership in the healthy set.

$$\text{healthy}(x) = \begin{cases} 0 & ; \text{ if } x < 18 \\ (x-18)/2 & ; \text{ if } 18 \le x \le 20 \\ 1 & ; \text{ if } 20 < x < 25 \\ (27-x)/2 & ; \text{ if } 25 \le x \le 27 \\ 0 & ; \text{ if } x > 27 \end{cases}$$

(a) Draw the graphic for the fuzzy set 'healthy'.
(b) What is the degree of membership to the set 'healthy' for a person having BMI of 26.2?

P6.12: In Example 6.8 in Chapter 6, $\underset{\sim}{R}$ represents the relation between height in cm and weight in kg for a 'healthy male adult'. Assume your height is 160 cm. Transform this crisp value to a discrete fuzzy set $\underset{\sim}{A}$, and then using *max-min* composition operator, determine a fuzzy set that gives the weight possibility distribution for you to be a healthy male adult.

P6.13: Consider the following statements:

Input	:	$\underset{\sim}{A}'$ is very small
Rule	:	IF $\underset{\sim}{A}'$ is small THEN $\underset{\sim}{B}'$ is large
Inference	:	$\underset{\sim}{B}'$ is very large

If $\underset{\sim}{R}$ is a fuzzy relation from X to Y representing the implication rule, and $\underset{\sim}{A}'$ is a fuzzy subset of X, then the fuzzy subset $\underset{\sim}{B}'$ of Y, which is induced by $\underset{\sim}{A}'$, is given by

$$\underset{\sim}{B}' = \underset{\sim}{A}' \circ \underset{\sim}{R}$$

where o operation (composition) is carried out by taking cylindrical extension of $\underset{\sim}{A}'$, taking the intersection with $\underset{\sim}{R}$, and projecting the result onto Y.

Define cylindrical extension, intersection, and projection operations that lead to *max-min* compositional rule of inference.

P6.14: Consider the fuzzy approximate reasoning problem of determining fuzzy set for humidity, given the fuzzy rule

"IF temperature (x) is high THEN humidity (y) is fairly high"

and the input fuzzy set "temperature is fairly high".

$\underset{\sim}{A} =$

x	20	30	40
$\mu_{\underset{\sim}{A}}(x)$	0.1	0.5	0.9

= 'temperature is high'

$\underset{\sim}{B} =$

y	20	50	70	90
$\mu_{\underset{\sim}{B}}(y)$	0.2	0.6	0.7	1

= 'humidity is fairly high'

$\underset{\sim}{A}' =$

x	20	30	40
$\mu_{\underset{\sim}{A}'}(x)$	0.01	0.25	0.81

= 'temperature is fairly high'

Determine fuzzy set $\underset{\sim}{B}'$ for humidity.

P6.15: Describe the structure of a fuzzy rule-based system. List and define the five basic components of this learning system.

P6.16: Given fuzzy inputs, a fuzzy inference system produces fuzzy output. Therefore, we require fuzzification of crisp inputs, and defuzzification of inferred fuzzy set to obtain crisp output from fuzzy set.

Describe widely-used methods for fuzzification and defuzzification.

P6.17: Figure P6.17 shows the fuzzy output of a certain control problem. Defuzzify by using the center of area method, to obtain the value of crisp control action.

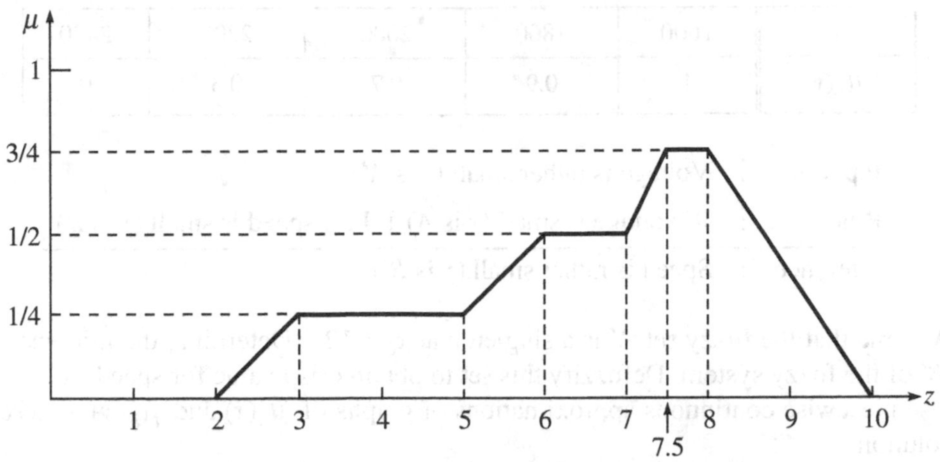

Figure P6.17

P6.18: In fuzzy inference system based on Mamdani model, the underlying nonlinearity for decision-making is shaped heuristically in the design process. Describe the steps involved in the design process, through an example.

P6.19: Find and present graphically the output fuzzy set for the single-input single-output system in Fig. P6.19, described by following two rules:

Rule 1 : IF x is *small* THEN y is *high*

Rule 2 : IF x is *medium* THEN y is *medium*

For the input $x_0 = 20$, find the crisp value y'.

(i) Give the results for different defuzzification methods.

(ii) Give the results for *min* and *product* implication operators.

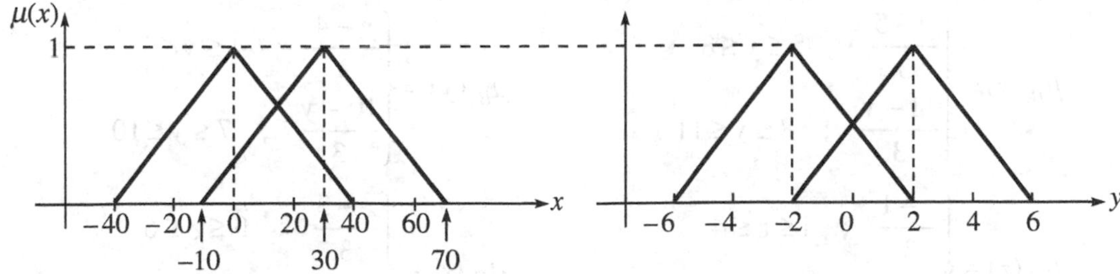

Figure P6.19

P6.20: Consider the fuzzy system concerning the terminal voltage and speed of an electric motor, described by the membership functions

x	100	150	200	250	300
$\mu_{\underset{\sim}{A}}(x)$	1	0.8	0.5	0.2	0.1

y	1600	1800	2000	2200	2400
$\mu_{\underset{\sim}{B}}(y)$	1	0.9	0.7	0.3	0

Input	:	Voltage is rather small (x is $\underset{\sim}{A}'$)
Rule	:	IF voltage is small (x is $\underset{\sim}{A}$) THEN speed is small (y is $\underset{\sim}{B}$)
Inference	:	Speed is rather small (y is $\underset{\sim}{B}'$)

Assume that the fuzzy set $\underset{\sim}{A}'$ is a singleton at $x_0 = 125$. Determine the inference fuzzy set $\underset{\sim}{B}'$ of the fuzzy system. Defuzzify this set to obtain crisp value for speed.
Use piecewise continuous approximations of graphs of $\mu_{\underset{\sim}{A}}(x)$ and $\mu_{\underset{\sim}{B}}(y)$ to describe your solution.

P6.21: Consider the two-input, one-output fuzzy system:

Input	:	x is $\underset{\sim}{A}'$ AND y is $\underset{\sim}{B}'$
Rule 1	:	IF x is $\underset{\sim}{A}_1$ AND y is $\underset{\sim}{B}_1$ THEN z is $\underset{\sim}{C}_1$
Rule 2	:	IF x is $\underset{\sim}{A}_2$ AND y is $\underset{\sim}{B}_2$ THEN z is $\underset{\sim}{C}_2$
Inference	:	z is $\underset{\sim}{C}'$

The fuzzy sets $\underset{\sim}{A}_i$, $\underset{\sim}{B}_i$ and $\underset{\sim}{C}_i$; $i = 1, 2$, have the membership functions

$$\mu_{\underset{\sim}{A}_1}(x) = \begin{cases} \dfrac{x-2}{3} & ; \ 2 \le x \le 5 \\ \dfrac{8-x}{3} & ; \ 5 \le x \le 8 \end{cases} \qquad \mu_{\underset{\sim}{A}_2}(x) = \begin{cases} \dfrac{x-3}{3} & ; \ 3 \le x \le 6 \\ \dfrac{9-x}{3} & ; \ 6 \le x \le 9 \end{cases}$$

$$\mu_{\underset{\sim}{B}_1}(y) = \begin{cases} \dfrac{y-5}{3} & ; \ 5 \le y \le 8 \\ \dfrac{11-y}{3} & ; \ 8 \le y \le 11 \end{cases} \qquad \mu_{\underset{\sim}{B}_2}(y) = \begin{cases} \dfrac{y-4}{3} & ; \ 4 \le y \le 7 \\ \dfrac{10-y}{3} & ; \ 7 \le y \le 10 \end{cases}$$

$$\mu_{\underset{\sim}{C}_1}(z) = \begin{cases} \dfrac{z-1}{3} & ; \ 1 \le z \le 4 \\ \dfrac{7-z}{3} & ; \ 4 \le z \le 7 \end{cases} \qquad \mu_{\underset{\sim}{C}_2}(z) = \begin{cases} \dfrac{z-3}{3} & ; \ 3 \le z \le 6 \\ \dfrac{9-z}{3} & ; \ 6 \le z \le 9 \end{cases}$$

Assume fuzzy sets $\underset{\sim}{A}'$ and $\underset{\sim}{B}'$ are singletons at $x_0 = 4$ and $y_0 = 8$. Determine the inference fuzzy set $\underset{\sim}{C}'$ of the fuzzy system. Defuzzify $\underset{\sim}{C}'$.

P6.22: The control objective is to design an automatic braking system for motor cars. We need two analog signals: vehicle speed (V), and a measure of distance (D) from the vehicle in the front.

A fuzzy logic control system will process these, giving a single output: braking force (B), which controls the brakes.

Term set for each of the variables (V, D, and B) is of the form:

{PS (positive small), PM (positive medium), PL (positive large)}

Membership functions for each term-set are given in Fig. P6.22

Suppose that for the control problem, two rules have to be fired:

Rule 1: If $D = PS$ AND $V = PM$ THEN $B = PL$

Rule 2: If $D = PM$ AND $V = PL$ THEN $B = PM$

For the sensor readings of $V = 55$ km/hr, and $D = 27$m from the car in front, find graphically

(i) the firing strengths of the two rules;
(ii) the aggregated output; and
(iii) defuzzified control action.

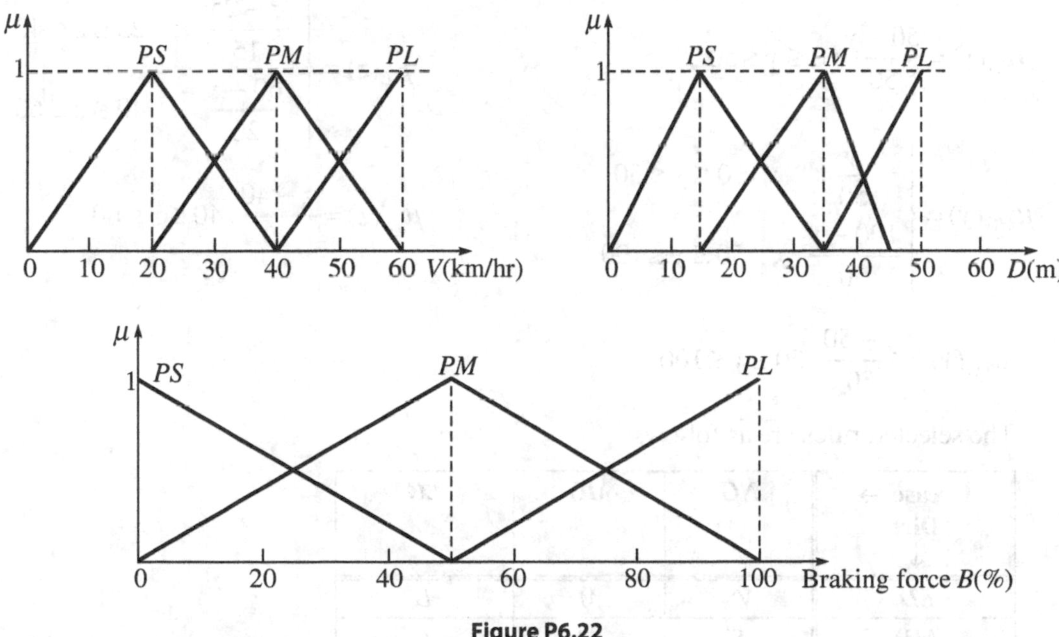

Figure P6.22

P6.23: The control objective is to automate the *wash time* when using a washing machine. Experts select for inputs *dirt* and *grease* of the clothes to be washed, and for output parameter the *wash time*, as follows:

Dirt $(x) \triangleq$ {SD (small dirt), MD (medium dirt), LD (large dirt)}

Grease $(y) \triangleq$ {NG (no grease), MG (medium grease), LG (large grease)}

Wash time $(z) \triangleq$ {VS (very short), S(short), M (medium), L (long), VL (very long)}

The degrees of the dirt and grease are measured on a scale from 0 to 100; wash time is measured in minutes from 0 to 60.

$$\mu_{\underset{\sim}{SD}}(x) = \frac{50-x}{50}; 0 \le x \le 50$$

$$\mu_{\underset{\sim}{MD}}(x) = \begin{cases} \dfrac{x}{50} & ; \quad 0 \le x \le 50 \\ \dfrac{100-x}{50} & ; \quad 50 \le x \le 100 \end{cases}$$

$$\mu_{\underset{\sim}{LD}}(x) = \frac{x-50}{50}; 50 \le x \le 100$$

$$\mu_{\underset{\sim}{NG}}(y) = \frac{50-y}{50}; 0 \le y \le 50$$

$$\mu_{\underset{\sim}{MG}}(y) = \begin{cases} \dfrac{y}{50} & ; \quad 0 \le y \le 50 \\ \dfrac{100-y}{50} & ; \quad 50 \le y \le 100 \end{cases}$$

$$\mu_{\underset{\sim}{LG}}(y) = \frac{y-50}{50}; 50 \le y \le 100$$

$$\mu_{\underset{\sim}{VS}}(z) = \frac{10-z}{10}; 0 \le z \le 10$$

$$\mu_{\underset{\sim}{S}}(z) = \begin{cases} \dfrac{z}{10} & ; \quad 0 \le z \le 10 \\ \dfrac{25-z}{15} & ; \quad 10 \le z \le 25 \end{cases}$$

$$\mu_{\underset{\sim}{M}}(z) = \begin{cases} \dfrac{z-10}{15} & ; \quad 10 \le z \le 25 \\ \dfrac{40-z}{15} & ; \quad 25 \le z \le 40 \end{cases}$$

$$\mu_{\underset{\sim}{L}}(z) = \begin{cases} \dfrac{z-25}{15} & ; \quad 25 \le z \le 40 \\ \dfrac{60-z}{20} & ; \quad 40 \le z \le 60 \end{cases}$$

$$\mu_{\underset{\sim}{VL}}(z) = \frac{z-40}{20}; 40 \le z \le 60$$

The selected rules are as follows:

Grease → **Dirt** **↓**	*NG*	*MG*	*LG*
SD	*VS*	*M*	*L*
MD	*S*	*M*	*L*
LD	*M*	*L*	*VL*

Find a crisp control output for the following sensor readings:

$$Dirt = 60; \; Grease = 70$$

P6.24: A fuzzy controller is acting according to the following rule base ($N = negative$, $M = medium$, $P = positive$):

R_1 : If x_1 is N AND x_2 is N THEN u is N

R_2 : If x_1 is N OR x_2 is P THEN u is M

R_3 : If x_1 is P OR x_2 is N THEN u is M

R_4 : If x_1 is P AND x_2 is P THEN u is P

The membership functions of the input and output variables are given in Fig. P6.24. Actual inputs are $x_1 = 2.5$ and $x_2 = 4$. Which rules are active and what will be the controller action u? Find u by applying standard fuzzy operations: *min* for AND, and *max* for OR.

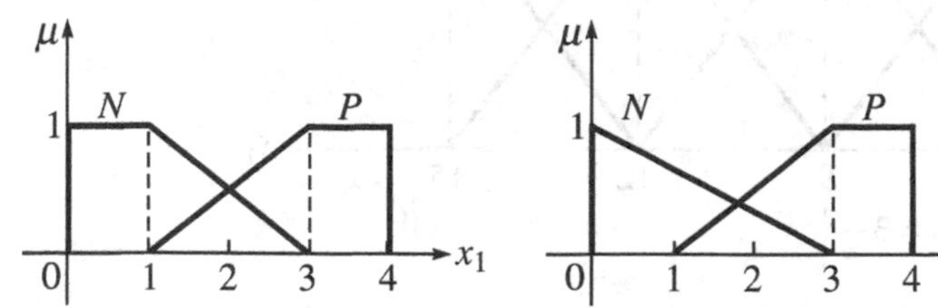

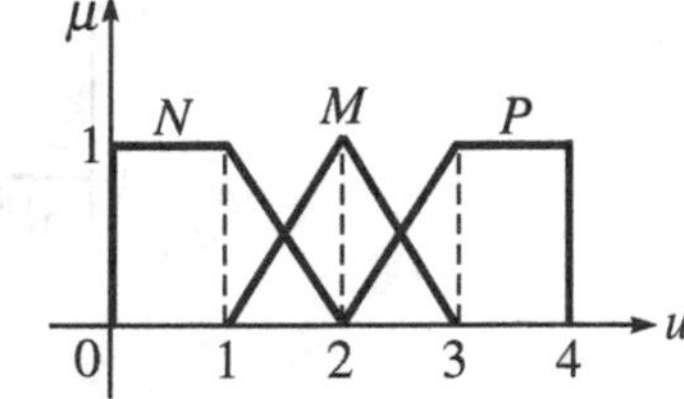

Figure P6.24

P6.25: Consider the following fuzzy model of a system with inputs x and y and output z:

Rule 1 : If x is A_3 OR y is B_1 THEN z is C_1
Rule 2 : If x is A_2 AND y is B_2 THEN z is C_2
Rule 3 : If x is A_1 THEN z is C_3

The membership functions of the input and output variables are given in Fig. P6.25. Actual inputs are x_1 and y_1. Find the output z by applying standard fuzzy operation: *min* for AND, and *max* for OR.

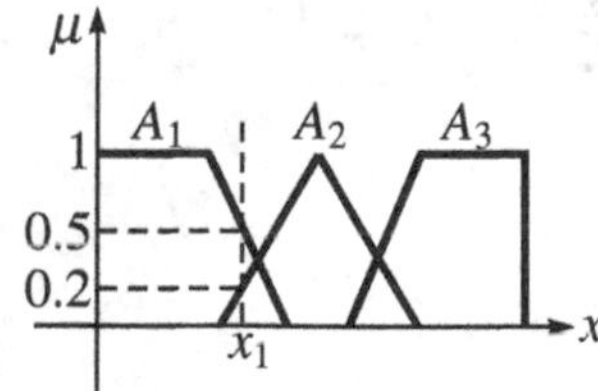

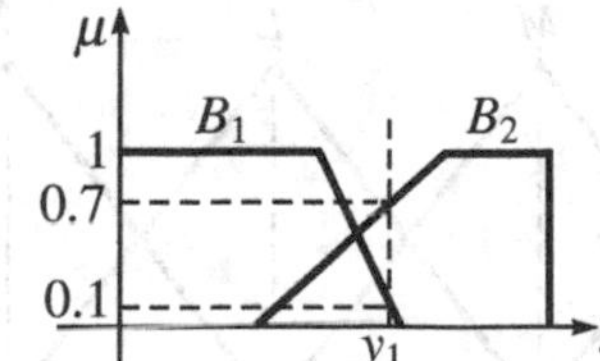

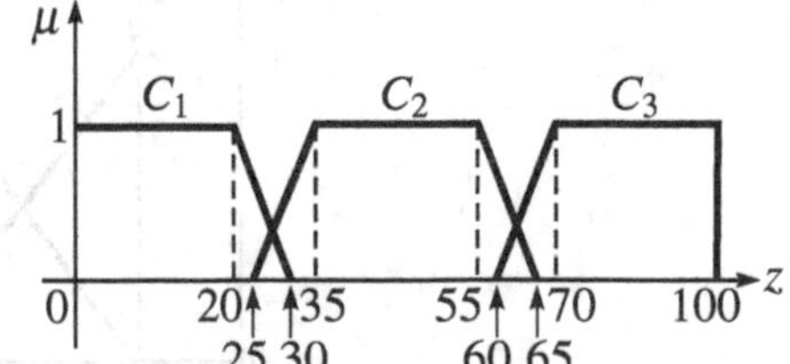

Figure P6.25

P6.26: A fuzzy controller is acting according to the following rule base (N = *negative*, P = *positive*):

R_1 : If x_1 is N AND x_2 is N THEN u is k_1
R_2 : If x_1 is N OR x_2 is P THEN u is k_2
R_3 : If x_1 is P OR x_2 is N THEN u is k_2
R_4 : If x_1 is P AND x_2 is P THEN u is k_3

The membership functions of the input variables are given in Fig. P6.24 and the membership functions of the output variable (which is a controller action) u are singletons placed at $k_1 = 1, k_2 = 2, k_3 = 3$. Actual inputs are $x_1 = 2.5$ and $x_2 = 4$. Find u by applying standard fuzzy operations: *min* for AND, and *max* for OR.

P6.27: Consider a Mamdani fuzzy model for a manufacturing process. The process is characterized by two input variables, x_1 and x_2, and one output variable y. The membership function distribution (isosceles triangles of base widths θ_1, θ_2, θ_3) of x_1, x_2, and y are shown in Fig. P6.27, and a rule base is given in Table P6.27. Determine the output of the model for $x_1 = 10$, $x_2 = 28$.

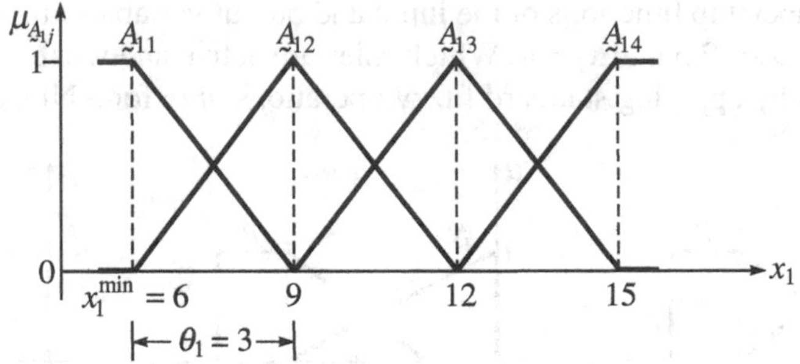

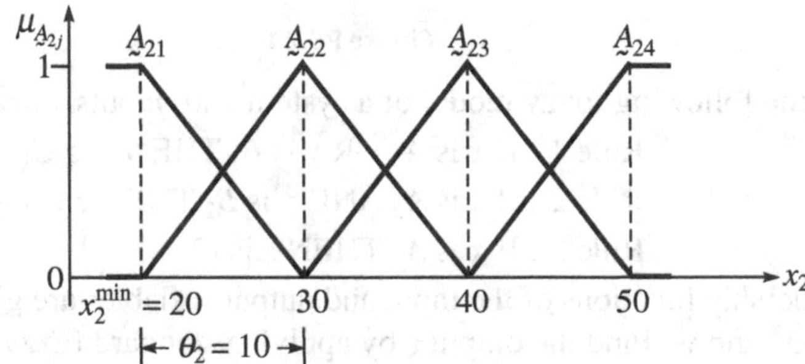

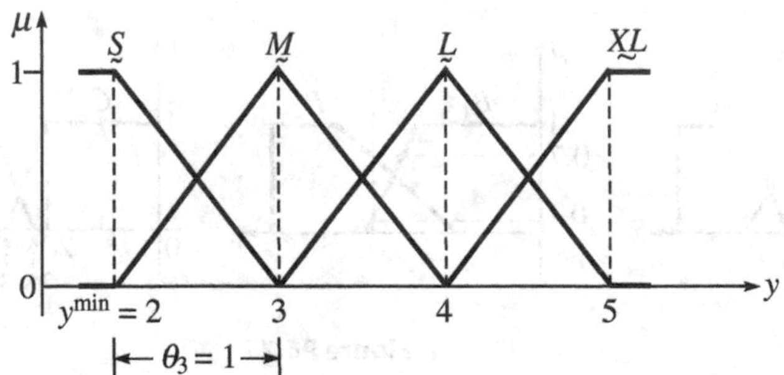

Figure P6.27

Table P6.27

$x_2 \rightarrow$ / $x_1 \downarrow$	A_{21}	A_{22}	A_{23}	A_{24}
A_{11}	S	S	M	L
A_{12}	S	M	L	L
A_{13}	M	M	L	XL
A_{14}	M	L	XL	XL

P6.28: We are given a thermal process shown in Fig. P6.28(a), where θ_i is the perturbation (from steady-state operating condition) in temperature of the liquid entering the insulated chamber, θ is the perturbation in temperature of the liquid leaving the chamber. The heating element has a knob to control the steam for circulation through the radiator. The higher the setting of the knob the hotter the fluid gets; the setting '0' of the knob indicating the steam supply for steady-state operating condition. The desired value for the variable θ is $\theta_d = 0$.

The control system we use is shown in Fig. P6.28(b). Fuzzy logic controller (FLC) inputs are error e and error-rate $\dot{e}$, and output is change-in-control Δu (change in heat knob setting); with input membership functions shown in Fig. P6.28(c) and output membership functions shown in Fig. P6.28(d), where *P, Z, N* denote Positive, Negative and Zero, respectively. We use nine rules in the rule-base given by Table 6.1 in Chapter 6. Furthermore, we use *min* operator to quantify the premise and implication; singleton fuzzification; and COG defuzzification.

Assume that the input variables have the following crisp values:

$$e = 15°\text{C}; \; \dot{e} = 4°\text{C/sec}$$

(a) Use Mamdani model and find the individual implied fuzzy sets. Show the overall implied fuzzy set graphically.
(b) Defuzzify using COG defuzzification method.

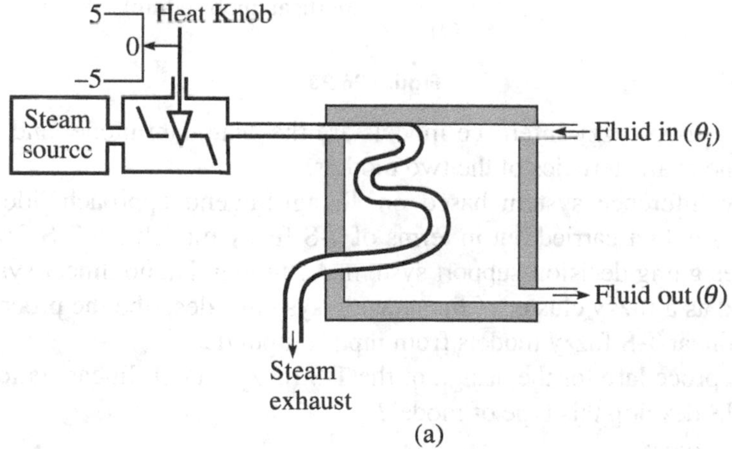

(a)

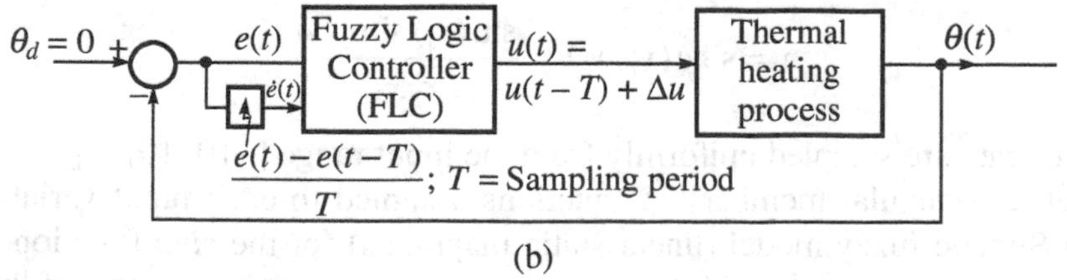

(b)

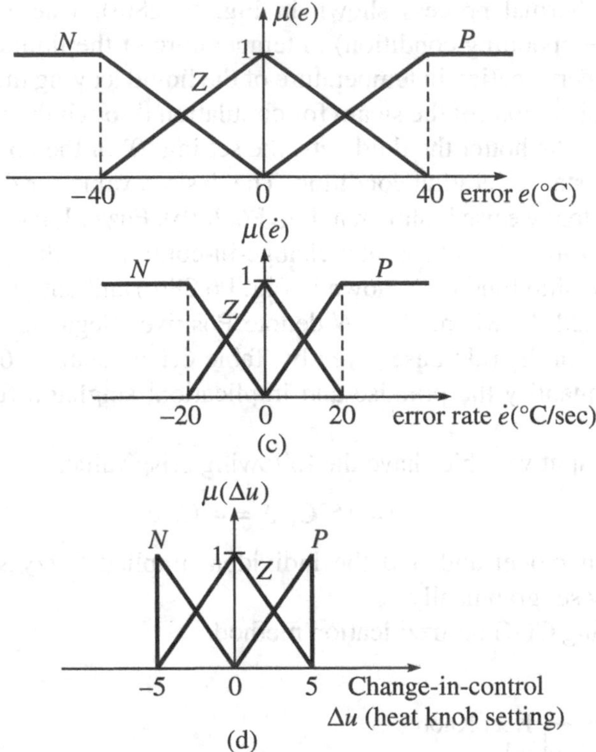

Figure P6.28

P6.29: Two widely-used fuzzy inference models are the Mamdani model and T-S fuzzy model. Describe the characteristics of the two models.

P6.30: In a fuzzy inference system based on Takagi-Sugeno approach, identification of the nonlinearity is first carried out in terms of T-S fuzzy model; the T-S fuzzy model is then used for designing decision-support system. Assuming that nonlinear system is effectively represented as a fuzzy cluster of linear static systems, describe the procedure of obtaining cluster of linear T-S fuzzy models from input-output data.

P6.31: Describe a procedure for the design of the T-S fuzzy model (linear static mappings). How does ANFIS develop this type of model?

P6.32: *Software exercise*

Consider a two-dimensional *sinc* equation defined by

$$y = \text{sinc}(x_1, x_2) = \frac{\sin(x_1)\sin(x_2)}{x_1 x_2}$$

Training data are sampled uniformly from the input range [−10, 10] × [−10, 10]. With two symmetric triangular membership functions assigned to each input variable, construct a Takagi-Sugeno fuzzy model (linear static mappings) for the *sinc* function. Give defining equations for determination of the premise and consequent parameters of the model.

P6.33: *Software exercise*

To identify the nonlinear system

$$y = (1 + (x_1)^{0.5} + (x_2)^{-1} + (x_3)^{-1.5})^2$$

we assign two membership functions to each input variable. Training and testing data are sampled uniformly from the input ranges $[1, 6] \times [1, 6] \times [1, 6]$ and $[1.5, 5.5] \times [1.5, 5.5] \times [1.5, 5.5]$, respectively. Extract Takagi-Sugeno fuzzy rules from the numerical input-output training data that could be employed in an ANFIS model.

P6.34: Consider a T-S fuzzy model for a manufacturing process. The process is characterized by two input variables, x_1 and x_2, and one output variable, y. The membership function distributions of x_1 and x_2 are shown in Fig. P6.34. Domain intervals of x_i are divided into $K_i = 3$ fuzzy sets. Therefore, there is a maximum of $K_1 \times K_2 = 9$ feasible rules. The output of the rth rule is expressed as

$$\hat{y}^{(r)} = a_j^{(r)} x_1 + b_k^{(r)} x_2$$

where j, $k = 1, 2, 3$; $a_1^{(r)} = 1$, $a_2^{(r)} = 2$ and $a_3^{(r)} = 3$ if x_1 is found to be $\underset{\sim}{A}_{11}$, $\underset{\sim}{A}_{12}$, and $\underset{\sim}{A}_{13}$, respectively; $b_1^{(r)} = 1$, $b_2^{(r)} = 2$, $b_3^{(r)} = 3$ if x_2 is found to be $\underset{\sim}{A}_{21}$, $\underset{\sim}{A}_{22}$, and $\underset{\sim}{A}_{23}$, respectively. Determine the output of the model if $x_1 = 6.0$ and $x_2 = 2.2$.

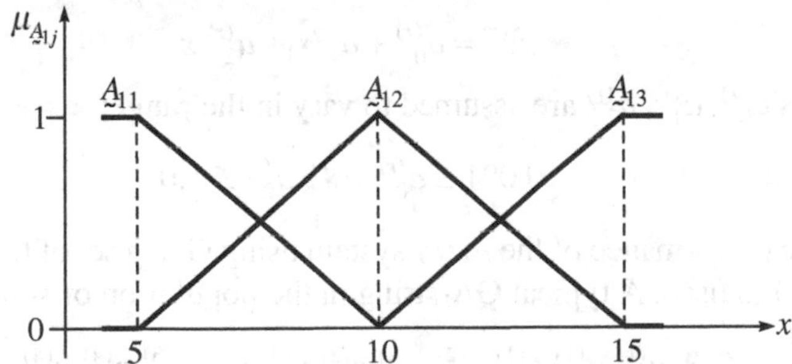

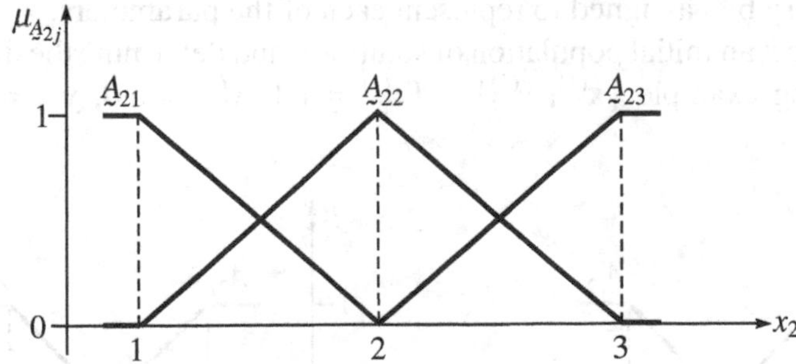

Figure P6.34

P6.35: Assume that a fuzzy inference system has two inputs x_1 and x_2, and one output y. The rule-base contains two T-S fuzzy rules as follows:

Rule 1 : If x_1 is $\underset{\sim}{A}_{11}$ AND x_2 is $\underset{\sim}{A}_{21}$ THEN $y^{(1)} = a_0^{(1)} + a_1^{(1)} x_1 + a_2^{(1)} x_2$

Rule 2 : If x_1 is $\underset{\sim}{A}_{12}$ AND x_2 is $\underset{\sim}{A}_{22}$ THEN $y^{(2)} = a_0^{(2)} + a_1^{(2)} x_1 + a_2^{(2)} x_2$

$\underset{\sim}{A}_{ij}$ are Gaussian functions.

For given input values x_1 and x_2, the inferred output is calculated by

$$\hat{y} = \frac{\mu^{(1)}y^{(1)} + \mu^{(2)}y^{(2)}}{\mu^{(1)} + \mu^{(2)}}$$

where $u^{(r)}, r = 1, 2$ are firing strengths of the two rules. Product inference is used to calculate the firing strengths of the rules.

Develop ANFIS architecture for this modeling problem, and derive learning algorithms based on least squares estimation and the gradient-descent methods.

P6.36: A fuzzy logic-based expert system is to be developed that will work based on T-S fuzzy model architecture to predict the output of a process. The database of the fuzzy system is shown in Fig. P6.36; x_1 and x_2 are two inputs with specified minimum values $x_1^{\min}$ and $x_2^{\min}$, respectively. The base-widths θ_1 and θ_2 are assumed to vary in the ranges:

$$0.8 \le \theta_1 \le 1.5; 4.0 \le \theta_2 \le 6.0$$

There are is a maximum of $R = 4$ feasible rules; the output of rth rule ($r = 1,2, ..., R$) is expressed as follows:

$$\hat{y}^{(r)} = a_0^{(r)} + a_1^{(r)}x_1 + a_2^{(r)}x_2$$

The parameters $a_0^{(r)}, a_1^{(r)}, a_2^{(r)}$ are assumed to vary in the range:

$$0.001 \le a_0^{(r)}, a_1^{(r)}, a_2^{(r)} \le 1.0$$

To optimize the performance of the fuzzy system using GA, a set of training examples $\{\mathbf{x}^{(i)}, y^{(i)}; i = 1, ..., N\}$ is used. A typical GA-string in the population of solutions is of the form:

$$\{\theta_1 \theta_2\, a_0^{(1)}\, a_1^{(1)}\, a_2^{(1)}\, a_0^{(2)}\, a_1^{(2)}\, a_2^{(2)}\, a_0^{(3)}\, a_1^{(3)}\, a_2^{(3)}\, a_0^{(4)}\, a_1^{(4)}\, a_2^{(4)}\}$$

with four binary bits assigned to represent each of the parameters.

Randomly select an initial population of solutions, and determine the deviation in prediction for the training example $\{\mathbf{x}^{(1)}, y^{(1)}\} = \{x_1^{(1)} = 1.1, x_2^{(1)} = 6.0, y^{(1)} = 5.0\}$ using the first GA-string.

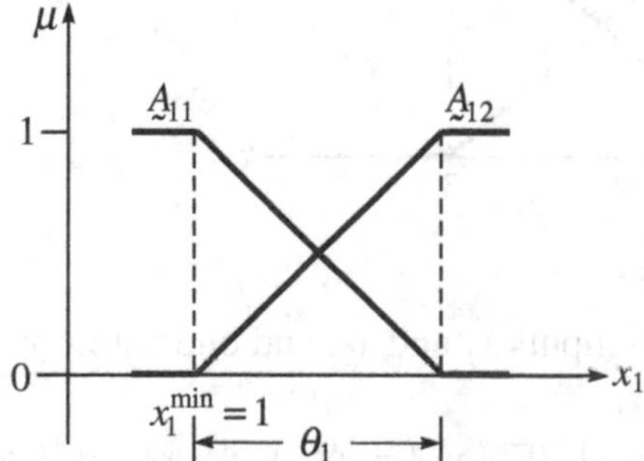

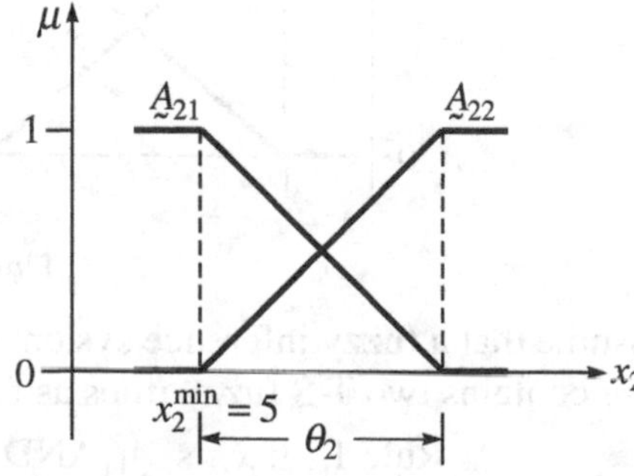

Figure P6.36

P6.37: A fuzzy logic-based expert system is to be developed that will work based on Mamdani model to predict the output of a process. The database of the fuzzy system is shown in Figs P6.36 and P6.37; x_1 and x_2 are two inputs with specified minimum values x_1^{min} and x_2^{min}, respectively, and y is the output with specified minimum value y^{min}. The base-widths θ_1, θ_2 and θ_3 of these isosceles triangles are tunable. The ranges of the tunable parameters are assumed to be

$$0.8 \le \theta_1 \le 1.5;\ 4.0 \le \theta_2 \le 6.0,;\ 0.5 \le \theta_3 \le 3$$

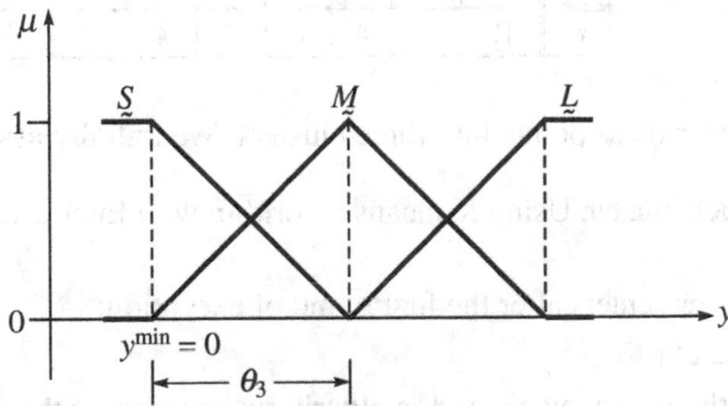

Figure P6.37

The Rule Base of the fuzzy system is given in Table P6.37

Table P6.37

$x_2 \rightarrow$ / $x_1 \downarrow$	$\underset{\sim}{A}_{21}$	$\underset{\sim}{A}_{22}$
$\underset{\sim}{A}_{11}$	$\underset{\sim}{S}$	$\underset{\sim}{M}$
$\underset{\sim}{A}_{12}$	$\underset{\sim}{M}$	$\underset{\sim}{L}$

To optimize the performance of the fuzzy system using GA, a set of training examples $\{\mathbf{x}^{(i)}, y^{(i)}; i = 1, ..., N\}$ is used. A typical GA-string in the population of solutions is of the form

$$\{\theta_1, \theta_2, \theta_3\}$$

with four binary bits assigned to represent each of the parameters.

Randomly select an initial population of solutions, and determine the deviation in prediction for the training example $\{\mathbf{x}^{(1)}, y^{(1)}\} = \{x_1^{(1)} = 1.1, x_2^{(1)} = 6.0, y^{(1)} = 5.0\}$ using the first GA-string.

P7.1: *K*-means clustering method is 'hard' in that it requires each object to belong exclusively to only one cluster. Describe two 'soft' clustering methods wherein *K*-means is a special case of the methods. Explain why *K*-means is a special case.

P7.2: Clustering is recognized as an important machine learning tool with broad applications. Give one application example for each of the following:

(a) Using clustering as a preprocessing tool for data preparation

(b) Using clustering as a major tool for decision-making

P7.3: Given the data points:

x_1	2	2	8	5	7	6	1	4
x_2	10	5	4	8	5	4	2	9

The task is to cluster these points into three clusters. We initially assign $\begin{bmatrix} 2 \\ 10 \end{bmatrix}$, $\begin{bmatrix} 5 \\ 8 \end{bmatrix}$ and $\begin{bmatrix} 1 \\ 2 \end{bmatrix}$ as the center of each cluster. Using *K*-means algorithm with Euclidean distance as similarity measure, find

(a) the three cluster centers after the first round of execution;

(b) the final three clusters.

P7.4: Describe each of the following clustering algorithms in terms of the following criteria:

(1) shapes of clusters that can be determined;

(2) input parameters that must be specified; and

(3) limitations.

(a) *K*-means

(b) Fuzzy *K*-means

(c) Gaussian Mixture Model

P7.5: Consider the dataset given in Table P7.5.

(a) The task is to cluster these data points into two clusters. We initially assign $[1\ \ 1\ \ 1]^T$, and $[-1\ \ 1\ \ -1]^T$ as the center of each cluster. Using *K*-means algorithm with Euclidean distance as similarity measure, find the three cluster centers.

(b) Now assign $[0\ \ 0\ \ 0]^T$ and $[1\ \ 1\ \ -1]^T$ as initial cluster centers and repeat part(a). Compare your result with that from part(a) and explain any differences, including the number of iterations for convergence.

Table 7.5

$s^{(i)}$	x_1	x_2	x_3
$s^{(1)}$	−7.82	−4.58	−3.97
$s^{(2)}$	−6.68	3.16	2.71
$s^{(3)}$	4.36	−2.19	2.09
$s^{(4)}$	6.72	0.88	2.80

(Contd.)

$s^{(5)}$	−8.64	3.06	3.50
$s^{(6)}$	−6.87	0.57	−5.45
$s^{(7)}$	4.47	−2.62	5.76
$s^{(8)}$	6.73	−2.01	4.18
$s^{(9)}$	−7.71	2.34	−6.33
$s^{(10)}$	−6.91	−0.49	−5.68
$s^{(11)}$	6.18	2.81	5.82
$s^{(12)}$	6.72	−0.93	−4.04
$s^{(13)}$	−6.25	−0.26	0.56
$s^{(14)}$	−6.94	−1.22	1.13
$s^{(15)}$	8.09	0.20	2.25
$s^{(16)}$	6.81	0.17	−4.15
$s^{(17)}$	−5.19	4.24	4.04
$s^{(18)}$	−6.38	−1.74	1.43
$s^{(19)}$	4.08	1.30	5.33
$s^{(20)}$	6.27	0.93	−2.78

P7.6: Repeat P7.5, but use instead fuzzy *K*-means algorithm with $m = 2$ (Table 7.2 in Chapter 7).

P7.7: Consider a problem in which the data $\mathcal{D}$ is a set of instances, assumed to be generated by a probability distribution that is a mixture of *K* distinct normal distributions. Each instance is generated using a two-step process. First, one of the *K* normal distributions are selected at random. Second, a single random instance $\mathbf{x}^{(i)}$ is generated according to this selected distribution. Consider a special case where the selection of the single normal distribution at each step is based on choosing each with uniform probability, where each of the *K* normal distributions has the same variance σ^2 and σ^2 is known.

Use EM algorithm to output a hypothesis that estimates the means of each of the *K* distributions in the Gaussian Mixture Model (GMM).

P7.8: *K*-means algorithm does hard partitioning but it is always better to do a soft partitioning so that instances in between two clusters can contribute to the parameters (the covariance matrix) of more than one cluster allowing a smooth transition between clusters.

Write the equations derived for EM algorithm, for estimating the means and covariances of the *K* normal distributions of Gaussian Mixture model (GMM). Show that this model uses Mahalanobis distance rather than Euclidean distance (which implies that features have the same scale and are independent), and hence taking care of differences in scale and dependencies.

P7.9: A numerical attribute *Temp* (°F) in a dataset, along with class labels of the dataset is given below.

75	80	85	72	69	72	83	64	81	71	65	75	68	70
+	–	–	–	+	+	+	+	+	–	–	+	+	+

For discretization of this data using entropy-based method, a cut-point at 75 ($\leq 75; > 75$) is to be examined. Calculate the Gain Ratio induced by this cut-point.

P7.10: Taking a suitable example of 10 data pairs (x_1, x_2), perform the following steps analytically:

Step 1: Subtract the mean

Step 2: Calculate the covariance matrix

Step 3: Calculate the eigenvalues and eigenvectors of the covariance matrix

Step 4: Choose principle components and form a feature vector

Step 5: Derive new dataset

P7.11: Consider now the dataset given in Table P7.11. Using an appropriate software tool, perform the five steps given in P7.10, and represent all the three-dimensional data in two dimensions. What are the eigenvectors and eigenvalues?

Table P7.11

x_1	7	4	6	8	8	7	5	9	7	8
x_2	4	1	3	6	5	2	3	5	4	2
x_3	3	8	5	1	7	9	3	8	5	2

P7.12: Reconsider the dataset given in P3.17. Use PCA for dimensionality reduction from two dimensions to one dimension. Confirm that classification performance on the reduced dataset is better using Fisher's Linear Discriminant analysis compared to PCA. Explain why so?

P7.13: A dealer of used cars has currently ten cars. The car dealer notes in his documents four features of each car: *number of doors, horsepower, color*, and *make*. A decision system built from this information system defines *car make* as the decision attribute, and other three features as condition attributes (Table P7.13).

(a) Using rough-set analysis, show that the decision attribute y depends on the condition attributes X to a degree $\gamma = 0.8$. It indicates that we cannot unambiguously infer on the membership of the objects of the space U to the condition-attribute set corresponding to class labels $q = 1, 2,$ and 3.

(b) A decision system by deleting objects 4 and 5 has now been created. Show that in this system, the decision attribute y depends on the condition attributes X to a degree $\gamma = 1$.

Table P7.13

Objects (U)	Number of doors (x_1)	Horsepower (x_2)	Color (x_3)	Make (y)
1	2	60	blue	Opel
2	2	100	black	Nissan
3	2	200	black	Ferrari
4	2	200	red	Ferrari
5	2	200	red	Opel
6	3	100	red	Opel
7	3	100	red	Opel
8	3	200	black	Ferrari
9	4	100	blue	Nissan
10	4	100	blue	Nissan

P7.14: The car dealer decided to add the *type of fuel used*, the *type of upholstery*, and *wheel rims*, to the features considered in P7.13. The new set of condition attributes is x_1, x_2, x_3, x_4, x_5, x_6 shown in Table P7.14.

(a) Obtain the dependency degree γ.

(b) Show that added attributes x_4, x_5, x_6 are of low significance.

Table P7.14

Objects (U)	Number of doors (x_1)	Horsepower (x_2)	Color (x_3)	Fuel (x_4)	Upholstery (x_5)	Rims (x_6)	Make (y)
1	2	60	blue	Ethyl gasoline	woven fabric	steel	Opel
2	2	100	black	Diesel oil	woven fabric	steel	Nissan
3	2	200	black	Ethyl gasoline	leather	Al	Ferrari
4	2	200	red	Ethyl gasoline	leather	Al	Ferrari
5	2	200	red	Ethyl gasoline	woven fabric	steel	Opel
6	3	100	red	Diesel oil	leather	steel	Opel
7	3	100	red	gas	woven fabric	steel	Opel
8	3	200	black	Ethyl gasoline	leather	Al	Ferrari
9	4	100	blue	gas	woven fabric	steel	Nissan
10	4	100	blue	Diesel oil	woven fabric	Al	Nissan

P7.15: Consider the dataset given in Table 7.4 of Chapter 7.

(a) Find the elementary sets *U*/IND(*Z*) for various subsets:

$$\{x_1,\}, \{x_2\}, \{x_3), \{x_1, x_2\}, \{x_2, x_3\}, \{x_1, x_2, x_3\}$$

(b) Find dependency degree of decision attribute *y* on these sets of condition attributes.

(c) Find the set *R* of all reducts, and the reduct $R_{\min}$ of minimal cardinality.

P7.16: Various methods have been reported in literature for data clustering. Out of them, an overview of widely-used ones has been given in this book.

Give an overview (your qualitative understanding) of the following clustering methods:

1. *K*-means
2. Fuzzy *K*-means
3. Probabilistic clusters
4. Hierarchical clustering
5. Spectral clustering
6. Clustering using Self-Organizing Maps

P7.17: Describe the steps involved in the design process (including relevant equations) for the following clustering methods:

1. *K*-means clustering algorithm
2. Fuzzy *K*-means algorithm
3. Gaussian mixtures clustering algorithm

P7.18: Data cleansing is an important data preprocessing requirement. Describe the major data problems that require cleansing operation.

P7.19: Discretization of numeric attributes in data matrix is a requirement for some machine learning algorithms. Describe various discretization methods known to you.

What are the merits of entropy-based data discretization? Describe this discretization method with the help of an example.

P7.20: Describe the basic features of the following attribute-reduction methods. Also outline the steps involved.

1. Principal Component Analysis (PCA)
2. Rough-set approach

P8.1: People decide to drive the car or take public transportation to go to work according to the weather and traffic situation. An example dataset is given in Table P8.1. The attributes are *Temperature* (x_1), *Wind* (x_2), and *Traffic-Jam* (x_3); and the target variable is *Car Driving Decision* (y).

Table P8.1

Day $s^{(i)}$	Temperature x_1	Wind x_2	Traffic-Jam x_3	Car Driving y
$s^{(1)}$	hot	weak	long	no
$s^{(2)}$	hot	strong	long	no
$s^{(3)}$	hot	weak	long	yes
$s^{(4)}$	mild	weak	long	yes
$s^{(5)}$	cool	weak	short	yes
$s^{(6)}$	cool	strong	short	no
$s^{(7)}$	cool	strong	short	yes
$s^{(8)}$	mild	weak	long	no
$s^{(9)}$	cool	weak	short	yes
$s^{(10)}$	mild	weak	short	yes
$s^{(11)}$	mild	strong	short	yes
$s^{(12)}$	mild	strong	long	yes
$s^{(13)}$	hot	weak	short	yes
$s^{(14)}$	mild	strong	long	no

The given dataset $\mathcal{D}$ has 14 examples altogether, including nine positive examples (car driving: yes), and five negative examples (car driving: no).

(a) Calculate the information gain for x_1, x_2, and x_3

(b) Choose the root node for the decision tree.

(c) Show a partial decision tree from the root node along with training examples sorted to each of its descendant node.

(d) You are required to continue with the decision-tree growing process till all nodes are pure (they contain examples that all have the same classification). Will it be possible to reach this situation for the given dataset? If not, why not? (**Hint:** Look for inconsistencies in the given dataset.)

P8.2: For the dataset given in Table P8.2, the following prior knowledge for fuzzification of data is assumed.

Attribute x_1 (universe of discourse: {–10, 50})

$$\mu_1\,(cool) = \begin{cases} 1 & ; \quad x_1 < 0 \\ 1 - \dfrac{x_1}{15} & ; \quad 0 \le x_1 \le 15 \\ 0 & ; \quad x_1 > 15 \end{cases}$$

$$\mu_2(mild) = \begin{cases} 0 & ; \quad x_1 < 0 \\ \dfrac{x_1}{15} - \dfrac{1}{3} & ; \quad 0 \le x_1 < 20 \\ 1 & ; \quad 20 \le x_1 < 30 \\ -\dfrac{x_1}{5} + 7 & ; \quad 30 \le x_1 \le 35 \\ 0 & ; \quad x_1 > 35 \end{cases}$$

$$\mu_3(hot) = \begin{cases} 0 & ; \quad x_1 < 25 \\ \dfrac{x_1}{10} - 2.5 & ; \quad 25 \le x_1 \le 35 \\ 1 & ; \quad x_1 > 35 \end{cases}$$

Attribute x_2 (universe of discourse: {0, 10})

$$\mu_1(weak) = \begin{cases} 1 & ; \quad x_2 < 3 \\ 2.5 - \dfrac{x_2}{2} & ; \quad 3 \le x_2 \le 5 \\ 0 & ; \quad x_2 > 5 \end{cases}$$

$$\mu_2(strong) = \begin{cases} 0 & ; \quad x_2 < 3 \\ \dfrac{x_2}{5} - 0.6 & ; \quad 3 \le x_2 \le 8 \\ 1 & ; \quad x_2 > 8 \end{cases}$$

Attribute x_3 (universe of discourse: {0, 20 kms})

$$\mu_1(short) = \begin{cases} 1 & ; \quad x_3 < 3 \\ 1.5 - \dfrac{x_3}{6} & ; \quad 3 \le x_3 \le 9 \\ 0 & ; \quad x_3 > 9 \end{cases}$$

$$\mu_2(long) = \begin{cases} 0 & ; \quad x_3 < 5 \\ \dfrac{x_3}{10} - 0.5 & ; \quad 5 \le x_3 \le 15 \\ 1 & ; \quad x_3 > 15 \end{cases}$$

(a) Give a graphical representation of the membership functions.

(b) Can we use fuzzy *K*-means clustering algorithm for fuzzification? If yes, describe the procedure. If not, why not?

P8.3: To build fuzzy decision tree for the dataset of Table P8.1, the prior knowledge on fuzzification of attributes x_1, x_2, and x_3 is given in problem P8.2.

(a) Choose the root node for the fuzzy decision tree.

(b) Show a partial fuzzy decision tree from root node along with training examples sorted to each of its descendent node.

(**Hint:** Membership functions for the data $\mathcal{D}$, $\mu_{\mathcal{D}}(s^{(i)})$; $i = 1, ..., 14$, are not given. Assume the dataset to be crisp: $\mu_{\mathcal{D}}(s^{(i)}) = 1$)

P8.4: Consider the dataset given in Table 8.5 of Chapter 8, with only two attributes x_1 (*height*) and x_2 (*weight*).

(a) Using fuzzy *K*-means algorithm, find membership functions of each attribute: Attribute $x_1(\mu_1(low), \mu_2(middle), \mu_3(high)$, Attribute $x_2(\mu_1(low), \mu_2(middle), \mu_3(heavy))$.

(b) Choose the root node for the fuzzy decision tree.

(c) Show a partial decision tree from root node along with training examples sorted to each of its descendent node.

You are advised to use an appropriate software tool (e.g., MATLAB) for fuzzification.

P8.5: Consider the training dataset given in Table P3.5.

(a) Construct a decision tree from the given data using "information gain" based splitting.

(b) For the given dataset, we want decision-tree induction based on " Gini index". We know that Gini index considers binary split for each attribute. Outline the procedure that need be adopted for construction of decision tree for the given dataset.

P8.6: A decision tree for the concept *buys-computer*, indicating whether a customer of a departmental store is likely to purchase a computer, is shown in Fig. P8.6. Each leaf node represents a class (either buys_computer = no or buys_computer = yes).

Convert the decision tree to classification IF-THEN rules.

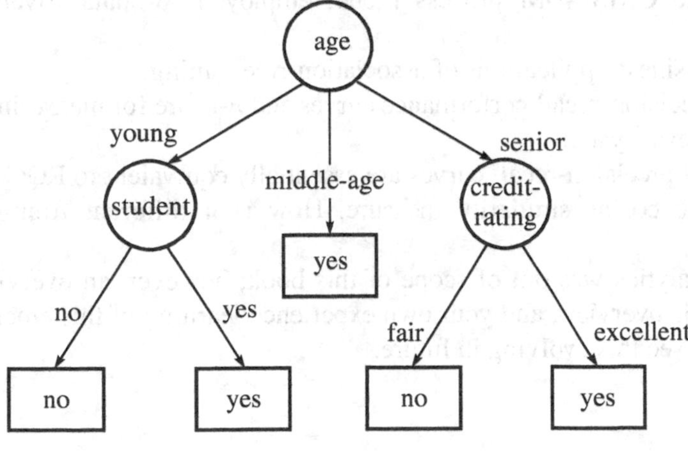

Figure P8.6

P8.7: Given a decision tree, we can (a) convert the decision tree to rules and then prune the resulting rules, or (b) prune the decision tree and then convert the pruned tree to rules. Describe relative merits of each option.

P8.8: Briefly outline the major steps of decision tree classification? How is tree pruning useful in decision tree induction?

P8.9: Describe the following popular impurity measures used for decision tree induction:
1. Information gain/entropy reduction
2. Gain ratio
3. Gini index

P8.10: Describe the procedure for fuzzy decision tree induction, highlighting parallels and differences with crisp decision tree induction.

P9.1: The fields: analytics, business intelligence, predictive analytics, machine learning, data mining; do not have precise boundaries for the spectrum of their functions. Their functions overlap. Understanding the essence of these tools for data-driven decision making is all that is important. Give your qualitative understanding of these terms.

P9.2: Data science is a collection of tools and techniques useful for data-driven decision making. With respect to this description of the term 'data science', list all the tools and techniques you have learnt from this book and other sources.

P9.3: Data exploration helps provide insights into the past, and is an essential ingredient for understanding the future. Summary statistics, data visualization, data clustering are some of the tools employed for data exploration

Many open-source/commercial software tools provide extensive breadth and depth in data exploration. Using software tools, learn the way how to get '360° view' of business through data exploration.

P9.4: On-Line Analytical Processing (OLAP) provides an easy-to-use GUI to query large data collections in data warehouses, for the purpose of data exploration. Describe the basic OLAP features provided in data warehouses.

P9.5: Describe the CRISP-DM process model employed for data-driven business decision making.

P9.6: Give two business applications of association-rule mining.

P9.7: Describe precision-recall performance curves and F-score for measuring accuracy of information retrieval systems.

Show that precision-recall curves are essentially equivalent to ROC curves.

P9.8: Describe the cosine similarity measure. How is it different from Euclidean distance measure?

P9.9: Big-data analytics was out of scope of this book; however, an overview has been given. Based on this overview, and your own experience/learning of this emerging field, describe how do you see this evolving in future.

REFERENCES

[1] T. Mitchel, *Machine Learning*, McGraw-Hill, 1997.

[2] S. Russell and P. Norvig, *Artificial Intelligence: A Modern Approach*, 3rd Edition, Pearson, 2015.

[3] V. Kecman, *Learning and Soft Computing*, The MIT Press 2001.

[4] R. O. Duda, P. E. Hart and D. G. Stork, *Pattern Classification*, 2nd Edition, Wiley 2002.

[5] M. Negnevitsky, *Artificial Intelligence: A Guide to Intelligent Systems*, 2nd Edition, Pearson, 2002.

[6] E. Alpaydin, *Introduction to Machine Learning*, 3rd Edition, The MIT Press, 2014.

[7] T. Hastie, R. Tibshirani and J. Friedman, *The Elements of Statistical Learning : Data Mining, Inference and Prediction*, 2nd Edition, Springer, 2009.

[8] C. M. Bishop, *Pattern Recognition and Machine Learning*, Springer, 2006.

[9] S. Shalev-Shwartz, and S. Ben-David, *Understanding Machine Learning from Theory to Algorithms*, Cambridge University Press, 2014.

[10] A. Webb and K. D. Copsey, *Statistical Pattern Recognition*, 3rd Edition, Wiley, 2011.

[11] K. P. Murphy, *Machine Learning: A Probabilistic Perspective*, The MIT Press, 2012.

[12] Y. S. Abu-Mostafa, M. Magdon-Ismail and H.-T. Lin, *Learning from Data: A Short Course*, AML, 2012.

[13] P. Flach, *Machine Learning: The Art and Science of Algorithms that Make of Data*, Cambridge University Press, 2012.

[14] J. Bell, *Machine Learning: Hands-On for Developers and Technical Professionals*, Wiley, 2015.

[15] S. Theodoridis, *Machine Learning: A Bayesian and Optimization Perspective*, Academic Press, 2015.

[16] B. Clarke, E. Fokoue and H. H. Zheng, *Principles and Theory of Data Mining and Machine Learning*, Springer, 2009.

[17] J. Han, M. Kamber and J. Pei, *Data Mining: Concepts and Techniques*, 3rd Edition, Morgan Kaufmann, 2012.

[18] I. H. Witten, E. Frank and M. A. Hall, *Data Mining: Practical Machine Learning Tools and Techniques*, 3rd Edition, Morgan Kaufmann, 2011.

[19] G. S. Linoff and M. J. A. Berry, *Data Mining Techniques: For Marketing, Sales, and Customer Relationship Management*, 3rd Edition, Wiley, 2011.

[20] G. Shmueli, N. R. Patel and P. C. Bruce, *Data Mining for Business Intelligence: Concepts, Techniques and Applications in Microsoft Office Excel with XLMiner*, Wiley, 2007.

[21] M. J. Zaki and W. Meira, Jr., *Data Mining and Analysis*, Cambridge University Press, 2014.

[22] P-N Tan, M. Steinbach and Vipin Kumar, *Introduction to Data Mining*, Addison-Wesley, 2005.

[23] D. Hand, H. Mannila and P. Smyth, *Principles of Data Mining*, The MIT Press, 2001.

[24] M. Kantardzic, *Data Mining: Concepts, Methods, Models and Algorithms*, 2nd Edition. Wiley, 2011.

[25] G. Xu, Y. Zong and Z. Yang, *Applied Data Mining*, CRC Press, 2013.

[26] J. M. Keller, D. Liu and D. B. Fogel, *Fundamentals of Computational Intelligence: Neural Networks, Fuzzy Systems and Evolutionary Computation*, Wiley, 2016.

[27] N. Siddique and H. Adeli, *Computational Intelligence: Synergies of Fuzzy Logic, Neural Networks and Evolutionary Computing*, Wiley, 2013.

[28] R. Eberhart and Y. Shi, *Computational Intelligence: Concepts to Implementations*, Morgan Kaufmann, 2007.

[29] A. P. Engelbrecht, *Computational Intelligence: An Introduction*, 2nd Edition, Wiley, 2007.

[30] R. S. Sutton and A. G. Barto, *Reinforcement Learning: An Introduction*, The MIT Press, 1998.

[31] C. Szepesvari, *Algorithms for Reinforcement Learning: Synthesis Lectures on Artificial Intelligence and Machine Learning*, Morgan and Claypool, 2010.

[32] D. P. Bertsekas and J. N. Tsitsiklis, *Neuro-Dynamic Programming*, Athena Scientific, 1996.

[33] M. Gopal, *Digital Control and State Variable Methods: Conventional and Intelligent Control Systems*, 4th Edition, McGraw-Hill, 2012.

[34] T. M. Mitchell, "Machine learning and data mining", *Communications of the ACM*, 42(11): 31-36, 1999.

[35] G. Strang, *Linear Algebra and its Applications*, 4th Edition, Cengage Learning, 2006.

[36] L. Breiman, "Bias-Variance, Regularization, Instability and Stabilization", *Neural Networks and Machine Learning*, 27-56, Springer, 1998.

[37] M. J. Kearns and U. V. Vazirani, *An Introduction to Computational Learning Theory*, The MIT Press, 1994.

[38] V. Vapnik, *Statistical Learning Theory*, Wiley, 1998.

[39] R. A. Caruana, A. Niculescu-Mizil, G. Crew and A. Ksikes, "Ensemble selection from libraries of models", *Twenty-First International Conference on Machine Learning*, 137-144, New York: ACM, 2004.

[40] L. I. Kuncheva, *Combining Pattern Classifiers: Methods and Algorithms*, Wiley, 2004.

[41] L. I. Kuncheva, Special Issue on Diversity in Multiple Classifier Systems, *Information Fusion*, 6: 1-115, 2005.

[42] T. Fawcett, "An Introduction to ROC Analysis", *Pattern Recognition Letters*, 27: 861-874, 2006.

[43] D. C. Montgomery, *Design and Analysis of Experiments*, 6th Edition, Wiley, 2005.

[44] I. Guyon and A. Elisseeff, "An introduction to variable and feature selection", *Journal of Machine Learning Research*, 3: 1157-1182, 2003.

[45] H. Liu and H. Motada (Eds.), *Computational Methods of Feature Selection*, CRC Press, 2008.

[46] S. M. Ross, *Introduction to Probability and Statistics for Engineers and Scientists*, Wiley, 1987.

[47] D. C. Montgomery and G. C. Runger, *Applied Statistics and Probability for Engineers*, Wiley. 2010.

[48] P. D. Hoff, *A First Course in Bayesian Statistical Methods*, Springer, 2009.

[49] F. Jensen, *An Introduction to Bayesian Networks*, Springer, 1996.

[50] Y. Chen, E. K. Garcia, M. R. Gupta, A. Rahimi, and L. Cazzanti, "Similarity-based classification: concepts and algorithms", *Journal of Machine Learning Research*, 11: 747-776, 2009.

[51] K. Q. Weinberger and L. K. Saul, "Distance metric learning for large margin classification", *Journal of Machine Learning Research*, 10: 207-244, 2009.

[52] P. D. Grunwald, *The Minimum Description Length Principle*, The MIT Press, 2007.

[53] B. Scholkopf and A. J. Smala, *Learning with Kernels: Support Vector Machines, Regularization, Optimization and Beyond*, The MIT Press, 2002.

[54] H. A. Taha, *Operations Research: An Introduction*, 8th Edition, Pearson, 2007.

[55] N. S. Kambo, *Mathematical Programming Techniques*, Revised Edition, Affiliated East-West Press, 1991.

[56] K. P. Suman, R. Loganathan and V. Ajay, *Machine Learning with SVM and Other Kernel Methods*, PHI Learning, 2009.

[57] K. P. Bennett, M. Momma and M. J. Embrechts, "MARK: A boosting algorithm for heterogeneous kernel models", *Proceedings of SIGKDD International Conference*, 2002.

[58] T. M. Cover, "Geometrical and statistical properties of systems of linear inequalities with applications in pattern recognition", *IEEE Transactions of Electronic Computers*, EC14: 326-334, 1965.

[59] T. G. Dietterich and G. Bakiri, "Solving multiclass learning problems via error-correcting output codes", *Journal of Artificial Intelligence Research*, 2: 263-286, 1995.

[60] E. L. Allwein, R. E. Schapire and Y. Singer, "Reducing Multiclass to Binary: a unifying approach for margin classifiers", *Journal of Machine Learning Research*, 1: 113-141, 2000.

[61] J. Platt, N. Cristianini and J. Shawe-Taylor, "Large margin DAGs for multiclass classification", *Advances in Neural Information Processing Systems*, 12: 547-553, The MIT Press, 2000.

[62] B. Scholkopf, A. J. Smola, R. Williamson and P. Bartlett, "New support vector algorithms", *Neural Computation*, 12: 1083-1121, 2000.

[63] J. Platt, "Sequential Minimal Optimization: A fast algorithm for training support vector machines", *Microsoft Research Technical Report* MSR-TR-98-14, 1998.

[64] S. S. Keerthi, S. K. Shevade, C. Bhattacharyya and K. R. K. Murthy, "Improvements to Platt's SMO algorithm for SVM classifier design", *Neural Computation*, 13: 637-649, 2001.

[65] S. K. Shevade, S. S. Keerthi, C. Bhattacharyya, and K. R. K. Murthy, "Improvements to SMO algorithm for SVM regression", *Proc. of International Joint Conference on Neural Networks* (IEEE-NN), 1999.

[66] A. Bordes, S. Ertekin, J. Weston and L. Bottou, "Fast kernel classifiers with online and active learning", *Journal of Machine Learning Research*, 6: 1579-1619, 2005.

[67] T. Jaochims. 2008: *SVMlight*, http://svmlight. jaochims.org

[68] C.-C. Chang and C.-J. Lin, "LIBSVM: A LIBrary for Support Vector Machines", *ACM Transactions on Intelligent Systems and Technology*, 27(2): 1-27, 2011.

[69] J. A. K. Suykens and J. Vandewalle, "Least squares support vector machine classifiers", *Neural Processing Letters*, 9(3): 293-300, 1999.

[70] G. Fung and O. L. Mangasarian, "Proximal support vector machine classifiers", *Proceedings of KDD*, 2001.

[71] O. L. Mangasarian and D. R. Musicant, "Lagrangian support vector machines", *Journal of Machine Learning Research*, 1: 161-177, 2001.

[72] E. Bredensteiner and K. Bennett, "Multicategory classification by support vector machines", *Comput. Optim. Appl.*, 12: 53-99, 1999.

[73] J. Weston and C. Watkins, "Multi-class support vector machines", *Proceedings of ESANN 99*, 1999.

[74] K. Crammer and Y. Singler, "On the algorithmic implementation of multi-class kernel-based vector machines", *J. Mach. Learn. Res.*, 2: 265-292, 2001.

[75] X. He, Z. Wang, C. Jin, Y. Zheng and X. Y. Xue, "A simplified multi-class support vector machine with reduced dual optimization", *Pattern Recognit. Lett.*, 33: 71-82, 2012.

[76] P. Thagard, *Mind: Introduction to Cognitive Science*, The MIT Press, 2005.

[77] G-B Huang, Q-Y Zhu and C-K Siew, "Extreme learning machine: theory and applications", Neurocomputing, 70(1): 489-501.

[78] Y. Bengio, "Learning Deep Architectures for AI", *Foundations and Trends of Machine Learning*, 2(1): 1-127, 2009.

[79] A. Itamar, D. C. Rose and T. P. Kamocoski, "Deep machine learning—a new frontier in artificial intelligence research—a survey", *IEEE Computational Intelligence Magazine*, 2013.

[80] J. Schmidhuber, "Deep learning neural networks: an overview", *Neural Networks*, 61: 85-117, 2015.

[81] Y. Bengio, L. Yann and H. Geoffrey, "Deep Learning", *Nature*, 521: 436-444, 2015.

[82] I. Goodfellow, Y. Bengio and A. Courville, *Deep Learning*, The MIT Press, 2016.

[83] S. Haykin, *Neural Networks: A Comprehensive Foundation*, 2nd Edition, Prentice-Hall, 1998.

[84] K-L Du and M. N. S. Swamy, *Neural Networks in a Softcomputing Framework*, Springer, 2006.

[85] L. Behera and I. Kar, *Intelligent Systems and Control: Principles and Applications*, Oxford University Press, 2009.

[86] K. Hornik, M. Stinchcombe and H. White, "Multilayer feedforward networks are universal approximators", *Neural Networks*, 2: 359-366, 1989.

[87] L. Behera, S. Kumar, and A. Patnaik, "On adaptive learning rate that guarantees convergence in feedforward neural networks", *IEEE Transactions on Neural Networks*, 17(5), 2006.

[88] C. M. Bishop, *Neural Networks for Pattern Recognition*, Oxford University Press, 1995.

[89] B. D. Ripley, *Pattern Recognition and Neural Networks*, Cambridge University Press, 1996.

[90] D. Michie, D. Spiegelhalter and C. Taylor (Eds.), *Machine Learning: Neural and Statistical Classification*, Overseas Press, 2009.

[91] D. K. Pratihar, *Soft Computing*, Narosa, 2008.

[92] T. J. Ross, *Fuzzy Logic with Engineering Applications*, 3rd Edition, Wiley, 2010.

[93] K. M. Passino and S. Yurkovich, *Fuzzy Control*, Addison Wesley, 1998.

[94] C. T. Lin and C. S. George Lee, *Neural Fuzzy Systems*, Prentice-Hall, 1996.

[95] E. H. Mamdani and S. Assilian, "An experiment in linguistic synthesis with a fuzzy logic controller", *International Journal of Man-Machine Studies*, 7(1): 1-13, 1975.

[96] E. H. Mamdani, "Application of fuzzy algorithms for control of a dynamic plant", *Proc. IEEE*, 121: 1585-1588, 1974.

[97] M. Sugeno, *Industrial Applications of Fuzzy Control*, North-Holland, 1985.

[98] T. Takagi and M. Sugeno, "Fuzzy identification of systems and its applications to modeling and control", *IEEE Transactions on Systems, Man and Cybernetics*, 15: 116-132, 1985.

[99] M. Sugeno and G. T. Kang, "Structure identification of fuzzy control", *Fuzzy Sets and Systems*, 28: 15-33, 1988.

[100] D. K. Pratihar, K. Dab and A. Ghosh, "A genetic-fuzzy approach for mobile robot navigation", *International Journal of Approximate Reasoning*, 20: 145-172, 1999.

[101] N. B. Hui, V. Mahender and D. K. Pratihar, "Time-optimal collision-free navigation of a car-like mobile robot using a neuro-fuzzy approach", *Fuzzy Sets and Systems*, 157(16): 2171-2204, 2006.

[102] H-P Gullich, "Fuzzy logic decision support system for credit risk evaluation", *EUFIT Fourth European Congress on Intelligent Techniques and Soft Computing*, 2219-2223, 1996.

[103] J. Zhang and A. J. Morris, "Fuzzy neural networks for nonlinear systems modeling", *IEE Proc. Control Theory Appl.*, 142: 551-561, 1995.

[104] L. Behera and K. K. Anand, "Guaranteed tracking and regulatory performance of nonlinear dynamic systems using fuzzy neural networks", *IEE Proc. Control Theory Appl.*, 146(5), 1999.

[105] J. J. Buckley and Y. Hayashi, "Fuzzy neural networks: a survey: *Fuzzy Sets and Systems*, 66: 1-13, 1994.

[106] J. S. R. Jang, "ANFIS: Adaptive Network-based Fuzzy Inference System", *IEEE Transactions on Systems, Man and Cybernetics*, 23(3): 665-685, 1993.

[107] K. Nigam, A. K. McCallum, S. Thrun and T. M. Mitchell, "Text classification from labeled and unlabeled documents using EM", *Machine Learning*, 39: 103-104, 2000.

[108] K. Nigam and R. Ghani, "Analyzing the effectiveness and applicability of co-training", *Proceedings of the Ninth International Conference on Information and Knowledge Management*, ACM Press, 2000.

[109] T.-Y. Liu, *Learning to Rank for Information Retrieval*, Springer, 2011.

[110] R. K. Herbrich and T. Graepel, "Large margin rank boundaries for ordinal regression", *Advances in Large Margin Classifiers*, 115-132, The MIT Press, 2000.

[111] T. Jaochims, "Optimizing search engines using clickthrough data", *ACM SIGKDD International Conference on Knowledge Discovery and Data Mining*, 133-142, ACM, 2002.

[112] O. Chapelle and S. S. Keerthi, "Efficient algorithms for ranking with SVMs", *Information Retrieval*, 11: 201-215, 2010.

[113] D. Freedman, R. Pisani and R. Purves, *Statistics*, 4th Edition, W. W. Norton and Co., 2007.

[114] J. L. Devore, *Probability and Statistics for Engineering and Sciences*, 4th Edition, Duxbury Press, 1995.

[115] W. Cleveland, *Visualizing Data*, Hobert Press, 1993.

[116] E. R. Tufte, *The Visual Display of Quantitative Information*, 2nd Edition, Graphics Press, 2001.

[117] V. J. Hodge and J. Austin, "A survey of outlier detection methodologies", *Artificial Intelligence Review*, 22: 85-126, 2004.

[118] V. Chandola, A. Banerjee and V. Kumar, "Anomaly detection: a survey", ACM Computing Surveys, 41(3): 15: 1-15:58, 2009.

[119] Halkidi and M. Vazirgiannis, "Clustering validity assessment: finding the optimal partitioning of dataset", *Proceedings of ICDM Conference*, 2001.

[120] R. Xu, and D. Wunsch II, "Survey of clustering algorithms", *IEEE Transactions on Neural Networks*, 16: 645-678, 2005.

[121] U. VonLuxburg, "A tutorial on spectral clustering", *Statistical Computing*, 17: 395-416, 2007.

[122] A. Y. Ng, M. I. Jordan and Y. Weiss, "On spectral clustering: analysis and an algorithm", *Advances in Neural Information Systems*, 14: 849-856, The MIT Press, 2001.

[123] M. Belkin and P. Niyogi, "Laplacian eigenmaps for dimensionality reduction and data representation", *Neural Computation*, 15: 1373-1396, 2003.

[124] T. Kohonen, *Self-Organizing Maps*, Springer, 1995.

[125] F. Hoppner, F. Klawonn, R. Kruse and T. Runkler, *Fuzzy Cluster Analysis: Methods for Classification, Data Analysis and Image Recognition*, Wiley, 1999.

[126] S. Marsland, *Machine Learning: An Algorithmic Perspective*, CRC Press, 2009.

[127] G. J. McLachlan and T. Krishnan, *The EM Algorithm and Extensions*, 2nd Edition, Wiley, 2008.

[128] T. Dasu and T. Johnson, *Exploratory Data Mining and Data Cleaning*, Wiley, 2003.

[129] H. Liu, F. Hussain, C. L. Tan and M. Dash, "Discretization: an enabling technique", *Data Mining and Knowledge Discovery*, 6: 393-423, 2002.

[130] I. Guyon and A. Elisseeff, "An introduction to variable and feature selection", *Journal of Machine Learning Research*, 3: 1157-1182, 2003.

[131] U. Fayyad and K. Irani, "Multi-interval discretization of continuous-valued attributes for classification learning", *Proc. 1993 Int. Joint Conf. Artificial Intelligence (IJCAI'93)*, 1022-1029, 1993.

[132] J. Mao and A. K. Jain, "Artificial neural networks for feature extraction and multivariate data projection", *IEEE Transactions on Neural Networks*, 6: 296-317, 1995.

[133] K. R. Muller, S. Mika, G. Ratsch, K. Tsuda and B. Scholkopf, "An introduction to kernel-based learning algorithms", *IEEE Transactions on Neural Networks*, 12: 181-201, 2001.

[134] Z. Pawlak, *Rough Sets: Theoretical Aspects of Reasoning about Data*, Kluwer Academic, 1991.

[135] W. Ziarko, "The discovery analysis, and representation of data dependencies in databases", *Knowledge Discovery in Databases*, 195-209, AAAI Press, 1991.

[136] K. Cios, W. Pedrycz and R. Swiniarski, *Data Mining Methods for Knowledge Discovery*, Kluwer Academic, 1998.

[137] A. Lenarcik and Z. Piasta, "Probabilistic rough classifiers with mixture of discrete and continuous variables", *Rough Sets and Data Mining: Analysis for Imprecise Data*, 373-383, Kluwer Academic, 1997.

[138] R. Swiniarski, "Rough sets and principle component analysis and their applications in feature extraction and selection, data model building and classification", *Rough Set Hybridization: A New Trend in Decision-Making*, Springer, 1999.

[139] A. Skowron and C. Rauszer, "The discernibility matrices and functions in information systems", *Intelligent Decision Support: Handbook of Applications and Advances of the Rough Set Theory*, 331-362, Kluwer Academic, 1992.

[140] J. R. Quinlan, "Induction of Decision Trees", *Machine Learning*, 1: 81-106, 1986.

[141] J. R. Quinlan, *C4.5: Programs of Machine Learning*, Morgan Kaufmann, 1993.

[142] L. Breiman, J. H. Friedman, R. A. Olshan and C. J. Stone, *Classification and Regression Trees*, Wadsworth International Group, 1984.

[143] T. Oates and D. Jensen, "The effects of training set size on decision tree complexity", *Proceedings of the Fourteenth International Conference on Machine Learning*, 254: 262, Morgan Kaufmann, 1997.

[144] O. T. Yildiz and E. Alpaydin, "Linear discriminant trees", *Seventeenth International Conference on Machine Learning*, Morgan Kaufman, 1175-1182, 2000.

[145] L. Rokach and O. Maimon, "Top-down induction of decision trees classifiers—a survey", *IEEE Transactions on Systems, Man, and Cybernetics*–part C, 35: 476-487, 2005.

[146] J. R. Quinlan, "Unknown attribute values in induction", *Proc. 1989 Int. Conference Machine Learning* (ICML' 89), 164-168, 1989.

[147] T.-S. Lim, W.-Y. Loh and Y.-S. Shih, "A comparison of prediction accuracy, complexity, and training time of thirty-three old and new classification algorithms", *Machine Learning*, 40: 203-228, 2000.

[148] S. K. Murthy, "Automatic construction of decision trees from data: a multi-disciplinary survey", *Data Mining and Knowledge Discovery*, 4: 345-389, 1998.

[149] T. Smith, B. Kao, K. Y. Yip, W-S Ho and S. D. Lee, "Decision trees for uncertain data", *IEEE Transactions on Knowledge and Data Engineering*, 23(1): 1-15, 2011.

[150] L. Rokach and O. Maimon, *Data Mining with Decision Trees: Theory and Applications*, 2nd Edition, World Scientific, 2015.

[151] A. Altay and D. Cinar, "Fuzzy Decision Trees", *In Fuzzy Statistical Decision-Making*, Springer, 2016.

[152] A. Kazemi and E. Mehrzadegan, "A new algorithm for optimization of fuzzy decision tree in data mining", *Journal of Optimization in Industrial Engineering*, 7: 29-35, 2011.

[153] Y. Yuan and M. J. Shaw, "Introduction to fuzzy decision trees", *Fuzzy Sets and Systems*, Elsevier, 69(2): 125-139, 1995.

[154] F. Provost and T. Fawcett, *Data Science for Business: What you Need to Know about Data Mining and Data-Analytic Thinking*, O'Reilly Media, 2013.

[155] T. H. Devenport, R. Morison and J. G. Harris, *Analytics at Work: Smarter Decisions, Better Results*, Harvard Business Review Press, 2010.

[156] T. H. Devenport and J. G. Harris, *Competing on Analytics: The New Science of Winning*, Harvard Business Review Press, 2007.

[157] J. D. Kelleher, B. M. Namee and A. D' Arcy, *Fundamentals of Machine Learning for Predictive Data Analytics: Algorithms, Worked Examples, and Case Studies*, The MIT Press, 2015.

[158] D. T. Larose and C. D. Larose, *Data Mining and Predictive Analytics*, 2nd Edition, Wiley, 2015.

[159] D. S. Putler and R. E. Krider, *Customer and Business Analytics: Applied Data Mining for Business Decision Making Using R*, CRC Press, 2012.

[160] C. Vercellis, *Business Intelligence: Data Mining and Optimization for Decision Making*, Wiley, 2009.

[161] N. Memon, J. J. Xu, D. L. Hicks and H. Chen, *Data Mining for Social Network Data*, Springer and Business Media, 2010.

[162] R. Agrawal, H. Mannila, R. Srikant, H. Toivonen and A. Verkamo, "Fast Discovery of Association Rules", *In Advances in Knowledge Discovery and Data Mining*, 307-328, The MIT Press, 1996.

[163] C. Zhang and S. Zhang, *Association Rule Mining: Models and Algorithms*, Springer, 2002.

[164] F. Sebastiani, "Machine learning in automated text categorization", *ACM Computing Surveys*, 34(1): 1-47, 2002.

[165] "The Porter Stemming Algorithm" http:// www.tartarus.org/-martin/PorterStemmer

[166] C. D. Maning and H. Schutze, *Foundations of Statistical Natural Language Processing*, The MIT Press, 1999.

[167] D. A. Grossman and O. Frieder, *Information Retrieval: Algorithms and Heuristics*, Springer, 2004.

[168] C. D. Manning, P. Raghavan, and H. Shutze, *Introduction to Information Retrieval*, Cambridge University Press, 2008.

[169] R. A. Baeza-Yates and B. A. Ribeiro-Neto, *Modern Information Retrieval*, 2nd Edition, Addison-Wesley, 2011.

[170] M. W. Berry, *Survey of Text Mining: Clustering, Classification and Retrieval*, Springer, 2003.

[171] A. N. Srivastava and M. Sahami (Eds.), *Text Mining: Classification, Clustering and Applications*, CRC Press, 2009.

[172] U. Fayyad and P. Smyth, "Image database exploration progress and challenges", *Proc. AAAI'93 Workshop Knowledge Discovery in Databases*, 14-27, 1993.

[173] A. Natsev, R. Rastogi and K. Shim,"WALRUS: A similarity retrieval algorithm for image databases", *Proc. 1999 ACM-SIGMOD Int. Conf. Management of Data*, 395-406, 1999.

[174] W. Hsu, M. L. Lee and J. Zhang, "Image mining: trends and developments", *J. Intelligent Information Systems*, 19: 7-23, 2002.

[175] E. B. Sudderth, A. Torralba, W. T. Freeman and A. S. Willsky, "Describing visual scenes using transformed objects and parts", *International Journal of Computer Vision*, 77: 291-330.

[176] L. R. Rabiner and B. H. Juang, *Foundations of Speech Recognition*, Prentice-Hall, 1993.

[177] F. Jelinek, *Statistical Methods for Speech Recognition*, The MIT Press, 1997.

[178] F. Camastra and A. Vinciarelli, *Machine Learning for Audio Image and Video Analysis*, Springer, 2008.

[179] M. Cord and P. Cunningham (Eds.), *Machine Learning Techniques for Multimedia: Case Studies on Organization and Retrieval*, Springer, 2008.

[180] C. Faloutsos and K-I. Lin, "FastMap: A fast algorithm for indexing, data-mining and visualization of traditional and multimedia datasets", *Proc. 1995 ACM-SIGMOD Int. Conf. Management of Data*, 163-174, 1995.

[181] Z. Zhang and R. Zhang, *Multimedia Data Mining: A Systematic Introduction to Concepts and Theory*, Chapman and Hall, 2009.

[182] H. Kargupta, A. Joshi, K. Shivakumar, Y. Yesha (Eds.), *Data Mining: Next Generation Challenges and Future Directions*, The MIT Press, 2004.

[183] M. Michael, C. Michele and D. Ambiga, *Big Data, Big Analytics: Emerging Business Intelligence and Analytical Trends in Today's Business*, Wiley, 2013.

[184] B. Franks, *Taming the Big Data Tidal Wave: Find Opportunities in Huge Data Streams with Advanced Analytics*, Wiley, 2012.

[185] T. H. Devenport, *Big Data @Work: Dispelling the Myths, Uncovering the Opportunities*, Harvard Business Review Press, 2014.

[186] C. R. Darwin, *On the Origin of Species by Means of Natural Selection*, London: John Murray, 1859.

[187] J. R. Koza, *Genetic Programming: On the Programming of the Computers by Means of Natural Selection*, The MIT Press, 1992.

[188] H. P. Schwefel, *Evolution and Optimum Seeking*, Wiley, 1995.

[189] J. H. Holland, *Adaptation in Natural and Artificial Systems*, University of Michigan Press, 1975.

[190] D. E. Goldberg, *Genetic Algorithms is Search, Optimization and Machine Learning*, Addison-Wesley, 1989.

[191] M. F. Akay, "Support vector machines combined with feature selection for breast cancer diagnosis", *Expert Systems With Applications*, 36(2; 2): 3240-3247, 2009.

[192] O. L. Mangasarian, W. N. Street, W. H. Wolberg, "Breast Cancer Diagnosis and Prognosis Via Linear Programming", *Operations Research*, 43(4): 570-577, 1995.

[193] M. D. Garris, J. L. Blue, G. T. Candela, D. L. Dimmick, J. Geist, P. J. Grother, S. A. Janet and C. L. Wilson, "NIST Form-Based Handprint Recognition System", *NISTIR* 5959, 1997.

[194] Y. LeCun, B. Broser, J. S. Denker, D. Henderson, R. E. Howard and L. D. Jackel, "Backpropagation Applied to Handwritten Zipcode Recognition", *Neural Computation*, 1(4): 541-551, 1989.

[195] S. Moro, P. Cartez and P. Rita, "A data-driven approach to predict the success of bank telemarketing", *Decision Support Systems*, 62: 22-31, 2014.

[196] H. A. Elsalamony, "Bank direct marketing analysis of data mining techniques", *International Journal of Computer Applications*, 85(7): 12-22, 2014.

[197] V. Rao and H. Rao, *C++ Neural and Fuzzy logic*, BPB Publications, 1996.

[198] E. W. Saad, D. V. Prokhorov and D. C. Wunsch, "Comparative study of stock trend prediction using time delay, recurrent, and probabilistic neural networks", *IEEE Trans. Neural Networks*, 9(6): 1456-1470, 1998.

[199] G. E. P. Box and G. M. Jenkins, *Time Series Analysis: Forecasting and Control*, 2nd ed., Holden-Day, 1970.

[200] S. Chen and S. Billings, "Neural networks for nonlinear dynamic system modeling and identification", *Int. J. Control*, 56: 319-346, 1992.

INDEX

M

N